# ANIMAL BEHAVIOR

# ANIMAL BEHAVIOR

## ITS DEVELOPMENT, ECOLOGY, AND EVOLUTION

**Robert A. Wallace**
DUKE UNIVERSITY

GOODYEAR PUBLISHING COMPANY, INC.
Santa Monica, California

**Library of Congress Cataloging in Publication Data**

Wallace, Robert Ardell, 1938–
Animal behavior, its development, ecology, and evolution.

Bibliography: p. 526
1. Animals, Habits and behavior of. I. Title.
QL751.W252 591.5 78–11611
ISBN 0–87620–043–9



Current printing (last digit):
10 9 8 7 6 5 4 3 2 1

ISBN: 0–87620–043–9
Y–0439–3

Printed in the United States of America

**Supervising Editor: Susan Smith**
**Designer: Christy Butterfield**
**Illustrators: Barbara Hack and Cyndie Clark**
**Photo Research: Linda Rill**

The quote on page 522 is from S. Milgram, "Einige Bedingungen von Autoritätsgehorsam und Seiner Verweigerung," *Z. Exptl. Angew Psychol.* 13:433–463. Reprinted with permission.

*For Bonni*

# CONTENTS IN BRIEF

# CONTENTS

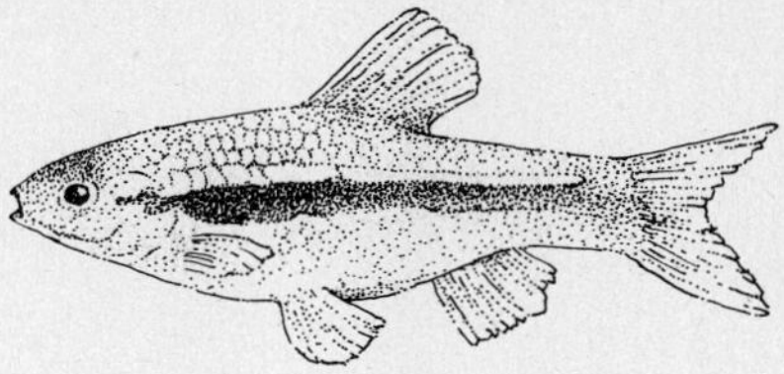

# PREFACE

Studying animal behavior, it seems to me, is a bit like being married. It has its good times and its not-so-good times, but if one stays on, it's probably because he or she simply wants to. In a sense, one can be wedded to the field in a very real sense so that printed words like *biology, behavior, animal, wilderness,* and *social* are as attractive and consoling as a loved one's visage. They have a pleasing sense about them. They catch your eye and subtly sharpen your attention.

But then there are those other times—times when one may become uncharacteristically silent, swathed in doubt. He or she may, at these times, become disconsolate and dissatisfied, willing, even, to junk the whole thing, whether a profession or a relationship. The animal behaviorist may, at such times, begin to ask new kinds of questions. Is this a legitimate field of study? It's too wide-ranging. We are too ill equipped. We haven't even answered the most basic questions. We are baffled at every turn, but here we are pretentiously trying to deal with esoterica. Who would build a roof without first having laid the foundation?

Of course, not everyone is so bothered. The short-sighted in either marriage or science are blessed by not continually noticing the alternatives. Everything seems in order. There is no reason to rock the boat or to be dissatisfied. Furthermore, they note, we have our operational paradigms. And we have our models and modellers.

Occasionally people like Tinbergen or Lorenz come along with particularly keen abilities and vision and build us a model, such as their explanation of instinct, and it fits what we know about the real world. Things seem to work more or less according to the model's predictions. But perhaps, the hand-wringers say, other models could be contrived that would explain even more, but that would be no closer to the truth. After all, many of our most sacred notions of behavior are difficult or impossible to test. We are limited in part by the paradox of an eye trying to see itself, and resorting, instead, to mirrors. For those interested in human behavior, other animals are very distorted mirrors indeed, but notice how many preconceptions about ourselves stem from observations of bright-eyed and bewhiskered beasties.

One gets discouraged at times, also, because it appears that we can't even study other animals by studying other animals. What motivates a crayfish may not even exist in the world of the rat, and one strain of rat may even behave differently from another. And we have other problems that involve trying to form unitary rules. For example, when we discuss territoriality in this book, we will find that it varies in almost as many ways as there are species. Does territoriality, then, really exist? Is it a unitary phenomenon? When someone says the word, will other people know what they mean? Or must wolf people talk to wolf people, and bird people to bird people? (This question is, in essence, the prickly problem of homology.)

The optimists, of course, will survey all the problems and cheerfully declare that they are of value because they stimulate interest and present challenges. They will note that precisely because the dust hasn't settled there is room for innovation. Furthermore, they might add that, in spite of any problems, the only way to ever learn about animal behavior is to study it. Our alternative is to not study animal behavior until we know enough about it to study it.

Dogmatism in science and in human relationships is boring, and the field of animal behavior thankfully has an abundance of disrespect for dogma. "Ah, but what about the . . ." has burst many an overblown idea. Animal behavior is not boring, but it does have the disconcerting quality of one solution leading to yet another problem. It must sorely try the souls of those who crave finality.

And perhaps there is no finality. I have the feeling that if we were ever to unravel the biochemistry of behavior, we would be left "sifting electrons," the essence having escaped us somewhere along the way.

I would like to make another point here. There is a certain snobbery afoot in the world of biology. There are those who continue to sight behavior along their noses, labelling it a "soft science" (that most pejorative of terms). It is not mechanistic or physiological enough, they say (ignoring reams of physiological data), and real "science" simply cannot be done within the fluid limits of the behavioral paradigm. Of course, we believers realize that much of our discipline is heavy with implicit or explicit assumption, and that our faith is great. We also know that data are cascading in at a tremendous and accelerating rate and that our footing grows ever firmer as the old premises are relentlessly tested by a growing army of young, hard-nosed, and well-trained ethologists who not only tromp about in muddied boots but who speak arithmetic and converse with terminals.

We wish no quarrel with the other sciences. It is interesting to know if a *Drosophila* intestine can regenerate and why cells adhere, but we believe it is also worthwhile to know where ducks go in winter. The ducks, in fact, are interesting at several levels. We can discover *where* they go easily enough. *How* they go presents problems of methodology, and *why* they go, problems of philosophy as much as physiology.

I began by mentioning marriage or, as some would have it, human bonding, and I might continue the analogy by saying that although we can't explain love, there are few who would deny its existence. Furthermore, this "unexplainable" entity must be one of the most compelling forces on the planet. We can examine, poke, sniff, and measure it to our heart's content, but in the final analysis there is no final analysis. It simply exists and must be dealt with and accounted for in our social equations. Once we accept it, homebuilders and esthetes alike can alter their attentions accordingly. And they will profit by so doing.

So it seems that the field of animal behavior, by its very breadth, must encompass some areas that have stronger support than others. I would argue with those who call it soft. I think there are now enough hard data to call it lumpy. The problem is simply, at this time, the lumps lie suspended in a rather amorphous matrix. Perhaps we can make bombykol, but we don't know what motivates moths, or what it feels like to be a moth, or what a moth senses and knows.

Anyway, it's too late. The study of animal behavior is alive and well. The whole study continues to be difficult, in part because some areas are so much more resistant to investigation than others. It is, admittedly and in a sense, a bit presumptuous for a species in which individuals cannot even predict what they, themselves, will do in a given situation and cannot define the love, hate, jealousy, or benevolence that so often drives them, to broach the great questions of animal behavior such as motivation. But we can, it turns out, offer suggestions for why love is adaptive and we do know where the ducks go. So, let's have a closer look, but let's stand prepared to alter our conclusions and tread roads not taken as the data grow ever harder and more precise.

# ACKNOWLEDGMENTS


This is a particularly difficult section to write, and I am tempted to simply refer you to the bibliography. The reason is, my debts are enormous. However, certain people have provided special encouragement, in some cases unwittingly, and I would like to thank them specifically. The first draft of this book was written on Ibiza, in the Spanish Baleares, and my library was limited by what I was able to bring with me. The miserable winter weather there gave me long days to pore over my material, and I often found myself engrossed not only in the works of people like Wilson, Lorenz, Tinbergen, but also Pianka, Emlen, Schoener, and a host of others who I had probably failed to fully appreciate until I took the time to see what they were really up to and the range of their thinking.

I am grateful for the formal reviews of Vernon Avila, George Barlow, Marc Bekoff, Constance Campbell, George Cox, Jerry Feldman, Charles Krebs, Melvin Kreitken, Richard Thompson, and Pauline Yahr.

I would also like to thank a number of people for personal encouragement and for their stimulating ideas, sometimes granted in something less than an academic context. Among these are David Barash, Joseph Connell, Lincoln Fairchild, John Goodman, H. B. Graves, David Gubernick, Ray Huey, Jack L. King, Neil Krekorian, Lee McGeorge, Denis Mitchell, William Murdoch, Gordon Orians, Stephen Rothstein, Daniel Rubenstein, Robert Selander, Delbert Thiessen, Niko Tinbergen, and Edward Wilson.

I would like to express my appreciation to the Animal Behavior Group at Duke University. In particular, I am grateful for having had the opportunity to work with Peter Klopfer, a most singular person, who turns out to be as formidable athletically as he is intellectually, and to whom I must reluctantly grant everything over twenty . . . for now.

Again, it is my great pleasure to have worked with Goodyear Publishing Company. My editor for these last several years, Clay Stratton, continues to give very special attention to our projects and shepherds them through their most difficult times with his usual finesse. I am also grateful for the expert assistances of Al Goodyear, Jack Pritchard, Bob Hollander, Gene Schwartz, Pamela Sullivan, Christy Butterfield, Linda Rill, Barbara Hack, Cyndie Clark, and Victoria Philp, and in particular I would like to thank two important new people in the Goodyear organization: Gerald Rafferty and Susan Smith, two people who care and are total professionals in every sense of the word. I can only say, if you plan to write and should ever get an opportunity to work with them, do it.

Finally, I suppose I should thank the residents of Ibiza for teaching me something about the range of human behavior.

# ANIMAL BEHAVIOR

# 1

# INTRODUCING ETHOLOGY

In one way or another, man has been studying animal behavior for thousands of years. The behavior of animals was important to him in earlier ages for some of the same reasons it is important to him now. The most skillful hunters and fishermen of any age were those who were able to make certain predictions regarding the behavior of their prey. It was important to know that when salmon are spawning they will not respond to the fisherman's bait, that many rodents escape toward darkness while most birds flee towards light, and that many kinds of animals will fight, some with awesome effect, if they are trapped.

The study of animal behavior may have occupied the fringes of man's consciousness for centuries on just such practical bases. Later, when animals were domesticated and put to work, it was necessary to understand new things about them. Dogs could easily be trained for the personal protection of individual humans, but any attempt at training watchcats was bound to fail. Although cats were stealthy, dogs were better trackers. Both dogs and cats were easier to teach than fish.

Recently, as we shall see, the study of animal behavior has taken on new dimensions. The goals as well as the techniques have changed. We no longer learn about animals simply to exploit them more efficiently (although this is certainly a major reason for our attention); now we sometimes study them because they present unsolved problems (and this is the "purest" motive). In other cases, we hope they will shed light on our own behavior. There are undoubtedly a host of less appealing reasons, but they won't be acknowledged here.

The techniques of the study of animal behavior have proliferated during the last seventy years. No longer are our observations recited anecdotally. Today, even the simplest records are subjected to the cold analysis of statistics. Field work is carefully designed—before one pulls on his boots—in order to maximize the yield of pertinent (or potentially pertinent) information. Experiments are often conducted in the field as well as in the laboratory, and, ideally, the environment of the animal is carefully controlled as variables are introduced singly in an effort to discover what component of the environment the animal is responding to. In other kinds of behavioral studies, the nervous system of the animal is dissected and analyzed in the minutest detail. Other researchers go one step further and attempt to describe, at the molecular level, how the animal interacts with its external and internal environment.

In other words, animal behavior is presently being investigated through a dazzling array of techniques. Hopefully, the variety of these techniques will become apparent as we consider the kinds of information that have become available to us in recent years.

## COMPARATIVE PSYCHOLOGY AND ETHOLOGY

Basically, the array of investigative techniques has sprung from two major philosophical camps. These major branches are generally called *comparative psychology* and *ethology* and they have not always viewed each other with benevolence. In the early days, the result of any disagreement between the two philosophies was that each rejected even the most valid tenets of the other, a most unfortunate confrontation.

Comparative psychology has traditionally been an American discipline. Its proponents have generally been grounded in psychology, and somehow this discipline came to focus on the Norway rat. Comparative psychologists have usually carried out their work in the laboratory, and their research has emphasized learning and the development of theories of behavior.

Ethology, on the other hand, was born in Europe, and the training of ethologists was traditionally in zoology. Generally, the research animals of the ethologists have been insects, fish, and birds, although, recently, more emphasis is being placed on mammals, especially primates. The ethologist has been concerned with the study of the evolution of behavior and, as a part of this concern, he has devoted much of his effort to the study of instinct. He usually acquires his information through field work—studying animals in their natural habitat—and through observation of the animals usually surrounding him wherever he lives.

Such categorical descriptions are, fortunately, not as valid as they once were. In spite of the sporadic and often vitriolic disagreements of the recent past, each group is now more accepting of the other's contributions. This acceptance should lead to a more complete and unified theory of animal behavior.

One important unifying matrix will probably prove to be neurophysiology, the study of how nervous systems function. In fact, as we shall see, the physiologists are already helping to thread the two great camps together. Even now, it is just not possible to seriously study ethology without dealing with neural mechanisms. Actually, the ethologists and physiologists haven't been as far apart as some might have believed. In some cases, the behavioral mechanism is explained by physiology, and ethologists have been concerned with mechanisms all along. In other cases, ethologists set the physiologists on the right track to begin with. Other bridges are being built as well. As the ethologist begins to be more involved with the physiology of his business, he may uneasily look around and find himself squarely in the turf of the psychologist. That uneasiness is disappearing, however, as psychologists are beginning to ask questions about the ecology and evolution of animal behavior. At the same time, the ecologists and evolutionists who study behavior are tightening their methodology and concerning themselves with hard statistical analysis. In the end, we can expect a more rigorous and meaningful ethology.

## ECOLOGY AND BEHAVIOR

Of course, it would be shortsighted to place all our hopes on the possible reconciliation of two fully developed disciplines, particularly since there are other important new advances occurring in the study of animal behavior, such as those relating to ecology. Ecology is sometimes defined as the study of the interrelationships of organisms to their environment, but to the biologist, the term has almost staggering connotations. The definition is simple, the concept is fascinating, and the practical problems of developing theory and gathering data are enormous. The key word is *interrelationship,* a term that implies reciprocal influence. The animal can influence the environment and the environment can influence the animal, which may cause the animal to change its influence on the environment, ad nauseum.

And, by the way, what comprises an environment? The term may encompass more than you imagined. It is not simply a habitat or substrate that lends itself to definition in physical terms. It also includes other organisms—organisms that can change in space and time and can adapt on an individual level to meet their own specific needs.

An animal's environment may include not only its parents and siblings but also other members of its own species with which it must compete and from among which it must seek its mate. It also includes its predators or its prey, as well as species of other adaptive types with which it might interact to one degree or another. Whereas for many animals the "physical" part of the environment may change slowly, or at least regularly, the biotic components may change rather quickly and unpredictably as sensitive cytoplasm responds adaptively to the changes in pressures placed upon it.

It also must be realized that interaction in a living system is not simply reciprocal or between two components; nor can it be described as the sum of such reciprocal interactions. Instead, in any such system an individual may be interacting on various levels with a number of kinds of organisms in an exceedingly complex feedback operation. As a hypothetical example, species A might act to increase the numbers of species B. Species B, in turn, might then reduce the numbers of species C. But the presence of species C is necessary for A to act effectively on B. And even this scheme neglects the operation on A, B, and C of yet another species and the effects of physical influences such as weather on the numbers of any of these (Figure 1.1).

As we begin to understand more about the basic principles of environmental systems, we can no longer be content to restrict the study of animal behavior to learning mechanisms or to the inheritance of behavior and its evolutionary descent. We must try to integrate our findings so that they might reflect more closely what is happening in the real world. We must form our interpretations in the light of new ecological information on the habitat and the niche, population dynamics, social structure, and the overall effects of altering any one of these. The real problem is how to integrate our expanding body of knowledge in new ways to produce more accurate descriptions of nature.

While increasingly detailed data are pouring in, the synthesizers among us are faced with ever greater problems in trying to make some sense of it. At the same time, the problem is, paradoxically, compounded by critical gaps in our levels of information. So, it seems that as long as we continue to investigate our world, we have basically two (not exclusive) options: First, we might continue to concentrate on gathering data whose real import we may not fully understand; or, second, we can try to put our information into some kind of ecological perspective. Admittedly, it is too early to expect much real success from the latter, but the effort at least focuses general attention on the need.

## My Aim

My aim in this book is to introduce the ethological concept of animal behavior, with attention to certain findings of comparative psychology and physiology. More importantly, once certain ideas are established, I will attempt to place them in some sort of ecological and evolutionary context. In other words, I want to try to answer a few simple questions about the overall advantage of certain behavioral patterns and how they came to be.

I will begin in this chapter by describing certain techniques and theories relevant to the observation of animal behavior. We will then consider the importance of the "ethogram" as a behavioral description of an animal. In the following three chapters, we will distinguish the concepts "ultimate" and "proximate" causation and, in so doing, we will establish some basic principles of evolution and physiology. We will then establish the ethological concepts of instinct and learning as bases for our following discussions. Additional principles of ecological and evolutionary theory will increasingly be introduced in later chapters as we explore specific areas of behavior. Hopefully, the result of such a format will be, first, to acquaint you with certain principles of behavior, and then to show you how these might operate in nature. I have mentioned the awesome nature of ecological studies, so our goals must remain modest. We cannot hope to solve great problems or even define with any confidence the important questions. We can only indicate how such a start might be made.

A note should be added here regarding my choice of certain phrases. Occasionally (although I hope it is kept to a minimum), I will refer to an animal's performance of an act *to* achieve something or other. Such a statement is not intended to imply intention, unless specifically stated; it is simply a shorthand way of saying that when a particular pattern is completed, a specific result usually follows. The account thus becomes more readable. It is a ploy often used when one biologist is addressing another or when one is certain the real meaning is understood.

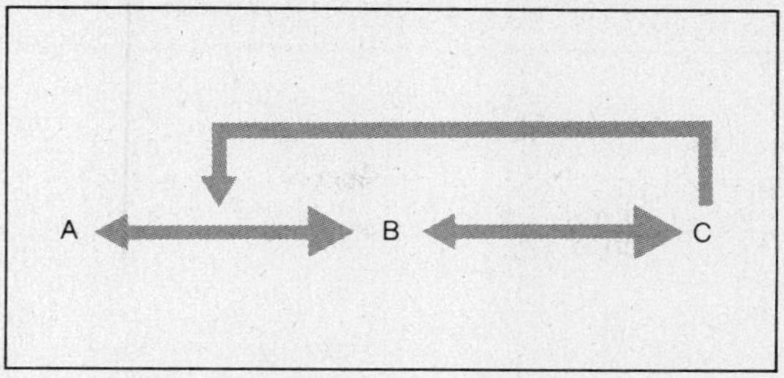


Figure 1.1

## THE DESCRIPTION OF BEHAVIOR

How does one classify a behavior? The question is deceptively simple. Obviously, one must first be able to know when a certain type of behavior has occurred. In some cases this is relatively simple. For example, we can be sure when we have witnessed certain displays that are used in communication. It is easy to count the "push-ups" of a lizard or the flashes of a firefly. But in other cases, the classification may be more difficult. Is the heron "bowing" or simply "looking down"? At what point has a dog changed from a belligerent to a submissive posture? Is a woodpecker excavating a hole for a nest or is it simply looking for insects?

What about the context of a behavior? A crouching dog may be crouching in fear or in preparation for a spring. We can usually ignore breathing patterns if we are describing nest building in terns, but we might wish to consider them if we are interested in how air-breathing Siamese fighting fish build their bubble nests.

These problems are less troublesome if we put the behavioral patterns into their proper adaptive context ("adaptive" refers to those patterns that produce a net benefit to the individual) and if we carefully design a system of description. The adaptive context, for example, can be described in terms of time. We might ask what immediately preceded a certain action, in an effort to determine what might have produced the pattern ("causal analysis"). In such a case, we would be interested in the physical and biotic (living) components of the animal's immediate environment and how they changed just prior to the behavioral action. If a male bird in the area of its nest shows aggressive behavior at the appearance of another male, one might deduce that the intruder *caused* the show of aggression. (We might then experiment to see what aspect of the intruder's appearance elicited the response.)

We might also approach such problems in terms of the individual's history. What, in the animal's experience, could have brought about the observed pattern? Are animals in parks more approachable because of some special character of parks or because the individual animals there are protected?

Then, we might ask about the evolutionary history of the animal. In other words, what has been the experience of the *species?* Many island species are so tame that they can be picked up by hand and show no sign of fear. Is this because no individuals among them have experienced harm by humans? It is more likely that they are fearless because, for generations, they have had nothing to fear and that there is now nothing in their genetic makeup that could trigger aversive behavior. In other words, there has been no "selection" for fear in these island species.

On the other hand, continental species show great fear of man. Their species have been exposed to humans and other large predators for untold generations. Whenever certain changes appeared in the genetic makeup of offspring that might cause them to tend toward tameness, these individuals were very likely caught by predators and they, therefore, left no offspring of their own to perpetuate the "tame" genes. So, in continental species, there is selection *for* "wildness," and "tameness" is selected *against.* (We may also say tame individuals that are caught are selected *out.*) More will be said about all this later.

### Description by Operation and Consequence

There are, basically, two ways to describe behavior: by operation and by consequence. *Description by operation* simply refers to what the animal does. A quick straightening of the leg may be called a "knee jerk." The description doesn't tell us whether the tendon of the knee was tapped first, and we don't know what neural events preceded the action. The description only includes the speed and the arc of the swing of the lower leg.

In description by operation, one simply places all behavioral patterns of a similar spatiotemporal (space-time) organization into a single group. Thus, sleeping patterns in animals may be grouped together, although there may be wide differences in the various animal groups. One may also utilize this kind of classification to place patterns of greater similarity into subgroups. For example, sleeping postures in arctic dogs are relatively uniform, as are such postures in many tropical birds. One reason for the uniformity is that the posture helps to regulate the body temperature.

In some cases, this kind of classification is rather simple, such as in describing knee jerks. In other cases, when a number of muscles are involved in a highly coordinated sequence, a range of description within a single class of behavior becomes possible and thus its classification becomes more difficult. In complex patterns at low intensity, certain components that are not so easily elicited may also be left out entirely, and, thus, the overall appearance of a pattern may be drastically changed.

The second method of describing behavior is called *description by consequence* (Hinde 1970). One might say that a bird is not simply performing "bending down and grasping motions," but is "picking up nest material." Such descriptions have their limits as well as their advantages. "Pressing down a lever" is a description by consequence, but it says nothing about whether the lever was pressed by a nose or a foot. "Approach behavior" does not specify whether the animal is tiptoeing or somersaulting.

In addition to its other limitations, description by consequence can be misleading. Is a mouse entering the hole or escaping from the cat? Also, such descriptions lend themselves to overinterpretation. Such problems usually arise when the consequence is assumed to be understood, or even desired, by the animal. "The rat is pushing the lever in order to obtain food" could be said of a trained rat, but not of a completely naive one in an experimental situation. At what point in the animal's training could the last five words of the description be added?

Often, describing by consequence may be clarified by modifying it with an operational description. "Approaching by hopping" is an example. In some cases, behavioral patterns grouped by consequence are extremely varied, such as is demonstrated in "hunting behavior." In other cases, they are exceedingly restricted, such as is demonstrated in "retrieval behavior" in geese (Chapter 8), where a goose always attempts to return eggs to the nest by reaching out from the nest and rolling the egg back with her chin. The goose's "retrieval behavior" is always performed with this peculiar chin movement in spite of the fact that it's obviously a tricky maneuver.

Description by consequence also has its advantages in that it reduces the room for error. For example, one can say with certainty that a dog did, in fact, bury a bone, or that a squirrel did open a nut, whereas the attempt to define the motions involved might lead to problems.

Another advantage of description by consequence is that it refers to environmental variables. A carnivore may pick up her young with the same motions she uses to crush the skulls of her prey. However, the adaptiveness (or advantages) of the two acts are quite distinct. The observer must distinguish the functions of the motion on the basis of the animal's relationship to its environment.

## THE ETHOGRAM

One of the key concepts in ethological studies is the establishment of the ethogram—the "behavioral profile" of a species. For example: How could escape behavior in rats be studied without knowing that rats tend to run along walls and other structures where their bodies are in contact with a vertical substrate? The interpretation of any experimental data is open to question without information regarding the *normal* behavior of the species. Supplying this information is the first task (if not the first love) of the ethologist. The need for this information gives him the opportunity to spend time with animals and to record their behavior without interfering with them as they go about their daily activities. It is not until their normal patterns and tendencies have been established that meaningful experiments can be devised that can yield insights into the causation and results of their behavior.

As an example of the establishment of an ethogram, consider Tinbergen's (1951, 1952c) description of the behavior of the three-spined stickleback, *Gasterosteus aculeatus* (Figure 1.2).

The stickleback is a common little fish, only a few centimeters long, found in the ditches and streams of Europe. When Tinbergen was a young lecturer at the University of Leyden, he was asked to organize an undergraduate course in animal behavior and he chose the stickleback as his research animal.

The choice was a fortunate one. He found that the fish was not only hardy and unafraid, but an excellent subject for studying innate behavior. Its displays were dramatic, its reproductive behavior was intriguing and unvarying, and it responded strongly and predictably to certain environmental cues.

Tinbergen found that the behavior of the little fish was the same whether it was in an aquarium or in its ditch. Sticklebacks begin to mate in early spring when the males, which have been in schools all winter, leave these groups and stake out individual territories from which they chase out any intruder, male or female.

Each male then builds his nest in his own territory. First he digs a shallow pit, carrying away mouthfuls of sand until the depression is about two inches square. Then he gathers up weeds, preferably thread algae, and places them in a heap over the pit. The weeds are then coated with a sticky material excreted by his kidneys. He then sets about shaping the weeds into a mound by repeatedly circling the heap and poking it

Figure 1.2 The sequence in courtship and mating in the three-spined stickleback, a small fish found in European ditches and streams.

with his snout. When the weeds have been fashioned into a mound, he bores a tunnel by wriggling through it. The tunnel is usually only a little shorter than he is. This, then, is the nest. After it is completed, the male changes color. He was drab, gray, and inconspicuous all winter long, but this is his time of breeding. Now his chin begins to faintly blush pink, and his back and eyes take on a greenish gloss. Then the pink becomes a bright red, and his back turns a bluish white. He is now resplendent in his nuptial colors. He is bold and ready to mate.

While the male has undergone his spring transformations, the female, too, has changed. Her colors remain drab, but her body is now shiny and swollen with up to one hundred eggs and she begins to roam the waters of her ditch. Whenever one of these wandering females enters a male's territory, he swims to her in a series of zigzags, first swimming away toward the nest and then turning and darting toward her, only to turn toward the nest again.

The female doesn't respond immediately. But finally she notices the displaying male and swims toward him. However, she moves strangely, in a curious headup position. The male swims quickly toward the nest and she follows. At the nest, the male makes a series of quick thrusts with his nose into the entrance. Then, he turns on his side with his raised dorsal fins pointing toward the female. At this, she quickly enters his "tunnel of love" and lies there, her body protruding from either end of the nest. The male then swims to the rear and prods the base of her tail with rhythmic thrusts, whereupon she lays her eggs and leaves. The male will stay with the eggs, fanning oxygen-laden water over them until they hatch. This, then, is the reproductive part of the stickleback ethogram. Such basic information was critical to the development of important experiments, as we shall see.

## ETHOLOGY AND HUMAN BEHAVIOR

It is fashionable, and perhaps safer, these days to develop the principles of ethology by discussing other animals, and then to end the book with a brief discussion of what might be called "human ethology." The trend has merit in that we know less about the ethology of our own species than we do about many other species. This is partly due to the difficulty of performing "experiments" on humans without harming them, violating their rights, or somehow offending them. Thus, such reticence is proper.

But our ignorance perhaps deters us more than it should. After all, our level of information varies between *all* the species we will consider. For example, we know much more about cichlid fish and perching birds than we do about loose-running city dogs or shrews. Yet we don't hesitate to discuss what we do know about dogs and shrews. Thus, hesitancy to discuss humans solely out of the fear that our data may not yet be complete is not entirely logical nor perhaps merited. In such cases, we should undoubtedly proceed with great care, phrasing our ideas in such a way as to discourage malicious speculation and stating our findings tentatively when all the data may not be in.

Another reason humans are usually considered separately and discussed in the most tenuous and delicate of terms, is that it's easy to offend people when they see themselves receiving the same treatment as other animals. The indignation can be based on any number of factors, from religion and political dogma to objections about methodology and premises. Some of the objections are unquestionably valid. Ethologists do often tend to overlook the incredible flexibility and compensatory abilities of humans. (But so do many psychologists, as has been shown by their inability to relate childhood experience with cross-cultural aggressiveness.) In our discussions, we will grant the behavioral flexibility of humans and try to keep it in mind as we consider our own species. It hinders the making of hard-and-fast rules, but then, the more one delves into ethology, the less one even hopes for any such "rules," no matter what the animal.

Perhaps considering the human condition along with that of other animals is irreverent, but it seems that ethology is fast approaching the point where a "principles" approach is merited. Probably at this time, the approach is premature, but the ground has been broken (or perhaps rent asunder) by writers such as Konrad Lorenz, Robert Ardrey, and Desmond Morris. Those who followed them have apparently been so chagrined by their boldness that they have been almost apologetic in their own heavily documented and carefully phrased writings (for example, Blurton-Jones 1972), and yet others have suggested that the study of human ethology is taboo, period.

I hope to take the eternally espoused and well-worn middle road. We will consider the human condition where the trait is discussed for other animals, instead of at the end of the book, as if humans were somehow not quite as subject to ecological and evolutionary influences as other animals. (That case could probably only be made for man's very recent history.) We will try to consider such matters circumspectly, keeping in mind that we are often dealing in speculation. As in any science, certain parts of the material will be "softer" than other parts and, in this book, some of the less well-documented material will deal with humans. Still, the speculation itself is often fascinating, and, probably, in time we may know more about its accuracy. It certainly provides food for thought, and at a time when we are holding conferences to see how many times over we shall hold the power to annihilate ourselves, perhaps introspection at any level is merited.

As we progress, just keep in mind that not all the data are in and that even now arguments are raging about much of the material we will consider. The choice of what falls into this category reflects my own biases as well as the prejudices of the "experts" I have chosen to cite. If arguments arise over these matters, so much the better.

# 2

# EVOLUTION AND ULTIMATE CAUSATION

Behavior evolves. A simple statement, but one that has provoked endless arguments over its veracity by its doubters and its mechanisms by its believers. Let's have a closer look.

Not long ago, the evolutionist Ernst Mayr wrote, "There are vast areas of modern biology, for instance . . . the study of behavior, in which the application of evolutionary principles is still in the most elemental stage" (1963). In the few years since that statement, there have been notable advances in theory and technique, but these have not been consistent in their assumptions and so their impact has been diminished. What we have is a proliferation of schools of thought, each scrambling for a toehold in the conceptual world (which actually means in the minds of the "synthesizers," those who are able to take the available data and fit them into the big picture).

How have the assumptions varied? Some groups held heartily to the idea of the inflexibility of an instinctive pattern and found their greatest challenge in deciding which patterns to consider instinctive. Others believed that almost all behavioral patterns are learned. And still others argued that any innate pattern can be greatly affected by environmental input; that only the propensity for performing a pattern is inherited; or that instinct and learning, while essentially distinctive, usually operate together to produce adaptive behavior. And there are almost as many points along the spectrum as there are behavioral scientists. The result is sometimes a revival of the old nature-nurture controversy—is a behavior "innate" or "learned"?—although wrapped in new syntactical garb. The question, of course, centers around whether we can treat behavior as we do morphology. Is any behavioral pattern inherited, variable only within limits and subject to natural selection? If heritable, is it "species-specific" (characteristic of a single species)? Can it be used to describe species and relationships between species? If these questions are valid, how can the problem best be approached?

We know, simply because some patterns persist in populations, that behavior is transmitted through the generations. The question is, *how?* It seems, on the surface, that such transmission could occur in two major ways: through cultural means or through genetics.

Our intuition tells us that culture, or learning, would be more important in some kinds of animals than others. For example, it could be an important means of perpetuating a behavioral pattern in mammalian carnivores or in primates, animals that are relatively intelligent and that have extended parental care. Each generation could learn from older generations, and thus behavioral patterns could be passed along by tradition. But learning can also be important in more "stereotyped" animals, such as birds. For example, a group of English titmice learned to open milk bottles left on porches, and each generation apparently taught the trick to the next (Hinde and Fisher 1951). However, such learning probably requires intelligence far beyond that of most species. Therefore, it has been suggested that behavioral tendencies survive in most populations by the operation of natural selection upon inherited variations (Manning 1971). It has also been suggested that relegating behavioral patterns as "learned" or "innate" is invalid, as we shall see.

## PROBLEMS IN STUDYING THE INHERITANCE OF BEHAVIOR

There are formidable problems in studying the inheritance of behavior. Some of these problems are similar to those confronting the comparative anatomist in the study of the evolution of physical traits. For example, some animals, such as birds, have changed rapidly while others, such as the horseshoe crab and the opossum, are virtually living fossils. Obviously, the intensity of natural selection on birds and opossums has been different. Today's opossum probably behaves rather like yesterday's, but we can only guess at the behavior of ancient birds. Also, there are specific traits that are more conservative than others. For example, the bones in the flipper of a porpoise are startlingly similar to the skeletal forelimb of a man or bat. Wristbones, then, are conservative. On the other hand, the bills of many species of birds seem to be evolutionarily labile, changing quickly in relatively few generations. As we shall see, behavioral characters also vary in their lability—some change within a few generations, while others are monumentally conservative. (And the adverb is deliberate.)

We must recognize that studying the evolution of behavior and the evolution of morphology are not entirely the same. The student of the evolution of behavior confronts special problems. For example, behavior itself cannot be fossilized, and the fossil records of the *results* of behavioral patterns are scarce. Such clues are usually limited to footprints or tracks, burrows, wounds, and tooth marks. As special momentos, there are animals fossilized in the act of copulation. Even rarer are fossil records of a behavioral pattern that indicate the development of the pattern. But they exist. For example, fossil burrows of wormlike animals of the Cretaceous and Tertiary periods called *Zoophycos* show a change in movement and orientation that probably led to increased foraging efficiency. Basically, they changed from tunneling randomly through the mud to excavating one large foot-laden area (Seilacher 1967).

### Comparative Behavior Studies and Evolution

As has been mentioned (also see the discussion of the empid flies in Chapter 8), one means of studying the evolution of behavior is through a comparison of living forms. For example, certain extant species of hymenopteran insects (bees, wasps, ants) show different levels of social organization that can be arranged from simple to complex. The arrangement is believed to indicate the probable evolutionary route of their social development. In these species, it seems that the steps were as follows: (1) the female remained with the eggs; (2) the female built a communal shelter and remained with the larvae; (3) the female began to survive successive generations of her offspring; (4) female workers became sterile and dependent on the group; and (5) distinct castes developed (Lindauer 1965).

As another example, the Meliponi bees show important interspecific differences in their "dance." The differences are believed to indicate the sequence in the development of these peculiar movements (Johnsgard 1967). (See Chapter 13 for details of the dance.)

One of the more fascinating studies, comparing living species in an effort to learn about the evolution of behavior, was done by Tom Gilliard (1963) on bowerbirds. He looked at all the birds with arena, or lek, behavior. As we shall see later, these are birds that court their mates on special places where the males gather at mating time and defend very small areas, which have no value except as mating places. The females then choose from among the gaudy males as each displays on its own small turf.

Gilliard assumed that arena behavior in all the various species is a relatively advanced characteristic with a common sequence of development. He said that once the stage is set by the proper environmental conditions, there begins rapid evolution to the highly distinct and specialized behavior seen in all the far-flung arena species. The "usual" system among birds, in which males and females share work loads, is considered to be the most primitive condition—and very conservative and resistant to change. In arena birds, however, the beautiful males no longer perform domestic duties. Such chores are left to the drab females.

There is wide variation in the twenty-four species of the brightly-colored birds of paradise that practice arena behavior. Some of these highly conspicuous males clear arenas by picking up debris in their bills or by violently fanning with their wings. This is probably the first step toward building bowers. From here, it is a rather small step to decorating the cleared arena with overturned leaves, bits of paper, plastic, or any other bauble. These odds and ends seem to attract the females of the species. Gilliard proposed that in bowerbirds—species in which the males are generally less colorful than the birds of paradise—the gaudy qualities that attract females are transferred from the individual to the bower (Figure 2.1). Since the lek is decorated with baubles of all sorts, the male is released from personal garishness and thus can become more cryptic, or hard to see. Thus, he is able to announce himself to females without advertising himself to predators.

It is interesting that in some species of bowerbirds, when the male approaches the female, he twists his head oddly as if to display a crest, which is no longer there—so, perhaps, there once *was* a crest there, which has disappeared, leaving the behavioral pattern. It is also theorized that such movements are simply a new use for his old nest-building motions and that the odd movement, itself, is sexually attractive.

The whole system involving conspicuous males and eventually arena building probably resulted from a shift in the work load to the female. Theoretically, if the females were somehow able to rear the young alone, the freed males could spend their energies attracting more females. Females could be expected to be able to rear the young alone, when the male's protection was not mandatory and when she could be expected to find enough food for them by herself. This last chore is relieved if the young are able to quickly feed themselves, as do, for example, pheasants or chickens.

In summary, arena behavior is the courtship behavior of emancipated males and is probably derived from nest-building behavior, and the cleared arenas are the first step toward building bowers. Generally, the duller the appearance of a bowerbird, the more elaborate its bower. These dull-colored bowerbirds that build the most conspicuous, complex, and ornate bowers are probably at the leading edge of avian evolution (Gilliard 1963).

Figure 2.1 The satin bowerbird, *Ptilonorhynchus violaceus,* is an avenue-builder. Avenues are decorated passageways through the grass into which the female must be enticed before mating can occur.

## Primitive Versus Advanced Behavior

When considering the evolutionary history of animals, it is tempting to oversimplify, calling some animals primitive and others advanced. However, such classifying is naive in that all animals show both primitive and advanced characters (Lorenz 1950). Thus, any animal may have some advanced characters, while possessing some primitive ones. For example, the mouth and teeth of "advanced" humans are much more primitive than those of a "primitive" monotreme, the platypus (Figure 2.2). But what sort of characters are regarded as "primitive" and "advanced"? Advanced characters are the more specialized ones. Human teeth are adapted to cutting, tearing, and chewing a variety of foods (thus, as a group, they are primitive and unspecialized), but the "bill" of the "primitive" platypus is a highly specialized organ for detecting and capturing food from murky stream bottoms.

Such inequities in characters result because evolution does not proceed in a simple linear fashion with all characters of any species changing at a constant rate, but rather it progresses as a branching tree, some branches (characters) evolving faster than others. "Advanced" species are usually considered to be those showing rapid evolution in many characters,

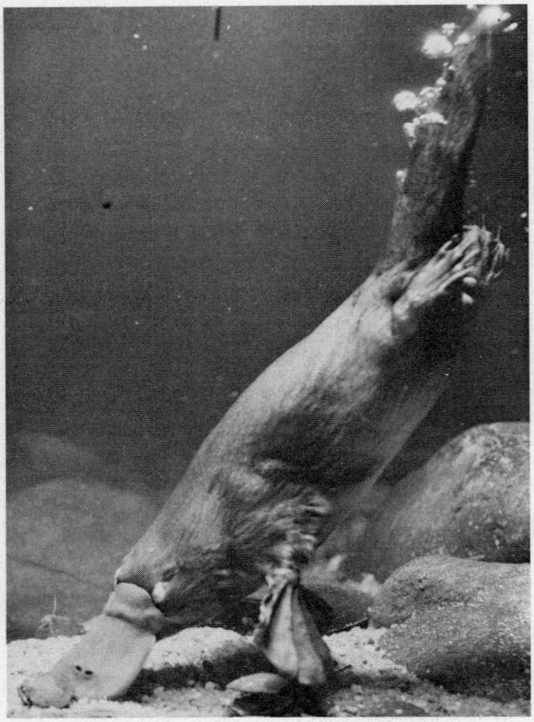

Figure 2.2 The platypus is "primitive" in many respects, but it has a remarkably advanced jaw and a leathery and extremely sensitive bill. Advanced characters in primitive animals indicate that specific characters may evolve at different rates.

whereas "primitive" lines are those evolving rapidly in relatively few characters. I will try to avoid using the terms *higher* and *lower*, an unfortunate and vestigial distinction implying that more humanlike creatures are higher on some natural scale. It ignores the fact that any animal may be well adapted to its own kind of environment.

There is also an unmerited tendency to assume that the nervous systems of less intelligent species correspond to earlier stages in the development of the human brain in keeping with an old, and discredited, notion of recapitulation (Figure 2.3).

Evolutionists are forever looking for trends—for unifying rules—but they don't come easily. And when they are formulated, they are often the targets of critical sharpshooters. The problem is, even when broad evolutionary progressions are considered, there are likely to be disturbing exceptions or disjunctions as one proceeds along the developmental line. For example, there seems to be a trend in some evolutionary lines toward larger brains. This tendency toward cerebration, or braininess, has been noted in vertebrates and especially in mammals. The rule is, the more advanced forms are generally more intelligent. Even among primates, the most intelligent of the mammals, the primitive insectivores apparently rely less on individual intelligence than do the advanced chimpanzees (Figure 2.4). But in birds, the progression toward cerebration is evident in only a few families, such as the crows and parrots. On the other hand, the octopus, which is related to the uninspiring clam, may be more intelligent than some vertebrates. No insect group shows the tendency toward the development of braininess, though insects show many advanced characters. This is not to say that all insect species are equally intelligent. Actually, there are important distinctions between the learning ability of insect species, even those at about the same stage of phylogenetic (evolutionary) development. For example, some hymenopteran insects, such as wasps, show a remarkable facility for certain kinds of learning, whereas the other "advanced" order of insects, the diptera (two-winged flies), show very little learning ability.

There is an understandable tendency in egocentric man to consider the characters he possesses as advanced and even to assume that all animals are ever progressing toward the attainment of humanlike characters. There are pronounced vestiges of this tendency among behavioral scientists who attempt to study our own behavior through the analysis of other "lower" animals, which are often, almost subconsciously, regarded as cheaper, shorter-lived, simpler, and expendable "men." This is a peculiar point of view, especially since there is no reason to assume that man is more "successful" than many other animals. After all, are our social systems more efficient than those of the hymenopteran insects? Perhaps bees do not agonize in guilt about their feminist societies, nor do they hold endless dialogues on the inequities within their system, but there are no bee slums.

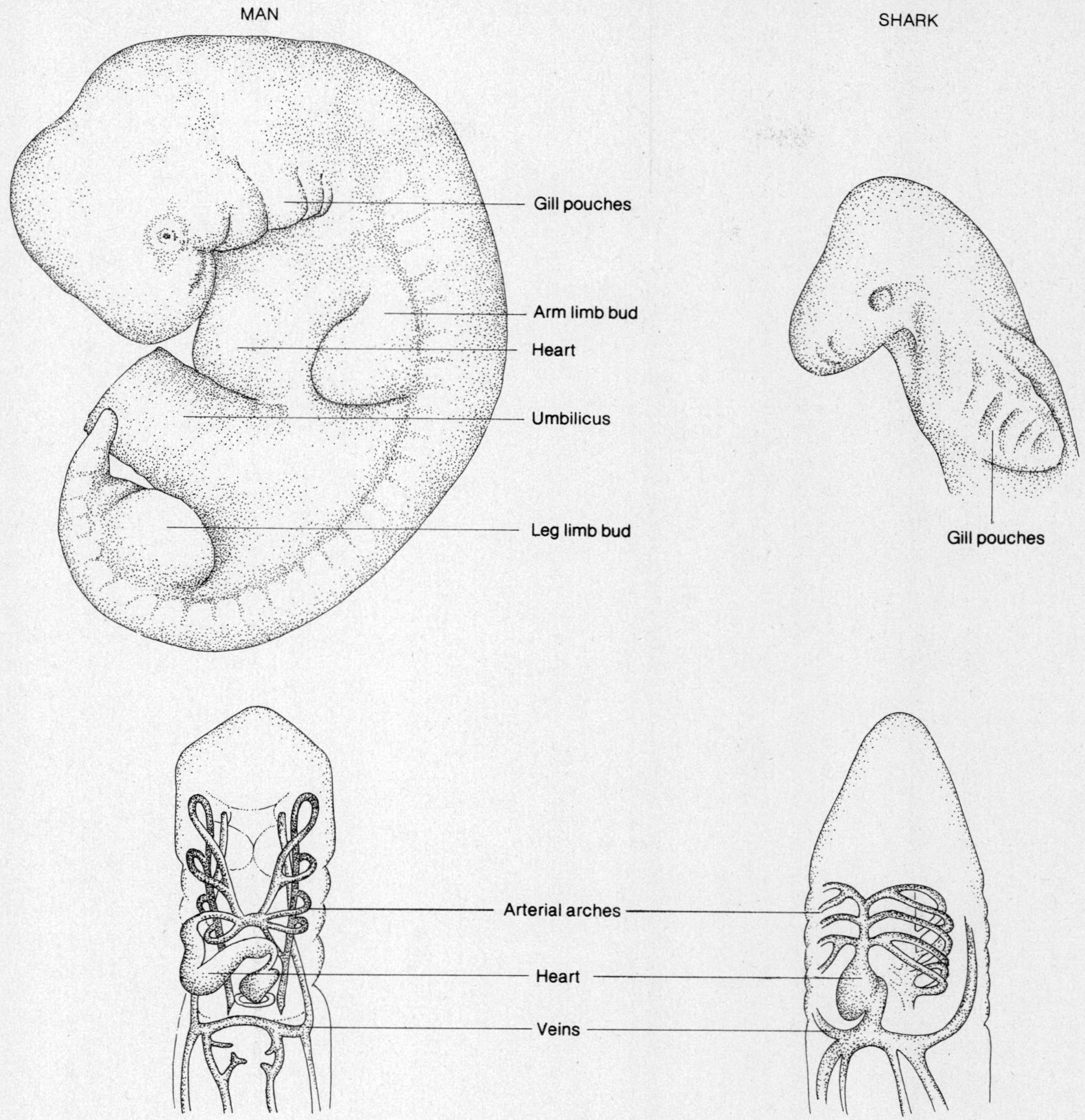


Figure 2.3 The theory of recapitulation. Developing human embryos go through stages in which they resemble other animals that are believed to be similar to the human ancestors. Note the similarities of gill pouches and blood vessels in embryonic sharks and humans.

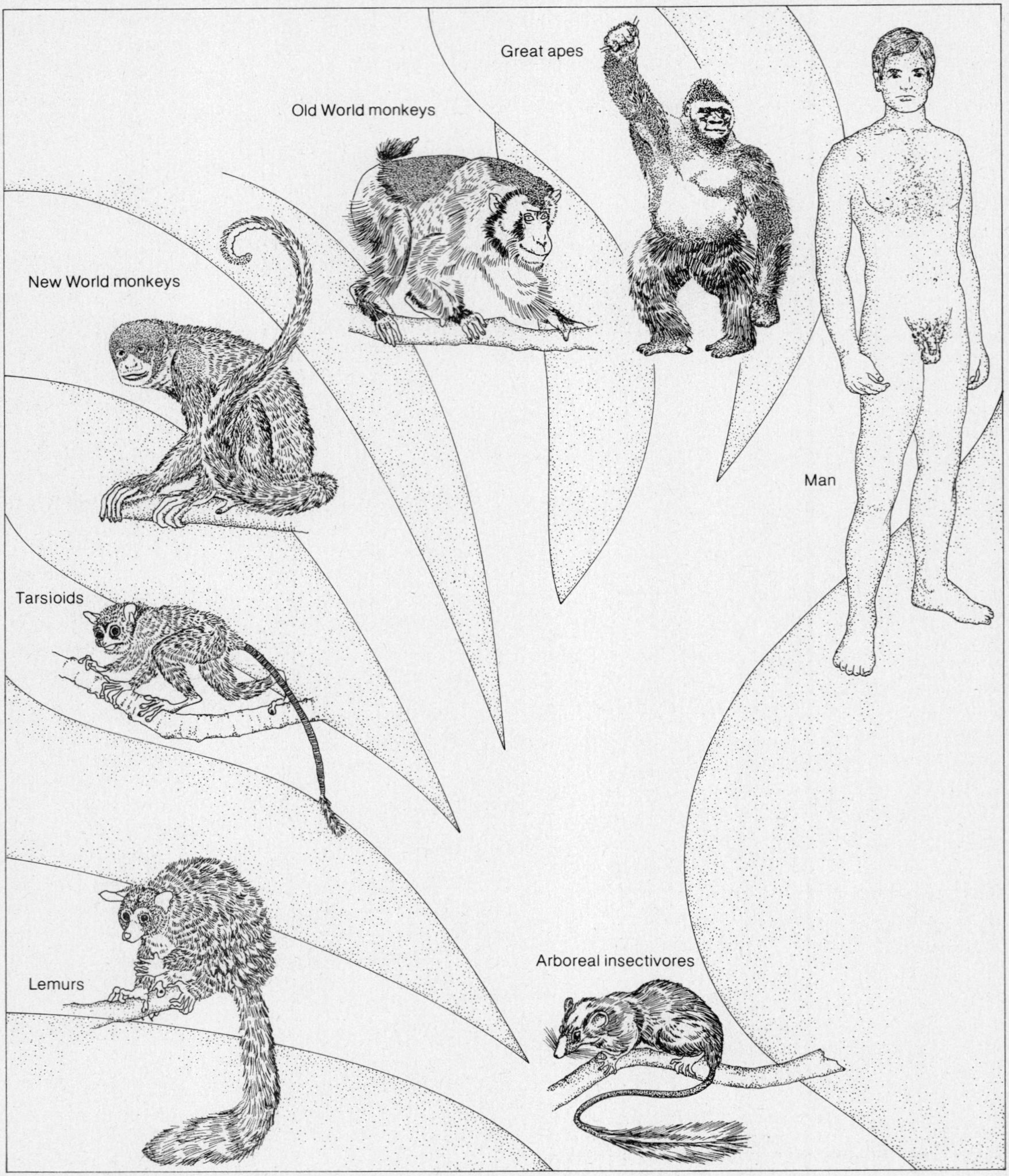

Great apes
Old World monkeys
New World monkeys
Man
Tarsioids
Arboreal insectivores
Lemurs

Figure 2.4 A simplified family tree of the primates. There are many inherent assumptions in any such evolutionary scheme, but the animals closest together can be expected to behave somewhat similarly.

## Extrapolation

One result of applying ecological and evolutionary theory to the study of behavior should be an increased reticence to extrapolate from one animal species to another, or from caged animals to wild ones. In the first instance, the bases for natural selection in the social organizations of ants and humans cannot be the same. Not only did vertebrates and insects separate early in their evolution (Figure 2.5), but their potentials and recent natural histories are vastly different. Likewise, birds and mammals diverged early from the reptilian stock (Figure 2.6). How valuable are experiments with artificially bred rats in explaining human learning? Even careful laboratory experiments on primates may not reflect the characters of *those* species, let alone man. Jane Goodall, after years of close association with wild chimpanzees, tells us that the behavior of caged laboratory chimpanzees bears little resemblance to that of their wild brothers.

Extrapolation, however, is not completely useless. The behavior of one organism could furnish important clues regarding why another species behaves as it does. In general, however, we are not yet in a position to make very many blanket statements. Instead we must consider each species separately. We must try to account for those influences that have a genetic basis, but we must consider environmental variables as well. These latter include those factors presently operating on any individual or population (those influences that we can see and measure). But we must also consider the historical influences of the environment on animals (those influences we can determine only through circumstantial evidence). As more evidence accumulates, we will be in a better position to make our extrapolations and our general rules.

## THE DEVELOPMENT OF THE MAMMALIAN BRAIN

Let's very briefly review a few ideas regarding the evolutionary development of mammals according to Romer (1970). The serious evolutionist might well argue with the generalized treatment we give the topic; however, not everyone is familiar with the precepts of biology, and so certain fundamental information on evolution might be useful. We will specifically consider the sorts of evolutionary influences that might lead toward the development of the higher learning centers.

First, we might note that the special group of reptiles that gave rise to mammals diverged from the other reptiles before the end of the Carboniferous period, about three hundred million years ago. The mammalian ancestors were the synapsids—a group that included the therapsids (Figure 2.7). They were somewhat doglike reptiles, but you would probably clear out a bar if you walked in with one on a leash. The therapsid skull suggests the animal had a well-developed sense of smell, and the skeleton is that of a swift-running animal. It probably ran down its prey like a dog, rather than stalking it, catlike. The brain was small, but it showed the beginnings of a mammallike cortex ("gray matter"—the part responsible for "higher" brain functions).

By the end of the Triassic period (about two hundred million years ago) the mighty dinosaurs appeared just as most of the therapsids died out (Table 2.1). The survivors, however, included a reptile about the size of a collie, which continued to coexist with the dinosaurs. This relatively small reptile was somehow able to survive in its dangerous world and ultimately it gave rise to the mammals. The period of coexistence with the dinosaurs—many of which were large, fleet-footed flesh eaters—very likely placed a high premium on intelligence and alertness in these premammals and their descendants. They were, after all, a very insignificant part of a world dominated by the Ruling Reptiles. About this time, probably due to intense pressure from predators, the early mammals began to develop larger and more complex brains, as only the more clever individuals of each generation escaped the dinosaurs long enough to reproduce and leave their "smart" genes.

As the brain changed, so did other systems in the early mammals. For example, they saw the advent of homoiothermy (warm-bloodedness—the condition in which body temperature is maintained within rather narrow limits). Homoiothermy meant the mammal could function at a certain level in a wide range of

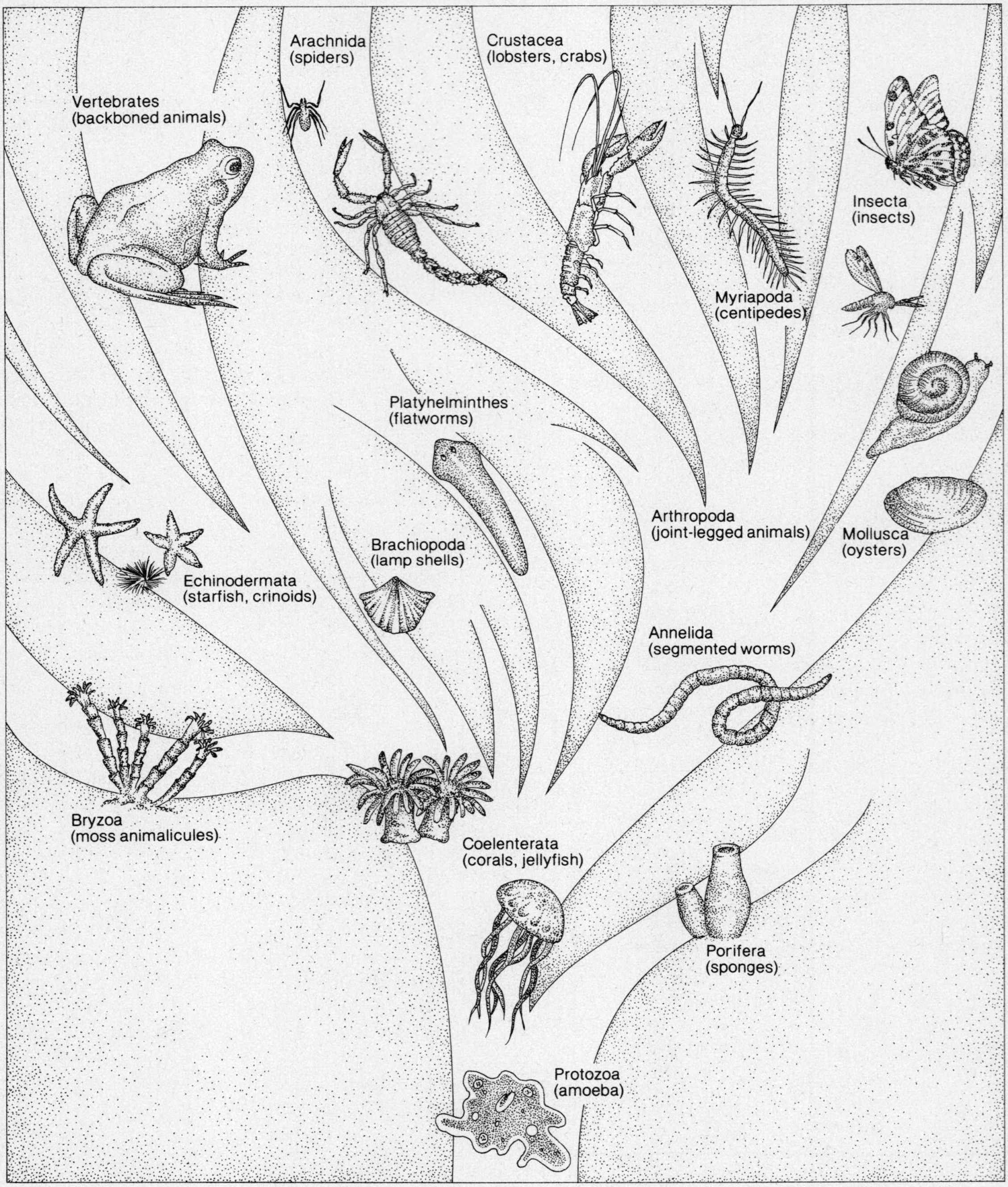

Arachnida
(spiders)
Crustacea
(lobsters, crabs)
Vertebrates
(backboned animals)
Insecta
(insects)
Myriapoda
(centipedes)
Platyhelminthes
(flatworms)
Arthropoda
(joint-legged animals)
Mollusca
(oysters)
Brachiopoda
(lamp shells)
Echinodermata
(starfish, crinoids)
Annelida
(segmented worms)
Bryzoa
(moss animalicules)
Coelenterata
(corals, jellyfish)
Porifera
(sponges)
Protozoa
(amoeba)

Figure 2.5 (Opposite page.) The family tree of animals. Any such tree is primarily useful in indicating roughly when a branch is believed to have arisen and in placing various animals in the proper lineage.

Figure 2.6 (Below.) A simplified family tree showing the vertebrate classes. Notice that birds and mammals are not closely related, but that they both apparently have arisen from the reptiles.

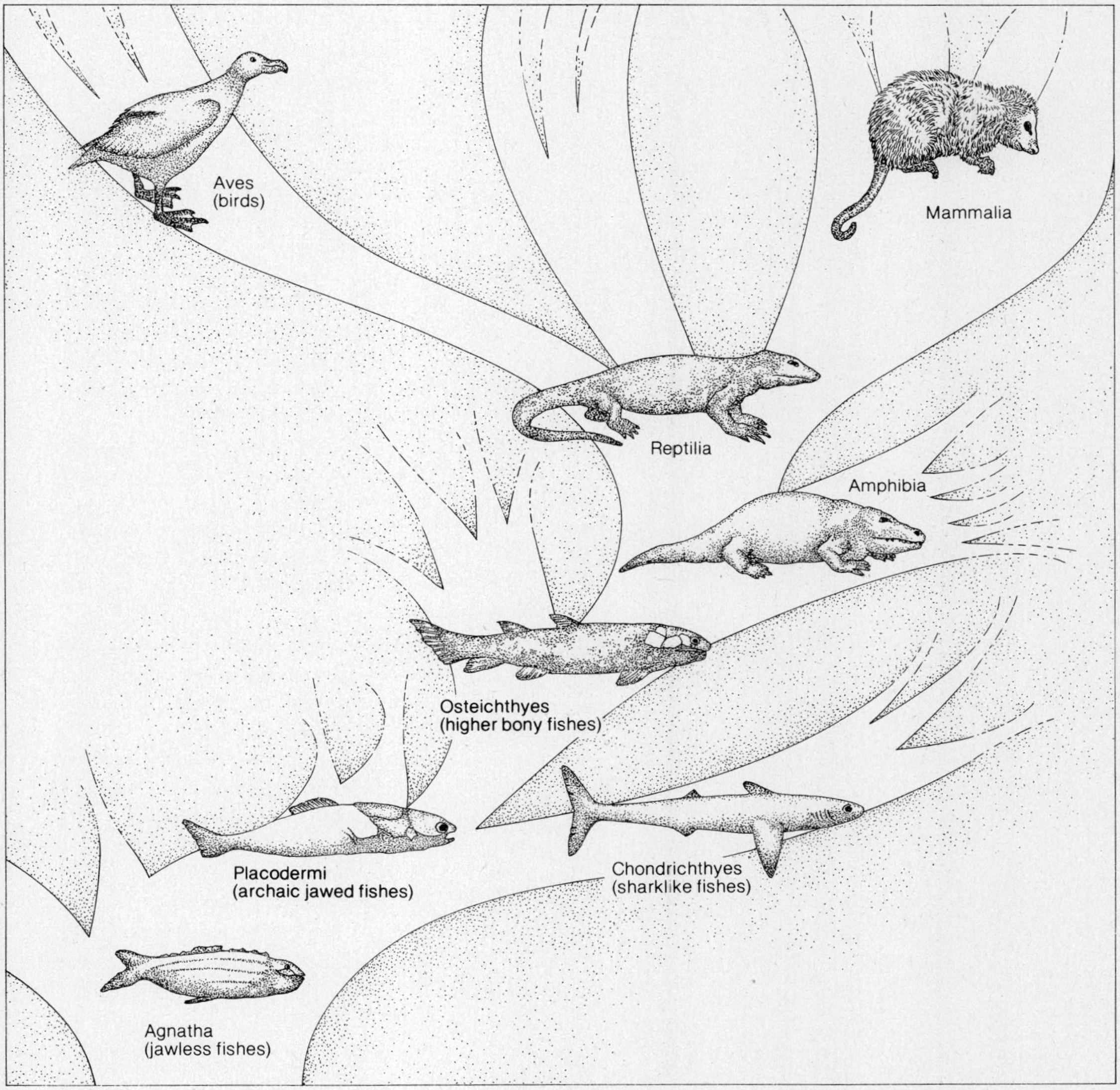

TABLE 2.1 Geologic timetable.

| ERA AND DURATION | PERIOD | EPOCH | YEARS BEFORE PRESENT (IN MILLIONS) | PRINCIPAL POINTS |
|---|---|---|---|---|
| *Cenozoic 75 Million Years* | *Quaternary* | *Recent* | *(11,000)* | *Age of man; end of last ice age; warmer climate.* |
| | | *Pleistocene* | *1* | *First human societies; large scale extinctions of plant and animal species; repeated glaciation.* |
| | *Tertiary* | *Pliocene* | *11* | *Appearance of man; volcanic activity; decline of forests; grasslands spreading.* |
| | | *Miocene* | *25* | *Appearance of anthropoid apes; rapid evolution of mammals. Formation of Sierra Mountains.* |
| | | *Oligocene* | *36* | *Appearance of most modern genera of mammals and monocotyledons; warmer climate.* |
| | | *Eocene* | *54* | *Appearance of hoofed mammals and carnivores; heavy erosion of mountains.* |
| | | *Paleocene* | *65* | *First placental mammals.* |
| *Mesozoic 165 Million Years* | *Cretaceous* | | *135* | *Appearance of monocots; oak and maple forests; first modern mammals; beginning of extinction of dinosaurs. Formation of Andes, Alps, Himalayas, and Rocky Mountains.* |
| | *Jurassic* | | *181* | *Appearance of birds and mammals; rapid evolution of dinosaurs; first flowering plants; shallow seas over much of Europe and North America.* |
| | *Triassic* | | *220* | *Appearance of dinosaurs; gymnosperms dominant; extinction of seed ferns; continents rising to reveal deserts.* |
| *Paleozoic 360 Million Years* | *Permian* | | *280* | *Widespread extinction of animals and plants; cooler, drier climates; widespread glaciation; mountains rising; atmospheric carbon dioxide and oxygen reduced.* |
| | *Pennsylvanian* | | *310* | *Appearance of reptiles; amphibians dominant; insects common. Gymnosperms appear; vast forests; great life abundant. Climates mild; lowlying land; extensive swamps; formation of enormous coal deposits.* |
| | *Mississippian* | | *355* | *Many sharks and amphibians; large scale trees and seed ferns; climate warm and humid.* |
| | *Devonian* | | *405* | *Appearance of seed plants; ascendance of bony fishes; first amphibians; small seas; higher, drier lands; glaciations.* |

TABLE 2.1 Geologic timetable (cont.).

| ERA AND DURATION | PERIOD | EPOCH | YEARS BEFORE PRESENT (IN MILLIONS) | PRINCIPAL POINTS |
|---|---|---|---|---|
| *Paleozoic 360 Million Years* | *Silurian* | | *425* | *Atmospheric oxygen reaches second critical level. Explosive evolution of many forms of life over the land; first land plants and animals. Great continental seas; continents increasingly dry.* |
| | *Ordovician* | | *500* | *Appearance of vertebrates, but invertebrates and algae dominant. Land largely submerged. Warm climates worldwide.* |
| | *Cambrian* | | *600* | *Atmospheric oxygen reaches first critical level. Explosive evolution of life in the oceans; first abundant marine fossils formed; tribolites dominant; appearance of most phyla of invertebrates. Lowlying lands; climates mild.* |
| *Pre-Cambrian 2100 Million Years* | | | *2700* | *Life confined to shallow pools, fossil formation extremely rare. Volcanic activity, mountain building, erosion, and glaciation. Photosynthetic life.* |

environmental temperatures. Hair probably developed about this time as an insulator of the body. The mammals were now able to incubate their eggs within their own warm bodies, and the young were born alive rather than encapsulated in eggs. Whereas the prehistoric reptiles probably left their eggs and young to make it on their own, as many modern reptiles do, the early mammals probably began to care for and, later, to nurse their young. This care meant that the young had time to develop complex nervous systems before they entered a world dominated by the stupid but voracious dinosaurs. Even today, extended parental care is characteristically found in the more intelligent species, presumably to provide time for the development of learning.

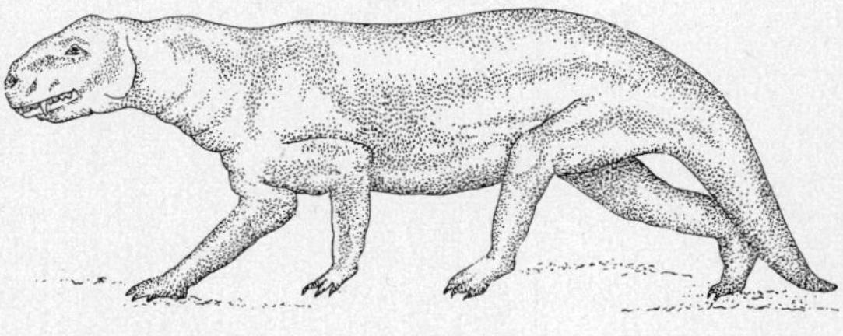

Figure 2.7 A mammallike reptile, *Lycaenops,* from the late Permian period of South Africa. Reptiles such as these are thought to be the forerunners of mammals.

Nursing, a convenient means of transforming nutrients that an adult was able to acquire into a form easily accessible to its young, could have also contributed to the development of family units—at least in the sense that mother-child bonds would have been encouraged. When the dinosaurs finally died out at the end of the Mesozoic (about sixty-five million years ago), the mammals, already equipped with well-developed brains born of a dangerous existence, began their rapid expansion over the earth.

Among mammals, the most intelligent are the primates, and most of these are tree-dwellers. Probably, in fact, the earliest primates were tree-dwellers, and the kind of specializations necessary for a complex arboreal life undoubtedly placed even greater emphasis on the development of the brain (Figure 2.8). As our ancestors moved from the forests to the plains, then, they were probably already relatively intelligent.

## CLASSIFICATION

Later in the book, I will refer to orders, classes, and families of animals, as well as genera and species. These terms are simply the most convenient way of sorting the different types of organisms according to their characters, to try to distinguish them from each other, and to show how they are related. For example, almost all living things are placed into one of about five categories, called kingdoms, according to whether they are considered animals, plants, or simpler organisms such as fungi, protozoa, or bacteria. Among animals, for example, those that show a number of characters in common are then placed in groups by themselves. Each of these groups may be more finely divided until finally, in the smallest groups, the members have so many characteristics in common that they can interbreed successfully. These are called species. From kingdom to species, the divisions are:

Kingdom
- Phylum
  - Class
    - Order
      - Family
        - Genus
          - Species

Of course, there are varieties within a species, and super- and subgroups of the classifications, but these will not concern us here. I only wish to show that, for example, the genera of a family are considered to be more closely related to each other than to members of other families of the order. Organisms are usually placed into their particular groups on the basis of their physical traits, or morphologies. In some cases, however, classification has been made on the basis of behavior, as we shall see later. But in most cases, behavior and morphology are tightly interrelated, each in a sense, reflecting the other.

## BEHAVIOR AND MORPHOLOGY

The relationship of morphology to behavior is in some cases obvious, and in other instances exceedingly subtle. A tiger has teeth and claws and big muscles. One can expect it to behave violently, at least at times, and it does. On the other hand, only careful experimentation revealed why sticklebacks with three spines were bolder than species with no spines (Hoogland, Morris, and Tinbergen 1957). It turned out that the spines offered protection from predators, and thus spined fish didn't have to be as careful as fish without such protection. Instead of spending valuable time and energy in hiding, they could be foraging or reproducing or engaging in some other useful pursuit. It seems simple enough, but the relation between spines and behavior was, at first, not entirely clear.

Species differences in morphology arise because of differences in selective pressures (i.e., different environments), and where there are morphological differences, behavioral differences can also be expected. The concept is a simple one. The questions surrounding it involve "why" any animal has a certain appearance, structure, size, or morphology. And then how does it employ its specializations? Why does a certain bird species have long legs? A related question centers over what nervous specializations must exist in order to utilize a particular structure. If a long-legged bird has a bill that is adapted for catching fish, one can expect that its neural activity will produce a certain affinity for water.

Such questions need not be solely applied to differences between species. For example, some species have more than one distinct morphology, and each "morph" may behave quite differently. Sexes, for example, are morphs. In humans men are generally larger, stronger, and swifter than women, whether we consider means or upper limits. On the other hand, women appear to be better "constructed," as determined by their resistance to many sorts of stress and their greater longevity. (Females of many species live longer than males, as we will see later, but it might be argued that this is because their roles have traditionally been less dangerous, or because they were protected by males.) Can we assume, on the basis of pronounced morphological differences, that the behavior (or anatomically, the neural complex) that utilizes the distinct qualities of each morph is specific to the sex? In other words, assuming there

has historically been an advantage to large size in males, should we assume the development of a particular mental set, such as aggressiveness, to go along with it? Or is it more logical to assume that size and strength are largely inherited, but the behavior that utilizes them is cultural or taught?

Such questions are being asked, and perhaps studies of the evolutionary development of behavior can help to shed light on these vexing problems. One of the first things we must consider, however, is how behavior and morphology are related.

The noted animal psychologist, Oscar Heinroth, supposedly once said, "Birds are so stupid because they can fly." The statement implies a great deal

Figure 2.8 There are several behavioral traits associated with the development of tree dwelling: (1) a trend toward a herbivorous or mixed diet; (2) a necessity for agility with a parallel development of the parts of the brain that control movement and coordination; (3) the development of the grasping hand; (4) a shift from nose to eyes as the most important source of environmental information; and (5) a trend toward increase in relative brain size (probably correlated with eye and hand development). Shown here is an orangutan from Malaysia.

about how a particular morphology might affect the development of behavior. The therapsids were forced to *deal* with the speedy, voracious, and formidable dinosaurs, and the result was the development of higher brain centers. The dinosaurs couldn't be outfought or outrun, so any animal that survived in their presence probably had to outthink them. Birds, on the other hand, developed lighter bones, air sacs, large pectoral muscles, and feathers. Thus equipped, they reduced the likelihood of having to deal seriously with terrestrial predators; they could simply leave danger behind. So while they became very specialized morphologically, they never did become very smart. In a sense, then, a morphological development may strongly influence certain behavioral patterns.

In other cases, a morphological pattern may act more directly in concert with a behavioral one. For example, some primates have ischial callosities—callouses on their rears—that enable them to sleep upright on limbs in trees that most predators are unable to climb (Washburn and Hamburg 1965).

Neither tree-sleeping nor ischial callosities would be much good without the other. Ischial callosities undoubtedly developed *de novo*. In other words, they probably arose anew and were not a modification of some preexisting external structure. But natural selection works this way. It is opportunistic, continually remodeling existing structures, as well as building anew. For example, it seems that the very earliest fish had lungs, and that these were changed into hydrostatic organs, the swim bladders (Romer 1970). The same sort of "conservation" is seen in some behavioral patterns. Rather than develop whole new patterns, new uses are made of the old ones, as we will see in the discussion on ritualization in Chapter 8.

It is often assumed that evolutionary changes in behavior also involve physical changes in the nervous system (Sperry 1958). Since the brain controls behavior in many species to a large degree, the rate of change in behavior might therefore be expected to be strongly influenced by the rate at which the nervous system can change.

On the other hand, perhaps the brain is capable of much more complex activity than it is normally called upon to employ (Harlow 1958). Thus, behavior can shift even between generations within the limitations of a single neural framework. In other words, pronounced behavioral changes may be possible without concomitant neural changes. To illustrate, some worms can learn a T-maze, the likes of which they would probably never encounter in nature. (The opposite, but probably evolutionarily less important, argument can also be made. Washburn and Avis in 1958 argued that the adoption of tools by early apelike creatures such as the australopithecines probably contributed to the development of brains that could better employ such tools.)

In the continuing attempt to associate behavioral patterns with their neural centers, some researchers have assumed that early mammals which had relatively small cerebrums or "thinking" centers, utilized primarily stereotyped or instinctive behavior (Simpson 1958). It is then assumed that in the evolutionary development of the brain, the bulk of the cerebrum was added secondarily. Thus the evolutionarily newer brain parts function in "higher" ways, such as in learning—or so goes the argument.

Before we leave this topic, I should mention that behavior may also be correlated with morphology on another level. Recent physiological studies have been based on electrodes implanted in different parts of the central nervous system that, when stimulated, elicit certain behavioral responses. Any such response is then assumed to be under the control of the area stimulated.

Other interesting but preliminary work is being done that indicates the importance of naturally occurring chemicals in the blood, such as metabolic by-products and hormones, which influence behavior by acting on the nervous system. Blood chemicals, after all, are part of our anatomical makeup. If it can be shown that certain normally occurring chemicals, such as lactates, influence a person's level of irritability or "mood," then perhaps altering the concentration of that chemical can produce a more desired mental state. If levels of nitrogenous chemicals in the blood are associated with "drive" or "ambition," then maybe we can be made to settle for that little farm in Pennsylvania by controlling their level in the blood, perhaps through diet. A result of such studies might be the reduction in the use of "unnatural" additives to the blood in an effort to change a mood or state of mind.

## The Sequence of Development

Since behavior is often directly correlated with a morphology or structure, we may ask which appeared first, the behavioral or the morphological pattern. The "mutationists" of earlier days lent support to the idea that a morphological change arises first and the animal then develops the behavior for employing it. We now believe the opposite is true. Of course, each case must be considered separately since no hard-and-fast rule is likely to cover all situations, but a general pattern does emerge that seems to include most cases: The behavior is usually considered older than the structure that utilizes it. For example, many species of birds raise the feathers on the top of the head when excited or alarmed, but only a few, such as the bluejay, have developed crests, which make the pattern obvious (Figure 2.9).

Figure 2.9 It is assumed that, in most cases, the behavioral pattern arose before the structure it employs. Many birds, such as white-crowned sparrows (left), raise their crests when excited. Some species, such as bluejays (right), have modified (enlarged) the crest with the result that when it is raised it is more conspicuous.

Lorenz (1937b) and Tinbergen (1940) stated that in displays (a particular form of communication that we will discuss later), the ceremony is almost always older than the structure being displayed. In other words, a bird does not bow, showing the top of its head, because it has a crest, but it developed a crest because it has a habit of bowing. Probably the "eyespots" on the wings of the butterfly *Precis almana* (see Figure 17.18 in Chapter 17) would have been ineffective if they had developed without the behavioral pattern (the wingspread) that exposes them (Manning 1971). The quick spreading of the wings could have developed because it had some startling effect on predatory birds. The "eyespots," which increased the effectiveness of the wing flash, probably appeared later.

Figure 2.11 Two views of a pregnant female stickleback in which her swollen appearance is emphasized by a behavioral pattern. It is possibly an unusual instance of a morphological trait arising before the behavior that employs it.

Figure 2.10 Sexual dimorphism in bills of the extinct New Zealand *huia*. Apparently, the males chipped holes in the bark of trees with their powerful bills while searching for insects, and the females followed probing in the openings with their long, curved bills. This may be an unusual kind of environmental exploitation by the sexes, since one does not normally prepare the environment for the other.

Also, as was mentioned earlier, in some cases a structure may arise on one adaptive basis, and then come to be employed in new ways. For example, in some birds the sexes have different bill sizes. The result is that they take different foods and do not strongly compete with one another. But the sexes could also come to recognize each other on the basis of bill size, thus enhancing mating, and so the sexual difference takes on a new importance (Figure 2.10).

In some cases, the structure may persist after the behavioral pattern has disappeared. For example, whereas an alarmed Guira cuckoo, *Guira guira*, raises its crest, its close relative, the smooth-billed ani, *Crotophaga ani*, has elongated crown feathers that are never erected (Davis 1942). It has been argued that when a structure or ornament is present with no ceremonial or display behavior associated with it, we may assume the structure is vestigial rather than a recently acquired morphological character awaiting its behavioral pattern (Armstrong 1965).

Tinbergen (1940) pointed out exceptions in which the structure may precede the behavior. He noted that the gravid three-spined stickleback positions her body so that her swollen belly is accented when she confronts the male (Figure 2.11). In this case, it is apparent that the structure did not follow the display. As another example, it has been suggested that the display might not precede the structure in "displacement patterns" (Armstrong 1965). (Displacement behavior is discussed in Chapter 8.)

It is important, then, to realize that behavior evolves. Its evolution is often correlated with changes in anatomy, but it is still unclear whether behavioral evolution can or should be studied with the same confidence we have or with the same assumptions we make when considering anatomical descent alone. We have only laid the groundwork for further discussions here. We will continue to approach the problem. Hopefully, the depth and ramifications of the basic question will become increasingly clear.

# 3

# BEHAVIORAL GENETICS

Is behavior inherited? Almost any hard-nosed biologist would reply, "Of course." But the simplicity of the question is deceptive. It turns out that the idea has been very difficult for certain behavioral scientists to accept. Their reticence stands in the face of evidence ranging from the appearance of patterns that are correctly performed with no chance of their being learned, to behavioral changes resulting from known genetic abnormalities. As an example of the latter, in some cases mutations produce gross behavioral changes that are associated with extreme morphological abnormalities. One type of mutation, for example, may produce microcephaly in humans—characterized by a small head and almost total absence of brain tissue. A mutation also produces "waltzing mice," which, because of inner ear abnormalities, spin when disturbed. These mutations produce great changes in the animal, and any shift of such magnitude in finely adapted organisms is likely to be deleterious and quickly selected out. We are more interested in small changes that may gradually accumulate through the generations, thus enabling selective forces to operate with less vigorous results. For a thorough discussion of the turbulent events leading to the acceptance of genetic influences on behavior, see Hirsch (1967).

## A SHORT COURSE IN GENETICS

I lied. The following discussion is not a short course in genetics. The field is far too complicated for such treatment and you probably have some knowledge about this area anyway. If not, there are a number of general biology, psychology, or anthropology textbooks that introduce the material in a simple manner. For a *beautiful* introduction, see Wallace (1978). Here, we will only briefly review some of the basic genetic tenets that are germane to following discussions. For a comprehensive introduction to the study of behavioral genetics, you might wish to refer to Fuller and Thompson (1960); Parsons (1967); Thiessen (1971); Hirsch (1967); and McClearn and DeFries (1973).

### Single Gene Effects

If a behavioral pattern is strongly affected by a single gene, the mutation of that gene can have devastating impact. Phenylketonuria, or PKU, is the result of a mutation in a single gene. The disease causes the accumulation of a single metabolic by-product in the blood. But this chemical produces severe abnormalities in the brain chemistry, resulting in serious mental retardation.

Other single gene changes may have much less severe effects, and these are, therefore, exceedingly difficult to detect (Benzer 1967). Some of the best examples of single gene mutations on behavior come from studies on the fruit fly *Drosophila*. For example, a single gene change is known to cause the *yellow* mutant of *D. melanogaster*, a variant that has less mating success than its normal ("wild type") brothers (Bastock 1956). It was found that its lack of success was partly attributable to the tendency of the yellow males to spend less time in the wing-vibration stage of courtship. It was a small change, but the evolutionary implications were enormous. We will see how small reproductive disadvantages can lead to rapid extinction of the organism suffering the disadvantage. Later, by the way, we will discuss hygenic and nonhygenic bees whose behavioral differences are attributable to differences in genes at *two* loci.

### Multiple Gene Effects

Most behavior that we have been able to trace genetically is due to the action of a whole set of genes. Actually, we know of only a few traits attributable to the action of one or two genes. The involvement of multiple genes means that the genetic source of a behavior is very hard to trace. If a behavioral trait is due to a number of genes, it can be expected to vary along a continuum within any population, and a change in a single gene among a complex of genes is simply not detectable. In the determination of multiple gene effects, statistical analysis of populations is necessary to describe the variance for any trait within the population. We will not go into the steps involved in such calculations, but if you're interested, see Parsons (1967) and Hirsch (1967).

### Problems in Studying Behavioral Genetics

The development of the cerebral cortex in "higher" animals has presented problems in studying behavioral genetics. Increasing cerebration has greatly increased behavioral potentials and flexibility at the cost of a reduced repertoire of "instinctive" or stereotyped patterns. As we will see, it is much more difficult to deal with genetic influences on learning than with such influences on instinctive patterns. Actually, at this point, our efforts in the field of behavioral genetics can only be viewed as preliminary, and our conclusions, because we have not yet clearly defined our paradigms, are tenuous.

Conceptually, the study of behavioral genetics demands that we consider all aspects of behavior, including the receptors, the central nervous system, and the action of the muscles. This means that we must consider all associated pathways, keeping in mind that genes can influence *any link in the chain.* The conceptual problems are enormous. Probably the greatest problem is that we usually have no idea of how genes can alter behavior. All we can do is note the behavioral change and admit that we have almost no idea of how it was really brought about.

Basically, the work has progressed along three major lines: (1) comparing behavior of animals whose genotypes are presumed to differ at only a single locus (in other words, by only a single gene); (2) comparing inbred lines whose behaviors have been rather thoroughly described (such as certain well-known laboratory mouse or rat strains); and (3) selecting for different behavioral characters from a single parent strain. The evidence all leads to the same conclusion (Manning 1964). Generally, innate patterns are very stable, but almost any genetic change in the chain of events that produces the behavior will cause a *quantitative* change, altering the frequency at which the patterns are performed. Later, I will have more to say about the significance of changes in frequency.

The point I would like to make now is that behavioral changes due to changes in genes are probably usually due to changes in the *threshold* for a particular action (an "action" here being a single, identifiable *part* of an overall instinctive pattern). Changes in threshold for the stimulus for an action can produce differences between strains in the frequency with which an instinctive pattern is performed, differences in the speed with which the pattern is performed, differences in the emphasis on various actions comprising a pattern, and differences in the relative importance of the various sensory modalities involved in perception. (For example, light affects the speed with which mating behavior may take place in various species of the fruit fly. One strain will not breed at all in the presence of light.) And the argument can be made that threshold changes can also account for changes in the sequence of actions that comprise an innate pattern (Manning 1964). For a remarkably clear dissection of the behavioral genetics of fruit flies, see Benzer (1967), and for a theoretical discussion, see Klopfer (1969b).

## GENETICS AND HUMAN BEHAVIOR

Genes can influence human behavior. The statement cannot be denied, but some of the implications are rife with dissension. Certain of the undeniable effects involve single gene changes, such as those that produce PKU or sickle cell anemia. And then, the human chromosome that has been given the number 21 (Figure 3.1) is occasionally duplicated in the continual genetic shuffling involved in reproduction.

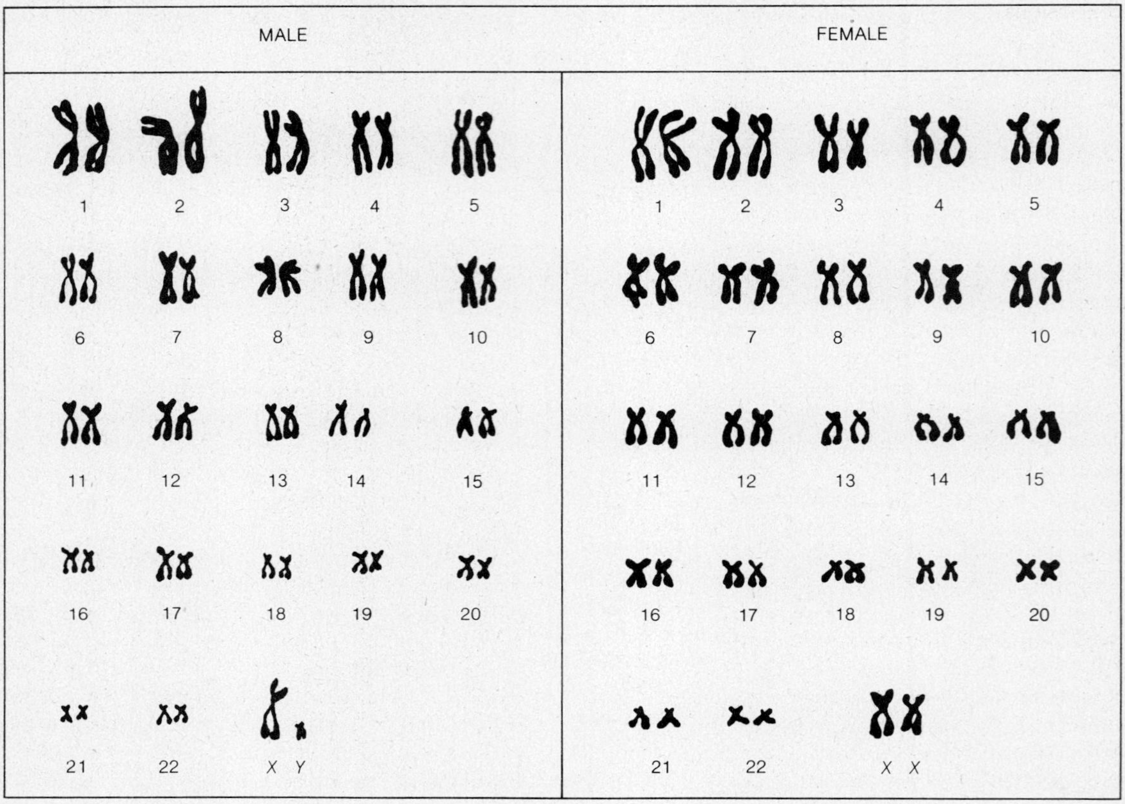


Figure 3.1 The normal number of chromosomes in human cells (note number 21). The pairing stems from one chromosome having come from the father and one from the mother. Genes are located along the chromosome like pearls on a strand. If a gene exists at a given site, or *locus,* on one chromosome, a gene coding for the same trait lies at that location on the sister chromosome, or *homolog*. Thus there are two genes (also called alleles) on different chromosomes at any locus. One may totally repress the effect of the other (a *dominant* acting on a recessive), or the two genes may act in concert to produce an intermediate condition.

The result is Down's syndrome, or "mongolism" (Figure 3.2), a condition whose most marked trait is severe mental retardation.

Genetic abnormalities may also produce more subtle changes that may simply be expressed as "tendencies." For example, men with an extra *Y* chromosome (an *XYY* genotype) are more likely to find themselves in a mental or penal institution than normal *XY* males. (Females do not carry a *Y* chromosome, but two *X* chromosomes instead.) It works out so that only one male in a thousand is an *XYY,* but about twenty out of every thousand institutionalized males bear this genotype (Hook 1973). An *XYY* male runs about one chance in twenty-eight of being in a jail or mental institution at some time in his adulthood. Interestingly, *XXY* males are likely to be mentally retarded, but they show no increased likelihood of being jailed.

This sort of information is easily quantifiable, and most people have no problem in accepting the conclusions. There is often a problem, however, for many people in accepting the notion that genes have anything to do with the likelihood of a person smoking or drinking. Or that I.Q. is inherited, or that genes may be responsible for different behavior in the sexes, or that genes may contribute to cheerfulness or schizophrenia. One can often find himself on thin ice by even bringing up the question.

You can rest assured that we will not enter into such emotion-charged debates. We will deal with material, however, that may touch directly or indirectly on such matters. We will not retreat from certain of these questions and we will consider specific approaches to the solutions.

We are almost compelled to do so. As we learn more about animal behavior, we are drawn inexorably toward applying our findings to man. The applications are perhaps premature, but almost always stimulating—even exciting. Also, because of our newborn technical abilities, the way we see ourselves is important. It's an unpopular turn of phrase, but we must begin to take ourselves more seriously. Our powers are too great for us to misunderstand ourselves. And if ethology provides us with one more mirror, so much the better. That mirror, however, is presently so hazy that dogmatism is folly. Whereas we cannot blindly extrapolate from mice and birds to men, we can see how selective forces can operate on behavior; and, keeping in mind that man is but a very special kind of animal, we may be able to use the techniques of ethology to learn more about ourselves.

Since most human characters are complex and malleable, we must keep in mind that many important behavioral differences in people are likely to be subtle and hard to detect. So, *is* there any information that indicates that "subtle" and complex behavioral traits may be inherited in man? There is. For example, we now know that such complex behavior as schizophrenia has a strong genetic component. If one parent is schizophrenic, there is a much greater risk of the children being afflicted with the disease than if neither parent has the trait (Heston 1970). The evidence is more compelling when the disease appears disproportionately in the offspring of schizoid parents when those children have been reared by foster parents.

Much of the evidence for the role of genetics in human behavior comes from comparing identical and fraternal twins. Identical twins result when the fertilized egg is divided at the two-celled stage, so that the twins have identical sets of genes. Fraternal twinning occurs when separate eggs are fertilized by different sperm. Thus, fraternal twins share no more genes than other brothers and/or sisters (in other words an average of fifty percent of their genes).

Figure 3.2 Down's syndrome. The disorder usually includes mental deficiency, a short, stocky body with a thick neck, and often heart abnormalities. It may occur as a result of an extra chromosome (number 21).

It has been theorized that identical twins reared apart, under different conditions, should develop more behavioral traits in common than fraternal twins reared apart. This seems to be the case, even in matters pertaining to I.Q. (Figure 3.3). It has also been found that in identical twins reared apart, if one smokes, the other one is likely to be a smoker too. But no such correlation is shown for fraternal twins. If one identical twin has a record of criminal arrest, the odds are greater that the other twin will, than holds true for fraternal twins (Dobzhansky 1962; Young 1971). This, of course, is not to say that there are genes for smoking or robbing banks, but that an individual may be particularly sensitive (vulnerable) to certain environmental stimuli.

The problem of investigating human behavioral genetics is compounded because the presence of genes for a trait does not insure that the trait *will* be expressed. We can deliberately and consciously override certain tendencies (perhaps at an expense, which we will discuss later). And because we are such

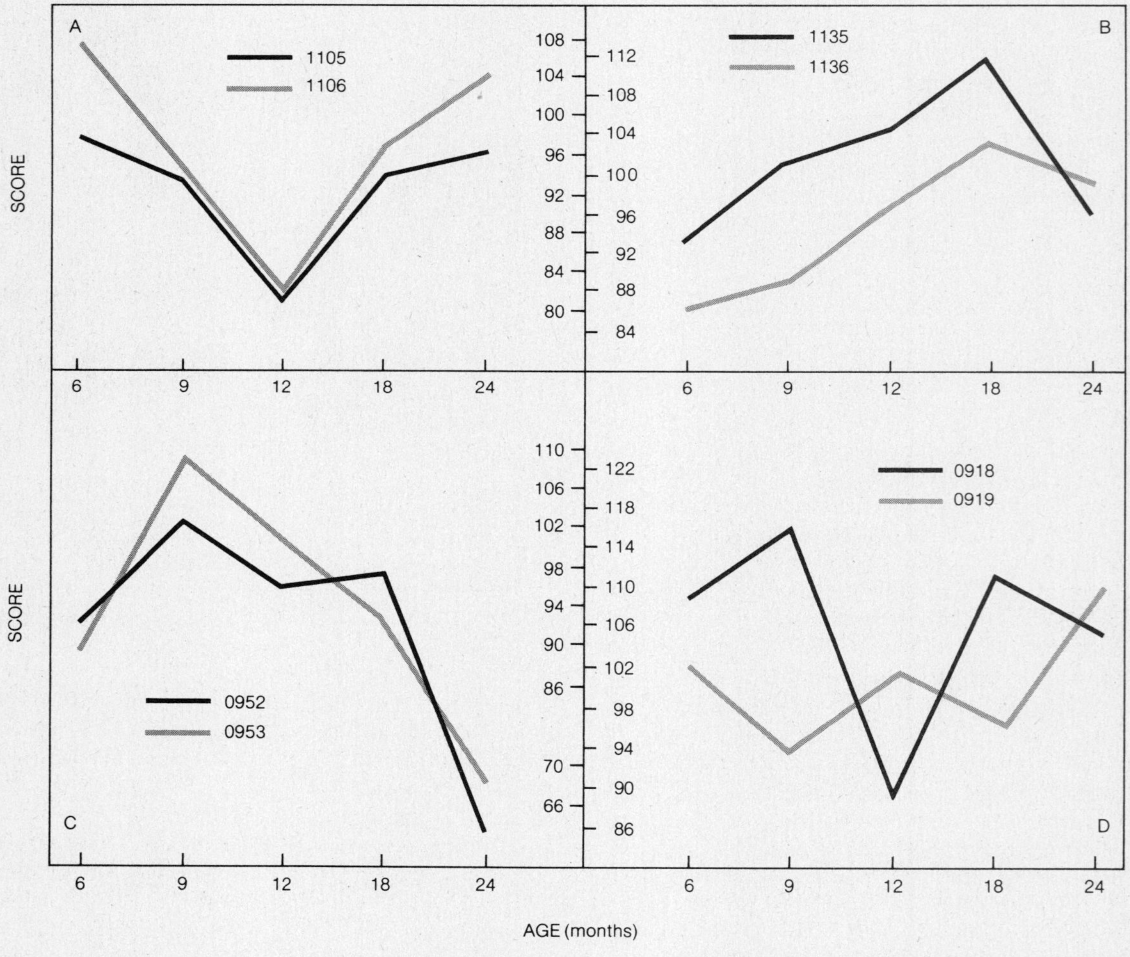


Figure 3.3 Mental development scores for maternal twins at ages 6 to 24 months. Twins usually show a high correlation in their mental ability (A–C). Rarely are twins markedly different in mental development (as in D).

adaptable animals, we are able to compensate in innumerable ways for genetically or historically-based dispositions.

### Units of Behavior

A serious problem in studying the genetics of behavior arises in trying to determine just what constitutes a unit of behavior. Some patterns are performed in a variety of contexts, some patterns gradually change into others, and some are subject to various levels of learning influences. Therefore, it is difficult to determine just when a given behavioral pattern has been observed. (Tinbergen, in 1959, described the usefulness of *displays* as behavioral units—as is discussed in the section on the homology of behavior in Chapter 5.) Then we have the problem of not knowing enough about animal behavior to be aware of the repertoire the animal might have at its disposal. For example, signals unrecognized by us may be used as modifiers of signals we do recognize. Also, we must be thoroughly familiar with the natural history of an animal in order to be able to interpret what we see. The baring of teeth in various primates may have entirely different meanings (Figure 3.4). As a further complication, the same behavioral change may be produced by a number of genetic alterations, thus making the search for the genetic control difficult. It is believed that several single-gene mutations in *Drosophila* all result in a reduction of the frequency of courtship patterns in males (Manning 1971).

In some cases, the problem of determining gene influences on behavior is made more difficult by the operation of "switch" genes. Here, a single gene may turn a set of other genes on or off. Lorenz (1941) described a number of apparently genetically based behavioral units that are held in common by several closely related species of ducks. One of the patterns involved courtship behavior of the drakes. He found that when two species of these ducks were crossed, the offspring showed a courtship pattern not found in the repertoire of either parent species. It seems that the pattern is controlled by a "block" of genes, which is, in turn, controlled by a "switch gene." The block has undoubtedly descended intact through the generations because the genes comprising it lie in close proximity along the chromosome and have been present, but "turned off," by the switch gene in the parent species. In the hybrid offspring, apparently the effect of the switch has been reduced, and the pattern, long hidden, emerges.

Figure 3.4 Grin face in the stumptailed macaque. This is a fear grin being given by a frightened individual. In humans, grinning may be used to indicate nonaggressiveness, if not fear.

Such information thus indicates the close relationship of the parent species. It is unlikely, however, that hybrid studies alone will provide much information about the inheritance of behavior, since closely related parent types must be used in the crosses, and their behavioral repertoires are likely to be similar.

As another example of the behavioral effects of genes, consider a case from honeybees. Some strains of honeybees are "hygienic"—that is, if a larva dies, the workers uncap the cell and place the corpse outside. Nonhygienic strains do not show this behavior. The classical crosses and backcrosses of the two strains were made, and it was found that the pattern was, in fact, a discrete unit of behavior (Rothenbuhler 1964 a,b). The genes controlling the behavior were apparently activated by switch genes. It has been shown that crosses between a number of related species or strains could result in behavioral changes in such behavior as fixed action patterns (Chapter 8), selective responsiveness, circadian rhythms, orientation, and learning ability (Fuller and Thompson 1960).

## VARIATION AND NATURAL SELECTION

It is apparent that all the members of some species do not look alike. You don't look exactly like your little brother, a fact that may please you no end. You can also tell your dogs apart. And individual pigeons are often easily distinguishable. Furthermore, you don't always react to a situation in the same way your kid brother might. Some dogs are much more aggressive or intelligent than others, and even pigeons may show strong behavioral differences. Two pigeons I hand-reared after they fell out of their nests behaved very differently. One would huddle in a corner of its cage when I tried to pick it up. The other would peck at my hand and then deliver startling and surprisingly strong blows with the "wrist" of its wing. The point is, and we will stress this later, members of populations—even closely related members—may differ strongly in both their appearance and behavior. Of course, the level of difference can vary from one population or species to the next. No matter how different the members of a population of gulls are, they are more alike than humans, at least from the human point of view.

Charles Darwin, in the last century, recognized the immensely important implications of such variation. In fact, he based a critical part of the development of his theory of natural selection on such variation. In particular, he was interested in variation in reproductive behavior.

Darwin noticed that the tremendous reproductive ability of living things is not always realized. The results of uncontrolled breeding has been calculated for a number of species—and the findings are almost always spectacular. For example, if a pair of common houseflies were to breed in April, and all their offspring survived, and all the offspring of each subsequent generation survived, by late summer, the entire earth would be thickly blanketed with some 191,010,000,000,000,000,000 flies (Keeton 1967)—a situation I would find quite intolerable. I can't even tolerate *one* fly. However, I am happy to report that the earth is not buried in flies. Nor elephants. Nor rabbits.

The fact is that the numbers of most species change slowly through time and usually fall within sustainable limits (although there are notable exceptions, as we shall see). Darwin was aware of all this and the conclusion was inescapable: some individuals reproduce, and some don't. Among those who do, some are better reproducers than others. But what determines which individuals will leave offspring?

In some cases, the answer is obvious. A chimpanzee that has a crippled arm as a result of polio contracted in its youth is more likely to be caught by predators than its healthier brothers, thus its reproductive life is shortened. An infant baboon that is feeble, or not very bright, may be caught and eaten by chimpanzees before it ever reproduces at all. An individual that would rather fight or eat than court is likely to leave few offspring. In any population, then, there are some individuals that will not survive to the reproductive stages, or if they do survive, they may leave fewer offspring than other individuals. The result is that each generation will be made up of the descendants of the most successful reproducers of previous generations. They can be expected to bear "good reproducer" traits, then, because they inherited these tendencies from their reproductive forebears.

Those things that influence the reproductive success of organisms provide the mechanism for the process known as *natural selection.* By the way, if humans are purposefully involved in selecting which organisms will be allowed to reproduce, the process is called *artificial selection.* High-yield milk producers were developed through applying the principles of artificial selection to dairy herds.

It should be pointed out that natural selection is not synonymous with evolution. Actually, evolution simply implies any change through time. Natural selection is the mechanism that brings about one kind of evolution. Populations may experience evolution in ways other than through natural selection, for example, through random "drift" (see Chapter 4), or through changes in learned patterns. As an example of the latter, traditions may change as they are passed along from one generation to the next.

Since we often regard evolution as occurring primarily through natural selection, we are often surprised at the number of instances when change occurs in other ways. There have been a number of interesting studies of nongenetic transmission of behavior. In one such study, it was shown that female rats handled in infancy produced *grandchildren* with altered behavior and weaning weight. The mechanism underlying these changes is not known, but it could have been brought about by changes in the maternal behavior of the grandmothers and mothers, or through physiological changes that might affect the developing fetus (the developing "mother") or alter her milk supply (Denenberg and Rosenberg 1967). There are also good studies of culturally transmitted vocalizations in sparrows (Marler and Tamura 1964).

Now let's consider variation and natural selection as they apply more directly to ethology.

## Variation in Populations

From an evolutionary point of view, there are two important opposing tendencies operating in populations. One favors the behavior that is "optimal" for most individuals of the species at a given time, and it therefore places a premium on uniformity of behavior throughout the group. The other favors variation between individuals. Such variation preserves evolutionary flexibility in that it provides something for natural selection to work on. Perhaps a fine distinction should be made here. Although variation may have its advantages and may persist in populations, I am not saying that there is selection for variation. In fact, I see no way it could operate. (See the discussion of genetic load at the end of this chapter.)

The uniformity of a species is enhanced by natural selection *strongly* favoring those members with certain qualities. The more any individual differs from that optimal type, the more it decreases the likelihood of its survival and/or reproductive success. But if this is the case, why doesn't natural selection simply tend to wipe out all those individuals not possessing a preponderance of the optimal traits? How is variation maintained in populations at all if natural selection tends to favor those individuals of a single adaptive type? In other words, if natural selection tends to reduce variation in populations, how is any variation maintained?

The original source of genetic variation is, of course, mutation. But here we're interested in the processes that might perpetuate those aberrant qualities. In some cases, variation may be due to the seemingly fickle nature of natural selection. Environments change. And whereas one type of trait may be favored for a while and continuously accumulate in the population, environmental conditions may change so that another trait is favored—perhaps one that was seemingly on its way out. The result of different traits being favored over a given time span is that no single trait completely dominates and variation is maintained.

Also, variation may be increased because there is such strong competition among individuals of the "optimal" type that those few variant individuals that are able to exploit marginal resources may enjoy an advantage (Pianka 1974).

Variation may be maintained in some populations through polymorphism, or the maintenance of distinctly different phenotypes that are not connected by a continuous series of intermediate types. Polymorphisms usually occur in heterogeneous environments composed of subunits of distinct types—in other words, in "patchy" environments. For example, an area may contain both live, green leaves and dead, brown leaves. Certain butterflies, called "leaf butterflies," which hide by mimicking leaves, may be green or brown. Green butterflies will only succeed, however, if they tend to sit among green leaves, and brown butterflies among brown leaves. Those individuals with different behavioral tendencies will be selected out of the population by hungry birds. Since each has its place, the two forms of butterfly persist (Pianka 1974).

Polymorphisms are maintained by *disruptive selection* (Figure 3.5) In our example, it occurs because intermediate types of butterflies blend with neither green nor brown leaves. Disruptive selection may also account for sexual dimorphism, where the male and female are markedly dissimilar. The ethological implications of sexual dimorphism are discussed in Chapter 26.

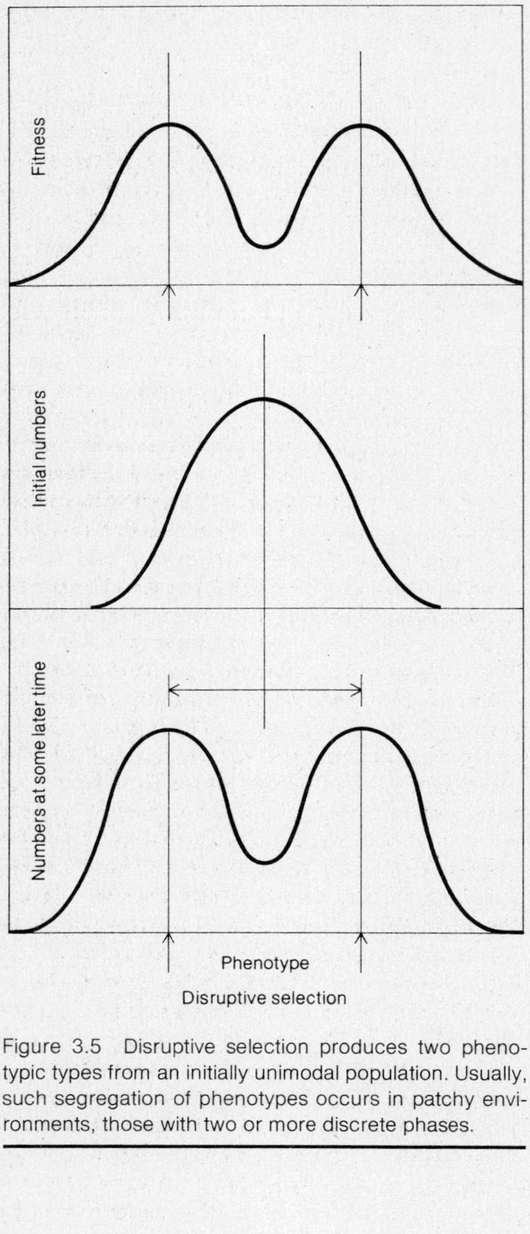


Figure 3.5 Disruptive selection produces two phenotypic types from an initially unimodal population. Usually, such segregation of phenotypes occurs in patchy environments, those with two or more discrete phases.

Finally, polymorphic variation may be maintained by *heterozygote advantage*. When an organism bears two different alleles for a character, it is said to be heterozygous for that character. It has been found that, in some cases, individuals that are heterozygous for a certain character are more reproductively successful than if they are homozygous (dominant or recessive) for the trait.

The best-known example of heterozygote advantage occurs in humans. Sickle cell anemia is presently a serious debilitating condition among American blacks. Individuals that are born homozygous for the disease quickly die, because their red blood cells are "collapsed" and sickle-shaped (Figure 3.6). Such cells can carry little oxygen to the body's tissues. However, individuals heterozygous for the condition may lead relatively normal lives. But their blood cells are not able to carry as much oxygen as normal cells, hence, they are "anemic" and may tire easily. Strangely enough, in parts of Africa and some other areas of the world where malaria is common, the heterozygous individual has an advantage over an individual with normal blood cells. The malarial parasite invades red blood cells, and normal individuals are highly susceptible to the disease. But the partially sickled cells of heterozygous individuals do not provide a suitable environment for the malarial parasite, and, hence, the anemic individual is conferred a high degree of immunity. Since those people with normal blood cells are often weakened or killed by malaria, the sickle gene will be maintained in the population by natural selection as long as malaria is exerting its selective effect.

The variation of any character within a population can be expressed graphically (Figure 3.7). If selection is strong as a result of, for example, a limited type of food or strong competition, variation will probably be lessened (B). If selection is weak (C), variation could be expected to increase.

Actually, in some cases, the norm may not be "optimal." (See the discussion of supernormal releasers in Chapter 8.) It is known, for example, that a black cardboard "bill" with a contrasting white dot elicits stronger begging responses in gull chicks of certain species than does the normal red bill of the parents. So why does the bill of the gulls retain its less effective color? Why doesn't it turn black, as darker colors are continually selected for? It is possible that the red bill color has a number of uses and that it is the result of a compromise between its various

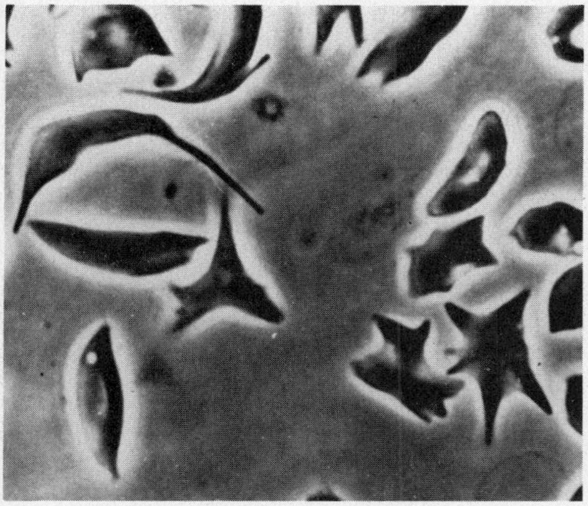

Figure 3.6 Sickle cell anemia, a disease caused by a mutation in a single gene and characterized by the collapse of fragile red blood cells, which may then clog blood vessels.

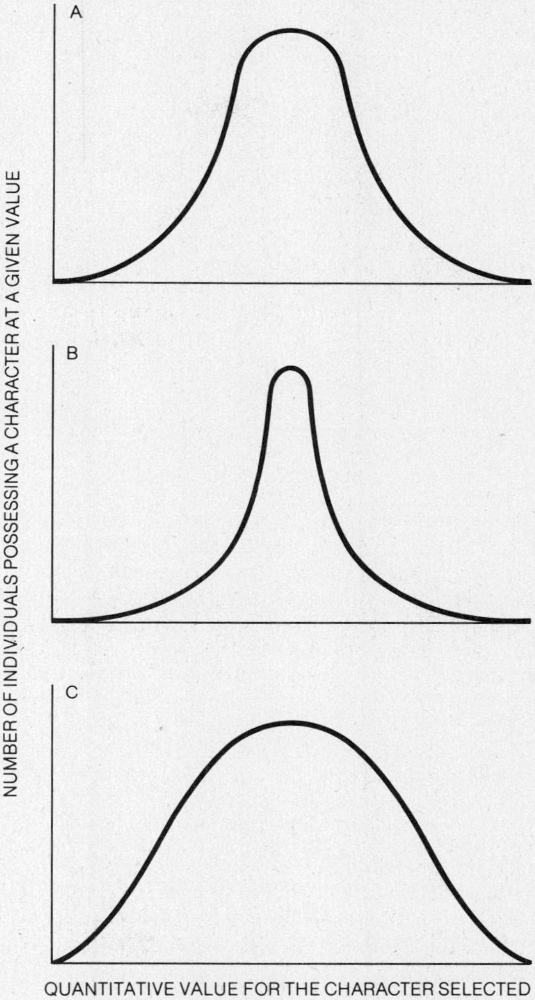


Figure 3.7 All populations vary (A) according to how strongly natural selection is operating on them. Strong selection (B) produces low variation. When selection is relaxed (C), variation is allowed to increase.

adaptive demands. Perhaps, for example, the red bill also functions as a threat to competitors. Perhaps, in fact, it would develop even more threatening qualities if it were not necessary that it also function in feeding the young. It can be argued, however, that the red bill *is* optimal in terms of fitness if not in terms of behavioral response.

But how much behavioral variation exists within populations? The question has never been answered very well, and no unitary rules have yet been drawn. We do know, however, that some species are highly variable. For example, mongrel dogs are usually immediately distinguishable even for humans comparing them for the first time. The variation could be due to artificially selected breeds ("subpopulations") interbreeding or to low natural selection operating on dogs cared for by humans. (Low selection produces high variation, in general. For an exception, see how variability is selected for in starfish in Chapter 17.)

Artificial breeding programs have also demonstrated that distinct behavioral characters can be developed in populations, as was done with laboratory populations of *Drosophila melanogaster* (Figure 3.8). As examples, we find that different strains of

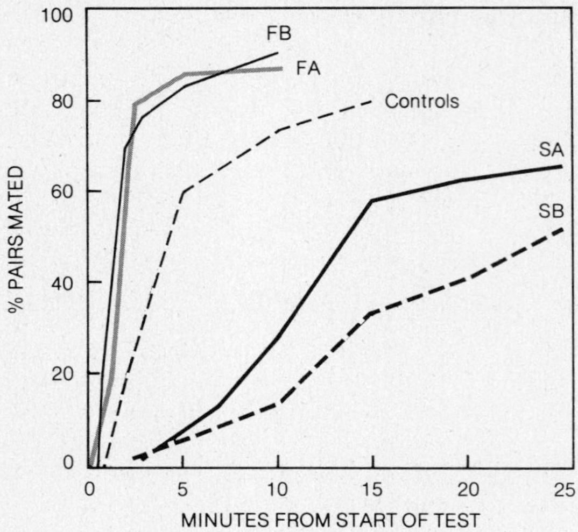


Figure 3.8 The results of selection for mating speed in the fruit fly, *Drosophila melanogaster.* The curves show the mating speed from two fast-mating lines, FA and FB, and two slow-mating lines, SA and SB, after the eighteenth selected generation. A comparison is made with unselected controls. About eighty percent of the fast lines have mated before the slow line has begun.

bees communicate in different ways, and various strains of laboratory rats behave differently (for example, some are wilder than others). Probably no one would deny the strong behavioral differences in domestic bulls and the famed "brave bulls" of Spain.

Human populations, of course, are notoriously variable. We come in a variety of shapes and sizes and we vary not only in our appearance, but in our behavior and abilities as well. It is not necessary for us to be able to judge the rate of a car's approach and then dart across the street with agility, fast cars removing slower or dimmer individuals from the population. We all simply wait for the light.

Those among us who are unable to provide for ourselves or our offspring are often provided for to some degree, thus individuals who are not "good providers" remain in the population. In other words, we have altered our environmental situation and reduced the level of selection operating upon us. The result is a species with highly individualized members and a wide range of abilities, appearances, and interests.

Variation in humans is also increased by our cosmopolitan dispersion. We occupy a wide range of habitats, and adaptations to local conditions have resulted in strong differences across our range. A short, stocky Eskimo retains body heat through a high ratio of body mass to surface area, and a tall, gangling Watusi loses heat through an increased surface area. I might add here that we know almost nothing about innate behavioral differences in human groups.

Great variation also may exist among wild animals—variation that is apparent even to the human eye. We have noted that in some bird species, such as gulls, individuals are often indistinguishable to us, but as we will see, they may be easily recognizable to each other. Among such birds, behavior also varies little from one individual to the next, or so it seems. The implication is that strong selection is operating on these birds.

Recognition of individuals is easier in less stereotyped animals. Jane Goodall was able to recognize her wild chimpanzees by their faces (Figure 3.9). But, importantly, they also differed in their size, vigor, aggressiveness, sexuality, parental care, and ability to form friendships. And even kinship cannot erase much of these differences. It has been demonstrated that wolf cubs from the same litter may be quite different behaviorally (Fox 1972).

Actually, one might well expect individuals of populations of "learning" animals, such as chimpanzees, to vary greatly in their behavior. Any variation (behavioral or morphological) might increase their ability to recognize each other. Thus, the variation would be adaptive. Since they are a wide-ranging and explorative species, these intelligent animals may also vary behaviorally because of differences in their experiences. It must also be admitted that perhaps primates do not drastically differ from each other in comparison to other kinds of animals after all. But since primates and humans have similar perceptive abilities, it could be that we can recognize subtle differences in these animals more easily than in other species. The rear-ends of dogs furnish all sorts of recognition signals to other dogs, but to humans they probably all smell alike. However, I have no data.

There is evidence that there may be great variation in the behavior of even the most stereotyped species. The Peckhams, in 1898, wrote about their observations of the female wasp, *Ammophila urnaria:*

*While one was beguiled from her hunting by every sorrel blossom she passed, another stuck to her work with indefatigable perseverance. While one stung her caterpillar so carelessly and made her nest in so shiftless a way that her young could only survive through some lucky chance, another devoted herself to these duties not only with conscientious thoroughness, but with an apparent craving after artistic perfection.*

In summary, behavioral variation exists within populations, and whereas we do know of some of the causes, there are undoubtedly others that we know almost nothing about. Perhaps the greatest question, and one that we have some hope of answering is, To what extent do populations of different adaptive types vary? Can we correlate population ecology and behavioral variation?

## Behavioral Variation in an Individual

It is probably true that parrots or chimpanzees show more variable behavior than do houseflies or tapeworms. There are surely great differences in the behavior of individuals in a population of chimpanzees, but randomly selected tapeworms of the same species (since "populations" of tapeworms essentially do not exist) probably all show very similar behavior. Let's consider for a moment variation in the behavior of the "average" tapeworm and the "average" chimpanzee.

We must first realize that we can't blithely compare their behavior in any meaningful way, other than to say chimpanzees seem to be more variable (or unpredictable). However, if we consider specific behavioral patterns in chimpanzees, we may find that some patterns within the repertoire of the species are far less variable than others. In fact, some behavioral patterns seem to approach the highly stereotyped reactions of the tapeworm.

Consider the clinging behavior in young chimpanzees. Just as all tapeworms attach by their hooked heads to intestines, so do all infant chimpanzees cling. Since chimpanzee mothers see to it that they

Figure 3.9 Some animals, such as chimpanzees, are variable enough that even humans can tell them apart. Such differences may help relatively intelligent members of a species to recognize each other.

are in close proximity to their newborn, it turns out that chimpanzee infants, when they cling, usually end up clinging to their mothers. The infant's clinging response is highly beneficial (or "adaptive"), since chimpanzees spend a lot of time in trees, and, at times, the mother may need both hands free.

Later on, when the chimpanzee infant is a bit older, it may climb to the mother's back or shoulders and ride there. At this point, we might expect a bit more variation from one infant to the next. The variation may be due to the fact that the mother is between the infant and the ground. Thus, the incessant pull of gravity tends to keep them in close association. And a well-balanced infant then has one or both hands free and can do all sorts of interesting things as he rides along. Riding infants, then, are able to make certain kinds of choices as the mother browses through the leafy canopy or dozes in the crotch of a branch. An infant that is well coordinated and strong enough to climb to the mother's back, may also be strong enough to hang onto a limb or to climb back aboard should he fall off. Therefore, it is probably less critical that he be as good a rider as he was a clinger.

Later, when the infant begins his first timorous explorations, at the first sign of anything "threatening" or unfamiliar, he may rush back to his protective mother. Of course, the "return to mother" response may be elicited more easily in the more timid individuals. The braver infant may only initiate a return, but before reaching the mother, he may decide the danger is not so great after all, and return to his explorations. And then when the infants begin to play with other youngsters in the group of the same age, there may be great variation in a single individual's behavior from one time to the next and in the way the various members of the group go about playing.

Now it should be noted that since most "threats" to exploring youngsters are likely to be more imagined than real, it is not as critical that retreating youngsters all behave in the same way. Another way to say it is, the *adaptive value* of returning all the way to the mother at the first hint of danger may not be as high as clinging to her during those first few days of life. And play behavior, while it may have great value in learning social patterns and improving strength and coordination, is not as important as returning to a protective mother at the sight of a threatening situation. In other words, perhaps, in chimpanzees, those traits that have the highest adaptive values have the lowest levels of variation.

On the other hand, in tapeworms, each behavorial pattern may have a very high adaptive value with very little leeway allowed in how it is performed. A tapeworm reaching an intestine *must* attach to the intestinal wall. Since it is hermaphroditic (male and female reproductive parts in the same individual), it *must* then self-fertilize. As the egg sacs break off from the lengthening worm, the eggs pass out with the feces to be eaten by an intermediate host. The new stage of the tapeworm *must* move through that host's tissue, then it *must* wrap itself in a cyst in the muscle tissue to await the time that the flesh is eaten by its new host. Each stage is critical. Should any one not be performed correctly, the cycle ends. Very little behavioral variation is permitted at any stage in the cycle. Thus, tapeworms are behaviorally rather inflexible and have never developed the mental facilities needed to deal with situations in which alternatives are presented.

Possibly, then, we might summarize the differences by correlating adaptive value and variation in two species with entirely different means of solving their problems (Figure 3.10). The diagram is obviously simplistic. For example, clinging, a single behavioral trait with possibly a relatively simple neurological basis, cannot be treated the same way as a complex behavior involving many patterns—such as play. However, this problem actually illustrates a point. The behavioral traits that are most complex and that may arise on a number of neurological bases (traits that may also interact in complex ways such as through reinforcement or compensation), may have a relatively low adaptive value. This may also be a way of strengthening an earlier point: natural selection reduces variation.

## Genetic Load

It was once assumed that variation was the price a species paid for genetic flexibility with which it could capitalize upon any change in the environment. This variation was termed *genetic load.* Genetic load may be defined as the extent to which a population departs from a perfect genetic constitution (Strickberger 1968). Given a certain environment, there is

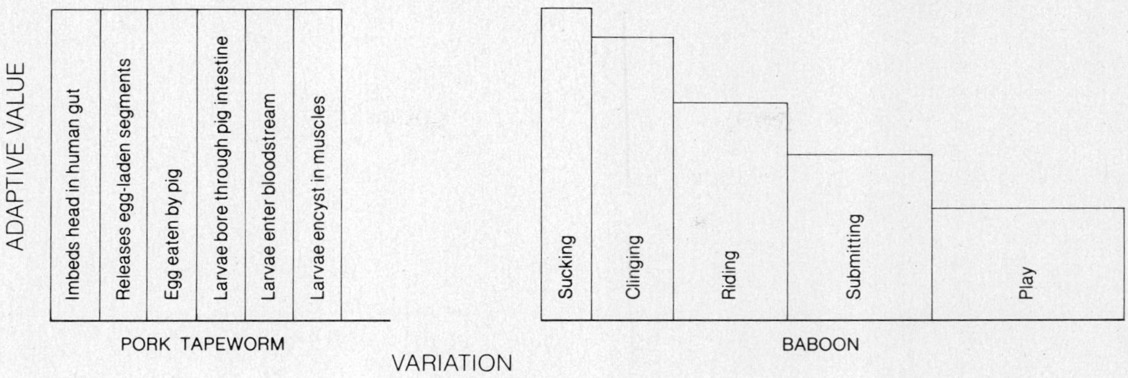


Figure 3.10 For some traits, variation is inversely related to adaptive value. In stereotyped species such as the tapeworm, each behavioral phase is critical and precise conditions must be met. More flexible species, such as baboons, also have low-variability phases of high adaptive value, and they show their greatest flexibility in areas that are less critical to immediate survival.

one optimal genotype for the population that is living there. However, variation in a population means that some individuals are less well suited to reproduce under the prevailing conditions. In fact, those varying too far from the optimum may die before they reproduce. The death toll has been called the "cost of natural selection" (Haldane 1957).

So, since variation implies inequality, some individuals survive, although they are different from those at the current optimum and therefore are at an immediate disadvantage. The original idea was that these variant members might come to be at an advantage should a change in the environment begin to favor individuals with their special qualities. This is not to say they accepted their lot graciously, however. The modern theory of natural selection does not assume that individuals sacrifice their own immediate reproductive output, even to save the population, as we shall see. However, as we know from our discussion of specializations of subgroups in a variable habitat, genetic variation may be important to the individuals in a population.

On the other side of the coin, we may ask how much a character may vary from the optimum before it no longer imparts an advantage to the bearer. The eyespot on a moth's wing can be crude indeed and still maintain a degree of its startling effect on birds, but at some point it becomes so crude that it loses its net advantage to the moth. It should be apparent that the degree of variation in a character is dependent upon a number of factors, including the life style of the organism that bears it and the nature and adaptiveness of the character in question.

It should be clear, then, that behavior may be genetically based. "Genetically based" is suitably vague because I do not want to imply that there are specific genes that control specific behavioral patterns. This may well be the case in certain instances, but the cases are probably sufficiently rare and the notion too pat and simplistic to be worth great consideration. There are too many extenuating factors operating in the development of most behavioral patterns. Nonetheless, certain patterns are undoubtedly influenced to some degree by genes, and to that degree they are subject to investigation by traditional genetics.

# 4

# NATURAL SELECTION AND ADAPTATION

Variation implies inequality. Thus, in a variable population, certain individuals are likely to be more reproductively successful than others. The characters that enhance reproductive success are likely to be propelled into succeeding generations, thus, these characters are said to be selected *for.* Other individuals possess characters that hinder the reproductive success of the bearer, and those characters are selected *against.* Ultimately, they may be selected *out* of the population. The process is called *natural selection.* To add to our jargon, we might mention that there may also be *selective pressures,* for example, for strength and speed among savannah-living baboons. But if there is greater *selective value* on speed than there is on strength, the baboons are likely to be leaner than they would be if the selective values were reversed. So, these are some of the terms we will use in the following discussions.

It should be made clear that natural selection does not operate on the *genotype,* or the genetic component, of individuals. It can only operate on the *phenotype,* or the expression of the genotype. Thus, it has no direct effect on genes for tallness, but it can operate on height—tall or short individuals being selected for, depending on the demands of the environment. It therefore cannot operate on recessive genes in heterozygous individuals. Again, natural selection operates on *expressed* characters. In behavioral studies, this idea is especially complex. Later, we will reconsider the inheritance of "tendencies," "predispositions," or "abilities" to behave in certain ways. If, in highly intelligent species, only some or parts of these tendencies are expressed, the rest being *consciously* suppressed, natural selection cannot operate on the entire inherited behavioral spectrum. This will be a particularly important point to keep in mind when we discuss human ethology. For example, if aggressiveness, or any of its components, *is* inherited in humans, and if it has come to interfere with our biological success on this crowded planet, it cannot be selected against if those tendencies are consciously overridden or expressed in subtle or socially acceptable ways. Here, then, the phenotype that is *present* may not be actually *expressed.*

## GREAT LEAPS OR SMALL STEPS?

It is sometimes assumed that sweeping changes in populations appear in a short time, perhaps a few generations, as species make great saltatory leaps, quickly changing from one form to another. We can't be sure at this point, but current theory suggests that this is not usually the case. In most cases, evolutionary change can be explained on the basis of the accumulation of small changes over long periods of time.

If behavioral patterns and the associated structures arose through a series of small steps, then what was the pressure behind them—the force pushing them along in these small, tedious steps? For example, we can see the advantage of reptiles suddenly becoming birds, but why would a reptile begin to develop tiny, useless wings? Why would it place its evolutionary eggs in a new basket, so to speak? It seems any running reptile would have reaped more advantage from becoming a better runner than a poor flier—and the first flights must have been poor attempts indeed if the traits necessary for flight arose in small steps.

Such questions cannot be answered in definite terms, but it has been suggested that, in insects, wings were probably rather well developed as courtship organs before they were used for flight (Alexander and Brown 1963). A comparison of advanced and primitive crickets among living species indicates that sound production originated from wing lifting and vibration during courtship. The probable sequence began with the development of winglike structures that were vibrated as a visual signal in courtship and that incidentally produced sound. Once the sound had become associated with the courtship, there followed an elaboration of auditory organs to better utilize the sound. Selection, then, improved both the producers and receptors of sound until, finally, sounds could have become more elaborate to carry more precise information (see Alexander 1966). Finally, the well-developed wings could have been used in locomotion, perhaps first in short hops, and later as organs of flight. Perhaps in birds, also, the wings developed on some adaptive basis other than flight, such as in display or for gliding—a habit that would accent the immediate advantage of a broader forelimb with its increased surface area, possibly without detracting too much from running or climbing ability.

One argument states that although a character has value now, it couldn't have developed piecemeal on the basis of its present adaptiveness; as an incipient character, it would have been selected out when it first appeared in an imperfect form. There is some evidence, however, that even the imperfect forms of adaptive characters have a selective advantage. As I mentioned, even very crude "eyespots" on the wings of moths had a repelling effect on birds (Blest 1957). Even *slightly* twiglike caterpillars have an advantage over those who are not disguised at all. Tinbergen (1965b) pointed out that our present experimental techniques are very crude and that probably a character would need to have enormous selective value before we could detect any advantage at all. For discussions of how minute selection advantage could produce evolutionary change, see Fisher (1930) and Haldane (1932).

### Generation Time and Evolution

As we know, there is a higher premium on some adaptive characters than others, and this is the basis of our concept of adaptive value. To illustrate the point in more specific terms, consider an oversimplified scheme in which a mutation is completely favorable (with no undesirable effects); its value would be designated as +1. A neutral change would be given 0. Neutral changes would be mutations that exerted no effect on the behavior or morphology of the organism. (These are generally considered rare, but a vigorous argument has developed; see Simpson 1958 and Cain 1964.) A −1 would designate those changes that are completely adverse or lethal. Now consider a gene with a selective value of 0.001, a very low figure, which means for every 999 individuals that survive each generation lacking the gene, 1 survives to carry it. Once a gene with this selective value appeared in 1 percent of the population, it would take 120,000 generations of selection for it to reach 100 percent (Manning 1971). Such predictions, of course, are predicated upon a stable environment that does not shift the selective value of the character. The time is described in generation time because this period differs so greatly among animals. A population of flies could therefore be expected to alter gene frequencies faster, in chronological time, than the same number of elephants. Changes in animals with long generation times are possible because of the great span involved in what is termed "geologic time."

### Stabilizing Selection and Directional Selection

Selective value for any character can shift with changes in the environment; furthermore, so can the *direction* of selection. A character can become less advantageous as time goes by, or it can even become detrimental. The peppered moth, *Biston betularia*, of England flies at night and rests on the bark of trees during the day where it is vulnerable to predatory birds. For ages, the population was comprised of speckled, gray individuals that blended well with the lichen-covered bark on which they rested. An occasional dark form would arise as a result of a rare dominant mutation in a single gene. These individuals were conspicuous and quickly captured by birds if their choice of a resting site did not change. (See Kettlewell 1956 for experimental evidence.) In the mid-1880s England began its conversion to an industrial economy and began burning soft coal in its factories. The greenery of the English countryside quietly submitted to its new cloak of soot and the gray, lichen-covered bark of trees turned black. Now the rare melanic or dark moths were hidden, and the peppered form was conspicuous to predators. In the industrial areas around Manchester, the dark form almost completely replaced its peppered colleagues. In rural western England, however, the environment has not changed, and the peppered moth still predominates. Where antipollution efforts have reduced the amount of soot in the air, the peppered moth is returning (Cook, Askew, and Bishop 1970) (Figure 4.1). (There is some evidence that moths tend to choose "matching" backgrounds by resting on surfaces that most closely resemble the reflectance of their forewings; see Sargent 1966.)

Most shifts in gene frequencies within a population are not so clearly documented in terms of selective advantage. The hamster, *Cricetus cricetus*, in Russia produces a mutant black form in addition to the normal gray form. In Russia, hamsters are trapped for fur, and records are kept of the distribution of its various forms. Thus, it was possible to trace the spread of the black form into certain areas such as the Ukraine (Gershenson 1945). The selection coefficients were rather high, indicating strong selection for these dark forms in some regions (Dobzhansky 1937), but just what the adaptive advantage might be remains a mystery.

We know that in some cases variation may be disjunct, variants falling into discrete classes, such as when a trait is due to the effect of a single gene. In other cases, variation is not expressed as an either/or

Figure 4.1 Light and dark forms of the peppered moth. The light form was more common in England until the industrial revolution blackened the trees with soot, thus shielding the dark form from hungry birds.

polymorphism, but along a continuum. The continuum may exist as a *gradient*, changing gradually from one extreme to the other. It may also exist along a "normal," or bell-shaped, distribution, with most individuals falling about midway between the extremes, as in Figure 4.2. The reduced number of individuals at either extreme of a normal distribution are usually a result of stronger selection operating against those that are most different from the optimum. Under stable conditions, the optimum can be expected to remain unchanged as *stabilizing selection* operates (A). However, as we know, the optimum can shift as the environment changes. In such a case, those at one extremity might come to be favored, and the curve representing the population might be shifted in that direction. The phenomenon is termed *directional selection* (B and C).

## Selection and Population Size

The likelihood of selection ever producing a change within any population is largely dependent upon the size of the population in which the change appears. In large populations, any genetic changes (mutations) that appear may be swamped by the other types (alleles) of that gene in the group, and therefore they never really catch hold. If, on the other hand, the population is small, the random death of any individual may have an exaggerated influence on the frequencies of genes that are found in that population. So, gene frequencies in small populations may shift rapidly, since any gene or gene combination may be lost purely by chance, such as by the deaths of the few members carrying it. Also, any new mutation can have an amplified impact in small populations because it is not strongly tempered by a multitude of "competing" alleles. The best examples of such rapidly evolving populations are found on islands, such as the Galapagos, the West Indies, and Australasia (Mayr 1963).

## Genetic Drift

All changes in small populations, it must be pointed out, may not be due to selection operating on variation. In small, permanently isolated populations, the rate and direction of evolution can be altered through the accumulation of random changes. The phenomenon is termed *genetic drift* (Wright 1951). Genetic drift is the chance fluctuation of the frequency of an allele within populations, and it is more important in small populations than in large ones. If

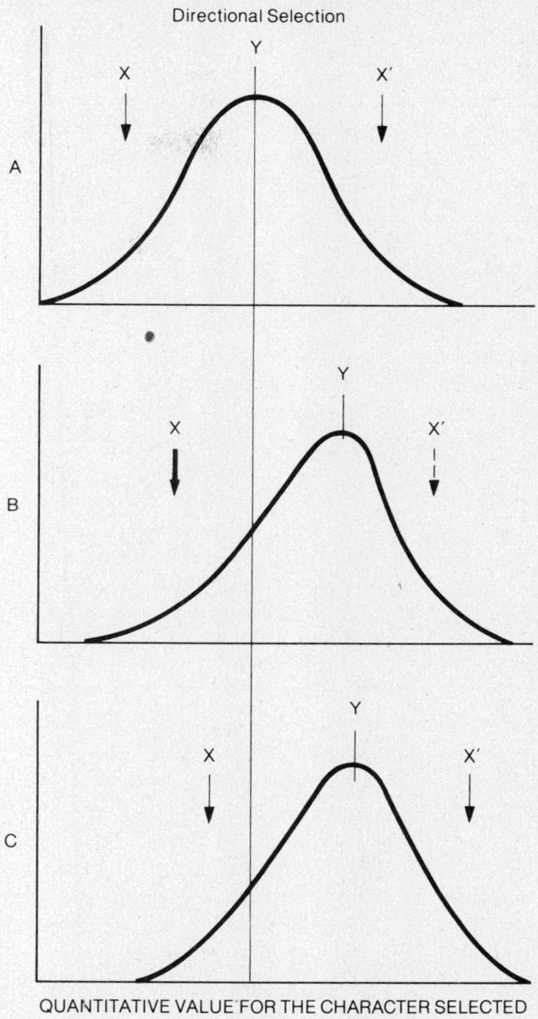


Figure 4.2 Directional selection. As long as the selection pressures (X and X′) remain stable, the value for any trait in the population will tend to remain stable. When the pressures of X and X′ change, the optimum Y may shift, causing the population to fall in line with the new optimum.

two alleles exist at a specific locus in a chromosome, and if no other force, such as natural selection, influences the relative frequencies of those alleles, one may ultimately be lost, while the other becomes "fixed," in that population, simply due to random changes. Genetic drift is, at present, not believed to be an important factor in evolution, but there is some disagreement over the matter (Wright 1951; Mayr 1963; Selander 1970).

Often, population traits that cannot be easily explained in terms of adaptation are said to be due to drift. The problem is, although drift may explain some evolutionary changes, it is probably invoked much too freely (Manning 1971). However, under certain conditions, drift could play an important part in evolution. Changes that accumulate in such a manner are, by definition, not acted upon by selective forces. However, if the environment should change, genes accumulated by drift could fall under the forces of natural selection. Those accumulated genes might then be selected out, or selection could increase their frequency, should they prove to be newly adaptive.

## ADAPTATION

As we have seen, adaptation is, in an evolutionary context, an adjustment of individuals to their particular sorts of environments—it is a result of natural selection.

### The Search for Adaptiveness

Tinbergen (1965b) objected to the free use of the concept of drift to explain what is apparently nonadaptive. (*Nonadaptive* will refer to an absence of beneficial effects. *Unadaptive* will imply a detrimental effect.) He stressed two points: First, the search for adaptiveness has to be made with respect to *all* the known properties of behavior mechanisms, *all* the known details of behavior control. He thereby emphasized the need for thorough understanding of the behavior of an animal in the *context* of the animal's natural history. We have actually employed such considerations in only a few instances, such as in Donald Griffin's (1958) astonishingly detailed description of the intricacies of echolocation by bats or in communication studies of some animals (Lanyon and Tavolga 1960). On the other hand, the adaptiveness of something as apparently simple as locomotor patterns has been investigated much less rigorously. In an early study that seemed to incorporate some of Tinbergen's notions, it was discovered that the way the godwit, *Limosa,* walks—by lifting and folding its feet—is an adaptation for walking through tall grass (Klomp 1954).

Secondly, Tinbergen pointed out that it is useless and even unscientific to made assumptions regarding the adaptiveness of any feature until one has tried to *find out* what the adaptiveness might be. Thayer's study (1909) is remarkable for its inclusion of worthless theorizing alongside important observations. In his paper, Thayer made the surprisingly naive statement that red coloration in flamingos enables them to disappear into the sunset, and in the same paper he pointed out the significance of countershading. (Countershading is the condition of animals having darker colors on the back than on the belly [Figure 4.3].) It was later found that countershading coloration was adaptive when employed with a certain behavioral set (correct position, scattering, and immobility) (De Ruiter 1955). It seems that a countershaded animal is less likely to be detected, since a view from the underside is likely to be against a lighted sky and from the topside, against a darker earth. (For other examples of apparently unadaptive characters that were found to have advantages, see Cain 1964.)

Tinbergen and his group (1962), as will be mentioned later, demonstrated the high adaptive value of the black-headed gull's habit of removing empty egg shells from the nest. It was found that shell removal made the nest less conspicuous to predators. Such "field experiments" can not only indicate the adaptiveness of a pattern, but can reflect what kinds of forces have been acting on a species. Tinbergen cautioned that the findings can only suggest what selective pressures were operating in the past to shape the present behavioral repertoire of the species.

### The Maintenance of Unadaptive Traits

It should be clear by now that we cannot expect that all observed characters, even when the data are in, will be found to be adaptive or at least neutral. There are undoubtedly unadaptive characters in any population. In some cases, these are being selected out, and we simply observe them on their way to oblivion. In other cases, though, unadaptive characters may be there because they are maintained. One way unadaptive traits can be maintained is through *pleiotropism,* multiple effects of a single genetic factor. As examples of pleiotropism, a certain gene in mice that produces short ears also produces fewer ribs; and a single genetic factor in *Drosophila* produces small wings, lowered fecundity, shorter life span, and other changes. If the net pleiotropic effect is beneficial, bad traits can be maintained in the population. Unadaptive characters may also continue in a population through *linkage.* In such a case, an unadaptive gene may be located very close to a highly adaptive one on the chromosome. So, when the genes undergo their continual "reshuffling," there is a high probability that the unadaptive gene will be carried along on the coattails of the adaptive one. But, again, we don't want to invoke pleiotropy or linkage, any more readily than we do drift, simply to account for behavior that we can't readily explain.

In the black-headed gull, which nests in dense colonies, the eggshell-removing behavior does not appear until after a delay in which the chicks become dry. It was at first theorized that the dangerous delay is a pleiotropic and unadaptive effect associated with the removal behavior. Tinbergen warned, however, that such theorizing is, in essence, a refusal to investigate. So, he developed an experiment to test for any adaptiveness of the time lapse. His results indicated that the delay is highly adaptive in that it gives the chick time to dry off, thus making it hard to swallow. Therefore, the pattern decreases the likelihood of the chick's being eaten by neighboring gulls while its parents are gone on their chores.

We should also be aware that even adaptive traits can have their unadaptive side. In fact, because of the complex nature of relationships in the wild, it can rarely, if ever, be said that any character is completely adaptive and without its drawbacks. For example, as a tree places its energy into growing taller, it may increase its exposure to sunlight, but, at the same time, its girth is reduced, thus increasing its susceptibility to high winds. As another example, the limbs of the horse are adapted for running fast. But, while the horse is indeed fleet-footed, its long, stiff limbs aren't very good for scratching. Tinbergen (1965b) pointed out that, in the black-headed gull, colonial nesting is advantageous as a means of predator defense. The drawback is that some birds lose their young to nearby cannibalistic neighbors. In sum, however, a trait is adaptive if the advantages outweigh the disadvantages.

Figure 4.3 Countershading is a common phenomenon among both terrestrial and aquatic animals. The darker back and lighter underside are believed to camouflage the animal against a lighter sky and a darkened ground. Shown here is a semipalmated plover.

## The Opportunism of Natural Selection

We should keep in mind that natural selection is opportunistic. It takes advantage of existing conditions. As an example, many species of freshwater fish build air-bubble nests that float at the surface among the vegetation. The male usually builds these nests by gulping mouthfuls of air and blowing out clusters of froth into which he puts the fertilized eggs. It seems that their nest-building pattern is an outgrowth of their peculiar habit of air breathing. Air breathing—gulping air from the surface—is, in turn, a behavioral result of low oxygen content in the tepid waters of many tropical lakes and rivers (Manning 1967).

When a character trait evolves in response to one set of selective forces and later turns out to have additional evolutionary possibilities, we can say that the trait (or the animal) is *preadapted* to take advantage of those new possibilities. As an example, behavioral flexibility is often a prerequisite for successful island colonization. Certain mainland conditions, such as high seasonal variability, or the conditions that exist at the periphery of a species' range, might select for behavioral flexibility. So, animals that live under such conditions are likely to be able to quickly exploit new situations and therefore might generally be superior colonizers.

There may also be more specific types of preadaptation—for example, for a certain *type* of niche. If a species attempting to colonize an island is not flexible enough to quickly adapt to the new conditions it encounters, it may still survive if it is already partially adapted through a compatible earlier evolutionary history. If, for example, a flycatcher colonizes an island, it may not find the same range or type of insects there that existed on the mainland. But it might be expected to be rather successful (after perhaps a few modifications in its flycatching behavior) in capturing the types of flying insects that *are* found in its new home. The flycatcher may be said to be preadapted to its niche on the island.

## Adaptive Routes

Adaptation, we learned, is life's answer to an environmental problem. But there may be more than one answer. The same environmental pressure may force different adaptive responses in different animals. Predation pressures may cause some mammals to become swifter and others to become stronger, as one responds by fleeing, the other by fighting. Animals that are threatened by desiccation may alter their behavior, for example, by becoming active only at night, but other species may alter their environment. Certain termites, for example, that are highly susceptible to desiccation are able to cross dry, open areas by building covered corridors over their paths.

In some cases, the adaptive response of an organism may set the stage for later responses. In other words, once an evolutionary direction is taken, there is no return, and there are subsequently fewer adaptive alternatives open to the animal. The animal, in a sense, has placed its bet on a certain strategy, and—as any gambler knows—some strategies are better than others. When the primitive and ancient fish, the *Sarcopterygii,* were faced with seasonal drought, one group apparently stayed in the mud and waited for the rains to return. These animals, in order to survive in their pools of mud, began to rely on oxygen from the air, and this is how the lungfish developed. Another group, however, left their drying pools and clambered overland in search of water. These gave rise to the earliest four-legged animals (von Wahlert 1965). The difference in the results of these two strategies may be seen as bipedal men stand gazing in curiosity at lungfish in an aquarium.

There is an interesting theoretical account of the development of the flatfish (such as halibut) that stresses the ethological component of evolutionary influences (von Wahlert 1965). It is assumed that the development of the flatfish stems from a change in the role of the ocean bottom in the lives of their ancestors. This role change was, in effect, a construct of a behavioral pattern. The progenitors of the flatfish, the argument goes, probably slept on their sides as do many modern fishes, such as wrasses and triggerfish. If, upon awakening, such early fishes tended to remain in the horizontal position for some reason, such as to escape predators in the area by remaining immobile, the stage might thus be set for a continued horizontal existence. Selection might then favor a shift of both eyes to the top and the development of the appropriate countershading. The flatfish could then come to use the bottom in different ways—as ambush sites by perch-flounders, hunting areas by flounders, or grazing sites by soles and tongues. In all cases, the complex morphological adaptations followed the shifts permitted by a key behavioral change. The ocean bottom didn't change—only the use to which it was put.

## Adaptive Interaction

When we speak of adaptation, we often are referring to an organism's adjustment to its physical environment. But, since other forms of life are also part of the environment, special problems arise in the adaptive picture for any animal. In many cases, an animal is subject to environmental influences over which it has no control at all. The weather heeds not the presence, whereabouts, or activities of animals, but exerts its effect independent of their influences. In the case of the biotic (living) component of the environment, however, a different situation exists. Animals interact. As an example, among prey animals, the

Figure 4.4 Nesting grounds of the herring gull are subject to predation because of their accessibility. Ground-nesting gulls must vigorously defend their nesting area by mobbing any prowler.

Figure 4.5 Herring gull nest. The nests are often hidden in clumps of grass in the nesting area. Thus the eggs are both concealed and shaded. A predator finding one nest must then actively search for the others.

slower are more likely to be caught. The result is a faster prey. None but the swiftest predators thus are able to catch the speedy prey, so the faster predators are those that survive and reproduce. The result is to place even greater premium on speed or evasiveness in the prey. So, while either predator or prey may be adapted to a regularly varying but slowly changing climatic system, each may be much more quickly evolving in response to the reciprocally adaptable living component of its environment.

The mutual adaptation of biota may also be seen in cooperative or competitive interaction. As we will see later, the results of competitive interaction are quite different from those of predator-prey interactions in which one species does not usually bring about the extinction of the other. In competing species, one may exclude the other from a habitat or a niche, thus forcing it to die out or adapt to a different set of environmental conditions. (For a discussion of an example of behavioral coadaptation, see Vander Wall and Balda 1977.)

## Multiple Effects of a Habitat Change

One notable study has been made of a change in the habitat of a gull species. Once the story was unravelled, the extent of the results of the change was startling. Most gulls, including the herring gull, nest in low-lying areas or on offshore islands (Figure 4.4) and are therefore subject to some predation. As a result of such predation they hide their nests (Figure 4.5), which are composed of bits of twigs and grass piled together among clumps of shore grasses. The eggs and young are also cryptic (Figure 4.6). The young have certain behavioral characters that maximize the effect of their camouflage. At their parents' warning call, they may quickly run under the nearest cover, and thus may escape any predator that has been attracted to the nest; or they "freeze" and rely on their markings to render them inconspicuous. Esther Cullen (1957 a,b), one of Tinbergen's students, investigated a small gull called the Atlantic kittiwake, *Rissa tridactyla*, which nests on cliff ledges (Figure 4.7). In addition to its selection of a peculiar nest site, the kittiwake displays a wide range of behavioral traits not found in other gulls. Apparently, many of these patterns have developed as a direct result of the change in location of its nest.

If one simply compared the behavioral characters of the kittiwake and other gulls, it would seem at first that the differences are so great that the species could only be distantly related. For example, the kittiwakes fight by grabbing each other's bills and twisting, thus causing the opponent to fall off the ledge, while ground-nesting gulls usually try to get above an opponent and peck down, or to pull at the body of the other. Probably because of their fighting patterns, ground nesters threaten by stretching into an "upright threat" posture. The kittiwake doesn't. And the kittiwake is different in other ways. Female kittiwakes sit down to copulate, thus reducing the risk of falling off their narrow ledges in a moment of unbridled passion. Also, the kittiwake builds much tighter and deeper nests than do ground nesters, since they would not be able to retrieve eggs that rolled out, as could their flatland fellows (Figure 4.8).

Figure 4.6 Cryptic young of the herring gull are mottled. The young respond to the parents' warning calls by running for cover or "freezing." Such traits are considered adaptations to nesting in vulnerable areas.

It is also interesting that when the kittiwake is on the nest one can almost reach out and touch the bird without eliciting an escape reaction; also, they don't defend their nests against predators. They have undoubtedly lost their fear response through generations of nesting in areas that were safe places because of their inaccessibility. However, when kittiwakes are on land, gathering nest material, they are vulnerable and, at such times, they move about in groups and are extremely shy.

Later on, when kittiwakes are nesting, conspicuous egg shells and droppings are left around the nest, a habit that in other gulls would attract predators. Young kittiwakes, of course, must remain in the nest until they can fly; therefore, all feeding is done at the nest. The young take the food directly from the parent, thus no partly digested matter is left lying around the nest as a potential source of disease. Ground nesters, on the other hand, feed their free-roaming young anywhere in the nesting area and they sometimes regurgitate food onto the ground near their young. Of course, ground nesters must be able to recognize their footloose offspring. The kittiwake, however, cannot recognize its young except by location—which is how the ground nesters recognize their *eggs*.

Also, the young of ground nesters "practice" flying by jumping high into the wind. It's a risky business living on a cliff ledge, so the young kittiwakes never jump very high and they always face the cliff when they do jump. There are other differences in the behavior of kittiwakes and ground-nesting gulls that can be related to the choice of nest sites, but these few examples show the range of repercussions that may stem from a single behavioral change.

Cullen believed that the kittiwake originated from ground-nesting stock and based her reasoning on certain similarities of kittiwakes to ground nesters: (1) the eggs are partly cryptic, like those of ground nesters, although the habit of removing conspicuous shells and droppings has been lost; and (2) young kittiwakes *can* run—an unusual ability in species that nest in precarious habitats.

Other studies have supported Cullen's work. For example, the nest site of the Galapagos swallow-tailed gull, *Larus (Creagus) furcatus,* is intermediate between the kittiwake and ground-nesting gulls. Not surprisingly, its behavior was also intermediate be-

Figure 4.7 Nesting area of the kittiwake. The simple shift of nesting sites from the ground to the cliffside has rendered the kittiwake nests inaccessible to predators and has triggered a whole gamut of behavioral changes.

Figure 4.8 The exposed nest of the kittiwake. The kittiwake need not take the trouble to conceal the nest since it is inaccessible to most predators.

tween the two types. Also, see a paper on the black noddy, *Anous tenuirostris* (Cullen and Ashmole 1963).

### The Search for Adaptiveness

The most fruitful searches for the adaptiveness of characters are likely to involve one of three basic approaches. First, we may make comparisons between species, such as the one involving differences in cliff-nesting and ground-nesting gulls and the earlier investigation of behavioral differences of predators toward the three-spined stickleback, *Gasteosteus aculeatus,* and unspined fish (Hoogland, Morris, and Tinbergen 1957). In the latter study, it was found that when a predatory perch or pike attacked a stickleback it usually spat it out and avoided the prickly species thereafter. Unspined fish such as minnows, roach, and rudd were more often attacked. (As we know, the three-spined stickleback is a bold fish that swims in schools, nests in open habitats, and has conspicuous nuptial colors.)

Secondly, we may compare individuals within a species. This was done in studies we'll consider later relating to nesting success of birds that nest at the center and at the edge of a colony and in the investigations of "roles" in primates. Both of these studies are described in Chapter 19. And thirdly, we may carry out ethological experiments of the sort pioneered by Tinbergen.

In his experiments, Tinbergen (1965b) delineated two operational approaches to the problem of studying adaptiveness. The first is to single out a particular environmental pressure, a "challenge to survival," and then to find out how the animal meets the challenge. As an example, he and his colleagues attempted to find out how the black-headed gull meets the problem of predation. They found that eggshell removal is part of the system and that patterns of habitat selection, daily timing of behavior, colony density, and synchronization of the breeding cycle are also involved. The diurnal (daily) rhythm of habitat selection was first described: In the breeding season, all birds spend the night on their roosts, which are located in wide, open spaces such as are found along the lower part of a seashore, open mud flats, or shallow lakes. Any approaching predator would be quickly spotted here. Only at daylight do they move to the breeding grounds where they establish territories and form pairs. Each night, the birds return to the open flats until the time that they lay eggs on the breeding grounds. The open flats would probably be unsuitable areas for raising young—perhaps because of periodic flooding. After the eggs are laid, of course, they continue to sit on them through the night.

The density of the nests seems to be a compromise between grouping together so as to be able to "mob" predators, and scattering so as to reduce the searching effectiveness of predators that are not deterred by mobbing, such as foxes. The advantage of synchrony in nesting was demonstrated by the high nestling mortality of those pairs that raised young out of phase with the peak nesting period of the group. Apparently, birds that nested earlier or later than the rest suffered high nest predation.

As a second general method, Tinbergen suggested starting from the other end; describe an observable behavioral pattern and then find out what it does. For example, what is the advantage, if any, in the "fanning" behavior of the male three-spined stickleback? Fanning is done mainly when there are developing eggs in the nest. The pattern involves a particular movement of the pectoral fins and the tail, which results in a flow of water around and through the nest. When the fanning was prevented, or when a watchglass was placed over the eggs to shield them from the current, the eggs died. The eggs survived, however, even in the absence of a male if they were fanned or aerated artificially. The tests showed that fanning is adaptive as a ventilation mechanism and that the rate of fanning is regulated so that the eggs always receive enough oxygen.

We should also remind ourselves that the search for the adaptiveness of any pattern is complicated by the fact that evolutionary directions may change. Anyone who wishes to compare a domestic strain of rat to the wild rat is making a mistake. *The* wild rat doesn't exist. Instead, there are multitudes of wild populations out there, each adapting to its own kinds of environments and following its own evolutionary pathways (Mitchell, Beatty, and Cox 1977).



## HUMAN EVOLUTION

Throughout this book you may be tempted to draw parallels from animal behavior to the human condition. And that's fine—*as long as* we have some idea of where humans fit in the scheme of things. Such knowledge will simply prohibit extrapolations in some cases, but will permit us to derive potentially valuable conclusions in other instances. It is important for us to know something about the descent of our own species, especially since learning about our own behavior motivated many of us to study animal behavior in the first place.

A number of popular authors in recent years, such as Robert Ardrey (1962, 1966, 1970) and Desmond Morris (1967), have burst upon the scene telling us what zoologists have always assumed. Humans are not simply the sums of their individual experiences. Rather, those experiences are superimposed upon the experiences of the species—experiences that have altered the genes of the group through the millennia. In other words, each new human mind is not a blank sheet, as some would have it, awaiting the inscription of its experiences—experiences that will then comprise its total behavioral construct or "identity." The formative effects of individual experiences are more analogous to the variations in the way a house can be built *after* the foundation has been laid. Any number of variations are possible, *within the limits imposed by that foundation.* So, let's take a look at our own evolutionary history, our foundations.

Perhaps our first impulse, in trying to learn what our ancestors were like, is to examine the living nonhuman primates around us. Almost anyone will quickly point out that our lines are closely related to those of the chimpanzees (in fact, our blood proteins are virtually indistinguishable from theirs) (Dayhoff 1972; King and Wilson 1976). But we behave quite differently from chimpanzees. An estrous female chimpanzee will copulate with almost any male she encounters, behavior untypical of human females. It has also been suggested that our social system is more like that of the social carnivores such as wolves (which it is). But, even here, the parallels are not complete. Human mothers don't go out on hunts, leaving their babies with aunts, and coming home to vomit undigested food for the aunt and children. Such behavior would be considered unseemly. And it has been suggested that we developed on the grassland and savannah, as did most baboons. But baboons rarely kill and eat game (Figure 4.9). On the other hand, the search for game is probably what drew our forebears to the grassy savannah in those early days. We are undeniably different. Our ancestors were probably unlike any living animal.

Figure 4.9 Modern savannah primates do not actually hunt, but they may kill and eat the young of other mammals that they come across. Shown here is a savanna baboon eating a young Thomson's gazelle.

## The Fossil Record

We probably separated from the line that led to the great apes about twenty to thirty million years ago. The prevailing theory is that this line changed for millions of years until it comprised the earliest known animal that humans will claim as kin, *Australopithecus africanus.* Fossils of this primitive hominid (manlike form) have been found in rock strata four million years old. The line continued to change as the eons passed, until about 500,000 years ago a form resulted that was different enough for taxonomists to place it in a different genus, *Homo erectus* (Figure 4.10). There were important differences between *A. africanus* and *H. erectus.* For one thing, *erectus*

Figure 4.10 *Australopithecus africanus* on the African savannah where they evolved. *Africanus* is generally not believed to be the ancestor of *Homo sapiens,* but the evidence is derived from scanty fossil data and so the human lineage is highly debated.

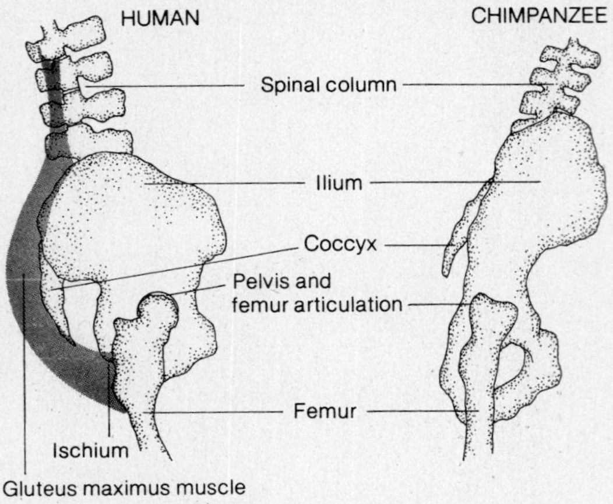


Figure 4.11 The human pelvis (left) and chimpanzee pelvis (right) compared. The expanded ilium in humans provides a surface for those muscles that aid in standing erect and in pulling the femur backward when running, as does the gluteus maximus.

tended to stand more upright, hence its name (Figure 4.11). It also had a larger brain (Figure 4.12) and a hand that was better able to manipulate things (Figure 4.13).

*Homo erectus* was also subject to natural selection and changed through time until the appearance of modern humans, *Homo sapiens,* about 250,000 years ago. The greatest difference between these two species was in the size of the brain, which increased from about 1,000 to 1,300 cubic centimeters. Interestingly, the increase in brain size is marked by significant changes in the remains of early human settlements (Binford and Binford 1969; Pfeiffer 1970). Obviously, our behavior was changing all the while.

## Tool Use

The notion of tool use has fascinated anthropologists for some time, and recently the ethological literature is filled with instances of animals using tools. Some cases were quite unexpected. For example, Egyptian

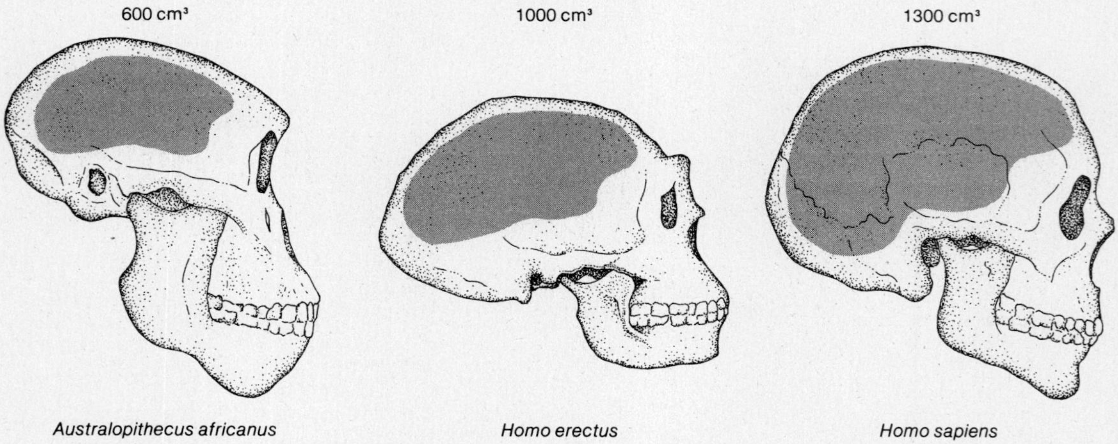


Figure 4.12 The skulls of two fossil hominids and that of modern man. The relative volumes of the brain cases are indicated. Many anthropologists believe *Homo erectus* to be in the human lineage.

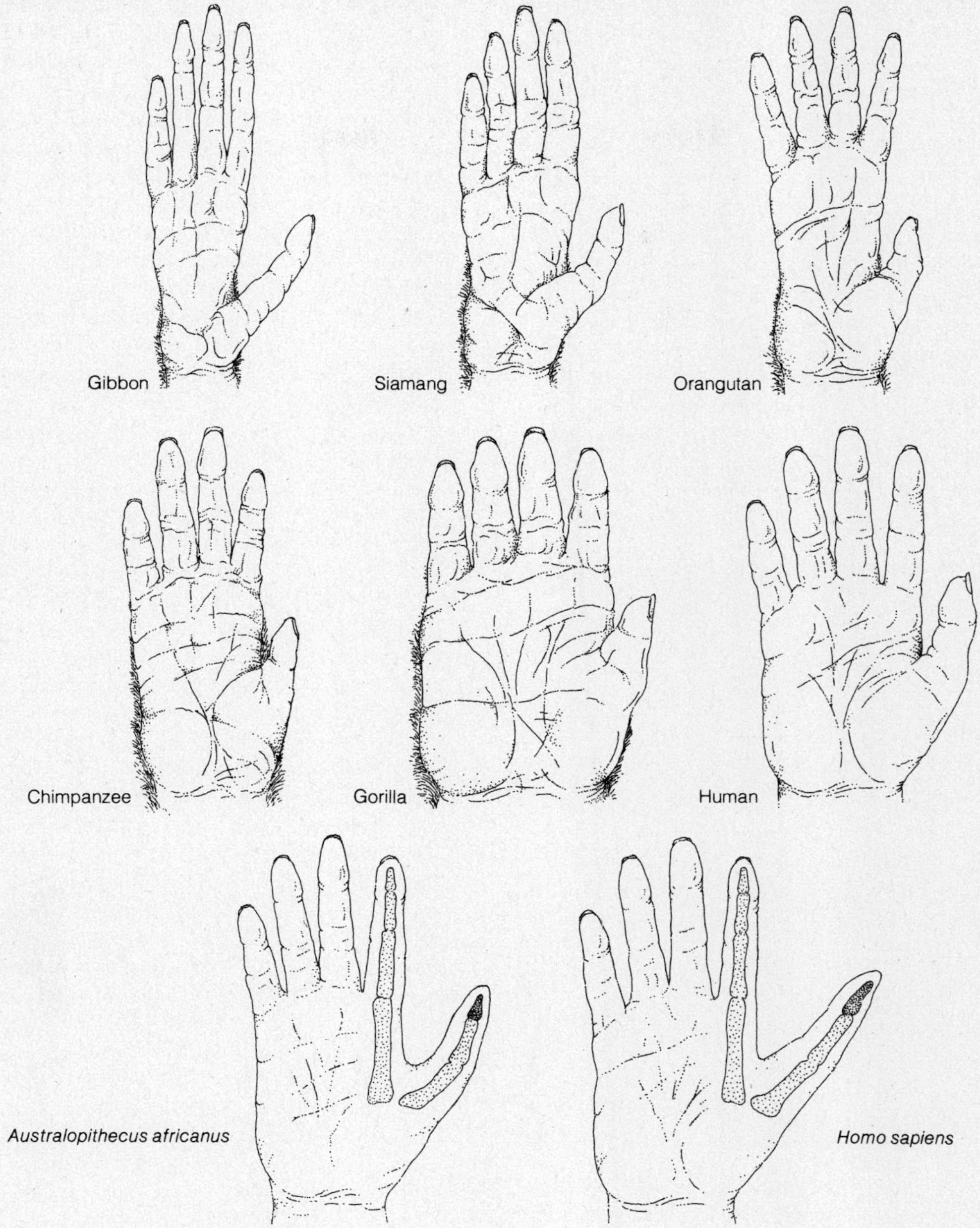


Figure 4.13 A comparison of the hands of six adult primates, adjusted to the same length, indicating the trend toward manual dexterity. The anatomical comparison below shows the longer, straighter thumb of *H. sapiens,* which permits a strong and precise grip.

vultures have been observed throwing stones at Ostrich eggs (Goodall and van Lawick 1966), a behavior so remarkable as to suggest that our list may not yet be complete (Figure 4.14). I personally am not very interested in tool using in nonhuman animals except for what it can tell us about natural history. I tend to agree with the rather remarkable primatologist K. R. L. Hall who said that tool use does not necessarily indicate a higher form of intelligence. A more interesting question is, What does early handling of objects have to do with using tools later in life? Consider the question again after you have dealt with our discussion of learning.

The historical progression of tool making and its use in human evolution is interesting, however. The remains of the small-brained *A. africanus* are found with only the most rudimentary sorts of tools (Figure 4.15). *H. erectus*, however, was somewhat more proficient at fashioning stones to use in ways that remain a puzzle. But early *H. sapiens* developed into a toolmaker par excellence. In fact, recently, some anthropologists have attempted to make stone tools and have been astounded at the difficulty in making anything even resembling the smooth and symmetrical forms of the late stone-age man. The smoothness and symmetry of their stones would have permitted them to be rapidly whirled or thrown with great accuracy. It would be most interesting to know just how good these people were with their tools and weapons.

Figure 4.14 Tool use was once considered a uniquely human trait. We now know several species use tools. For example, the Egyptian vulture drops rocks over the shells of ostrich eggs to smash them. Its aim is terrible.

In time, man's tools became more complex. He built bows and spear-hurlers. He also built boats since he is naturally a poor swimmer (like the gibbon, whereas the quadripedal macaque swims quite easily since it doen't need to learn any new movements). He also began to hunt with dogs, and, even today, bushmen who use dogs are more successful hunters than those who don't (Washburn and Lancaster

It is remarkable that nearly all estimates of the height of fossil men fall between five feet two inches and five feet ten inches. The uniformity in size is surprising. It is obvious, then, that throughout almost all of man's evolution there were strong selection pressures keeping his size within rather strict limits. It is only recently that we have been freed from these contraints, and drastic size differences have resulted (but see Pilbeam and Gould 1974). Tinbergen (personal communication and see 1974) believes that there may be an inherent risk in our sudden surge in size in the industrialized countries. We spent millions of years developing within rather narrow size limits, but the improved diet and medication of the industrial revolution have suddenly propelled us to new physical dimensions. Tinbergen suggests that our new height and bulk are placing new kinds of stresses on a system never intended to withstand it. Perhaps, also, the result is physical or psychosomatic stress. It would be most interesting to see correlations of various sorts of stress diseases and body size. (By 1985, the United States Air Force estimates the average airman will be a little over 5 feet 10 inches and will weigh about 177 pounds. In 1967, the figures were 5 feet 9¾ inches and 173 pounds. The average union soldier in the Civil War was 5 feet 7½ inches and 139 pounds [*Newsweek* 1975]. You may have been surprised at the diminutive sizes of knights' armor you have seen in museums.)

## Hunting

Although the genus *Homo* has existed for about a half a million years, agriculture has been important only in the last five thousand years or so (Washburn and Lancaster 1973). In fact, the entire evolution of the earliest forms of *H. erectus* into its different races (of which there were a number) took place while he was a hunter. As is discussed in Chapter 30, these primates came out of the jungle, onto the savannah—a move that was associated with a switch to meat eating. Thus, the progession in tool use was very likely associated with increasing proficiency in bringing down large prey. The important point is that the factors that produced *Homo sapiens* were preagricultural. Agriculture, as a matter of fact, has dominated less than one percent of human history, and there is no evidence of major biological changes during that time (Washburn and Lancaster 1973).

What agriculture did do was to provide man with a stable and storable food supply. Thus, he no longer was dependent on the environmental whims that dictated the presence of large game. Agriculture thus enabled man to reduce the risks of starvation and was a major factor in the rise in the human populations, as we will see later.

So, in spite of certain similarities with other animal groups, *Homo sapiens* has had an evolutionary history uniquely its own. We are derived from social, sexually oriented, hunting primates. We have exaggerated learning abilities that are the results of millennia of natural selection. But we are more than naked apes. We are a different kind of animal. The question is, how different? And from what?

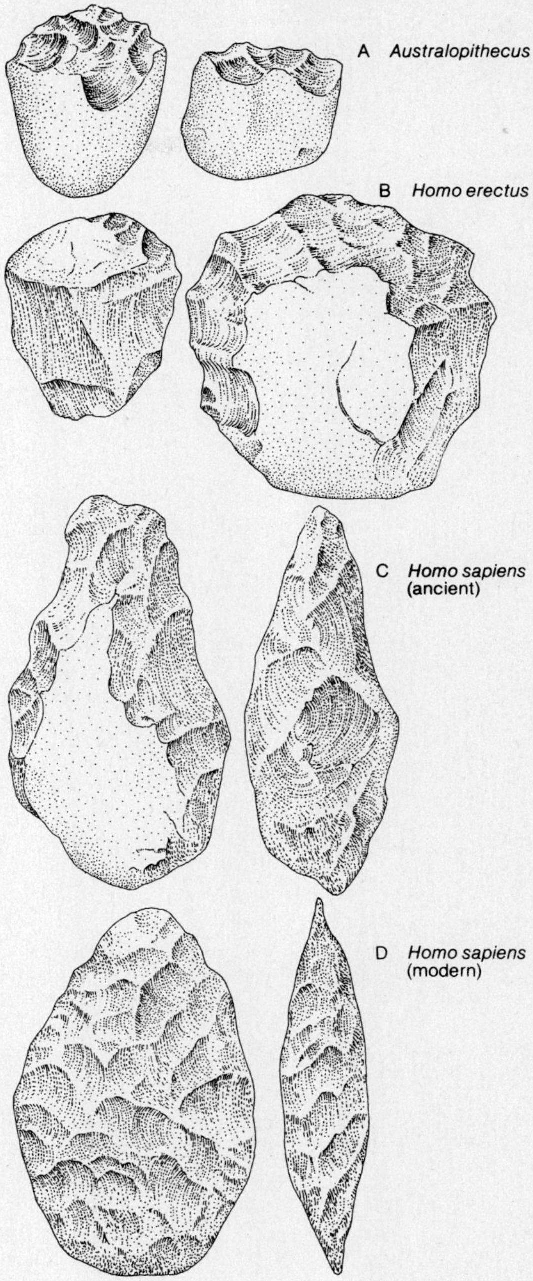


Figure 4.15 The evolution of stone hand tools. The chipped rocks attributed to Australopithecus are so roughly made that they scarcely look like tools at all. *Homo erectus* made better tools, but the art gradually culminated in the work of *Homo sapiens*.

# 5

# THE HOMOLOGY OF BEHAVIOR

"Is it not possible that beneath all the variations of individual behavior there lies an inner structure of inherited behavior which characterizes all the members of a given species, genus, or larger taxonomic group—just as the skeleton of a primordial ancestor characterizes the form and structure of all mammals today?"

The question was asked by Lorenz in 1958. The concept, however, was born much earlier. The American psychologist C. O. Whitman (1899) laid the groundwork by stating that instinct and structure are to be studied from the common viewpoint of phyletic descent. Whitman was not at all bothered by the special difficulties in such a proposal, such as in following the development of an instinct the way one can trace the development of an organ. Those who attempted such studies, however, met problems beyond the ones encountered by other evolutionists who used the tools of anatomy and paleontology. A number of researchers have pointed out the critical nature of the lack of paleontological evidence in behavioral studies (Blest 1961; Hinde and Tinbergen 1958; Mayr 1958; Simpson 1958; Tinbergen 1951; Wickler 1961).

## FORMING CRITERIA AND OTHER PROBLEMS

What is actually involved here is whether or not behavioral studies can tell us much about evolutionary relationships between animals. In other words, can behavior be homologized? *Homology* has been described as resemblance due to inheritance from a common ancestry (Simpson 1961). To give an anatomical example, the bone structure of the flipper of a whale is more similar to that of the arm of a man than to the pectoral fin of a fish (Figure 5.1). The reason is assumed to be because the whale, the descendant of a land vertebrate that returned to the sea, is more closely related to man than to fish. In other words, the dichotomy separating the ancestors of whales and men occurred later than that which separated the fish from land animals.

Today, the homologizing of behavior is more likely to be attempted by ethologists than by comparative psychologists, but there is some argument over whether the attempt is valid at all (Atz 1970). It has been argued that the principle of homology is in some sense tautological in its reasoning: evolutionary relationships are based on homologies, and homologies, in turn, are explained on the basis of phylogeny (relationships). In a practical sense, the student of behavior can avoid the problem to a degree by studying animals that have been already classified on some other basis, such as through the usual comparative morphological studies. In addition, the behavioral scientist should proceed with inordinate caution. It seems that, perhaps too often, we tend to jump to conclusions about behavioral homologies. There is often a disconcerting tendency to label a behavioral character as homologous before such a judgment is warranted and to ignore contradictory evidence. Besides stressing the need for objectivity, I hope here to describe the limited value of attempting to homologize behavior and to discuss what criteria should be met before homology is indicated.

Generally, there are four criteria for determining morphological homology: (1) position—homology is indicated if structures occur within comparable anatomical areas; (2) special quality—characters might be homologous if they agree in several unusual characters; (3) constancy—dissimilar structures may be homologized if intermediate connecting structures are found in related species; (4) distribution—the probability of homology increases with the frequency of occurrence of a pattern among related species and decreases with its appearance in unrelated forms (Remane 1961; Atz 1970). It may be said that the more continuously distributed and the more similar a display is throughout a group of animals, the more probable is its common origin in the group.

The criterion for behavioral homology is basically similar to that for morphological homology: two patterns must appear in two related forms and must have been present as a single behavior in their common ancestor (Wickler 1961). However, the above stringent criteria for morphological homology cannot be rigidly applied to behavior. For example, how could one determine "position" in a behavioral pattern? Such a criterion would demand precise knowledge of the location of the neural center of the pattern and such information is, at present, generally not available. So, already we see that behavioral homology involves more assumptions than that based on morphology.

## BEHAVIORAL HOMOLOGY AND LEARNING

The problem of behavioral homology is conceptually more difficult in some types of animals than others. For example, mammals present special problems because of their behavioral flexibility. They have a great ability to learn and thus to alter their behavior on the basis of individual experience. The investigator must be sure that the differences between such flexible species are based in genetic diversity and not in environmental dissimilarity. (For example, a chimpanzee from a forest group might have a different behavioral repertoire than one from a savannah group where trees are sparser and vision is perhaps more important. But such differences are probably learned.) The investigator must also have a good idea

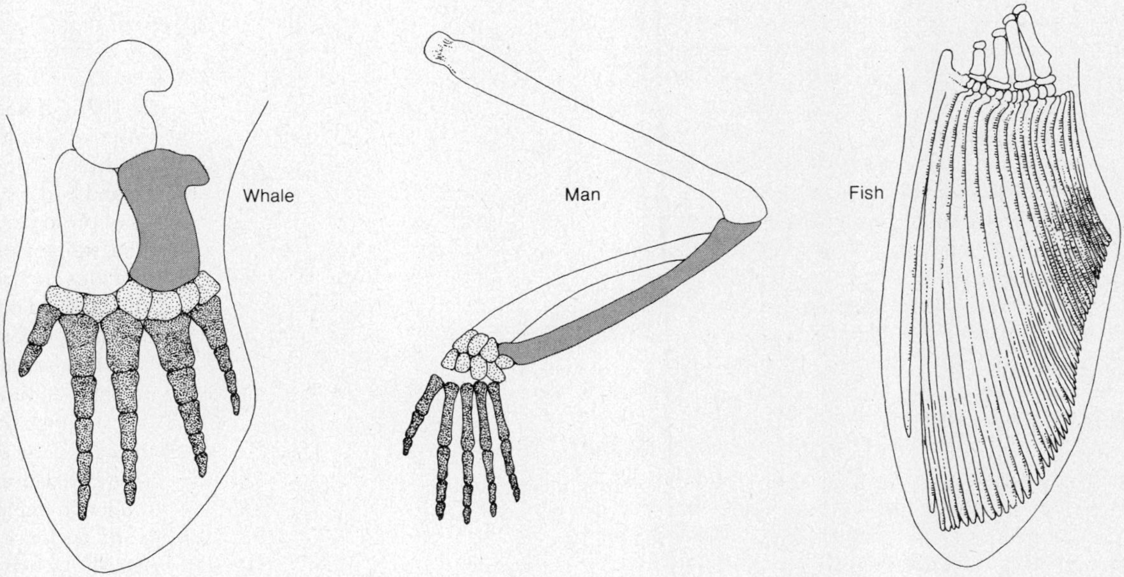


Figure 5.1 Vertebrate forelimbs. The whale's limb is functionally similar to the fish's fin, but in fact it is evolutionarily more closely related to the human arm, demonstrating morphological homology.

of the degree to which learning has affected an innate pattern or what part of the entire adaptive pattern is due to learning. (For example, an inexperienced ringdove must *learn* to walk up to its chicks in order to perform an *innate* pattern—feeding; see Lehrman 1955.)

Tinbergen (1951) maintained that only the "innate basis" of behavior could be homologized (see below); others disagreed and argued that either learned or innate patterns could be treated in this way (Wickler 1961). One reason sometimes given for homologizing learned patterns is that learning *abilities* of animals are inherited anyway. However, such an argument neglects the fact that under natural conditions animals may learn far less than is actually possible, and members of the same species living under different environmental conditions may learn very different things. So, in general, learned behavior is probably not a valid subject for homology.

## CONSERVATIVE AND LABILE CHARACTERS

Some behavioral characters are considered labile in that they may evolve rapidly. Conservative characters are those that change little, maintaining a degree of constancy even through drastic evolutionary changes of its bearers. Some characters are actually so conservative as to be useless in homologies, and others lose their value by their very lability. Morphologically, you might consider the paired eyes of vertebrates as a conservative character and the marking of tricolored dogs as a labile pattern. Neither tells us anything about relationships. In other cases, environmental pressures can alter a rather conservative character so that it loses its taxonomic value. For example, many types of animals are born with an escape reaction to fast-approaching objects. But then, they can *learn* to ignore such stimuli as skittering leaves or passing automobiles. So, just because a pattern changes, it doesn't mean that it could not be innate. The hoarding tendency in some rats may be reduced or obliterated by training, but this does not mean they lacked the genetic tendency to begin with.

Some patterns stubbornly resist any environmental modification of their expression and show a large degree of independence from environmental stimuli. For example, dogs and some birds both scratch their heads by reaching the scratching foot in front of the shoulder (Figure 5.2). So do almost all legged reptiles. In the case of the bird, it would be much simpler to move its foot directly to the head without moving the wing at all. It has been suggested that the motion in birds is a relic of their reptilian ancestry (Heinroth 1930). Lorenz (1958) interpreted the performance of the clumsier action as an indication of an extremely conservative inborn behavior. He recommended such conservative patterns as tools for evolutionary study. It must be admitted, however, that such an invariable pattern as the scratching behavior has little to recommend it beyond indicating the relationship of broad classes of animals—so in this case behavioral homology is a rather blunt instrument. In general, however, conservative factors are indeed more useful than labile ones in the reconstruction of phylogenies (Mayr 1958).

Even within a single adaptive pattern, the various components may vary in their evolutionary lability (Mayr 1958). For example, in birds, courtship move-

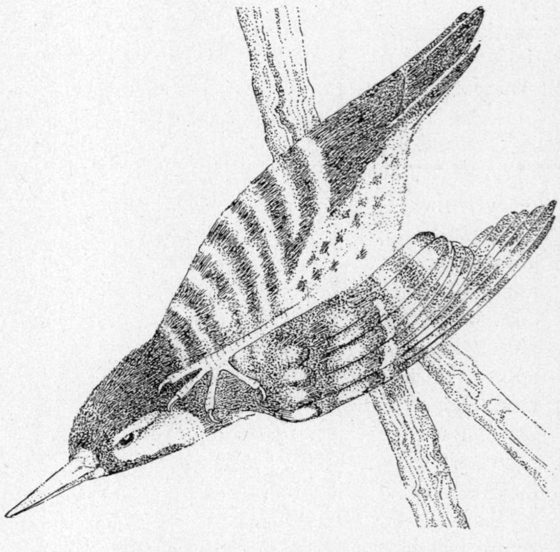

Figure 5.2 Many bird species scratch in the manner of reptiles and mammals in that the scratching leg is brought up behind the forelimb. This is an awkward movement and is considered to be due to inheritance of the pattern from reptiles.

ments seem much more conservative than the neural centers that respond to the appropriate environmental stimuli. In other words, changing environmental stimuli continue to elicit the same behavioral responses.

Researchers using behavioral homology have stressed the importance of selecting the appropriate character (Hinde and Tinbergen 1958). We have seen that strongly labile or extremely conservative characters are not suitable for studying relationships between closely related species, but it was pointed out that conservative characters gain in value as the relationship between the species decreases. For example, all species of tits in the genus *Parus* (family: Paridae) nest in holes and use moss for nests. The behavior differentiates *Parus* from others in the Paridae family, but does not distinguish between species of *Parus*. (Probably, in general, nest building itself is not suitable for homology for reasons discussed below.) As another example, the precise "tail flicks" made by passerines when moving through the foliage are conservative within families and can help assign genera to families (Andrew 1956). Threat postures and some courtship displays vary between closely related groups, and so are useful for studying relationships between genera or species. However, these patterns are often too divergent to be used to distinguish between families (Hinde 1952).

## ANALOGOUS BEHAVIOR

As with morphological characters, one cannot use behavioral comparisons to determine taxonomic relationships if the behavioral similarity is due to analogy (similar characters acquired independently in "unrelated" groups). An example of an analogous pattern is the high-pitched warning call given in many species of birds. Such a call may have developed independently in several distantly related species because it carries well and its source is difficult for predators to detect. Thus, warning calls cannot be used to indicate relationships.

Since there is always the possibility that character similarity in different groups is due to "convergence," the "special-quality" criterion discussed earlier is increasingly important. Lorenz, by the way, once proposed that "display behavior" is unlikely to show convergence, but this has been shown not to be the case (Marler 1957a; Tinbergen 1959). For other discussions on the convergence of behavior, see Blest (1961); Cullen (1959); Remane (1961); Mayr (1958).

Social systems also do not lend themselves to homology because of their complexity and because similarity may be due to convergence rather than relationship. We will see in the discussion of "grades" among primates, later on, that there are strong environmental effects on social structure that could lead to convergence, or analogous behavior, in animals living in similar habitats. In addition, social systems may be exceedingly labile. For example, in times of high food supply, a baboon troop may swell because nonbreeding males join breeding units at this time. But with the onset of seasonal food paucity, some males may move off in bachelor herds to forage separately. The social structure of these troops, then, would be partly a function of the time of year they were observed.

The assessment of behavioral development is made particularly difficult by the interdependence of various patterns. For example, among passerines (perching birds) territoriality, song, sexual dimorphism, and aggressiveness may all be linked in a complex feedback system so that a change in one may shift the expression of the others (Hinde and Tinbergen 1958).

Since similarity between two species does not necessarily mean common descent, the student of evolution is forced to decide whether a certain pattern is due to homology or analogy. As we have seen, the decision is more difficult for the behaviorist than for the anatomist. The anatomist can always fall back on such rules as "Owen's criterion" of homology, which stresses the usefulness of general anatomic position and the relationship of adjacent organs. There are those who would say that since the behaviorist lacks this knowledge, homologizing of patterns is beyond his realm. But considering the advances in critical areas of biology, such temerity may not be merited.

## TOWARD A MORE RIGOROUS APPROACH

We may ask, then, what criteria can be used to establish homology? Actually, the method is essentially the same as that used by taxonomists even before Darwin: conclusions must be based on the sum total of behavioral characters. Since the study of evolution by behavioral description faces so many special problems, it is important that the attempt to create phylogeny (relationships) is first postulated on morphological grounds, at least until the data are found to be contradictory. In addition, the use of behavioral homologies in the study of evolution will most likely avoid the problems mentioned here if only special movements called fixed action patterns are compared. (The idea is not new; in fact, it was employed in some of the earlier work of Tinbergen [1942] and Lorenz [1950].)

Fixed action patterns are those movements that seem to be innately based in that they are performed much the same way among all the individuals of a species. Threat postures and courtship displays are examples. Fixed action patterns are not only rather invariable within populations, but are not often greatly affected by learning or practice. However, a problem arises in that these patterns may appear as parts of more complex behaviors. Thus, the fixed action pattern must somehow be separated from the overall behavioral sequence (Whitman 1919; Heinroth 1910). A complex behavioral sequence, such as that which is involved in nest building, may be made up of a *series* of fixed action patterns. In some cases, the performance of one pattern may even serve to initiate the next so that the proper sequence is maintained. Learning could operate in many ways on any such complex pattern, but it might function primarily in altering the timing of any fixed pattern. Learning could also operate, of course, to improve the performance of a motor pattern—smoothing out the rough edges of the innate behavior.

In any case, however, complex adaptive sequences are far too intricate, and too little understood, to be the subject of homology. We just can't deal with overall adaptive patterns that are composed of multiple fixed action patterns and that can be altered by learning. We must concentrate on a single, genetically based component of the overall pattern. So, if nest building or other complex acts may not be homologized, yet we wish to utilize, for example, nest-building behavior, we might concentrate on a fixed component of the behavior—particularly one that could have no other use than as a very special nest-building action. For example, we might wish to homologize the peculiar sideways sweep of the head that is used to insert material into the fabric of a nest once the nest has reached a certain stage of completion. (Quantitative or qualitative differences in such patterns, or differences in the sequence in which equivalent fixed patterns are employed, may, of course, result in different kinds of nests.)

As a final note, because there may be problems in identifying fixed patterns, one might wish to employ only those patterns that appeared in out-of-context behavior, such as we will see in play (Chapter 11) and in "vacuum" behavior (Chapter 8).

## The Search for Neural Controls

A fixed action pattern, it must be remembered, is also composed of component parts, and all those parts must be rather rigidly controlled in order for the pattern to have adaptive value at all. For example, Figure 5.3 shows the sequence, duration, and

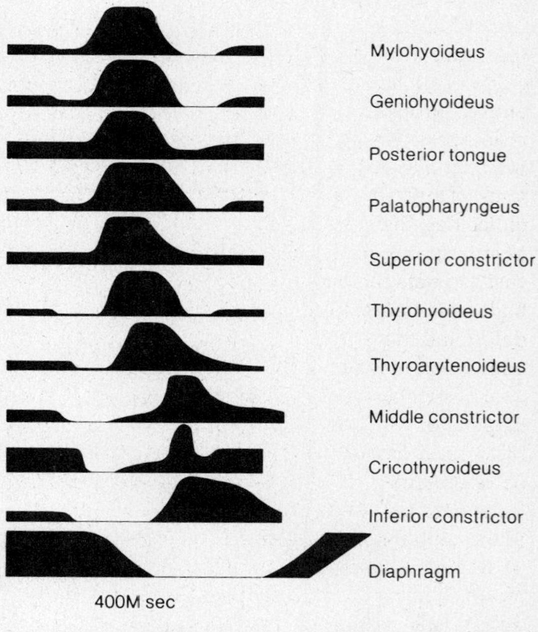


Figure 5.3 Diagrammatic summary of the sequence, timing, and intensity of muscular contraction for swallowing in the dog. The height of the shaded area shows the strength of the contraction for each muscle involved.

strength of contraction of a set of specific muscles involved in the fixed pattern of swallowing in dogs. The contraction of each individual muscle must occur within rather narrow limits. A major change in the action of any one of these muscles would probably be strongly selected against since the swallowing "action profile" is so highly coordinated. In other words, even a "simple" fixed action pattern, such as swallowing, might be maintained within tight limits by the complexity of its component interactions. It doesn't matter whether such a fixed action pattern is under the control of one neural center or several, a great change in any of its components is likely to render the resulting pattern unadaptive and useless.

Ideally, one would wish to compare all fixed action patterns in terms of its components. We would, of course, need action profiles for these behaviors. Such descriptions would greatly increase the validity of comparisons of behavior, but the problem is, we just don't have enough such action profiles to permit comparisons. So what can we do? Realistically, what should be our goal?

To start, field and laboratory studies might help us to select the appropriate characters for comparison. Such studies, of course, require great knowledge of the natural histories of the animals, and few field workers are familiar enough with the behavior of any animal group. (One of the great imperatives of ethology is "Know Thy Animal.") In addition, laboratory experiments present formidable technical problems (Doty and Bosma 1956). In spite of such difficulties, the comparison of a behavior at the level of specific muscle interaction would help to close the vast gap between our knowledge of adaptive patterns and their neural control. One might then seek to determine the action of the next largest component of muscles, muscle bundles (Figure 5.4). The ultimate goal in such a study would be to discover the precise innervation of the motor components and then to locate and describe the associated centers of these patterns in the central nervous system. This, however, is the hard part. A review of the complex interactions of the central nervous system should dispel any hope of finding a discrete control center for a specific motor function. For the present, then, we must work within our available methodological framework, gleaning what is essentially circumstantial evidence.

Ethologist Peter Klopfer (1969b, 1973) warned that associating motor patterns with their neural centers may be very difficult because the same behavioral pattern might be the result of any number of neural and muscular interactions. If he is correct, the likelihood of ever unequivocally homologizing behavior becomes even more remote. (See also Marler and Hamilton 1966; Barlow 1968.)

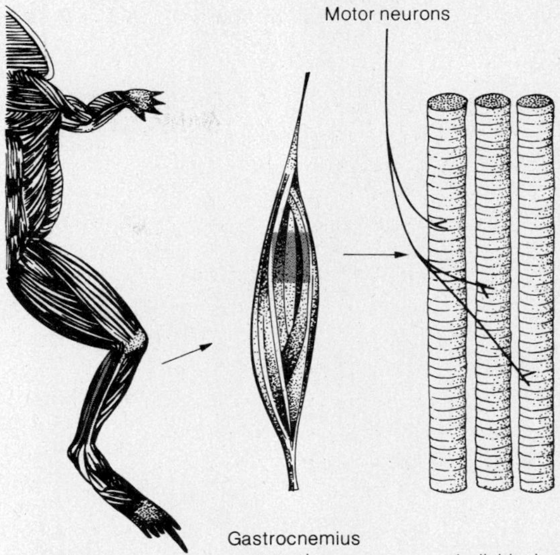


Figure 5.4 The gastrocnemius (shown here in the frog) is a muscle composed of muscle bundles and is innervated by nerves stemming from various neural centers. Behavioral homology ultimately must include descriptions of the neural roots of the neurons stimulating each muscle bundle.

## HOW PATTERNS MAY VARY

In some cases, the same fixed action pattern may appear throughout a group of related species. Any such patterns may be considered to be truly homologous, having descended to each species from a common ancestor (Manning 1971). Even homologous patterns, of course, are not identical in one species to the next. They can be expected to vary somewhat between species due to variations in

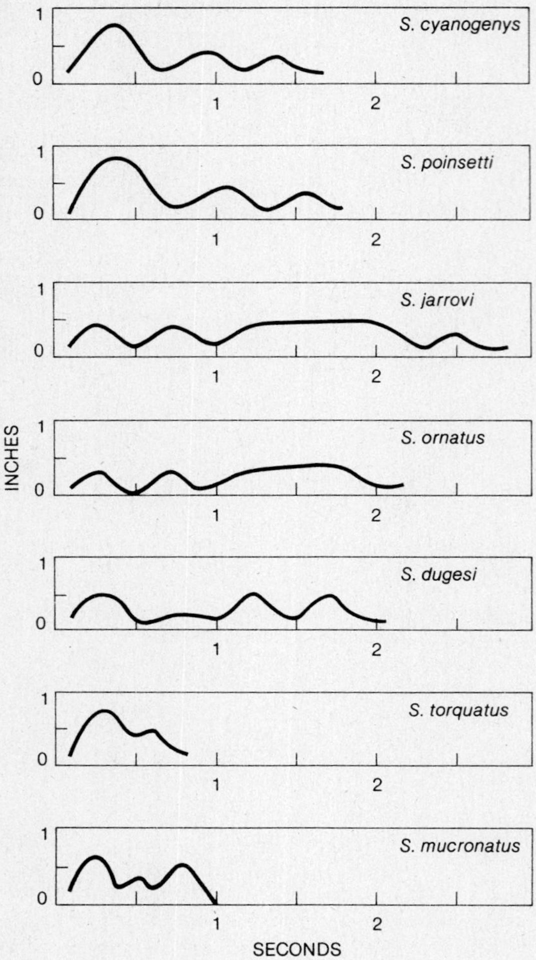


Figure 5.5 The head-bobbing patterns of some species of *Sceloporus* lizards. The height of the head movement is shown in the vertical axis. Variations in speed, amplitude, and pattern of movement are shown on the horizontal axis.

natural selection operating on different species. The differences may arise in a number of ways. For example, patterns composed of the same component parts may vary due to differences in the emphasis with which each part is expressed. Also, they may vary in how rapidly they are performed, as is demonstrated in lizards that perform "pushup" displays at different rates (Figure 5.5). In addition, Tinbergen (1959) showed that a group of gulls may have the same behavioral repertoire, with each species bearing a characteristic probability that any pattern will appear.

Drastic changes in the probability of a pattern appearing can cause qualitative changes in behavior. Two species of fruit flies both showed "scissoring," in which both wings are opened and closed slowly, and "vibration," in which one wing is extended and vibrated rapidly in the vertical plane. Scissoring was far more prevalent in one species, *Drosophila simulans*, and vibration appeared more frequently in the other, *Drosophila melanogaster* (Manning 1959). In some cases, a pattern can become so infrequent as to disappear, thus causing a *qualitative* change in the repertoire.

Changes in the frequency with which a pattern is performed may reflect shifts in the *threshold* for the stimulus of that pattern (Manning 1971). It is possible that changes in threshold are responsible for some of the behavioral differences between related species. The threshold changes could be brought about by alterations in the central nervous system, the sensory system, the motor system, the "motivational system," or any combination of these. In addition, changes in threshold could directly be brought about by genes acting on the nervous system itself, or indirectly by their affecting motor rate or hormone secretion, which in turn affect the motor system.

While proof is not well established, there is some indication that frequency alterations are likely to be produced by changes in the nervous system itself. Experiments on two inbred lines of guinea pigs illustrated that differences in the frequency of reproductive patterns of the two lines remain constant even after one is injected with sex hormones (which would probably alter motivation and receptor threshold), and thyroid hormones (which would increase the rate of metabolism and general level of activity). These results indicate that the centers controlling theshold for reproductive patterns lie in the brain, not in peripheral systems such as the receptors or the muscles. Of course, the assumption here is that peripheral controlling mechanisms are more likely to be affected by hormones than central ones; and there is, after all, evidence that the brain itself can be altered by hormones (Manning 1971).

Fixed action patterns may be altered by even small changes in threshold. Each fixed action pattern is believed to have a coordinating "center" in the central nervous system that is inherited. This center, when activated, apparently produces a stereotyped pattern of neural signals to lower centers (see Chapter 8) that control muscles and groups of muscles. As the lower centers activate each muscle group in a particular order, at the correct time, and for the proper duration, the result is the fixed action pattern. The pattern may be completely stereotyped, as is probably the case in swallowing, or it may be altered by environmental feedback operating on the orientation component of an instinctive pattern (see the egg-retrieving experiment on a nesting goose or the variable orientation in inciting behavior in female ducks in Chapter 8). It has been suggested that genetic changes that affect thresholds (either within the coordinating centers or in one of the subordinate muscle groups) will alter the fixed action pattern so that certain muscles might be activated sooner or later in the sequence, held for a longer or shorter time, or changed in the intensity or speed of their response as more or fewer motor fibers are activated (Manning 1971).

So, we see that because of the numerous components (muscles) involved in any fixed action pattern, each with its own specific characteristics, changes in the patterns undoubtedly occur very slowly. In such a coordinated system, any change in a component is likely to be "dissonant," or strongly selected against, since it is unlikely that other components would be simultaneously adjusted to provide a harmonious whole. Thus, changes in component parts of fixed action patterns are not likely to be frequent or pronounced. Threshold changes for the *performance* of the fixed pattern, however, occur more easily. In regard to this latter point, threshold for a stimulus may be affected by the animal's early experience (see reviews in Hinde 1970; Marler and Hamilton 1966). For example, when young male rats are exposed to different rearing conditions, social experience, and early handling, the age at which sexual behavior can first be elicited varies widely, but the sexual behavior itself remains unchanged (Manning 1971).

It can be shown that the threshold for some fixed action patterns is dependent upon other adaptive characteristics of the animal. For example, noxious moths with warning markings move at the slightest touch (low threshold), whereas palatable species are much harder to budge, since it is to their advantage to remain unnoticed (high threshold). As a different sort of example, consider the "intention movements" of birds. A bird intending to fly crouches, raises its tail and wings, and draws in its neck. In some species, this intention movement has become incorporated into display, or signaling, patterns, particularly in shy species. Shy birds, such as those that serve as prey species, are more inclined to "take off," and thus they might be more familiar with the flight-intention movements than bolder birds. These shyer species are more likely to incorporate flight intention movements as a signal. In some finches, for example, the sexes are very wary of each other and tend to fly away at the slightest threat by the opposite sex. Thus, we might expect flight intention movements to be a part of their mating behavior (Hinde and Tinbergen 1958).

Any behavior may become more "fixed" if it is used in communication, especially as a sexual display, to elicit an innate pattern in another individual. It doesn't matter whether the signal maintains its original adaptiveness (such as do flight intention movements in birds, when the movement is read as a signal by another individual, but the signaler is, in fact, ready to take off). But patterns may also be released from their original adaptiveness and come to be employed *only* in signaling (such as the dancing with "veils" in empid flies discussed in Chapter 8). Once a fixed action pattern becomes used in communication, it is subject not only to the constraints against change within the nervous system of the signaler, but also to the constraints imposed by the *recipient* of the signal. After all, any great change in a signal might render the pattern unrecognizable, and the behavior would thus lose its function in communication.

## WEIGHTING CHARACTERS

Tinbergen (1959) advocated "weighting" characters when determining phylogenetic relationships. The procedure is sometimes not rigorously employed in morphological comparisons, but it is essential in behavioral studies. The weighting is done as follows: when species are believed to be related, based on a number of similarities, the *differences* between them are considered to be recently acquired. These differences can then be "peeled away" (or ignored) to permit examination of the phylogenetically older "core." The core behaviors of the species can then be compared. If such cores are more similar between species than their total complex of characters, affinity is indicated. If the cores are more different than would be suggested by the total complex, then common descent of the species is doubtful. To illustrate, consider Esther Cullen's work in Chapter 4 showing that the differences between the behavior of kittiwakes and ground-nesting gulls are largely due to differences in their nesting sites. The nesting differences were considered to be superimposed on a common adaptive theme. The core behavior of the two groups *is* similar; the differences are due to relatively recent evolution. Cullen's detailed ethological study was able to show which behavior was to be "peeled away" and what parts were subject to valid comparison in evolutionary studies. When the kittiwake's peculiar mouth-open "choking" display is stripped of its mouth-open quality, the result is a display *more* similar, not less, to that of other gull species.

As another example, facing away or head flagging in black-headed gulls is performed more strongly than in gulls with less distinctly marked heads. The more extreme turn of the head hides the mask and bill, which are used in aggressive displays. A similar behavior is seen in kittiwake chicks. But why? It turns out that the behavior is a corollary of their peculiar nesting habitat, since they must appease neighbors nesting nearby (within striking range). The pattern is performed a bit differently in black-headed gulls and kittiwakes. But if those differences are ignored, the adaptive appeasement role of the pattern becomes obvious. Once such morphological and ecological influences are understood, it becomes easier to see the strong homogeneity of the pattern throughout gulls as a group. So, "peeling off" the obviously *special* adaptiveness of a pattern (such as the more pronounced turning of the black-headed gulls) makes it easier to compare the *cores*.

I might jump ahead of our discussion here to say that the necessity of stripping to reveal the core behavior becomes apparent when attempting to homologize ritualized behavior (Chapter 8). Once a common pattern has been ritualized, it may be altered to such a degree that its original basis is no longer easily recognizable. Only careful comparison of a ritualized pattern throughout a group of species can reveal what the basic pattern of a ritualized behavior might be and which aspects of it have been added secondarily (Tinbergen 1962).

In any homology study, one would ideally search for a "progression" of characters—slight character differences from one species to the next—so that when the species are arranged from "primitive" to "advanced," a trend or cline becomes evident. The criterion of continuity mentioned earlier deemphasizes the stress on anatomical similarity. But this criterion is difficult to meet even in anatomical studies. After all, structures that are undoubtedly homologous may differ greatly anatomically, such as do the pineal eyes of reptiles and mammals (de Beer 1938, 1958). There are also certain theoretical problems in attempting to place various characters in their proper evolutionary order. Lorenz (1952) showed how a "progression" of characters may be devised from rather unrelated forms if certain evidence is lacking (Figures 5.6 and 5.7).

As a final note, I should add that homologous characters need not have similar developmental (embryological) origins as well as phylogenetic continuity. It has been suggested that: (1) homologous structures are not necessarily controlled by "identical" genes, and that characters controlled by similar kinds of genes are not necessarily homologous; and (2) homologous characters do not necessarily arise from similar parts of the egg or embryo (de Beer 1958).

## A PROCEDURE NOT INVOLVING THE SEARCH FOR NEURAL CONTROL

British ethologist Robert Hinde, with Niko Tinbergen (1958), reviewed the procedure for homology studies and described what we need to know for homology studies and what we can expect them to tell us. First, they said, one should look for similarities between species for whatever behavioral character has been chosen. Since the species are assumed to be related *on other grounds*, one could assume common an-

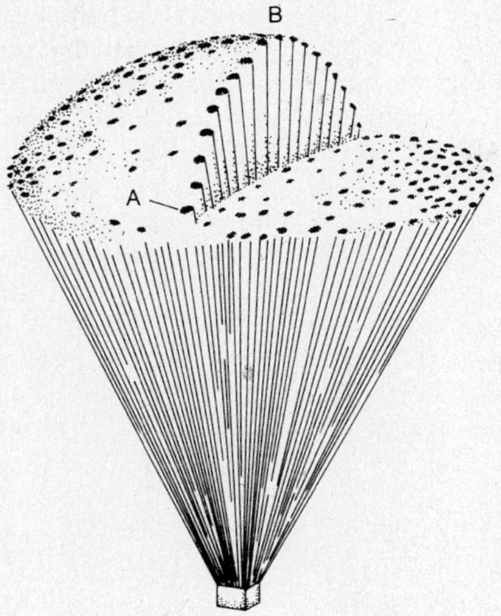


Figure 5.6 Diagram of a misleading "series" of animal forms from simple (A) to the more complex (B). In spite of the apparent sequence, no real phylogenetic series exists.

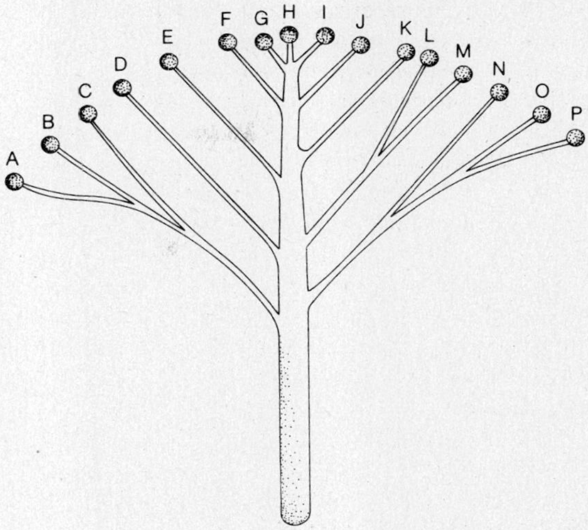


Figure 5.7 Similarities of animal forms may be based on true phylogenetic relationships. Here animals in rows A through P hold part of their evolutionary history in common.

cestry of the pattern. Second, the possible causation and function of the differences should be examined. Hypotheses would then be formed regarding which form of the pattern is older, thus establishing a basis for setting up an evolutionary sequence. Third, hypotheses could be developed regarding the probable evolutionary origins of the behavioral pattern. An effort might then be made to explain the differences between the hypothetical original and the present form. It could then be asked if the change has been adaptive, and therefore the result of natural selection, rather than due to some other phenomenon such as drift. Here, experimental studies might be employed in order to determine adaptiveness. Tinbergen (1959) added that to show adaptiveness, the comparative method should be used in conjunction with motivational analysis (which indicates the neural origin of a pattern) and studies of function (which indicate the selection pressures). The final result of such an effort might be a tentative description of the descent of a behavioral pattern, and this is about the most that can be expected from any comparative study.

There is a need for comparative ethological studies in order to understand more about those traits that appear in related species as well as why some traits are variable between species, and to what degree. For examples of such studies that have been made to date on a wide range of animals, you might wish to refer to: Crane (1952, 1957) on mantids and fiddler crabs; Blest (1957) on moths; Spieth (1952) on fruit flies; Crook (1963, 1970b) on weaverbirds and primates; Goss-Custard (1970) on wading birds; Aldrich-Blake (1970) on forest monkeys; Dilger (1960, 1962) on lovebirds; Lorenz (1941) and Johnsgard (1965) on ducks; and Tinbergen (1959) on gulls.

## Some Examples of Homologous Behavior

So, let's review some examples of behavioral patterns that have been considered homologous. In spite of the recent tendency to tighten the rules for homologizing behavior, some of the early work has remained relatively unscathed.

Whitman, in 1919, found that while no distinguishing morphological character is peculiar to all pigeons and doves, they all drink water by sucking it up through pumping the esophagus, rather than by tilting the head back as other birds do. The sandgrouse (Pteroclidae) has downy young that greatly resemble young grouse of the family Tetraonidae. For years, the sandgrouse was considered closely related to other grouse, until it was found that they drink by pumping water up. Comparative anatomists had found some similarities allying the sandgrouse with the pigeons, and so the behavioral work reinforced the anatomical findings. The result was that the sandgrouse was then placed into the order Columbiformes with other birds showing the peculiar drinking behavior (but not without argument; see Cade, Willoughby, and Maclean 1966). For another early study, see Friedmann's (1929) study of cuckoo relationships.

Courtship displays or species-specific rituals are sometimes considered as valid subjects for homologizing (Tinbergen 1959). Figure 5.8 shows, for example, the homologous "grunt-whistle" display in different species of ducks. Lorenz (1939) looked at sixteen courtship displays that appear in many species of ducks. By considering two of these—the chin

Figure 5.8 Drakes of three closely related species exhibiting a homologous fixed action pattern, the grunt whistle. Although the patterns are essentially the same, the timing and emphasis of each movement vary among each species. (A) is the mallard *(Anas platyrhynchos);* (B) the crested duck *(A. specularoides);* and (C) the gadwall *(A. strepera).*

lift (a pattern derived from drinking motions) and the over-the-shoulder inciting (Chapter 8)—he determined that the gadwall should be placed between the mallards and the wigeons. Later, it was found that the mallard-wigeon hybrids showed the same mingling of the fixed action patterns of the parent species as appears in the gadwall, and, thus, his hypothesis was supported.

Members of Tinbergen's group covered the globe, describing gulls at a number of sites around the world, and they found that, in spite of species differences, the signaling systems were very similar. Morphological similarity suggested a common ancestry, and Tinbergen (1959) used behavioral traits, primarily displays, to try to find out how the ancestral group was divided into the thirty-five or so distinct species that now exist. He was successful in that he was able to distinguish between families on the basis of such behavior.

Display behavior has certain qualities that recommend it for homology studies. Since displays are basically signals, they must be obvious to other animals, and therefore there is a good chance that a human observer will also be able to detect them. The primary advantage, however, is that such patterns are usually remarkably distinct and discrete. A displaying animal generally moves rapidly from one display to the next, diminishing the chance for intermediate activity that might confuse the other animal. Also, there is likely to be strong similarity of displays throughout a population.

All the large gulls have similar signaling systems that were apparently derived from adaptive patterns already in the gulls' repertoire. Thus, a current of similarity runs through the behavior of all the gull species (Figure 5.9). However, this is not to say that there are not differences between them. For example, as we have seen, the displays of the hooded gulls are distinctly their own, as are those of the kittiwake.

There is considerable variation in the displays, plumage, and fishing behavior in the families of pelicans, boobies, and cormorants. (Their families are Pelecanidae, Sulidae, and Phalacrocoracidae respectively, but they are all in the order Pelicaniformes.) However, there are similarities in the behavior of the three families that indicate their relationship. For example, all three use similar postures in signaling. In addition, the "flying-up" posture of the gannet is similar to the "nest-relief" ceremony of the brown pelican. It should be noted here that while the displays may be homologous, the conditions that elicit the displays may be quite different. Thus, homologous patterns may have different adaptive roles.

The accepted view, based on morphology, is that the pelicans are relatives of the ibises and spoonbills (both of which are in the order Ciconiiformes and the family Threskiornithidae) and this is bolstered by descriptions of their flight patterns. They all fly by intermittently flapping and then planing over the surface of the water. Cormorants, also placed with gannets and pelicans in the order Pelicaniformes, sometimes plane, but rarely. Interestingly, they adopt the flap-plane rhythm of pelicans whenever the species fly together, thus showing that the flight mannerism is held in common, but is more latent, or harder to elicit, in the cormorants.

Environmental factors may influence displays, and thus diminish their usefulness in homology. For example, it has been found that males of the shag, the North Atlantic cormorant, and the southern cormorant all attempt to attract females by gaping and showing off a brilliant yellow-lined mouth. But the Pacific species, Brandt's cormorant, doesn't gape. Instead, it distends a blue gular (throat) pouch when advertising for a female (Williams 1942). Interestingly, other displays of these four species are performed very similarly (Figure 5.10). The main difference, then, is that the Brandt's relies less on the open mouth and more on the colored throat. The evolutionary explanation may be indicated by comparing the habitat of the southern and Brandt's cormorants.

Because Brandt's lives among rocks that are pounded by noisy Pacific swells, it is theorized that any sounds it might make would be drowned out. Since vocalization never was an important part of its displays, then, there was no reason for it to develop the habit of opening its mouth to attract females. Gaping was left to those species who could be heard, at least at one time. In fact, the southern cormorant lives in a less noisy environment and not only gapes, but has retained its vocalizations as well.

And then there are the titmice. Ornithologists have not been able to subdivide the family of titmice, Paridae, on a morphological basis (Mayr 1948). Yet the nest-building habits of members of the family suggest that four, perhaps rather distantly related, groups of birds have unfortunately been lumped together. The first group includes the true titmice—the genus *Parus* and its relatives, which always nest in hollow trees or other cavities. The second includes the long-tailed tit, *Segithalos,* the bush tit, *Psaltriparus,* and related forms, which build an oval nest with a lateral entrance. These are very social birds, all with essentially the same call notes and habits. The third group consists of the penduline titmouse, *Remiz,* and its relatives, which build a peculiar retort-shaped nest consisting of plant material worked into a feltlike consistency. Lastly, the bearded titmouse, *Panurus,* builds a stick nest with a lateral entrance that, along with certain other habits, unmasks the little bird as belonging to the babbler family Timalidae.

These differences indicate that the four genera are not very closely related. The information that was accumulated also provides clues for placing them in their proper relationship. Nest building in the Paridae should probably be looked at again, considering only fixed action patterns, although it is probable that no reclassification would be merited, since each type of nest undoubtedly requires highly specific and distinct fixed patterns. Such studies might, however, provide new insight into the development of the patterns.

Figure 5.9 Sequence of the "oblique long call" in two species of gulls. In (A) the head of both species is in the oblique position, but is jerked down lower in the herring gull in (B). It is then thrown back as in (C). In (D) the call dies away as the head is gradually lowered.

TABLE 5.1 Behavioral differences in two species of wasps (*Ammophila*).

| BEHAVIOR | SPECIES: A. CAMPESTRIS | SPECIES: A. ADRIAANSEI |
|---|---|---|
| *Material used to fill nest hole* | *Carried from nearby "quarry"* | *Carried from distant source* |
| *Food* | *Supplies* | *Caterpillars* |
| *Provisioning sequence* | *Egg, then prey* | *Prey, then egg* |
| *Breeding season* | *Early, to August* | *Late, to mid-September* |

*Source: Adriaanse (1947).*

As another case, consider a study of how grasshoppers clean their antennae. The behavior has been used to justify the separation of various grasshoppers into families (Jacobs 1953). Gryllidae, Tettigoniidae, and other groups with long antennae, clean them with their maxillae mouthparts. The Acrididae, on the other hand, step on an antenna and pull it through between the leg and the ground. And then the Tetrigidae clean their antennae by stroking them with their legs, and then cleaning the legs with their mouths. In a study such as this, the behavior is undoubtedly a series of relatively simple fixed action patterns of a species not strongly affected by learning, and therefore one has more confidence in the validity of such comparisons.

Behavioral characters have, in some cases, proven to be particularly useful in distinguishing morphologically similar species, the so-called "sibling" species (Mayr 1958). As an example, consider the wasp genus *Ammophila*. It was noticed that some individuals in the species *A. campestris* had at least one peculiar behavior pattern not seen in the others. Close examination revealed that these aberrant individuals were actually a different species (Table 5.1). Sibling species have also been revealed by other behavioral studies (Barber 1951; Evans 1953). In one such study, morphologically indistinguishable birds were separated into two families on the basis of their behavior (Table 5.2).

Figure 5.10 Mating display of Brandt's cormorant *(Phalacrocorax penicillatus)* and the Southern cormorant *(P. carbo sinensis).*

## The Validity of Behavioral Homology as a Tool

In spite of such demonstrations of the usefulness of behavior in systematics and evolution, there continues to be some argument over the validity of the studies. Basically, the question of the usefulness of

**TABLE 5.2 Separation of two groups of morphologically similar birds into different families based on behavioral characteristics.**

| CHARACTER | ESTRILDIDAE (ZEBRA FINCHES AND GRASS FINCHES) | PLOCEIDAE (WEAVER FINCHES AND WAX BILLS) |
|---|---|---|
| 1. *Nest* | *A globular structure of small twigs or grass stems with lateral entrance* | *Usually a finely woven structure, often hanging from twigs* |
| 2. *Pair bond* | *Tightly knit, often lasting through years* | *Either no pair formation, or polygamy, or a pair bond of short duration* |
| 3. *Parental care* | *Nest building, incubation, feeding of young jointly done by both parents* | *Most parental duties performed by the female alone* |
| 4. *Tail movements* | *Lateral* | *Vertical* |
| 5. *Courtship posture* | *Stiff upright, with wings pressed against body; stereotyped repetition of song strophe* | *Excited courtship dances with wing fluttering and occasional display flights; noisy chatter* |
| 6. *Incubation* | *From first egg on; young hatch with daily intervals* | *After completion of clutch; young hatch simultaneously* |
| 7. *Tongue, gape, and throat of nestlings* | *With peculiar species—or genus-specific pigment spots and papillae* | *Without markings or papillae* |
| 8. *Feeding of nestlings* | *Take regurgitated food from the crop of the parent* | *Normal food begging; parents feed with the bill* |
| 9. *Begging of young* | *Without wing flutter* | *With wing flutter* |
| 10. *Nest hygiene* | *Droppings of young not removed* | *Droppings of young removed by parents* |

*Source: Mayr (1958).*

behavioral homology stems from the fact that there is a problem in homologizing neural centers due to the sketchiness of our information regarding the relationship of morphological structures and behavioral patterns.

In addition, evolutionist G. G. Simpson argued some time ago that in order for one behavior to be called homologous with another, the morphological structures used to perform the pattern must also be homologous. He said that the behaviors should be considered as abstractions, and are useless without a consideration of the structures that perform these functions. Of course, the "structure" need not be as obvious as an appendage; it may be an enzyme, a hormone, a part of the central nervous system, or any other character of the physical phenotype.

Tinbergen (1962) said that movements employing homologous structures in a similar way may be considered homologous. But structural homology is not enough. Some have pointed out that, in fact, nonhomologous behavior may arise from homologous structures (Lehrman 1953). It has also been noted that, although homologous structures are involved in a behavioral pattern and may *indicate* the patterns are homologous, such a relationship cannot be taken as proof. The anal fin of the males of some species of the guppy family Poeciliidae is modified into a gonopodium with which to fertilize the female. The anal fin and gonopodium are homologous, but the elaborate and specific motions of the latter cannot be homologized with the limited types of movements of species with the unmodified fin (Rosen and Gorden 1953).

In summary, it is argued that behavioral homology cannot have definitive value until we are able to associate the behavior with its neural center (Atz 1970). However, if we had information regarding the specific part of the nervous system that was responsible for each behavioral pattern, then we would probably choose to homologize the neural centers rather than the behavioral manifestations of their activity. So, in a sense, the demand for finding the neural centers would ultimately result in behaviorists comparing morphologies, and the "art of the science" would have advanced not at all. Anyway, we do not have this information, and herein lies the crux of the problem. Perhaps our understanding of behavior and its physiological basis is the widest gap between disciplines in science (Bullock 1965). Although homologous "nuclear regions" are known in the brains of animals as diverse as fish and mammals, this information has yielded nothing in terms of the homology of their behavior (Atz 1970). Even if more were known about which nervous center is related to a given behavior, still we might not be able to relate these correlates in evolutionary time. The problem lies in our not being sure that any behavior has been continuously associated with any particular part of the central nervous system.

There is no question that the homology of behavior presents more fundamental problems than such treatment of morphological characters. (For a criticism of the logic involved, see Cracraft 1972.) However, it has been demonstrated in certain cases that carefully designed behavioral studies may be used to indicate relationships. In spite of the inherent problems in methodology and certain important questions regarding the value of the results, our best approach is probably to primarily utilize fixed action patterns, to describe these in terms of their action profiles, and to attempt to relate these to their neural centers. Until our level of information, particularly regarding the nature of neural "centers" of behavior, increases to the point that such an approach might justify real expectations, we should continue to phrase our findings in tentative terms. More confidence would be merited, of course, when any anatomical or physiological studies are preceded by careful and extended fieldwork across the ranges of the species in question. And this has traditionally been the interest of the field ethologist.

# 6

# NERVOUS SYSTEMS, HORMONES, AND PROXIMATE CAUSATION

While the courting male wolf-spider is still a safe distance away from his deadly paramour, he stops and waves his specially adorned appendages (Figure 6.1). Should the female fail to signal back he will come no nearer. To do so would mean certain death.

If the female responds appropriately, however, the male will continue his cautious approach. When he reaches her, he moves behind her, quickly rakes his sperm-laden arm across the platelike covering of her receptacle, and, without further ceremony, clears out. Thus, he survives to breed another day, and the female will have found the mate she needed in order to leave her own offspring. Now, spiders are not very bright creatures, so it is not likely that any conscious decisions were made. The male could not be said to have been hopeful, nor was a coy female deciding whether she would be seduced by his advances. More likely, the male was equipped with receptors that detected the presence and position of the female, and he was "compelled" to move toward her. At a certain distance he began his semaphore signaling, and, in the absence of the appropriate response, it might be said that he was literally *unable* to continue his advances.

The female, on the other hand, would have "automatically" responded with the proper signal upon perceiving his waving, *if* she were in a reproductive condition—in other words, if the time were right for her. Whether the time was "right" might depend on whether the amount of daylight striking her light receptors had been sufficient to trigger the formation of hormones—chemical messengers—in her body, which prepared her physiologically and psychologically to receive the male. Daylength, of course, would be dependent on the season, and, considering how long it takes her eggs to hatch and the requirements of the young, the season could be critical to her reproductive effort. Many generations of natural selection would have produced females that only respond appropriately to the male when the time is right. The point is, behavior may be dependent on ecology, evolution, and the animal's internal state.

In this chapter, we will briefly deal with aspects of nerve anatomy and physiology. We will concern ourselves primarily with the *mechanisms* of behavior. I can only lay the groundwork here since physiological psychology is one of the fastest growing disciplines in the behavioral sciences. My goal is to review what nervous systems and their components look like, then to discuss how they work. Following this, I will discuss what are called the external and internal environments and how they are related to perception. I hope not only to introduce the fundamentals of neuroanatomy and physiology, but to show that specializations in the perceptive abilities of different species cause them to quite literally live in different kinds of worlds. Once we understand how different the neural mechanisms are in the various species, we should become increasingly hesitant about blithely extrapolating from one species to another.

## NEURONS

Neurons are nerve cells. Groups of neurons sometimes travel along the same channels and may form thick, white fibers called nerves. Neurons (and often those who study them) are irritable. In the case of the neurons, it means that they respond in specific ways

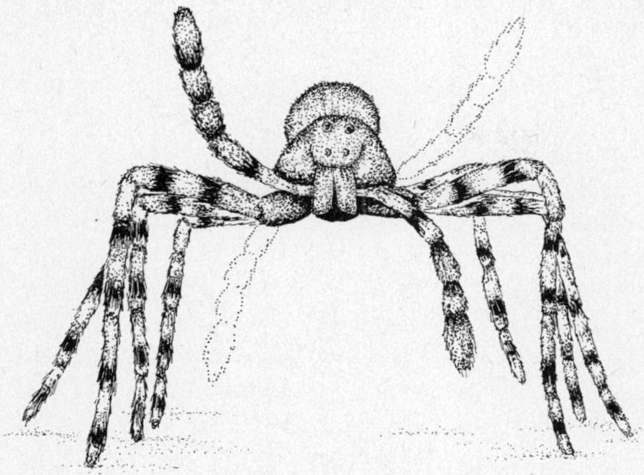

Figure 6.1 The male wolf-spider *(Lycosa amentata)* waves appendages as he approaches the larger female, thus blocking predatory behavior on her part.

Figure 6.2 The neural and muscular components of a bird, indicating schematically the responses of the various parts in order to produce an adaptive behavior.

to environmental stimuli. Furthermore, that response can move as an *impulse* from one place to another along the neuron. While some neurons are involved in transmitting such impulses, others function in the *detection* of environmental stimuli. These are called *receptors*. Some receptors react to a number of various stimuli, while others are more specific in that they register, for example, only cold or pressure. The stimulation of any receptor usually results in other neurons being activated, such as those that carry the message to the spinal cord and brain. Once the brain, or in some cases the spinal cord, has received the message, other neurons are involved in *decoding* it ("that was a sound," "that was light"). Then, other neurons may be brought into play as the brain or spinal cord acts on the information it has received and transmits messages, for example, to the muscles that move the organism toward light and away from noise, as is the case in many birds engaged in escape behavior (Figure 6.2). Keep in mind that a stimulus may be anything that elicits an impulse. Thus, it may be environmental, or one neuron may provide a stimulus for another.

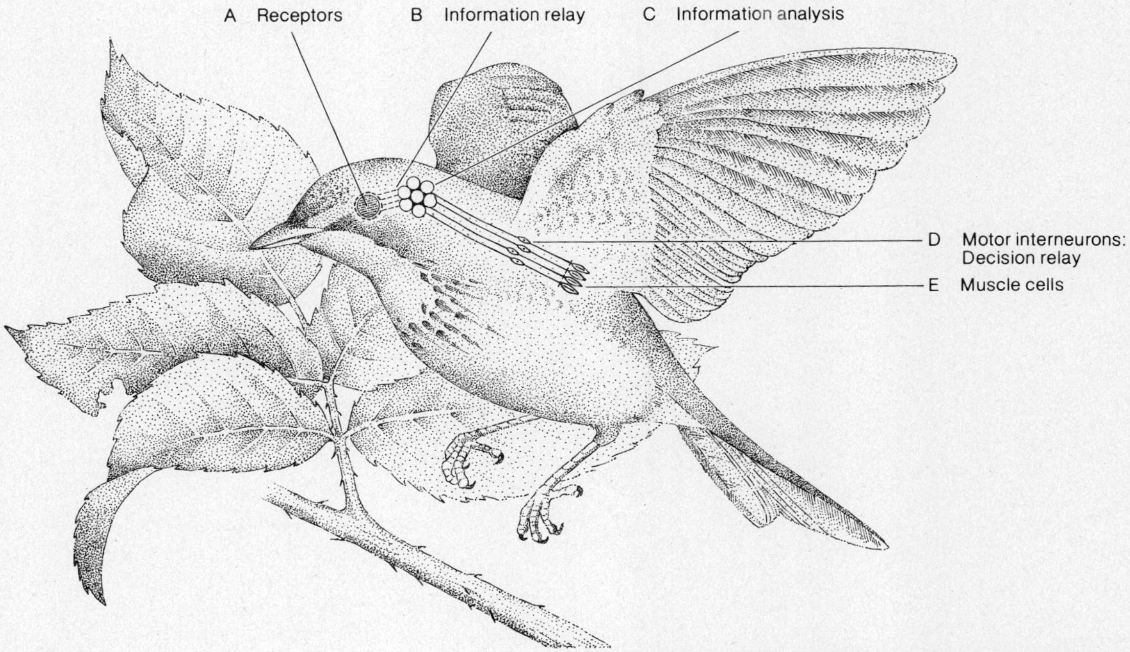

What we've been talking about, of course, are vertebrate nervous systems. But not all animals have brains and spinal cords. For example, the *Hydra* has only a nerve net running throughout its saclike body; there is no nerve center at all.

Nerve cells in most animals, however, basically behave very similarly, so let's consider in general terms how neurons work.

## HOW NEURONS WORK

It is important to know that neurons come in an incredible array of sizes and shapes. A few representative types are shown in Figure 6.3, and you will note that they have certain features in common. Basically, they all have a *dendrite* region, the part that receives the stimulus. ("Dendrite" comes from "tree"; this part of the neuron is usually highly branched.) The *cell body,* the part that harbors the nucleus, may or may not lie adjacent to the dendrite. The *axon* is the "trunk" of the nerve. Axons may be very long, for example, those that run from the base of the human spine and down the leg may be three meters long. The axon terminates in the *synaptic region.* The synaptic region joins, or *synapses,* with other neurons; maybe one, maybe several. The impulse, then, travels along the neuron from the dendrite toward the synapse at the end of the axon.

Neurons operate by being able to pump sodium ions out of the cell, while retaining potassium ions on the inside. (An ion is a molecule with an electrical charge—positive or negative.) Both of these ions are positively charged, but the result of so many sodium ions on the outside coupled with inequalities in the distribution of other ions, results in the inside of the cell bearing a net negative charge with respect to the outside. Thus, there is a net difference across the membrane of about −70 millivolts (the *resting potential*). This difference is maintained as long as the cell is at rest. Should a stimulus of sufficient strength be detected in the dendrite region, however, a sequence of changes will begin. First, at the point where the stimulus reaches the dendrite, the membrane increases its permeability to the charged particles at that point. Next, since opposites attract, the positively charged sodium molecules that have been kept out rush in toward the negatively charged interior, as some potassium molecules leak out. One result of such a shift is that the part of the membrane immediately adjacent to the area of "depolarization" also becomes permeable to the charged particles, and as the same sorts of changes take place here, the next area, in turn, becomes permeable. The result is that a wave of depolarization (the *action potential*) sweeps along the axon, but behind it, the resting condition is quickly restored. This happens because the membrane behind the wave of depolarization becomes very permeable to potassium ions, so they quickly leak out, carrying their positive charges with them. Thus, the combination of these ions and the sodium ions that are being expelled, produces a net positive charge outside again—in fact, a stronger charge difference than existed before. The difference reaches about −80 millivolts before shifting ions quickly restore it to its resting potential of −70 millivolts (Figure 6.4).

When the impulse reaches the end of the neuron, the synaptic region, a chemical is secreted into the space that separates that neuron from the dendrites of the next neuron (or neurons). The chemical (acetylcholine or some other chemical "messenger") is held in the bulbous terminals of the axon in small packets (Figure 6.5). Upon receiving the action potential, these packets move to the surface and release their chemical—which may then cause the next neuron to fire (or, under some conditions, they may inhibit it from firing; see Figure 6.6).

This is, of course, a brief and generalized account of neural transmission. The actual complexities are staggering. But we will settle for this oversimplified description, since it is probably sufficient for our purposes, and lengthy treatises on the subject may be found elsewhere (for example, Roeder 1963). It is important to know, however, that the scheme we have outlined here holds for all types of neurons. A light-sensitive neuron in the retina responds in the same way as does a neuron coming from the auditory apparatus of the ear.

Also, you should keep in mind that the action potential for any nerve remains exactly the same, no matter how strong the stimulus. The stimulus needs only to be strong enough to initiate the action potential. The *intensity* of the stimulus is determined in other ways. For example, if we place our hand on a hot stove, rather than a warm one, the firing rate of a neuron in our hand that responds to heat or pain may

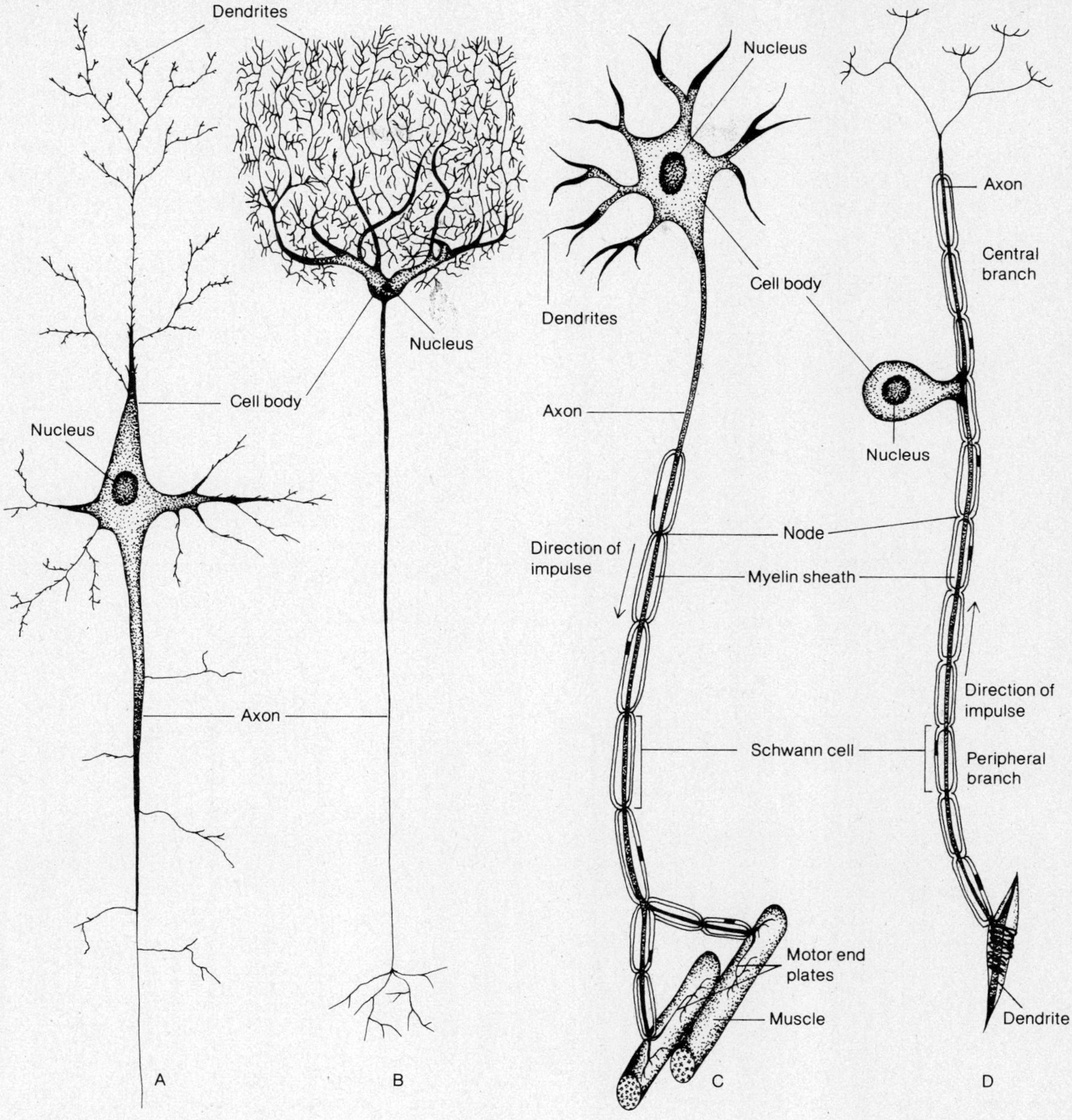


Figure 6.3 A variety of neurons from different parts of the human body. (A) and (B) are cells with different kinds of cell bodies, axons, and dendrites; (C) is a motor neuron; and (D), a sensory neuron. Although they are all structurally different, their basic structure and mechanisms are essentially the same.

be increased. Also, a very hot stove will activate more neurons since the threshholds of various neurons that register the same sort of stimulus (heat) may vary.

One might ask, then, if all neurons fire the same way, how are we able to register so many sensations (salt, heat, light)? One reason is that some receptors are sensitive only to certain stimuli. And then different kinds of neurons use different pathways in getting to the brain (or other centers where the messages are decoded). A series of neurons that register "salt" follow a different route to the brain than those that register "light," and thus stimulate different parts of the brain. If a taste neuron were to be artificially synapsed with a light neuron, presumably we would be able to taste what we were seeing.

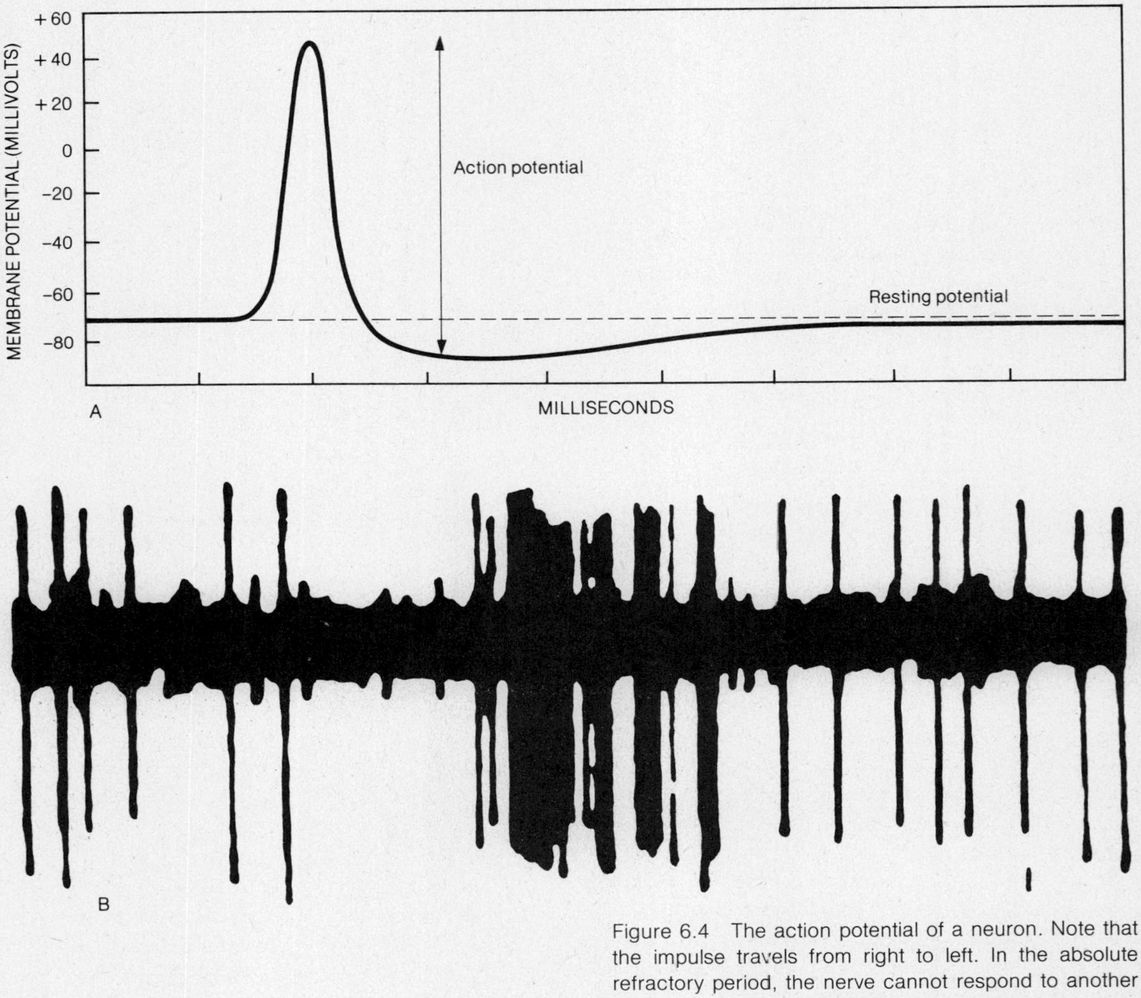


Figure 6.4 The action potential of a neuron. Note that the impulse travels from right to left. In the absolute refractory period, the nerve cannot respond to another stimulus; in the relative refractory period, only to an unusually strong stimulus (A). The pattern and frequency of impulses vary, depending on the intensity of the stimulus (B), as seen in this recording from a cat's single neuron.

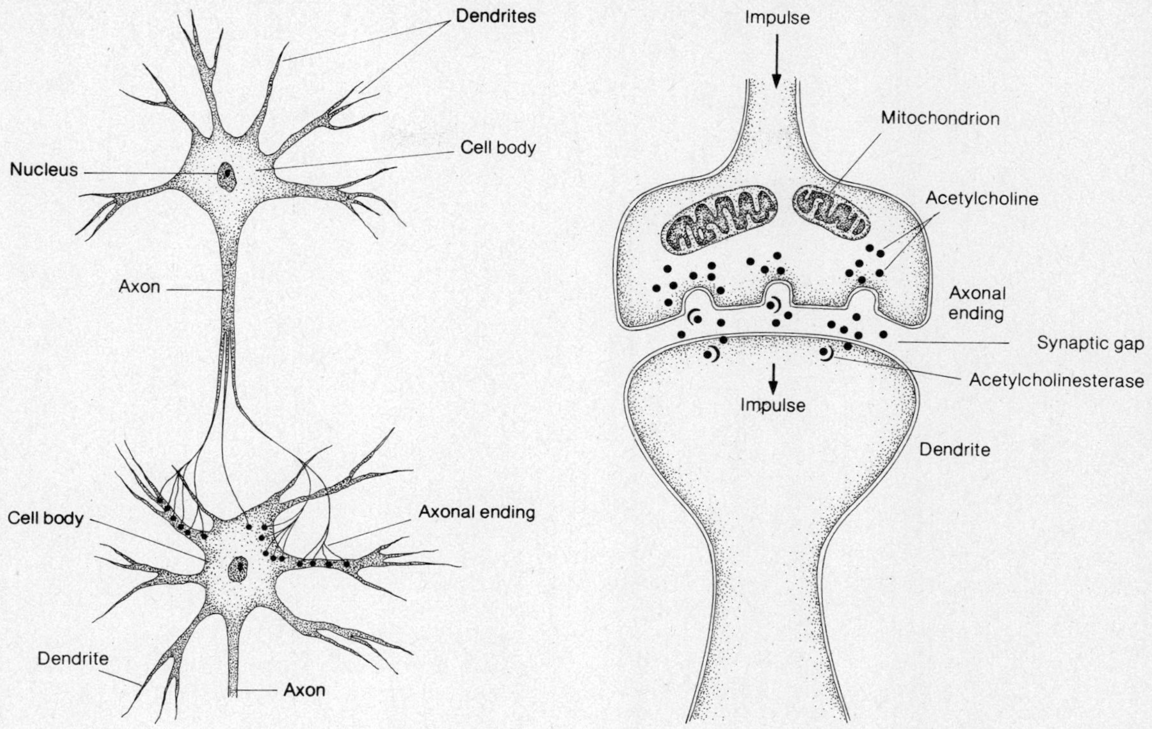


Figure 6.5 The synapse is formed where the axonal endings from one neuron stimulate the dendrites of the next by releasing chemicals into the gap between the two neurons. The chemical is quickly deactivated by an enzyme; here, acetylcholinesterase.

Actually, as undoubtedly fascinating as all this is, the firing of individual neurons is probably not particularly important to the performance of a behavior. Instead, it is the intricate interplay of millions of neurons that results in the adaptive functioning of the nervous system. Multiple synapses at a single neural junction may cause impulses to spread, fanlike, throughout the nervous system, triggering new impulses or waning by having triggered inhibitory neurons. The body is a seething mass of impulses as it seeks to adjust and regulate itself in this essentially hostile world. The brain, or its equivalent, is constantly monitoring, integrating, suppressing, ignoring, and coordinating the various signals coming from millions of neurons all over the body.

Since neurons may activate muscles and produce behavior, then, we may ask what determines whether muscles will be activated. If a man sitting alone notices an attractive woman watching him from across the room, the distant, pensive gaze he effects as he lights a cigarette may be purposeful. The slight tapping of his foot under the table may not be. Notice the behavior of someone studying for a test when he believes he is well prepared, and someone hastily cramming for an exam. Why the differences? What does clenching one's hair have to do with preparing for an exam? Why was *that* behavioral pattern generated?

This is the hard part. We know more about receptors and the mechanisms of neurons than we know about their integration. The problem is, we are dreadfully short on information (in spite of a *vast* body of data) regarding what goes on in the central nervous system. But let's have a look at the system in some selected species.

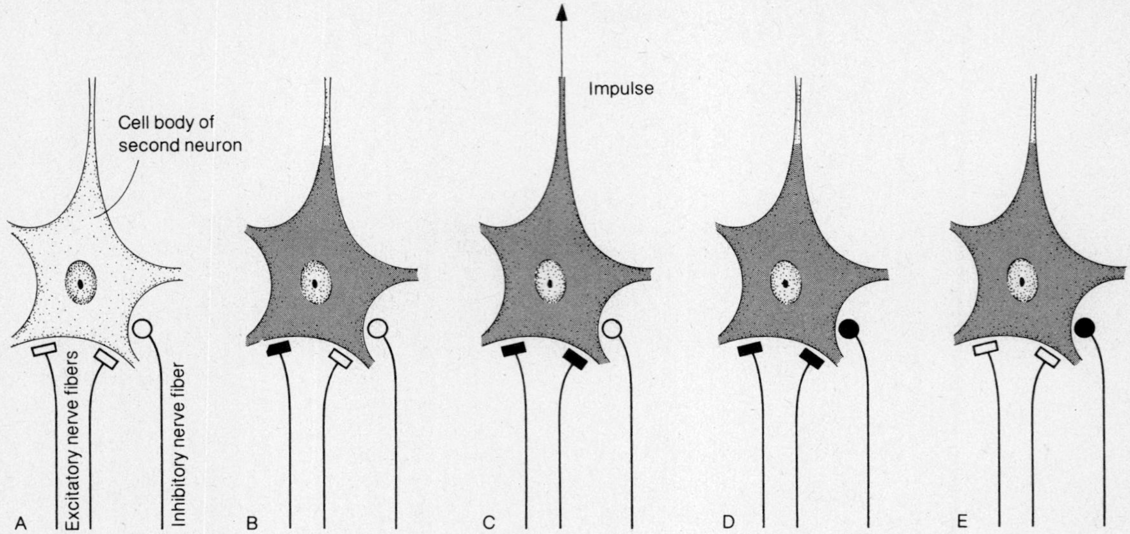


Figure 6.6 The opposing effects of excitatory and inhibitory neurons on the dendrite of the next neuron: (A) no firing; no impulse; (B) subthreshold firing by one neuron; no impulse; (C) threshold reached by two neurons firing; impulse; (D) inhibitory neuron fires and negates excitatory neurons; no impulse; (E) inhibitory neuron firing alone; no impulse.

## THE CENTRAL NERVOUS SYSTEM

The *central nervous system* in higher animals is composed of the brain and spinal cord. The presumed evolutionary pathway is outlined in Figure 6.7. As for most systems, the fossil record is too scanty to provide clear evidence of evolutionary development, so we often try to rely on living species, arranging them in some kind of order that we hope traces their evolution by moving from primitive to advanced. From the coelenterate's netlike system with no "brain" or centralized nerve area, paired longitudinal cords appear in such animals as flatworms. In the flatworms, there is also no real brain, but there are more nerves concentrated in the head region than in the rest of the body. In the segmented worms, the paired nerves have fused into a single ventral cord with bulbous swellings or "ganglia" in the head region. The nervous system of insects is a modification of this scheme, with insects showing marked specialization of segments. In the vertebrates, the brain is highly developed and very large with respect to the dorsal spinal cord. The longitudinal nerve, which has moved from the ventral, or belly, area (as we find in simpler animals) to the dorsal, or back, area, has become encased in a protective column of bones. The delicate brain is encased in a thick skull (some reputedly thicker than others).

### Cephalization

Some "primitive" animals have *radially* symmetrical bodies (Figure 6.8), while others have *bilateral* symmetry, or two roughly identical sides, and also a head and "tail" region. Cephalization, the development of a "head" region, probably was first associated with animals that had formed longitudinal body plans. Once a species began to develop nerve centers, it would have been to its advantage to localize the bulk of those centers in the area that would have been at the leading end as the animal moved into new environments. With most of the sensory mechanisms located in the head end of mobile animals, they could more quickly sense what sort of environment they were moving into. Their behavior, of course, would have had to correspond to the cephalization since, if

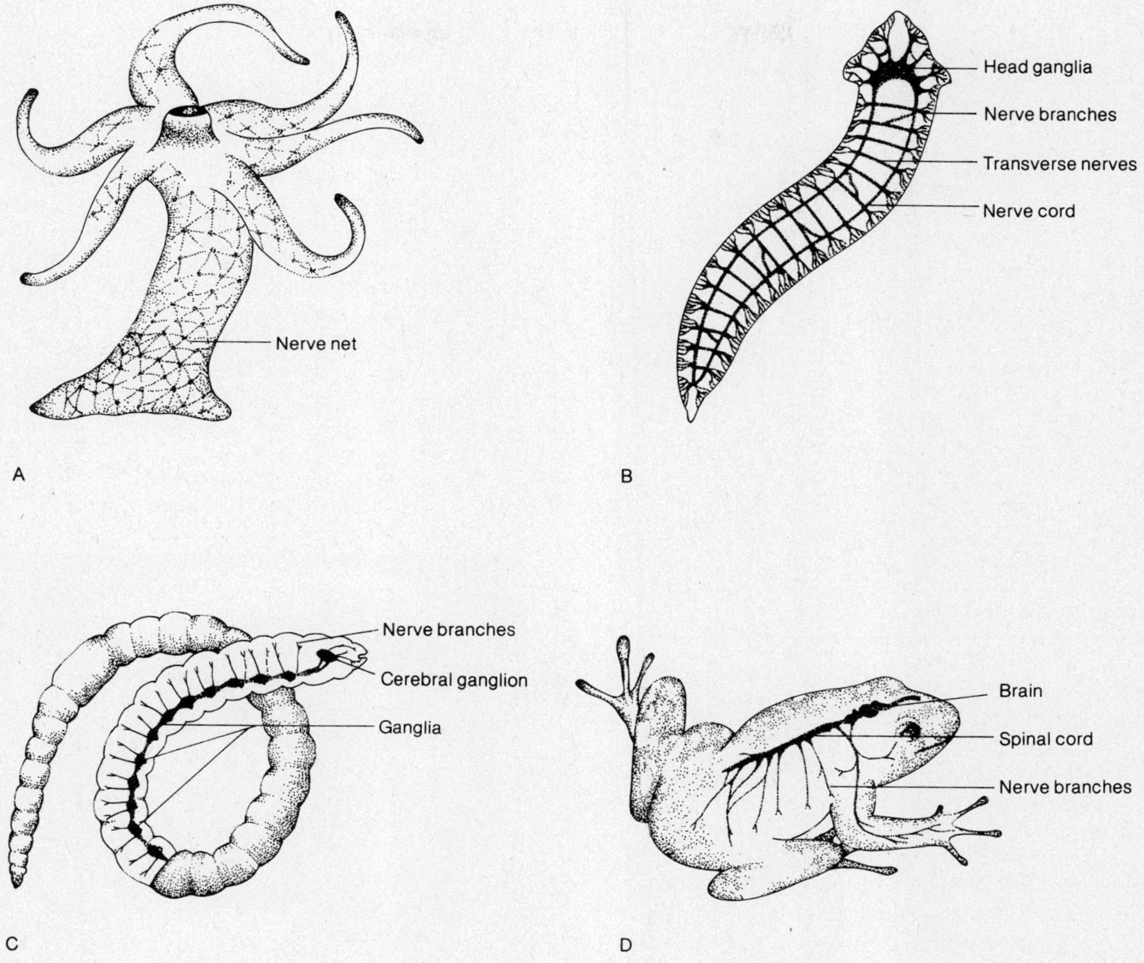


Figure 6.7 Examples of animal nervous systems. It is likely that the evolutionary route of vertebrate nervous systems was much the same. (A) the *Hydra,* with its nerve net composed of connecting neurons; (B) the planarian, or flatworm, with its longitudinal nerves connected by transverse nerves and its nerve concentration in the head; (C) the earthworm, with its single ventral nerve cord and well-defined cerebral ganglion; and (D) the frog, with its dorsal hollow nerve cord and the well-developed brain protected by bone.

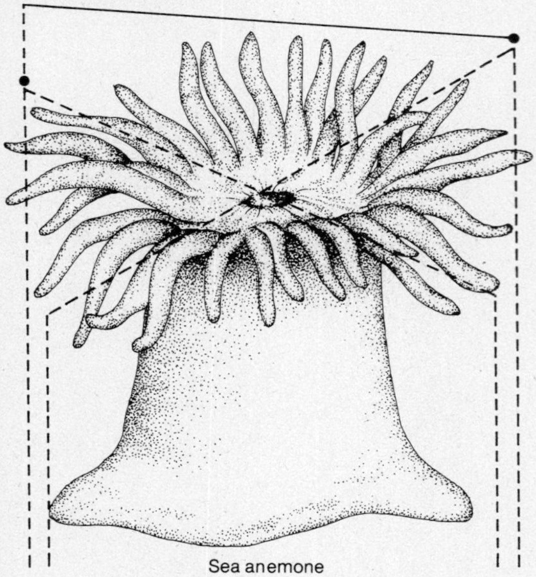


Figure 6.8 Some rather sedentary animals, such as this sea anemone, have radial symmetry. Bilateral symmetry may have set the stage for the evolution of intelligence by concentrating neural tissue at one end.

they *backed* through their environments, they might find themselves in an inhospitable situation after it was too late.

Another effect of cephalization is that the brain develops relatively early in many kinds of embryos (Figure 6.9), and can then exert a directing influence, through chemical mediators, over the rest of the developing body. Many of the chemicals that function in growth patterns are produced in brain tissue, as we shall see in the discussion of hormones.

## THE PHYLOGENY OF NERVOUS SYSTEMS

Let's now consider the nervous systems of a selected group of animals. An awareness of the range of nervous systems in the animal kingdom should develop a hesitancy to extrapolate from the behavior of one kind of animal to another, a crime of which ethologists often stand accused.

### The Coelenterates

The simplest (that is, most primitive) true nervous system appears in the coelenterates, such as the *Hydra* (Figure 6.10). It is a netlike system that was once thought to be continuous and unbroken, but now it seems, at least in some coelenterates, that the net is actually composed of separate neurons. These neurons come close together as they overlay each other, but, in many cases, do not actually touch, a distinction with important implications. If the system is not one continuous net, an impulse could take alternate pathways, permitting a certain amount of facilitation and inhibition at the various synapses, and making possible a more complex behavior than could ever occur in a truly netlike arrangement (Mackie 1960).

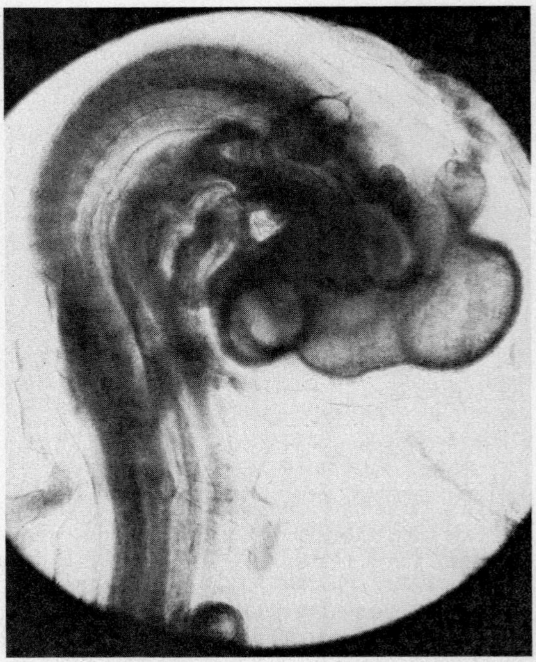

Figure 6.9 At seventy-two hours, the chick embryo has rolled onto its side and flexed its back, revealing swollen lobes of the developing brain overlying the beating heart.

Coelenterates may have slow conducting neurons, such as those in which the impulse travels at about half a centimeter per second. And they may have fast conducting neurons that transmit action potentials at about two meters per second. In the sea anemones, the twitching movements you may have seen are due to the action of their fast neurons, and the sustained contraction that results when you get the nerve to touch one in a tide pool is due to the activation of slower neurons (Passano and Pantin 1955).

## Flatworms

The most primitive flatworms have nervous systems very similar to those of the coelenterates, but others are more neurally specialized. For example, the relatively advanced planarians (Figure 6.11) have a bilaterally symmetrical body plan and a nervous system with a concentration of neurons in the head region. Here, in the anterior end, the rate of cell activity is higher and sensitivity is greatest. It is in the planarians that we first see *ganglia*, clumps of nervous tissue that contain the cell bodies of the neurons in that area. Evolutionarily, the localization of ganglia probably set the stage for the development of a brain, as we saw in our discussion of cephalization.

Also, the muscles of the flatworms make an important departure from the simpler animals. Unlike the coelenterates, which have simple "contractile elements" running in only two directions, circularly and longitudinally, the flatworms have real muscles running in three directions; circularly, longitudinally, and dorsoventrally (from the back to the belly) (Figure 6.12). Thus, they are capable of a wide range of complex movement.

You will notice in Figure 6.7 that the paired nerve cords give rise to nerve branches at regular intervals. This condition, in the evolutionary history of animals, probably set the stage for the next important development: segmentation.

## Annelids

The annelids are segmented worms, the most familiar example being the earthworm (Figure 6.13). In the earthworm, large ganglia surround the esophagus in the head and each segment has its own pair of fused ganglia emanating from the ventral nerve cord. If the head of a flatworm is removed, it will simply grow a new one, but if an earthworm is debrained, its repair is less complete and its behavior is affected accordingly. For example, it will dig more slowly than a

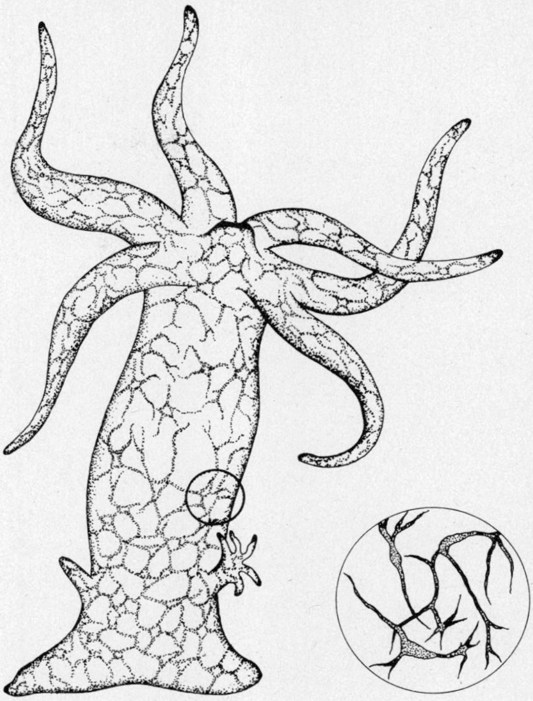

Figure 6.10 The simple nervous system of the *Hydra.* It was once believed the system was an anastomosing net, but we now believe that in some cases separate neurons simply lie in close association.

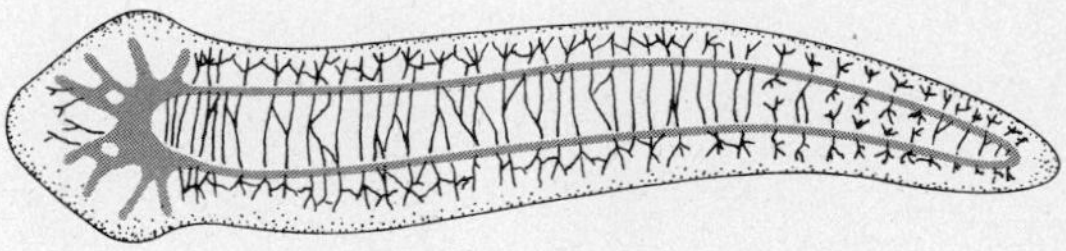

Figure 6.11 The bilateral nervous system of a planarian flatworm showing the concentration of nerves in the head.

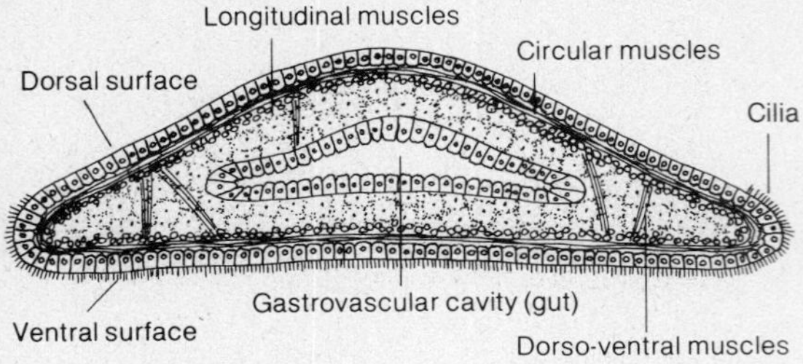


Figure 6.12 The three muscle systems shown in the planarian flatworm. Longitudinal muscles permit shortening, the circular permit rounding, and the dorso-ventral permit flattening.

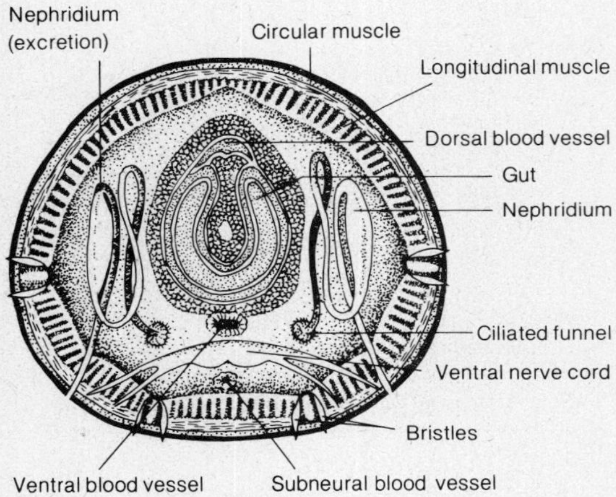


Figure 6.13 Cross section of an annelid showing the single ventral nerve cord.

normal worm, but it *will* dig. In polychetes, a more advanced kind of annelid, debrained worms become insensitive to light and cannot dig at all (Prosser and Brown 1961).

## Molluscs

Molluscs display a wide variety of neural specialization, so we are forced to ignore snails and clams for the moment and concentrate on the octopus because of the complexity of its nervous system. The elaborate ganglia surrounding the esophagus permit a surprising range of coordinated activity in this peculiar and intelligent invertebrate. The ganglionic complex of the octopus shows marked subdivision and specialization to the extent that it may actually be called a brain (Figure 6.14). A distinct and identifiable part of that complex serves as a respiratory center; another part controls the rapid color changes of the animal; other parts deal with associations and are hence important in learning; and other parts control eye movements (Prosser and Brown 1961). Because of the complexity of its brain, the octopus has been an important research animal in the study of learning and memory.

## Echinoderms

Adult echinoderms, such as sea urchins and starfish, have "reverted" to radial symmetry. Basically, their nervous systems consist of a nerve ring around the mouth opening from which extend five, or groups of five, nerve trunks. These, then, are connected to a peripheral nerve net (Figure 6.15). Echinoderms are relatively sessile animals and hence do not require complex nervous systems. Their radial symmetry means they have equal sensory and motor abilities on all sides. But it also means there is little central control, and thus complex, coordinated behavioral patterns have never developed.

## Arthropods

Arthropods, such as insects, crayfish, and lobsters, are interesting from an evolutionary point of view because they have taken segmentation one step further: the various segments have become different from each other. As neural specializations developed along the nerve cord, so did the morphological parts with which they were associated. Thus, in a crayfish, the specialized appendages that are the mouthparts are controlled by one set of ganglia; another set controls and coordinates the walking legs; and yet another set direts the tiny swimmerets under the abdomen.

The brain of a typical arthropod such as a grasshopper is composed of three ganglia above the esophagus, and one below (Figure 6.16). In arthropods, each brain part controls very specific patterns. For example, in the most dorsal ganglion there are two centers, one for vision and one for integration of information. Interestingly, if an arthropod has only a simple eye, which for the most part registers light, the visual centers will take up only about three percent of the brain. On the other hand, if the animal is a highly visual species such as the housefly, the visual centers may occupy as much as eighty percent of the brain (Maier and Maier 1970).

Arthropods may show complex and coordinated behavior because the brain can inhibit or facilitate the activity of another part. For example, the most dorsal ganglion inhibits activity in the subesophageal ganglion, which, in turn, facilitates or enhances the activity of ganglia along the ventral nerve cord (Prosser and Brown 1961).

### Vertebrates

Vertebrates are backboned animals with a hollow, dorsal nerve cord. Their central nervous system is composed of this spinal cord and its bulbous anterior end—the large, complex ganglionic mass that is the brain.

Among vertebrates, there is a tendency toward increased cephalization. That is, the more recently evolved classes generally have larger brains. (Fish, amphibia, reptiles, birds, and mammals are classes of vertebrates. Birds and mammals are considered to have more complex, or advanced, systems than the others.) In a relatively simple group, the Chondrichthyes (fish with cartilaginous skeletons such as sharks), the "thinking center" of the brain, the cerebrum, is so small and unspecialized that if it is removed, the shark will, for a time, show little or no change in behavior. It may even still attack. Most of a shark's brain, in fact, is devoted to its sense of smell. The same is true of the bony fishes (the next most complex group of vertebrates), but their brains are more differentiated, and removal of the cerebrum leads to more severe alteration in behavior.

There are basically three parts to the vertebrate brain: the *forebrain, midbrain,* and *hindbrain.* The cerebrum develops from the forebrain. Notice in Figure 6.17 that the fish brain is dominated by the midbrain (composed of the cerebellum) and hindbrain (composed of the medulla). In many animals, the midbrain functions in analyzing visual and olfactory stimuli, but in humans the midbrain is greatly reduced and simply serves as a bridge between the forebrain and hindbrain.

There are actually a number of differences in the central nervous systems of vertebrates. For example, in fish, the spinal cord is about the same diameter

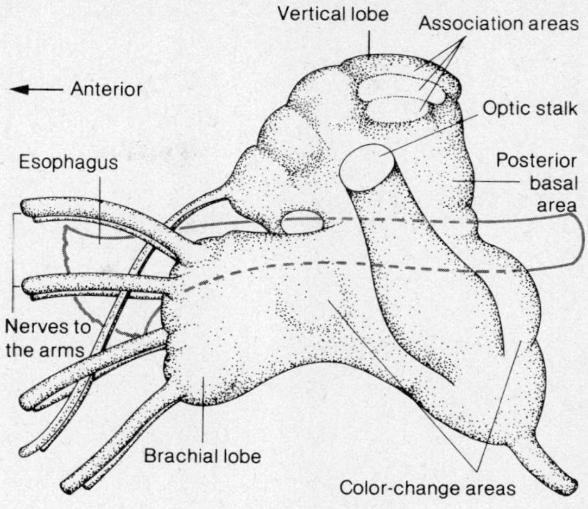


Figure 6.14 The brain of the octopus showing the large lobes surrounding the esophagus.

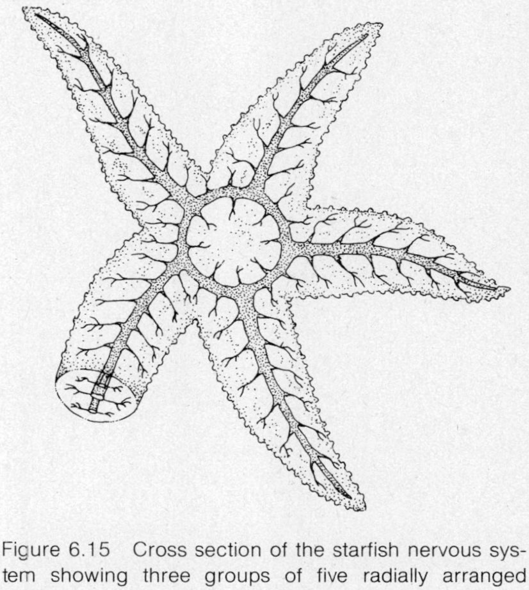

Figure 6.15 Cross section of the starfish nervous system showing three groups of five radially arranged nerves.

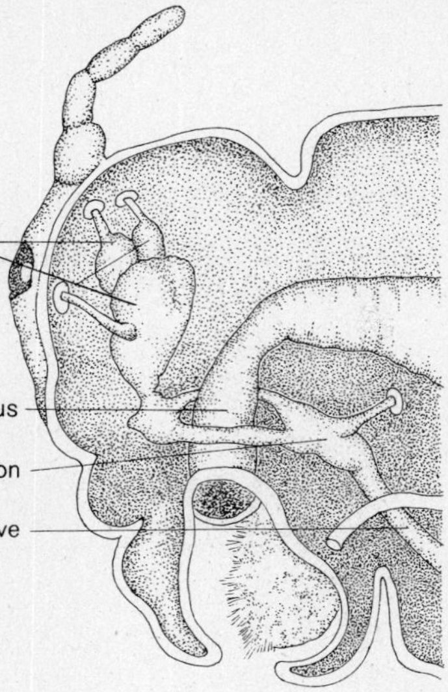


Figure 6.16 Grasshopper brain showing well-defined ganglia above and below the esophagus, connecting to a ventral nerve.

along its entire length, but in the four-legged amphibia and reptiles, there are two enlargements from which the nerves to the two sets of limbs arise.

In amphibians and reptiles, the thalamus is large and well differentiated and serves as a center for several of the senses, while also integrating sensory input. Another well-developed structure, the corpus striatum, also has certain integrative functions. Finally, the forebrain of amphibians and reptiles is much more developed than that of fish. The primitive three-layered covering of the forebrain of fish is abandoned for new forebrain tissue with four or five layers of cells in the amphibia, and in reptiles there appears a suggestion of a six-layered cortical area. Multilayers are considered an advanced condition (Grossman 1967).

The *autonomic nervous system* is found only in the vertebrates. The autonomic system, in its more developed forms, is composed of two divisions: the *sympathetic* and *parasympathetic*. Generally, these have opposing functions. Autonomic systems of various complexities are found throughout the vertebrates, but we can illustrate the basic roles of these systems by noting how they operate in humans (Figure 6.18). If a large, or even a medium-sized, bear rushes into the room where you are reading, your sympathetic system will be activated. Your heart rate will increase (carrying blood and oxygen to the running muscles), the pupils of your eyes will dilate (the better to see the bear), blood will leave the peripheral vessels (minimizing bleeding if the bear should beat you to the window), and blood will be shunted away from the digestive tract to areas where it is needed, such as the legs. Actually, these are only a few of the changes that would result from the coming of the bear.

If the bear turns out to be only your roommate dressed in a bear suit, your parasympathetic system will take over, and, through an entirely different set of nerves, all the earlier responses will be reversed as your system returns to normal. You might then wish to activate the sympathetic system of your roommate.

The relationship between the autonomic and central nervous systems is quite close. In fact, in some cases, autonomic centers may be located *within* the central nervous system. Also, commonly, autonomic fibers synapse with motor and sensory fibers going to and from the spinal cord. The cell bodies of sympathetic neurons are clustered in ganglia near the spinal cord; the ganglia of the parasympathetic nerves are usually located near the organs they innervate.

Just as the central nervous system tends to be more developed in more complex animals, so does the autonomic nervous system. Very primitive fish, such as lamprey, may have only a few parasympathetic neurons associated with the heart, and sympathetic ganglia scattered through the body (Johnels 1956). The autonomic nervous system is only somewhat more elaborate in sharks. In amphibians, at least five organs are innervated by both divisions of the autonomic nervous system: the esophagus, stomach, rectum, heart, and bladder (Hoar 1966). Dual innervation, of course, permits an organism to more finely attune itself to its environment on a short-term basis. In mammals, a number of organs are innervated by both types of autonomic nerves. In general, the phyletic trend is toward the development of specific nerves, sympathetic and parasympathetic, in the autonomic nervous system, and an increase in the number of organs innervated.

**Higher Vertebrates** As unlikely as it seems, the nervous systems of birds are structurally similar in many respects to those of reptiles. In birds, however, the midbrain gives rise to a large, complex cerebellum, the structure associated with the coordination of the body (Figure 6.19). The size of the cerebellum in birds is undoubtedly related to the complexity of flying. In addition, from the midbrain of birds spring great optic lobes whose size should not be surprising, considering the incredibly sharp vision of many of these creatures.

The forebrain of birds also shows development beyond that of reptiles. For example, the *hypothalamus* is well differentiated, and the cerebral hemispheres (the left and right halves of the cerebrum) are also relatively well developed. In addition, the *corpus striatum,* which seems to be involved in a number of complex behavioral patterns, is highly differentiated and functions as a kind of coordinating center of the brain (Stettner and Matyniak 1968).

The autonomic nervous system of birds is more developed than that of the lower vertebrates. For example, it is only in birds and mammals that the sympathetic and parasympathetic nerves enter the spinal cord in different regions along its length. The sympathetic division enters the spinal cord along the middle area; the parasympathetic nerves, at the most anterior and posterior ends.

The most complex neural development occurs in mammals. There are more neurons in the mammalian brain, and they interact in more ways, than in any other group. Also, mammals show a rather marked tendency towards increased cerebration at the expense of the lower brain centers (Figure 6.20). In addition, the *hypothalamus* and *limbic system* are well developed (Figure 6.21). The limbic system lies underneath the cerebral surface and is composed of many individual structures, including the hippocampus, amygdala, and septum, which are believed to operate as a single functional system in certain kinds of learning and emotions, and, perhaps, memory. The limbic structures interact in complex ways with the hypothalamus. The *corpus callosum* surrounds the limbic system and is a very large fiber tract joining the two halves of the brain. Thus, in the more behaviorally complex animals there is a reduction in the proportion of the brain matter devoted to simple sensory registration and motor control.

The hypothalamus of mammals is an exceedingly important structure, its critical role not suggested by

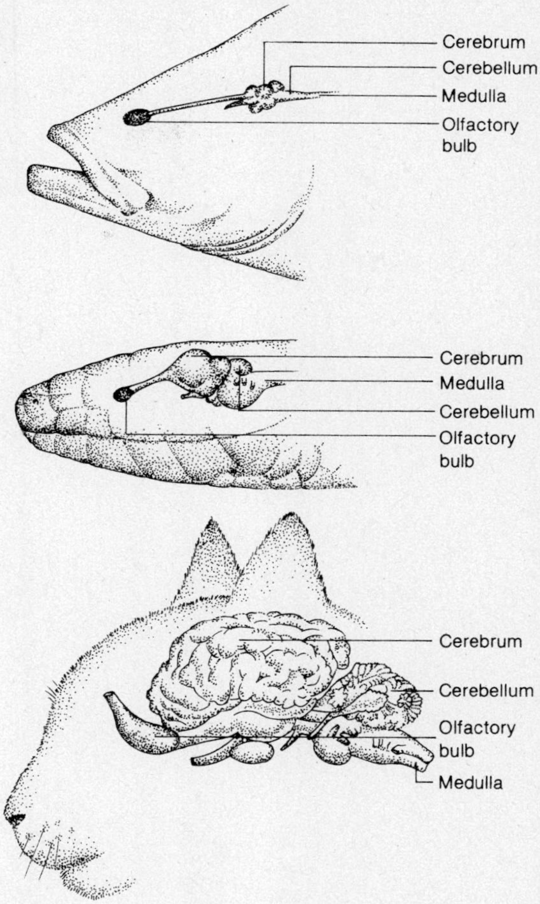


Figure 6.17 A comparison of the brains of a fish, a reptile, and a mammal. Notice the diminutive size of the cerebrum relative to the olfactory bulb in the fish. Also note the relative mass of the lower brain (here, the cerebellum and medulla) compared to that of the cerebrum. Reptiles have a somewhat larger, but still smooth, cerebrum and a reduced olfactory area. The brain of the cat is dominated by the convoluted cerebrum. The cerebellum, involved in coordination, is well developed in the cat, as is the olfactory bulb.

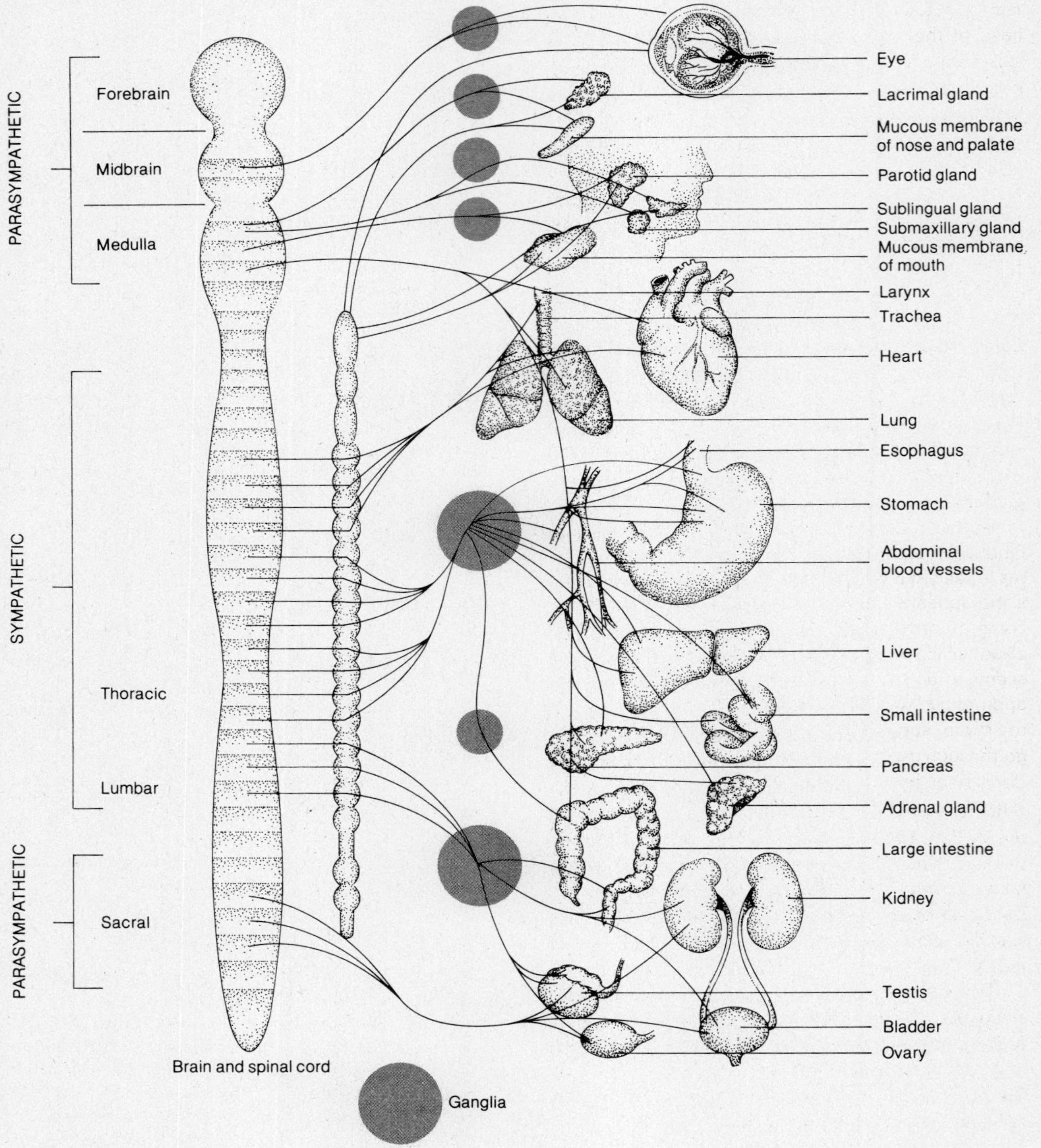


Figure 6.18 The autonomic nervous system showing the sympathetic and parasympathetic components and the organs they innervate.

the fact that it is but a tiny structure, buried at the base of the brain. It is composed of a number of subdivisions, or nuclei, each with a specific role. Certain nuclei are associated with eating, drinking, sex, sleep, body temperature, and fluid balance. If the hypothalamus is removed, the animal becomes incapable of regulating these activities and soon dies. The hypothalamus is not only associated with drives such as hunger and sex, but it harbors what are loosely called "pleasure" and "pain" centers. Electrical stimulation of the pleasure center apparently generates enormously good feelings of some sort in the experimental animal. An animal that has been trained to press the lever that activates the electrodes in the pleasure center of its own brain, may sit and doggedly press the switch until it collapses.

The mammalian hypothalamus is also intimately connected with the *reticular system,* which lies at the base of the brain in close association with the *medulla,* the swollen end of the spinal cord, and which, with the *pons,* comprises the mammalian hindbrain. The reticular system is a peculiar apparatus indeed. It has a netlike or weblike structure, giving it the appearance of felt, and it boasts an extremely complex array of billions of interconnected nerves about which we know very little. The reticular system seems to be associated with arousal and attention. It apparently "wiretaps" the impulses heading toward the brain, suppressing some and allowing others to go through. All the little sounds around you that you don't hear are apparently being filtered out by the reticular system. The reticular system is highly selective in that it can be directed to let certain impulses through. You might at times find yourself engaged in some activity that could take a *general* stimulus on the order of an earthquake to distract you, while the *specific* sound of a turning doorknob immediately gets your attention.

The reticular system may also be important in sleep. As it shuts down various nerves, you fail to register the sensory input being picked up by the millions of receptors in your body. On those nights when you just can't sleep, it may be because the reticular system is continuing to allow all that disturbing sensory input to reach your brain.

The incoming impulses that the reticular system allows to pass are directed to a large structure called the *thalamus.* Like the hypothalamus, it is composed of a number of specific nuclei, and, as each is stimulated, it sends the signal to the appropriate part

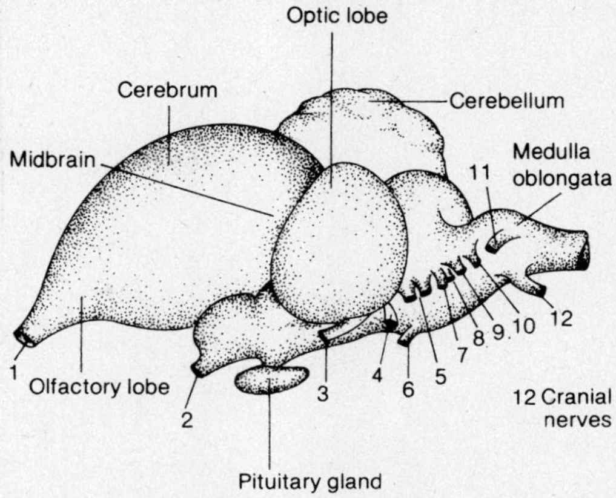


Figure 6.19 The brain of birds is well developed compared to that of reptiles. Note the pronounced midbrain area and the small, unconvoluted cerebrum.

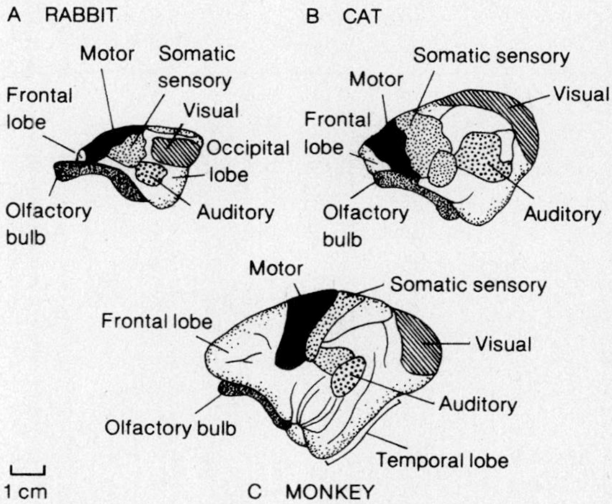


Figure 6.20 Side view of the brains of the rabbit, cat, and monkey showing increasing cerebration and reduced emphasis on areas related solely to sensory and motor control.

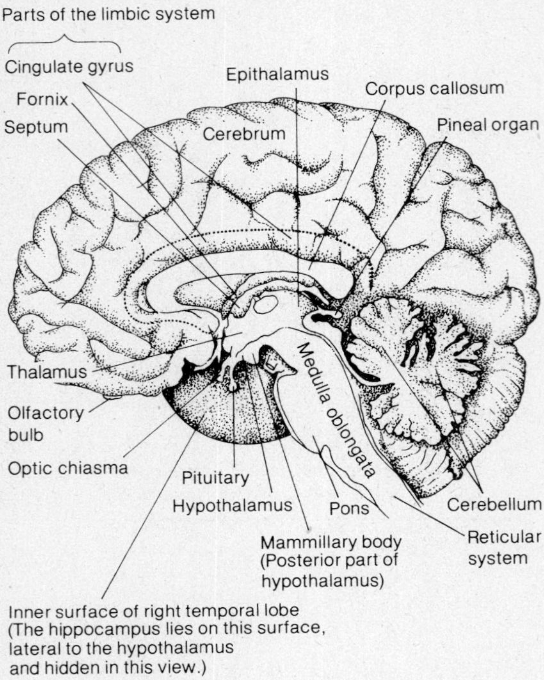


Figure 6.21 Section through the human brain showing the enormous cerebrum, the corpus callosum, hypothalamus, and the limbic system.

of the cortex. It also receives information being sent down from the cortex. While the nuclei of the thalamus are basically involved in sending sensory input to specific areas of the brain, there are also nonspecific neurons emerging from the thalamus to permeate all parts of the cortex. These seem to parallel the reticular neurons in function, helping to arouse the organism and focus its attention.

The *cerebral cortex,* the thin, outer layer of the cerebrum, consists of six layers of cells. Amphibians and fish have no cortex, and birds and reptiles have very little. Mammals have more, with their learning ability roughly proportional to its size. There are great differences in the cerebral cortex of more intelligent mammals, such as monkeys, and those less so, such as mice and rats. The cortex is larger and more convoluted (folded) in the smarter mammals. Also, they have a larger "association area." These areas receive innumerable horizontal fibers from other parts of the cortex, and it was once believed that they "associate" or coordinate the diverse sensory input. Association areas are now defined as those cortical areas that do not function specifically in sensory or motor activities.

The cortex is divided into a number of parts, the most obvious of which are the right and left *hemispheres.* Since the hemispheres are virtually identical, many parts of the brain are duplicated. The hemispheres in humans are not totally identical, however; certain skills (for example, mathematical) are seemingly localized more in one half than the other. The hemispheres are subdivided further, more or less separated by particularly deep sulci or grooves, but these subdivisions need not concern us here.

The more intelligent mammals (for example, primates) show greater functional differentiation in the cortex than do other mammals (for example, rats). Thus, if part of a rat's cortex is destroyed, other parts may be able to compensate for the loss since the various areas are roughly equipotent. Behavior is altered according to how *much* of the brain is destroyed. Destruction of certain parts of a monkey's brain, however, has very specific effects that cannot be compensated for by other parts of the brain (Morgan 1965).

The cortex is also more important in motor activities in more cerebrated mammals than in less intelligent ones. If motor areas of the brain of a rat are destroyed, there is little lasting effect on its motor abilities. It is still able to move about, scratch, feed, etc., after it recovers. Destruction of the motor areas of monkeys or apes, however, results in serious motor defects (Thompson 1967). The implication is that, as the cortex becomes more highly developed, it takes over certain functions that were under the control of lower (noncortical) brain centers in less cerebrated mammals.

## HORMONES

Closely associated, physiologically, with an animal's neural system is its hormonal, or *endocrine,* system. Endocrine glands release chemicals into the blood and are then carried to other places, "target sites," where they cause various sorts of changes. Hormonal effects are usually long-lasting. For example, some hormones help direct sexual development in vertebrates. The development of sexual systems may then secondarily affect the animal's behavior. Among vertebrates, hormones may also influence the rate of metabolism, growth, and even color changes. Other hormones may increase or decrease the sensitivity of receptors. For example, high levels of sex hormones may cause extreme sensitivity of the genitalia so that almost any touch is arousing.

Hormones may affect an animal's response to environmental variables as well as its motivation. After a male ringdove has courted and copulated, and his mate has laid the eggs, the sight of those eggs stimulates the production of the hormone called progesterone (Lehrman 1955). This hormone, in turn, motivates the male to help incubate the eggs. In the absence of the hormone, the sight of eggs has no influence on the male whatever. The neural centers that mediate nesting behavior are simply not activated, and the bird is therefore not motivated to nest.

Among invertebrates, there is not much evidence that animals with poorly developed circulatory systems produce hormones, but endocrine systems seem to be well developed in annelids, molluscs, and arthropods (Gilbert 1963; Wells 1960; Jenkin 1962; Hoar 1966).

Among vertebrates, a group of important hormones are secreted by the pituitary, adrenal glands, thyroid, and gonads (testes and ovaries). (Figure 6.22 shows the location of the endocrine glands in the human.) The anterior part of the pituitary releases six hormones that are actually manufactured in the hypothalamus and then carried to the pituitary by the bloodstream: (1) the *adrenocorticotrophic hormone* (ACTH) stimulates the cortex of the adrenal gland to secrete cortical hormones, which adjust the metabolism to various forms of stress; (2) the *somatotrophic hormone* (STH) stimulates growth of the skeleton; (3) the *thyrotrophic hormone* stimulates the growth and function of the thyroid gland; (4) the *follicle-stimulating hormone* (FSH) stimulates the growth of the ovarian follicles (which contain the egg); (5) *prolactin*, or the lactogenic hormone, is related in very complex ways to parental behavior; and (6) the *luteinizing hormone* (LH) facilitates the development of the ruptured follicle, the corpus luteum, after the egg has been released (it induces the gonads to secrete hormones and, in some species, it causes the release of a new egg).

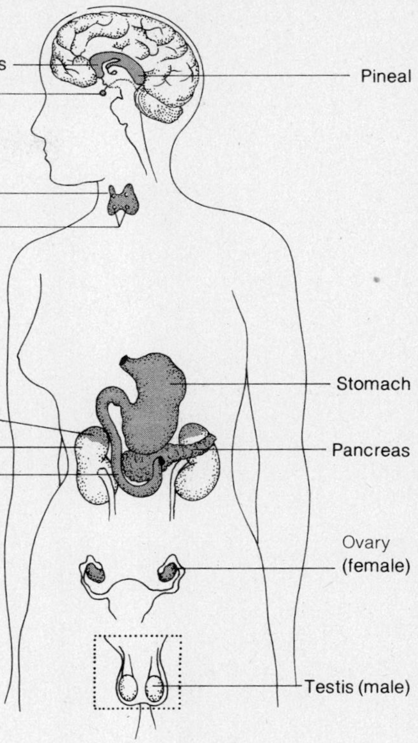


Figure 6.22 The human endocrine system and certain associated structures.

The *posterior pituitary* releases only two hormones, *oxytocin* and *vasopressin*. The former stimulates uterine contractions during labor. The latter regulates water absorbtion in the kidneys and constricts small arteries throughout the circulatory system. Again, both these hormones are synthesized in the hypothalamus and are simply released by the pituitary. But, strangely enough, they reach the pituitary by the axons of nerves (Maier and Maier 1970).

The intermediate lobe of the pituitary secretes *intermedin*, a hormone that functions in color change in lower vertebrates such as some fish and lizards.

*Thyroxin*, secreted by the *thyroid gland*, regulates metabolic rate. In some lower vertebrates, such as amphibians, thyroxin also regulates metamorphosis, directing, for example, the changes of a tadpole to a frog.

The *adrenal cortex* secretes *cortical hormones,* which have very diverse and complex roles. For example, they function in the adaptation to long-term stress and help to develop male sex characteristics.

The *adrenal medulla,* or the more interior part of the adrenal gland, secretes *adrenaline* and *noradrenaline,* both of which function in sympathetic reactions of the autonomic nervous system. Adrenaline seems to be associated with fear reactions, however, and noradrenaline with anger (Maier and Maier 1970). Interestingly, the adrenal medulla is under neural control, while the adrenal cortex is controlled by the hormone ACTH.

The *gonads,* logically enough, secrete the *sex hormones.* In males, certain cells of the *testes* secrete *testosterone,* which causes the development of male secondary sex characteristics (for example, beards, deep voices—the traits that are superimposed on the primary sex characteristics, or the plumbing) and the typical highly impressive, masculine behavior. The *ovaries* in females secrete *estrogen,* which has a comparable role in female development (for example, enlarged breasts, widened pelvis) and maternal behavior.

Once an egg is released from a follicle and becomes fertilized, the remaining corpus luteum secretes *progesterone,* a hormone of pregnancy that functions in influencing parental behavior.

It is important to realize that hormonal systems function in complex ways, interacting with each other and with other physiological mechanisms in an incredibly coordinated orchestration of responses that produces the behavior that we see.

Also, the role of hormones in the *development* of behavior must be recognized. One dramatic story will illustrate the point. It seems that, at least in rodents, the sexual behavior of the organism can be altered through the presence of hormones before birth. No matter what the genetic sex of the embryo, the hypothalamus can be directed to cause the animal to behave as a male or a female later in life, the response being triggered simply by the presence of certain steroid hormones (the class of hormones that includes the sex hormones) about the time of birth. If the steroids are present, the hypothalamus-pituitary complex will develop so as to cause masculine behavior. One specific effect is that the pituitary will produce the luteinizing hormone (LH) and the follicle-stimulating hormone (FSH) in a nonrhythmic way, characteristic of males. If the steroids are not present, development and future behavior will proceed along female lines. FSH and LH will be produced rhythmically so that estrous cycles occur (Manning 1971).

It seems that, normally, the testes of male embryos become active briefly around birth. The testosterone they secrete gets the male started down the proper sexual path. The ovary does not secrete at this point and the result is a feminine course of development. A tiny injection of testosterone into a female embryo just before birth will result in a female that shows no regular reproductive cycles, is sterile, mounts and thrusts often, and is much more aggressive than normal (Bronson and Desjardins 1968). We will say more about this later.

In this chapter, we have learned something about the anatomy and physiology of nerves, the evolution of nervous systems, the specific traits of receptors, and basic endocrinology. The wide-ranging "catalogue" treatment is designed to, first, introduce basic principles and, second, to show the variety of neural apparatuses with which different animals must deal with their worlds. We must keep in mind not only what the neural specialization of any animal might be, but also what kind of environment it deals with. Is it highly restricted, limited to a narrow niche? Or is it wide-niched, living under a variety of conditions? Does it stalk the earth at night, or does it go about in the hard light of day? What *can* it do, and what *does* it do? In other words, we must seek to understand animal behavior by maintaining a proper duality in our viewpoint. We must consider the kinds of worlds in which animals live and the kind of equipment they have with which to perform their roles.

# 7

# PERCEPTUAL SYSTEMS

Let's now briefly consider the ways in which animals perceive their environments. In the consideration of why an animal responds to a situation in a certain way, such information is critical. Much of this information will not be intuitive because we are able to perceive only a small part of our environment. The world around us is richer and far different than we, as humans, know. There are deep violet colors in what seem to us to be simple white flowers. We can't be expected to react to the violet, because we can't see such short wavelengths. My skunk, Flower, squints myopically at everything. But he lives in a world of thundering odors. He can find a breadcrumb that I can't even see on the floor. Our limitations are, in a sense, then, the limitations of our receptors, our perceptual systems. All animals are thus limited, but in different ways and to different degrees. Let's get some idea, now, of the particular perceptual systems present in various kinds of animals.

## THERMORECEPTORS

We begin with a type of perception not usually given much attention by humans. Thermoreception, or the detection of heat, is apparently possible because of temperature-dependent changes in the speed of certain chemical reactions. However, since the rate of firing in several types of receptors is temperature-dependent, most animals never had to develop heat receptors per se.

### Invertebrates

While most invertebrates do not have thermoreceptors, certain insects such as the locusts (Acrilidae) have heat-sensitive cells scattered throughout their bodies. And cockroaches *(Periplaneta americana)* have heat receptors on their legs that apparently help them select an appropriate habitat (Carthy 1958).

### Vertebrates

Pit vipers (Crotalidae) and pythons (Pythonidae) have pits between their eyes and nostrils that serve as heat detectors and enable the snakes to detect their prey.

Regarding mammals, there has been some disagreement over whether specific receptors detect cold or heat, or whether simple, free, nerve endings act as thermoreceptors. One group argues that two specific kinds of receptors are needed in order to perceive heat and cold. But opponents of this idea point out that the same neuron could register both heat and cold if it fired in different temporal patterns when stimulated by different temperatures. In any case, neurons that register warmth are located deeper in the skin than are cold receptors. Whether the brain registers "warmth" or "cold" also depends on the immediate previous experience of the receptors. If the skin has been very warm and touches an object of a cooler temperature, it will seem cool. If the skin has been cooler, the same object will register as warm.

## TACTILE RECEPTORS

Tactile, or touch, receptors are extremely sensitive, fast-firing neurons. They respond to any force that deforms or alters the shape of their cell membranes.

Figure 7.1 Many arthropods, such as insects and spiders, are covered with hairs that, when moved, stimulate sensory cells. This hairy spider has captured a cockroach.

In some cases, bristles or hairlike structures protrude from the surface of the cell membrane so that objects in the environment are detected by these dead "feelers" before living tissue touches them.

### Invertebrates

In invertebrates, tactile receptors are scattered over the body and may give rise to bristles or hairs. For example, arthropods usually detect touch by hairlike processes (Figure 7.1). Web-building spiders have hairy legs that respond to vibrations set up in the web—an indication of captured prey. Cockroaches have tiny hairs protruding from their abdomen that, when stimulated by very light air currents, send them scurrying for safety, since those currents may indicate the approach of something that gets after cockroaches. Many aquatic insects have bristles on their heads that are sensitive to water currents. R. Stimson Wilcox at the State University of New York at Binghamton has accumulated fascinating evidence on how certain aquatic insects use water vibration as a communication device.

### Vertebrates

In vertebrates, there are basically two kinds of tactile receptors. Distance receptors, such as the lateral line organs of fish and aquatic amphibians, are sensitive to water motion. Such sensitivity can be used to detect the presence of predators or prey, or schooling fish can use the information to keep track of each other's whereabouts. One species of fish, *Aplocheilus lineatus,* even has modified lateral line organs on its head with which it can locate surface-swimming insects (Schwartz 1965).

The second type of receptor is the contact receptor. These are of two types: pressure and touch. The pressure receptors are, not unexpectedly, located deeper in the skin and seem to utilize encapsulated nerve endings called Pacinian corpuscles. Light touch is probably registered in rather unspecialized free nerve endings. The sense of touch in most mammals is particularly great around the hairy areas and genitals, as you may have noticed. Among primates, however, tactile sensitivity is actually greatest around the lips and eyes and the fingertips (Altman 1966).

## AUDITORY RECEPTORS

Audition, or the sense of hearing, is similar in some respects to lateral line reception in that both are long distance receptors. However, auditory receptors are sensitive to higher frequencies and they function in both air and water. You may have noticed how well you can hear when swimming under water, and beginning scuba divers may be surprised at the noise level of their bubbles.

### Invertebrates

The simplest group of animals with specialized sound receptors is the insects, although other invertebrates may be highly sensitive to sound-associated vibrations (Maier and Maier 1970). Insects have several kinds of sound receptors. Most have sensory hairs that respond to low-frequency vibrations of air, while others have a specialized organ, located in a leg, that is sensitive to movements in whatever the insect is standing on. Yet others, including grasshoppers and crickets, have tympanic organs that respond to high-frequency vibrations. The tympanic structures operate in a manner roughly similar to the human eardrum. Let's take a closer look at the auditory mechanism of a remarkable insect and see how it evolved and why it is important to the insect's survival.

Noctuid moths are a favorite prey of certain bats. The bats capture their prey on the wing, locating these flying insects by emitting high-frequency sounds that bounce back to the bat from any structure in the environment.

Kenneth Roeder has provided us with a fascinating account of how the relatively simple auditory apparatus of the moth is used to detect an approaching bat and how the moth then tries to avoid the bat (1963). He found that these moths have two ears, one on either side of the thorax (the insect midsection), but each ear, it turns out, has only two receptors! One, called the A1 cell, is sensitive to low-intensity sounds. The other, the A2 cell, responds only to loud sounds. Surprisingly, neither neuron can distinguish frequencies—a low sound of 20,000 hertz elicits the same neural action pattern as a high pitch of 40,000 hertz.

The A1 can detect sounds becoming louder, however, and fires more frequently, with a shorter lag time between receiving the stimulus and firing. Interestingly, the A1 fires more frequently in response to pulses of sound than to continuous sounds. It will fire increasingly slower if a pulse changes to an uninterrupted sound. It just so happens that bats emit sound pulses.

The very sensitive A1 neuron is able to detect bat sounds long before the bat is aware of the moth. The moth can not only determine the distance of the bat by the intensity of its sound, but whether it is coming nearer, because the sound of an approaching bat grows louder.

The moth is also able to detect the direction of the bat. If the bat, for example, is on the left side, the left thoracic receptors of the moth are exposed to the sounds, while the receptors on the right are shielded (Figure 7.2). Therefore, the left receptor fires sooner and more frequently upon receiving each bat sound. If the bat is directly behind, both neurons will fire simultaneously. Thus, the moth can determine the distance and direction of the bat. But what about its *altitude?*

If the bat is above the moth, the bat sounds will be deflected by the upward beat of the moth's wings. But if the bat is beneath the moth, the wingbeats will have no effect on the pattern of neural firing. The moth, then, is able to decode the incoming data, probably in its thoracic ganglion (from which the auditory neurons emerge), so that it is aware of both the presence and precise location of the bat.

But what does it do with this information? It turns out that if the bat were passing some distance away, the moth would simply turn and fly directly away, thus decreasing the likelihood of detection by increasing the intervening distance and by exposing less surface area to the bat. The moth probably turns until the A1 firing from each ear is equalized. When the bat changes direction, so does the moth.

Bats fly faster than moths, though, and if a bat should draw to within eight feet of the moth, the moth would be detected and caught if it tried to outrun its pursuer. If the bat and moth are on a collision course, that is, if the moth is about to be caught, the sounds of the onrushing bat will be very loud. At this point the A2 fiber is roused and begins to fire. These messages are relayed to the brain, which then apparently shuts off the thoracic ganglion that had been coordinating the antidetection behavior. Now the jig is up, and more drastic measures are in order. The moth's wings begin to beat in peculiar, irregular patterns or not at all. The insect itself probably has no way of knowing where it is going as it begins a series of loops, rolls, and dives. But it is also very difficult for the bat to plot a course to intercept the moth. If the moth crashes into the ground, so much the better. It is safe here because the earth will mask its echos.

The noctuid moth's evolutionary response to the hunting behavior of the bat serves as a beautiful example of the adaptive response of one organism to another. Also, it shows clearly that the sensory apparatus of any animal is not likely to respond to elements that are irrelevant to its well-being. It is not important for moths to be able to distinguish frequencies of sound, but it is important that they are sensitive to differences in sound volume. Anyone who tried to train a moth to respond to different sound frequencies could only conclude that moths are untrainable.

### Vertebrates

Vertebrates usually detect sound by registering its vibrations in the inner ear. Among most fish, sound vibrations in water are conducted *directly* to the inner ear, but in the order Cypriniformes (the minnows, catfishes, and suckers), a set of tiny bones, the Weberian ossicles, connect the swim bladder to the

Figure 7.2 The relationship of sound pulses from a hunting bat and auditory neural firing in the hunted moth. (A) When a hunting bat, emitting its high-pitched sounds, approaches a noctuid moth from the side, the receptors on that side fire slightly sooner and more rapidly than those on the shielded side. (B) When the bat is behind the moth, the moth's receptors on both sides fire with a similar rapid pattern. (C) When the bat is above the moth, the moth's auditory receptors fire when its wings are up, but not when its wings cover the receptors on the down stroke.

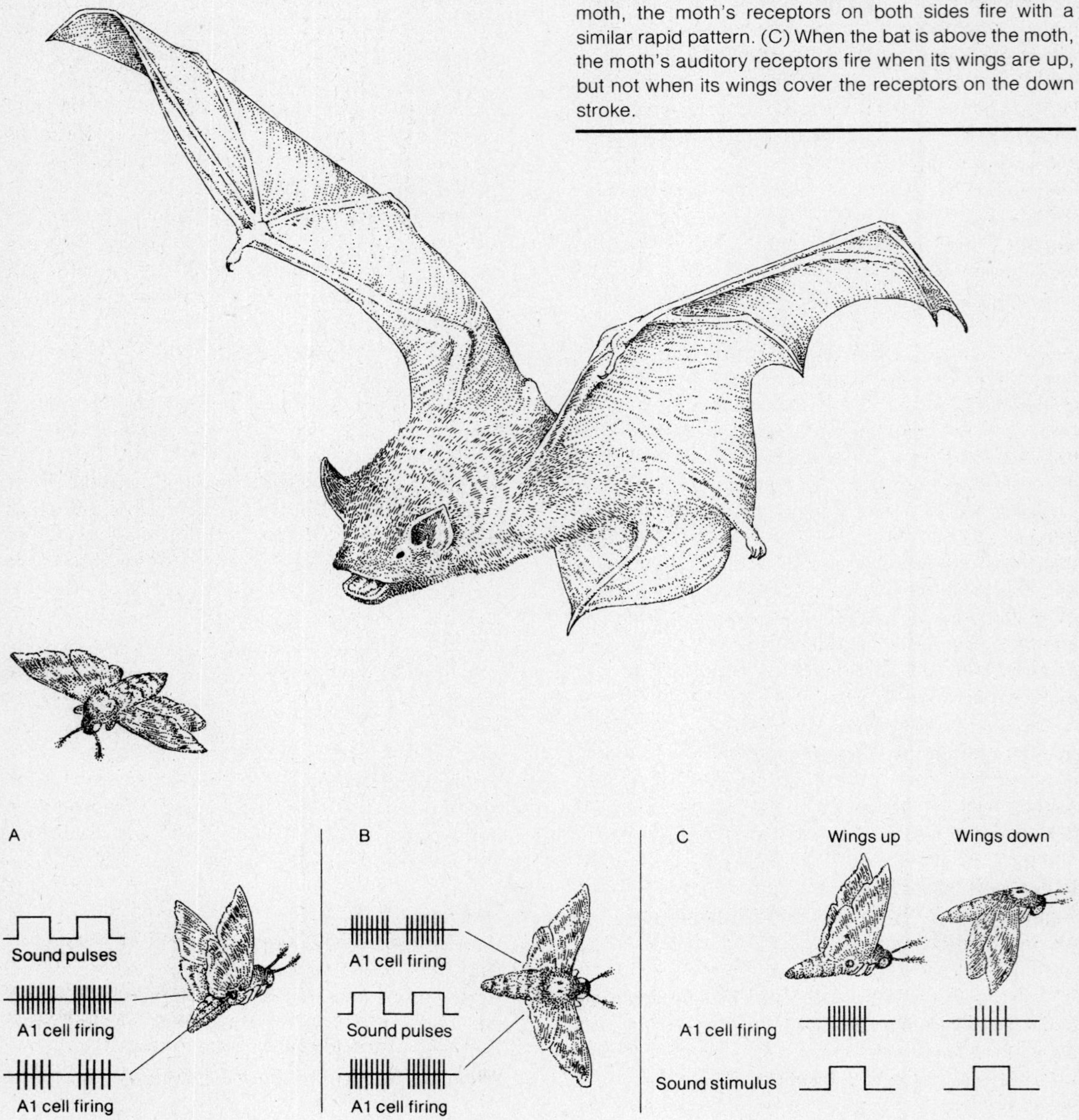

inner ear. The air-filled bladder resonates with sound, moves the bones, and receptors in the inner ear are stimulated. A fish with such an apparatus can detect a much wider frequency of sound than those lacking it.

Land vertebrates have an *auditory canal* and one to three *middle ear* bones that vibrate the tympanic membrane (eardrum) of the *inner ear*. The tympanic vibrations usually stimulate auditory receptors that then carry the impulses to the brain. The sounds may be amplified by the auditory canal or the middle ear bones.

Birds have a much more sensitive auditory system than fish, amphibians, or reptiles. Small birds are particularly sensitive to high frequency sounds (Schwartzkopff 1955). Birds such as the owl are able to locate the source of sounds with great accuracy. This is apparently because their ears are relatively far apart, and one means of locating sound sources is by detecting the difference in time it takes sound to reach each ear. This means, for example, that the source of any sound reaching both ears simultaneously is located in the body's midplane. Also, in the barn owl (Figure 7.3), its peculiar facial disc may act as a focusing apparatus for sound.

It has been discovered that barn owls can catch a mouse in a completely dark room in spite of the fact that they have no *external ears* (the part that sticks out, for example, the conspicuous part of a dog's ear, or a bat's) (Payne 1962). The barn owl has only auditory canals leading to the bones of the middle ear. Furthermore, these two canals are, oddly enough, placed asymmetrically. But a barn owl is able to capture a skittering leaf pulled along the floor by a string in a dark room as quickly as it would a scampering mouse (indicating that sight and smell are not involved), and, if unable to see, it will leap into the middle of an expensive loudspeaker from which emanate mouse sounds. By careful experimentation with microphone recorders placed in the auditory canals, it was discovered that the design of the canals and facial feathers block out sounds coming from most directions. So, to locate a sound source, the bird simply turns its head until the sound waves are maximized and equal in both ears. The prey is then dead ahead. This places the object directly in the bird's line of sight, so on moonlit nights the bird's vision can be brought into play.

We can see, then, that when analyzing perceptual abilities, one must consider not only the stimuli necessary to evoke the firing of a neuron, but also the morphological and behavioral adaptations that operate in conjunction with the perceptive apparatus.

Figure 7.3 The barn owl, a night hunter, has a facial disc of densely packed feathers that may gather sounds and aid in detecting their source.

In mammals, the auditory apparatus (Figure 7.4) differs in three basic ways from that of lower animals. There is an external ear, which most mammals are able to move so as to maximize sound input and to localize sound. The middle ear has three bones, the malleus, incus, and stapes (as opposed to one in lower vertebrates). These bones are linked in such a way that the movement of the tympanic membrane in response to sound causes them to resonate the oval window of the snail-shaped cochlea. The result is that fluids inside the cochlea are shifted about. (The shifting is possible because a part of the cochlea wall is flexible—the oval window.) The cochlea is divided longitudinally by complex membranes, forming the vestibular and tympanic canals. The actual organ of hearing is located on the basilar membrane separating them. Receptors there have hairlike appendages that reach upward to contact an overhanging tectorial membrane. The vibrating fluid of the cochlea causes the basilar and tectorial membranes to draw together and pull apart, thus distorting the hair cells, causing them to send impulses over the auditory nerve. The basilar membrane is relatively thick and stiff at the base near the tympanum, but at the other end, it is thin and flexible. This thinner end lies in what is the center of the coiled cochlea. Frequency discrimination is possible because high-pitched tones stimulate the hairs at the thicker base of the membrane; low-pitched tones are registered at the apex (Maier and Maier 1970).

Intensity of sound in mammals is apparently a function of the number of nerve fibers stimulated and the number of nerve impulses generated in each fiber (Davis 1961).

## VISUAL RECEPTORS

Humans are highly visual animals, and hence we seem to have a special appreciation for sharp vision in other animals. Some of those other animals see well, indeed, by our standards; some see poorly, and some see a world that would be totally alien to us.

Visible light is a particular part of the spectrum of electromagnetic energy. Its wavelength ranges from about three hundred to ten thousand millimicrons, but no animal can see more than a part of this range. The shortest wavelengths, such as Xrays, beta rays, and gamma rays, are not seen by any animal. Nor are the very long ones, such as radio waves and infrared waves.

Visual receptors contain *pigments* (colored substances) that absorb certain wavelengths of electromagnetic energy and transform that energy into a neural stimulus. Shorter waves contain just too much energy for these pigments to capture, and, in fact,

Figure 7.4 The human ear. The vibrating eardrum moves a series of bones that cause fluid to stimulate sensory areas in the cochlea (A), as described in the text. (B) Cross section of the cochlea. Sound waves move the basilar membrane and distort receptor cells that are anchored to the tectorial membrane. The distortion causes action potentials in the auditory nerve. (C) The cochlea, shown straightened. The areas of sensitivity to various frequencies are shown. Those near the oval window register the lowest sounds.

they may even destroy the pigments. The longest waves contain so little energy that they simply can't be absorbed easily, and thus they can't be detected. It seems that photoreception in many animals is mediated by a photochemical process in which light temporarily breaks down some substance in the animal's visual cells. It is this chemical reaction, then, that activates the receptors.

As we know, it is possible to detect differences in light intensity or brightness. A number of physiological studies have been directed at discovering just how this is done. Theory has it that radiant energy comes in packets called *photons,* and it has been suggested that the detection of intensity may simply be the perception of the number of photons that are bombarding a receptor at any given time. Also, *relay neurons,* which are neurons stimulated by more than one receptor, may fire at different rates according to how many receptors are stimulating them. Intensity might also be detected by some sort of "decoder cell" that measures the frequency at which relay neurons fire and translates this into yet other neural patterns, which then register the sensation.

Many animals are also able to discriminate between different wavelengths; in other words, they have color vision. Color vision has developed in a wide range of animals including insects, fish, reptiles, birds, and mammals, and apparently, it has evolved several times. This notion is supported by the fact that the mechanisms for detecting color may be quite different in various animals. For example, in reptiles and birds there is but a single pigment. But the light striking this pigment must pass through oil droplets of different colors, each color filtering out a different wavelength. In the higher primates, including man, there are no oil droplets in the fluid of the eye; instead, there are several pigments in the retina, each

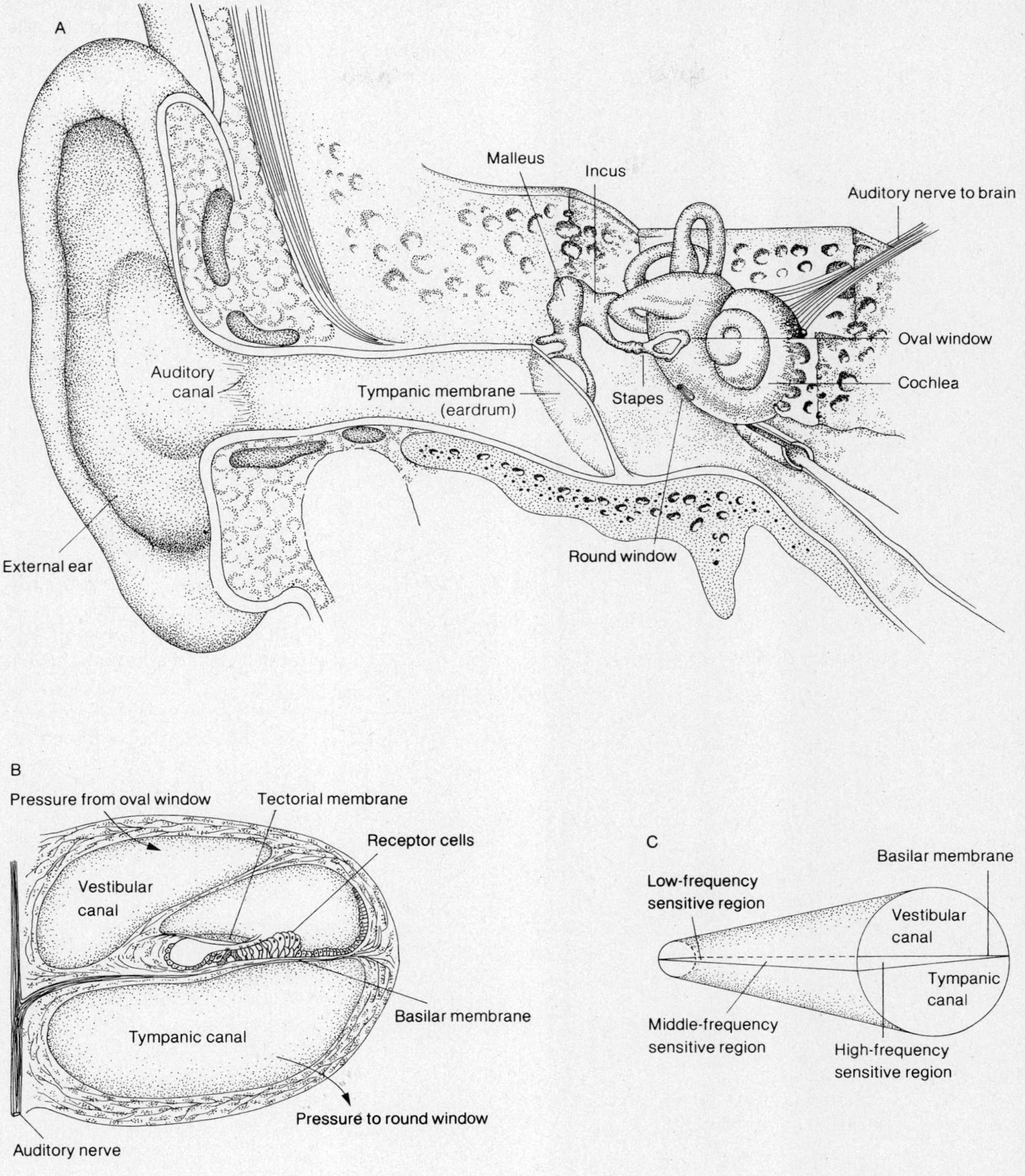

A
Malleus
Incus
Auditory nerve to brain
Oval window
Cochlea
Stapes
Auditory canal
Tympanic membrane (eardrum)
Round window
External ear
B
Pressure from oval window
Tectorial membrane
Receptor cells
Vestibular canal
Basilar membrane
Tympanic canal
Pressure to round window
Auditory nerve
C
Basilar membrane
Low-frequency sensitive region
Vestibular canal
Tympanic canal
Middle-frequency sensitive region
High-frequency sensitive region

of which absorbs a specific wavelength. Even within related groups of animals, there may be wide differences in color vision. For example, New World monkeys have color vision roughly equivalent to that of partially color-blind humans (those with protanopia). But Old World monkeys and apes see color about as well as we do. Interestingly, other mammals may not fare as well. For example, ground squirrels and cats have very poor color vision (Sechzer and Brown 1964; Crescitelli and Pollack 1965), and bulls probably can't see red. Now, let's consider vision in some specific cases among invertebrates and vertebrates.

### Invertebrates

The most primitive sorts of photoreceptors are found in the one-celled organisms called *Euglena* (Figure 7.5). The *Euglena* has a simple light-sensitive eyespot that is shaded on one side so that the animal is able to detect roughly which way the light is coming from. Another flagellated protozoa, *Pouchetia,* has an eyespot covered with a material that functions as a kind of lens in that it crudely focuses light.

The planarian flatworm has an improved directional system in its two eyespots that are shaded on opposite sides. This directional ability is, of course, important in shade-seeking animals.

Among invertebrates, one of the highest levels of visual acuity is found among cephalopod molluscs such as octopus and squid. In fact, there is a remarkable similarity between the cephalopod eye (Figure 7.6) and the vertebrate eye (see page 101). The octopus is particularly sensitive to the "pointed" end of other animals since, in the water, this is usually the head area and indicates the probable direction of the object's movement. By the way, the octopus may not have learned this; its response may be due to "stimulus filtering" (discussed later in the chapter), in which the receptors register only certain aspects of the environment.

Interestingly, the octopus is able to easily discriminate between how "wide" and how "tall" an object is. For example, it can tell a vertical line from a horizontal one. But it reacts the same to diagonal lines that are tilted to the right or to the left, since they are

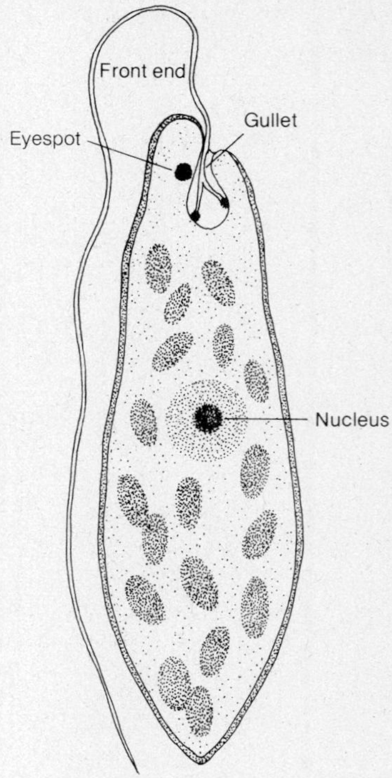


Figure 7.5 The *Euglena,* a protozoan, has one of the simplest light receptors known, a primitive eyespot.

equally tall (Young 1961). The ability is undoubtedly partly due to the fact that the pupil of the octopus eye is a pivoting slit that remains horizontal, regardless of the position of the animal.

In arthropods, such as spiders, crayfish, and insects, visual ability is well developed. Spiders have eight eyes (two rows of four), but most spiders don't have enough photoreceptors to form clear images. The jumping spiders (Salticidae), however, have a relatively large number of photoreceptors. This is undoubtedly beneficial since they leap onto their prey, and it wouldn't do to leap onto a hungry bird.

Crustaceans, such as crayfish, have both median eyes and compound eyes. The median eye is simpler, lacks a lens, and is characteristic of larval stages. The compound eyes may be on stalks and are the first eyes we've considered that are especially sensitive to movement. Compound eyes are actually composed of numerous "eyes," or *ommatidia* (Figure 7.7). The ommatidia cannot move to follow an image, so they stare blankly in their separate directions until something moves into their field of vision. When an ommatidium is stimulated, it fires an impulse to the central nervous system. So, as an image moves across the compound eye's field of vision, it fires a number of receptors in turn. Thus, even very slight movements are detected. The convex surface of such eyes may permit a visual field of 180 degrees, but the world is apparently perceived as some sort of mosaic.

Mantids have two types of ommatidia in their compound eyes. The more central ones are in sleeves of pigment so that an impulse cannot "spark across," thus, the stimulation of one cannot affect the others. This means the deadly little predators are very sensitive to an insect slowly crawling on the wall. Even the slowest movement will cause new ommatidia to fire, with no confusing effects caused by diffusing stimuli. However, since threshold intensity for each ommatidium must be reached, these receptors function best in good light. The more peripheral ommatidia, on the other hand, have transparent sleeves, so light passes from one ommatidium to another and can have an additive effect. Thus these areas have high sensitivity, but perhaps produce fuzzy outlines as light dances about the thresholds of various receptors simultaneously. Because light that stimulates one receptor is not blocked from reaching others as well, these function especially well in dim light (Roeder 1963).

Visual perception is well developed in the honeybee *(Apis)*, which is particularly sensitive to form and color. Interestingly, among bees, some kinds of forms are more preferred than others. For example, they seem to prefer a form with many parts, that is, a form that is strongly subdivided or that has an uneven outline. This may be because such forms produce numerous changes in the firing of the ommatidia, causing a "flickering effect" that bees find attractive. It may not have escaped you that flowers are subdivided and usually have uneven outlines.

In an experiment by Karl von Frisch (1953), the color vision of humans and bees was compared. It was found that bees can see ultraviolet light (which is invisible to us), and, in fact, some flowers use ultraviolet colors to attract bees (Figure 7.8). On the other hand, bees are blind to red light, which we see easily, and it turns out that red flowers are not normally pollinated by bees, but by wind, birds, or moths. Thus,

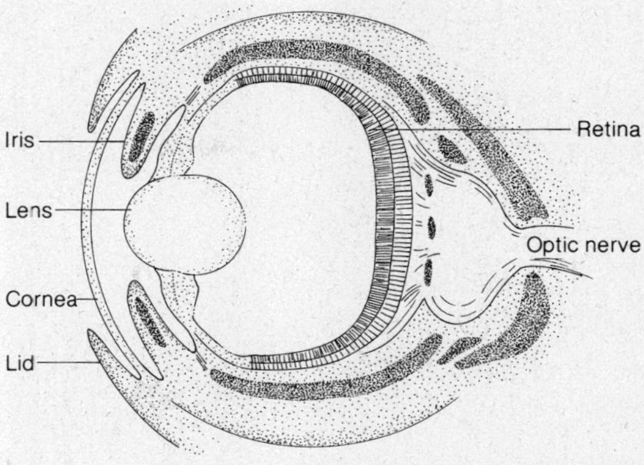


Figure 7.6 The eye of the cuttlefish, a squidlike cephalopod. Light is focussed by a lens onto a sensitive retina in a manner similar to the visual system of vertebrates.

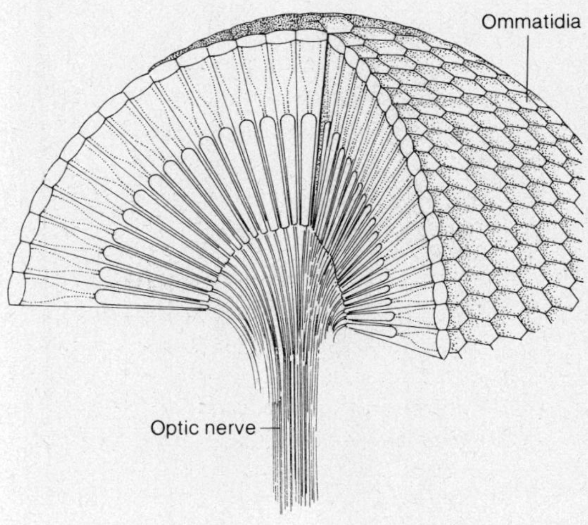


Figure 7.7 The compound eye of an arthropod. Each ommatidium fires separately and with full force when stimulated. Note the extreme field of vision.

the color of flowers and the sensory abilities of bees have evolved together to the mutual advantage of each.

Also, bees and some other arthropods are sensitive to polarized light. The plane of polarized light in the sky (see Chapter 13) is related to the position of the sun, and some animals are able to use this information to orient by the sun, even when it is obscured by clouds (Waterman 1961).

## Vertebrates

As has been mentioned, the vertebrate eye is unexpectedly similar to that of cephalopods, such as the octopus. In each, light entering through the pupil is focused by a lens onto a sensitive pigment layer at the back of the eye, the retina. The major difference between the two eyes is that their retinal layers face in opposite directions. In mammals, the photoreceptors are oriented away from the lens, and in cephalopods, the photoreceptors are oriented toward the lens.

Two types of photoreceptors are found in the mammalian retina: highly sensitive *rods*, which function in dim light and, where they are concentrated in the retina, lend strong light sensitivity; and *cones*, which are highly sensitive to color, but are not stimulated in dim light. There is not a clear line of distinction between rods and cones, however (Figure 7.9). Some light receptors have qualities associated with both.

As you sit reading this fascinating account, out of the corner of your eye you might be distracted by a slight movement. You look up; it's only a fly walking along the wall. It's a good trick, but you've seen it before, and eagerly return to your text. You noticed the fly because the vertebrate eye is extremely sensitive to movement, although the mechanism is different than that employed by arthropods.

The optic nerve of vertebrates is comprised of neurons (fibers) that respond to one of three types of stimuli. For example, some fibers only respond when incoming light stops or is turned off ("off" fibers). The second type fires *either* when a light appears or when it stops ("on-off" fibers). And others continue to fire as long as a light source is present ("on" fibers). Remember that although we're speaking of "light" in its broadest sense, light carries information regarding color, texture, and form in the vertebrate visual system. There is a preponderance of "off" and "off and on" fibers in vertebrate eyes, and so these fibers are highly sensitive to anything moving across the

Figure 7.8 Humans may see white or yellow flowers (top), but bees see shorter wavelengths and, to them, those same flowers appear as violet, especially near the nectar-laden centers (bottom).

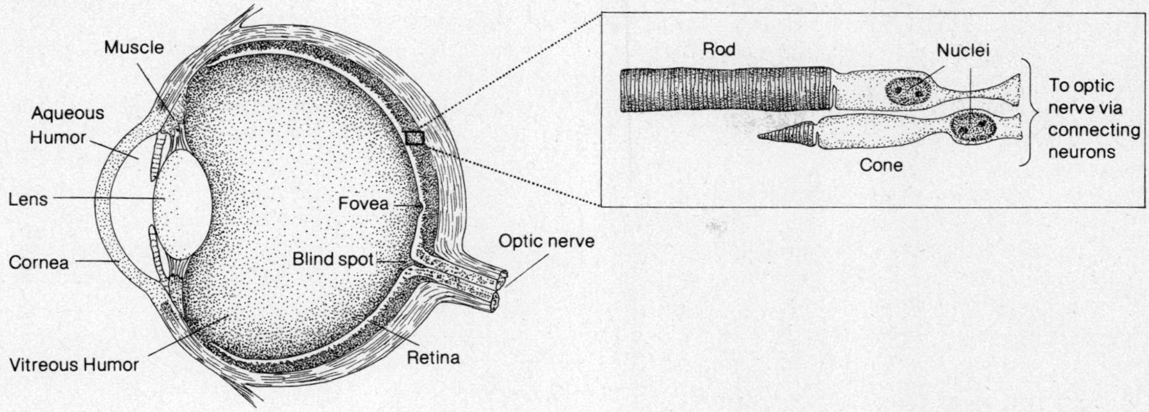


Figure 7.9 The human eye. Light is focussed through a lens and falls on a retina composed of color-sensitive cones and light-sensitive rods. The fovea is a high-density area of receptors.

field of vision. This has a number of ecological consequences.

In most vertebrates and some invertebrates, there is a "visual alarm" reaction to anything that rapidly increases in size. If very young and inexperienced animals are shown a small circle projected onto a wall that becomes quickly larger, they will duck or show evasive behavior. There is obviously some mechanism in their eyes and associated neural apparatus that triggers an evasive response to certain cues. Through the long history of the animals' evolution, rapidly enlarging objects could have meant that something was after them—and gaining. The response, then, could have become genetically influenced so that each animal didn't have to learn the proper response the hard way.

Many vertebrates have an inborn ability to perceive depth. For example, many land vertebrates will not cross a "visual cliff" (Figure 7.10), although they have had no experience with cliffs. Apparently, the stimulus that triggers the behavioral response is associated with parallax, or the relative apparent movement of things near and those farther away. As the head is moved from side to side, nearer objects seem to move more. It is this distance-associated discrepancy that provides the environmental cue. The central nervous system seems to react to the parallax cue by causing the animal to "stop," "jump," or "back up" (Gibson and Walk 1960).

There may be wide differences in perceptive abilities, even within classes of animals. For example, the vision of fast-moving fish, such as bass and trout, in clear water is particularly well developed, and shows color discrimination, depth perception, and some form discrimination (information useful to fly-tiers). Fish that live in dimly lit environments tend to have less discriminating vision in spite of having larger eyes. If they have bioluminescent organs (light producers), they may have unusual numbers of rods to detect that kind of light in other animals of the same species (as do, for example, *Bathylagus* and *Zenion*) (Maier and Maier 1970).

While amphibians occupy both land and water habitats, the amphibian eye is adapted to terrestrial environments. For protection against the drying atmosphere, amphibians have eyelids and glands that keep the eyes moistened. Conservation of moisture over the eye is a problem with almost all terrestrial vertebrates, and the amphibian's adaptations are not unusual. But the amphibian eye *is* unusual in some respects. For example, the frog has well-developed muscles that are able to distort the shape of the eyeball and thus to adjust the distance from the lens to the retina. This ability aids in distance perception (and focusing) in animals that are not able to judge distance by focusing two laterally directed eyes on the same object. (Apparently, animals that can "fix" an object with both eyes can then extrapolate its distance from the angle thus formed.) Since many frogs eat small insects, their retina is especially adapted to register movement. In fact, some ganglia only respond to photoreceptors that have been stimulated by small moving objects passing close to the frog (Barlow 1953). Movement of the object is important in visual discrimination in some amphibian species. Frogs of the genera *Rana* and *Hyla* can differentiate between a circle, a star, and a cross, but only if the forms are moving (Thorpe 1963).

Among reptiles, many terrestrial lizards and turtles have a third eyelid (frogs have two of these), which further moistens and protects the eye. They have well-developed tear glands, are able to move their eyes in their sockets more easily than amphibians, and can adjust the shape of the lens. Changes in the shape of the lens aid in focusing—as does the distortion of the entire eyeball in frogs.

Snakes are also reptiles, but snakes don't blink, because they have no eyelids. Their eyes are protected, instead, by a transparent skin over the eyes. They have no tear glands and their eyes are relatively

Figure 7.10 The day-old goat could not have learned the dangers of cliffs, but it avoids them even when the cliff is harmlessly covered with glass.

fixed in their sockets, lending them their sinister stare.

Birds have well-developed vision, although their eyes are structurally similar to those of reptiles. Their eyes are relatively large and they have well-developed third eyelids *(nictitating membranes).* These membranes are apparently closed during flight to protect the eyes (Weichert 1959), just like a barnstormer dons his goggles.

High-flying birds are well known for their visual acuity, and some are real eagle-eyes. Part of their visual ability is due to the density of their receptors. In fact, in the European buzzard *(Buteo buteo),* the most sensitive part of the retina, the *fovea,* has about one million cones per square millimeter (Walls 1942).

Not surprisingly, night-hunting birds have more rods than cones in their retinas, and day-flying birds have more color-registering cones. In some birds, there is a horizontal streak across the retina that is heavily laden with cones. In order for a hawk to take a good look at something it has noticed on the ground, it must tilt its head so that the image falls on this band. However, because of this streak, it is able to scan the horizon with great clarity as it flies along.

Color vision in birds seems to be facilitated by colored oil droplets within the receptors themselves. Every receptor has its own color of oil. The result is that colored objects seen by the bird are strongly contrasted. Some receptors have red droplets that act as haze filters, cutting out certain of the wavelengths that produce glare. Therefore, "early birds," those species that are up and about when there is likely to be haze, have the greatest number of cells with red droplets (Wald 1959). As an aside, birds are able to learn form discrimination (the difference between circles, squares, and so forth) easily, but, for some reason, only if the forms are of very nearly the same size and brightness (Bingham 1922).

Among mammals, vision is particularly good in primates and carnivores. As would be expected, nocturnal mammals have eyes that are highly sensitive to light and so they are able to easily discriminate vague forms in the darkened world in which they live. However, animals more active during the day have the sharpest vision, because they are more likely to need it.

Mammals are very sensitive to movement, as has been mentioned, but in some species, only movements in a certain *direction* elicit neural impulses. For example, rabbits—whose eyes are located on the sides of the head, have certain ganglia that respond only if an object is moving from the posterior to the anterior, but not in the reverse direction (Barlow and Hill 1963). Also, rabbits have special ganglia in their retinas that respond only if a small or distant body moves into the receptor field and then *stops* (as would a stealthy predator), or if such an object moved *out* of the receptor field. Grazing rabbits often stop and raise their heads and remain perfectly still for a few moments. Thus, the rabbit's visual images remain stable, and most cells stop firing. However, should something in the rabbit's visual field move slowly and then stop, the "predator-finding" ganglion will fire, and the sneaky lynx will have been discovered while it is still a safe distance away (Levick 1967).

Form-discrimination is accomplished rather easily in most mammals. In one test of rats, though, they failed to learn a simple visual discrimination in one thousand trials. It turned out that the failure was in the experimental design. Rats apparently prefer to rely on other senses than sight when making discriminations, so they would walk up to the images, sniff them, find no differences, and therefore fail to discriminate. However, if they were forced to leap from a pedestal toward one of two images, and if they were then punished for the wrong choice, they quickly learned to discriminate visually (Maier and Maier 1970).

Visual discrimination is most highly developed in primates. Some chimpanzees are apparently even able to recognize pictures of common objects, as has been discovered in the ongoing experiments to attempt communication with them.

Primates are unusual among mammals in that they can see color. It has been suggested that this is how chimpanzees, for example, can distinguish ripe red or yellow fruit from unripe green fruit. It might be added that color vision in primordial man may have enabled him to rather easily change from his two to four million years of being a hunter-gatherer, to being a formidable and effective hunter, since color vision is immensely helpful to predators. Ethologist George Schaller found that during the fawning season he could rather easily find newborn antelope on the African plains. The defense of the young antelope was to "freeze"—remain motionless. The effort would be rather effective against a color-blind hunter looking for movement or contrast, or a predator hunting by scent (in which case he must be downwind), but is of very little use against animals with color vision. (My cat-hating German shepherd will walk right past a motionless downwind cat, unless that cat is white, or in rarer instances, black—colors that usually contrast with the environment.) Interestingly, animals that are preyed upon by color-visioned animals such as birds are often cryptic in coloration as well as markings (Figure 7.11).

## CHEMORECEPTORS

Basically, there are three kinds of chemoreception. The simplest type is a general ability of relatively unspecialized receptors to detect certain chemicals. Taste, or *gustation,* is a more specialized ability. It operates best when the receptors are in direct contact with the stimulus, a requirement that is not demanded of the third, and most sensitive, type of chemoreception—olfaction, the sense of smell. Olfactory neurons can react to very low concentrations of chemicals in air or water, and hence they can respond to molecules that have diffused or been swept from the stimulus source. The source, then, can be detected some distance away.

### Invertebrates

Some invertebrates have the greatest powers of chemical detection yet known, but others make do with relatively unspecialized abilities. Planarians, for example, have generalized receptors located in pits on their bodies, over which beating cilia move water currents. On the other hand, insects have distinct taste and smell receptors, sometimes of astounding sensitivity. Taste receptors may be found in the mouthparts, antennae, or forelegs—and, in some cases, the ovipositor, or egg-laying apparatus (since

Figure 7.11 Those species that are hunted by color-visioned birds, are often cryptic in color and markings, such as this viper from the Foz Desert in Angola.

many insects lay their eggs on certain plants and in other very specific places). Olfactory receptors, however, generally lie at the back of minute tubules that penetrate the insect body. As a molecule of a chemical diffuses in through the tubule, it becomes dissolved in the fluids that bathe the receptor (Schneider 1969).

The silk moth (*Bombyx*) smells with its antennae. Adult males have enormous antennae with thousands of sensory hairs (Figure 7.12). About seventy percent of the adult male receptors respond to only a certain complex molecule called *bombykol* and, astoundingly, only one molecule is necessary to generate an impulse (Schneider 1962, 1969). (A human receptor would need to contact several million of these molecules in order to detect anything at all.) We will discuss this in more detail in the chapter on communication, but, to summarize, a female moth in reproductive condition need only stay put and secrete bombykol. As the molecules drift downwind and touch the antennae of a male, they enter the tiny pores of a "hair" that houses an olfactory receptor, dissolve in a solution covering the receptor, and interact with its membrane. The male then heads upwind, a behavior that, sooner or later, should lead him to the siren that released the chemical.

The female, it should be noted, has smaller antennae with much lower sensitivity. She is not particularly good at detecting males, but then, why should she be? The male uses his energy in seeking her out, while she, on the other hand, is better off placing energy into egg formation. This is not to say she is generally insensitive, however. She is very sensitive to other environmental cues, such as those that indicate a suitable site for egg laying.

In contrast, the honeybee is sensitive to not only one scent, but can differentiate a total of forty-three different odors (von Frisch 1950). This fine power of discrimination is essential to both of the leading notions of how bees communicate food sources to other bees. (See the discussion of bee orientation in Chapter 13.)

Figure 7.12 The male silk moth *Bombyx* has one of the most sensitive chemical detectors in the animal kingdom in its great, brushy antennae.

### Vertebrates

Chemoreception in vertebrates may occur by any of the three major means of detection. Receptors generally sensitive to a number of chemicals are scattered over the body surfaces of many vertebrates. It has been argued that they are the same as pain receptors, but actually the two sensations have been separated experimentally (Morgan 1965). In one test, acid could still be detected on the skin after the pain receptors had been deadened by cocaine.

Olfaction is usually accomplished by moving chemical-laden water or air into a canal or sac that contains the chemical receptors. The chemical cannot be detected, however, until it goes into solution. You may be aware that your nose is full of a solution, and, in fact, nasal passages are moist in all terrestrial vertebrates, since in this entire group olfactory receptors are found in the nose (Figure 7.13). Fish and aquatic amphibians need not manufacture a nasal liquid, however, since any chemical is likely to already be in solution when the animals encounter it. In these groups, water is moved into a nasal sac that is lined with a folded membrane covered with receptors.

In land vertebrates, the best sense of smell is found in mammals, especially carnivores and rodents. It is obvious why olfaction is important in carnivores, for example, in wolves that must track their prey. It is less clear why rodents have evolved such strong olfactory ability, since their food sources are not usually widely dispersed and usually do not run away. However, rodents discriminate among them-

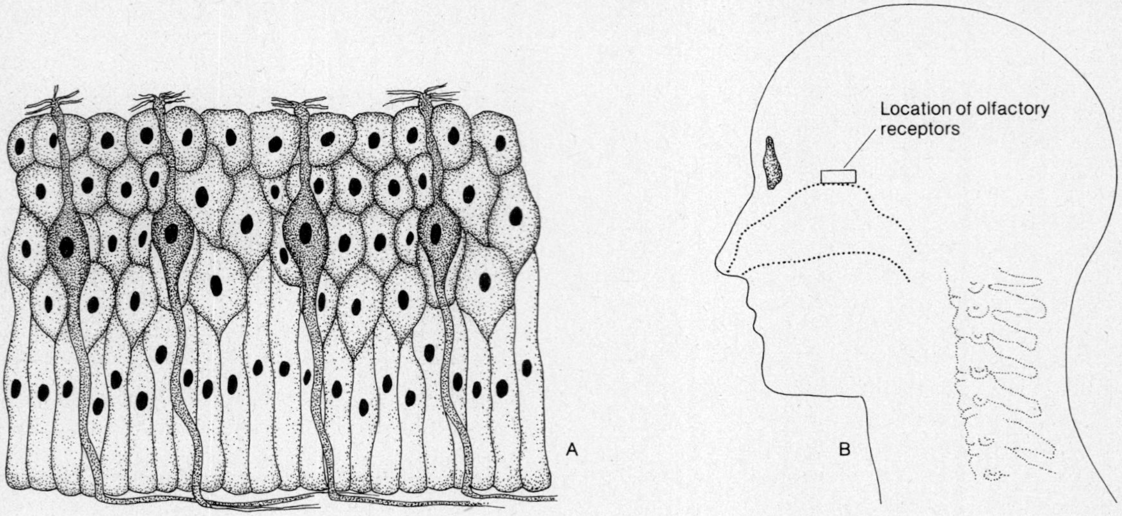


Figure 7.13 The structure (A) and location (B) of olfactory receptors in humans.

selves on the basis of families, or "tribes," and hence olfaction may be important in social interactions.

Olfaction is especially well developed in the dog family (Canidae). Wild canids, and some breeds of domestic dogs, have long muzzles that accommodate a large area of folded nasal epithelium. The increased surface area holds more olfactory receptors and hence lends greater powers of smell.

Olfaction is rather poor in primates. Chimpanzees smell about like you do (no offense), but olfaction is undoubtedly more important in mating behavior among the other primates than among humans.

Interestingly, toothed whales of the suborder Odontocetes have no sense of smell at all. Apparently, whales are derived from terrestrial vertebrates that returned to the sea and whose descendants were unable to adapt their noses to the new environment.

There is certainly no accounting for esthetics in smell. You may have noticed that your dog is fascinated by something you were chagrined to find on your shoe. Of course, much of our esthetic values may be acquired. Americans have been sold the idea that human body odor—*any* body odor—is bad, and must be removed, disguised, or overpowered at all costs. Preliminary evidence, however, suggests that certain body smells can be attractive to the opposite sex in humans. There are indications that women secrete odorous vaginal substances at the time they are most likely to become pregnant. Some women also describe being attracted to the normal body scents of men—as long as it is not overwhelming. Perhaps, to our forebears, human smells were not "bad."

Smell may play a more important part in human sexual relations than was previously believed. For example, a certain odor derived from musk glands of other mammals, which is important in their mating, can hardly be detected by men and women before puberty and after menopause. However, young women in reproductive condition can easily smell these substances, especially about two weeks after menstruation, a time when they are most likely to be fertile. Men have difficulty in detecting the scent, but can do so after injection of the female hormone, estrogen (LeMagnen 1952).

The sense of taste depends upon certain chemicals initiating impulses in the neurons of *taste buds.* Among fish, these receptors may be scattered over the body's surface, but are often more concentrated in the head region. In aquatic animals, the sense of taste is sometimes very difficult to distinguish from that of smell or general chemical detection, but it cannot be denied that the chemical senses are well developed in these animals, as we will see in our discussion of chemical communication (Chapter 16). As an example of chemoreception in fish, the bullhead (or catfish) *Ictalurus*, has high concentrations of taste buds on its long, sensitive "whiskers" and around the mouth. Of course, such sensitivity is important in an animal that feeds on the dark bottoms of murky rivers or lakes.

In all terrestrial vertebrates, taste receptors are confined to the mouth area. And, as you already know, there are four basic tastes: sweet, sour, salty, and bitter. In humans, receptors sensitive to each are located in specific areas of the tongue (sweet on the tip, sour on the sides, bitter on the back, and all parts of the tongue are sensitive to salt). Try to distinguish salt from sugar by simply touching the tip of your tongue to a few grains of either substance. The human taste bud is shown in Figure 7.14. Physiological studies indicate that each taste bud does not contain receptors that are specific for a certain taste, but rather that a single cell responds to three or more tastes and is simply more sensitive to one than to others. It should not be surprising, then, that each taste has its own neural code, producing distinctive firing patterns in different kinds of receptors (Pfaffman 1959).

Interestingly, animals are variously sensitive to the taste of sweet. You may have difficulty in trying to get your cat to eat a cube of sugar that a rat might taste and a guinea pig might devour in a flash. The reason is, cats can't taste sweets, while rats are moderately sensitive, and guinea pigs are highly sensitive to the taste (Beidler 1961). And we cannot discount individual differences within groups. Your dog might not deign to touch a cube of sugar. But many Spanish dogs that beg at sidewalk cafes eagerly accept such an offering. It might be because these dogs are hungrier, and thus have a higher tolerance. But many of the ones with the most attentive, pleading expressions are very fat. The Spanish dogs may also have been fed sugar as puppies, and thus acquired the taste. It is obvious that individual differences must be taken into account before generalizing about how any species will react to a specific taste.

Taste ability is obviously important in determining diet, and so has undoubtedly been molded by natural selection. For example, cats are almost exclusively flesh-eaters and would therefore have little use for sweet-sensitive taste buds. On the other hand, humans are omnivorous. That is, we'll eat almost anything, plant or animal. It has undoubtedly been important to us, through our long eons of development, to be able to finely distinguish between a variety of tastes and smells. Those things that smelled and tasted good were probably good for us. Something rotten probably lost its appeal for us (but increased its value in the eyes of other animals, such as vultures). We have recently, through sophisticated chemical manufacturing, been able to fool our traditional chemical warning system so that we now happily consume small cans of "lemon pudding" that are dyed chemical mixtures containing no lemon whatsoever. Read the label next time. Or note the ingredients in your "non-dairy" creamer.

From an adaptive viewpoint, it is interesting that while we are quickly satiated by salty and sour foods, and may even avoid them, we have no upper thresh-

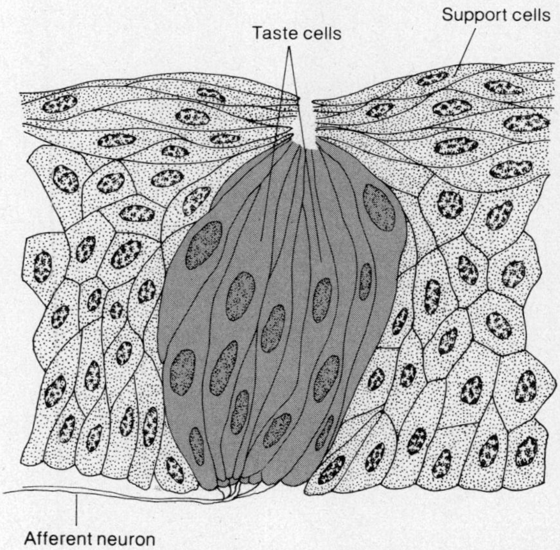


Figure 7.14 Structure of a taste bud from the human tongue.

old for sweets. There is no point at which a sweet taste would be avoided. This may be because, until recently in our history, it was impossible to get enough sweets. Sweet is the taste of carbohydrate, an important type of food (Nordsiek 1972), and food has traditionally been hard to come by for many species, including our own. Therefore, when food was found, it may have been to man's advantage to be able to glut himself on it without becoming tired of the taste. Recently, as we know, the situation has changed. It's easy to get sweets. And many of us are finding to our sorrow that we *still* don't seem to get tired of the taste.

On the other hand, sour is the taste of acid, and a certain amount of acidic food is good. Some acids are associated with important vitamins such as vitamin C (citric acid, ascorbic acid), and these are needed in small amounts, so we enjoy the taste of acid—up to a point. Too much acid can be bad for digestion, and it is hard on the gastric lining as it is added to the acid already present in the stomach. One reason that a glassful of pure lemon juice doesn't hold much appeal may be related to our evolutionary history. Sour is the taste of unripe fruit. In our early hunter-and-gatherer societies, and perhaps well into our agricultural period, it may have been to our advantage to avoid eating unripe fruit—to wait. By waiting, if alternate foods were temporarily available, the fruit would have time to grow to its maximum size and carbohydrate content. As it ripened and its tissues changed to high carbohydrate levels, the taste would change to sweet—the "ready" signal. Thus, the avoidance of high-level sour tastes could maximize the food-yield from certain sources such as fruits. It might also be to a fruit tree's advantage to render its fruit unpalatable until the seeds are ready to be dispersed. At that time, when the fruit becomes desirable, it is pulled down and carried away, the seed to be dropped in a new and possibly hospitable area.

Regarding salt, low-level dosages are generally agreeable, and thus we tend to eat things that contain important salts such as sodium, calcium, magnesium, and potassium. Highly salty substances are often avoided though—and with good reason. To eat too much of them could easily result in an imbalance of critical ions in our bodies. Finally, bitter substances are perhaps roundly rejected by the palate because bitter is the taste of so many poisonous foods. Some insects eat alkyloid plants and thus render themselves unpalatable to birds.

## PROPRIOCEPTION

Proprioceptors respond to displacement and pressure changes caused by the relative positions of different parts of the body. Thus, you may not be able to see your foot under the table, but you could probably point to it; your proprioceptors tell you where it is. Proprioception is most well developed in mammals—such as the primates, which often amaze us with their agility (Altman 1966). In vertebrates, such receptors are located in muscles, in tendons, and in muscle coverings, all of which stretch to various degrees depending on the position of the limb. The receptors fire according to how much they are stretched, in other words, according to the position of the limb.

Several types of proprioceptors are found in arthropods. For example, cockroaches—which are running animals, as you know if you have ever chased one—have hairs on the inside of their legs that respond to flexion of the knees and that probably function in coordinating running patterns. The closely related crabs and lobsters, on the other hand, have receptors located within the muscles—a condition similar to that of humans (Carthy 1958).

Another type of proprioception gives information regarding the position of the body with respect to gravity. In most invertebrates, this information is conveyed by *statocysts,* which are openings containing sensory neurons and a granule of some sort. In crayfish, for example, as long as a granule is resting on the sensory hairs at the bottom of the statocyst, the animal will remain in an upright position. If the animal is turned over so that the granule falls away from the hairs, it will immediately right itself. If, on the other hand, the granule, which is usually sand, is replaced by a bit of iron, the animal will respond normally unless the iron is lifted by a magnet, in which case the crayfish will flip onto its back.

Equilibrium receptors in vertebrates are located in the inner ear. Usually, three fluid-filled loops, or semicircular canals, and two saclike chambers register the movement of the animal (Figure 7.15; see also Figure 7.4). The semicircular canals connect to two bulges at their base that are lined with sensory hairs. The fluid moves these hairs if the animal changes speed or direction. Since the three loops lie at right angles to each other, movement in any direction is detected. The two bulges, or chambers, the utricle and saccule, contain granules or small bones and function very much like statocysts.

In humans, there may be a lag in time between the cessation of an impulse and its sensory reception and decoding. Thus, pilots have great difficulty "flying blind" in small planes. After a turn has been completed, the cessation of the movement may not register with the pilot for a moment and he tends to overcompensate (John Willister, flight instructor, personal comment).

## SELECTIVE PERCEPTION

In spite of the wide variety of impulses bombarding an animal from its environment, an animal is able to detect only a limited range, and of these impulses, it may ignore all but a few key stimuli, as we shall see in our discussion of instinct. We must, therefore, distinguish between *perceived* and *effective* impulses (Tinbergen 1951). Ethologist Eibl-Eibesfeldt, as late as 1970, stated that the former are studied by physiologists, the latter by ethologists. Fortunately, the validity of that statement is waning, as ethology is incorporating an increasingly wide front.

The process of distinguishing between perceived and effective impulses by an animal is called *stimulus-filtering*. For example, when a toad spies a fly walking along a wall, it begins to orient itself in preparation for the tongue flick that will capture the insect. But if the insect stops, the toad shows no further reaction. The fly is still clearly visible, but now it elicits nothing in the toad's central nervous system that would compel the toad to continue its orientation. Somewhere between the eye of the toad and its muscles, the sight of the immobile fly is filtered out. The toad sees the fly, certainly, but not as an orienting stimulus. Should the fly begin to move again, however, the stimulus apparently either shifts neural pathways so that it stimulates a different ("orienting") brain center, or the movement causes a different chain of neural events in the same center that was being stimulated by the immobile fly. The frog moves and the fly is nabbed.

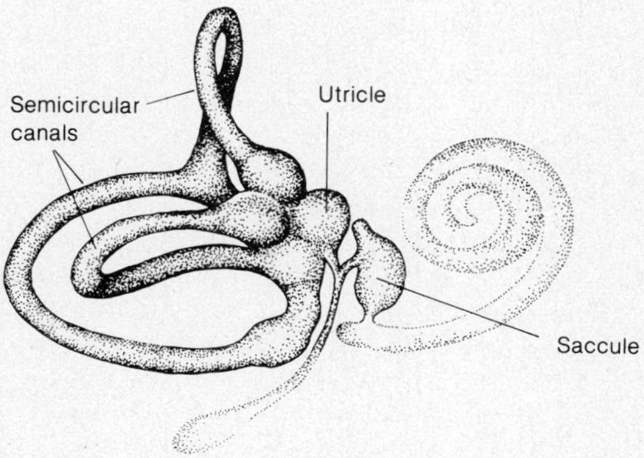


Figure 7.15 The semicircular canals and the two chambers that function as equilibrium devices in mammals. (See also Figure 7.4.)

One of the central questions in modern ethology is how do animals selectively filter incoming stimuli? A frog sitting in a marsh that is resounding with an evening chorus of all sorts of frog species may "hear" only the call of its own species. Why? Perhaps the tympanum ("eardrum") is of such a size that it will only vibrate in response to sounds of a certain frequency. Or perhaps the auditory neurons, or their associated ganglia, will fire only in response to certain frequencies and/or volumes of sound. Or perhaps, once the neural impulse enters the central nervous system, some are ignored, while others stimulate the proper brain center. So, perhaps the brain registers all incoming stimuli and selectively ignores certain signals, while responding to others by sending out impulses to the appropriate muscles. Probably all of these ideas are too simplistic, and the real answers are much more various and complex. Robert Hinde (1970) gives an excellent review of the selective reception of stimuli.

One last thing: if the "coding" (filtering) is done at the level of the receptor, it is called *peripheral coding;* if it takes place in the central nervous system, it is called *central coding.* An advantage of peripheral coding is that useless or irrelevant information is discarded early so that no neural activity is wasted on its transmission. Such economy may be important in lower animals in which the number of nerve cells is relatively small. On the other hand, peripheral coding is disadvantageous in that there are fewer neurons operating at the peripheral level than in the central nervous system. Therefore, any peripheral coding, or filtering, is likely to be far less specific than central coding. The complexity and coordination of any resulting behavior will therefore be reduced.

It would seem, then, that peripheral coding would be more useful in animals that live in a stimulus-poor, or homogeneous, environment where the complex integration of stimuli is not necessary. It might also be advantageous in animals with relatively few brain cells. Central coding, on the other hand, might be expected in animals living in rich or varied environments with high levels of potential stimuli that could best be sorted in the brain. The generalization seems to hold. Animals, such as the frog, that rely more on "genetically-based" and rather inflexible patterns tend toward dependence on peripheral coding. Mammals such as the cat and the monkey, on the other hand, show comparatively little peripheral coding. For references and notable exceptions to this trend see Muntz (1971).

In this chapter, then, we have seen that the nature of the world is partly defined by the nature of the beast. Organisms are able to perceive only parts of the vast spectrum of potential stimuli provided by the environment. Some kinds of animals perceive more kinds of stimuli than others, and perception can vary from one time to the next. The number and kinds of stimuli any organism can perceive is determined by what sorts of information the organism needs in order to survive and reproduce. Some need a great deal of information, some need very little. This brief survey of the perceptive abilities of animals should have underscored one more thing: it is unwise to blithely extrapolate from one kind of animal to another without considering differences in the kinds of information they have available to them.

# 8

# THE CLASSICAL INSTINCT THEORY

A young long-tailed tailorbird has managed to acquire a mate and is in the process of building its first nest. The quick, little bird first finds two large, hanging leaves and then proceeds to sew them together in a most curious fashion, first by punching holes in their margins with its sharp beak and then threading strands of spider webs, cocoons, or bits of string through the holes. It manages to tie stopknots in each piece of thread and to pull it tight so that the leaves form a sort of pouch. It then searches the area for twigs and grass, which will make the floor of the nest, and on this cushion it deposits its eggs.

The tailorbird has not witnessed nest building by older birds and it has had no previous experience in such a task. But on its first try it builds a nest—a nest that may not be perfect, or even as good as those of older birds, but it is adequate, and the tailorbird is able to raise its young (Figure 8.1).

Among animals, there are many such instances of elaborate behavior that show adherence to a "plan" that is relatively constant throughout a species and is not acquired through previous experience or learning. These types of behavioral patterns are considered to be "instinctive," a term that, through the years, has had many meanings.

It is unfortunate that even in the sciences, where objectivity is of prime importance, certain ideas may be temporarily fashionable, while others become unpopular and are largely excluded from serious consideration. The study of instinct has suffered from several such swings of the prejudicial pendulum. A brief review of the history of the study of instinct will point this out.

## THE MODERN DEVELOPMENT OF THE IDEA

Those who study animal behavior once thought that innate patterns stemmed from certain inborn neural "seats" that, when activated, produced instinctive behavior. But about the middle of this century, ethologists began to see that an innate behavioral pattern could not simply be ascribed to the existence of an "instinct" for that pattern. The idea was overly simplistic. The hard evidence just didn't suggest the presence of a neurological "seat" for an instinct—a single center that produced a particular behavior when it was somehow turned on. So, instead of attributing complex innate behavior to a single neurological event, a multilevel model was developed, primarily by Lorenz and Tinbergen. It is called the "classical model," since it has served as the basis for recent concepts. In this chapter, we will see what the original model was all about. Just keep in mind that the concept has undergone changes as we will see in later discussions. However, the idea is more than a historical oddity; it has served as a framework upon which later notions have been superimposed, even by those who most strongly denounce the old idea. Let's begin with an example.

### Appetitive and Consummatory Behavior

Hunting, in some birds, involves instinctive behavior. As a hunting peregrine falcon (Figure 8.2) begins searching for prey, its behavior is highly variable. It is

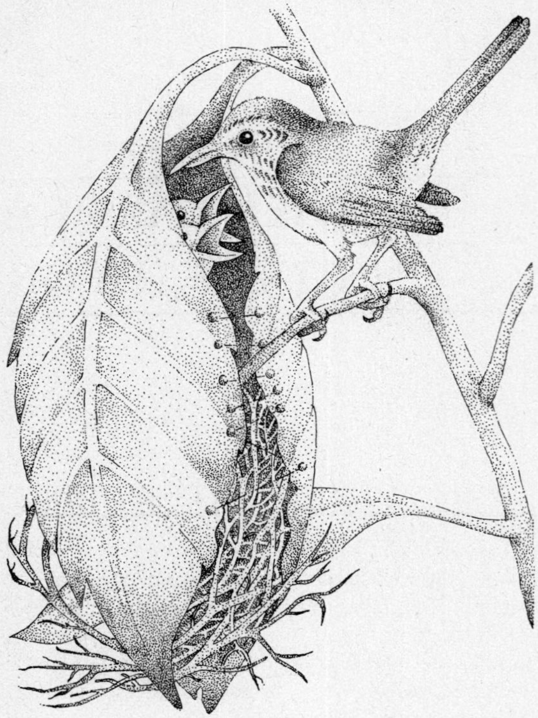

Figure 8.1 The nest of the long-tailed tailorbird, made by its sewing two large leaves together. The bird is able to perform this complex task correctly on its first attempt.

hard to predict which way it might bank, or whether it will rise lazily on air columns or suddenly drop toward the ground. At such times, the bird is hungry and might be equally pleased at the sight of a flying bird or a scampering mouse. Then our falcon sees a group of teal flying below. Its behavior suddenly becomes less random as it swoops toward them in the first stage of "teal-capturing behavior." The swoop is performed in somewhat the same way throughout the species, but it may vary a bit from one individual to the next. The falcon might hit at the front or the back of the group and might strike at birds to the right or the left of the flock as it passes. This is likely to be a sham pass to scatter the flock so that an individual bird may be selected for pursuit, and there is less variation in its performance than in the random searching. Once an individual teal is selected, the falcon goes through a series of maneuvers to keep the hapless bird separated from the flock as it closes in. The falcon's behavior becomes increasingly less variable as it enters a final high-speed dive.

The early part of the dive is still variable to a degree and can be altered in response to changes in the teal's escape behavior; if the teal changes direction, so does the falcon. However, there is a point at which the falcon's behavior can no longer be altered. It is committed to a specific pattern. Its actions now are performed in very much the same way as they have been at this stage in previous hunts. This is the brief moment at which the falcon's feet are tightly clenched into a fist as it swoops at almost unbelievable speed. Unless the teal makes a last-second change in its pattern, it will be knocked from the sky, and the falcon's hunt will have been successful (Figure 8.3).

The performance of the strike, a so-called *fixed action pattern,* ends that part of the instinctive sequence, but it usually triggers another one that also begins with a variable pattern. The falcon may follow its stunned prey to the ground through a number of routes and it may glide, flap, or walk to the body. Once there, it grasps the body in a number of ways (which may depend partly on what it has learned about teal), but its behavior is again becoming increasingly stereotyped. It tears food from the body in much the same way it has in previous kills and in much the same way other falcons treat their kills. Finally, though, when it swallows, a series of muscles are called into play that contract for the same duration and in the same sequence under all conditions. Swallowing, then, is a fixed action pattern.

Now let's reexamine what has happened. The behavior leading to the capture of the teal began as a highly variable pattern that became more and more stereotyped once the falcon began the sequence leading to the kill. In the instinctive model, the more variable actions are considered "higher" on some sort of hierarchical scale—which we will say more about shortly. Recall, by the way, that the final part is completely stereotyped and can't be changed by environmental cues. The entire variable part of the pattern is dependent not only upon the presence of a stimulus (the teal), but also upon the changing relationship of the stimulus to the environment (the teal's position), and is called the *appetitive* stage. The completely stereotyped action that ends a sequence is called the *consummatory* stage.

The appetitive behavior is considered to be a searching (the result of an "appetite") for a *situation that contains a stimulus* that could release the consummatory behavior or fixed action pattern. The objective of appetitive behavior, according to this scheme, is not the attainment of an object (here, food), nor does its performance have any survival value per se. Instead, the objective is simply the performance of a consummatory act. This undoubtedly sounds strange, but there is, in fact, some supporting evidence as we shall see later. Of course, if all goes well, the performance of the consummatory act does have survival value (here, it provides nourishment).

There may be an appetitive aspect to each *part* of an instinctive sequence. The German ethologist Leyhausen (1965b) noted that cats usually catch mice through performing a specific sequence of behavior patterns: (1) lying in wait, (2) sneaking up, (3) pouncing, (4) pawing, and, finally, (5) eating. He argued that each of these patterns has its own appetitive motivation. For example, if a cat is allowed only to (3) pounce, (4) paw, and (5) eat mice, it will become more "anxious" to perform the (1) hiding and watching, and the (2) creeping-up patterns. The animal will search for a situation that will enable it to perform the missing part of the sequence. Thus, each phase in the complex art of mouse-catching has its own appetitive and consummatory phases. Other experiments have shown that if the ultimate goal (filling one's stomach with a mouse, in this case) is not achieved, the mouse-catching sequence will be performed again much sooner than if the cat is allowed to eat the mouse.

In another set of experiments (Janowitz and Grossman 1949), dogs were surgically altered so that the esophagus emptied, not into the stomach, but onto the floor. Tubes were also attached to the dog's stomach so that it could be filled without the dog actively eating. It was thus possible to dissociate eating behavior from the results of having eaten. One question was, What happens if a dog with the esophageal opening is allowed to become hungry, and then his stomach is filled without allowing him to perform eating behavior? Would he still behave as if he were hungry, even though he was not allowed to perform the various stages of consummatory behavior? It turns out that the artificially fed dogs did not

Figure 8.2 A hunting peregrine falcon. The first stages of its hunt may be very variable, focussing only later when prey is spotted.

Figure 8.3 When a peregrine falcon spots its prey, it may show increasingly stereotyped behavior as it ends each instinctive sequence with a consummatory pattern.

behave as if they were still hungry. Thus, it was theorized that receptors in a distended stomach can switch off the motivation to perform normal eating sequences. When thirsty dogs were allowed to drink, the water never reached their stomachs, but they drank only a little longer than normal dogs. The greatest difference was, that since the water deficit was not replaced, they returned to the water pan sooner than normal dogs. However, if their stomachs were artificially filled, they lost all thirst behavior. Obviously, the behavior was at least partly dependent on the condition of the stomach. For eating and drinking behavior in dogs, then, it is not necessarily the simple performance of a consummatory pattern that reduces the appetitive behavior or "drive." So, already we can see that the classical model does not satisfy all conditions.

Most of the recent evidence supports the idea that the period of rest following goal-oriented, or appetitive, behavior results from sensory feedback of some sort that indicates "goal achieved." This makes good sense from a biological standpoint. After all, it takes about an hour for digested food to enter the bloodstream and replenish exhausted food reserves. If an animal continued to eat until that happened it would eat too much. But a simple system of sensory feedback based on receptors in a distended stomach solves the problem. The animal stops eating before its actual physiological needs are met (Manning 1971).

The ethologist Eckhard H. Hess (1962), at the University of Chicago, pointed out that appetitive behavior may be characterized in three ways: by (1) a motor pattern (usually locomotion), (2) an orientation (or spatial arrangement), and (3) a search for stimuli to which the animal is receptive. For example, a hungry squirrel (1) climbs (2) up trees (3) looking for food. It had earlier been suggested that many of the acts performed during the appetitive phase are not completely innate, but that some acts, especially the more variable ones, may be done in association with patterns learned through experience. So a squirrel may be more inclined to forage in a tree because it found an acorn there once before. It should be clear that appetitive patterns may be more important to some kinds of animals than others. For example, a hungry lioness must get up, find, and stalk her prey (obvious appetitive behavior), whereas a horse may be standing knee-deep in its food most of the time. It seems clear that appetitive behavior plays different roles in the lives of lions and horses.

**Purposiveness** In appetitive behavior, then, the animal may be consciously striving to attain some end (*purposive* behavior). An example is a frog turning toward a fly in order to correctly release his fixed action pattern, the darting tongue (Figure 8.4). An experienced frog, it may be said, wants to turn toward the fly and has a succulent fly "in mind" as he turns. Tinbergen (1951) stated, "The consummatory act is relatively simple; at its most complex, it is a chain of reactions, each of which may be a simultaneous combination of an orienting pattern (a taxis) and a fixed pattern. But appetitive behavior is a true purposive activity, offering all the problems of plasticity, adaptiveness, and complex integration that baffle the scientist in his study of behavior as a whole." However, it should be pointed out that appetitive behavior may also include nonpurposive action as is demonstrated, for example, in seasonal migration. Even an experienced bird doesn't have in mind the lush vegetation and warm winds of Mexico when autumn arrives. To oversimplify, when something in its nervous system is affected by the shortening days, it becomes increasingly restless and one day it just goes.

Figure 8.4 A fly-catching frog. The frog first orients itself with respect to the fly and then releases its fixed action pattern, the tongue flick.

**Orientation Movements** Lorenz and Tinbergen (1938) described in detail the last level of appetitive behavior to occur before the consummatory act. The animal at this stage continually adjusts itself spatially to its environment. This adjustment is usually a locomotion or a turning of all or part of the body so that the animal is oriented correctly. The orientation is necessary for the consummatory fixed action pattern to be effective. The change of position, which is steered by a taxis mechanism (Chapter 13) is, in some cases, completed before the consummatory act is initiated. When the taxis is thus performed before the fixed action pattern, the two types of action may be easily distinguished. The snapping reaction of the sea horse (*Hippocampus leach*) illustrates such a sequence (Figure 8.5). The fish must assume a very precise posture in which the prospective prey is "exactly in the central plane of its head, in a definite direction and at a definite distance obliquely in front of and above its oral slit" (Lorenz and Tinbergen 1938). The sea horse may follow a small crab back and forth for several minutes. Then, when it finally attains the proper spatial relationship with the crab, it snaps at the prey. The snapping action, like the tongue-darting action of a frog, is the consummatory act. Once this reaction has begun, the sea horse no longer responds to spatial changes in the environment. If the crab moves a bit to one side after the final motion is begun, no adjustment is made and the sea horse misses.

Figure 8.5 The sea horse moves so that its prey comes to lie in the central plane of its head as determined by the angle of the converging lines of sight. The snapping behavior is then released.

In other instances, the appetitive orientation and consummatory patterns are performed simultaneously rather than sequentially. If the two types of actions are performed simultaneously, it is difficult to determine what part of the observed behavior is due to each factor. Lorenz and Tinbergen, however, were able to separate the two components in an experiment that was remarkable for its simplicity. They found that if an egg is removed from the nest of a graylag goose, she will retrieve the egg by rolling it back under her bill (Figure 8.6). She continually makes small adjustments as the eccentric egg rolls off course. But if an experimenter removes the egg during this retrieval process, she will continue the egg-rolling motions until the fixed pattern is completed (or until the egg *would* have been retrieved if it hadn't been removed). Furthermore, if the egg is removed while she is "chin rolling," she will continue to tuck her head, but will stop the small side-to-side adjustments. So, here, the orientation and the consummatory act are normally performed simultaneously. Without experimenting, the two separate phases of the pattern would not have been distinguishable to an observer. Again, the fixed action

Figure 8.6 Brooding goose retrieving an egg. The chin-tucking movements are separable from the orientation motions that keep the egg on track.

pattern (egg-rolling motion) is independent of the environment. Once the pattern is begun, it continues to completion. The orientation movements, however, are dependent upon what is going on in the environment.

**Action-Specific Energy** Lorenz, in 1937 and in 1950, advanced the concept of *action-specific energy*, a notion that at this point has not gained wide acceptance, partly because of a lack of neurological evidence to support it. In presenting his case, Lorenz noted that the ease with which a stimulus released its consummatory act was dependent upon how long it had been since the act was released. ("Released" is the proper choice of words in this case.) He postulated that action-specific energy, or energy for a specific action, is constantly being produced in the animal's central nervous system. This energy is held in check by an inhibitory mechanism until the appropriate stimulus releases it. The released energy activates certain muscle systems and a particular behavior pattern results. If the proper stimulus is not found, the action-specific energy continues to build up until it can be contained no more, and then it goes off on its own, without being triggered by a stimulus. This is called *vacuum activity* and has been described in a variety of animals. Lorenz noticed it when one of his hand-reared starlings flew to a perch on a statue in his house and began to "catch insects." The professor climbed up on the furniture several times to see what the bird was catching until he was finally satisfied that no insects were there. Lorenz's explanation is that in the absence of a stimulus, the innate motor pattern was discharging due to the buildup of its own energy.

Lorenz's notion of action-specific energy also accounted for another observation. It had been noted that there is a progressive increase in the number of situations that will act as releasers as time goes by. For example, nest-building behavior in tits is normally released when the bird finds an ideal site. However, if the bird fails to find an ideal site, its threshold of acceptance becomes lower and lower, presumably due to a buildup of reproductive action-specific energy, until finally the tit will make do with a less-than-perfect site (Hinde 1952). One of the most important ecological implications of this phenomenon is its effect on animal dispersion. Animals may be forced to occupy a previously unoccupied geographical area after being displaced by storms or being driven out of a better area by territory holders. In the absence of ideal nesting sites, inferior ones are accepted, and thereby a new area is colonized. In time, the population may adapt itself to its new habitat, to the extent that it becomes "preferred."

Progressive lowering of the threshold for a fixed reproductive motor pattern was also described by Lorenz. In an experiment, a male blond ringdove was deprived of a female of its own species. After a while, Lorenz found it would court females of another species when they were presented. If no stimulus was presented as time went by, it showed a willingness to court a stuffed pigeon, then a rolled-up cloth, then a corner of its cage (the linear perspective of the converging lines at least presented a point at which the bird could focus its gaze). The lowering of threshold is a property of instinctive acts that is not found in purely reflexive acts, in partial answer to those who wish to call instinct an elaborate reflex.

It should be pointed out that the lowering of thresholds in the absence of a stimulus, or the performance of vacuum behavior, does not necessarily suggest a buildup of any kind of "energy." The notion of action-specific energy is primarily useful in predicting certain kinds of behavior, but should not be uncritically applied. It should also not be used as an "explanation" of behavior in the absence of some sort of supporting neurological evidence. To further complicate matters, we will see that those factors that switch on appetitive behavior and switch off the consummatory act may be entirely different processes. Thus, at this point, conceptualizing a buildup and flow of some sort of "energy" is probably not merited.

**Against Oversimplifying the Model** Subsequent work has shown that the relationship of appetitive and consummatory behavior is actually much more complex than it has been presented here. For example, what seems to be appetitive in one species may be consummatory in another. Also, environmental conditions that initiate one instinctive act may also serve to end another one, even when its stimulus is still present. Such conditions are, in the words of Thorpe (1963), "consummatory stimuli." He adds, "The goal of an animal is not always the performance of an action; it may sometimes be the perception of the environmental situation; the achievement of such a situation may in fact be as effective in bringing a particular piece of behavior to a close as is the satisfaction of what is more crudely thought of as a philosophical need."

## The Behavioral Hierarchy

We have seen that generalized appetitive behavior leads to increasingly specific behavioral levels until, finally, the inflexible fixed pattern of the consummatory level is released. Tinbergen (1951) proposed a hierarchical model that defines these behavioral levels and their relationships. To show how his model may be used, consider the reproductive behavior of the male stickleback described in the ethogram in Chapter 1. In spring, the gradual increase in the period of daylight brings the male stickleback to an internal physiological state of readiness to perform reproductive behavior. The fish migrates into shallow water, where the temperature is warmer, and here establishes a territory. When he settles on a territory, certain skin cells expand, turning him a bright red; he reacts to strangers in a hostile manner and he begins to build a nest. It is difficult to predict which component of the reproductive drive (say, aggression or nest building) will be manifest at any particular time. Each is released by its own special set of stimuli. For example, fighting may be released by the sight of a red male intruding into the territory, and nest building by the presence of an adequate site.

Increasingly specific stimuli are necessary to elicit behavioral patterns lower on the hierarchical scale. In Figure 8.7, it may be seen that there is more than one type of fighting behavior, and the intruding red male may provoke any of these. If he bites, the defending male will bite back; if he threatens, the defender will threaten back; if he flees, the defender will give chase, and so on. Thus, when environmental conditions cause our male to prepare physiologically for reproduction, the visual stimulus of a suitable territory

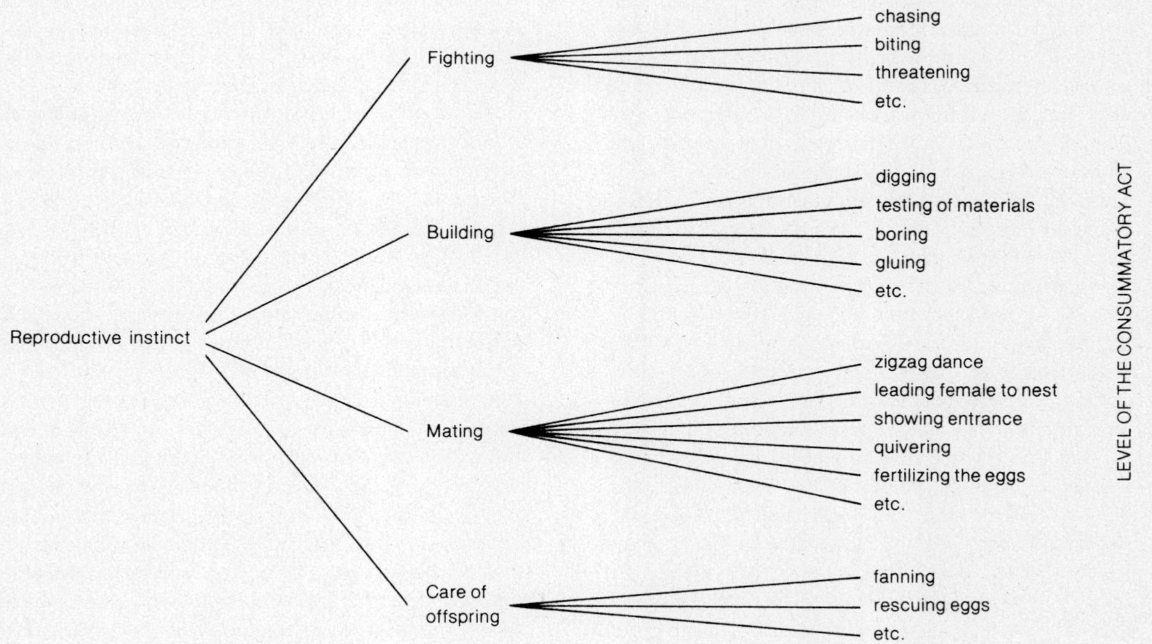


Figure 8.7 The hierarchical organization of the reproductive instinct of the male three-spined stickleback. See the text for an explanation.

elicits the propensity for either fighting or nest building (thus these are placed at the same level in the behavioral hierarchy). An intruder releases fighting behavior, but the type of fighting depends on even more specific stimuli provided by the intruder. (Chasing, biting, and so forth, are at the next lowest level of the hierarchy.)

Tinbergen's overall hierarchical model (Figure 8.7) involves a mechanism by which each center passes the energy of the general reproductive drive to more specific centers when a neural block is removed. The "block-remover" is the part of the nervous system that reacts to specific environmental stimuli (in the classical scheme, called the innate releasing mechanism, about which more will be said later). Tinbergen coupled his model with a hierarchical theory proposed by Weiss (1941) showing the pathway of motor patterns that are activated by the consummatory act. In the integrated scheme (Figure 8.8), general behavior patterns (such as reproductive behavior) are released by certain environmental stimuli. As the "energy" flows to lower levels (its path being determined by which blocks are removed by particular stimuli that remove specific blocks), each act is less flexible than the one before until the consummatory level is reached. The consummatory act activates a motor center that is the neural source of a specific activity, for example, a particular nest-building motion in fish. This center activates a subordinate center, such as the one that controls the fin. This, in turn, activates the center controlling a specific set of fin rays, then the center governing the individual muscles in the fin rays, and finally the center for the motor units (motor end plates and the fibers they activate). It has been pointed out that certain imperfections exist in too strict an application of the Tinbergen-Weiss model. For example, upon close examination, it is not clear just how appetitive behavior could fit consistently into this scheme.

And this shortcoming points up basic problems with such physical models. First, flow charts and diagrams purporting to explain such complex phenomena as behavior are likely to be simplistic. They probably won't cover all the cases. And if they do, they are likely to be too broad to be useful. Second, they tend to give us a false sense of security. They

Figure 8.8 The theoretical organization of the instinctive centers and their relationship to the motor patterns. Motivational impulses (shown as straight arrows) "load" the centers (shown as circles). These impulses may begin in the external environment as well as in higher centers. The heavy lines indicate inhibitors that prevent a steady discharge of motor impulses. Releasers activate the innate releasing mechanisms that remove the blocks. This enables the animal to show a specific appetitive behavior until more specific releasers trigger the next lowest center with its increasingly specific appetitive behavior. The curved arrows between centers of the same level show mutual inhibition and the presence of displacement activities, discussed later. Below the consummatory level, several motor centers are activated at the same time.

usually make us think we know more than we do. In reviewing these, or any of the other diagrams we will encounter, a good dose of skepticism will probably help. Such physical descriptions are primarily useful in putting across ideas that would be difficult to explain verbally. Just keep in mind that not all aspects of a complex model are equally grounded in real evidence, and that such descriptions do not usually leave room for the exceptions.

Although every facet of complex innate behavior is not accounted for in Tinbergen's model, his scheme was important because it did furnish a conceptual bridge between observed instinctive patterns and their physiological interpretation. In 1943, W. R. Hess began a series of physiological experiments that basically supported the model. He implanted tiny electrodes in the brains of cats and found that stimulation of certain parts of the brain could cause the animals to execute instinctive patterns. By administering a slight shock to a proper brain center, a cat could be made to fight, eat, or sleep. It was interesting that the cats showed normal appetitive behavior in these cases; in other words, they would actively search for food if the "eat" center was stimulated. Specifically, the entire coordinated instinctive pattern (with its normal appetitive and consummatory parts) appeared only when parts of the hypothalamus were stimulated. If the electrodes were implanted lower, say in the medulla or spinal cord, disjointed behavior resulted—*parts* of instinctive patterns. And these parts were, interestingly, lower on the hierarchical scale. It would be expected intuitively

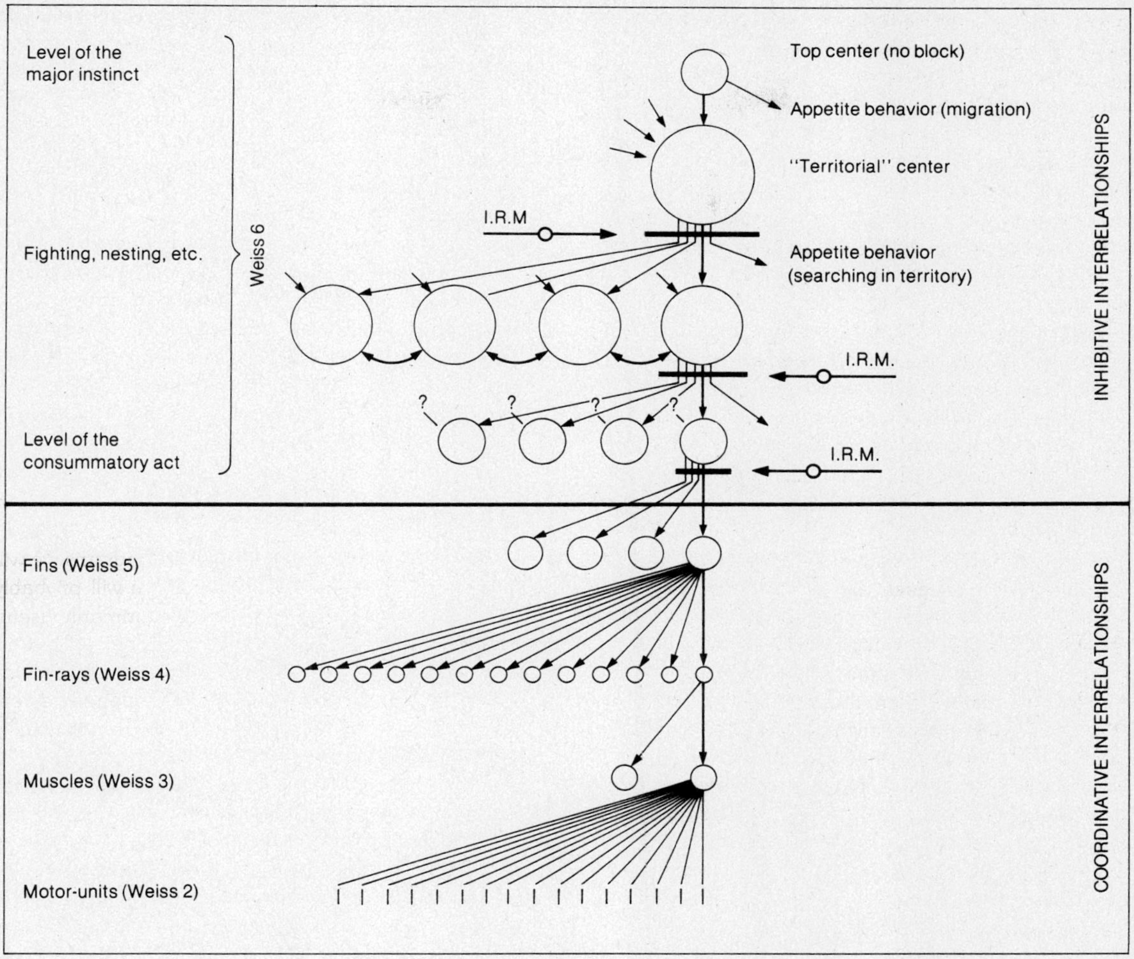


that the more variable, higher centers would be located closer to the brain. For reviews of electrode implantation experiments, see Marler and Hamilton (1966); Corso (1967); and Hinde (1970).

In the hierarchy model, then, the higher levels control the purposive and flexible behavior. The lower levels are involved in increasingly simpler and more stereotyped movements until, at the consummatory level, the orienting taxic component operates closely with the rigid fixed action pattern. Hormones and internal stimuli are thought to operate on the higher centers as do the environmental stimuli that "motivate" behavior rather than release it. (See the discussions of releasers at the end of this chapter and drives in Chapter 9.)

## Releasers

Now, let's consider *releasers,* those parts of the environment that trigger instinctive acts. In our previous example, the diving patterns of a falcon could be said to have been released by the sight of the teal below.

**Selecting Parts of the Environment** A hen will not attempt to rescue a chick under a bell jar even though the chick is visibly in distress. But the same hen will very excitedly try to seek out her chick when a recording of its distress call is played. The releaser of rescue behavior in this case is auditory, not visual. The indifference of the hen to the *sight* of her distressed young compared to her immediate reaction to its *call* shows us two things. First, insight is not involved. The hen is not reacting to the chick's dilemma. She is responding to a particular sound. Second, only part of the situation that the hen perceives reaches through to that part of the hen's nervous system that says "rescue."

Tinbergen (1951) described a variety of models that would release aggressive behavior in male stickleback fish (Figure 8.9). A realistic model of a male stickleback, but one with a white belly, elicited little response from males in reproductive condition. However, a motley assortment of crude figures with the underparts painted red were attacked by those same males. Obviously, the red underpart provided a stronger attack stimulus than did the lifelike configuration of another fish. In the world of the male stickleback, a smallish object with a red underpart is very likely to be another male in reproductive condition—a competitor. Animals do not always respond to all aspects of any environmental situation, then. Instead, only certain parts are selected for response by the nervous system. These parts are reliably associated with overall environmental conditions so that their presence usually signals those conditions. The parts are called *releasers*.

It is important to understand that each instinctive act is not summoned by a single releaser. Instead, a particular response may be elicited by a number of releasers. For example, it has been found that fighting behavior in the male cichlid fish is elicited by five stimuli: (1) silvery blueness, (2) dark margin, (3) highness and broadness, (4) parallel orientation to opponent, and (5) tail beating (Seitz 1940). It was found that any one of these stimuli would elicit hostile behavior, and that any two would elicit about twice the reaction of one (although the relative strengths of reactions would be very hard to measure precisely). Therefore, responses to releasers appear to be additive, so that the whole is equal to the sum of its parts (not greater than the sum of its parts, as is the case in gestalt perception, discussed in Chapter 11). The additive effect is called the *law of heterogeneous summation.*

**Orientation and Environmental Relationships** In some behavioral patterns, different stimuli may be needed to release the consummatory act and direct its orientation. *Daphnia* in a dish of water will swim to the top when the level of $CO_2$ in the water is high. The response is probably adaptive, since surface water usually has a higher oxygen tension than deeper water. However, if the dish is lighted from the bottom, and $CO_2$ is added to the water in which *Daphnia* are swimming, the little arthropods will dive to the bottom. So, apparently, high levels of $CO_2$ *release* swimming behavior, which is then *directed* by light.

The spatial (orientation) response, by the way, is likely to be less stereotyped than is the fixed action

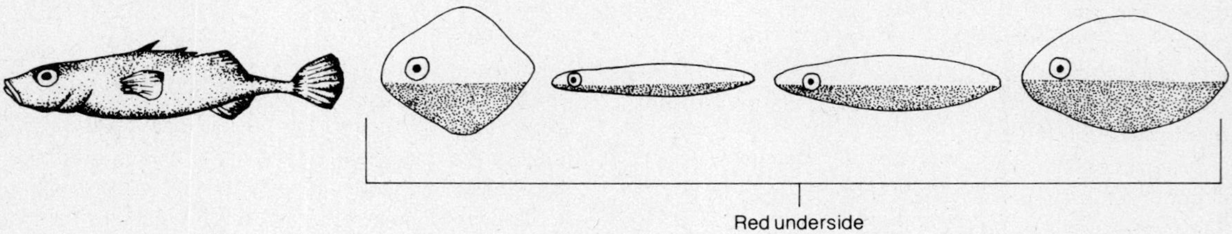


Figure 8.9 A lifelike model of a stickleback and four crude models. The crude models with red undersides elicited more responses than the lifelike model lacking red.

pattern, due to the number of adaptive alternatives open to an animal making a spatial change. Sometimes this freedom of choice in orienting gets the animal into trouble. Once, when Tinbergen was spotted in his blind by a herring gull he had been watching, the startled gull gave the alarm call to its young and they scrambled into what appeared to be the nearest safe place, his tent!

In other cases, the same stimulus may serve as both releaser and director. Tinbergen, in one of his early experiments with releasers, found that young herring gulls, when begging for food from a parent, peck at a red spot on the mandible of the adult. So he made some artificial heads and rearranged the features a bit. With such models, he found that if the red spot is shifted to a place above the eye, the model elicits only one-quarter the number of pecking responses as it does in its normal position (Figure 8.10), but of the responses elicited, about half are directed to the mandible and half to the new site. It has since been found that the pecking responses are due partly to other factors, such as the height of the model and its direction of movement. Such experiments provide examples of the orienting influence of some releasers and they illustrate the importance of configuration and spatial arrangement in visual releasers.

Consider another classical experiment that illustrates this principle more clearly: the reaction of young ducks and geese to a silhouette model of a flying bird (Figure 8.11). The wings were symmetrical; at one end there was a short protuberance, at the other end a longer one. When it was pulled along a guide-wire in one direction, it appeared to have a long neck and a short tail, like a goose. Flown the other way, it resembled a short-necked bird of prey. In the first case, it was ignored by young chickens, turkeys, ducks, and geese on the ground. When they viewed it being pulled along the other way, however, they tried to escape. Keep in mind that the configuration hadn't changed, only the relationship of directional movement to configuration.

There has been quite a bit of argument over this experiment, but when all is said and done, it appears that some wild birds do possess an innate tendency to escape when they perceive the hawk configuration, even for the first time. The "recognition" of the pattern is apparently centrally mediated, peripheral filtering probably not being involved. Also, the key elements that release escape behavior evidently are a

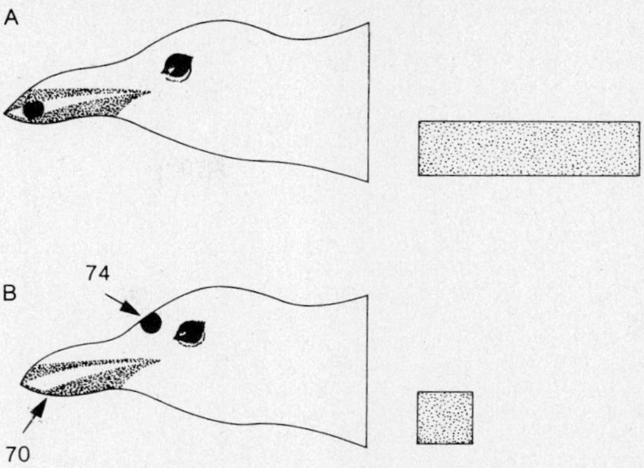


Figure 8.10 The bars show the relative number of pecks aimed at (A) and (B) when both were presented. The numbers in (B) show that pecks at this model were about equally divided between the misplaced red spot and the place where the spot normally is.

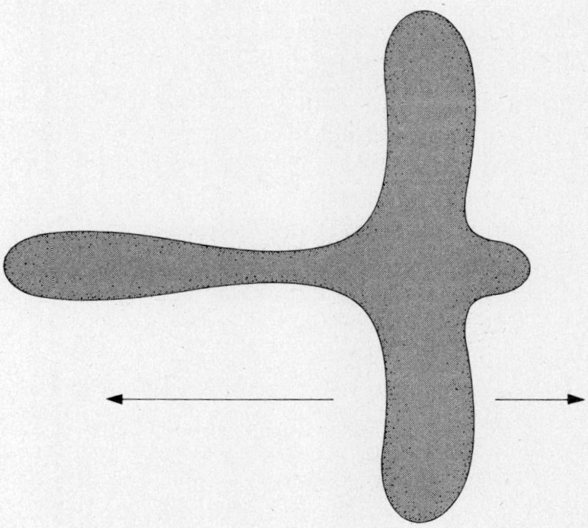

Figure 8.11 The silhouette model resembles a long-necked goose when pulled from right to left. It resembles a short-necked bird of prey when pulled left to right, and in this direction it elicits escape reactions in some ground-dwelling birds.

short neck and the appropriate, apparent speed of the model.

Tinbergen performed other experiments with gulls that showed the importance of relative sizes in releasers. Nestling herring gulls will gape at any part of a model that breaks the contour. This break is apparently regarded by the chicks as the head of the parent. Thus, consider the models represented in Figure 8.12. In (A), the chick gapes at the smaller "head." In (B), it gapes at the larger. It apparently prefers those heads that are of the appropriate size *relative to the body.*

These experiments point out a difficulty in designing and interpreting laboratory experiments with animals. Releasers in the wild are undoubtedly complex and highly interrelated, and so valid measurements of isolated factors are extremely difficult to obtain. Experiments must, therefore, be designed very carefully, with an ethologist's basic familiarity with his subject.

The problems of studying releasers, or their specific components, called "sign stimuli," are also compounded by other factors. For example, in some cases, a *set* of releasers is necessary to elicit an entire adaptive response. Bees may initially be attracted to a colored paper flower, but they will not land unless the flower scent is also there, so in this case, the visual releaser alone will not initiate a behavioral pattern that has survival value. We can also see here that perceptive abilities play complex roles in eliciting behavioral patterns. Let's explore this idea further.

**The Role of Perception in Instinct** The male of the grayling butterfly (*Eumenis semele*) is attracted to virgin females on the basis of their darkness (the darker the better), their nearness, and their fluttering movement. Color seems to play no part. Normal females are brown, but a male would as soon court a red or blue female if they were equally dark. It would appear, then, that the butterflies are color-blind. But they aren't. In tests using paper flowers of various colors, *Eumenis* showed real color discrimination in feeding, being attracted primarily to blue and yellow flowers (Tinbergen 1951). So, here is a case of different senses being brought into play in specific behavioral contexts.

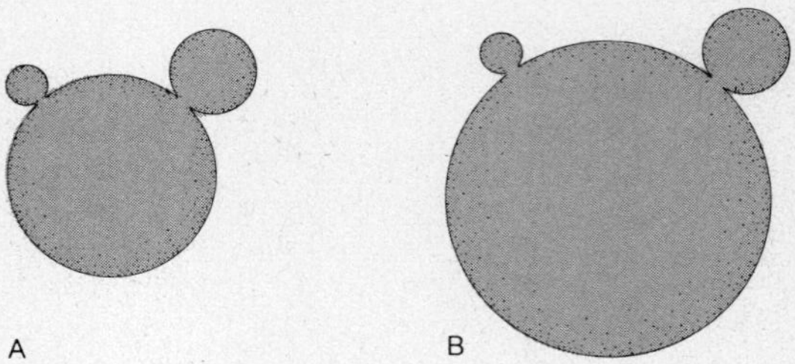


Figure 8.12 Models with two "heads." When model (A) is presented to a nestling, gaping responses are directed toward the small "head." When (B) is presented, the large "head" elicits gaping, indicating that the relative size of the head is important.

We can see, then, that a knowledge of the perceptive capacities of an animal will not tell us what part of its environment is actually triggering a behavior, since, as we know, a certain behavior may be stimulated by cues from only a small part of the environment. As another example, *Dysticus marginalis,* the carnivorous water beetle, has very well-developed compound eyes, but a clearly visible and succulent tadpole suspended in a sealed test tube does not stimulate an attack. However, if meat extract is poured into the water, the beetle will capture *any* solid object, edible or not. And it will do it again and again. (The failure to learn from such mistakes is one of the most conspicuous characteristics of purely instinctive behavior.) The point here is that while an animal may be able to perceive an environmental factor, it may not react to it unless it is "programmed" to do so. It may fail to utilize certain facilities under a given condition, and some species can never learn to bring other perceptive devices into play.

**Environmental Influences on Perception** Some types of patterns such as capturing behavior in *Dysticus* can be elicited at any time, but we have also learned that there are other types of behavior that can be released only at special times. Let's reexamine the principle in terms of releasers. Courting behavior in the male stickleback is released by the sight of the female's belly swollen with eggs and the peculiar posture of the "ready" female. The female, for her part, reacts sexually to certain releasers in the male's behavior, such as his zigzag dance. However, none of these stimuli has any effect at all in autumn or in winter. It would be most unfortunate if a pair were to mate and waste time and energy on a brood that would surely die! It turns out that since the reproductive behavior of both males and females is under the control of seasonal factors, they are likely to perform, and be susceptible to, mating patterns at the same time of year. Should a misguided male perform his zigzag dance in winter, a female may see him, but since the seasonal daylength has not caused her to produce reproductive hormones, she will remain totally unimpressed by his efforts. The environment, then, may dictate the nature of the hormones coursing through the bloodstream, and these, in turn, may influence whether an animal responds to a certain environmental cue.

**Sequencing and Mixing Parts of Patterns** One of Lorenz's students found that male and female cichlid fish possess the same behavioral repertoire. Thus, the question arose, How does a male recognize a female, or how is homosexual mating prevented? It turned out that although the repertoires were identical, the behavior was not. It was found that cichlids performed three types of behavior upon meeting: (1) aggression, (2) escape, and (3) sexual behavior. But the sexes differed in how they *mixed* the three. Successful pairing demanded an "awe-inspiring" male that showed mainly (1) and (3). The submissive and "awed" female showed mainly (2) and (3). It may be assumed that there is a sexual difference in the thresholds for the elicitation of each act. Also, of course, there may have been subtle signals, characteristic of each sex, that went unrecognized by the human observer.

In a sequence of instinctive reactions, an animal must know when to stop each act and begin the next. In some situations, the transition must be accomplished when the releaser of the first act is still present. How does the first releaser come to be ignored so that the sequence can continue without the animal performing the first step over and over again? The answer may lie in the relationship of the releasers to the consummatory act. Following a consummatory act, there is a period of rest or of satiation. During these periods, the threshold for the release of the behavior that leads to that pattern may be raised above the level that was necessary to elicit the reaction the first time, so that the second act is then more likely to be released. (Lorenz might say that the action-specific energy for the first pattern had been reduced.) Any such explanation would be even more feasible if each act were to serve as the releaser for the next act. Tinbergen (1953b) described the apparent operation of such a system in the nest-building behavior of the long-tailed tit (*Aegithalos caudatus*), and Thorpe (1963) analyzed such behavior in terms of sequential releasers. Thorpe described the bird putting moss on the tree, the first act. "Moss on site" then releases a search for spider silk. "Silk on moss" releases a stretching motion that is employed in binding the nest together, and so on, until the nest is completed. So, in those species for which learning is not the basis for proper sequencing, it may be that as each phase is consummated the threshold for that act is raised, thus permitting the release of the next phase. And, in some cases, the completion of an act may, itself, release the next phase of the pattern.

**Supernormal Releasers** It has been found that, in fish, the numerous sign stimuli composing a releaser could be segregated and analyzed independently (recall the law of heterogeneous summation) and that a deficiency of one part could be compensated for by an increase in another part (Seitz 1940, 1942). This obviously means that the releasing value of any sign stimulus is not fixed. If blue is good, bluer is better. Since the releasing value of each sign stimulus is modifiable, then, a question arises regarding the extent to which a releaser may increase its effect. Interestingly, sign stimuli or releasers can be artificially "boosted" so that they are preferred over their normal condition by the animal. Oystercatchers prefer eggs larger than normal and will try to sit on "eggs" that are normally marked but too large to incubate (Figure 8.13). Ringed plovers (*Charadrius hiaticula*) prefer a more strongly spotted egg over their own eggs (Figure 8.14). Exaggerated releasing stimuli that are "preferred" over normal stimuli are called *supernormal releasers.*

As another example, the butterfly *Argynnis paphia* L. reacts to supernormal releasers (Magnus 1958). The fluttering of females' wings alternately shows light and dark wing parts. When yellow-orange (normal wing colors) and dark strips were placed on a rotating cylinder in view of a male, his excitement

Figure 8.13 An oystercatcher (*Haematopus ostralegus*) shows a preference for a giant egg rather than a normal egg (foreground) or a herring gull's egg (left).

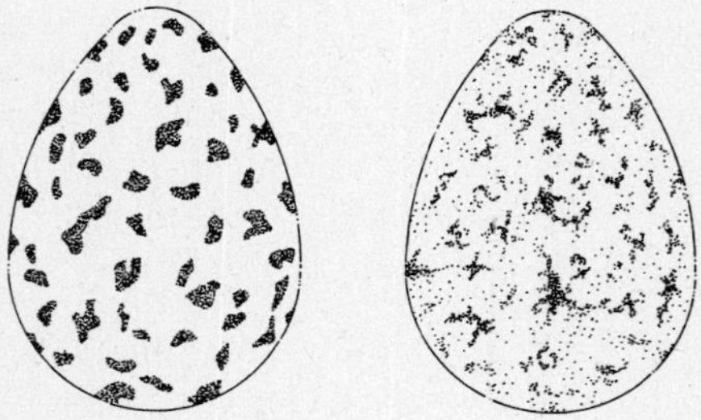

Figure 8.14 A supernormal egg that is unusual only in the size of its spots (left) is preferred by the ringed plover (*Charadrius hiaticula*) to a normal egg (right).

increased as the flash frequency exceeded that shown by normal females. So, in this case, the trait (darkness) that was important to the male grayling butterfly we discussed earlier is ignored by *Argynnis* in favor of the flashes of wing colors, and the flashes of the color increase in appeal with their frequency, even beyond that which is normal. One might ask, then, why haven't females developed faster and faster wing flashes through natural selection? The most obvious answer is that the wing flashes are derived from flight movements and the locomotive action predominates over the signaling function (Barlow 1968).

Certain parasitic birds, such as some cuckoos and cowbirds, have young that are larger and more "babyish" than the host's. This may be the reason why the host may attend to the young parasite in preference to its own young. There are undoubtedly good examples of supernormal releasers among the mimicking animals also. In the cases of both parasitizing and mimicking, the supernormal releasers are directed toward members of *other* species.

Releasers have been compared to a key that unlocks directives for a behavioral pattern in the animal's neuromotor system. The occurrence of supernormal releasers and the evidence for heterogeneous summation, however, show that the lock-and-key analogy is not airtight. But it does serve to point up the idea that an inherited mechanism (the lock) seems to exist within the nervous system. We really don't know very much about this theoretical lock, but that hasn't stopped us from giving it a name, the innate releasing mechanism.

## The Innate Releasing Mechanism

As we have seen, only a small part of the environmental situation serves as a releaser for each instinctive act. How and where are these specific incoming environmental signals decoded in the nervous system of the animal? How is a coordinated motor response achieved? Some ethologists have assumed the existence of an inborn neural center, an innate releasing mechanism (IRM) that responds in a specific way to the releasers.

**The Development of the Concept** Originally, ethologists considered the IRM to be a central filtering system for incoming stimuli—stimuli that the peripheral receptors had allowed to pass (Lorenz 1937a). Tinbergen (1951) defined an IRM as a "special neurosensory mechanism that releases the reaction and is responsible for its (the reaction's) selective susceptibility to a very special combination of sign stimuli." In terms of the hierarchical model (Figure 8.8), the IRMs act to remove inhibitory blocks in the nervous system. These blocks, of course, are essential in preventing the inopportune discharge of a behavioral pattern. It certainly wouldn't do for a minnow to attempt to mate with a pike—even in the breeding season. And this illustrates another relationship: the intimate association of releasers and IRMs. A pike just doesn't have the size or configuration to release reproductive behavior in a minnow. The sight of a pike is much more likely to stimulate the minnow's IRM that removes the block that inhibits escape behavior.

The association of the IRM with central filtering, in the minds of ethologists, has led to certain problems. It has given rise to the notion that there is a special device in the central nervous systems of animals that lies waiting to be switched on or off. Such a concept is undoubtedly simplistic. Perhaps the word "system" should be substituted for "mechanism," since, because we really know so little about IRMs, the more vague we are, the better—for the present.

The concept of the IRM is useful, however, if we keep in mind that we don't know how central filtering is accomplished, that it may (and probably does) involve a multitude of complex neuronal interactions at various levels (from peripheral to central), and that the systems may differ in various kinds of animals. The concept also has its value if it serves to stimulate research into the nature of central filters.

**Experiments in Search of the IRM** Studies of the pecking responses in gull chicks have provided some insight into the nature of the IRM in one group of birds. Tinbergen and Perdeck (1950) found that young herring gulls responded differently to various cardboard models of gulls' bills. The appearance of the head seemed to be unimportant. The results are shown in Figure 8.15. They found that the mandible spot was the critical element in releasing begging

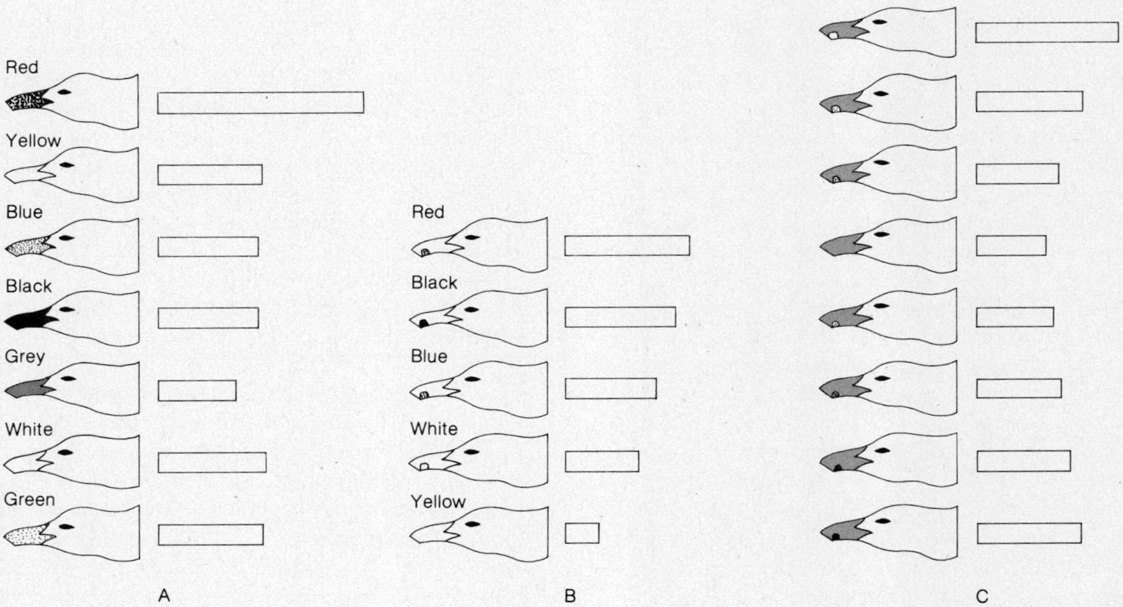


Figure 8.15 Three tests to elicit peck responses in gull chicks. The bars indicate relative numbers of responses: (A) measured bill color; (B) spot color (all bills were yellow); (C) contrast between spot and gray bills.

responses in the chicks. The degree of contrast between bill and spot was important, as was the redness of the bill or spot in eliciting responses. The IRM, it seemed, reacts to contrast and redness, but ignores other bill color and the appearance of the head.

The preference for redness, it turns out, could be due to the fact that the visual receptors in gulls have high levels of red, yellow, and orange oil droplets. Any incoming light has to pass through the droplets before reaching the light-sensitive area. Thus, the droplets would tend to filter out colors from the blue end of the spectrum, allowing red light to pass through unhindered. The result is that blue will appear darker (or blacker) than an equal intensity of red light. Thus, it seems that the filtering could be operating, in part, at the periphery.

But the story grows more complex. In two species of terns that have black bills (the sandwich tern, whose bill has a yellow tip, and the wideawake tern), black is responded to about as well as red (Cullen 1962; Weidmann 1961). Since there is no conceivable way the black could be filtered out by oil droplets, or somehow altered in its perceived wavelength properties by the eye so that it appears red, the filtering must be done at another level. Thus, the IRM, in this case, apparently functions partly in the central nervous system.

From Tinbergen's principle of heterogeneous summation, it seems that some IRMs may be stimulated by more than one sign stimulus. It isn't particularly important *which part* of a set of stimuli (or a releaser) is missing, but it is important *how much* is missing, since one part can be compensated for by other parts, up to a point. Although the various sign stimuli for a certain response may emanate from entirely different parts of the environment, they do not differ in their qualitative effects on the motor response as a whole. In other words, the releaser may be composed of diverse components, but the response is specific. The important principle to be gathered from this is that apparently somewhere in the central nervous system sign stimuli are added and, if their *total* effect is strong enough, the IRM sends an impulse to the effectors.

Certain changes may occur in the IRM, or the filtering system in general, during the postnatal life of the animal. In some cases, its selectivity may increase. In other words, the range of stimuli that will affect it may become narrower. An escape reaction may be released in a young bird by skittering leaves, but later they cease to elicit any response. A young toad may snap at any insect within range, including wasps, but in time, it becomes more selective. Apparently, learning has altered the expression of the IRM.

At the same time, there are many warnings in the ethological literature against too simple an interpretation of what, at first, appears to be modification of an instinctive act by learning. Lorenz has repeatedly emphasized that since instincts are inherited, just as organs are, they must mature just as organs do. A newly fledged bird does not suddenly begin using his sex organs in mating. Instead, sexual organs and associated instinctive patterns develop together as the bird becomes sexually mature. Such behavioral maturation might easily be confused with learning.

As an animal matures, the IRM may also show a stronger response to a certain releaser. In animal socialization, this effect may be associated with a phenomenon known as imprinting, discussed in Chapter 11. Basically, a gosling will follow almost any moving object until, at some point early in its life, it "selects" a certain type of object, usually an adult goose. After this, it will attempt to associate only with the selected object. For a detailed discussion of how ethology and neurophysiology may be integrated to answer questions associated with response changes, see Ewert (1974).

## RELEASERS AND FIXED ACTION PATTERNS IN HUMAN INFANTS

If a human infant is touched on the mouth in the first few days following birth, it will begin to move its head rhythmically, turning it to the left and right in what is apparently a search for the mother's nipple. The movement continues until the baby closes its mouth around the nipple or some substitute (Prechtl and Shleidt 1950) (Figure 8.16). In a few days, these blind responses will be replaced by an active searching for the breast. This is only one of a number of functional behavioral patterns found in newborn humans. Because of the universality of the nursing patterns and the presence of appetitive (searching) and fixed consummatory (sucking) components, such behavior

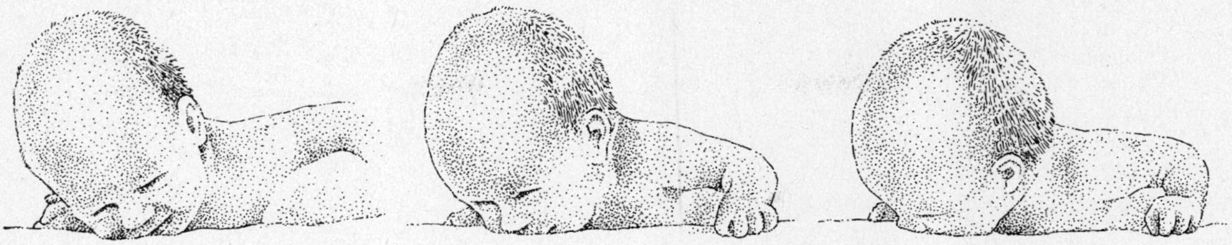

Figure 8.16 Human infants, touched near the mouth, search for the mother's nipple in a stereotyped fashion.

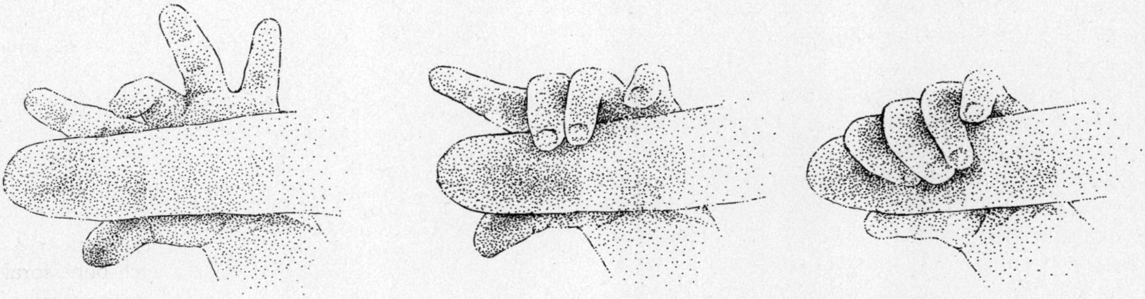

Figure 8.17 Human infants innately grasp with the hands, closing the middle finger first and the thumb last.

has been described as instinctive. For a detailed discussion of fixed patterns in humans, see Eibl-Eibesfeldt (1970).

In spite of the fact that certain of the infant's feeding patterns are fixed, after a time, they, or their degree of fixity, may change. Not only do the breast-searching patterns of the human infant change to more "purposeful" (goal-oriented) searching patterns, but the motor patterns involved in sucking also change. At first, sucking is accomplished by the infant producing a partial vacuum in its mouth (the way one would suck through a straw). Later, however, the sucking is accomplished by simply depressing the tongue as the mouth remains open. It is not known whether experience has altered the original pattern or whether one fixed action pattern has replaced another.

One of the most characteristic types of infant behavior is the grasping reflex of the hand. The response is elicited by touching the palm (Figure 8.17). Interestingly, the fingers close in a certain sequence. And what is more, the response is more strongly elicited by touching the palm with hair than with a solid object. The first thing that springs to mind, of course, is that this is a vestigial response left from the time when the mother had hair to which the infant had to cling. (Babies still often hold tightly to their mother's clothing, so perhaps the pattern still has some value.) It turns out that the hand grasp is more pronounced in premature babies than in those born later, and in fact, a premature baby can be hung on a clothesline. Surprisingly, they can also grasp strongly with their feet (Figure 8.18). Much of this ability is lost later, a factor that indicates that the tendency is now only rudimentary and vestigial and is disappearing from our behavioral repertoire.

Swimming movements (which, as we will see later, are the same as those employed by primates that walk on all fours with their palms down) can also be elicited in babies. In fact, they can actually swim in a coordinated fashion as long as their chin is held up. But this innate ability also disappears with age. They are unable to swim at all by the age of three or four months.

Walking is another innate behavioral pattern. If a newborn baby is held up, it will perform, in a rough fashion, the very complex patterns of walking (Figure 8.19). And if placed on its stomach, a newborn baby will perform crawling movements by moving diagonally opposed limbs. Crawling and walking patterns, of course, do not disappear, probably because we are now a terrestrial species, hanging and clinging less, but crawling and walking more.

It has also been noticed that the law of heterogeneous summation has less and less validity as an infant grows older (Bower 1966). Babies were trained to turn their heads when the image at the upper left of Figure 8.20A was shown, the reward being a fifteen-second peek-a-boo by the experimenter. When the baby was considered "trained," the entire stimulus and then all its parts (the rest of Figure 8.20A) were shown alternately with no reward given, until finally the baby failed to respond at all.

As shown in Figure 8.20B, with eight- and twelve-week-old infants, the law of heterogeneous summation held firm. The parts produced the same response as the integrated whole. At sixteen and twenty weeks, however, the whole produced more responses than the sum of the responses that were produced by showing the infant only parts. In a second experiment, infants were trained on the parts until they would react to any of the three parts, and then they were tested on the parts and the whole (which they had not seen before) as in the previous set of experiments. At ten weeks, there was again little difference, but at twenty weeks the mean number of responses to the whole was *smaller* than the sum of the responses to the parts. Thus, younger infants react more to disassociated parts of any situation, much as some animals react to sign stimuli. Later on, however, the integrated whole becomes more important. In this experiment, of course, learning obviously plays a role. Previously, you will recall, we discussed heterogeneous summation only with regard to innate patterns, ignoring learning input.

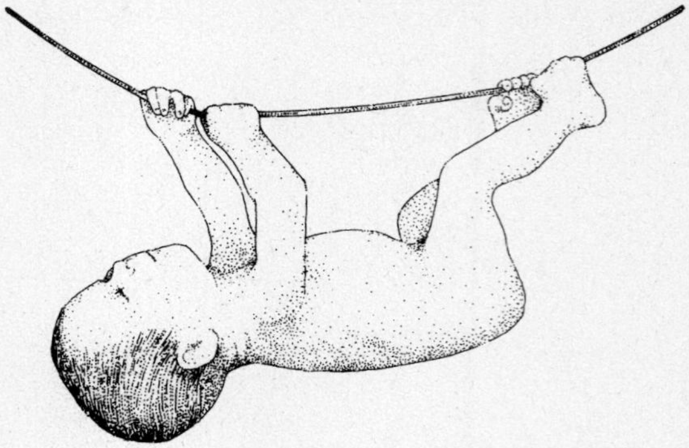

Figure 8.18 A premature (seven months) human infant grasping with both its hands and feet.

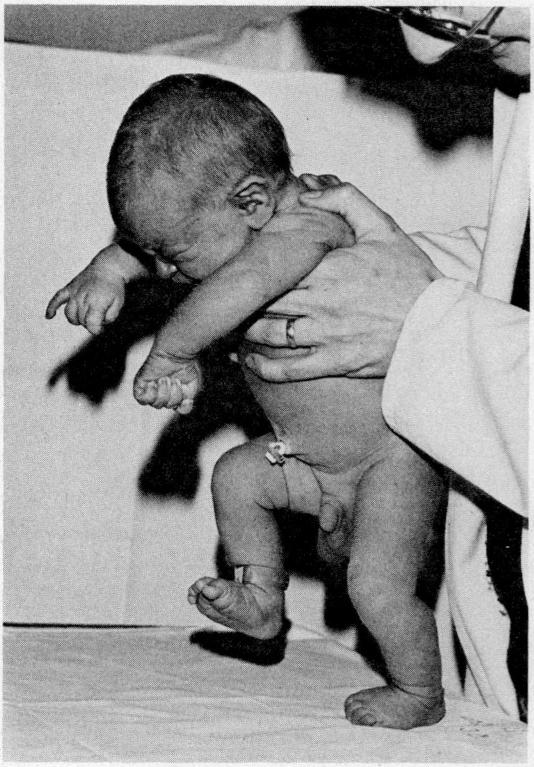

Figure 8.19 If a newborn baby is held erect it will automatically perform the very complex muscular movement associated with walking.

The infant smile (Figure 8.21) may be a highly adaptive innate behavior. It almost universally elicits delight in the mother, and this may be very important at a time in the infant's life when it is so often a royal pain in the neck. One wonders at times how anyone could love anything so small, helpless, demanding, inarticulate, inexperienced, dirty, and broke. But then the infant looks directly at you and smiles, and all those doubts vanish amid coos of delight. The smile is no doubt an important bond builder at a time when the infant is otherwise likely to elicit rejection behavior.

It is interesting that an infant is able to "fixate" (look directly at) anything (such as a mother's face) so early (at four to five weeks). Even blind infants will fixate by staring directly ahead (where the mother's face is likely to be) as they smile broadly (Eibl-Eibesfeldt 1970). Infants also smile in their sleep or at other times that have nothing to do with their pleasure at seeing their mother. This "spontaneous" smile that elicits maternal behavior will later be replaced by a "true" smile, in other words, one that appears when a person smiles after being smiled at.

It has been found that three- to six-month old infants will smile at practically anything (Spitz and Wolf 1946), from scarecrow faces to normal and grimacing human faces. In fact, until the onset of the second month, eye-sized, highly-contrasting spots on a square or round "head" elicit smiles more frequently than a painted face or a rectangular bar on the same kind of head (Ahrens 1953). It makes no difference whether the spots are in a parallel or

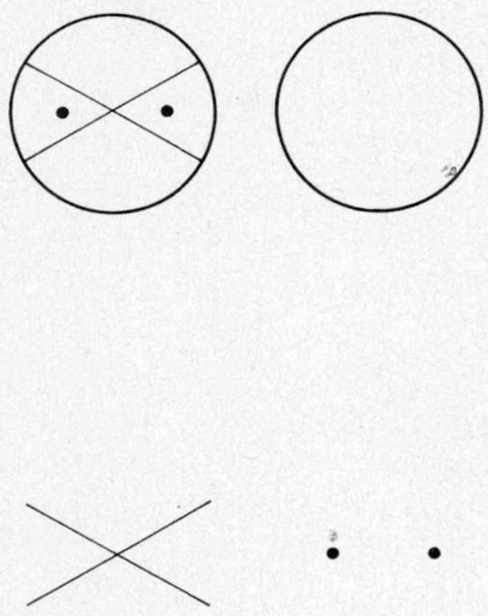


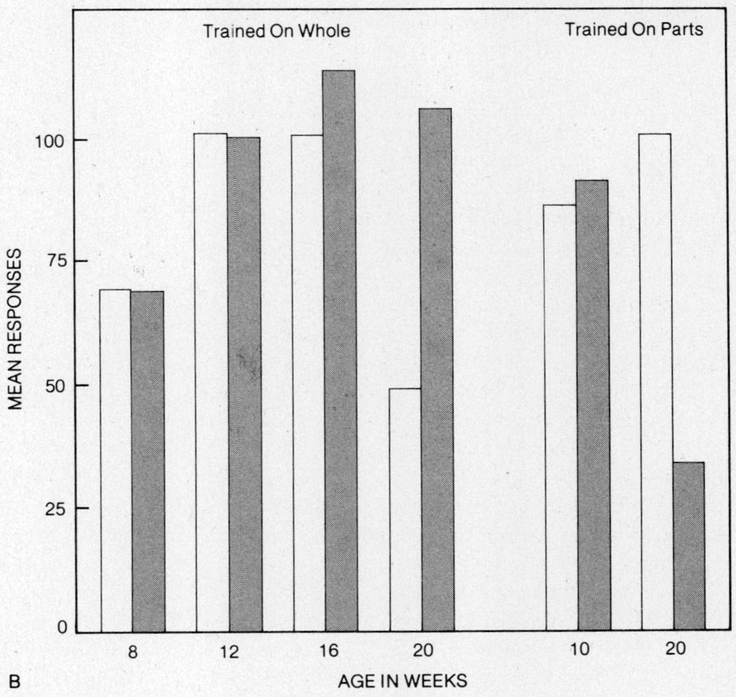


Figure 8.20 (A) The stimuli used for assessing the validity of the law of heterogeneous summation in human infants. The three parts are put together at upper left. (B) The relative effectiveness of whole and partial releasers on infants who had been trained to respond to one or the other.

Figure 8.21 The infant smile may be an adaptive behavior that acts to turn away wrath in adults.

horizontal position or whether there are three pairs of spots. The baby smiles. The baby won't smile at one spot, however. There's no percentage there. Mothers generally have at least two eyes.

At about the second month, spots placed horizontally get more smiles than if they are presented vertically (as if the mother were leaning over from the side). The child pays closer attention, at this point, to the eyes than to other parts of the face.

At about four months, the child seems to become aware of the mouth, and in the fifth month, broadening the mouth (or a line representing the mouth) elicits more smiling. It is apparently responding to a "smile" (a broadened line). At about six months, the effectiveness of models decreases as actual faces begin to elicit more smiles.

At fourteen months, babies respond to forehead wrinkles. Vertical frown wrinkles caused the children to cry, scream, or run away. In these experiments, some of the infants were institutionalized children who had never seen a "threat-frown" before.

There are many instances of stereotyped actions in humans. Eibl-Eibesfeldt (1970) described laughing and crying responses in children born deaf and blind. He also found that the "eyebrow flash" seems to be a pan-cultural phenomenon (Figure 8.22). For other examples, you might refer to Eibl-Eibesfeldt (1970) and the included references. It seems clear, in any case, that humans are born with certain "fixed" behavioral patterns, and that a frown, smile, or grimace has essentially the same meaning in any culture.

## RITUALIZATION

Julian Huxley, Lorenz's teacher, in his pre-World War I study of the behavior of the great crested grebe, found that, in the course of evolution, certain behavioral patterns lose their original functions and become purely symbolic, useful only in communication (Blest 1961). He called the process *ritualization*. In the evolution of ritualization, two coordinated events must occur: the first is a behavioral pattern performed by the "actor" (the one performing the ritual), and the second, the response of the "reactor" (the one observing the ritual).

In the development of ritualization, if a movement that is well established in the behavioral repertoire of a species is traditionally followed by another behavioral pattern, the first movement may come to be regarded as a signal, a harbinger of the second. The signaling value could arise in two ways. Relatively intelligent animals, such as birds and mammals, could individually learn those movements that enable them to predict what another animal is likely to do. Secondly, if it is advantageous to respond "appropriately" to such movements, and if conditions remain stable, in time the proper response to another animal's movement could come to be genetically controlled.

Any benefits to such a signaling system can be reciprocal. An actor may derive a benefit from its actions being accurately "read" by its colleagues. Thus, for example, threat signals could cause another animal (the reactor) to move away, saving the actor a fight. In fact, the incorporation into the innate behavioral repertoire will only occur if the signal is beneficial both to the actor and observer. Probably if such a signal were not mutually beneficial the observers would generally fail to read it or it would be suppressed or continually changed as the population evolved.

It has been suggested that ritual displays are not often derived from the *entire* original behavior sequence, but from only a small part of it such as the *intention movements* (Daanje 1950). Intention movements are incipient preparatory motions that are, in themselves, and performed alone, of little adaptive value. They are, instead, part of an overall adaptive pattern. Examples are the quick crouches made by a small bird just before flight or the over-the-shoulder glances of a woodpecker that is about to fly. Each of these is part of an adaptive pattern—flight. The small bird must crouch to launch himself, and the woodpecker must be sure he isn't going to crash into a tree trunk as he departs. Intention movements may occur just before the entire pattern is carried out or, as in some cases, they might appear in immature animals before the complete pattern has developed. (Conceivably, intention movements may be reactions to weak environmental stimuli that are not strong enough to initiate a complete response.)

If ritualization developed from such intention movements, then natural selection could develop the display gradually since the adaptive pattern itself (the model) would not be affected (Klopfer and Hailman 1967). The adaptive model (for example, crouching before flight) could be retained while developing the display from its nonadaptive (or adaptively neutral) intention movement (for example, crouching without launching into flight). It should be pointed out that not all ritualized behavior develops from nonadaptive models. In some cases, a pattern might become incorporated as a signal while it is serving its original function.

A behavioral pattern is recognized as a ritual act when it is apparent that it is no longer a flexible behavior, composed of many separate parts, that is performed in response to a number of environmental stimuli. Instead, the various parts of a ritualized behavior become welded into a stereotyped pattern that is performed in response to relatively few cues.

The evolutionary development of a ritualized pattern is easier to imagine if you keep in mind that innate patterns are evolutionarily conservative (not easily changed). Innate patterns, unless selected against, may persist in the behavioral repertoire long after they have ceased to be useful. Dogs still turn

Figure 8.22 The eyebrow flash is a cross-cultural human behavior, here seen in Caucasian and Polynesian women. It has been suggested that the omission of the flash may be taken as a sign of hostility.

around several times before lying down on a rug. Their ancestors probably performed the act to drive out insects or spiders from their beds, and since there is no strong selection against it, the act is still performed. The point is, once an act is innate, if it were not strongly selected against, there would be plenty of time for it to come to be used in new ways.

When an animal has retained the original adaptive behavior with its ritualized derivative, it may be possible to distinguish the two. Consider the *inciting behavior* found in certain species of ducks, such as the European sheldrake. In the pattern's original form, if the female felt the territory was being threatened by, for example, an intruding pair of ducks, she responded by running toward her protective drake. She usually stopped in front of him and looked back over her shoulder at the intruders (Figure 8.23), causing the male to attack the object of her concern. In the original pattern, the angle of her body and neck was a function of the position of the two males. Her body directed toward her male, she looked over her shoulder at the intruders. So the head-body angle varied depending on where the males were. As the pattern became ritualized, the female looked over her shoulder in the same way, forming the same angle with her neck, no matter where her male and the intruder were. The peculiar curve of her neck acted as a clear signal to incite her male to attack any intruder he could see. Adaptive orientation is still seen in the female's eye, which remains fixed on the strange male, but she is apparently unable to resist moving her neck in what has become an innate pattern.

There is an interesting example of ritualization among the insects. In empid flies (*Hyperborean empis*), the male brings a dead fly for the female to eat while he mates with her. Such behavior may be adaptive in keeping the cannibalistic female from eating the male when he approaches her. In a North American species, the courting male spins a white balloon with a few small insects trapped in it. In the southern empid (*Hilara maura*), little veils, which may or may not hold food, are carried between the legs of the courting male. The male Alpine tailor fly (*H.*

Figure 8.23 Inciting behavior in ducks. The female at right is inciting her male by holding her neck at a sharp angle. Since the pattern is ritualized, this sometimes results in her looking at the intruders and sometimes not.

*sartor*) spins a little veil that is carried between the middle and hind legs, but never carries food in it. At the sight of the veil, the female fails to attack and responds sexually instead. So, the simultaneous existence of original and varying degrees of ritual behavior in different species indicates the evolutionary development of certain patterns. As Lorenz (1963) asked, Who would have guessed the origin of the "lovely little veils"?

This chapter, then, has been a review of the classical concept of instinct. The model is flawed, and I have pointed out several examples of its inadequacies. However, many of its premises have held firm in the face of new information, and some premises have even been strengthened by incoming data. While the model cannot be uncritically applied in every case, it forms the basis for what we have to work with at the moment, and we should not yield to any pseudointellectual trend to dismiss it as unfashionable. In subsequent chapters, we will repeatedly see both its specific usefulness and its often tacit acceptance by those at the leading edge of ethological research.

# 9

# DRIVE AND MOTIVATION

Why do we do the things we do? Why do we do things we wish we wouldn't do? Why are birds singing in my yard this snowy spring morning? Why will quarreling bluejays wake me up too early next month? Why is some behavior so predictable? We will have the answer to many such questions when we learn more about something called drive.

The concept of drive is currently undergoing major changes based primarily on neurophysiological studies. The term has been variously interpreted and defined so that at present it may mean many things. However, we will settle for the definition that is probably accepted by most ethologists.

## THE DEFINITION OF DRIVE

Drive has been generally regarded as the force that motivates an animal to seek a consummatory stimulus. It may thus be considered the "motivation" behind appetitive behavior. Hinde (1956a, 1959, 1960a) reviewed the concept and decided that drive should be considered simply a description of behavior without physiological implication. It is apparent that such a broad definition places serious limitations on the usefulness of the concept as a research tool. However, even an operational concept of drive can provide a useful way of looking at appetitive behavior. For example, theoretically, the importance an animal places on the performance of a consummatory act (and hence, presumably, the importance of an instinct) may be determined by measuring the force of the drive.

One of the problems in studying drive stems from the fact that in spite of its theoretical utility, it is difficult to quantify. One attempt centered over the observation that cold weather inhibits songbirds from singing (Nice 1937). Temperature was correlated with song frequency (Figure 9.1) in order to determine "how badly the birds wanted to sing." The general usefulness of this sort of analysis was questioned, however, on the basis that some drives do not lend themselves to this general experimental technique.

In another attempt to get at the problem, it was suggested that there are five ways of measuring a rat's motivation to satisfy its hunger (or thirst) (Manning 1971). (1) How much does it eat when given the opportunity? (2) How bitter can food be made before it is refused? For example, a hungry rat presented with pellets increasingly saturated by quinine would finally refuse pellets with a certain level of bitterness. Of course, care would have to be taken to make sure the hunger drive was not reduced by the rat filling himself with low quinine dosage pellets. (3) How strongly will it move toward food? A rat fitted with a harness attached to a spring scale would presumably pull harder to reach food if it were hungry. (4) How much pain will it accept? Once a rat has learned that he must cross an electric grid in order to reach food, one can measure the voltage he is willing to accept to reach the food. (5) How fast will it perform a task? After a rat in a Skinner box has learned that bar pressing is followed by food, the rate of bar pressing can be taken as a measure of motivation.

But these are only a few suggestions of how motivation can be measured in one kind of animal. Undoubtedly many other means could be devised. Interestingly enough, when a variety of tests is used to measure motivation, the results may vary (Figure 9.2)

(Miller 1957). We really don't know yet what causes such discrepancies, but apparently the most reliable figures would be those that involve a number of such tests. One disturbing point comes through clearly enough from all this. We don't know what we're measuring.

### Timing of Drives

Some drives appear in sequence, each dependent upon the successful performance of a specific preceding act. For example, Tinbergen (1951) found that the male stickleback will not start to build its nest, although material is available, until he has defended his territory for awhile. Nor will he court females until he has built the nest. If his pit is filled in after he has dug it, he will dig it again—and again, if necessary—but finally the stimulus of a filled-in pit will have no further effect on him; the next drive takes over, and he will build the nest without completing the pit.

An interesting characteristic of some drives is their rapid appearance and disappearance, a trait that may result in alternately performed acts. The zigzag dance of the male stickleback (Chapter 1) is considered to be an example of such alternation. He swims away in an apparent attempt to lead her to the nest. Then he turns and charges what he suddenly deems to be an intruder into his territory. As he nears her, the aggressive drive is replaced by the reproductive, and he turns and begins leading her again.

Some animals, such as birds, may show longer-term alternation, producing a "waxing and waning" of drives (Howard 1929). Because they are of a relatively long duration, they may be considered to be higher on the hierarchical scale than rapidly alternating drives. An example is the seasonal appearance of reproductive behavior. The internal physiological changes that are correlated with seasonal behavior are due to the influence of hormones. In fact, the very appearance of these hormones is triggered by some seasonal stimulus such as changing day length. The hormones alter the internal milieu of the animal and affect the nervous system in a general way. These changes may make the animal responsive to yet different environmental stimuli and cause the annual fluctuations in behavior.

Hormones are very important to some types of motivation, such as those related to the reproductive drive, but are less important to other drives, for example, hunger. Reproductive hormones may tell an animal when it is time to reproduce, but other

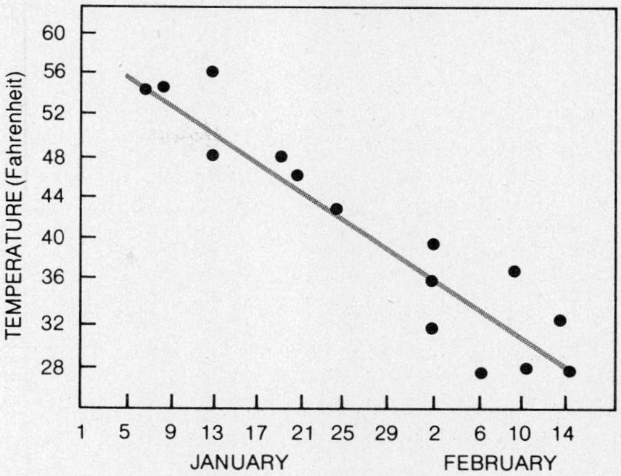


Figure 9.1 Measuring drive by noting the occurrence of song (the scattered dots) in the song sparrow *Melospiza melodia* at decreasing temperatures as the breeding season draws on.

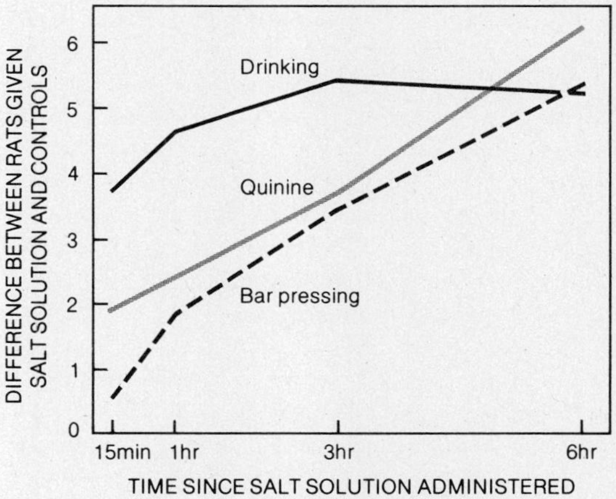


Figure 9.2 Differences in the results of three tests attempting to measure the same drive: thirst. Five cubic centimeters of strong salt solution were injected into the stomach of water-satiated rats before they were tested.

factors, such as blood sugar, tell him when it is time to eat. However, in some cases, behavior that is not directly under hormonal control may be indirectly influenced by changes in hormone levels. For example, an animal in which the sex drive is strong is less disposed toward flight or eating. Witness the gaunt stallion that is penned within sight of a herd of mares. Hormonally-induced behavior does not always supercede other behavior, however. In most animals, when "escape" behavior is elicited, the thresholds of both foraging and reproduction are raised so that the animal is not as likely to respond to either of these drives. I should add that drives may also interact at lower hierarchical centers. For example, nest building renders the stickleback less susceptible to stimuli that release fighting.

Drives, as we have seen, may play a great part in determining the behavior of an animal, and when a drive is strong the animal may have few behavioral alternatives. For example, a drive may be strong enough to cause a brooding hen to fight a predator she would normally flee. I was once attacked by a ptarmigan with chicks as I stepped into a clearing in an Alaskan forest. My cat Mary Elizabeth, who apparently had certain objections to any form of voluntary birth control, once leaped upon Bear, a friendly and curious German shepherd that had inadvertently entered a room where her kittens were. Mary Elizabeth would have ignored or avoided Bear under most conditions, but she was compelled to risk her life by a kind of molecule that happened to be in her blood at the time.

Some drives can build to much higher levels than others. As we have seen, perhaps some drives can become so strong that they "break through," in the absence of a stimulus, to produce vacuum behavior. All drives, however, are not capable of building to such levels. Weak drives can be defined, in the absence of testing, as those that rapidly wane upon performance of the behavior. Dogs or wolves, even those that are being continually fed, apparently gradually develop an urge to hunt. After being allowed to "hunt," even a little and under contrived conditions, they show almost no interest in this activity for a while. The hunting drive in wolves appears to a rather weak one that responds to some slowly accumulating internal factor—one that is easily fatigued.

### Interactions of Drives

Lorenz (1963) pointed out that drives can interact in a number of ways: one can accelerate the other; one can strengthen the other; they can be superimposed on one another and give simultaneous and identical expression; and, rarely, one can negate the other. In this last case, it has been noted that the escape drive usually, but not always, subjugates the other drives.

A closer examination of simultaneous drives will reveal more of the nature of drive interaction. For example, four types of simultaneous interactions have been described (Hess 1962). Alternation between two incompletely expressed drives, such as in the zigzag behavior of the stickleback, is called *successive ambivalent behavior.* It may also be seen in the quick, fluttering attacks and retreats of territorial birds fighting at their common borders. *Simultaneous ambivalent behavior* is the result of multiple drives occurring simultaneously. A rather remarkable example has been suggested in the threat posture of the cat: perhaps the back paws advance more rapidly than the more "timorous" front paws, causing the cat to "hunch" and to appear larger to enemies (Leyhausen 1956). *Redirected behavior* occurs when the object of a behavioral pattern is changed. An animal, feeling aggressive toward its mate, may attack another animal. A nesting bird may swoop at an intruding human, but, at the last minute, overtaken by fear, he swerves and attacks another bird instead. His aggressive act has not changed, only the object of the aggression. The last type of drive interaction is called *displacement activity.* Here, when two drives are in conflict, a third and completely unrelated pattern may appear. Tinbergen and Kortlandt reported this behavioral phenomenon independently in 1940. Hostile gulls, which are disposed toward both fighting and flight, may suddenly begin preening. A man who can't decide between two things he wants to do may scratch his head. The responses appear to be nonadaptive in that they are unrelated to the problem at hand. However, the *type* of displacement activity that appears may be directly related to some other immediate stimulus. In other words, the displacement activity may not be irrelevant, but may be adaptive in that it solves another, although unrelated, problem. For example, nesting terns, if approached, are torn between impulses to stay on the nest and to escape. In such cases, they may show displacement behavior, such as feeding motions; but, if their feathers happened to be wet, their displacement pattern is likely to be "preening," the usual response to wet feathers (Van Iersel and Bol 1958). In other cases, displacement behavior might have a lower adaptive value. For example, in fighting roosters, it might produce either eating or drinking behavior—depending on whether food or water is nearby—even though the birds may have just finished eating or drinking and therefore have no physiological need for food or water at that time. The point is, displacement patterns may be influenced by environmental cues.

Displacement patterns may not be performed as they normally are in their usual (adaptive) context; in other words, the irrelevant displacement pattern may differ somewhat from the normal functional pattern. For example, in fighting sticklebacks that suddenly begin displacement digging, the digging is a bit different from the normal "model" in that the spines are erected. A digging stickleback with flattened spines is simply digging, but one with erect spines has come to be understood by other sticklebacks to be in a threat position. So, here, the displacement activity now has a ritualized role in communication.

Some promising research on drives is being done with electrophysiological studies of the central nervous system. These studies indicate that displacement activities may not be due to a conflict of drives at all, but may be more properly described as the result of heterogeneous control in the central nervous system. Because of such complexities it has been suggested that the overused term "drive" is not an appropriate term for the intricate neural activity it attempts to describe. Although, in this discussion, we have considered drive to be a measurable motivating force, it should be remembered that it is being considered as such only because the concept aids in explaining some behavior. For the time being, the ethological definition of drive is an operational theory only; one that sometimes works in helping us to understand *why* animals do what they do.

## The Role of the Hypothalamus in Drinking

Probably the hypothalamus has been studied more than any other part of the central nervous system in the search for the neural centers of drive and motivation. Teitelbaum and Epstein (1962) and Manning (1971) have summarized much of the evidence concerning the physiological bases of motivation, and the following discussion is based largely on their descriptions.

Cells in the outer parts of the hypothalamus are extremely sensitive to the concentration of circulating body fluids. In other words, they can measure how thick the blood is. If the blood is "thick," the hypothalamus causes the pituitary to secrete a hormone called the antidiuretic hormone (ADH) that causes the kidneys to reabsorb more water into the blood. If the link between the hypothalamus and pituitary is damaged, ADH may not be secreted, causing the kidneys to pass enormous quantities of water. The animal therefore compensates by continually drinking. The condition is called *diabetes insipidus.* (Alcohol also suppresses ADH secretion, so that on those rare nights when you have imbibed heavily, you may have noticed you had to make a disproportionate number of trips to the facilities. And the next day you were thirsty.) On the other hand, if the blood is "thin," the kidney passes more water. So we see, the animal with thick blood is motivated to go out and seek water. In the first instance, the problem is solved by purely physiological means; in the second, by behavioral activity. But both are mediated by the same part of the brain.

Physiologists have been able to stimulate very specific areas of the brain with microelectrodes, or by inserting incredibly tiny hypodermic needles into various neural structures with which they inject specific chemicals. The results have shed light on the mechanisms of motivation. For example, if the lateral areas of the hypothalamus of an animal are stimulated either electrically or by injecting a salt solution (which registers as "thick" or concentrated blood), the animal will drink enormous quantities of water. After such treatment, animals will even drink foul water, which they would normally disdain.

If the lateral hypothalamus is damaged, an animal may stop drinking altogether, even to the point of death. It may die of severe dehydration, although surrounded by water. It doesn't simply ignore water, even while dying of thirst; it *avoids* it! If water is placed in such a rat's mouth, it will spit it out, acting as if it's bitter.

It could be argued that "motivational centers" are not being stimulated at all, but that the chemicals or electrodes are only triggering motor patterns. In other words, the artificial stimulus is only altering the action of certain muscles, producing a drinking pattern. If this were true, motivation or desire might not be involved at all.

It would be helpful to see whether animals artificially stimulated in such a way show normal appetitive patterns, and some such studies have already been started. In experiments with goats (Andersson and McCann 1955), artificially stimulated animals walked over to the corner of their pen and searched out their water bowls, thus showing normal appetitive behavior. There are a number of experiments, however, that indicate that artificially stimulated animals (which normally show only the response while the stimulus is actually being applied) are more easily distracted from their goals than are normally stimulated animals (Valenstein et al. 1970). In experiments on rats, slight amounts of quinine in water that they are artificially stimulated to drink will stop the drinking. Also, there is another puzzling aspect to all this. Normally, rats lap water at a very constant rate, thirstier rats simply drink longer. But artificially stimulated rats change their rate of lapping depending on the strength of the hypothalamic stimulation. So, apparently, motor centers *are* being stimulated somehow.

As we can see from these few examples, the problem of associating a specific nerve center with a particular behavior is a difficult one. It is apparent that the hypothalamus is involved in motivational states, but we can't say just how or what its mechanism might be. Also, it undoubtedly works in concert with brain centers other than the pituitary. For example, there is ample evidence that other parts of the forebrain are involved in drinking, as well as in feeding, sexual, and aggressive behavior (see Grossman 1968 for a review). Also, recently, questions have been raised about whether such centers are actually very specific (centers are here defined as a relatively small group of cells that act together in directing a behavioral pattern) (Valenstein et al. 1970). For example, it was found that the same electrode in the lateral hypothalamus could stimulate either eating or drinking behavior, depending on the situation in which the animal found itself at the time. So, it was suggested that hypothalamus stimulation provides only a general stimulus and facilitates any responses that the animal might be predisposed to perform. Other researchers favor the notion of specific drive centers, suggesting that feeding and drinking stimulation by the same electrode probably occurs because the centers are located very close to each other and the stimulus "diffuses" from one to the other. And, finally, to leave you with an appreciation of the complexity of the problem (which may be the same as ensuring total confusion), the various parts of the brain can compensate for each other so that a rat that has stopped eating due to hypothalamic lesions can learn to eat normally again by careful training. The eating center is apparently shifted to a different part of the brain (Teitelbaum and Epstein 1962).

The study of drive and motivation, then, is fraught with dissension and interpretation. In fact, the entire problem has decreased in popularity in recent years partly because it is so thorny. Unitary rules have been hard to formulate because of the exceptions, and the prospects for greater understanding seem remote. At worst, however, the problem can be only temporarily shelved because if we have any hope of ever understanding behavior, we must know about its sources.

# 10

# LEARNING, MEMORY, AND REWARD

We have all, at one time or another, been asked to believe startling accounts of animal intelligence, insight, and understanding. One of the great disadvantages of studying animal behavior, in fact, is that once your cover is blown at a cocktail party, you are invariably subjected to countless stories that begin, "I had a dog once who could. . . ." In time, you learn to fix a smile and allow your eyes to glaze over and your mind to wander. You must be prepared, however, to be brought to earth by data that, in some cases, support some of the most farfetched of these stories. In time, you may even learn to listen with an open mind, realizing that we really don't know much about animal intelligence. There are, even now, great differences of opinion among scientists as to what certain animals can learn, how much they can learn, and how important it is to them. There is also some argument as to just how one should go about investigating the phenomenon in animals.

Clever Hans was a horse owned by a European some years ago. He amazed the scientific world with his ability to solve a number of types of problems, the answers to which he communicated by tapping the ground with his forefoot. For years, Clever Hans withstood the most rigorous tests of skeptics until it was finally determined that the perceptive horse was reading extremely subtle clues given inadvertently by his owner. (The man would breathe a slight sigh of relief when Hans had tapped enough.)

Today, even preliminary tests would rule out such an error. We now rely strongly on what we like to call the "scientific method," and while the scientific method itself is seldom the subject of argument, the formation of hypotheses and the methods of testing them can lead to sharp disagreement. One of the most volatile areas involves learning. There has been vigorous debate over our assumptions and over the importance of learning in the world of animals. But we will not concentrate on the arguments here; we will assume the ethological approach.

As we have seen, certain patterns of animal behavior may be, in the final analysis, deeply seated in the genetic constitution of the animal. Such patterns are the product of patient and unprejudiced, yet blind and compromising, influences of natural selection. The patterns that survive this selection are the products of severe trials and are indelibly stamped into the nervous system of the species. In a sense, then, the members of any species are, to some degree, "programmed" to behave in a certain way.

However, we also know that in a complex and ever-changing world, it is not reasonable to assume that all possible problems may be anticipated genetically so that an immediate and proper response is already programmed into the animals' behavioral repertoire. Instead, some degree of individual behavioral flexibility is important. The degree of this flexibility, of course, could be expected to vary widely between and within species. Some species show very little variation in their behavioral patterns and, as might be expected, they are usually very specialized for a particular life-style. Intestinal parasites, for example, probably are not able to profit greatly by individual experience. They live under very constant conditions as adults and, rather than depend on their individual cleverness to insure new generations, they simply lay thousands upon thousands of eggs and leave the rest to chance. Other animals may show greater flexibility of behavior. Many vertebrates live under a variety of conditions during their lifetime, and their behavioral patterns are more affected by their individual experience. The success they achieve in solving their problems is in part dependent upon how well they are able to make associations based on past experience—how well they learn.

And, then, even among these more flexible "learning" animals, there are certain patterns that are more flexible, more free to vary, and to be influenced by learning, than others. For example, the thrusting motion of a copulating male monkey is probably innate, but he must *learn* how to mount the female. So learning is more important to some animals than others, and some patterns are more labile than others.

Learning has been notoriously difficult to define. In light of the great biases and schisms in the study of behavior, it is not surprising that there are a number of definitions being bandied about. However, we will develop the concept by slightly modifying the general framework set out by Thorpe (1963). The concepts we will encounter are the basis for many current biochemical and neurophysiological studies, and the provision of such foundations might well prove to be one of the most important functions of ethology.

## WHAT IS LEARNING?

There have been many attempts to define learning. Thorpe (1963) defined it as "that process which manifests itself by adaptive changes in individual behavior as a result of experience." According to others, learning occurs when the probability of certain behavior patterns in specific stimulus situations has been changed as a result of experience with this, or other similar, stimulus situations (Hilgard 1956). In this latter definition, changes due to maturational or fatigue processes are ruled out. Hinde (1970) focused on the exclusive part of the definition and called learning "changes which cannot be understood in terms of maturational growth processes in the nervous system, fatigue, or sensory adaptation." "Adaptive behavioral modification" is a well-worn definition, but is too broad to be very useful, and "a relatively permanent change attributable to practice" is perhaps too narrow. The increasing number of indications of interaction between neurological centers make forming any definition very difficult. But we shall use the above definitions to form one of our own. We shall define learning as adaptive changes in behavior that are due to experience and that are associated with changes in the central nervous system. These patterns may function with innate patterns, but are probably ultimately distinguishable from them. In the following discussions, it should become eminently clear that although learned and innate patterns are theoretically distinguishable, *adaptive* patterns are usually comprised of elements of both, and at our present level of knowledge, debate over the relative contributions of each is not usually fruitful.

### Memory and Learning

*Many years had elapsed during which nothing of Cambray had any existence for me, when one day in winter my mother offered me some tea. . . . I raised, to my lips a spoonful of the tea in which I had soaked a morsel of cake. No sooner had the warm liquid, and the crumbs with it, touched my palate than a shudder ran through my whole body. . . . And suddenly the memory returns. The taste was that of the little crumb of madeleine which on Sunday mornings at Cambray my aunt Leonie used to give me, dipping it first in her own cup of real or of lime-flower tea. . . . Once I had recognized the taste all the flowers in our garden and in M. Swann's park, and all the water lilies on the Vivonne and the good folk of the village and their little dwellings and the parish church and the whole of Cambray sprang into being.* [*Marcel Proust,* Remembrance of Things Past]

We are continually learning, sometimes purposefully, sometimes inadvertently. Not only are the momentous lessons of life tucked away in our minds, but also multitudes of small bits of information that we may bear for years, only to carry them unsummoned to our graves. Let's now look at this mysterious storage and retrieval system we call memory.

When learning occurs, it theoretically brings about changes in the central nervous system. There is a more or less permanent record of the learned thing, and it can sometimes be consciously recalled. At other times, it can be unexpectedly "jarred" into consciousness when an experience triggers a memory, perhaps involuntarily, as occurred in Proust's remembrances. The hypothesized physical change that occurs when something is learned is called the *memory trace* or *engram.*

A number of surprising findings have led many researchers to believe that virtually everything we encounter is learned—stored away in the brain. Wilder Penfield (1959) and his group, working in Montreal, probed the conscious human brain (which has no pain receptors) with electrodes and asked the patients to describe their sensations as various parts of the brain were stimulated. The results were startling. Some patients "heard" conversations that had taken place years before, some heard music, and some seemed to find themselves with old friends, long deceased. They seemed almost to "relive" the experiences, rather than simply remember them. The powers of recall under such circumstances are phenomenal.

Hypnotism may also help us recall far beyond our normal abilities. In one case, a hypnotized bricklayer described markings on every brick that he had laid in building a wall many years before. His recollections were carefully recorded, then the wall was found and examined. The markings were there! (But see also Hunter 1964 for a description of our abilities to embellish partially recalled events.)

Although we may have entered every jot and tittle of our lives into our mind's ledger, we are not able to consciously recall much of it. Sometimes we can't even recall what we have *tried* to memorize—as you well know. But apparently the information is there, just the same. The fault lies with our recall techniques. The implications of this finding are enormous. If our research ever enables us to recall or "relive" earlier events, we might "reread" novels on long train trips—if trains still exist—and all our exams would be virtually open-book tests. In addition, perhaps terminally ill and pain-racked patients could be stimulated to relive happier, youthful days with loved ones now gone, while life dwindled in a changed body.

**The Advantage of Forgetting** There are all sorts of strange abilities associated with memory in humans. Idiot savants, a special class of retardates, have very low I.Q.'s, but are able to multiply two five-figure numbers in their heads. Others can immediately tell you the day of the week on which Christmas day fell in 1492—or any day in any year (although this is probably not a memory feat). A normally intelligent Russian man made his living giving stage performances in which he would memorize lines of fifty words. Once, in just a few moments, he memorized the nonsense formula:

$$N \cdot \sqrt{d^2 \cdot \frac{85^3}{vx}} \cdot \sqrt[3]{\frac{276^2 \cdot 86x}{n^2v \cdot 264}} \cdot n^2b = SV \cdot \frac{1624}{32^2} \cdot r^2s$$

Fifteen years later, he was asked, with no warning, if he could remember it. He repeated the entire formula without a single mistake (Sinsheimer 1971). And what about those people with "photographic memories"? They exist, and they are called *eidetikers.* Proof of their abilities has been demonstrated with stereograms. These are apparently random dot patterns on two different cards that, when superimposed by use of a stereoscope, produce a three-dimensional image. In one such experiment, a woman was asked to look at a ten-thousand-dot pattern with her right eye for one minute. After ten seconds, she viewed the other "random" dot pattern with her left eye. She then recalled the positions of the dots on the first card and conjured up the image of a "T" (Stromeyer 1970).

So, if such abilities are possible for our species, why haven't they been selected for so that by now we can all, more or less, perform such feats? On the surface, the advantages seem enormous. The reason we can't is, in part, because there are serious drawbacks to such abilities. Both the Russian "mnemonist" and the eidetiker could look at a barren tree, "recall" its leaves, and then when they looked away, they might be confused over whether the tree was leafy or not. The mnemonist could watch the hands on a clock for a long time without knowing their position, remembering (or "seeing") only where they *had* been, so he was late a lot. Also, what about all those insignificant events that are of no advantage to remember? The energetically expensive neural apparatus would be wasted in retaining such information for recall (assuming that retention takes energy). Another problem with remembering everything concerns time-lapse. The mnemonist recalled everything so well that it seemed to him as if every event of his life had just occurred. Forgetting, then, gives a sense of time.

Obviously, the reason we can't remember as well as the people in these examples is that we, as a species, have not found it necessary. We generally don't need to recall every stone on the path to the place where we found food yesterday. We only need to remember the location of the food. In fact, remembering too much about our physical environment might cause us to relate to the environment on the basis of remembrances. Thus, any change in the environment might not be adjusted to as quickly as if we were never quite sure what to expect, because we knew we couldn't rely too strongly on our memories.

**The Consolidation Hypothesis and Octopus Memory** Apparently, there are two separate mechanisms by which information is stored, as evidenced in the differences in *long-term* and *short-term memory.* The existence of these two types of memory was first noticed in humans with brain concussions who were unable to recall what happened just before an accident, but could remember what happened earlier. On the basis of such findings, two memory centers were postulated whose information could be affected independently.

Information stored in the long-term memory "center" is relatively permanent (subject to very slow decay). Once information is processed into the long-term system, it is not easily lost and therefore may be recalled with minimum confusion after long periods of time. Short-term memory, on the other hand, is rather easily disrupted and is subject to rather rapid decay. (This is the reason "cramming" for exams is not a good idea. If the examination is postponed, all is lost; or if the student is suddenly disturbed or steps in something on the way to class, he may forget parts of the material. So, walk slowly and be careful.) Also, there is evidence that the short-term memory can be overloaded, whereas the long-term cannot.

Apparently, short-term storage may be necessary before long-term storage can occur. In a series of experiments with the octopus, it was found that if an inedible object is repeatedly presented to one arm of an octopus, the arm will come to reject it after learning that it is inedible. But if the object is then immediately presented to other arms, they will accept it. However, if some time is allowed to elapse before the second presentation, all the arms will reject it (Wells 1959). Apparently, the information acquired by one arm is only locally adaptive and must filter from a hypothesized short-term system to a long-term system before a general adaptive (total rejection) pattern can appear. The experiment actually only shows that *something* filters from neural centers associated with one arm to general centers. The short- and long-term phenomena are postulated because the learning arm ends up with no more "information" than the other arms once the information has become integrated.

Such observations have led to the development of a *consolidation hypothesis* that suggests the pathway of information storage. First, any registered experience enters a short-term memory system where it causes changes that last for about an hour, but then the system returns to normal (forgets). But, second, while information is still in the short-term system, and subject to recall by memory, it is being shifted into the long-term memory system, where it forms an *engram.* Actually, the hypothesis does not suggest that the engram must be formed from the short-term memory system; it could be formed through a parallel mechanism. In any case, the transfer from short-term to long-term systems is called *consolidation,* and during this process, the memory is susceptible to all sorts of disturbances or even obliteration.

Although the consolidation hypothesis was developed from research on a lowly invertebrate, the octopus, it may have wide applicability for vertebrates. The reason is, the octopus brain is remarkably similar to that of vertebrates. In fact, an octopus apparently solves some of its problems in the same way as the esteemed laboratory rat. This should not be surprising, since the visual receptors, brain structure, and neural physiology in the two animals are highly similar. The octopus brain, you recall, is composed of large lobes, such as the optic lobes that are connected to the brain by large optic tracts, and the subesophageal lobes that control the mantle covering (see Figure 6.14 in Chapter 6). It turns out that not only is the octopus ideal to work with anatomically, but its predictable behavior renders it especially valuable in learning experiments. Let's consider a few of the studies in the search for the formation of the engram.

An octopus has a legendary appetite and it will voraciously attack a crab on sight (Figure 10.1). But if the attacking octopus is shocked every time it touches a crab that is presented with a horizontal rectangle near it, it will quickly fail to respond to any crab associated with a horizontal rectangle. However, if a crab is then presented near a vertical rectangle, it is out of luck. The octopus doesn't associate vertical rectangles with unpleasantness and will attack.

It turns out that if the vertical supraesophageal lobe of the brain is removed from an octopus trained in this way, it will attack a crab accompanied by any kind of rectangle. Obviously, it has forgotten what it had learned. It would seem, then, that the engram for this task is stored in the vertical lobe. But, as you might suspect by now, it's not that simple. If the same octopus is given a choice between a crab presented with a vertical rectangle and one with a horizontal rectangle, it will usually choose the one that was never associated with a shock. So, apparently, the vertical lobe has something to do with attack inhibition, but, then, where is the engram?

It turns out that the engrams of visual learning are apparently located in the optic lobes themselves. Large nerve tracts run from the optic lobes to the supraesophageal lobes. It has been suggested that the optic lobes and vertical lobe may have a relationship similar to that of the cortex and limbic system in mammals. The vertical lobe permits the octopus to "concentrate" on the results of each training session and helps to consolidate the engrams in the optic lobes. Strangely, but not unexpectedly (since the same is true in the cortex in mammals), the engram is spread rather evenly throughout the optic lobe. In other words, if ten percent of the optic lobe is removed after training, the octopus will do only about ninety percent as well as it had. Forty percent removal results in a forty percent loss of efficiency, and so on.

Apparently, the vertical lobe (like the limbic system in mammals) is also associated with registering the results of an action. It answers the question, Is crab-grabbing associated with food or shock? Fibers that seem to record pain and food sensations, by the way, enter both the optic and supraesophageal lobes.

**The Mammalian Memory** Some of the most surprising results in the area of mammalian memory have come from experiments involving what is called the "split-brain" technique. It would seem better, somehow, to say that the term is a misnomer. But it isn't. Brains are split.

The technique was developed by Sperry (1961) and his colleagues. It involves splitting the cerebral hemispheres apart by cutting them down the midline through a tract, the *corpus callosum,* which links the hemispheres on the underside of the brain.

One of the things that led Sperry to perform his experiments was a rather startling discovery in brain-damaged humans. The halves of the brain could be severed, cutting through the corpus callosum, all the way back to the cerebellum with apparently no ill effect. It took a long series of careful experiments to determine the function of the corpus callosum and to

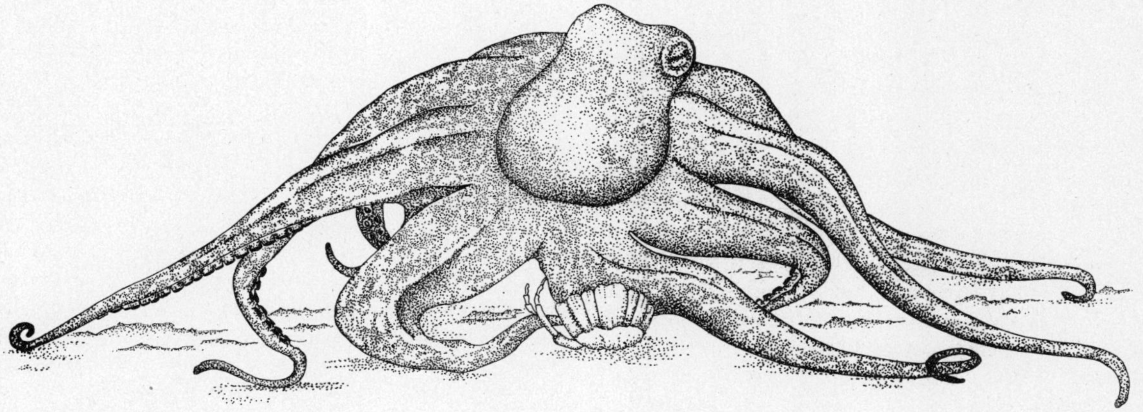

Figure 10.1 The octopus lends itself to experimental laboratory work because of its voracious and predictable appetite. It will normally attack a crab on sight.

solve the puzzle of how brains could be split without causing death. Let's review some of the key experiments.

If we learn a visual discrimination with one eye covered, we can make the same discrimination with the other eye. Anatomically, the reason is that fibers from the right and left halves of each eye cross over the optic chiasma to the other side of the brain, as shown in Figure 10.2. Thus, the visual centers in both halves of the cerebrum receive information from both eyes. Now, if the optic chiasma is split, the right eye can still make the same discriminations as the left, and vice versa. However, if the corpus callosum is also split, a task learned by using one eye cannot be repeated if the other eye is used. In fact, anything learned by one half of the brain is not transferred to the other half. This means, in effect, that mammals have two brains that can act independently. It is even possible to train both halves of the brain separately. With one eye covered, a split-brain monkey can be taught to approach a certain symbol that it has learned to avoid if it is seen only with the other eye. If that monkey sees the object with both eyes, however, usually one side of the brain will take over and he will respond without hesitation by either approaching or withdrawing.

The split-brain research shows that one function of the corpus callosum may be to transfer engrams, or their precursors, from one cerebral hemisphere to the other with the result that most information is stored bilaterally. However, this assumption is not universally accepted. Robert Doty of the University of Rochester implanted electrodes in a monkey's brain and stimulated it electrically in a series of learning trials. He found that if his animals learn the events associated with shock in the left hemisphere, they "remember" the shock's significance when the corresponding region of the right hemisphere is tested. But he also found that the memory is isolated in the "trained" hemisphere if the corpus callosum is cut some time after training. Doty interpreted this to mean that, whereas each half of the brain can "tap" the other half, the engram is not actually transferred from one half to the other. Instead, the engram stored in one side is simply available for use by the other side. The advantage of such a system, according to Doty, is that the brain can store twice the number of memories that it could if both halves stored the same engram.

The hemispheres are also connected by a bundle of nerve fibers, about the size of a dandelion stem, called the *anterior commissure*. Doty's split-brain experiments indicated that if engrams are transferable between hemispheres, the anterior commissure may be involved (Libassi 1974). In our discussion here, we will assume that engrams exist and that they *are* bilaterally transferable.

Since an animal can function normally with only one cerebral hemisphere, the other hemisphere can

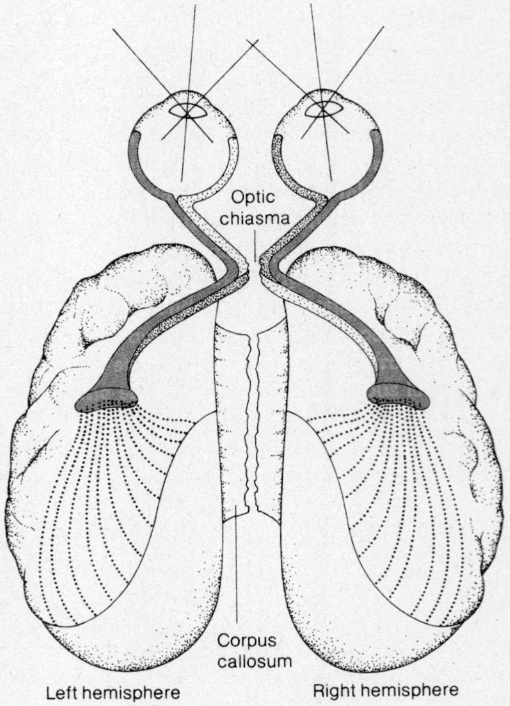


Figure 10.2 Dorsal view of a split brain, both the optic chiasma and corpus callosum being cut. The fibers from the median half of each retina go to the cerebral cortex of the opposite hemisphere. When the optic chiasma is cut, a hemisphere can only register the things seen by the lateral retina from the eye on its own side. When the corpus callosum is also cut, no visual information can pass from one side of the brain to the other.

be experimented with rather drastically without immediately endangering the health of the animal. For example, most of one hemisphere can be removed to see what the function of the remaining tissue might be. Through such harsh techniques, it was found that the various areas of the cortex are not equipotential. Certain areas have certain propensities or abilities that cannot be assumed by other areas.

Also, specific areas of the brain can only transfer certain kinds of information to the other hemisphere. The areas of the hindbrain, it turns out, can transfer only very simple kinds of information, and only certain types at that. In split-brain cats, lower hindbrain centers, for example, can transfer brightness-discrimination learning, but not pattern discriminations.

It is hoped that we will soon have answers to a series of important questions through split-brain experimentation. For example, if engrams pass through the corpus callosum, how long does it take? If only some of the corpus callosum's fibers are cut, does the engram become imperfect or just take longer to get through? What connections are necessary between cortical areas and subcortical areas in order for the former to store engrams?

As was suggested earlier, removal of parts of the higher brain centers affects mammals and the octopus in much the same way. It doesn't matter so much which part of the cerebrum is excised, but how *much*. Again, fifty percent removal results in about fifty percent memory loss. Thus, the engram seems to be diffused throughout the cortical tissue.

We still don't know whether consolidation takes place in the cortex itself, or if it occurs in some lower brain center and is then transferred to the cortex. But the lower brain centers undoubtedly play a role in consolidation. For example, the *hippocampus* is part of the limbic system, and one hippocampic area lies along the lower inside edge of each cerebral hemisphere (see Figures 6.21 and 28.1). The hippocampus connects directly to the hypothalamus. Damage to both lobes of the hippocampus results in a person being unable to recall anything that had happened just prior to the brain damage, but has no problem in remembering what happened much earlier. Apparently, in the latter case, the person is drawing on a fully formed engram.

The hippocampus is also attached by broad tracts to the reticular system, that mysterious net that functions in arousing certain parts of the cortex. Damage to the reticulum itself retards recall of recent events; thus it would seem that continuing *attention* is important in forming an engram. Attention is impaired by damage to either the hippocampus or reticulum.

So, here we are, speculating about how engrams are formed when we don't even know if engrams *exist,* at least not in the sense that we have been discussing them. There is a disturbing tendency among only casually interested people to hear a name or description and to construct an imaginary picture that accounts for what is heard, and then to assume that this imaginary thing exists. It's a trap. If we are ever to gain some insight—some real understanding—of the state of the behavioral sciences, we must keep in mind the level of supportive evidence for each assumption we encounter. It might be added that the evidence is usually much more substantive than is presented here, and for details you should go to the original material.

**How Engrams May Be Formed** Now, with this reminder, let's consider a hypothesis describing how an engram is formed. First, the sense organs register events that occur in the environment. Second, the information is passed to specific parts of the cortex and, at the same time, to the reticular and limbic systems. If the events have any importance, the reticular system activates certain neural centers and the engram begins to form in the cortex. The limbic system registers the "significance" of these events by recording their results—do they reduce thirst, sex needs, hunger? Recall that pleasure centers are associated with the hypothalamus and limbic systems, but stimulation of different areas in these same systems produces aversive behavior. The limbic system then passes signals to the hypothalamus, which, as we know, has a regulative function in "bodily needs." The results of all this cause the continued development of engrams in the cerebral cortex, which then react with what might broadly be called "approach" or "withdraw" motor channels, and thus the engram functions in influencing how the animal will behave when next presented with those particular environmental events (Manning 1971).

As you might guess, this scenario is not universally accepted. Neurophysiologists, for example, might object to the notion of a "connection" between engrams and motor centers, since there is no physical evidence of such connections. That is why I phrased the last sentence in the above paragraph as I did. You have to walk on eggs around neurophysiologists.

### Cells and Learning

The search for the engram has, of course, involved cellular approaches. Does learning alter the morphology of brain cells? That is, do they look different after learning has occurred? Are new synapses, or connections, formed? Are new cells formed in the learning process? Does the permeability of neural membranes change? Does the transmitter substance change? The problem is, the brain contains billions upon billions of neurons. And to complicate matters, they are extremely small ones at that.

While no one knows just how memory works, various hypotheses have been proposed. It has been proposed that memory consists of reverberating circuits, which, when once activated through sensory experience, continue. It has since been suggested that the reverberating units require too much energy to be maintained indefinitely, so that perhaps they function in short-term memory, while the engram itself is a stable, if diffuse, structure in the central nervous system. Yet others have shown that morphological changes appear in nerve cells thought to be associated with learning, such as changes in the number of microsomes (tiny bodies found in the neuron), the size and shape of apical dendrites, and changes at the synapses themselves (Gerard 1961). Also, biochemical hypotheses have been proposed that are based on a fascinating set of experiments with planarians (*Dugesia*). In these tests, after a planarian learned a task, its head was cut off. When a new head was generated, the animal was still able to do the task (McConnell, Jacobson, and Kimble 1959). The changes that occurred, then, must have been distributed throughout the animal's body and must not have been restricted to its primitive "brain." But if the deheaded sections were regrown in solutions containing an enzyme that destroys ribonucleic acids (RNA), no memory survived the regeneration (Corning and John 1961). So, attention has been focused on RNA as the key "memory chemical." (RNA, you may recall, is involved in translating genetic directions into specific enzymes in the cell.)

In other experiments, trained planarians were fed to untrained ones, and the cannibals learned the same task faster than those on a diet of untrained planarians (McConnell 1962). It has been argued that the experiments are not conclusive, since it was found that even if a planarian ate an untrained brother, it learned somewhat faster than planarians that ate other things (Hartry, Keith-Lee, and Morton 1964).

Similar experiments were attempted on mammals. When trained rats were killed and RNA extracted from their brains and injected into untrained rats, the latter acquired the learned pattern in a significant number of instances. Other experiments indicated that not only was a specific response transferred, but also the tendency to react to a specific stimulus (for example, light, sound). These experiments have all been severely criticized, and the findings, for now, must remain tentative.

In another experiment, right-handed rats were trained to reach into a test tube with their left paw. Since each paw is controlled by only one side of the brain, it was reasoned that the cellular results of such training should appear in only one of the cerebral hemispheres. The rats were then killed and the neurons from both hemispheres analyzed. The results showed the level of the kind of RNA that codes for specific enzymes to be much higher on the "trained" side (Hyden 1969).

A somewhat cruder set of experiments has been done on goldfish. Puromycin, which inhibits protein synthesis (enzymes are proteins), was injected into the brains of goldfish. Those fish receiving their shot before they learned to associate a light with an electrical shock, continued to avoid the light, but by the next day, they had forgotten what they had learned. If the fish were given their shots an hour *after* the training period, they showed no unusual memory loss the next day (Agranoff 1967). It seems, then, that injections before training permitted short-term memory to operate, but did not permit the development of the engram. So, apparently in these animals, it takes less than an hour for the engram to form.

At this point, there is no conclusive evidence that learning alters the structure or function of existing neurons or causes new ones to form. Nerve cells change as the individual undergoes new experiences (Figure 10.3), but such changes could simply be due to the growth process and have nothing to do with learning. It has been shown, however, that baby rats that are stimulated with electric shocks or other treatment develop more dendrites in their neurons than rats not stimulated (Shapiro and Vukovich 1970). It would be interesting to measure dendrite and RNA differences in animals raised with as little environmental stimulation as possible, those raised with mildly unpleasant experiences such as periodic shock, those raised under extremely agreeable conditions (perhaps with electrical stimulation of pleasure centers), and those in "enriched" environments, continually exposed to novel situations.

Most studies of memory are still in the exploratory stages, and as more data appear, the picture grows increasingly complex. We must certainly avoid attempting to formulate an oversimplified model at this time (Hinde 1970).

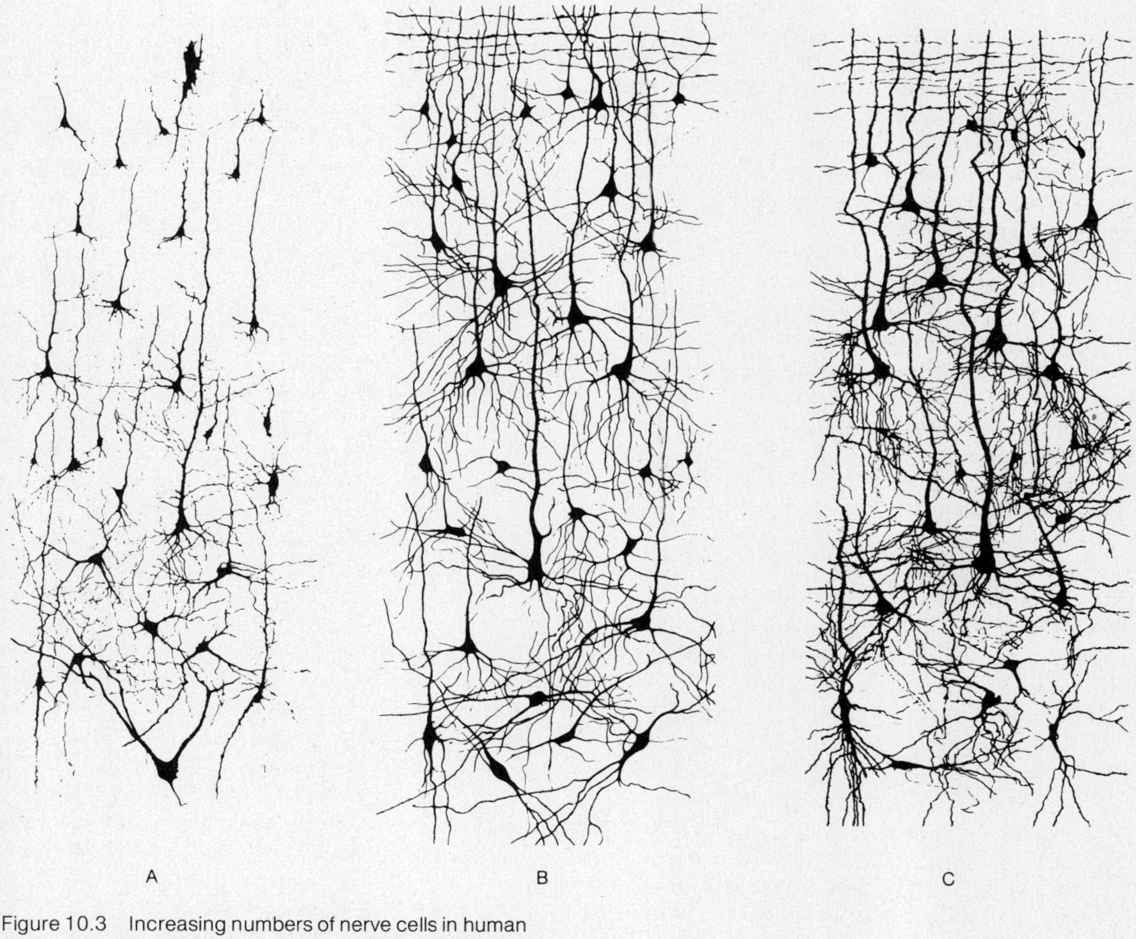

A B C

Figure 10.3 Increasing numbers of nerve cells in human infants with age. Sections of the cerebrum in infants that died at (A) one month, (B) six months, and (C) twenty-four months.

## REWARD AND REINFORCEMENT

In evolution, changes are preserved through their survival value. If their net effect helps the organism to survive, they are "good" and are maintained. But how does an individual ascertain which behavioral patterns are good? Thorndike (1911) said that if an act fulfills or relieves body needs, it is good, and that act is thereby reinforced so that it will be likely to be repeated when that body need appears again. Others have postulated that "good" should be defined as that which reduces a drive (such as the hunger drive), in other words, that which relieves physiological tension. It should be pointed out, however, that the reward need not actually be of biological advantage (as in the case of addicting drugs). It should also be pointed out that reward may occur in the form of reduced psychological tensions, as we see in anxiety-reducing, nonadaptive habits or superstitions. Although there is no real biological advantage in these acts, when performed they may relieve psychological tensions—perhaps simply because of their familiarity.

The drive-reduction hypothesis runs into trouble from experiments with avoidance responses. A rat placed on an electric grid will run off if he is shocked. If a light is presented within five seconds before the shock, after a few trials he will run off whenever the light comes on. So, here is a case of learning occurring in the apparent absence of a drive-reduction. Drive-reductionists like to point out that perhaps the traumatic experience of the shock caused an increased level of tension that came to be associated with the light, and when the light came on, the rat hauled tail in order to reduce its anxiety. Therefore, they say, it is reduced anxiety (called a "secondary drive"), after all, that is rewarding. We will see later that the drive-reducing hypothesis ran into trouble again when it was found that animals learn by merely exploring. (Is there an exploration drive?) Or that monkeys will learn to open a window to be able to watch an electric train run in circles. (Is there a curiosity drive?) It will become apparent that unless we wish to widen the usual definition of drive, we must admit that learning can occur without it.

However, we must not neglect the importance of reward in normal learning situations. Most rewards probably do reduce "anxiety" or "drive," and rewards have been shown time and again to be a critical factor in association learning. It's just that the exceptions do not permit drive-reduction to be considered synonymous with reward.

Interestingly, drive may even interfere with the attainment of the reward. For example, food may act as a reward for a hungry animal. The hungrier he is, the faster he will learn—up to a point. He may reach a level of hunger where his motivation is so high that it interferes with learning. In other words, he is so desperate for the reward that he can't put his mind to the problem of how to get it. He makes all the wrong moves and ends up with nothing. Perhaps you have experienced the same problem in a different context.

Another question is, At what level does reinforcement operate? Lorenz, basing his ideas on Craig's (1918) work, proposed that the mechanism that directly brings about the good or adaptive effect is the fixed action pattern, or consummatory act. (We have already discussed some problems arising with too strict an application of the idea.) Lorenz stated that the completion of an act (for example, eating) has a reinforcing effect through *reafference* or sensory feedback. As an example of such a learning-instinct interaction, consider nest building in crows. A crow, standing on a suitable nest site with nest material in its beak, performs a downward and sideward sweep of the head that forces the material against the substrate and later against the partially completed nest. When the twig meets resistance, the bird shoves harder, thrusting repeatedly, as would a man trying to push a pipe cleaner through an obstructed part of the pipestem. When the twig is wedged in, its resistance is increased, and the efforts of the bird are heightened until, when it sticks fast, the bird consummates its activities in an orgiastic maximum of effort. And then it loses interest for a time. Some species of crows possess no innate mechanism to guide their selection of nest material, and so young birds will attempt to build with anything they can carry. Most of the unlikely material, which may include light bulbs or pieces of ice, will not wedge firmly into the nest, and so they are prevented from reaching the consummatory stimulus situation that means both relief and biological success. Such failure quickly extinguishes the bird's acceptance of inadequate material until, finally, the birds become true connoisseurs of twigs. In such situations, then, the reinforcer of a learning situation is the performance of an innate consummatory act (Lorenz 1969).

It has also been shown that learning may place innate patterns into their proper adaptive sequence through reinforcement mechanisms. Nest building in rats is accomplished by the performance of three motor patterns (Eibl-Eibesfeldt 1970). First, the rat brings nest material to a selected site. Second, it stands in the center of the material and, turning to and fro, stacks the material into a roughly circular wall. Finally, it pats the wall, tamping it down and smoothing the inside. Naive rats, offered paper strips, will get into a frenzy of all three activities—running about with nest material, patting, and building all at once. Each act is performed well—but no nest results. The naive rat, after carrying a few paper strips, may stand above them performing heaping movements in the air and tamping down a wall that doesn't exist. The failure of the pattern to provide a rewarding reaffirmation by its resistance to the rat's paws teaches the rat not to attempt heaping before the nest material is carried in, or patting before the wall is built.

Lorenz (1969) pointed out that the structure and functional properties of the consummatory act were never fully understood until their teaching function was realized. He also noted that there does not seem to be any case in which conditioning by reinforcement could be demonstrated in a behavioral system not including appetitive behavior. However, rats that are allowed to press a bar and trigger an electronic impulse in their own pleasure centers will learn a task to be able to press the bar. On the face of it, this seems to refute Lorenz's notion that appetitive behavior must be involved, since there is not likely to be an appetitive pattern for bar pressing or pleasure-center stimulation. However, we don't know what pleasure-center stimulation "feels like"—does it feel like sexual relief? Eating sumptuously? We only know the rat will frantically press the bar up to seven thousand times per hour, until it collapses in exhaustion. Also, since learning plays such an important part in appetitive behavior, can the goal be an acquired one in certain cases, not associated with some innate determined "goal"? At this point, we simply don't know enough about what appetitive action really is to be able to support or deny Lorenz's assumption.

Reward, according to Thorndike (1911), is the major factor through which the stimulus-response (S-R) bond is strengthened. Of course, the reinforcing effect of some commodity, such as food or water, is not a function of properties of the commodity, but rather the properties of the animal. Water may reinforce the behavior of a laboratory rat, but not some species of desert rodents that derive water from their food and never drink.

### Social Reinforcement in Humans

Not long ago, B. F. Skinner caused another stir—this time with his book *Beyond Freedom and Dignity* (1971). In it, Skinner surmises that the utopian society might be reached through a system of positive rewards. Those individuals that acted in accordance with some grand design would be rewarded. Aberrant individuals would go unrewarded. And, in time, almost all members of the society would be constrained to go along the prescribed route. Of course, a wide variety of humanists, from behavioral scientists to drunks, attacked the idea. They feared that such a system would result in reduced freedom and a loss of dignity.

One might argue, of course, that such a system of rewards has been shaping our society for centuries. In some cases, the coercion has been dramatic, as in the response to civil unrest in the 1960s or, on a different scale, the American or the Spanish Civil War. The results of the Spanish Civil War of the 1930s are still visible today in the scarred bodies and buildings. This says nothing of the battered psyches of people who managed to survive.

Of course, these were rather drastic sorts of social molding. Usually, it occurs much more subtley. Usually you don't notice it. For example, here you sit, reading a book, when perhaps you would rather be outside. So why are you here? Obviously, your friends and parents approve of your being in school. You, yourself, feel good about it. So, you are rewarded. Also, if you do well, make good grades, and stay out of jail, you can expect to be further rewarded with a good job, money, friends, and perhaps power. If you are *not* in school and do *not* have a job, you may receive a lot of sidelong glances from society's more "respectable" and "productive" members. They may make you feel uneasy in a multitude of little ways. You are not meeting society's expectations for you.

It is probably erroneous, however, to speak of "society's" expectations. The idea implies a "big brother," or some other consciously manipulating entity. It is more likely that the society or the "system" that molds us all, is actually nonexistent. At least in the sense of its being a body, aware of itself and its decisions. It may simply be what it is often called—a *system*—not led, not controlled, but effective and self-perpetuating nonetheless.

Any such system could have developed through selective processes. It's here because it works. Cultures in which the individuals produced and/or acquired have historically been the ones that prospered and survived. Present societies are descended from successful earlier systems. As individuals in a society became aware of what "worked," those things could have been consciously incorporated into the codes, laws, ethics, or traditions of that culture. And the citizens of that culture would have been encouraged to go along. The encouragement has sometimes been the threat of dire punishment, but it is likely that the long-term, persistent, and gentle effects of peer approval have probably played a much more important part in social patterning. This is not to say that any individual is aware of his own role in such modeling. The hostile and curious stares that scruffy and strangely dressed people receive from "decent citizens" are probably not delivered with any intention of social modeling, but the citizen doing the staring is, nonetheless, a part of the great nebulous "system" if he makes the scruffy one feel uncomfortable.

To see how easy such modeling is, try an experiment on your instructor. He or she won't be aware of it, of course, since it's well known that instructors never read the text. In your next lecture, smile approvingly at him every time he touches his glasses, or whatever, and stop smiling as soon as he takes his hand away. See how long it takes him to talk continuously without removing his hand. Children have apparently used the system to direct their parent's behavior. When they are criticized, they say nothing or as little as possible. Whenever they are treated well or rewarded, they never fail to let the parents know how much they approve of them. The parents soon behave in such a way as to gain as much approval as possible.

Much of our discussion on human social reinforcement has been pure speculation, as you know. However, it seems important to recognize that the principle of subtle reward may be an important factor in human social systems and that its role deserves an increasing amount of research attention.

# 11

# KINDS OF LEARNING

Learning, the great benchmark by which we often choose to measure and compare ourselves, remains a mysterious process. We still don't know much about it. Our efforts to discover its mechanisms continue to disappoint us. They have, when all is said and done, yielded precious little. So we have vindictively attacked it with our weapon of last resort: labels. Unfortunately, every group of behavioral scientists seems to have its own labelling system, and for that reason even this weapon has lost its effectiveness. This may be a good thing. Our confusion should serve to remind us that the battle is not won.

So, now let's consider a labelling system generally acceptable to ethologists. Although in ethological theory learning is generally divided along rather precise lines, first proposed by Thorpe (1963), keep in mind that no such distinct divisions actually exist in nature. Individual adaptive modification of behavior may involve several of these "types," and, in fact, they may all eventually be proven to be based in the same neurological mechanism, or in neurological events not yet imagined.

Roughly in keeping with Thorpe's classification of learning classes, we will consider the following kinds of learning:

*(1)* Habituation
*(2)* Associative learning
  *(a)* *Classical conditioning (also called Type I, or the conditioned reflex)*
  *(b)* *Operant conditioning (also called Type II, or instrumental conditioning)*
  *(c)* *Trial and error*
  *(d)* *Gestalt perception*
  *(e)* *Learning sets*
*(3)* Latent learning
*(4)* Insight learning
*(5)* Imitation
*(6)* Play
*(7)* Imprinting

## HABITUATION

We discuss habituation first because, on the surface, it is the simplest kind of learning. However, it is undoubtedly of great adaptive value since it is found in such a wide range of animals. But the diversity of habituating species also indicates that the mechanisms involved in this kind of learning may also be varied.

*Habituation* may be defined as the relatively persistent waning of a response as a result of repeated or continuous stimulation that is not followed by any kind of reinforcement (Thorpe 1963; Hinde 1970). Persistent, here, means that the effects of habituation are rather long lasting, once they are acquired. In its widest sense, habituation is simply learning *not* to respond to specific stimuli that tend to be without significance. A bird must learn not to take flight at the sight of windblown leaves. So, habituation is simply dropping responses instead of adding or changing them.

One finds habituation everywhere. *Hydra* living in a shallow dish of water will quickly retract their tentacles and shorten their bodies if the glass is tapped. But after a few such taps, their reaction slows and may stop. A snail crawling across a pane of glass will quickly retract if the glass is tapped, but if the tapping continues, it will finally be ignored. Birds will at first be startled by a scarecrow, but will come to ignore it or sit complacently on its arms, preening their feathers after a large meal of the farmer's grain. In certain eastern countries where water flows through the fields, the problem has been solved. The running water gradually fills ballast buckets on a pivoting scarecrow. When the buckets are heavy enough, a mechanism is released that causes the supine scarecrow to spring erect. The birds are scared away, the buckets are emptied by the scarecrow's action, and the scarecrow falls back into the grass until the buckets become full again. If the cycle is slow enough, the birds never habituate and the crops are saved.

Timing is critical in habituation. The marine ragworm *Nereis* will withdraw into its burrow in response to a wide range of stimuli, from touches and shadows to bright light and electrical shocks. Habituation to a bright flash of light occurred in less than forty trials when the stimuli were given at thirty-second intervals; but eighty trials were necessary for habituation to occur if the stimuli were given five minutes apart. Also, it was found that, for the ragworm, each type of stimulus—light, electricity, shadows, touch, or whatever—had its own characteristic rate of habituation at any trial frequency (Clark 1960).

Figure 11.1 Nesting common murres. Neighbors habituate to each other and therefore save time and energy that would have been spent in needless squabbling.

## The Adaptiveness of Habituation

The adaptiveness of habituation may readily be seen in social or colonial animals. If colonial birds such as seabirds, which often nest within a few feet of each other on rocky coasts (Figure 11.1), did not habituate, a great part of their time and energy would be spent quarreling with neighbors. Instead, they come to reduce their aggressive responses to any neighbor as long as that neighbor keeps a certain distance. In many species, this defended area is defined, quite simply, by how far a bird sitting on the nest can reach with its bill. Lorenz (1963) noticed that in "nonanonymous" communities, in which each individual animal is recognized by the rest of the community, animals may only habituate their immediate neighbors, which are, of course, the ones most frequently encountered. In such communities, an animal from another neighborhood will receive harsh treatment from residents into whose turf it wanders.

Lorenz (1969) theorized that in habituation, an innate releasing mechanism becomes associated with an individually acquired perception. This interesting notion seems believable when one realizes that the stimulus that is habituated to is a group of perceptions rather than a simple sign stimulus. So, the changing reactions to a perception (which is, after all, habituation) is dependent upon a number of variables occurring simultaneously—variables that are characteristic of a certain situation. In other words, the context is important; a stimulus that has been habituated to must appear in its usual context if it is to be ignored. Out of context, it may arouse the original reactions. Thus, ducks will mob a well-known dog if he should appear along a stretch of shore where he is not usually seen.

Habituation may also be important in interspecific relationships among animals in the wild. There is some evidence that harmless animals are encountered much more often than predatory ones. Habituation, therefore, may occur to harmless species, but not to the harmful ones. The difference in the reaction of birds to Tinbergen's long- or short-necked model (Figure 8.11 in Chapter 8) has been explained on this basis. Geese may be seen more often than hawks and, therefore, the latter are not habituated to. The difference in the numbers of times the two types of animals are encountered does not imply that there are more geese than hawks. Selection for adaptive behavioral patterns may result in potential prey avoiding situations in which predators are likely to be found, and may cause predators to move about in a more cryptic manner than innocuous species.

Lack of habituation to short-necked or "predator-type" birds does not entirely explain why animals seek to escape them. On several occasions, I noticed that the Jamaican woodpecker *Centurus radiolatus*, immediately after uttering a warning cry, drops from high limbs into dense undergrowth at the approach of the large turkey vulture *Cathartes aura*. There are no reports of the woodpeckers being attacked by the slow-flying vultures. Also, the vultures usually fly at low altitudes, just above the treetops, so there is little possibility of a high-speed dive that would be necessary to catch a wary woodpecker. Although the woodpeckers may have to take evasive measures several times in a single day, they apparently do not habituate to the harmless vultures. They seem to be just as alarmed after the fifth encounter as they were after the first. The pattern may be a vestigial response from the period when the woodpecker's ancestors lived among avian predators; or it may be a useful pattern that, although it wastes some energy, is adaptive since real predators with a similar image, such as hawks, also live in the area. Both explanations for the pattern suggest that the woodpecker's present habitat is a relatively new one for them and they haven't yet adapted to it very closely.

Habituation may play an important role in habitat selection (Klopfer and Hailman 1965). An animal placed in new surroundings may show a general wariness and fear response. Once habituated to an area, however, it may feel at ease only in that familiar surrounding where novel or unusual stimuli are rare. If the habitat is rewarding in any way, such as through supplying food or warmth, the animal's positive response to it, or areas like it, may become firmly established.

### The Mechanisms of Habituation

There are some fascinating puzzles associated with the study of habituation. Habituation is, in some cases, a selection (perhaps a selecting out) of releasers that elicit behavioral responses. In other words, it is a mechanism that reduces the instances in which a certain IRM (such as that associated with escape behavior) is activated. In experiments, a number of innate responses have been found that, in spite of their great survival value, wane rapidly if released a number of times in quick succession. This is true, for example, in the owl-mobbing reaction of the chaffinch (Hinde 1960b). In a series of experiments, not only was the mobbing reaction reduced in birds that were allowed to repeatedly mob a stuffed owl, but the response did not regain its intensity even after a rest of several months. Furthermore, no reinforcement could restore the mobbing behavior, not even being chased by a real owl. Certainly, such an effective defense mechanism as mobbing is not made to function only once or twice, losing strength after that. Why, then, does it die out? Such puzzles are scattered throughout the ethological literature and await new interpretations or redesign of experiments. Lorenz (1969) suggested such anomalies indicate experimental error.

Habituation, by the way, must be considered as distinct from sensory adaptation and fatigue. Sensory adaptation is due to repeated stimulation of the receptors until the receptors fail to respond. For example, the sense of smell quickly adapts to odors so that they go unnoticed. And then, an effector, such as a muscle, may fail to respond simply because of fatigue. Both sensory adaptation and fatigue are very short-term phenomena, a factor that, in part, distinguishes them from habituation.

In sum, it is not known whether different mechanisms function in habituation to different environmental factors, but it is believed that the neural mechanisms probably differ widely among different groups of animals. Habituation is probably one of the more widely occurring types of learning responses, and it has been considered to be among the simpler types of learning. However, in the final analysis, it cannot be shown to be the result of a "simple" mechanism; in fact, new evidence indicates increasingly complex facets of the problem. For example, there are a number of factors that can increase and decrease response strength when chaffinches mob a predatory owl (Hinde 1970).

## CLASSICAL CONDITIONING

Pavlov's widely heralded experiment involved ringing a bell just prior to feeding a dog. The dog came to associate the sound with food so that, for a while, even when the food was not present, the dog would salivate upon hearing the bell. The phenomenon is called the *conditioned reflex.* "Conditioned" is an unfortunate translation from the Russian. The term should be "conditional," implying that the behavior is conditional upon something happening (Figure 11.2).

In Pavlov's experiment, the food is called the unconditioned stimulus, and the bell the conditioned stimulus. Because the unconditioned stimulus is necessary in the training process, it is said to "reinforce" the conditioned response (salivation).

It is interesting that there may also be a conditioned and unconditioned response. In other words, the response to the food and to the bell are not identical in all respects, usually because of an incomplete response to the conditioned stimulus (the bell). Another difference is that the conditioned response may be anticipatory; the animal may look around as if expecting the unconditioned stimulus (the food) to appear. In fact, Pavlov's dogs soon came to recognize the experimental situation and were eager to be put into their restraining harnesses so that the experiment could begin. The point is, the conditioned response is not simply a duplication of the unconditioned response.

The conditioned response is formed by the *association* of a strange stimulus with a reward. Thus we have *positive reinforcement.* If the stimulus is associated with a painful or unpleasant experience, it will produce *negative reinforcement.* Thus, if a dog comes to associate a particular sound with an electrical shock to its left paw, it will begin to raise the paw in response to the sound alone. Birds will recognize certain noxious insects as unpalatable after only one or two episodes with them and will avoid them like the plague (Figure 11.3).

Thorndike (1911), in testing a variety of different animals, found that when an animal learned a response through conditioning, a new stimulus only somewhat like the original one could elicit the same response to some degree. The phenomenon is called *stimulus generalization,* and usually the more characteristics the new stimulus has in common with the original stimulus, the stronger will be the response.

*Irradiation* is a theoretical construct Pavlov advanced to explain stimulus generalization. Irradiation involves the acquiring of conditioning effects by receptors not involved in conditioning trials. If the conditioned stimulus is touching a certain area of the skin, touching other areas of the skin may produce a partial response. It is not an unexpected rule that the farther the stimulus is from the conditioned area, the weaker the response will be. (It is assumed that the same sorts of receptors exist in both places.)

## OPERANT CONDITIONING

In 1938, B. F. Skinner (Figure 11.4) published his landmark work, *The Behavior of Organisms.* The basis for his approach to the study of learning is called *operant conditioning.* Thorpe (1963) also called it Type II, or instrumental, conditioning, as opposed to Pavlov's Type I, or classical, conditioning. Type II differs from Type I in several important respects. In Type II, the conditioned response is a *voluntary* action of an animal and is a result of appetitive behavior. Also, in establishing Type II conditioning, both the stimulus and the response must precede the reward (Thorpe 1963).

In some of the earliest experiments demonstrating operant conditioning, Thorndike used "problem boxes." If a cat is placed inside a problem box, it will seek to escape and will roam around, sniff and poke and push, trying to find an exit. Finally, if it happens to make the proper move, for example, pull a ring, the door opens and the cat is free. If the sequence is repeated again and again, the cat comes to associate the ring with the door opening, until finally, when it is put into the box it promptly pulls the ring and leaves.

The basic differences in classical and operant conditioning are shown in Figure 11.5. Skinner, by the way, preferred to call the reward "reinforcement." He argued that terms like *reward* and *satisfaction* imply sensations that might be intuitively inferred but are not measurable. But reinforcement is defined operationally; it alters the probability of a response.

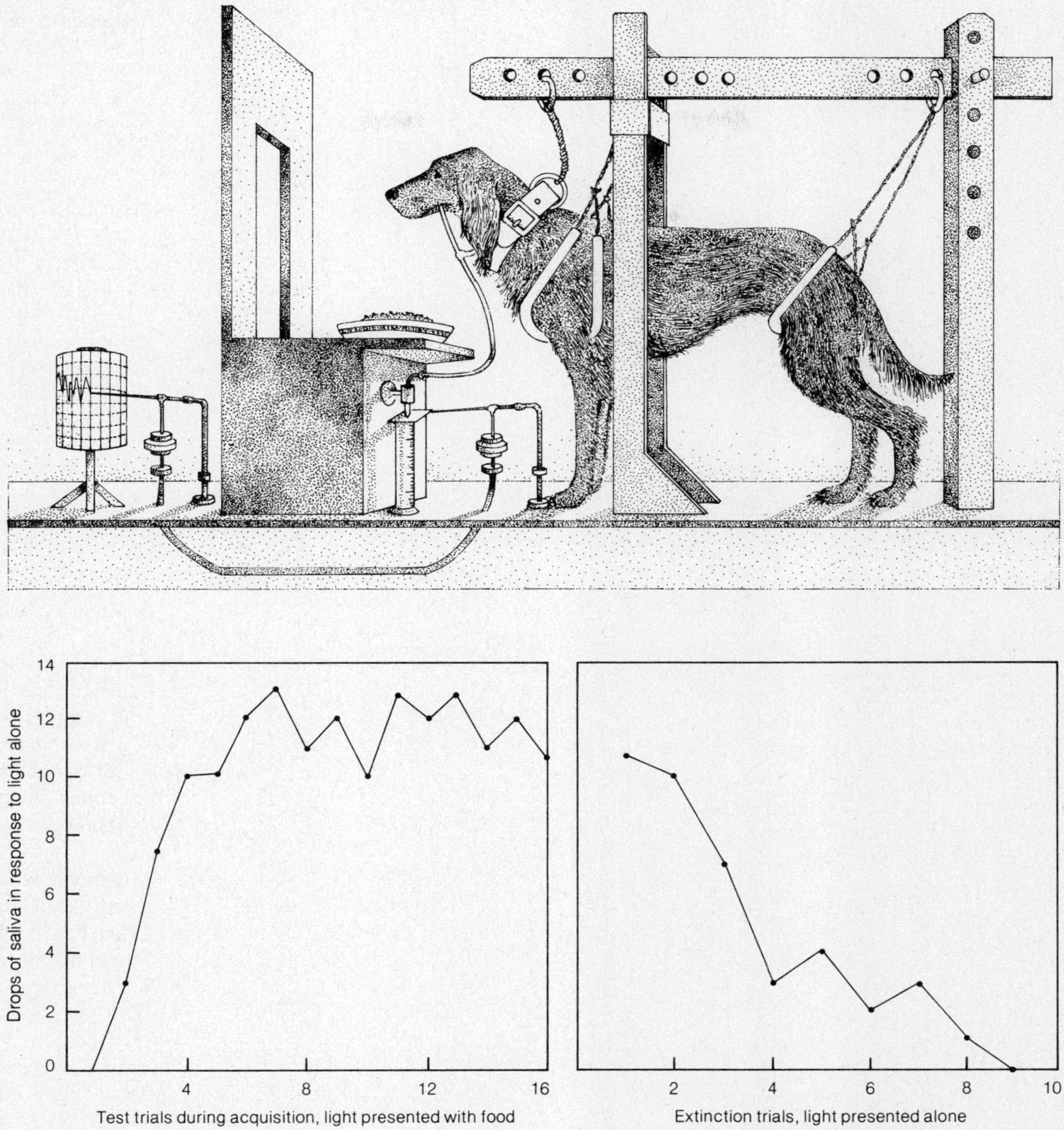


Figure 11.2 Pavlov's apparatus with which he demonstrated classical conditioning in the dog. The graphs show the number of trials necessary to learn the association (left), and in the absence of reward, the number of fruitless trials before extinction (right).

Figure 11.3 Animals can learn through unpleasant experience. The bluejay eating a poisonous monarch butterfly and then vomiting has learned a lesson this way.

Skinner's entire approach to science has been called "unusual," among other things. Rather than set up and test hypotheses, he suggested testing all possible stimulus-response situations, analyzing the results, and inductively arriving at scientific principles. Many psychologists do not accept Skinner's approach, arguing that most types of behavior are far too complex for associations and correlations to be tested randomly.

The mechanical test situations Skinner devised, on the other hand, have been widely used to test operant (Type II) conditioning. Testing this type of learning might involve animals running mazes or opening boxes. The important thing is that upon cue, the animal must *do* something to receive his reward. A typical test situation might take place as follows: the researcher wants the animal to learn to press a bar while a buzzer is sounding. The hungry animal moves randomly, at first, until it accidentally pushes the lever while the buzzer is sounding, and, lo! food is delivered. The fact that the reward appears under a certain set of circumstances each time increases the probability that the proper sequence will be repeated until, finally, a dependable stimulus-response chain is established.

Actually, the effects of classical and operant conditioning can be hard to distinguish experimentally. One reason is that both types of learning may occur simultaneously (Mitchell, Scott, and Williams 1973). For example, in classical conditioning, there may be skeletal responses (perhaps undetected) accompanying the conditioned response. On the other hand, when operant conditioning is achieved, a classically conditioned response may be occurring at the same time. Researchers, however, have been able to separate the two kinds of learning in some experiments. In one, a dog was taught to respond to one stimulus by performing an action in order to receive a second stimulus, which was, in turn, rewarded by food. The first stimulus evoked an instrumental response without salivation; the second evoked salivation, thereby separating the two types of conditioning. (Thorpe separated the two in trial-and-error learning, as is described next.)

## TRIAL-AND-ERROR LEARNING

Although probably neither classical nor instrumental conditioning, alone, is important in natural learning situations, the two processes occurring together may be important to learning that occurs in the wild. Thorpe (1963) referred to this kind of simultaneous learning process as *trial-and-error learning.* Other researchers, however, consider instrumental conditioning and trial-and-error learning to be one and the same, and Thorpe admitted that the greater component of trial-and-error learning, even according to his scheme, is instrumental Type II learning. His ponderous definition of trial-and-error learning includes only an element of instrumental learning: "The development of an association, as the result of reinforcement during appetitive behavior, between a stimulus or a situation and an independent motor action as an item in that behavior when both stimulus and motor action precede the reinforcement and the motor action is not the inevitable inherited response to the reinforcement." It's a bit complex, but useful.

Although Types I and II conditioning are sometimes an integral part of trial-and-error learning, the two have been experimentally separated. A guinea pig was placed in a special harness so that when it turned its head in a certain way, food would be released from a container. The appropriate head movement would deliver food only if the movement occurred directly after a bell was rung. At first, the guinea pig moved in a highly distorted fashion. But when it finally moved its head in the correct way, following the bell, food was delivered. After several repetitions of the sequence, the irrelevant motions were eliminated and, upon hearing the bell, the guinea pig immediately moved its head in the appropriate way. It was also found that the guinea pig experienced anticipatory Type I conditioning (such as salivation in response to the bell). This reaction was distinct from the gradual selection of the proper motor pattern that demonstrated instrumental, or Type II, conditioning (Thorpe 1963).

The general nonspecific "trials" that are first attempted in such situations should not be considered as random movements, but rather as appetitive attempts to solve a problem of bodily need. The problem is realized by the animal in the form of a drive (here, hunger), and the trials are attempts, however simple, to cope with the problem at hand. Appetitive motivation, we see, is necessary before trial-and-error learning can begin.

In the study of wild animals, such separation of the components of trial-and-error behavior is probably not possible. However, the same sort of conditioning experience in nature is not hard to imagine. A young animal could be expected to come to associate certain environmental signals with the presence of food and to react with an increasingly fruitful pattern of motor responses. Consider a young quail. In the period just after hatching, the chick will peck at anything until it hits upon something edible. It gradually learns to associate the sight of certain objects with food until, when it locates an object that has the "food image," it pecks at it and then performs the consummatory act of swallowing. It may show anticipatory reactions at the sight of the food. The reinforcement, then, is preceded by the stimulus (hunger), the expectation (based on experience) that it will soon be satisfied, and the active response (pecking)—a necessary sequence in trial-and-error learning. The reward is retroactive and influences future pecking. This process, together with habituation to inedible objects, results in increased foraging efficiency.

Figure 11.4 B. F. Skinner, one of the most important twentieth-century psychologists. He is responsible for the discipline known as behaviorism.

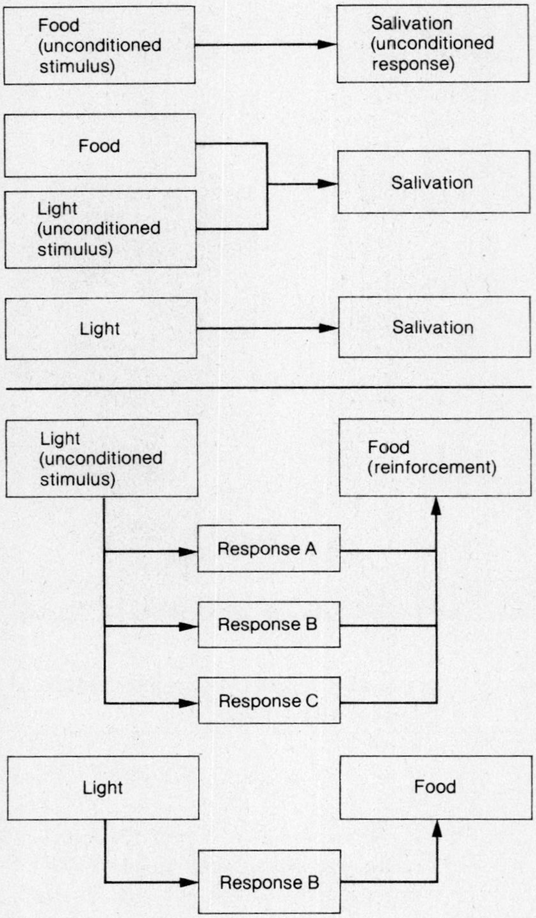


Figure 11.5 The organizational differences between classical conditioning (top) and operant, or instrumental, learning (bottom). Note that in classical conditioning, an unconditioned stimulus becomes paired with a conditioned stimulus. The conditioned stimulus then becomes a substitute for the unconditioned stimulus to produce the unconditioned response. In instrumental learning, when a conditioned stimulus occurs, there is an opportunity for the animal to respond in various ways. However, only the "correct" response is reinforced. Finally, the stimulus produces a specific response.

In nature, trial-and-error learning may also be important in escape reactions when environmental signals are associated with an attacking predator. Flocking or grouping animals, in particular, could learn the signs of impending danger through the alarm behavior or bad luck of their fellows and learn to respond properly without being personally attacked.

A central problem with most of the research on conditioning is that it has been performed under such artificial conditions that little light has been shed on its adaptiveness. It is likely that conditioning is an important part of individual accommodation to a changing environment. Thorpe (1963) theorized about the specific behavioral level at which conditioning could be operating. Conditioning is, he said, "a method by which the innate releasing mechanism is elaborated or modified to fit the environmental situation more exactly." Note that this differs from Lorenz's concept of how habituation occurs: by altering the releasing effect of environmental stimuli. However, Thorpe added that conditioning also apparently plays an important part in adapting the receptor to specific stimuli as a result of experience. So, here again, we see the close interrelationship of learning and instinct, and how experience may modify the entire adaptive response.

There is a variety of information that indicates that trial-and-error (or operant) learning is an extremely complicated phenomenon, and perhaps should not be thought of as a single type of neurological event at all. In an experiment, a cockroach was killed and its leg removed. The part that included only a ventral ganglion and a pair of legs with their muscles was arranged so that when the leg was extended, an electrical circuit was completed, thus shocking the animal. After this happened a few times, the leg remained flexed and showed signs of "remembering" for several hours. Controls receiving random shocks showed no such flexion (Pritchatt 1968). One wonders how this finding can be reconciled with the usual assumption that instrumental learning takes place in the central nervous system.

To further complicate matters, trial-and-error learning may be accomplished by a different sequence of neural events in different kinds of animals (Schneirla 1959). For example, insects learning a maze first learn the important turns individually and later integrate the information. Rats, on the other hand, learn and integrate the patterns at the same time. These sorts of findings should stand as a warning against any tendency to develop simplistic learning models and as encouragement to continue cross-species examinations, with an eye toward finding the common, as well as diverse, principles of learning.

In general, then, we may regard learning as having two major roles. One is in establishing individually-acquired adaptive patterns. The second is in "fine-tuning" phyletically acquired patterns by altering the IRM and/or the response to releasers.

## Rules of Associative Learning

There are certain general principles common to all associative learning (Type I, Type II, trial and error), as we have seen. However, there are also some notable exceptions to these principles that should discourage us from hastily drawing up hard-and-fast rules. Let's review a few of these general principles and some of the important exceptions.

We know that if the conditioned stimulus (such as a light) and an unconditioned stimulus (such as food) occur closely in time, learning occurs faster. In fact, it has been shown that it is most rapid when the conditioned stimulus begins before the unconditioned stimulus and then goes on to overlap it (Skinner 1938). On the other hand, learning is greatly impeded if the conditioned stimulus ends more than about a second before the unconditioned stimulus begins, or if the conditioned stimulus continues after the unconditioned stimulus has stopped (Figure 11.6).

Timing is also important in other conditioning patterns. Delays of only a few seconds between a rat pressing a bar and the delivery of food greatly impeded the rat's learning to press the bar (Skinner 1938). Such phenomena have presented thorny problems for animal trainers for movies or circuses. In many instances, the reward just couldn't be given fast enough. Because of this, a system of "secondary reinforcement" was developed. Secondary reinforcement occurs, for example, when a rat learns that lever pressing does not bring food unless the light is on, so it learns to turn the light on.

The wild cousins of those rats have given us something else to think about. Wild rats are hard to get rid of and it isn't considered chic to have them in your house. But they seem to be able to survive almost anywhere, and some are large enough to give pause to any marauding cat. Also, they are hard to

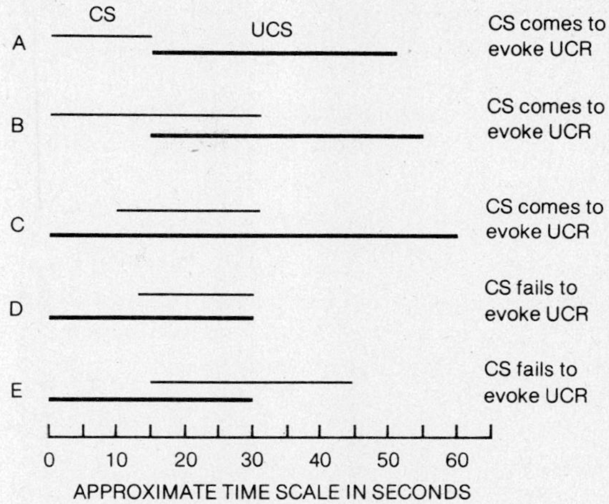


Figure 11.6 The importance of timing in a conditioned stimulus (CS) evoking an unconditioned response (UCR). Note that the CS must not end with or extend beyond the unconditioned stimulus (UCS).

poison. (An exception is the one that, when I was a graduate student, my landlord managed to poison. It took its revenge by crawling between the walls to die. After a few days, I almost moved out, and the people next door did. But this only shows that in any given population there may be a few rats that are not very bright. I'm not sure what it says of landlords or graduate students.) In any case, why are rats so hard to poison?

An answer was suggested when it was discovered that wild rats (who must subsist on a variety of foods) only *nibble* at any new food (Barnett 1963). Then, if nothing happens, they may eat a bit more the next time they encounter it until, finally, it becomes a normal part of their diet. Since they only nibble a little at first, if it is poisonous, they usually survive and then they avoid it from then on. The point is that some time must pass in order for the rat to feel ill after nibbling poisoned food. Almost all poisons take at least an hour to work. So, apparently, the wild rats learn even if the effects are delayed for an hour. Denis Mitchell and his colleagues (1975, 1976, 1978), however, have data indicating that the delayed-learning is vastly more complex than was believed, as will be mentioned later.

If rats have such an ability, it may be important to them because of their close association with humans. Over the years, in our waste, we have provided them with increasingly exotic and bizarre types of potential food. As dietary specialists tell us, much of our leftovers may be unpalatable or dangerous for consumption. Thus the learning mechanisms of rats may have adapted to living off suspicious human refuse. It seems less likely that such learning ability would be found in those animals that have traditionally encountered a relatively limited and stable range of wild food types.

It was also found, by the way, that rats deprived of the vitamin thiamine would select thiamine-rich food, even though the effect of eating the vitamin wouldn't have been felt until some time later. It is hard to relate this finding to any sort of learning mode. Perhaps it falls better under some appetitive-control system. But it does serve to point up that sometimes animals really do eat the things that are good for them, as is stated so frequently in popular animal lore.

Another characteristic of associative learning is that it improves with *repetition.* A rat learning a maze reduces its number of errors with each trial until it runs directly to the food box with no wrong turns at all. After this, of course, its performance cannot be improved upon. However, if it is continually permitted to run the maze after it consistently achieves perfect scores, it remembers the route longer.

If a conditioned response (performing an action for a reward) should begin to go unrewarded, the correct response will gradually disappear. This waning is called *extinction.* For some reason, classically conditioned responses are apparently much easier to extinguish than instrumentally conditioned responses. But the extinction may not be permanent. Even after a response has been completely extinguished in a series of trials, if the animal is allowed to rest for a few hours and is then given the conditioned stimulus again, the conditioned response may appear through what is called *spontaneous recovery.* Such recovery can be elicited several times before the extinction becomes permanent.

In the wild, spontaneous recovery may have some adaptive value. For example, suppose a certain behavior has traditionally yielded food or other reward predictably enough for the behavior to become entrenched in the animal's repertoire. Then, suppose an unusual, but temporary, condition should come about so that for a time the behavior does not yield a reward. After a time, the conditioned response would be extinguished. However, if environmental conditions should then return to normal, the animal would quickly begin to respond appropriately through spontaneous recovery; a new conditioning process would not be necessary. It seems that probably the adaptiveness of spontaneous recovery would be highest in animals living in rather changeable or variable habitats, or those that travelled through various types of habitats on a recurring basis.

## LATENT LEARNING

The term *latent learning* is, to some extent, unsatisfactory in that it refers to only one phase of what is a broad and rather complex learning process. For reasons that are discussed below, Thorpe (1963) preferred the term *exploratory learning* and defined it as the association of indifferent stimuli, or situations, those without patent reward.

It has been found that rats will learn a maze even when there is no obvious reward such as food or water. A monkey will perform work, such as pulling a lever, simply to be able to look out of the box in which it is kept. The motivation in latent learning situations seems simply to be a desire to get to know the surroundings. It is such an unspecific sort of motivation that, at first, animal psychologists were at a loss to explain why their animals were willing to do work when the only reward seemed to be the chance to look around. It was not believed to be simple curiosity because curiosity is considered a private domain of the more intelligent species, particularly mammals, and latent learning has been described in a wide range of animals, including insects. Also, while curiosity implies a very general condition, the exploratory drive might be considered to be the application of that condition in developing an adaptive pattern.

Let's consider an example. If rats are allowed to repeatedly run a maze without any sort of reward at the end, they will learn the maze anyway, but they seem to learn it very slowly. However, if food begins to appear at the end of the maze, these same rats will be able to run the maze correctly after only a very few trials. Apparently, they were learning all along, but it was latent learning and not evident until it was activated by a patent reward. So, why did they learn without a reward? Was learning the maze a kind of reward in itself? Also, it would be interesting to know if extinction occurred after the food was again withheld.

Critics contend there is no difference between latent and trial-and-error learning, doubting that reinforcement is necessary in order for trial-and-error learning to occur. But Hinde (1970) suggested that trial and error involves learning the characteristics of a situation *and* putting this knowledge into action, thus gaining a reward; latent learning, however, involves only the former. The reevaluations of Thorpe's (1963) divisions serve to point up the close interrelationships (and the arbitrary separations) of the various kinds of learning.

### The Adaptiveness of Latent Learning

Latent learning is undoubtedly very important in many kinds of wild animals. Many young mammals, in particular, undoubtedly benefit greatly from exploring the area around their protected den (Figure 11.7). At the time that they are learning about their home area by just poking around, they are being cared for by parents, so they need not learn with a reward in mind.

Figure 11.7 Young coyotes venturing from their den. As they explore their surroundings and gain confidence, they explore farther away, always gaining information about their surroundings. At any sign of danger, it is important that they be able to quickly find their den and scramble into its safety. They may learn the location of the den relative to other parts of their environment through latent learning.

The result of such exploration is that when the young animals start out on their own, their chances of survival and finding food are increased because they are operating in familiar surroundings.

Latent learning may also be important in adult animals. A mongoose, for example, may range several hundred yards from its burrow. It is important that it learns where food is likely to be found so that it will be exposed to predators for the shortest possible period of time. It must also know where to find holes into which it can dart at the approach of a predatory cat. In a laboratory experiment, two groups of mice were turned loose in a room with an owl. One group was given a few days to familiarize themselves with the room before the owl was introduced. The second group was put into the room at the same time the owl was. The mice that were familiar with the room fared much better—that is, fewer were caught (Metzgar 1967). So, the survival value of an exploratory drive may not be obvious at first, but its adaptiveness becomes apparent when a need arises and is quickly met. So, we see that after establishing the location of a commodity, such as food or a hole, the animal can reasonably expect it to be there when it returns. This principle of *expectancy* is important in the next type of learning to be considered.

## INSIGHT LEARNING

The lines separating the various types of learning are, of necessity, arbitrarily drawn because of (1) imperfect knowledge of how learning is really accomplished, (2) the natural intertwining and intergrading of the various types of learning, and (3) the complex interactions of learning and instinct that produce an adaptive response. As an example of the latter, consider that innate appetitive behavior may constitute the "trial" part of trial-and-error learning, and that any trials are innately restricted so that only certain ones are attempted. In an experienced animal, a trial may be nonrandom, to some degree, because of the principle of expectancy. In this case, the trial is based on an "if . . . then" hypothesis. In its highest form, insight learning may be a type of trial-and-error process, except that the trials are performed mentally. I will say more about this shortly.

### Expectancy

Reinforcement in trial-and-error learning is, in essence, a confirmation of an expectancy. At first, an animal may not know what a certain behavioral pattern will lead to, but trial-and-error learning may cause the animal to expect a certain result to follow a specific type of action. This is the basis of insight—a "what-leads-to-what" expectancy (Tolman 1937). In what ways might expectancy be adaptive? Expectancy might be a device that cuts down the "cost" of adjustment (Berlyne 1960). The system would work best when the events (action, result) appeared in a regularly occurring sequence so that the element of predictability could develop. Anticipation on the part of the animal arises when a reward comes to be associated with an action.

Expectancy is also adaptive in that it opens the way for a prior selection of certain responses to impending situations. Therefore, the likelihood of choosing the proper response is increased. Expectation, too, may reduce stress by making adjustments less abrupt. If an animal has an idea of what may be about to happen, the actual occurrence may be less traumatic so that it is able to act from a more normal psychological set than if it were completely surprised.

At times, complete environmental cues may be lacking or the cues may be incompletely learned. In either case, the result may be "anticipatory arousal." Such cues may indicate an impending situation without indicating exactly what it is. The animal may then brace for the event and react to it more successfully than if no warning had been given. Anticipatory arousal may explain the fabled sixth sense claimed by the more dramatic adventurers.

How common is expectancy among wild animals? We don't know. However, it is probably fairly prevalent among vertebrates and is present in at least some invertebrates. In intelligent animals, such as the primates, expectancy has been clearly demonstrated. In one experiment, monkeys were allowed to watch various foods being placed under inverted cups. Before they were allowed access to the cups, their favorite food, banana, was secretly replaced by less desirable, but perfectly edible, lettuce. When the monkeys were allowed access to the cups, they lifted the ones under which they had seen the bananas placed. Upon finding lettuce, they refused to eat it. They had obviously expected bananas and were not about to accept lettuce instead. They were more likely to eat lettuce when they were expecting it.

The principle of expectancy, then, is the basis for insight learning. But let us make a distinction. *Insight* is that part of the mental process that relates to understanding relationships; *insight learning* is demonstrated in the behavioral solution of the problem (Klopfer and Hailman 1967).

### Gestalt Perception

The concept of insight originated with the Gestalt psychologists. It has been called the "ah-ha erlebnis," a term that conveys the sudden understanding of relationships. Put simply, gestalt is the consideration of the whole rather than its parts. Perception, of course, may depend on how a stimulus is interpreted in the central nervous system, and gestalt perception allows the nervous system to "interpret" stimuli on the basis of past experience. A broken circle may be seen as an entire circle (Figure 11.8), and we tend to supply the missing lines in symbolic drawings (Figure 11.9). (Ah-ha! This is a horse.) The entertainment industry counts on our gestalt abilities. We have all seen children's books in which, when the pages were flipped rapidly, figures appeared to move in a continuous motion. And while we laugh at early films in which the frames pass at a low enough frequency for us to distinguish motion as disjointed, the 32-mm films currently in use in theaters also produce disjointed images. However, they pass at a high enough frequency that Hollywood can count on us not to notice. These motions are disjointed, not continuous, but we have learned through experience that motion is continuous, and therefore we supply the missing parts.

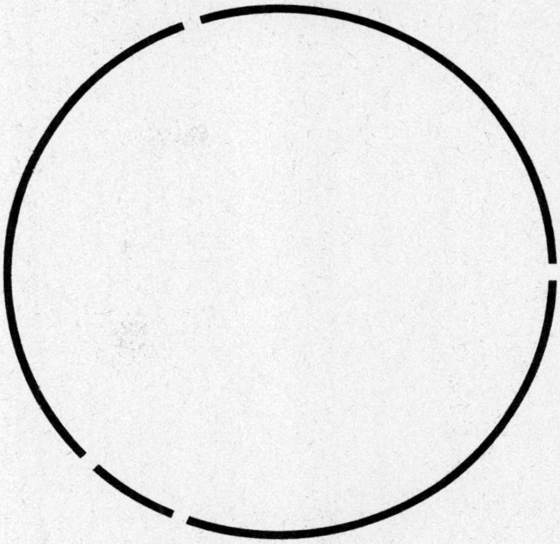

Figure 11.8 A glance will allow us to believe that this incomplete circle is an entire circle because our eyes "supply" the missing parts.

Gestalt psychology was born in Germany about the time of Pavlov and was based on the work of Wertheimer (1912). The Gestaltists believed that insight involves the perception of relationships and is a high order of mental ability. Trial and error, according to the theory, would only be employed as a last resort in problem solving.

Gestalt psychology assumes that psychological organization moves in a given direction: toward integrating what is learned and perceived. In other words, animals tend to organize their world into a good gestalt.

Not only do different species depend, to various degrees, on the perception of wholes (gestalten) or parts (sign stimuli), individuals may also vary in these respects at different times or in different contexts. Some birds, as we have seen, recognize a biologically important object by the perception of a small part of its total observable characteristics. A territorial robin will attack a tuft of red feathers placed at a certain height as readily as it would a stuffed male (Lack 1939). The sign stimulus is "red at a certain height." On the other hand, as Lorenz has pointed out, caged birds may show alarm when approached by their keeper who is wearing a hat for the first time. The "keeper" configuration has changed and he is not recognized. So, perception may be based on sign stimuli for some types of environmental objects and on gestalten for others.

There is some evidence (see Hailman's discussion in Klopfer's 1962 book) that the relative importance of gestalten and sign stimuli may change as an individual matures. Young gulls may, at first, respond only to specific sign stimuli such as the spots on the parents' bills, but as they gain experience and better vision, they begin to relate more to the entire parent configuration. It has been shown experimentally that

Figure 11.9 We see these lines as a horse, although no such configuration is actually shown. Again, our mind is willing to furnish the missing parts.

as young birds mature, they show greater response to models that are more normal in their proportions.

It has been suggested that habitat selection in birds is based primarily on gestalt perception (Lack 1933). In a complex habitat, probably no single clue or set of clues is a clear indicator of habitat conditions (Pielowski 1961; Klopfer and Hailman 1965). (But ethologist Gordon Orians is accumulating evidence that blackbirds may use very specific cues in locating suitable territories.)

Of course, any study of habitat preferences in animals must be preceded by an analysis of the animals' sensory abilities. We must keep in mind that man's perception of the world is uniquely his own. We do live side by side with animals who live in a different world. If a planarian were asked what the world is like, he might describe it in terms of vague sensations of heat and cold, changing chemicals, and light and dark accented by hazy shadows. An insect, able to perceive ultraviolet rays, might describe the beautiful colors in what to us is a white flower, and it might be taken by the shimmering and shifting mosaic of its environment. We know no world of ultraviolet colors and mosaic patterns. The perceived environment is the animal's *Umwelt,* and we must have some knowledge of it before we can effectively discuss what elements in it might be influencing the animal.

Insight, as described by the Gestaltists, is distinguishable from trial and error in that, using insight, not all the possible solutions to a problem are attempted. In one test where food was placed outside the arm's reach of a chimpanzee (true insight is best demonstrated in primates), the animal used short sticks, which were placed within its cage, to reach longer sticks placed outside, and then used the longer sticks to reach the food. If the food were suspended well above its reach, the chimpanzee sized up the situation and then stacked boxes (through considerable trial and error), so that a short stick would reach (Köhler 1927). It appears that a chimpanzee in such situations tries a number of possibilities mentally, until it hits upon a likely one, and then tries it physically.

For a long time, it was believed that insight learning is not possible below the primate level. This is probably wrong. For example, some birds can apparently learn this way. An example might be the wily crow, which has survived in the face of human efforts to eradicate him. We really haven't taken a good look at insight learning in various species, so who can say, at this point, where the ability ends.

## Learning Sets

Insight behavior has been defined as "the sudden production of a new adaptive response not arrived at by trial behavior," and "the solution of a problem by the sudden adaptive reorganization of experience" (Thorpe 1963). The key word here is *experience,* which is only implied in the first definition. In problem solving, any factor that reduces randomness and lowers the number of unfruitful trials is adaptive, and one such factor is experience. In fact, the experience need not be directly associated with the problem at hand. Sometimes, when presented with a new problem, an animal will solve the problem on the first attempt because of experience in dealing with the component parts of the problem, although it has never seen them put together in just such a way before. So, unlike trial-and-error behavior, insight learning is strongly dependent upon experience.

By using trial-and-error learning and insight learning, animals can develop more complex ways of solving problems (Harlow 1949). In an experiment, a monkey is shown two boxes of different shapes, one with food under it. He can lift either or both, but is only rewarded in one case. He is then presented with several more pairs of boxes with the same shapes as the first pair, and with the food always under a box of a certain shape. At the end of this set of experiments, a new set is begun in which he is shown another pair of boxes with different shapes from those used in the first set. In several more trials, the food is again always under a box of a particular shape. After several such experiences, the monkey begins to make a steady improvement in the percent of correct choices in the successive trials until, finally, once it learns which box the food is under, it chooses the right box every time. In effect, the monkey has learned to learn. This *learning set,* as Harlow called it, is very important to any consideration of learning and experience in the wild because it describes how experience, during the lifetime of an animal, facilitates its ability to solve new problems; and, in nature, few environmental situations will be similar in all respects.

## Imitation

Among some animals, learning need not be directly related to personal experience. Thorpe (1963) referred to insight learning acquired through observing the behavior of other animals and called the process *imitation.* He warned that this is a rather nebulous classification that may encompass a variety of learning phenomena. Among these is *social facilitation,* which is the inclusion of an established behavioral pattern into a new situation. In this case, the pattern is already in the behavioral repertoire of the learning animal; only the timing of its employment is altered as a result of watching another animal. A dog may learn to "speak" after watching another dog being rewarded for simply barking on command. The observing dog could bark to begin with; the learning just altered the probability of when the bark would occur. The phenomenon is evidenced in human behavior in the contagion of yawns (try it).

Behavior may also be altered by *true imitation,* which is the copying of an act otherwise unlikely to appear in the behavioral repertoire at all. Thorpe believed this sort of behavior is limited to birdsong and some primate behavior.

Another type of imitation involves *tradition learning.* Here, there is a tendency to alter the usual response to parts of the environment after watching another individual. Milk bottles left on porches had been ignored for years by English birds, including populations of tits (*Parus* spp.). Then some members of the population learned to open the bottles to get to the cream after watching other birds who had managed to do so (Hinde and Fisher 1951). Of course, somewhere along the line some ingenious bird must have employed a different learning process in order to remove the first lid. As another example of tradition learning, gregarious birds may learn to avoid predators by observing the flock's behavior toward certain potentially dangerous species. (I have mentioned how a bird may learn the clues of impending danger by watching actual attacks on other birds.)

Species may differ in how well they are able to learn through tradition. For example, Lorenz has described how jackdaws pass down information from one generation to the next. The birds travel in flocks, and it is likely that at least one bird will be able to recognize just about any form of danger. When it gives the alarm and the flock takes wing, there is probably an increased likelihood that others in the flock will show an aversion to that same stimulus the next time it is encountered. On the other hand, birds that live alone or in pairs, such as thrushes, must rely to a greater degree on an inherent system of recognition and they are, therefore, probably less likely to be able to learn through tradition. It should also be pointed out that there are important differences in learning abilities of species of various adaptive types. (*Adaptive type* is a general term that loosely describes the behavioral spectrum of the animal and includes foraging patterns, social behavior, habitat preference, learning abilities, and so forth.) Under experimental conditions, the inquisitive and exploratory tits showed no adverse effect when they are trained in pairs, but in the greenfinch (*Chloris chloris*), single birds learn food-discrimination tasks much more rapidly than members of pairs. Peter Klopfer (1961) maintained that such predilections as that of the greenfinch can only be adaptive in species with relatively conservative food habits or those of a solitary nature.

Probably the best known instance of tradition learning involves the Japanese macaques (*Macaca fusca*) on Koshima Island. In 1952, scientists began feeding them sweet potatoes. A year later, a young female named Imo was observed washing the sand off her potatoes by holding them in one hand and rinsing with the other. The washing pattern spread throughout the community over the next few years—first through the closer families, then through playmates. By 1962, three-fourths of all monkeys who were at least two years old were potato washers (Figure 11.10). Some had even specialized to the point of salting their potato by dipping it in seawater before each bite.

These same monkeys were also fed wheat, which they had to sort laboriously from the beach sand where it was spread. Then the enterprising Imo discovered that by throwing handfuls of both sand and wheat into the water, the lighter wheat would float and could be quickly separated out. This technique spread through the group in much the same way as washing potatoes had (Itani 1958; Kawai 1965).

Such learning can also operate in the other direction. For example, the presence of naive animals has been shown to reduce the effects of learning in trained animals. A greenfinch, trained to avoid unpalatable food, will begin to make mistakes if it is allowed to watch an untrained companion.

As I emphasized earlier, the various types of learning are not as clear-cut as is suggested by the classifications. Thorpe (1963) noted that complexities can arise because the categories may broadly intergrade and because each category contains examples that may differ greatly in their complexity. There are other complicating factors as well. For example, each category may involve an element of *perceptual learning*. The importance of perceptual learning is widely debated, but basically it involves information that has been inconspicuously gathered during the development of the animal. A certain experience may produce no effects at the time, but may affect subsequent learning. A rat, given early nonreinforced experience with a triangle and a circle, will later be able to discriminate between geometric figures more easily than a rat with no such experience. An animal (or a baby) given an enriched environment full of playthings with various properties may learn things associated with such properties more easily later in life. If perceptual learning influences other kinds of learning, we must obviously take our categorizing with a grain of salt.

Figure 11.10 Long-term studies of Japanese macaques have provided fascinating evidence of culturally transmitted behavior. These macaques learned to wash sweet potatoes by watching a highly intelligent young female in their group.

## PLAY

Play has been observed in many species of animals and it probably has been subjected, more than any other type of behavior, to the greatest anthropomorphic interpretations. Part of the problem in interpretation lies in the fact that since we do it, we think we must understand it, and because play is associated with the more intelligent species—those with which we feel we have more in common. As a matter of fact, we really don't know much about play.

Part of the problem of investigating play arises because it is hard to define, in spite of the fact that we all believe we can recognize it when we see it. Mason (1965), in describing play in primates, states, "In spite of the difficulties in arriving at an acceptable and comprehensive definition of play, the fact remains that observers show considerable agreement in judging 'playful behavior.'" However, in the less intelligent of the playful animals, it may very well appear in forms that are not recognizable to the human observer. For the present, we can only discuss some of the results of studies of play in a few groups of animals and, essentially, we are forced to view these subjectively and with human behavior as the basis.

### Who Plays?

Playlike behavior has been described in fish, but a bit of imagination is required. Play, for example, is suggested in low intensity nest-building actions in some fish (Van Iersel 1953). "Shooters" (*Jaculator*) may squirt water at their aquarium keeper (Thorpe 1963). But this could be because they associate him with food and they tend to shoot at food. The argument for fish play is not completely convincing, but enough evidence has been presented to warrant further investigation (Beach 1945).

It was believed for a time that birds do not play, but in recent years, people have reported play, or something like play, in some birds. Eider ducks, for example, have been seen "shooting the rapids" in Iceland (Roberts 1934). They walked up the bank of a stream, jumped in, careened through the rapids, got out, walked back up, and jumped in again. An Anna's hummingbird repeatedly floated down a small stream of water that was flowing from a hose (Stoner 1947). Also, it is said that some passerines (perching birds) may sing out of sheer exuberance (Hartshorne 1958), and some have interpreted this as a form of play.

Lorenz described playful behavior in his birds, but probably play is not an important learning technique in most bird species. It would be extremely difficult to imagine anything as starkly humorless and unplayful as a chicken! There may be an explanation for the unplayfulness of precocial birds such as chickens. Precocial species must run about and forage successfully for themselves a few hours after hatching and have no time or energy for behavior that is not immediately productive. On the other hand, playful behavior might be expected in altricial species, such as crows, which are cared for by parents as they face their first weeks of life. Also, crows are omnivorous (eating a wide variety of plant and animal food) and are relatively wide-ranging, and, therefore, the kind of behavioral flexibility that permits play may be selected for. Play might give animals that have to be adaptable the opportunity to become familiar with their surroundings and to develop the ability to deal with various conditions by experimenting with their environment in ways that seem to have no immediate adaptive value.

There is good reason to believe that play does not exist in animals below the vertebrates. It can only be expected, to any great degree, in those species for which learning or socialization is important and in those which are behaviorally flexible. Also, play should only be expected in those species in which appetitive behavior is not always strongly bound to a consummatory act. In the behavioral patterns of many animals, appetitive behavior must lead to a consummatory act. In more intelligent animals, such as birds and particularly mammals, however, appetitive behavior is flexible enough to be broken off without leading to the consummatory act.

One of the characteristics of play is that it is not necessarily relevant to any environmental situation existing at the time it is performed. It can appear without environmental cues and without being preceded by the behavior's normal appetitive behavior. A young kitten may suddenly arch its back, spit, and appear terrified of something that is familiar and presents no threat at all. Then, just as suddenly, it abandons its defensive posture and rolls in mock predatory behavior. Then it stops to delicately sniff a mysterious spot, with no trace of its former behavior. Then it . . . .

### The Adaptiveness of Play

Since the goal of play is not the performance of an adaptive consummatory act, one might expect it in domestic animals whose needs are provided for by man. Adult domestic dogs, for example, have been labelled "neotenous" (physically developed but at a permanently immature stage of social development) because of their continuing propensity for play. As a matter of fact, it has been demonstrated that play in a number of animals occurs most frequently when physiological needs are met (Baldwin and Baldwin 1972).

In ordinary appetitive behavior, the animal learns how to overcome obstacles. Thus, the consummatory act becomes more easily reached as time goes by. But if the consummatory "goal" (such as food) is already provided for an animal (for example, by a parent), the appetitive search may become an end in itself. When this happens, the learning that takes place may not be primarily adaptive in helping to reach the consummatory situation, but instead may be more useful in encouraging the exploration of the environment. So, the adaptiveness of play may not be in helping the animal to learn to reach a goal, but in helping to expand his horizons, in giving him a better overall understanding of his world. This understanding, of course, is important to another behavioral process, namely, insight.

One of the most persistent and obvious interpretations of play behavior is that it provides practice. Puppies play-fighting and kittens stalking a ball of yarn certainly support the notion. Puppies and kittens in play-fighting often grab each other by the throat in the same manner they will later use in making quick kills of their prey (Figure 11.11). Meerkats (*Suricata*) employ the same sort of bite when playing, copulating, or killing prey (Eibl-Eibesfeldt 1970). These young animals are obviously learning motor patterns that will be important later. However, what about those patterns that appear *only* in play? Why would a bird practice shooting the rapids? There are many accounts of otters racing uphill just to be able to slide down again. To include such behavior, the practice theory would have to be modified to include patterns that improve coordination or some other general faculty.

**Play in Primates**

Harry Harlow (1963) described five stages in the development of play behavior in laboratory rhesus monkeys: (stage one) presocial, when the infant may

Figure 11.11 The cheetah has a tight hold on the throat of a male reedbuck. Many predators learn their deadly neckbites by practicing in play.

manipulate any object in its environment; (stage two) rough-and-tumble wrestling with other infants; (stage three) chasing; (stage four) rough-and-tumble play with chasing; (stage five) aggressive play, which begins at about the age of one year. Aggressive play is rarely harmful in itself, but seems to set the stage for real aggression later on.

Play, in addition to affording practice, has a socializing function in some species (Poirier and Smith 1974). The play group, in primates, may provide a situation in which youthful experiments may be tried and mistakes go largely unpunished. Since small, playful monkeys usually don't have big teeth, their mistakes usually present no threat to others. While playing, young primates can also learn about each other's temperaments and physical abilities, thus setting the stage for formation of dominance hierarchies with a minimum of violence when the monkeys do become able to inflict damage.

Harlow's primate studies pointed up interesting sex and age differences in primate play. Gender differences appear in the first year in rapidly developing primate species, but somewhat later in species that mature slowly. These sexual differences are usually evident in the percentage of time spent in each type of play. For example, juvenile male baboons, both in the laboratory and field, play more roughly and more frequently than do females. Females groom more frequently and spend more of their time with newborn infants. Such sexual differences are not restricted to baboons. Among most nonhuman primates, young males show a definite preference for rough-and-tumble play, an activity not particularly popular with females. Also, young males tend to range farther from the mother than do females. When older juvenile and young subadult males are off playing somewhere, females of the same age are usually associating with adult females, especially those with newborn young. Juvenile female vervet monkeys seem to develop their social relationships in mutual grooming sessions with other females, while holding and exchanging infants. It has also been shown that the young females "pretend" motherly behavior with infants, a type of play generally lacking in males.

Among baboons (Figure 11.12), the play groups are mostly composed of males, since females leave such groups when they are one to three years old, depending on the species. Later, among hamadryas baboons, the females will stay together as males move about freely and independently, and thereby may form the stable nucleus of the troop.

In some wild baboons, there are sexual differences in play by the age of two or three months, and the differences increase with age. Males spend most of their time with males of their own age; females tend to associate with the mother subgroup. Also, male juveniles tend to interact with more individuals than do females, who mainly associate with only one other female. Interestingly, a similar pattern has been found in human children (Knudson 1971; see Poirier and Smith 1974 for details and extensive references).

In many primates and other animals, the incidence of play decreases with age, but the trend is especially evident among male primates. Reduction in play may reduce the risk of misinterpreted signals among increasingly dangerous peers. Adults can be expected to play less for another reason. Since young animals are the most explorative, curious, and innovative, they are more likely to "test" the environment than the adults. Young Japanese macaques, for example, tend to be much more adaptable to new situations than older monkeys (Itani 1958). Hans Kummer (1971), a field ethologist working in Africa, made the rather surprising suggestion that the conservatism of adults may serve as a safety reservoir of behavioral patterns that are known to work, while a species may ensure its survival by allowing its younger members to experiment. The result would be that new behaviors, foods, or routes may be discovered, and if the experimentation should prove risky, the group has less invested in the young—in other words, they are more expendable. Kummer also suggested that the reason young males play more than young females is that, for reproductive purposes, males are more expendable (only a few males can keep all the females pregnant). Such suggestions invoke spectres of "group-selection," a process that most biologists believe is nonexistent or unimportant among animals.

## IMPRINTING

In 1910, Oscar Heinroth described a surprising behavioral pattern in goslings. He found that they tend to follow a large moving object soon after

Figure 11.12 Two 3-month-old baboons at play. Note the open mouth and wide-open eyes of a play face.

hatching and to relate in specific ways to that kind of object later in life. In the wild, of course, the first big moving thing they would see would probably be a parent, so the adaptiveness of the pattern is obvious. Not only do they tend to follow their parents when young, they later will tend to mate with animals that resemble their parents. Lorenz (1937b) called the process *imprinting*. It has been described in many types of animals from insects to mammals. Imprinting, as we shall see, may have a wide range of advantages in various species. For example, young insects do not follow their parents, but they may imprint on a certain kind of edible plant and direct their foraging attentions to it in later life.

Imprinting has been described as a rapid learning of certain general characteristics of a stimulus object. It apparently occurs at a very specific time early in the animal's life. And, evidently, it occurs without reinforcement. This, and the fact that the adaptiveness of the imprinting may not be apparent until long after the original exposure, give it certain similarities to latent learning.

Imprinting is probably most easily evidenced in the young of precocial birds (you will recall that these are species, such as ducks and chickens, in which the young can run around soon after hatching). These species form an attachment early in life with a mother figure, which, as luck would have it, is usually their mother. (But it might be something else. When I was a graduate student, one of our laboratory chickens we named Spot seemed to have imprinted on people in white lab coats and would follow anyone wearing one, occasionally walking over with whomever was going to check the mail.)

Most of the work on imprinting has been done with birds because they are convenient research animals. They will imprint on practically anything—balloons, toothpaste boxes, or the "blind" in which the experimenter is hiding so as not to influence the behavior of his charges. Color and shape do not seem to matter, but the imprinting takes place faster if the object contrasts with the background and if it makes some sort of noise. In fact, young wood ducks actually imprint on sound (Klopfer 1959). In an experimental situation, a young wood duck will approach any sound, but later it will follow only familiar sounds. It turns out that wood ducks nest in dark holes, so that during the imprinting period they may never even see their mother. Also, their mother looks very different from their father so there could be mating problems later on if both the young male and female wood ducks were to imprint on the female image. The problem is partially circumvented in species in which the sexes have different appearances by only the male ducklings imprinting sexually on the mother (Schutz 1965). In these species, ducklings exposed to adult males during their imprinting period will later form homosexual bonds, even in the presence of females.

Lorenz (1952) once found that one of his male jackdaws (a crowlike bird) had become sexually imprinted on him. In fact, the bird insisted upon "courtship feeding" the professor. He described his dilemma: "Remarkably enough, he recognized the human mouth in an anatomically correct way as the orifice of ingestion and he was overjoyed if I opened my lips to him, uttering at the same time an adequate begging note. This must be considered as an act of self-sacrifice on my part, since even I cannot pretend to like the taste of finely minced worm, generously mixed with jackdaw saliva." But he found that if he *didn't* open his mouth he would suddenly find himself with an earful of worm pulp!

Imprinting, unlike other forms of learning, has been described as an irreversible process. Birds that have been sexually imprinted on humans will court other birds normally if no humans are around, but will break all bonds and begin courting humans if one should appear (G. Hellman, cited in Eibl-Eibesfeldt 1970). It has been suggested, however, that since the birds will sometimes accept a sexual partner other than the one they imprinted on, perhaps the process is not so irreversible after all. Actually, though, reversibility is not suggested so much as flexibility. The preferred object has not changed.

So, there are several phenomena associated with imprinting, each of which has received some criticism from researchers. (1) The animal forms a social attachment to another animal or object. (2) Sexual attachments are formed, although they may not be apparent for several months. (3) There is a critical period when imprinting is most likely to occur. Regarding this last point, the period may vary from one individual to another, even within the same species. It may be quite brief, but it is always very early in the individual's development. The exact length of the period depends on what one wishes to use as an index. The period is relatively long if one simply includes the tendency to follow a model on the first exposure. But if one measures how well mallard ducklings discriminate and stay close to a model five to seventy hours after a single thirty-minute exposure to it, the period turns out to be much shorter (Figure 11.13).

Also, Eckhard Hess (1964), a pioneer in the study of imprinting, reported that the more difficulty a duckling has in following a model, the stronger it will imprint on that model (Figure 11.14), a notion that has provoked some argument. Hess, by the way, is a strong proponent of the notion that imprinting occurs through the mechanisms of learning, although many other researchers disagree, stressing that innate mechanisms other than learning must be involved. They believe that the imprinting phenomenon has too many special qualities for it to be considered just a special form of learning.

For discussions and criticisms of the imprinting paradigm, see Klopfer (1962); Klopfer and Hailman (1967); Bateson (1971); and Hess (1973).

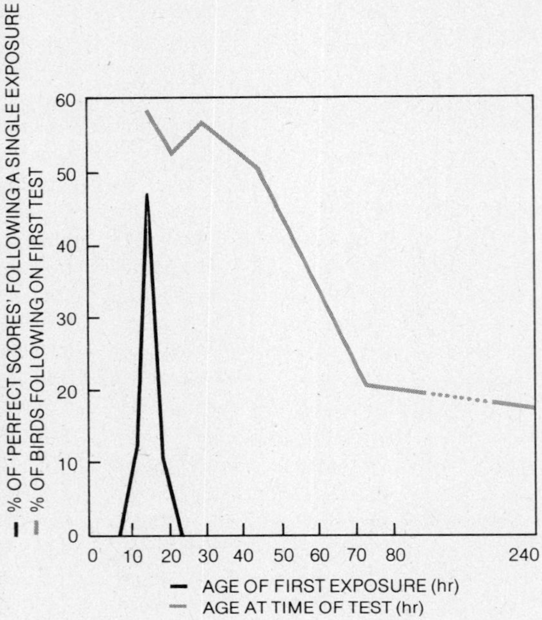


Figure 11.13 One of the problems in studying imprinting lies in investigative technique. For example, two different curves were generated when ducklings (left) were tested at the ages shown after a single training exposure, and when birds (right) were simply tested to see if they would follow a moving object on their first exposure to it.

## AUTONOMIC LEARNING

There is one last type of learning that should be mentioned—one that is just beginning to be explored, but that may have enormous implications. It is called *autonomic learning.*

It has long been agreed that visceral responses (those under the control of the autonomic nervous system) could be classically conditioned, as Pavlov clearly demonstrated. But even Pavlov did not suspect that the dog could be instrumentally conditioned to salivate *in order to receive* food. Most people believed that classical and instrumental conditioning operate by different neurophysiological mechanisms and that visceral responses could only be brought about through classical conditioning. Of course, there were those learning theorists, all along, who argued that the mechanism for learning is the same for both types, and that if an arm could consciously be trained to bend, an iris could consciously be trained to contract. They are now being taken more seriously.

L. V. DiCara (1970) described a set of experiments that he performed with N. E. Miller of Rockefeller University in which they were able to produce, through instrumental conditioning, increases and decreases in heart rate, blood pressure, intestinal contractions, control of blood vessel diameter, and even rate of urine formation. Basically, their experiments involved using rats that were paralyzed with a curare derivative (the arrow poison of certain South American Indians). The poisoned animals remained fully conscious, but were unable to voluntarily move their skeletal muscles. Positive reinforcement was accomplished by stimulating an electrode implanted in the "pleasure center" of the brain, and negative reinforcement, by a mildly unpleasant electric shock. The minor physiological variations that occur naturally throughout the body were the starting points for the experiments. For example, if they wished to train the animal to slow its heart rate, as the heart temporarily slowed naturally, they stimulated the pleasure center; whenever the heart speeded up, they administered a shock. Soon the heart rate remained slow. The reverse effect could be achieved also. When a rapid heart rate was continually rewarded, the heart rate increased.

Some recent experiments by the Canadians have shown that even brain-wave patterns can be altered by training. For example, rats were trained to raise or lower the voltage of their brain waves, and dogs were trained to alter the activity of one kind of brain wave, the theta wave. (Yogis, in practicing certain types of meditation, usually increase the number of certain brain waves, particularly alpha waves. In this country, novices in "biofeedback" programs are sometimes tested with simple electroencephalographs to determine whether they are properly controlling their thought patterns. And in recent years, the conscious control of alpha waves became another American fad as people sought to escape the stress of the rat race.)

To rule out the possibility that such learning was really due to some sort of general changes in the autonomic nervous system, a variety of tests were performed. For example, when heart rate and intestinal contraction were trained separately, it was found that changes could be brought about in one without altering the other. Also, it was found that visceral changes can be brought about in specific areas without affecting other parts of the body. To amaze your friends, you can even learn to dilate the blood vessels of one ear more than those of the other ear.

The question arises, Is this sort of learning ability useful? Does it have a role in helping the animal adjust to the rigors of the real world? DiCara suggested that just as skeletal muscles respond according to what happens in the *external* environment, so, possibly, the autonomic system responds to the *internal* environment, fine-tuning the animal to its world.

Another possible adaptiveness of visceral conditioning is in maintaining homeostasis (the steady state, or constant internal condition). In the event of injury or disease to normal homeostatic mechanisms, the body could learn to compensate for these functions.

In the wild, the implications of autonomic learning are enormous. Perhaps such learning plays a far more important role in animal ecology than has yet been imagined. For example, perhaps gulls swimming in frigid North Atlantic waters learn to respond to unusually cold temperatures by constricting the vessels in their feet, thus reducing the amount of fluid in that area and rendering the foot less susceptible to freezing. Perhaps an arctic fox can learn to withdraw blood from his ears on the same basis. Such learning, of course, would function only as part of an overall adaptive pattern that involves morphological specialization as well. For example, the gull has circulatory "shunts" that permit blood to bypass extremities under certain conditions, and the ear of the arctic fox is small with a minimal surface area exposed to the elements. I need only point out here the implications of autonomic learning in habitat selection. The ability would permit an animal to exist in highly variable or changing habitats by increasing its flexibility by expanding its physiological tolerance.

What are the implications of such learning for humans? Aside from the benefits extolled by the practitioners of Yoga, and the bizarre feats accomplished by certain cultists (some of these phenomena may be based on autonomic conditioning), there are certain very practical therapeutic benefits of the process. Already, researchers have shown some promising results in helping heart patients to control their heart rate, and in training epileptic patients to

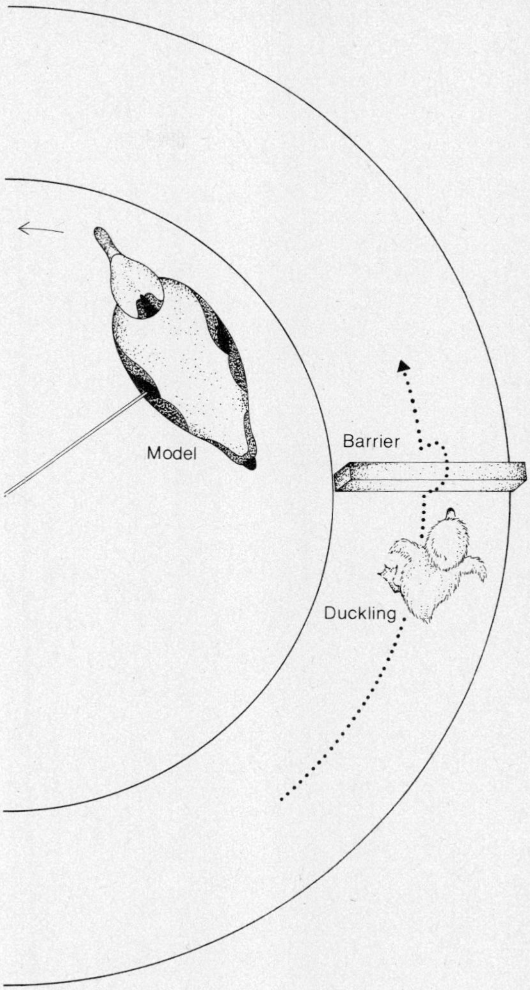


Figure 11.14 In Hess' experimental setup, a duckling is forced to cross barriers in order to follow a model. He believed such difficulties produce a stronger imprinting.

control a type of abnormal brain wave, the paroxysmal spike.

On the other hand, visceral learning may be at the root of certain psychosomatic illness. If prolonged tension, for example, brings about headaches, the appearance of the discomfort may be rewarded by rest, a round of ale, or even (for some) a gratifying feeling of martyrdom. The appearance of the headache is therefore reinforced until finally certain internal changes may be brought about (such as the constriction of certain blood vessels) in order to cause a headache that will be rewarded. For further information on autonomic learning, you might refer to Miller (1961, 1969); Miller and Carmona (1967); Trowill (1967); and Miller and DiCara (1967).

It should be apparent, now, that the separation of the different kinds of learning into tidy categories is not merited. In the real world, animals may be confronted with a highly variable environment, and their challenges certainly cannot be met by one or even a few kinds of learning. We may well find that what we now call a *kind* of learning is not a discrete function after all, but instead a part of an overall adaptive response that is comprised of different but intergrading "kinds of learning," which can shift in importance according to the immediate demands. In any case, we should be wary of assigning any behavioral pattern to a single slot, labelling it, and considering the task finished. It seems, in fact, that our task has just begun.

# 12

# THE EVOLUTION OF LEARNING

It is often easy to forget that while instincts have been molded and forged over the eons, so have learning abilities. Not only have some lines become more intelligent, some have become more specialized in that whatever they are able to learn, they are able to learn it well. It must be remembered, however, that acumen in one area does not indicate intelligence across the board. Animals are best able to learn those things that are important to them.

## RESTRICTIONS ON LEARNING

Perhaps you recall seeing on some late-night, TV talk show demonstrations "proving" that pigs are smarter than horses. However, horse lovers, take heart. Pigs are not necessarily smarter at everything, and besides, the stirrups drag.

The often-asked questions regarding relative animal intelligence are, for the most part, nonsensical; and the concept of an "animal I.Q." is unrealistic. Probably no test will ever be devised that would measure intelligence without somehow being biased by not accounting for species differences in innate tendencies or physical abilities. A cat is not dumber than a bird because it can't learn to chirp. Nor is a cat smarter than a dog because it sneaks up on its prey, since cats tire quickly and their hunting by stealth is an innate pattern, while longer-winded dogs have traditionally run down their quarry.

It is interesting that an animal can learn more about some things than others. For example, squirrels can learn very little when hiding food, but they can learn a great deal about how to open nuts (Eibl-Eibesfeldt 1970). (See also the discussion of song development in the white-crowned sparrow in Chapter 18.)

Another example of action-dependent learning occurs in the female digger wasp, which carries caterpillars to a burrow for her larvae to feed on. She first drops the caterpillar near the opening, goes inside, inspects the burrow, then reaches out and drags the caterpillar in. If the caterpillar is moved away some distance from the burrow while she is inside, she will come back out, locate the caterpillar, and repeat the sequence. She will not vary her pattern for thirty or forty trials, but then she will finally enter the nest *with* her prey and stop the nonsense (Baerends 1941). Perhaps she has learned that it won't do to leave the prey unattended. However, her very strong tendency is to first enter the hole alone.

In other situations, however, this same wasp shows remarkable learning abilities. For example, she can care for as many as fifteen nests at a time. Before she begins hunting for the day, to replenish the nests, she checks all the nests on a morning inspection tour. If caterpillars have been experimentally removed before she gets there, she will add more that day than she normally would. If caterpillars have been added, she will bring fewer. She remembers the condition of each nest as it was *during* her morning tour; any changes made after she has begun her work do not affect her behavior. She is only influenced by the conditions of the nests on her first visit of the day.

Other animals depend on learning in more ways. The more intelligent species usually have extended parental care so that there is time to learn specific behavioral patterns and proper responses before having to meet critical situations alone. In some cases, it seems that the parents actively teach their offspring. Female meerkats (*Suricata suricata,* a South African viverrid related to the civet and mongoose) present food to their young, and the young thereby develop food preferences. The behavior, here, has two advantages—it teaches the young and it feeds them. In other cases, teaching behavior may be more specific. In one instance, a mother gorilla was observed removing an inedible hagenia leaf from the mouth of her offspring (Schaller 1963). Of course, to argue that she was teaching the young gorilla, one would need to determine that she caused the infant to avoid those leaves as a result of her behavior.

"Learning animals" usually occupy a heterogeneous habitat or "niche." Some researchers believe that chimpanzees are more intelligent than gorillas because they are more threatened in their habitat. Gorillas have few enemies other than guess who. They also live surrounded by food in the deepest parts of the African jungles. Chimpanzees, on the other hand, had to confront the big cats (until cat skins recently became fashionable for coats) and were, in general, more vulnerable than the powerful gorillas. Also, their habitat is "marginal," or highly variable, thus they must be rather wily beasts in order to survive.

Species differences in learning abilities can sometimes be revealed only by careful experimentation. For example, the sunfish (Centrarchidae), which live only in the northern hemisphere, can orient by the sun upon their first exposure to it. provided the sun moves from left to right (as it does when viewed from north of the equator). But they are unable to orient to a sun moving in the other direction, as it would if they were in the southern hemisphere (Braemer and Schwassmann 1960). On the other hand, the cichlid (*Aequidens portalegrensis*), whose range includes both hemispheres, can orient by either a northern or a southern sun depending on the direction the sun was moving when it was *first* seen (Lorenz 1965).

## THE FALLACY OF INTELLECTUAL CONTINUITY

M. E. Bitterman (1965, 1975) at Bryn Mawr described experiments that showed inherent learning differences in various species of animals (monkeys, rats, pigeons, turtles, and fish) and caused us to reassess some of our earlier assumptions. It had often been suggested that as we go "up" the phylogenetic tree, there is a continuous increase in learning ability. For example, Thorndike, writing at Harvard in the early 1900s, assumed no intellectual uniqueness anywhere in the evolutionary hierarchy and said that species differed in intelligence only in degree. This theory of intellectual continuity has been based largely on the results of curves called learning functions. To get these results, experimental psychologists give an animal a choice between two alternative courses of action, one consistently rewarded, the other never. After a number of trials, the animal tends to choose the rewarded alternative. The trials are then plotted against the errors. The results for monkeys and fish, for example, are similar. They show a relatively steady increase in the number of correct responses in every set of trials. The main difference is that monkeys learn faster than fish.

The theory of intellectual continuity, stated simply, says that an animal is born with certain tendencies to react in certain ways to certain stimuli. These tendencies are functions of inherited neural connections between sensory and motor systems. Learning ability, according to this scheme, is simply the ability to make or break these connections as environmental circumstances dictate. The assumption in the first half of this century was that more intelligent animals can simply form more of these connections because of a more complicated central nervous system. This idea means, of course, that the evolution of intelligence just involves refining the old learning equipment—that differences are quantitative, not qualitative. The theory itself discouraged tests on different animals and resulted in our choosing convenient animals, such as the rat.

Bitterman, however, was able to illustrate important qualitative differences in learning ability between several species. For example, he jarred the old continuity argument with his "habit reversal" experiments. In these kinds of experiments, once an animal learns to discriminate between two alternatives for a reward, the reward is transferred to the opposite choice. As these reversals continue, learning function is determined by how long it takes to make the switch. Monkeys and rats show a steady improvement in performance; fish do not. In fish, later reversals are accomplished no more easily than earlier ones. Turtles show an intermediate ability by progressively improving in spatial (for example, high versus low, choices), but not in visual (for example, red versus green, circles versus square) problems. Therefore, habit reversal requires an intellectual ability that is present in higher animals, but is not present in fish, and is manifested in the turtle in only a restricted *class* of problems.

Bitterman also found that different species show differences in probability-learning experiments. In these experiments, the rewarded alternative is changed *within* a given trial session (not from session to session). For example, a given choice would be rewarded randomly seventy percent of the time, the alternate choice thirty percent of the time. In these experiments, both spatial and visual clues were used. Rats and monkeys tended to "maximize" by choosing the most frequently rewarded alternative in every case and, therefore, collecting seventy percent of the time. Fish, on the other hand, "match." This means that if an alternative is rewarded seventy percent of the time, they will choose it seventy percent of the time. In some cases, the rats and monkeys did not maximize, but matched, although in a different pattern from fish. Fish matched randomly, but rats and monkeys, when they matched, tended to choose the alternative that produced the reward in the preceding trial. Or in some cases, they chose, rather consistently, the opposite ones. Both mammals and fish solved both spatial and visual probability-learning tests in their particular manner. The turtle, on the other hand, matched in visual problems and maximized in spatial problems. Therefore, in both habit reversal and probability-learning, the turtle was ratlike in solving spatial problems, but fishlike in solving visual problems. Bitterman concluded that the modes of adjustment evolved by the higher animals appear earlier in spatial than in visual contexts. (Of course, one cannot—must not—speak of "the" turtle or "the" monkey as representative of all species of turtles or monkeys, since species and individuals within these groups are likely to behave differently, and subpopulations may differ as well.)

The ability to maximize, it should be mentioned, produces a higher percentage of correct choices than does matching. For example, in an experiment where the reward ratio is seventy to thirty, the probability of success is seventy percent if the subject is maximizing, but only fifty-eight percent $(.70 \times .70) + (.30 \times .30)$ if the animal is matching.

In some cases, there may be sexual differences in what constitutes a learning situation. An estrous female, for example, is reinforcing to male rats; the males will learn a maze to be able to reach a receptive female. Females, on the other hand, respond equally to sexually active and passive males. It has been concluded, therefore, that mating does not act as a reward in the female rat. Sexual differences in behavior may also be a function of anatomy and unrelated to reproduction. The female Hispaniolan woodpecker (*Centurus striatus*) has a shorter, thinner bill than the male (Figure 12.1), and she gleans food primarily from the surface of trees, whereas the male more often uses his strong bill to excavate burrowed larvae (Wallace 1974). Whether behavioral differences arise as a result of learning to employ different anatomical specializations, or whether the morphologies are associated with innate behavioral differences that allow the senses to use different equipment, is not known.

Other learning constraints may be a function of maturation. For example, young chaffinches will hop over the ground when they are about two weeks old,

Figure 12.1 Males and females of the Hispaniolan woodpecker forage differently, but it is not known whether they are simply utilizing different bill sizes or whether their nervous systems are also different. Would a male with a smaller bill forage like a female?

occasionally pecking at it. But they will starve to death in the midst of food unless they are fed by an adult bird. It is not possible to train them because if food is gradually lowered to the ground, at some point they will stop looking at it. If it is held in front of them, they do not take it, but only gape, waiting to be fed (Sevenster, reported by Hinde 1970).

Since there are limits to what certain kinds of animals can learn, it may be assumed that there are also differences in what members of various species teach their offspring. The differences may be great, even within a single order of animals, such as birds. There is some evidence, for example, that fish hawks (ospreys) teach their young to fish. Among woodpeckers, the young learn to forage on their own, but the parents continue to feed them while they learn the technique. In some pelicans, on the other hand, the parents abandon the nest before their offspring leave, so the young aren't taught anything about foraging and must perform the complex art of fishing on their own.

For discussions of the impact of the traditional assumption of intellectual continuity in comparative psychology, see Hodos and Campbell (1969) and Lockard (1971). For summaries of the theoretical implications of various learning experiments and extensive bibliographies, it would be useful to consult Estes et al. (1954); Hilgard (1956); Koch (1959); Hilgard and Marquis (1961); Klopfer and Hailman (1967); Pribram (1969); Hinde (1970).

## THE NATURE-NURTURE ARGUMENT

I have stressed throughout our previous discussions the interaction of innate and learned patterns in producing an adaptive response. The reason I have given the matter so much attention is that, for years, researchers argued over whether a certain behavior was due to instinct (nature) or learning (nurture). It would seem, on the basis of what we have considered, that the nature-nurture argument is dead. However, it turns out that it is only severely wounded. In defense of the antagonists, it should be mentioned that although the argument has been a dead end, it has not proved unproductive. In fact, it has stimulated some of the most illuminating arguments in modern science. Let's review the major positions.

The operational model of instinct, as we have considered it, was laid out in its basic structure by Lorenz and his co-workers in 1937 and 1950, and amended by Tinbergen (1942, 1951) and others. What they proposed was an explanation of the operation of inborn behavioral patterns. The model, in spite of its imperfections, demanded serious consideration, and the arguments began immediately. The result was a strong polarization of views in the scientific community based partly on misunderstandings of position and subscription to dogma, but partly on honest differences in the interpretation of the same material.

The argument centered over not only whether certain behavioral patterns are innate or learned, but also whether there is any value in attempting to winnow out and identify the specific contributions of these two phenomena to an adaptive pattern.

As is often the case, the principal proponents of each side probably disagreed less than the vigor of their supporters' arguments would have indicated. Neither side had any information that the other lacked, and the beginnings of a realistic model for the basis of adaptive behavior could probably have been designed in those early days (1950s) that incorporated the best information from each camp. However, shortly after Lorenz's (1950) paper, which was powerfully buttressed by Tinbergen's (1951) book, harsh criticisms were raised that began an unfortunate crystallization of positions.

The first volleys were intended to be warnings against arbitrarily attempting to place behavioral actions into one category or the other (Hebb 1953; Lehrman 1953; Schneirla 1956). These were rather vehement attacks upon Lorenz's assumption that if behavior could ever be broken up into its individual elements, it would then be possible to name each element as either "innate" or "learned." Lorenz's critics argued that the dichotomizing was artificial and that the more important question involved the overall ontogeny of behavior.

Lorenz tended to dismiss his critics as "American psychologists"—rat-runners—who were not prepared to ask the important questions. The charge might have been directed with more justification at others than the principals who were involved, but the gauntlet had been thrown and the response was in kind.

There has been much discussion of the argument in the ensuing period (Lorenz 1961, 1965; Eibl-Eibesfeldt 1961b; Thorpe 1963; Tinbergen 1963;

Schneirla 1966; Lehrman 1970; Wilson 1975). The very persistence of the argument for over twenty years seems to indicate that (1) it is a problem outside the scope of science, (2) our investigative technique is inadequate, or (3) we are treating the available data in a biased or irrational manner. It is far too early to assume the first. Our problem is probably the second with the added influence of the third, more or less.

It is currently popular to decry the existence of the hoary argument and to maintain that there is really no argument at all. However, the positions of the major proponents changed little. Lorenz, as late as 1965, stated, "It would be hard to exaggerate the importance attributed by ethologists to the distinction between the innate and the learned." Lehrman (1970) proved no less apologetic when he said of his 1953 criticism, "I do not now disagree with any of the basic ideas expressed in my critique."

Lehrman (1970) attempted to discuss rationally the problems of semantics that have too often led to unnecessary disagreement, and it is all too clear that such clarification is eminently necessary. There is one case of a confusion of terms, however, that is actually more conceptual than semantic, and the difference should be made clear. Throughout the book, I refer to "experience" when I mean learning situations, and in the contexts of the usage, there is hopefully no problem regarding my intention. However, T. C. Schneirla (1966), one of the more important theorists of the time, considered experience to include a wide range of stimulative effects, from biochemical influences on the developing nervous system to learning. If this meaning is accepted, then one must certainly admit the influences of experience on instinct. Considering the developmental interaction of cells, not even one neuron could develop normally without the experience of its proper milieu. Such an interpretation of the term is supported by certain experiments. For example, flight in insects is considered to be an innate pattern, but fruit flies (*Drosophila melanogaster*) can fly normally, erratically, or not at all, depending on the temperature of their environment during incubation. However, in the interest of clarity, the term experience should normally be reserved for processes in which an organism has the opportunity to learn.

## Two Experiments

In spite of the tainted atmosphere, there have been several studies that attempt to analyze the innate and learned components of adaptive patterns. One such study by Jack Hailman (1969) was boldly entitled "How an Instinct Is Learned." In this report, Hailman described his work with the feeding behavior in gull chicks wherein he repeated and extended certain of Tinbergen and Perdeck's (1950) earlier experiments. He argued that, in gull chicks, the feeding instinct is not completely developed at birth, but that its normal development is strongly dependent on the chick's experience. As evidence, he showed that experience is necessary in order for the gull chick to properly peck the bill of the parent, a pattern that induces the parent to regurgitate food. He also found that the chicks quickly learn the visual characters of the parent so that increasingly "natural" models must be used in order to release the pecking responses.

At least part of the improvement in Hailman's chicks is attributable to increased coordination of the muscles, which results in more accurate pecks. Such coordination might be a result of learning to hit their target through practice—and this is the most obvious explanation. However, the improvement, in other cases, might also be due to the "maturation of an instinct" described by Lorenz.

An interesting experiment conducted by H. R. Hess (1956) demonstrated the maturation of the aiming mechanism in domestic chicks. Newly hatched chicks tend to peck at objects around them, but their aim is not very good. The head of a nail pushed into soft clay provided a target in the experiments, and imprints around the nailhead showed how well the chicks were doing. At first, the imprints were scattered around the nailhead (Figure 12.2A), but by the fourth day they were clustered closely around the target (Figure 12.2B). Hess was able to show learning was not involved. He fitted his chicks with prismatic goggles so that the objects they viewed were displaced to the right (Figure 12.2C). Figure 12.2D shows that their aim had improved by the fourth day, but it had little relationship to the position of the nail, since the cluster of imprints was still displaced from the target. The birds with goggles were never able to learn to hit the nail.

In view of such findings showing how both innate and learned patterns are often necessary to produce an adaptive response, it has been suggested that the classical instinct theory is strictly applicable to inver-

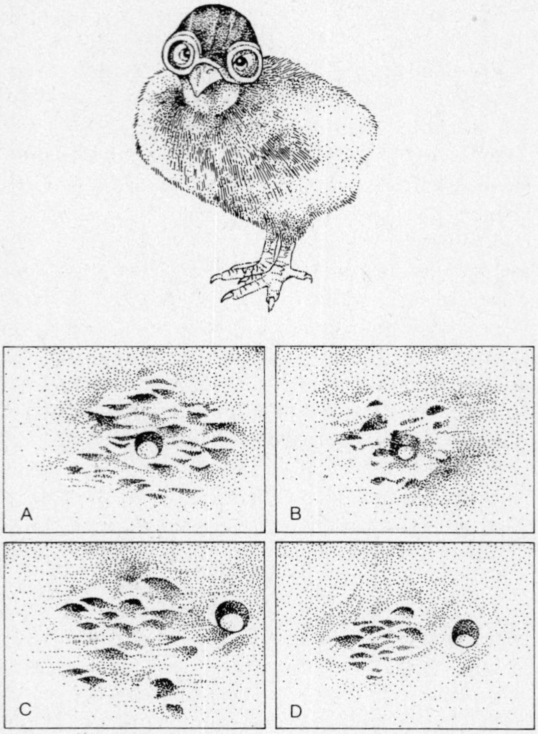


Figure 12.2 Chick wearing prism goggles as part of a test to demonstrate the maturation of an instinct. (A), the marks made by one-day-old chick, and (B), the marks made by four-day-old chick, show normal improvement as the scattered impressions around the nail tighten. When goggles shifted the appearance of the target to the left in other chicks, the pattern also tightened from (C), the one-day-old chick, to (D), the four-day-old chick, but the chicks continued to miss the target.

tebrates and lower vertebrates, animals that can't learn much (Crook 1970b). This is not to suggest that instinct operates on a different principle in higher vertebrates. Instead, we should simply remind ourselves that in such animals, the strong influences of learning on any "innate" pattern would be likely to obstruct objective analysis of the separate components.

It is not probable that evidence of any sort will suddenly end the enduring argument. Rather, its escort to its long-awaited rest will probably be the slow accumulation of experimental data, carefully analyzed, gradually yielding bases for generalities. Even these generalities will need to be carefully qualified, however, since we already know that certain types of animals are more likely to rely on innate patterns than others, and that in various types of animals the normal interaction between innate and learned patterns may be quantitatively and qualitatively quite distinct.

## BEHAVIOR AND EVOLUTIONARY STRATEGIES

Darwin started all the trouble. He not only upset conservative England at the very time the horrors of the French revolution were occurring, just when the pious British were denouncing the "French atheists," but he also forced the reluctant reexamination of many of the foundations of scientific theory. With Darwin, also, came new questions regarding the evolution of behavior. How can selection operate on behavior? What can comparative studies tell us? Which components of behavior are genetically controlled and to what extent? We have seen the results of the attempts to answer these questions. And almost none can yet be answered in definite terms.

Ernst Mayr (1974) provided a plausible explanation for at least part of our problem and suggested ways to begin a new search for answers. He agreed that the problem, indeed, lies in asking how much of a pattern is innate and how much is acquired. The trouble with this dichotomy, he said, is that "innate" refers to the genotype and "acquired" to the phenotype, and therefore one *cannot* be the opposite of the other. One can find grounds for disagreeing with too strict an application of this statement, but Mayr's suggestions for closing the conceptual gap certainly merit strong consideration. His suggestion is that we view behavior as the result of either open or closed programs.

### Open and Closed Programs

Instinctive behavior, of course, is never directly controlled by the genotype (Klopfer 1969b), but rather by a genetic program being translated into a behavioral program in the neuromuscular system. This is another way of saying that genes do not code for behavior, but only for particular developmental sequences in the nervous system that, when activated, result in a particular behavior or tendency toward a behavior.

A genetic program that does not allow appreciable environmental modifications in the process of translation into the observable phenotype is called a *closed genetic program*. Closed programs are common among the "lower" animals. For example, if one isolates a newly hatched *Drosophila* (fruit fly) female until she is ready to mate and then offers her a choice of males of several species, she will choose the male of her own kind. She seems to have an "innate" knowledge of the qualities of her own species (Mayr 1958). Experience will not alter her preference much.

As another example, cowbirds are nest parasitizers. They lay their eggs in the nests of other birds and then let the foster parents worry about raising the young. The young cowbird, then, never sees its own parents or even its own species until it leaves the nest. But when it says goodbye to its foster parents, it seeks out the company of other cowbirds. The association with the host species has not altered its ability to recognize its own species in any way. This, again, must be a closed genetic program.

There may be great advantages to closed genetic programs, especially for species that do not tend to learn much, or do not tend to learn certain things. And they may be particularly important in species that are short-lived or that only reproduce once, such as many insects who have no time to learn. A mating mistake for them means the loss of one's genes from the population.

The situation is radically different in species that have an extended period of parental care. Here, the offspring have time to learn, to have their genetic programs altered by experience. They provide examples of *open genetic programs*.

Open genetic programs can be illustrated by a reconsideration of imprinting. Take, for example, imprinting in African and Australian estrilid finches (Immelmann 1972). If an egg from species *A* is transferred to the nest of species *B*, and the *A* chick is then reared by species *B*, it becomes imprinted on species *B*. Later in life, if such an imprinted adult male is put in a cage with a female of his own species *A*, the two will mate, but only with extreme reluctance. After the female lays the eggs, and they are hatched, the male will help feed the young in normal fashion; but if a female of species *B* is then put into the cage, the male will abandon his own family and his own species and mate with the *B* female. Species recognition in such finches, then, is clearly largely controlled by an open program. Open programs are also found among the more intelligent vertebrates, such as primates. In these animals, parental care and a long life span permit the incorporation of learned patterns into the behavioral repertoire—patterns that are influenced very little by genetic constraints.

## Evolutionary Implications of Genetic Programs

Now, we might ask, Under what evolutionary conditions is a closed genetic program favored over an open one? To answer this, we must consider three types of animal behavior: (1) the communication within species, (2) the communication between species, and (3) noncommunicative behavior.

**Intraspecific Communication** In intraspecific communication, the signals must be precise. They are too important to be weakened by ambiguities. Courtship displays, for example, may have several functions, which include stimulating the sex partner, synchronizing courtship (Figure 12.3), and serving as an *isolating mechanism*. The latter functions in keeping members of two different species from mating since any surviving offspring will almost surely be inviable or weakened. The sexual signals must be unmistakable, and stablizing selection will continually weed out aberrant behavior until a rigid and stereotyped pattern is achieved, possibly through ritualization.

Intraspecific interactions also occur between parent and offspring (these are discussed in more detail in Chapter 17, but here we are leading up to a special point). Again, the communication patterns are highly specific. In several species of wild pheasants, for instance, the young have characteristic markings. If an egg of another species is placed into the nest, when it hatches and the hen sees that the chick lacks the appropriate markings, she will kill it immediately. Here is a case of a closed program; she can't love a strange chick. As an example of an open program, a domestic hen will cheerfully raise whatever comes out of the egg she has been sitting on. In general, however, the communication between mother and offspring in birds must be precise. For example, a young partridge must freeze when the mother gives

Figure 12.3 Courtship of the great crested grebe. In the head-shaking ceremony (top), the birds face each other, raise their crest feathers and shake their heads from side to side. The male then approaches the female with his head submerged (center). Just as he reaches her, the male suddenly lunges out of the water. The birds then present weeds to each other (bottom). Such elaborate ceremonies insure intraspecific lines of communication.

the warning call. They cannot give it their own interpretation. They must do it right the first time they hear it.

Signals between members of the same species are also rather specific. Among carnivores, when an individual loses a fight with a group member, it tends to adopt a largely stereotyped submissive posture (Figure 12.4). In wolves, for example, the loser virtually immobilizes the aggressor by throwing himself on his back and exposing his abdomen (Mech 1970). Generally, hostile behavior between members of the same species includes numerous ritualized displays, most of which are controlled by a relatively closed genetic program (Mayr 1974).

**Interspecific Communication** Interspecific communication involves signaling between members of different species (see later discussions of competition, symbiosis, parasitism, and predator-prey relationships). Interestingly, the behavior of the prey toward the predator may clearly involve interspecific communication, but the predator very often treats the prey as if it were an inanimate part of the environment. In other words, the predator doesn't signal to the prey, while the reverse may not be true. For example, birds don't signal to moths, but when young moths of certain species are prodded by a curious foraging bird, they respond by quickly opening their wings, thus startling the bird. The young moth has not learned its response, but it signals appropriately to an animal with which it shares its habitat. And such responses usually result from the action of closed systems.

As an example of an open interspecific signaling system wolves must learn that a rolled-up porcupine is associated with pain in spite of an abundance of signals that suggest a vulnerable, succulent prey.

**Noncommunicative Behavior** Ecologists have long been concerned with noncommunicative behavior such as food selection and habitat selection, areas that, for some reason, have been rather neglected by ethologists. One reason may have been Lorenz's traditional emphasis on dyadic (two-member) systems, and plant food or a habitat cannot respond as part of such a system. Thus, in these studies there are no displays, those behaviors which have been the grist for the ethologist's mills for years. But since the environment is not "sensitive" to the behavior of the animal, the animal is more likely to be able to respond to the environment through open programs. This means the animal's behavior toward its environment can be expected to be more variable and flexible than if it were interacting with another life form. Phenotypic flexibility rather than genetic precision, then, becomes increasingly important.

But while much of the animal's behavior toward the environment is modifiable through learning (Mayr 1974), it should be made clear that there are genetic components even in habitat and food selection. For example, among parasites there may be a rigid host specificity. Even mammals may show such closed patterns. The panda lives on bamboo, the koala on a few species of eucalyptus. They have almost no flexibility in food sources. Generally, though, food and habitat selectivity probably functions through open programs.

So what does all this mean in terms of the evolution of behavior?

### The Evolutionary Consequences

Mayr (1974) contended that courtship and intrasexual aggression probably has had little evolutionary significance compared to the development of parental behavior. After all, extended parental behavior may be a key factor in the development of open genetic programs. It is believed, for example, that the unique development of humans is strongly related to greatly extended parental care, which gave learning time to occur. Sociality, which is often based on learning, has also had a great role in evolution, particularly among social insects and higher primates, as we shall see later.

Interspecific communicative behavior seems to have contributed rather little to evolutionary change. Mayr has even maintained that predation, "cleaning behavior," and mixed herds and flocks seem to have had little impact (Mayr 1974). An exception might be when predation pressures forced species members to

Figure 12.4 Stereotyped, submissive behavior in a wolf. Such behavior turns off aggression in other wolves.

aggregate as a protective measure, thus encouraging sociality.

Instead, by far the most important factor in evolution has been noncommunicative behavioral shifts of the animal in its utilization of the environment. Such shifts are involved in all instances of adaptive radiation, when several types spring from a single parent stock as each adapts to its own environment. Shifts in habitat and food may have been important in our own evolution as we changed from a mostly vegetarian diet to a more carnivorous one and began preying on large, mobile ungulates.

Thus, since intraspecific communication consists of formal signaling in which little leeway is allowed, it must show low variability. The program must be essentially closed. Generally, natural selection should favor the evolution of a closed program when there is a reliable relationship between the stimulus and its "correct" response. On the other hand, noncommunicative behavior that leads to the exploitation of natural resources should be flexible and variable. Natural selection can then operate on this variability in the course of evolution. The expansion of the niche or the exploitation of new habitats would be hindered by closed or "genetically fixed" behavior.

Much of the controversy in animal behavior could probably be resolved if we were more keenly aware of the differences in the results of closed and open genetic programs. Ethologists have traditionally been more interested in closed systems, comparative psychologists in open ones. It is no wonder the conclusions of the two groups were so different. One is reminded of the proverbial blind men's description of an elephant. Behavior, then, cannot be treated in a one-sided typological manner, as a single, unitary phenomenon. Behavioral development may vary according to its adaptiveness and the nature of its formative pressures.

# 13

# ORIENTATION AND THE EVOLUTION OF MIGRATION

Perhaps you have been strangely moved while standing on a crisp autumn day shrouded in the reds, yellows, and browns of the season, watching formations of ducks or geese flying strongly against a steely sky. You may have noticed that if it is early in the day, the flocks may be heading almost due south. If it is nearing dusk, you may see the group swing toward a grain field for the night. But the next day they will head southward again.

The following spring you may stand beside a swift-moving river in the Pacific Northwest and watch salmon struggling up a fish ladder. As they rest in the deeper pools before their next powerful surge, which will carry them one step nearer the spawning ground, they all face one way—upstream.

Both the birds and the fish are increasing their likelihood of success by positioning themselves correctly with respect to their environment. So, let's now concern ourselves with the ways animals meet their very specific spatial problems. Although we will discuss *space* here, and *time* later, in terms of the adaptive responses of animals, they are really not easily separable. The separate treatment is admittedly artificial and is done primarily for the sake of simplicity. It should also be pointed out that the study of animal navigation is one of the most dynamic areas in modern biology, and the body of evidence is growing at a prodigious rate. Therefore we can only hope to review some of the most general tenets of a fascinating phenomenon.

## KINESIS

Animals may adjust themselves spatially in a number of ways and with a variety of adaptive effects. Perhaps the simplest type of adjustment is kinetic. *Kinesis* may be defined as a type of movement that is related to the intensity of the stimulus, but is independent of its spatial properties. To illustrate, consider the humidity kinesis of the wood louse. If these arthropods are distributed over the floor of a chamber that has both humid and dry areas, they will soon be found congregated in the humid part of the floor. The explanation is rather simple: they are more active in the drier areas. This means they scramble about over the dry patches and when they reach the more humid parts, they slow down or stop altogether (Edney 1954). The result is, they tend to end up where it is moist. Their receptors need only to be able to register variations in humidity. So the response, in spite of its appearance, is nondirectional.

## TAXIS

Other movements are influenced by the direction of the stimulus. The simplest of these is the *taxis* (*plural* taxes). Taxic responses are becoming increasingly important in behavioral studies, especially those by the behavioral ecologists. Actually, little is known about the degree to which taxes function in the behavior of most animals, and among vertebrates the problem becomes more complex, because we tend to think in terms of their "preferences."

There are almost as many lists of specific taxes as there are authors, and there is little need to confuse the issue further here. Suffice it to say that taxes are described in terms of whether an animal moves toward or away from certain environmental characters. Thus, an animal that moves toward light, away from gravity, and toward water is positively phototaxic, negatively geotaxic, and positively hydrotaxic.

A few examples will illustrate the complexity of taxic responses. First, consider the behavior of the grayling butterfly (*Eumenis semele*). To escape from pursuing predators, it flies upwards toward the sun. If blinded in one eye, it will "escape" in circles as it attempts to equalize the stimulus. But the same half-blinded butterfly will follow a female in a straight line. So not all movements, even within the same individual, are dependent on the same type of mechanism (Dethier and Stellar 1961).

Animals may also change their taxic response to different intensities of the same stimulus. The peculiar little unicellular flagellate *Euglena*, which requires light to carry on its photosynthesis, moves toward a dim light, but away from a very bright one (strong sunlight will disrupt its chlorophyll molecules). So, here, the taxic response is dependent on the intensity as well as the direction of the stimulus.

An animal may also change its taxic response with time. Newly hatched blowfly larvae are positively phototaxic—a response that facilitates its dispersal from the dark hatching area. As the larvae mature, however, they become negatively phototaxic and begin to hide in darkened areas. Of course, along with changes in taxic responses, changes in habitat preference and morphology may also be expected. Hawkmoth caterpillars keep their lighter colored dorsal surface toward the sunlight when they are young, but as they grow older, their ventral surfaces become lighter, and they move to the undersides of leaves, thus coordinating their behavioral and morphological changes.

The taxic response of any organism may also change with the motivational state of the animal. For example, water beetle larvae (Dytiscidae) tend to swim toward the illuminated surface of their water, a behavior that normally takes them toward air. After they have replenished their oxygen supply, however, they are no longer positively phototaxic. In other words, they reverse their response to the stimulus.

Most taxic responses are considered to be innate, so one might wonder to what degree such responses can be altered by learning. Actually, it has been found that some taxic responses can even be *reversed* by conditioning. For example, a cockroach that is fed under light and given a mild electric shock when it approaches darkened areas, will begin to avoid its usual habitat. There are even reports of a similar type of conditioning being accomplished in the lowly planarian.

Investigations of taxes must be controlled very carefully because the responses are often not as simple as they may appear. Poorly designed experiments may, at best, be inconclusive. For example, an experimenter should be aware that taxic interactions may occur, and these must be separated by experimental design in order for a single component to be analyzed. To illustrate, the upright position of a fish may depend upon dorsal light *and* the effects of gravity. When these two are altered in relation to each other, the results may indicate their individual effects (Figure 13.1). For an example of how two taxic responses that are usually expressed in the same way may come to have a special adaptive significance, see the discussion of the honeybee in this chapter.

The last point regarding taxes is that orientation is not always simply directed toward or away from a stimulus. Ants and bees may move on a course at a certain angle from a stimulus, such as the sun. An innate timing mechanism enables homing ants to change the angle of movement with respect to the sun as the sun moves across the sky. There is some basis for arguing that this sort of movement should not be considered a taxis, but rather a compass orientation of the type discussed later.

Some animals are capable of much more complex orientation than simple kineses and taxes. *Piloting* is the ability to find a goal by referring to familiar landmarks. The animal may either search randomly or systematically for the landmarks. *Directional compass orientation* is found in those animals that are able to head in a geographical direction without the use of landmarks. And, lastly, some animals show true *navigation*, which we will define as the ability to maintain or establish reference to a goal without the use of landmarks. Our simple definition of navigation purposefully neglects the abundant exceptions and qualifications that often result in an admission that no definition is actually possible. (Such fate has befallen other words, such as "hormone" or "species.")

Investigations of animal orientation have involved a wide variety of animal species, and it would be impossible to attempt to summarize all the work here. Instead, we will consider a few species that show a variety of sensory capacities and that utilize various environmental signals as their directional clues. We will direct the greater part of our attention to orientation in birds because of the wealth of information available in the literature and because the experimental techniques themselves have often been fascinating examples of scientific imagination.

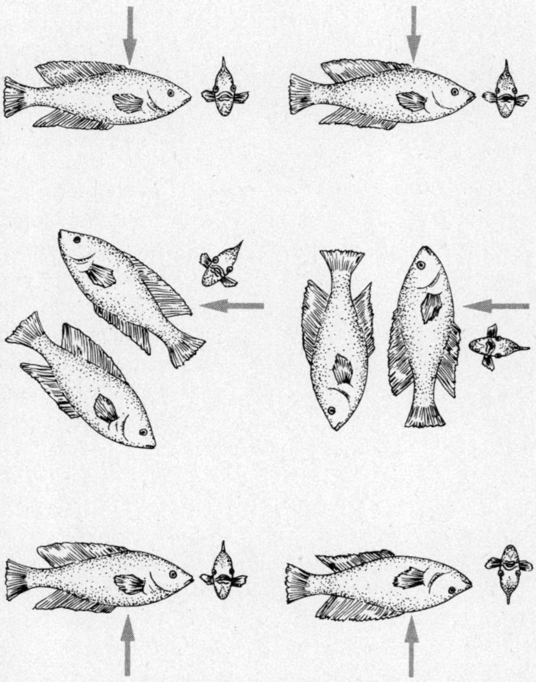

Figure 13.1 The orientation of *Crenilabrus restratus* to light coming from different directions with respect to gravity. Gravitational pull is downward. The direction of the light is shown by the arrows. The fish on the left is normal and orients correctly with regard to both light and gravity. The fish on the right has had the ear labyrinth removed and is unable to sense the direction of gravity. It responds only to light.

## Arthropods

We begin with arthropods because less complex invertebrates generally show only very weak directional abilities. For example, no one is very impressed by the tendency of earthworms to crawl to the surface at night. Arthropods, on the other hand, show a wide variety of migratory abilities, some of the most pronounced movements occurring in marine arthropods. To illustrate, the tiny crustaceans that help make up the ocean's zooplankton show daily vertical movements. They rise to the surface soon after dawn, stay for a few hours, and by midday they move back into the relative safety of the ocean depths, only to rise again in late afternoon. There is more food near the surface, but there are also more predators, and the midday sun's piercing energy is dangerous to their frail bodies.

Marine crustaceans may also move back and forth between shallow inshore waters and deeper areas, but these movements mostly occur in large, bottom dwellers, such as the spring lobster (*Panulirus*). They move in groups, sometimes single file, holding onto each other's tails in a long conga line, as they wend their way to warm, shallow waters where they will mate. They will spawn, or deposit their eggs, in the darkness of deeper water, only then to return towards shore. When winter comes, the lobsters again return to the briny depths (Bainbridge 1961). Perhaps the stability of the cold ocean depths is less traumatic than the variability of the shallow water in winter. So here, apparently, two separate migratory movements are related to two different factors, reproductive and physiological.

Some insects are famous for their migratory ability. The monarch butterfly (*Danaus plexippus*) apparently flies from the northern United States and Canada to Florida and Southern California each winter and returns north in the spring (Johnson 1964).

And then there are the locusts in Africa. For some reason, in certain years, enormous numbers of these insects emerge from their eggs to sweep over the African countryside, stripping bare the vegetation, including crops (Figure 13.2). Also, these aren't your average locust; they differ in color, size, shape, and level of activity from the usual variety. These swarming insects move with the prevailing winds, so thankfully most do not return to their place of origin (Johnson 1964). Since they move with the winds, either their orientation ability is weak, forcing them to take the easiest route, or they respond taxically to the direction of the prevailing winds, a response that could have been selected for because of the rather constant wind direction.

**Bees** One of the most fascinating stories to emerge from the annals of behavioral research is that of the dancing honeybees described by Nobelist Karl von Frisch (1950, 1953).

Honeybees (*Apis mellifera*) are, of course, highly social animals that behave in a very coordinated and cooperative manner. They work together to build a hive, care for the young, keep the hive at the proper temperature, defend it, and gather food. The surprising thing is that they apparently tell each other where food is to be found.

According to von Frisch, scout bees, upon finding a good food source, bring back a full load of pollen and nectar to the hive. Even if they have found the food by a very roundabout course, they may fly directly home on a beeline. When they return to the hive, they communicate their find to the other bees by "dancing" on the vertical combs inside the dark hive. The dance describes the direction and distance of the food source. Distance may be communicated by several methods. The round dance (Figure 13.3, left)

Figure 13.2 African locusts periodically emerge in overwhelming numbers, moving with the winds and stripping the countryside bare. Note that no grass or leaves are left and that some bark has been eaten off the limbs.

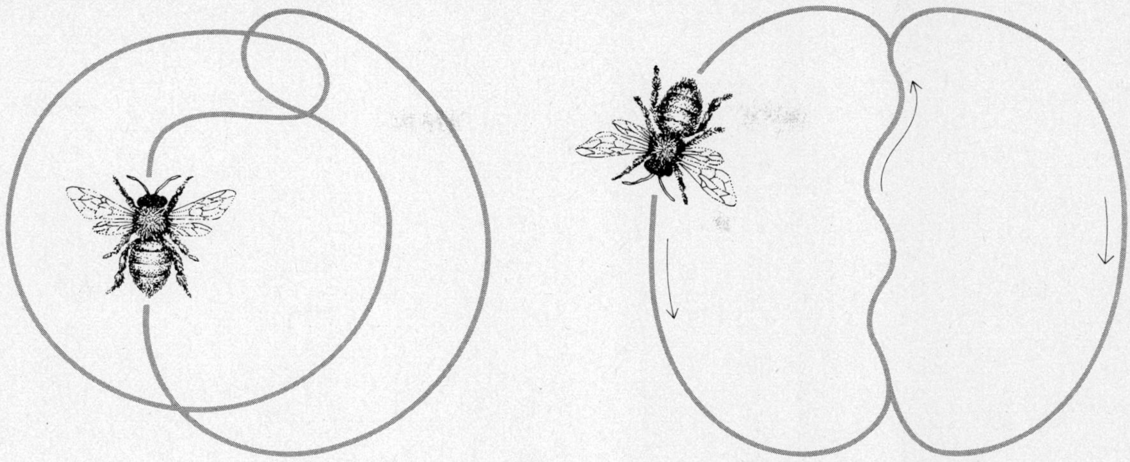

Figure 13.3 The dances of the bees. Successful foragers show the position of food by these dances. If food is less than one hundred meters from the hive, the round dance (left) is performed. If the food is farther than that, they dance the waggle-tail figure-eight (right).

is performed when the food is less than one hundred meters from the hive. The waggle-tail dance (Figure 13.3, right) is done when the food is farther away. In the latter case, the distance is communicated by the time spent in "waggling," by the tempo of the dance, and possibly, by sounds produced by the bee, using either pulse rate or period of sound production. The pulse rate is about 2.5 times faster than the waggling rate. Substrate vibrations set up by the dancing bee may also function in relaying information.

Distance is apparently described by bees in terms of the energy expended in covering it. This means that if a bee must fly into a headwind or uphill, or if it is forced to walk, it will report that the food is farther away than it is.

The direction of the find is described by the angle made by the "waggle" part of the dance and a vertical line. This angle indicates the position of the sun relative to the food. Of course the angle changes as the sun moves across the sky.

The bee's system of communication appears even more complicated when it is realized that the dance is performed in the darkness of the hive and that the other bees apparently pick up the information by touching the dancer. The information gained tactilely is employed visually when the bee emerges from the dark hive and sets out at a certain angle to the sun.

Even more complex, theoretically, is how bees communicate the angle of the food relative to the sun, since the dance is performed inside the dark hive, out of sight of the sky. What happens is this: since the waggling part of the dance cannot be performed with direct reference to the sun, it is performed with reference to gravity on the vertical face of the comb (Figure 13.4). An angle of forty degrees from vertical means the bees should fly at a horizontal angle forty degrees from a line running from the hive to the sun's position.

The evolutionary basis for the transposition of the angle was explored in studies of ants (Vowles 1954). It was concluded that the raw material for natural selection in this case was probably positive phototaxis (normally upward toward the sun) of disturbed arthropods, operating with negative geotaxis, which would produce the same adaptive response. An alarmed bee, it is suggested, flies toward the sky and away from gravity in what might be considered a redundant system of escape orientation. The idea is that the association of the two taxes may have served as the evolutionary basis for one coming to be described in terms of the other.

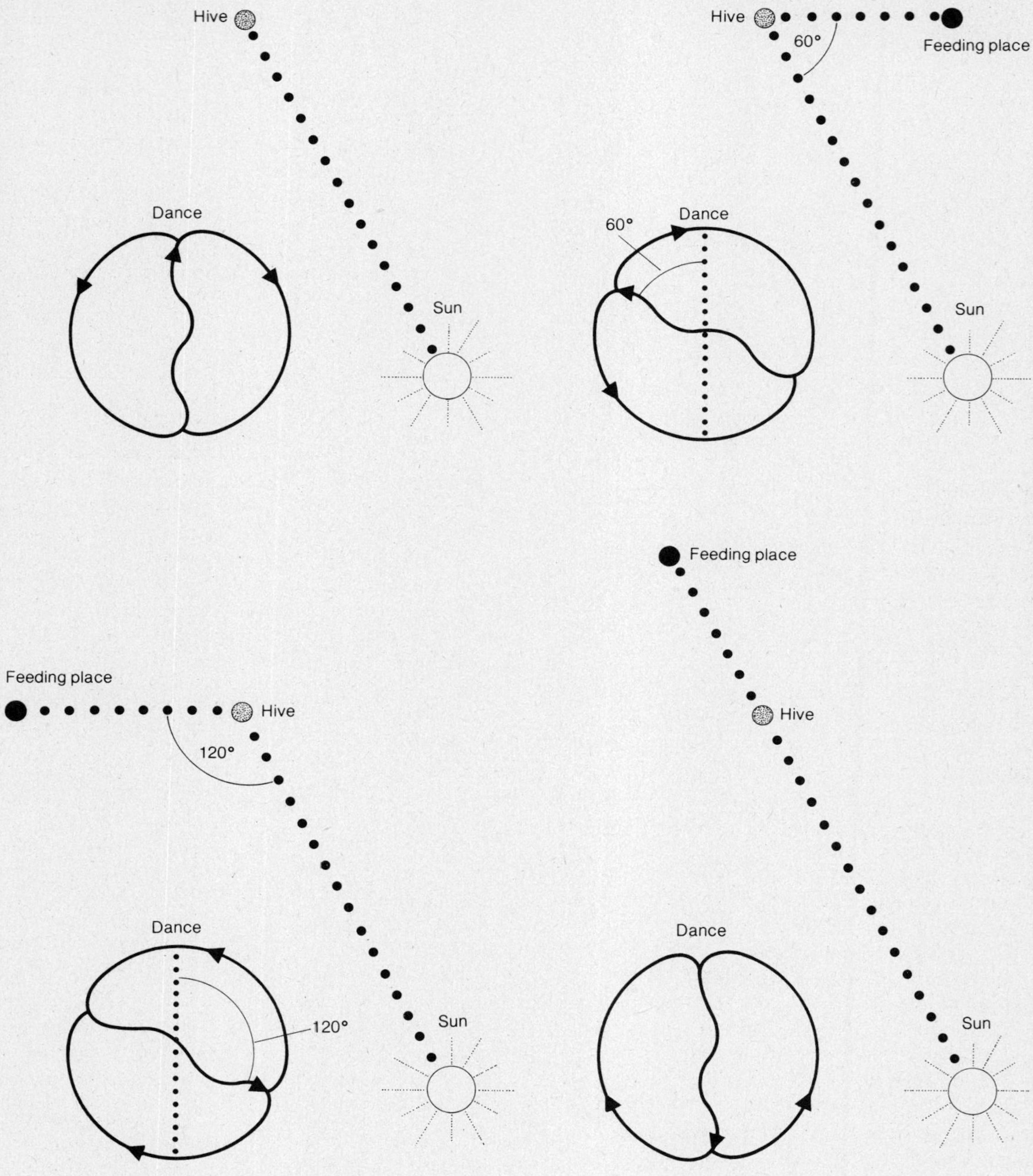

Hive
Dance
Sun
Hive
Feeding place
60°
Dance
60°
Sun
Feeding place
Hive
120°
Dance
120°
Sun
Feeding place
Hive
Dance
Sun

Figure 13.4 The waggle dance of the honeybee is performed on the vertical face of combs inside the dark hive. The direction of the waggle across the diameter of the circle made by the bee shows the direction of the food discovery. In the hive the direction is relative to the vertical, in flight it is relative to the sun. Therefore, the other bees in the hive must transpose the angle of the dance and a vertical line to an angle relative to the sun before they set out to locate the food that the scout has announced.

It might be asked if, in such a complex arthropod system, learning can be involved, or if the entire scheme is under innate control. It turns out that learning is important. Experiments have shown that bees must learn the sun's track, as demonstrated by the fact that only experienced individuals compensate correctly for the apparent movement of the sun (Marler and Hamilton 1966). It is interesting that bees can learn the sun's course after being exposed to less than one complete journey of the sun across the sky. But, if insects can learn, they can also forget. In the ant *Formica rufa*, the ability to compensate for the sun's apparent motion is greatly reduced when they emerge from hibernation. It seems the ants lose some of their navigational ability during their long winter's sleep.

It should be pointed out that von Frisch's description of communication in bees has been criticized by Adrian Wenner (1971). Wenner accumulated evidence that indicates that among bees, communication may take place through odor; he recommended the rejection of the bee-language hypothesis. Wenner and his colleagues (Wenner 1974; Wells and Wenner 1974) list the following reasons for their rejection:

*(1)* Bees newly recruited to a crop preferentially orient toward the center of an odor field.
*(2)* Recruit success at finding food is dependent upon odor concentration, but not upon the Nasanov gland (abdominal scent gland) secretions at the food source or the dance frequency in the hive.
*(3)* With no odor cues, recruits cannot find a food site.
*(4)* Recruits attending dances without direction information can still find the food.
*(5)* Even with artificial odor in the food or at the food site, most recruits do not find the site visited by foragers.
*(6)* Most recruits require several reconnaisance flights before they locate food previously found by foragers.
*(7)* Most recruits take much longer to find food than they would if they took a direct flight.
*(8)* Recruits apparently interact in flight and perhaps in the hive.
*(9)* Recruit success is more closely correlated with the cumulative number of foraging trips than with the number of foragers involved.
*(10)* After foragers begin making regular trips to a food site, recruits do not arrive until about thirty minutes later.
*(11)* When bees are fed scented sugar solution at a location away from the hive, recruit success increases proportionately with time.
*(12)* Crop-attached bees are redirected to their foraging areas by a conditioned response to the odor of the crop, and no dancing is required.
*(13)* Recruits always arrive at a food source from downwind.

For references and specific information, you may wish to consult Wells and Wenner (1974). Their criticism of the bee language hypothesis has generated a great deal of discussion and strong polarization in certain circles, and one is not always convinced that the arguments are unbiased. One might wonder, if directional information is contained in the dance, but is not utilized by recruits, why do the bees dance? The student of the philosophy of science might like to carefully analyze these dissenting viewpoints. Is there a problem with the question?

Many of the discrepancies between Wenner and von Frisch have recently been resolved by James Gould and his co-workers (1970). By experimentally manipulating the bees, Gould has been able to duplicate the results of both Wenner and von Frisch and to produce combinations of behaviors. He has argued that the differences in results are due to small differences in experimental technique.

## Fish

It has been established that some fish can navigate long distances, but our knowledge is extremely limited regarding which species possess this ability and what types of mechanisms are involved. Fishermen are aware that certain species of fish appear in certain areas at a given time of year, and in many cases, this is virtually the extent of our knowledge. Where do the fish go at other times? And how do they find their way?

Migration of fish is somewhat different from that of birds. A fish spends the early part of its life in the waters where it was born and then journeys to another place, where it reaches breeding condition, only to return home to breed, perhaps years later. As an example of migration in fish, let's consider the fascinating case of the salmon.

Salmon hatch from eggs in the redds, or spawning areas, of North America and Europe. The redds are usually in the cold, clear headwaters of gravel-bottomed rivers. The young salmon may spend two years in these home waters where they grow large and mature. Then they swim to sea, fanning out in all directions once they reach the ocean, many of them eventually appearing in well-known deepwater feeding grounds far from land. They spend two to seven years in the ocean (depending on the species) where they reach their breeding condition. Salmon in breeding condition are large, glistening, beautifully colored creatures and, in this state, they head from their feeding grounds back through the trackless sea to the very river from which they came. When they reach their river, they swim upstream, always turning up the correct tributary until they reach the same riverlet where they spent their youth. It is a remarkable journey from the depths of the open ocean back to the home redds, and it is performed flawlessly. They return to their own grounds with such consistency that streams near each other may hold strains of fish that are following separate evolutionary lines.

Once home, the females shed their eggs into the water to be fertilized by the males. The journey is over. They are no longer the beautiful creatures they were. Their fat stores are gone and their color has faded. Those of the Pacific species (*Oncorhynchus* spp.), completely spent, die. But the European Atlantic species (*Salmo salar*) may feed for a while after breeding, gain some strength, and return to sea to make the breeding journey another year as even larger specimens.

A fascinating set of experiments conducted in British Columbia shed some light on the timing of salmon migration. Two branch rivers, the Upper Columbia and the Willamette, converge into the Columbia River, which flows into the Pacific Ocean. The king salmon of the Upper Columbia are autumn-run fish, while those of the Willamette swim upstream to spawn in the spring. Eggs were collected from the Willamette fish and hatched in the Upper Columbia. The fish grew well in their new home and eventually made the journey to the sea. When they returned from the sea, they swam up the Columbia with the Willamette fish in spring; but when they reached the fork in the river, they left the Willamette fish and turned up the Upper Columbia, although this tributary was not suitable for breeding at this time of year. Their behavior was interpreted to mean that the time of movement is innate, but the characteristics of the home waters are learned.

One of the pioneer workers in fish navigation, Arthur Hasler (1960), pointed out that the fish learn, or are imprinted to, the odor of their home waters in the first few months of their lives. The evidence is convincing. It has been shown that many fish have phenomenal abilities to distinguish certain types of molecules in water. Salmon can be halted in their spawning by certain organic smells, such as the scents of predatory seals and otters. A fisherman who dips his hand into the river upstream may cause migrating salmon below to reverse their direction. In tests, salmon and other fish have shown the ability to detect very small amounts of various soluble chemicals. Other tests have shown that when the salmons' sense of smell is destroyed, they are unable to find their way home. So salmon apparently use their sense of smell to find the waters where they spent their youth.

Hasler said the fish do not follow a gradient because they would adapt to constant exposure to the odor. Instead, he argued, they crisscross the scent, much as a trailing dog might, and constantly move upstream in taxic response to the direction of the current. Of course, this means that different tributaries must be distinctively scented.

So, do streams have their own odors? We know that in some places there are strong differences in stream solutes because of geological differences in the stream-bed material. Other scent differences could arise through the type of plant material that has decomposed in the water or by various effects of its physical characteristics, such as its depth, its speed, and its degree of oxygenation (which might vary according to whether large rocks break the water surface and cause rills or bubbles).

Even if fish do find their way up the proper streams to their homes, olfaction cannot be the answer to the question of how salmon navigate from far at sea to the mouth of the river. Marked Pacific salmon have been caught at sea as far as 2,375 miles from their home river. European salmon are known to travel hundreds of miles out to sea after leaving their rivers. River effluent would simply be too diluted by the salt water to be used in navigation.

So, how do the fish manage to navigate so unerringly back to their river mouth? There are strong indications that they may use the sun as a compass (Hasler 1960). In tests with a device similar to a Kramer cage (see Figure 14.5), white bass (*Roccus chrysops*) were conditioned to swim into an enclosure that lay in a certain compass direction. Under a clear, daytime sky they were able to perform correctly, but under overcast skies they were unable to orient. In conclusive tests with mirrors, it was shown that the white bass is able to orient by using the sun's position in association with a biological clock. Although the ability to *orient* under test conditions should not be equated with the ability to *navigate* in the wild, the possibility of sun navigation is very real.

## Frogs and Salamanders

Some species of frogs and salamanders can orient by using celestial clues in conjunction with biological clocks. It would seem that celestial clues would only be useful to animals with keen vision. However, displacement experiments with newts (*Taricha rivularis*) showed that they were able to home over relatively long distances even after they had been blinded. So, what receptors were they using? It was thought that they may have used olfaction, but in experiments designed to preclude olfaction, they were still able to return. Then a remarkable set of experiments revealed some rather startling information.

A group of blinded southern cricket frogs (*Acris gryllus*) were able to orient with respect to their learned shoreline at home (Taylor and Ferguson 1970). When their biological clocks were altered by changing the light-dark regime to advance six hours ($\frac{1}{4}$ day), the animals shifted their direction ninety degrees ($\frac{1}{4}$ circle). The blind frogs were obviously reacting to light! So a system of extraoptic photoreceptors (EOPs) was hypothesized. However, such a system would need to be sensitive not only to the presence of light, but to its direction. Then it was found that when an area of the skull between the eyes was covered with opaque teflon, sightless frogs lost all directional ability.

The part of the skull that was shielded lay over a peculiar little structure that has been the subject of much debate by comparative anatomists and physiologists through the years: the *pineal complex*. The pineal is generally thought to be the vestige of a third eye that was functional in the distant ancestors of many extant vertebrates, including man. In some modern lizards, a small lens and retinalike receptors are still present at the tip of the structure (Weichert 1965). The pineal complex develops embryonically as an outpocketing of the forebrain, just as do the eyes. Both photoreceptive and secretory cells have been found in the pineal, but very little is known about the function of the structure. Many vertebrate evolutionists dismiss it as an interesting, but no longer useful, body, but these experiments indicate that its present adaptiveness in some species should be reconsidered.

Recent experiments have demonstrated that orientation by frogs and salamanders may be accomplished by the detection of polarized light in the overhead sky, and that the detection does not involve the eyes (Adler 1976). Polarized light, by the way, may be an orienting clue in a wide variety of organisms from arthropods to mammals.

## Green Turtles

Ocean-going green turtles (*Chelonia mydas*) (Figure 13.5) have presented some of the most intriguing problems in studies of animal navigation (Carr 1965). They are apparently blue-water navigators, crossing hundreds of miles of open ocean to find small islands. Female green turtles normally inhabit the coast of Brazil where they graze on the submerged and gently swaying "turtle grass," common in shallow and sheltered tropical waters. But every two or three years, they suddenly leave these peaceful environs and begin an incredible journey to tiny Ascencion Island, only 5 miles wide and 1,400 miles away (Figure 13.6).

Figure 13.5 The green turtle. Its life history and travels largely remain a mystery.

Here, they crawl ashore, find an appropriate place, and lay their eggs. Then they begin the long trek back to their feeding areas. Other turtles in other places make similar sorts of migrations. Another nesting area has, for years, been the research site of Archie Carr and his students at the University of Florida: Tortuguero in Costa Rica, which is the only green turtle nesting area in the western Caribbean.

These turtles have fascinating life histories. Once the little males enter the sea, they never come out again. But the nesting females crawl ashore three to seven times each nesting season, depositing about one hundred eggs each time. The female must lay large numbers of eggs because so few will make it. The hazards are enormous. Many of the young hatchlings, scrambling across the sand toward the surging breakers, will be caught by circling seabirds. Those who make it to the water face an alarming array of marine predators.

No one has any idea where the little turtles go after they enter the sea; we know only that for the first year they will be carnivorous, probably probing the murky sands for small bottom dwellers. But their first trick is making it to the water. The birds circling overhead are ravenous and time is precious. So the tiny turtles, only seconds old, do not delay. They immediately run toward the surf, even if they can't see or hear it. So, how do they know which way to go? Various sensory techniques have been experimentally ruled out, and now it seems that they are able to detect subtle differences in the quality of light over the land and water. Orientation is not accomplished by compass direction, since baby turtles flown from Caribbean nests to Pacific beaches and allowed to emerge from artificial nests, scurried straight to the water, just as the Pacific group did. Females that nest out of sight of the sea also use the apparently brighter light over water to get back to the sea. Turtles don't see very well, but their power to distinguish light is acute. When one eye of a turtle was covered with a filter that cut out only one percent of the light, the turtle deviated toward the uncovered eye (Ehrenfeld 1968).

Tagging experiments indicate that the turtles actively navigate from feeding to nesting areas, but no one knows how. Turtles are hard to follow and their sensory abilities aren't understood very well. Can they imprint on their nesting grounds by smell? How can they get their longitudinal bearings without being able to get a bearing on the horizon? At present, the questions are many and the answers are few regarding the mysterious abilities of the green turtle, but a variety of very sophisticated devices, such as tracking by satellite (which has been used successfully with large mammals), are being suggested. Hopefully, more sophisticated techniques will force the information we seek from these lowly reptiles.

And now, let's discuss a few points regarding mammals, saving birds for last, since they will occupy most of our attention.

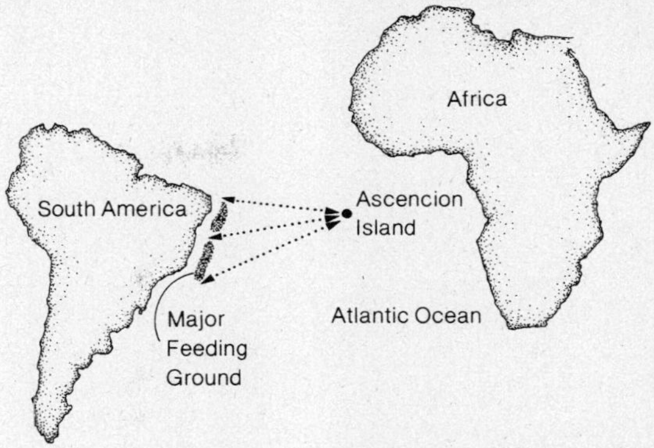


Figure 13.6 Migratory routes of green turtles between shallow feeding waters off Brazil and the nesting grounds on tiny mid-Atlantic Ascencion Island.

## Mammals

One of the objections of environmentalists to building the huge oil pipeline across the delicate arctic tundra was that it would interfere with the migration of caribou as they moved between their northern wintering area to the southern woods, some four hundred to five hundred miles away. The pipeline was built anyway, but the argument focused our attention on a little-understood phenomenon: the migration of mammals.

Further south, each fall, people head to sea in chartered vessels to watch one of the greatest migratory spectacles of all—the southward movement of the gray whale (*Eschrichtius glaucus*). The gray whale follows the North American coastline, possibly over alternate routes (Wolfson 1977), as it moves from the frigid North Pacific and Arctic Ocean waters to breed in the placid lagoons of Baja California, Mexico. Pregnant females lead the way, with other females and males following close behind. (The males are ready to inseminate any female not already in a delicate condition.) Although the actual breeding takes place in the sheltered Mexican waters, the females will not give birth until some months later on another southern sojourn. The whales do not retrace their path northward when their breeding season is over. Instead, they take a more seaward route and are therefore much harder to observe (Norris 1967). We have known about the migration habits of some whales for years, but we still understand almost nothing about the physiological mechanisms or the environmental cues involved.

The migratory ability of northern fur seals is also poorly understood, but each year males that spend the winter in northern waters move to the Pribilof Islands off Alaska. After the dominant males establish beach rights through fighting, they are then joined by females coming up from the California coast. Also, each year, a group of Alaskan Indians, which the Interior Department claims needs the jobs, drive the terrified seals landward by shouting and beating on cans. Then, while the seals huddle together for protection, the men club them to death with long, heavy bludgeons. The bloody spectacle is witnessed each year by busloads of tourists, and the skins are sent off to a furrier in the southeast to be made into coats and other apparel for people who wish to demonstrate their class—such as it is. The regularity of the seals' return furnishes a dependable supply of pelts, but again, the mechanisms remain a mystery.

We also know very little about migration in flying mammals. We do know, however, that some species of bats may migrate as far as eight hundred miles. For example, the European pipistrelle (*Pipistrelus*), migrates between southeastern Europe and central Russia each year, and some bats can home. One female pallid bat (*Antrozous pallidus*), returned to her home after being displaced eight times, once as far as sixty-eight miles (Davis and Cockrum 1962). We don't know how any bat navigates.

In some instances, both the ultimate and proximate factors behind mammalian migration present interesting questions. For example, some African elephants regularly migrate to areas where the umganu tree grows. It seems that the umganu fruit makes them drunk, and they stagger about, trumpeting, and looking for fights (Maier and Maier 1970). It would be interesting to know if they go there for food or alcohol. Some animals are born with their migration routes as fixed as any physical trait, but elephants are long-lived and social animals, so it is very likely that they learn their migratory routes.

There are many other instances of migrating mammals. We have all heard of the lemmings of Norway that sporadically move in great herds to the sea where many hurl themselves into the water and drown. However, most of us perhaps are not aware that a similar drama occurs in the United States. Periodically, legions of gray squirrels in the northeast set off on mysterious cross-crountry treks on which many are killed attempting to cross highways and rivers and others fall to predators. What is their destination? What triggers the movement? The last century witnessed great migrations of the tiny antelope, the African springbok. Millions died when they reached the inhospitable African coast, some by wading into the ocean and drowning. No one knows why they started or why they suddenly stopped.

## Birds

It has been known for centuries that birds disappear from certain areas every year, only to reappear the following year. Over the centuries, many theories have been developed to account for the phenomenon. For example, it has been suggested that the birds had been to nether areas, or to the moon, or had temporarily changed into another kind of animal. These suggestions have been largely discredited. It now seems that they migrate.

**Migration** One of the most startling aspects of bird migration is its sheer volume. The numbers are overwhelming! On some autumn or spring nights, the sky is virtually *full* of birds on their migratory routes. On radar screens they look like a snowstorm as species after species moves past in response to some primordial urge. We can only guess at *why* they go, but we have some rather substantial evidence on *where* they go. *How* is another problem (Emlen 1974).

The best answers as to where birds go when they migrate have come from banding experiments. When a bird is caught, a metal band with the address of an agency and a request for notification is fastened to the leg, and it is released. Later, when the bird is found, caught, or shot, the finder sends the band, or the information on it, to the proper address. In this way, information begins to accumulate, indicating the species' range, its seasonal habitat, and something of the rate at which it can travel. Some of the results have been startling. For example, a lesser yellowlegs was banded one August day in Massachusetts and was shot in Martinique six days later. Since the distance was 1,930 miles, the bird reached *at least* 322 miles per day. Since many members of the world's enormous army of birdwatchers have become involved in bird banding, a very large body of evidence has accumulated regarding the movements of many species.

Of course, there are many ways to learn about migratory movement in birds. Daytime (diurnal) migrants may be seen with the naked eye or with the use of telescopic aids. Nighttime (nocturnal) migrators have been counted by focusing a telescope on the moon and recording the birds that crossed the field of vision. Also, it was found that during peak migratory periods, a patient observer could count the bodies of birds (often many thousands) that crashed into radio towers or other tall structures on nights of low visibility. One of the most valuable sources of information, however, is radar. The "angels" that plague military men or flight controllers are often groups of migrating birds. "Surveillance" radar is now augmented by "tracking" radar, a device with which single birds can be tracked and sometimes identified as to species.

The incoming data from a number of sources, primarily banding experiments, have furnished many surprises, and some of the feats performed by migratory birds seem incredible. If there were not indisputable evidence in many of these cases, they would be denied on the basis of common sense. As an example, consider the Arctic terns that nest within ten degrees of the North Pole in the barren Canadian Arctic. It may be easier to raise a family in the far north where there are fewer parasites and predators, but then where do they go to spend their winter? Some tropical retreat? Actually, some fly via the west coast of Africa to winter on the *Antarctic* ice packs. (This is harder to explain, except that there are still fewer predators than in tropical or temperate zones, and the birds would have to undergo less physiological adjustment to changes in temperature. The south pole *is* warmer when the north is having its winter, but it is still very cold.) In the spring, they return to the same nest sites in the Arctic. The distance, of course, is equal to circumnavigating the globe, and it's done by the fragile birds once each year.

Another migrator *par excellence* is the greater shearwater, which ranges over the vast expanses of the entire Atlantic Ocean and somehow returns to the breeding areas of the isles of Tristan de Cunha, tiny islands only thirty square miles in area and fifteen-hundred miles from the nearest land. Also consider the bristle-thighed curlew, which nests in coastal Alaska and winters on Pacific islands six thousand miles away. Each leg of the journey necessitates sea crossings of two thousand miles. The sheer physical power involved in these feats seems astounding and has been the subject of much research of a histochemical nature in order to determine where the cellular energy reserves are and how these are utilized so efficiently (Wallace 1966a). However, we are less concerned with the physical than with the mental. How do they navigate?

**Advantages of Migration—Ultimate Factors** At this point, we might ask why migration arose to begin with. One almost totally unsubstantiated answer is that migration was initiated by birds fleeing southward before advancing ice age glaciers. The genes of those individuals that did not have inherent tendencies to retreat southward before the cold were eliminated from the population by severe weather. Other theories include the effects of continental drift and adaptations that permit the exploitation of temporary or moving food supplies. Ornithologist J. C. Welty, (1963, p. 449) writes:

*Without question, ecological stresses have been the most frequent causes of the migratory habit. Birds are driven out of their breeding areas by periodic climatic cycles, which bring on food shortages that are aggravated by the annual increase in bird populations. Although the original impetus to migrate may be entirely external, the ceaseless repetition of seasons of famine, brought on by arctic frost or tropic drought, may eventually, through natural selection, winnow the sensitive protoplasm of a species until the urge to migrate at a particular time of year becomes innate, and comes into play at the appropriate season even in the absence of the original impelling threat of starvation.*

For a review of the theories of how migration may have developed, see Dorst (1962).

What are the ecological implications of migration? Certainly it must have significant advantages in order for it to have evolved in so many animal groups. After all, the energy expenditure is great; large areas of inhospitable terrain must be passed; and the weather is unstable in the migratory periods. In addition to such dangers, many migratory birds must annually relinquish their rights to hard-won territories and compete vigorously to become reestablished the following year. So what rewards could possibly override such disadvantages?

We can intuitively understand the advantages of animals moving from impending arctic winters to the sunny tropics. We can even see why birds might move shorter distances, for example from Nebraska to Oklahoma, for the winter. Only a cursory familiarity with the elements of nature could also convince us of the advantages of simply moving from a mountain top to a valley as winter draws on. In each case, the birds are trading a less hospitable habitat for a more hospitable one.

However, we must remember that there are factors other than weather that make migration advantageous. For example, during the northern winters the food supply drops low and forces any resident species into increasingly severe competition with its neighbors for such commodities. Some species have very narrow food preferences, and their food may even disappear from parts of their range each year. The pennant-winged nightjar (*Semeiophorus vexillarius*) breeds during the rainy season each fall in southern Africa. Then, in February, it follows the rains across the equator to winter in the area from Nigeria westward. Thus it traces the appearance of its preferred food, flying termites (Streseman 1927–1934).

If there is so much food in the warmer winter habitats, why don't migratory species just stay there? Why do they return to their summer homes at all? Well, there are certain important advantages to rearing broods in temperate or even polar areas in the spring. For example, days in the north are long, and so the birds' working day can be extended; they can bring more food to their offspring in a given season. Food productivity is high in polar regions in summer because of increased sunlight for photosynthesis, optimal temperatures for photosynthesis, and nutrient-rich upwellings in polar oceans. And since, effectively, more food is available, more young can be raised. It is known that, generally, the farther north from the tropics a species breeds, the larger is its brood (Welty 1963). And, of course, with more food, any brood can be raised faster.

Another advantage in returning to the temperate zone to breed is in escaping the high level of competition that exists in the species-packed tropics. The annual flush of life in the temperate zones provides a predictable and otherwise underutilized supply of food that can be exploited readily by mobile species such as birds.

Farther north in the polar region, the breeding period is very short because of the weather cycles, and this "inconvenience" works to the advantage of nesting birds that are in danger of falling prey to predators. The short season means that a great number of birds must nest simultaneously; thus, the likelihood of any single individual being taken by a predator is reduced. Also, since the birds come and go, their predators are denied a stable food supply. And then, by leaving certain geographical areas each year, migratory species deprive many parasites and microorganisms of permanent hosts to which they can closely adapt. Also, the long harsh winters in the frozen north act further to reduce the numbers of parasites in that area. By the same token, predators that are unable to escape the rigorous northern climes might also be expected to be fewer in number the following spring.

Migratory movement may also be advantageous to the populations of the migrators. The "scrambling" of the population each year might cause some individuals to end up in new habitats where intraspecific competition might be low and where they might thus be able to establish themselves. Because of the strong differences in summer and winter habitats, migrating populations might develop individuals with high behavioral flexibility—an important quality for colonizers, as we shall see later.

Dispersal may result in the isolation of a small population of birds, and such pockets may then rapidly increase their evolutionary rate. Mutations have a greater effect on smaller populations than on larger ones. The reason, as we have seen, is that gene changes in smaller populations are not subject to the swamping effect of alternative genes, as is the case in larger populations. Not only are such small-population effects important in quickly adapting an isolated group to its new surroundings, but new genes or gene combinations thus established in a population may furnish important new genetic material should interbreeding with other populations occur.

**What Influences Migration—Proximal Factors**
When first-year birds of certain migratory species, such as starlings (Perdeck 1958), were displaced experimentally, they flew in the same direction of the parent group and, importantly, they flew for the same distance. In other words, they paralleled the parent group. In some cases, this meant that they ended up in ecologically unsatisfactory places. In another experiment, storks were taken from nests in the Baltic region, reared in West Germany, and released after the local storks had left (Schüz 1949). The experimental group had a strong tendency to head south-southeast, the direction of their parent population, even though the local West German storks were heading southwest. It seems, therefore, that for some species the tendency to move in a certain direction is

innate and continues for a set length of time. When the drive is fatigued, the animal stops. Apparently the continuance of the drive is a function of time and not of distance. As support for this idea, it has been found that the restless activity in caged migrators diminishes at about the time they would have reached their goal (Perdeck 1958).

Ecologist R. H. MacArthur (1959), studied migratory birds of the United States by making censuses in "virgin," or climax, forests and grasslands. He found that there is no simple correlation between climate and the proportion of migrants (Figure 13.7). In fact, he found that fewer species migrate from the prairies, with their severe winters and hot summers, than from the west coast forests, with their mild winters and cool summers. In seeking to explain his correlations, MacArthur suggested that migration is greatest where the difference between the winter and summer supply of food is the greatest. He based his estimates of food supply on the level of foliation. Eastern forests, which are largely deciduous, have great outbreaks of defoliating insects, and these insects provide food for birds. When the leaves fall, the food supply of the birds also dwindles, so the net food level for birds is lowered. On the other hand, there is much less seasonal change in the insect populations of the western habitats, and these areas harbor few migrants. The seasonal depletion of insects in forested areas could well have provided the impetus for the birds to leave the area. They may have initially moved in all directions, but those that tended to move southward probably enjoyed the greatest foraging success. The intense competition provided by the numerous resident tropical species could have driven the birds out in the spring. And those that tended to move northward found fewer competitors and a burgeoning food supply. In time, over thousands of years, the pattern could have become stamped into the nervous system of the migrants so that when the proximal cues (for example, changes in food level, competition, light, or temperature) appeared and the physiological conditions were right, they moved in the "correct" directions.

**Continuing Puzzles** There are several puzzling facets of the general problems of bird migration. For example, some individuals within a single population may be sedentary, while others are migratory, and some may change their migratory status. Also, in some cases, stock that does not ordinarily migrate at all can be induced to do so. Hatchlings from nonmigratory English mallards were allowed to mix with Baltic and Finnish migratory mallards and, of 116 young, 19 were recovered up to fifteen hundred miles away in the wintering grounds of the migrators (Matthews 1968). Other tests were run to discover whether migratory *direction* in young migrators can be influenced by experienced birds of another population. It was found that, in some species, displaced first-year birds abandoned their normal migratory direction and flew with their foster-group in a different direction.

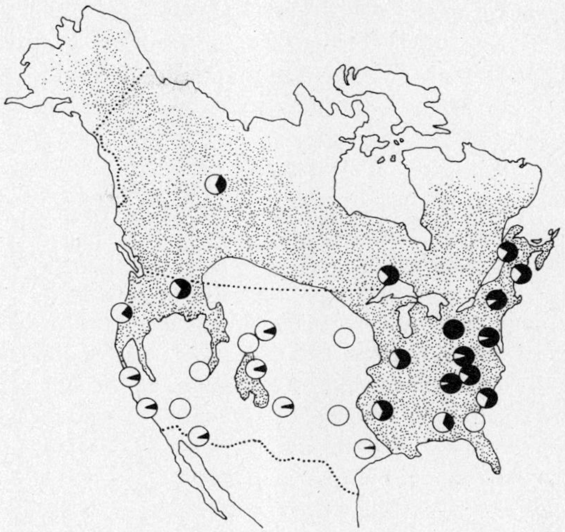

Figure 13.7 Proportions of migrant species that will leave various undisturbed vegetation areas in this Nearctic Zone in winter. Migrators are shown as black areas in the circles that represent populations of breeding birds. The stippled area represents forests.

It should not be assumed, however, that the migratory route is learned from older birds. The urge to stay with others of the same species is a strong one and may overcome normal directional tendencies in the young; however, the migratory route has been demonstrated to be innate in several species.

The English mallards just mentioned apparently followed their foster-group, but in other species, the young make the trip alone. In Europe, as in other places, the parasitic cuckoo (*Cuculus*) lays its eggs in the nests of other birds, which then care for the young birds. The cuckoo thus grows up without ever seeing its own species. When the fall season approaches, the adult cuckoos migrate even before the young leave their foster-parents' nests. As soon as the young cuckoos can fly, however, they leave their foster-parents and migrate two thousand miles to the cuckoo wintering ground, and they do it alone. As another example, the young Atlantic puffin (*Fratercula arctica*) is deserted by its parents in the last days of its nestling period. Before it has even learned to fly well, it tumbles from its nest, flutters to the sea, and begins to swim toward the adult birds already at the wintering area a thousand miles across open ocean.

If a bird has directional ability and a sense of distance, it should theoretically be able to migrate successfully between wintering and breeding grounds, provided no mishaps occur. In the wild, though, mishaps do occur. Violent storms may take birds far off their migratory course. If they were only able to continue on a set compass direction for a certain distance, they might be doomed to running courses parallel to the parent population, never to return to the group. However, some birds have appeared back in the home area after being experimentally displaced. Thus, something other than directional and distance abilities is involved: true navigation. There are indications that navigational ability improves with age so that birds can correct more easily for such disasters. For example, some birds that had successfully completed a round-trip migration were experimentally displaced, and it was found that they would ignore the movement of local migrants as well as their own normal direction, and would fly in the direction of their proper home for that time of year. Apparently, learning increased their navigational ability.

The central question here is, How do they navigate? What directional clues do they use? A great amount of time and energy has been spent in trying to unravel the secrets of the avian navigators, but the problem is far from solved. Some of the evidence is contradictory, and in some cases we must rely on sheer conjecture. However, some of the most promising evidence has come from the analysis of navigation in homing species, as we shall see in the next chapter.

# 14

# HOMING AND CELESTIAL CUES

We have known for centuries that some species of birds are able to return to their nests after they have been taken away. The ancient Greeks and Romans were aware of this ability in their pigeons and they used the birds to carry messages home as they journeyed across the land. Julius Caesar's armies communicated by pigeon. So did the armies of the ancient Persians, Assyrians, Phoenicians, and Egyptians. It was not until about 1825, however, that homing ability began to be selected for by Belgian breeders. Since that time, the breed has changed a great deal from its rock dove stock. Pigeons are rarely used as messengers today (although they saw some service in both World Wars; in fact, there is a monument to the carrier pigeon in Brussels). Now, however, they are bred primarily for racing, as pets, and as subjects for navigation experiments (Figure 14.1).

## HOMING

Homing is the ability to return to one's home area when displaced. It is known in many different kinds of organisms including, as we have seen, fish and amphibians. Some of the most interesting stories concern dogs and cats, but these are mostly anecdotal, so we will ignore them here. A great deal of work has been done on pigeons, however.

### Pigeon Studies

There is much folklore and anecdotal material regarding the abilities of homing pigeons. However, the truth is that the homing speed accomplished by racing pigeons is rare and partly a result of intensive training. That training has traditionally been accomplished by taking the birds longer and longer distances from their nests, usually in the same directions. Normally, they have reared young in these nests and, presumably, are strongly attached to them. A pigeon trained to home in one direction can find its way if released in another direction, but it takes longer. By the way, it should not be assumed that lengthening the training distance in the same direction simply permits the birds to accumulate knowledge of landmarks. Actually, while landmarks may be important in some cases, trained homing pigeons released well out of sight of any known landmarks are also able to navigate homeward.

Strangely enough, homing pigeons differ greatly in their homing ability. Some birds can barely get home at all, while others (a very small percent) are able to home long distances at high speeds on the very first attempt. Only five to ten percent of pigeons are *ever* able to achieve long-distance, high-speed homing, but perhaps ninety percent can manage to navigate homeward, even if a bit slowly.

Inconsistencies have even been shown in the homing ability of individual birds in different trials. Of 122 pigeons tested on 549 sorties (a sortie is a single flight by a single bird), fifty-seven percent vanished into the horizon in the correct direction within 2½ minutes of release at least once, but *not one* had a consistent record of such swift starts (Matthews 1968).

Older pigeons are better homers than younger ones (Matthews 1968.) Whether the older birds have learned to navigate better, or whether their homing ability has simply matured, is not known. Young trained birds are able to get their initial heading just as well as older birds, but older birds generally reach the nest faster. One partial explanation is that they are more familiar with the landmarks around the home. The importance of landmarks was indicated in experiments in which birds were reared in a cage so that they were unable to see the horizon or surrounding area; when these birds were displaced and released, some were lost for good.

In comparing homing over different distances, it was discovered that highly experienced pigeons can home well if released 50 to 100 miles away and that they are successful from short distances of 2½ to 18 miles away, but their homing ability is impaired from intermediate distances of 25 to 35 miles (Matthews 1955, 1963). This information indicates that the navigational mechanisms are particularly functional when the animal is displaced over great distances, but not as sensitive when detecting minor (25 to 35 miles) displacement. One explanation is that very different coordinates would be easier to compare than more similar ones. And short-distance homing, it seems, would be aided by landmark recognition.

The importance of landmarks is not clear, however. Charles Walcott and Robert Green (1974) at Cornell reported that their homing pigeons flew at altitudes of only a few hundred feet (just over treetop level). At such heights, the birds would not have been able to see many landmarks. And, in fact, they found

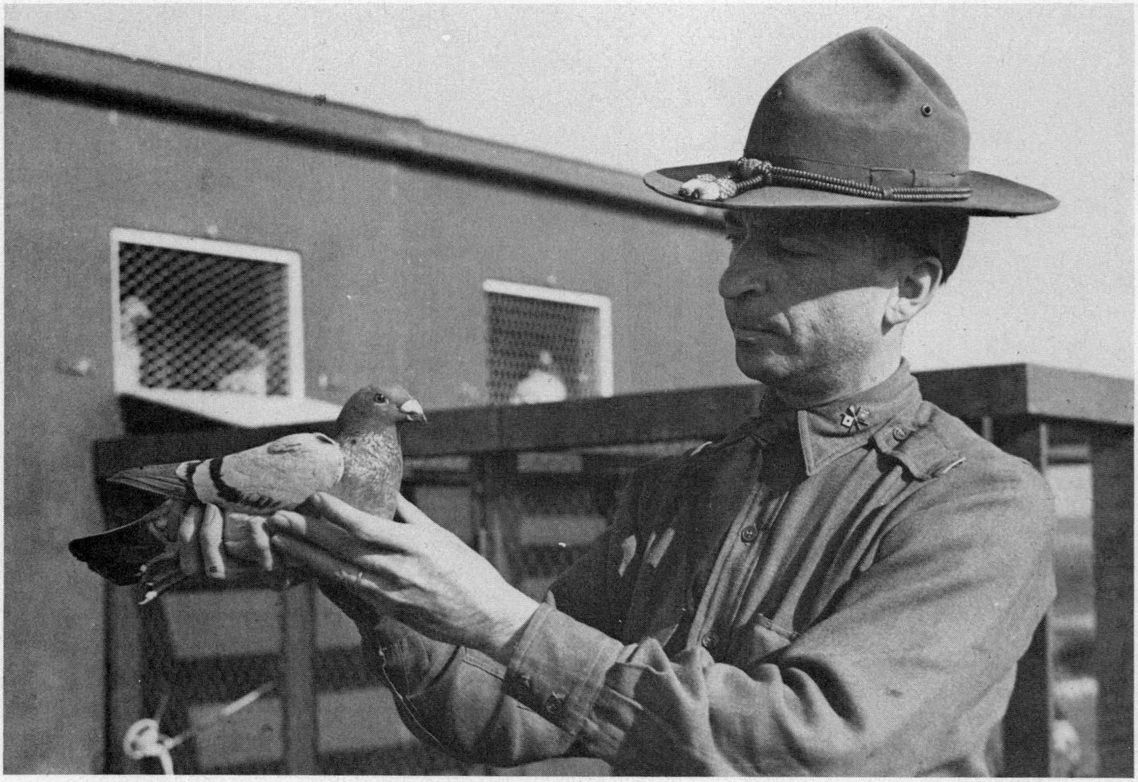

Figure 14.1 Two hundred fifty American soldiers in World War I owed their lives to this homing pigeon who carried a distress message to its handler. Reinforcements were sent to rescue the beleaguered battalion.

that any single bird might take a number of routes, even when homing from the same place, a behavior that indicates landmarks are not very important. On the other hand, one of their pigeons had been trained over a course that had a prominent mountain. Every time *this* pigeon was released, it first headed for the nearest mountain and only later regained its directional sense. (But other pigeons with the same training failed to show the same fascination with the mountain.) Walcott and his group suspected that in one group of experiments, the pigeons were cueing on a large white building, Harvard's William James Center for the Behavioral Sciences, but they were unable to persuade the university to remove the building so that they could test their hypothesis.

Homing pigeons undoubtedly use other clues besides landmarks. In one notable experiment (Schmidt-Köenig and Schlichte 1972), pigeons were fitted with translucent contact lenses so that they could see about as well as you can through a ground-glass door. Obviously, they couldn't see distant objects at all. But some of them came home—flying higher than usual and fluttering helicopterlike to the ground, but there they were. A few passed the roost or landed some distance away; but others were right on target. They could be picked up by hand, since they apparently couldn't see anyone coming. Obviously, the landmark hypothesis wasn't enough.

A number of people have shown that pigeons, as well as a variety of other animals, can get their compass direction by using the position of the sun. But the sun "moves." Therefore a bird can't simply maintain a constant angle to it. If the bird wished to fly east, it would have to fly into the sun at morning, keep the sun over its right wing at midday (in the northern hemisphere), and fly away from the sun at evening. By changing its position with reference to the sun as the day wears on, a bird could maintain a constant direction. Biological clocks are discussed later, but, for now, keep in mind that pigeons have internal clocks and *can* get their compass heading from the sun (see Walcott and Green 1974 for references).

It might be pointed out that knowing compass direction is not the same as being able to use that information to go home. Suppose you were left in the middle of a desert with a compass. You might wonder whether you were in the Gobi, the Kalahari, or the Mojave. You would need to know which way civilization and water lay with respect to your position. Being able to walk due north with a compass would be no trick, but being able to find San Bernardino would. The first would employ compass orientation; the second, true navigation.

There is tentative evidence that there may be two systems working in pigeon orientation: one, a sun compass that depends on an internal biological clock, and the other, a navigational system that doesn't (Walcott and Green 1974). In one experiment, heavy water ($D_2O$—water with a neutron in the nucleus) was put in pigeons' drinking water. Heavy water is known to upset biological clocks in a variety of organisms, including birds. In these pigeons, it retarded the onset of their "active period" by forty-five to seventy-five minutes a day. Most animals operate on roughly a twenty-four-hour schedule and gear their activity periods and their sleep with the clock. The clock, as we shall see later, is normally "set" by the light period each day. If the animals are kept in a place where their light regimen is altered so that it occurs earlier or later, the animals' activity periods will shift in accordance with the light shift, a process called "clock shifting." No one understands how heavy water shifts clocks, but six pigeons were fed heavy water and, at the same time, they were exposed to random light-dark periods so as to throw their timing off as much as possible. Then the birds were taken to Ithaca, New York, 260 miles away, and left there for a week. During this period, the birds reset their clocks—but on Ithaca time. The birds were then taken to a place somewhere between Ithaca and the home loft and released. Surprisingly, they flew home as fast as would birds taken directly from the loft. Thus, although their timing was upset (which normally alters compass direction accordingly), their navigational abilities remained unscathed.

It should not be assumed that clocks are unimportant, however. When birds were clock-shifted six hours ($\frac{1}{4}$ day) and released under a visible sun, they flew off course at ninety degrees ($\frac{1}{4}$ circle) to the correct bearing (Keeton 1969). But clock-shifted birds released under *overcast* skies flew home correctly, showing that clocks may be used only under some conditions. The experiment also showed that the sun may be important at times, but that pigeons can navigate without it. Redundant systems are also suggested by the fact that clock-shifted birds fly off course under sunny skies, but not under overcast conditions.

The place where the bird is released has something to do with how well it gets home. Some places seem to have built-in biases. For example, when William Keeton at Cornell released pigeons at Castor Hill, New York, they mysteriously flew seventy degrees to the right of the proper direction, no matter *which* way home was. But even more astonishing is that birds released at a place called Hornell, New York, almost never get home! They fly first in this direction, then in that, but almost none make it. No one can explain the strange "Hornell effect."

What about the earth's magnetism? Certainly it would be unaffected by clouds, and since there are local magnetic anomalies over the earth's surface, it could also help account for the many anomalies in navigation. There has been a lot of argument over magnetism as an orientation cue, but an undergraduate, Robert Green, at the State University of New York at Stony Brook, answered some important questions (Walcott and Green 1974). Electrical coils were placed around a pigeon's head (Figure 14.2). They were powered with a mercury battery strapped to its back (control animals had the same pack with a dead battery). Two types of coils were used, but both generated a field strength of 0.6 Gauss, a magnetic strength that equals that of the earth's field. One coil was arranged so that it would cause a compass needle to point upward. The other would deflect a compass needle downward. Green found that under overcast skies, the birds with the first type of field headed predominantly toward home. But those with the second type of field headed in just the opposite direction! Under sunny skies, there was only a small but perceptible difference (eleven degrees to thirty-three degrees) in the two groups. So, it seems that pigeons *can* sense magnetic fields somehow, and that their receptors are located in the head. It also seems that if the sun is present, they use it. If it isn't, they rely on other cues.

Finally, there is some evidence that pigeons may use olfaction in homing (Papi et al. 1972), and although this finding is not very well substantiated yet, it is being given increasingly serious attention.

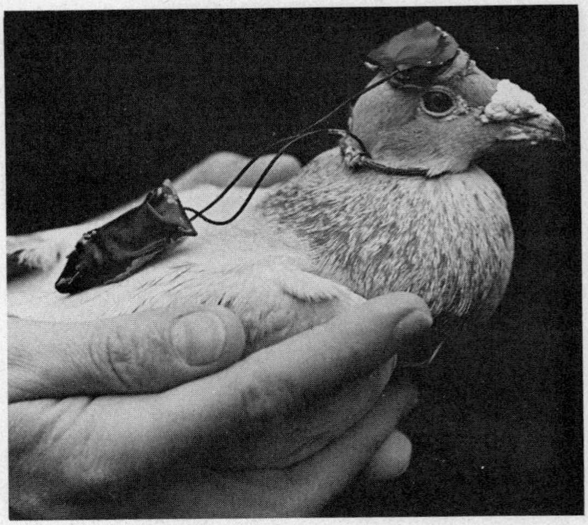

Figure 14.2 Pigeon with electrical coils on the head. Birds carrying coils with reversed polarity tended to home in opposite directions, indicating that pigeons can detect slight magnetic fields and that magnetism can influence their orientation.

The upshot of all this is that pigeons apparently use a variety of orientation and navigation devices in homing. But the complexity only adds to the mystery. How do they navigate from long distances under overcast skies? As yet, we do not even have *one* satisfactory explanation.

### Wild Bird Studies

Homing experiments are often done with domestic pigeons because they are easy to work with, but the best homers are wild seabirds. Homing in wild birds has probably been the result of strong natural selection, since they are often displaced from their homes by natural forces, such as windstorms. There have not been as many experiments with seabirds as with homing pigeons, but in some cases, the data that have been collected are impressive. For example, a Manx shearwater (*Puffinus puffinus*) was taken from its burrow nest and sent by air to Boston, over three thousand miles away, in a "black box" (a carrying case that does not permit the animal to see out). It was back on its nest $12\frac{1}{2}$ days later. There are a number of other remarkable records of shearwaters' homing (Figure 14.3).

The longest homing flights yet recorded were made by two Laysan albatrosses (*Diomedea immutabilis*) that were taken from Midway Island in the central Pacific and released in Washington state. They were home in ten and twelve days. In another experiment, a Laysan albatross returned to its nest after being released four thousand miles away in the Philippines, but that trip took thirty-two days.

Since, regrettably, it is difficult to follow birds on their homing flights except by radio tracking, little is known about what they do en route. Regarding the third Laysan albatross, why did its flight take so much longer? Actually, its flight time should not be surprising, since most birds home at rather slow speeds. Most days they fly distances that could be covered in only a few hours of flying time. The delay is not necessarily related to navigational problems, since they may be spending time in feeding or resting or even soaring.

The route would, of course, influence how long it took to get home. Perhaps some species are inclined to fly over certain types of familiar surfaces such as land or sea, or even along coastlines, even if it meant a slower journey. However, in experiments with the strictly marine Leach's petrel (*Oceanodroma leucorhoa*), it was found that the birds would take the shortest route homeward even if it meant flying overland, but strangely enough, they still homed at the widely different speeds of 40 to 217 miles per day (Billings 1968).

**Nonsense Orientation** There is an interesting trait in some populations of birds that has confused studies of navigational ability. Certain populations show a directional tendency that is called *nonsense*, or *fixed, orientation.* We see it in some bird populations in which all the individuals head the same direction no matter where they are released in reference to their home. The direction is consistently the same under various experimental conditions. However, after an initial nonsense heading, the bird may reorient itself and head toward home.

Nonsense orientation is a population, or "subgroup," tendency rather than a species characteristic. Populations within a given species may or may not show nonsense orientation, and different populations may show different nonsense tendencies.

Nonsense orientation is found in a wide range of animal species from penguins (Emlen and Penney 1964; Penney and Emlen 1967) to arthropods. On the other hand, some species, such as bank swallows and Manx shearwaters, which are excellent navigators, apparently have no such orientation at all. Although there seems to be no reason for nonsense headings, the tendency is so regular and predictable that, once determined, it can be a useful tool in orientation experimentation.

So, although it is not known what transpires en route in migrating and homing flights, the navigational abilities of some species of birds are firmly established. Now let's examine some of the evidence regarding the mechanisms that might be involved. For further discussions of navigational mechanisms, you might refer to Kramer (1957, 1961); Adler (1963); Lindauer (1964); Schmidt-Köenig (1965, 1975); Lockley (1967); Emlen (1974).

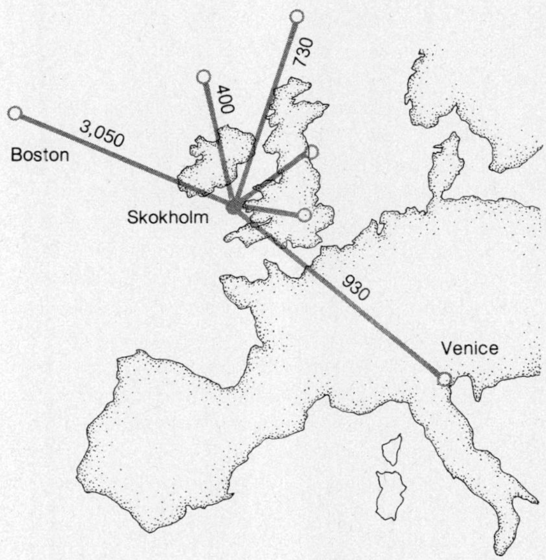


Figure 14.3 Results of homing experiments with Manx shearwaters. The map shows the distance over which individual birds successfully returned to their Skokholm breeding ground. The numbers indicate the distance (land miles) travelled.

## DIURNAL NAVIGATION CLUES

Gustav Kramer (1949), a German biologist, found that caged birds show directional tendencies in their restlessness at migration time. In other words, they huddle in that part of the cage that lies in the direction they would normally take. Also, in their fluttering about, they launch themselves most often in the proper migratory direction. In noting these tendencies, Kramer set the stage for a series of experiments that has yielded fascinating evidence in the quest for the navigational mechanisms of birds.

Kramer set up experiments with caged starlings (*Sturnus vulgaris*) (Figure 14.4) and found that these daytime migrators oriented in the normal migratory direction unless the sky was overcast, in which case they lost their directional ability and moved about randomly. When the sun reappeared, they oriented correctly again. Apparently they were using the sun as a compass. Kramer tested his "sun-compass hypothesis" by blocking the sun from the birds' view and using mirrors to make the sun appear to lie in different directions. In each case, the birds oriented according to the position of the new sun.

In other experiments, Kramer and his students trained their birds to expect food in a box that was no different in appearance from other food boxes that encircled the enclosure (Figure 14.5), except that it lay in a certain compass direction. As long as the birds could see the sun, they would approach the proper food box. Even when objects in the enclosure were moved about, the birds were not confused. When the feeding boxes were rotated and the wire cage enclosing the birds was lifted, the birds simply walked over to the box that lay in the proper compass direction. However, on overcast days, the birds were often disoriented—further evidence of a sun-compass. When they placed an empty box in the direction in which the birds had been trained to expect food, the birds would try the box several times before they gave up. Kramer said it was amusing to see their puzzlement at the empty box, but they were not easily persuaded that they had made an orientation error.

Kramer noticed an interesting pattern emerging in his artificial sun experiments. If the sun remained stationary, the birds would continually adjust their angle with reference to it as though it were moving. This reaction became much more pronounced when the sun was raised and lowered at the same rate of the apparent rise and fall of the real sun. He had noticed earlier that when the real sun was used, the birds were always able to orient in the proper direction even as the sun moved across the sky. They were able to compensate for the sun's apparent movement, and the compensation meant they possessed some sort of independent timing mechanism (Figure 14.6).

You will recall the experiments in which the internal clock of pigeons was shifted six hours ($\frac{1}{4}$ day), and when the birds were taken from home and released under the sun they departed ninety degrees ($\frac{1}{4}$ circle) from the home direction. In another experiment, starlings were reared with their clocks retarded by six hours under artificial light. They were then placed outside when the sun was at the same altitude (3:00 P.M.) as the artificial sun inside (set at

Figure 14.4 Starlings are daytime migrators and were the subject of Gustav Kramer's pioneering work in bird navigation.

9:00 A.M.). It was found that their orientation changed ninety degrees clockwise. Their rate of change was fifteen degrees per hour, or the sun's rate of apparent movement (Hoffman 1954).

A converse test involved moving Adelie penguins (*Pygoscelis adeliae*) that showed a northward nonsense orientation to a different longitude (at the poles, small lateral shifts result in large time changes). Their nonsense orientation changed, as expected, by deviating counterclockwise when transported eastward, and clockwise when transported westward. Nonsense orientation may be changed in birds, by the way, by resetting their biological clocks.

Just how do birds utilize the sun in navigation? The sun actually can provide several kinds of information, but it isn't clear which clues are used. For example, the sun may be used as a simple compass; but knowing which way is north does not solve all navigational problems, as I mentioned earlier.

Human navigators describe the compass direction of any celestial body by taking the number of degrees, measured clockwise, from a known compass direction—usually north. This reading is called the *azimuth*. A second reading may be taken of *altitude*, or the vertical distance from the horizon to the celestial object (Figure 14.7). Simply observing the altitude of the sun only places the observer on a *circle of Sumner*. This circle includes all those points on the earth where, at a given time, the sun is at a given altitude (Figure 14.8). Since the circle may extend for thousands of miles, it is of little use by itself. Of course, a bird could restrict its location to a part of the circle if it knew which way the sun was moving; but the sun moves slowly. One wonders, then, can a bird actually detect the movement of the sun along its arc? The movement is so slow that it might be easier to see movement in the minute hand of a watch. But it turns out that pigeons *are* able to detect such movement (Meyer 1964). They can distinguish a stationary artificial sun from one moving at fifteen degrees per hour, the sun's normal rate. (Humans are unable to distinguish movement slower than sixty degrees per hour.) The necessity for determining the direction of the sun's movement may account for the delay some birds exhibit before they begin orienting properly (fifty minutes or more for some species [Walcott and Michener 1967]). It may also explain why even the best homers often start courses sixty degrees to ninety degrees from the correct course.

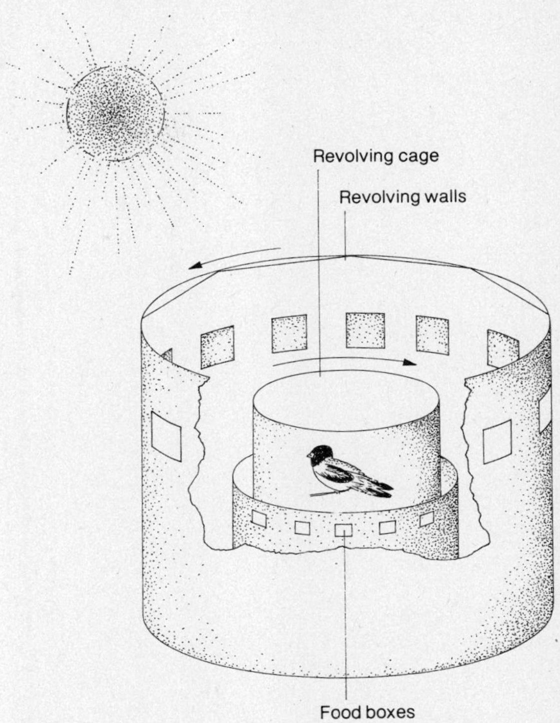


Figure 14.5 Kramer's orientation cage. The bird can see only the sky through the glass roof. The apparent direction of the sun can be shifted with mirrors.

Theories of sun navigation involve assumptions that the bird knows where the sun should be at a certain time of day in the bird's home area. Therefore, it is suggested that homing is an effort to move so as to restore the sun to its proper position for that time of day. If a bird were able to extrapolate, from a brief glimpse of the sun's movement, just where the sun will be at its peak (or, in the Northern Hemisphere, when the sun will arrive at due south in its arc), it would be possible for the bird to determine its own latitude (Matthews 1968). If the sun were going

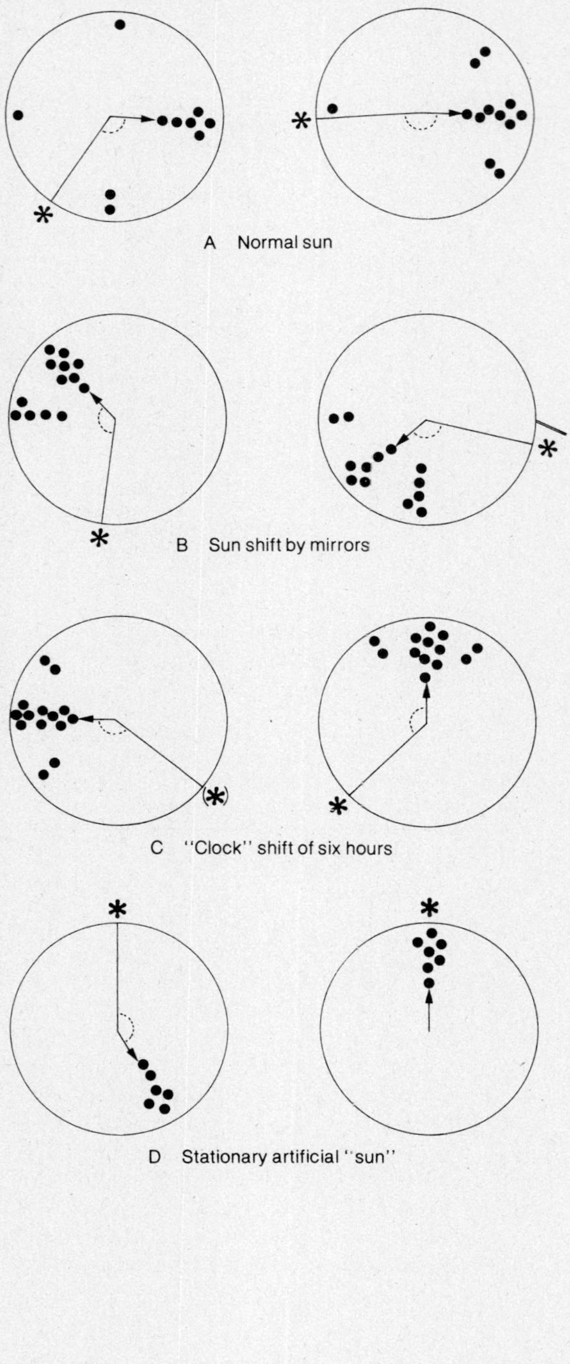


Figure 14.6 The analysis of sun-compass orientation in starlings tested in Kramer cages. The results in (A) show that the birds maintain the same heading even as the sun changes position. In (B) orientation is shifted as the apparent position of the sun is changed by mirrors. In (C) the birds' orientation is shifted by altering their internal clocks. In (D) the birds change bearing if the artificial sun is held at the same compass direction. The dots represent the primary directions of activity in (A) and (B) and food box choices in (C) and (D). The sun's apparent direction, real or perceived, is indicated with an asterisk.

to be lower at noon than it would be at home, then the bird would be too far north (Figure 14.9). There are indications that birds, as well as bees and lizards, have the ability to project the sun to its highest point, but the ability has never been conclusively demonstrated in vertebrates (Lockley 1967).

Most researchers now agree that the sun is used only as a simple compass, but it has also been suggested that birds might detect longitudinal displacement by use of the sun's arc (Matthews 1968). For example, if the sun were on a course such that at noon (home time) it would be lower in the eastern sky than if observed from home, the bird could correct for this by flying eastward. However, there are those who argue that extrapolation to the sun's highest point in its arc is beyond the ability of birds. Actually, they point out, the correct bearing can be taken from azimuthal position alone. Even this simpler idea, however, is complicated by the fact that the rate of change in azimuth varies with the time of day. The sun changes its compass direction more slowly in the morning and evening than at midday at almost all points on the earth.

Are there possibly other diurnal celestial clues that could be used in navigation? In Keeton's (1969) experiments with pigeons, resetting the biological clocks caused the birds to orient incorrectly in homing tests when the sun was visible. Yet when he tested these birds under overcast skies, they flew home. Keeton interpreted this as evidence of a redundant mechanism built into the bird's navigational system that is not time dependent and that is not employed when the sun is visible. Walcott and Green (1974) talked about two systems, a time independent navigational system and a time dependent compass.

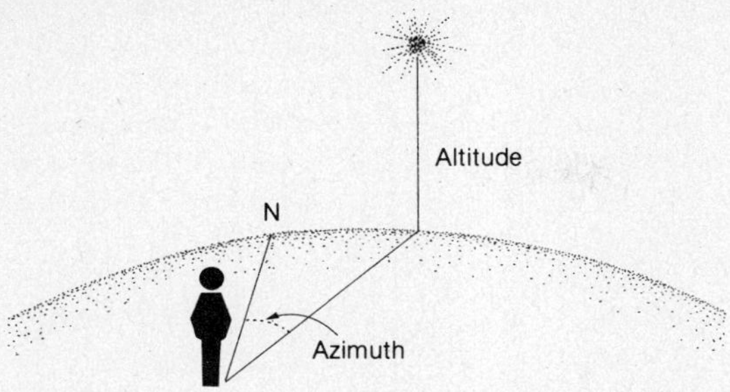


Figure 14.7 The position of a celestial object as determined by azimuth and altitude readings.

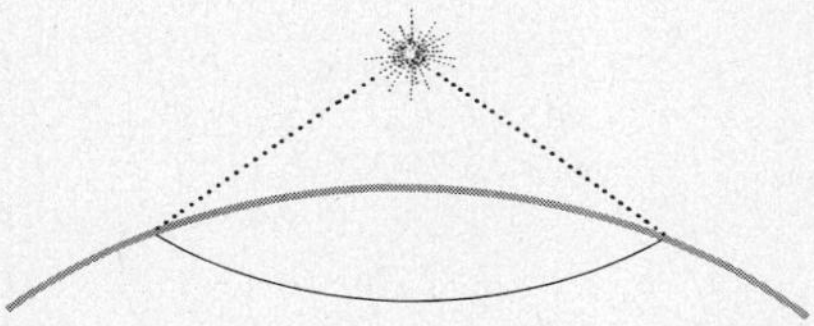

Figure 14.8 Circle of Sumner. The sun appears at the same altitude with reference to the earth's horizon from any point on the circle.

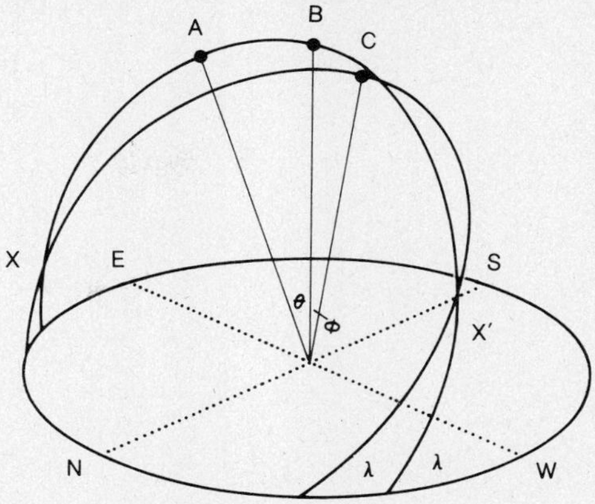


Figure 14.9 Perceived changes in the sun arc due to a move in a southwest direction at noon (home time). The altitude of the highest point of the arc *B* is greater (by Θ) than at home *C*. The inclination of the arc is also greater. The observer could be aware of the shift since the sun *A* would not have moved as far along its arc as it would have at home. In order to avail itself of such information, a bird would have to be able to extrapolate a new arc from observing movement of the sun after the bird had changed geographic positions. The bird would also have to memorize the home arc and be able to use it as a reference.

Implicit in Keeton's interpretation of his experiments is the possibility that birds use diurnal celestial clues other than the sun. Two potential clues exist: the planets and the daytime moon. It is true that either Mercury, Venus, or Jupiter is above the horizon most of the time, and although the positions of the planets are not nearly as stable as those of the stars, the planets do not change position much at a given hour over a period of a few days. Actually, there are reports that bird's eyes do show adaptations for reading celestial clues (Matthews 1968), but there is no good evidence that even these specialists are able to see the planets in a daytime sky. Although the daytime moon has been suggested as a possible celestial clue, there has been no direct evidence to support the suggestion. Also, many problems arise in attempting to explain how the daytime moon might be useful. The stars, moon, and planets would be expected to best be utilized in nighttime navigation, and there is evidence indicating that the night sky does, in fact, hold navigational clues.

## NOCTURNAL NAVIGATION CLUES

It has been known for many years that birds migrate both by day and night. Some species are more inclined to fly at one time than the other, but many species fly at all hours. After the evidence for daytime celestial navigation began to appear, it was assumed for a while that nocturnal migrants took their bearings at sunset and muddled about as best they could through the night. Then the German ornithologists Franz Sauer and E. M. Sauer (1955) reported that birds can navigate at night. They put two sylviid warblers (blackcaps and garden warblers) in Kramer cages and exposed them to the autumn night sky of Germany. The cages were variously modified with devices to record the direction of the birds' activities

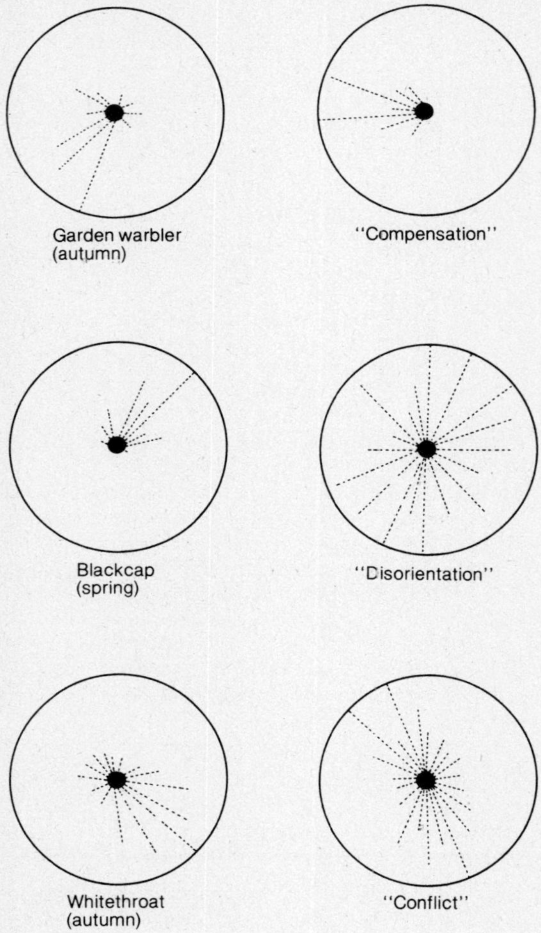


Figure 14.10 The migratory orientation of warblers in cages under the dome of a planetarium. At left shows what happens when the dome is aligned with the outside sky. At right shows the directional activity of the warblers when the dome is rotated. Compensation occurs when the birds orient correctly to the new pattern. Disorientation occurs when the birds show no directional activity. Conflict is the result of a multidirectional phenomenon. Note that the longest ray indicates the direction most often taken. The remaining rays are proportionate.

without the necessity of continuous observation. The Sauers' birds had been hand-reared indoors and had never seen the sky, but after one glimpse of the stars they oriented southward. When the test was repeated in the spring, they oriented northward. So, here was evidence that the night sky could provide navigational clues.

Franz Sauer then began a series of experiments aimed at discovering just which objects in the nighttime sky the birds use as clues. He moved his Kramer cage inside the Bremer planetarium so that the nighttime sky could be controlled. He first lined up his planetarium sky with the sky outside and found that the birds oriented themselves in the proper migratory direction for that time of year. When the "sky" was advanced from its normal position for that time of year, the birds became disoriented and moved about randomly. In some of these experiments, the moon and planets were not projected, but the birds oriented correctly anyway. Apparently, then, they were taking their bearings from the stars (Figure 14.10). Sauer concluded that the birds were not relying on the stars themselves, but on the altitude and azimuth of the starry sky. They were able to derive a "fix" by determining longitude from the azimuth and latitude from the altitude of certain stars, a rather complex set of calculations, as skeptics have consistently pointed out. Other critics have maintained that Sauer's work suffers from a lack of direct observation, a lack of quantitative analysis, and small sample sizes.

Of course, human navigators have long been aware of the usefulness of stars in direction finding; constellations may provide rough information even to the amateur navigator. For example, when Orion's belt is perpendicular to the horizon, it is in the eastern sky. The angle is forty-five degrees when it is in the southern sky, and it is parallel to the horizon when it is in the western sky. And you are probably already aware that the North Star is to the north.

The apparent motion of stars presents serious difficulties in our attempt to understand nocturnal navigation. There has been some argument over whether birds navigate by constellations or individual stars, but simplicity would dictate that they use those stars that show the least apparent movement. Stars do not show the seasonal rise and fall of their arcs as the sun does, but they rise four minutes earlier each night (two hours earlier each month). So the stellar day is twenty-three hours and fifty-six minutes long. This means the star sphere continuously slips westward until a given star returns to the position where it was a year ago at a given time. This also means a bird would have to fly at a different angle relative to a certain star, depending on which night of the year it flew. As the seasons change, new stars appear and others disappear below the horizon. The only groups of stars that are always above the horizon are those within thirty-five degrees of Polaris, the relatively stationary North Star. There is evidence that this star group is the most important one as a celestial cue for birds (Emlen 1967).

Many birds cross the equator in their annual migrations, and in so doing an entirely new sky with its myriad celestial patterns comes into view. Out of this strange sky the bird must pick its navigational clues, and it must do it right the first time. Of course, the North Star is lost from view as the Southern Cross comes into view. But the Southern Cross is not as stationary as the Polaris group and is therefore not as good a navigational clue. So the theory of navigation by the stars becomes increasingly complex.

If birds are able to compensate for the apparent motion of the stars, they must have an internal clock. As is the case with diurnal navigation, a clock would permit the bird to alter its direction relative to any single changing celestial clue. A clock allows the bird to know where a regular moving object should be at any time. A night clock is much more demanding than a diurnal one that uses a single celestial clue (Emlen 1967). If individual stars are used in navigation, the rate of compensation for apparent movement must differ depending on which star is used. A star appearing near Polaris moves through a smaller arc, due to the earth's rotation, than one forming a concentric circle farther away; therefore, since both stars complete their circles in the same period of time, the star nearest Polaris seems to move more slowly, just as the hands of a clock appear to move more slowly near the pivot point and faster near the tips. Actually, the rate is the same at both positions: in the case of celestial cues, fifteen degrees per hour. So, if a bird uses stars of different declinations, its compensation rates must vary accordingly. The same sort of difficulty arises if star groups are used. Groups nearest the North Star describe a smaller arc.

Some of the evidence implicating biological clocks in nocturnal navigation has been puzzling and contradictory. For example, some experiments indicate that timing is irrelevant to nighttime orientation. Matthews (1968) reset the clocks of mallards (*Anas platyrhynchos*), a species known to migrate both by day and night. Three groups of birds were tested by setting group I six hours ahead, group II six hours behind, and group III twelve hours out of phase. When released in daylight, the controls flew northwest as expected, group I flew southwest, group II northeast, and group III flew southeast. When released at night, however, all three groups went northwest with the control group. One interpretation is that birds can get more reliable information from stars than from the sun and therefore cannot be thrown off course by clock resetting (Griffin 1964). Matthews observed on some occasions that his mallards flew northwest when thin clouds obscured the stars, but not the moon. So, possibly, mallards as well as other birds show "moon-compass orientation." It should be pointed out, however, that Matthew's conclusions in celestial navigation have generated quite a bit of argument.

Apparently, then, there is great diversity of strategies used by navigating birds. Some evidently use circumpolar stars and therefore do not need to compensate for apparent movement. Others may use other star patterns, and some may reset their star compass every few days by referring to some sort of magnetic compass. The questions grow almost daily and it appears that the likelihood of developing universal laws grows dim as we unveil the rich spectrum of animal navigation. For good measure, let's consider a few more examples indicating the complexity of the puzzle.

On the basis of experiments with nocturnal migrating indigo buntings (*Passerina cyanea*) (Figure 14.11)), Emlen (1967) suggested that time compensation for stellar motion is not an essential component in the nighttime orientation mechanism. He arrived at this conclusion by shifting planetarium skies three, six, and twelve hours ahead of and behind local time and finding that twenty-five out of twenty-eight buntings tested maintained their original heading. Subsequent experiments showed that the birds were orienting by clues from the planetarium itself, but the findings still indicated that the birds were not relying on a time clock. These results conflict with

Figure 14.11 Indigo buntings are nighttime migrators that have presented us with puzzling and contradictory evidence regarding their orienting mechanism.

those of some of the experiments performed by Sauer that indicated that clocks are important in nighttime navigation.

Emlen (1967) also found that if some of the stars in the planetarium are turned off, indigo buntings can still orient correctly. Even if some of the brightest lights are turned off, they do not become disoriented. This evidence serves as further support for the notion that the birds are using the overall configuration—the gestalt—of star patterns.

Theoretically, there are two ways a bird could determine simple compass direction ("which way is north?") from the stars. One requires a sense of time; the other doesn't. In the first instance, a bird could locate a specific star, or group of stars, and then fly at a particular angle to its star reference. Since stars move, the angle would have to change through the night in order for the bird to maintain a constant course. The bird would need to possess both a clock and a means of pattern recognition in order to utilize stellar references.

The second way a nighttime migrator could gain directional information is to utilize the configuration of the stars. For example, humans can easily locate the Big Dipper and use it to find the North Star, thereby gaining directional information without the use of a timing mechanism. Star patterns such as the Big Dipper remain constant, even as they move across the sky, and they all maintain a distinct relationship to the North Star.

Emlen (1975) found that when indigo buntings are exposed to stars on a planetarium ceiling that are out of phase with the birds' internal time sense (advanced and retarded, three, six, and twelve hours from local time), the birds were unaffected. They maintained their normal migratory direction. Apparently, then, indigo buntings obtain directional information from star patterns, much as humans do. He also found that when various patterns of stars were turned off in the artificial night sky, the birds were still able to orient correctly. So, the ability to read a number of constellations provided a redundant orientation system that would be valuable under some conditions, such as spotty cloud cover.

But why do birds orient southward in the fall and northward in the spring? Simply being able to ascertain directions at a time when migratory restlessness appears wouldn't be enough to send a bird off in the proper direction. Of course, the fall sky is different from the spring sky. So, do birds have a built-in response to each stellar configuration? Emlen (1975) altered the annual clocks of one group of buntings by exposing them to artificial light periods so that their spring and fall molting patterns were reversed. When he compared them under the same sky to a group of buntings that had not been subjected to clock-tampering, the two groups oriented in opposite directions. He concluded that directionality is partly under the influence of hormone levels that are stimulated by certain light regimes. (His conclusions were supported by A. H. Meier and D. D. Martin of Louisiana State University who reversed the orientation of another nighttime migrant, the white-throated sparrow, by altering the levels of two hormones, prolactin and corticosterone, in the birds' bodies.)

One of the stumbling blocks in developing a theory of an innate mechanism of navigation in animals has been an unwillingness on the part of many biologists to believe that animals are born with some sort of star map. Emlen (1975) has relieved us of the necessity of our developing a "willful suspension of disbelief." He found that if young indigo buntings are never permitted to see a night sky or any point of light, but are reared in a windowless room with diffuse lighting, they will develop migratory restlessness at the proper time, but when exposed to a planetarium sky that is normal for the season, they will flutter randomly, unable to read the stars. Buntings that are allowed to view a "normal" planetarium sky every night, on the other hand, are able to orient correctly. Thus the map is not inborn; experience with the stars is necessary for the normal maturation of stellar orientation.

Emlen also found that there is a critical period in the lives of very young birds when they are able to learn about sky patterns. This time window is brief, and the young birds will learn any sky pattern presented to them at this time, even nonsense star patterns. We see, then, that an innate time window is coupled with a learning pattern to produce an adaptive response.

These experiments revealed another interesting fact in this continuing puzzle. In the planetarium sky, a new pole star was selected. All the other stars were made to revolve around Betelgeuse, a bright star in Orion, rather than the North Star. Buntings that were exposed to the new pole star on alternate nights as they matured oriented away from Betelgeuse, not the North Star, when placed under a normal autumn night sky. Thus, the young birds determined the north-south axis by the movement of the stars, those around the pole star describing smaller arcs. The axis, then, provides one reference system, and star patterns are of no use until a reference system has been established. However, once the reference system and star-configuration information have been coupled, the buntings can derive the north-south axis from the star patterns alone.

For further discussion and criticisms of the planetarium experiments, see Adler (1963) and Schmidt-Koenig (1965, 1975). Unfortunately, much of the experimental evidence is contradictory, but these discrepancies will doubtless be cleared up as experimental techniques are perfected and more sensitive statistical procedures are developed.

## OTHER TYPES OF NAVIGATIONAL DEVICES

It is possible that birds can orient by clues other than visual ones. In fact, it has been reported that birds can orient in the absence of any celestial clues (Gerdes 1962; Fromme 1961).

Once a bird has begun a migratory flight, should it encounter overcast skies, it might conceivably use a number of types of clues to maintain the proper course. There are theories involving the use of Coriolis force, or the sensing of latitudinal changes in subtle forces due to the earth's rotation. Inertial receptors of the inner-ear labyrinth, which might allow the bird to feel slight changes in its speed or direction, have also been hypothesized, but, they are undiscovered as yet. It has been suggested, also, that birds that call to each other in flight might detect wind movement by differences in the sound produced by members of different parts of the flock. These theories are almost entirely unsubstantiated by experimental evidence.

One of the interesting problems of migratory flight centers on how the birds maintain the proper altitude even in completely overcast skies and when the ground is not visible. Radar observations show that they continue to move inside and between cloud layers and that their flight path varies only about forty vertical meters for distances of two miles at altitudes of seven hundred meters to eleven hundred meters. It has been suggested that perhaps they listen for sounds or echoes from the ground, such as those reported by balloonists. Some cave-dwelling birds are able to use echolocation to fly in the dark in much the same way a bat does (see the discussion on bats at the end of this chapter), so the idea of such a system being available to migrating birds is not unreasonable. Another theory is that the birds may be sensitive to small changes in atmospheric pressure, such as we feel in our ears when driving over mountains. It has been suggested that the perception of changes in atmospheric pressure may be the function of the paratympanic organ (Griffin 1955). Recent evidence indicates that birds are indeed highly sensitive to air pressure changes.

## EMITTED ENERGY ORIENTATION

Most orientation and navigation is accomplished by organisms perceiving environmental features that are independent of the organism, such as sunlight. Another type of orienting mechanism, however, is the clue that is derived from the activity of the animal itself. Remember the theory that night-flying birds that continually call in flight may be listening for deflection of the sound waves to determine their altitude or their relationship to the rest of the flock. Also, recall the discussion of echolocation in bats. The general mechanism of "reading" self-produced energy patterns is called *emitted-energy orientation* (Marler and Hamilton 1966) and it has been found to be a feature of a widely diverse group of animals.

At present, there is no evidence that emitted-energy orientation functions in any other way than (1) simply giving an animal information about the topography of its immediate environment; and (2) making known the presence of other objects in the air or water through which the animal is moving. There is no indication that this type of orientation functions in navigation other than in merely permitting the organism to move successfully through its immediate environment.

As would be expected, the species that have evolved such systems are generally those that cannot utilize light in orientation or those that live in places where light energy is not present in sufficient amounts to be useful. As examples, we will briefly consider water beetles, electric fishes, bats, and porpoises.

### Water Beetles

We have all seen groups of the whirligig beetles (*Gyrinus*), spinning crazily over the surface of freshwater ponds, and may have wondered how they keep from crashing into each other. It has been demonstrated that as long as the beetles keep moving, they are able to avoid contact. But if one stops, he may be bumped. It seems each beetle is aware of the location and direction of the others by the vibrating waves they set up. The impulses are registered by a modified second antennal segment.

### Electric Fish

The dark, turbid waters of the rivers of South America and Africa are the homes of two unusual groups of fish that have evolved separately, but that are remarkably similar in some ways. The mormyrid and gymnotid fish swim with equal ease either frontward or backward by simply reversing the direction of their undulating fins. The undulations, by the way, are closely regulated to maintain a precise alignment of the electroreceptors. The fish avoid obstacles when swimming either way and when they enter a crack or crevice, they enter it backwards. Backward movement should not be surprising in animals that do not primarily utilize sight in orientation, and these fish do not.

It has been found that orientation, obstacle avoidance, and recognition of the presence of other animals are a result of sensitivity to changes in electrical fields, which the fish produce in a tail organ (Lissmann 1951). Conducting objects, such as other animals, distort the electrical field differently than do inanimate objects, such as jutting rocks. Other electric fish distort the field in a characteristic manner also, and so the animal is able to recognize the presence of members of its own species.

### Bats

In 1793 and 1794, Spallanzani performed a set of experiments that showed that blinded bats could find their way about a room. Then it was found that if the ears of blinded bats were plugged, that ability was lost. The subject was essentially one of conjecture until, in this century, the mechanism was found to be echolocation (Griffin 1958). Several species of bats simply bounce sound waves off any object in their immediate environment and by noting the time taken for the echo to return, they calculate the distance to the object. The size of an object may be determined by the distortion of the sound wave when it is deflected. The bats are thereby able to successfully move through the elaborate darkened mazes of their home caves without crashing headlong into the walls, as would a visually oriented creature.

Not all bats forage on airborne creatures. In addition to the fruit-eating flying foxes (Megachiroptera), and the fabled vampire (*Desmodus*), there are the fishing bats. The best known of these is *Noctilio leporinus*, the bulldog bat. This bat takes small fish near the surface of the water. Its toenails are elongated and sharply curved to act as gaffs, and the bats often trawl by flying along dragging their hind claws in the water. The search is not always random, however, for it seems they are able to locate fish by echolocation. Surface-feeding fish often break the surface of the water, causing ripples. The small deviations in the pattern of the echo caused by the ripples aid the fish-eating bats in locating their prey.

### Porpoises

The development of antisubmarine devices during World War II led to a great amount of interest in underwater sounds. The early workers in sonar were surprised to find the ocean a rather noisy place, with much of the sound produced by animals. This finding should not have been surprising, however, since the physical properties of water cause sound to attenuate less and to travel more rapidly than in air.

Porpoises (*Phocaena*) are among the marine species that are believed to use echolocation. Captive porpoises are able to move about and locate food in very turbid water, which renders vision almost useless. It is now known that porpoises continually emit high-frequency pulses at intervals of fifteen to twenty seconds and that these pulses function in orientation and in location of other bodies in the water. Interestingly, porpoises have no vocal chords and apparently produce sound by transferring air from one nasal sac to another (Kellogg 1961; Norris 1969; Erulkar 1972).

Now, let's consider the phenomenon of biological clocks in animals. It will become apparent that orientation and timing in animals are tightly interrelated. The subject of internal clocks is an intriguing one, even when considered alone, however, and it provides some fascinating examples of scientific detective work.

# 15

# TIMING AND BIOLOGICAL CLOCKS

Perhaps you have noticed that the small carnivore sharing your home becomes restless each evening. The eyes of the cat are dark-adapted, and the claws are sheathed in silent pads. If you were not feeding it, it would find its greatest foraging success at night when its stealth and quickness would enable it to catch smaller animals. For ages, its ancestors hunted at night, and now the pattern is firmly fixed in the entire group. We can guess how a tiger knows when night comes, but the question arises, how does an apartment-dwelling cat know when it is night? In this chapter, we will investigate timing in living things.

It is important for most animals to know the time of day or to behave as if they did. Just as living things must arrange themselves spatially, they must also arrange themselves in time. Since time itself is apparently not segmented so that it can be measured by the rate at which the units pass, it is usually measured in terms of regularly occurring events instead. (Notice that in our attempts to conceptualize time, we think in terms of space, thus accenting the arbitrariness of the division.) Because the earth revolves around the sun with regularity, rotates at a constant rate on its tilted axis, and has a predictable and recurring position in relation to the moon, life on earth is subjected to cyclic fluctuations in its environment. Of course, the life that exists in the changing environment must change so as to behave in the most adaptive manner according to the conditions.

The rhythmical nature of life comes as something of a surprise—even to many biologists—but this should not be the case in the light of what we know about evolutionary principles. After all, life developed under extremely cyclical conditions. The differences in the phases of the cycles are often so pronounced that most life forms would be expected to show cyclical behavior so as to be able to accommodate themselves as specifically as possible to each phase. Even very subtle environmental rhythmicity may elicit specific responses in populations that are exposed to it over long periods of time. We are all somewhat taken aback, from time to time, as new data appear that expand the list of species showing cyclic behavior, and as the rhythmic nature of life expresses itself in myriad ways. Such revelations only serve to point up our level of ignorance on the subject. Admittedly, this level is somewhat higher than might be expected since the first reports appeared over two thousand years ago.

## A BRIEF HISTORY

About 300 years before Aristotle, men wrote about plants that raise their leaves to the sky during the day and fold them at night. But it wasn't noticed for another 2,400 years that the leaves would rise and fall even when the plants were indoors under artificial light and apparently deprived of environmental information regarding the time of day. Then it was found that even in plants that were kept in total darkness and at a constant temperature, the leaves would continue to rise and fall with a startling regularity (Figure 15.1). This meant, of course, that the organisms were somewhat able to keep time, and so the search for the "biological clock" was on.

The first reports of biological clocks in this country came from John Welsh at Harvard (working with crustaceans) and Orlando Park at Northwestern (working with insects) in 1934. The response to their work was unfortunate. The scientific community greeted their reports with distinct coolness, and their evidence was almost ignored. For the most part, biologists simply found it very difficult to accept either of the two proposed hypotheses: independent internal timers, or the ability to read exceedingly subtle environmental clues. It was generally thought that the fault lay with experimental technique. Then, in 1948, Dr. Frank A. Brown, feeling his status secure enough in the scientific world to approach an unpopular problem, took up the sword. After analyzing his own work, he announced his opinion: organisms time themselves by environmental clues. These clues, he theorized, are not the obvious ones such as light and temperature, but subtle environmental influences that are able to pervade man's best attempts to provide constant conditions. His statements generated discussion, and since biologists are rather testy sorts, the banner of the opposition was quickly raised. *Their* position was that timing is by internal mechanisms and is independent of the environment. Unfortunately, we are not able to announce a decision in the battle. The issue is unresolved to this day. So we shall consider evidence for the two arguments, but before doing so, it would be useful to discuss a few general aspects of what is now known about the phenomenon.

## THE ADAPTIVENESS OF RHYTHMS

Just as there are many environmental cycles (such as those imposed by the moon and sun), there are also a number of biological rhythms. Such rhythms may vary from the rapid staccato of firing neurons to slow yearly seasonal changes, such as we see in those animals that build up fat or store food before winter. There are even rhythmic aspects of events that occur only occasionally or just once in populations. For example, most people are born in the very early morning. And, if you die of natural causes, you will probably also expire in the wee hours. However, probably the most obvious and best-known cyclical changes occur on about a twenty-four-hour basis, and most of our discussion will draw on these daily changes.

It should be pointed out that rhythms may not be solely related to abiotic (unliving) influences such as temperature and light. They may also be coordinated with the activities of other organisms, which may be, in turn, locked in step with the environment. Bees time their visits to flowers so that they arrive at the time when they are beginning their daily period of nectar secretion. This means the bee can gather the maximum amount of food with a minimum effort, and it means the flowers, which open with the sun, need not waste energy by continuously secreting their enticing nectar. There are numerous examples of rhythms of one species coordinating its activities with another group. Owls hunt mostly at night when the mice are out, and mice come out at night when they can't be seen.

*Differences* in the rhythms of two species occupying the same range may be important in reducing competition. Fly-catching birds and bats are not in direct competition because of the time differences in their feeding behavior. The swifts of Spain work mainly in the late afternoon, but I have observed that they yield the air to bats as dusk draws on.

Some animals show an interesting age-dependent rhythmicity. They may operate on one time schedule early in their lives and on another one later. For example, badgers play in the sun near their den when they are young. As they grow older, they gradually shift their period of greatest activity toward evening and night. Such shifts are accompanied by other

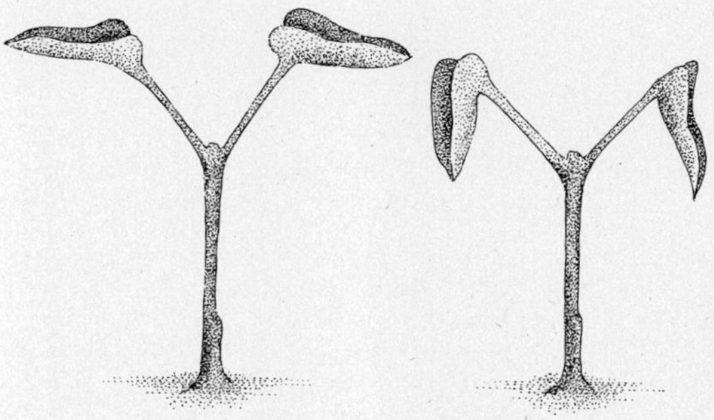

Figure 15.1 The normal sleep movements of plants. The leaves rise in the morning and fall at night. Such movements persist even when the plant is placed under constant artificial light (or in total darkness) and even temperature.

changes in behavior. The younger badgers are quite calm during the day, but as they grow older, they become more uneasy during the day and increasingly bolder at night (Figure 15.2) (Eibl-Eibesfeldt 1950). In some cases, nighttime activity may be a function, not so much of photoperiod, but of the amount of moonlight that is illuminating the environment. Prey animals can be seen far more easily in moonlight than starlight and therefore some of them tend to restrict their activities to periods when the moon is down (with apologies to Steinbeck) (Lockard and Owings 1974).

It is not entirely correct to speak of *the* rhythm of an organism, since several cyclic events may be taking place simultaneously within a single organism. Furthermore, such rhythms may be entirely out of phase with each other. Michael Menaker, of the University of Texas, for example, pointed out that in mice, body temperature, glycogen, and DNA content of a regenerating liver show daily rhythms but that each may reach their maximum values at different times of the day.

Proper coordination in timing (called *phase relationships*) among the various internal physiological processes are sometimes critical (Menaker 1969). Substrates, enzymes, energy sources, and information-carrying molecules must appear in a properly spaced sequence if the body is to function as a coordinated whole. For example, in chick embryos the importance of the phase relationship of cellular constituents is indicated by the rather sudden appearance of a great many enzymes on about the eleventh day of incubation. After their rapid formation in the watery embryonic tissues, they begin functioning in proper phase relationship to each other.

As an indication of the complex nature of the biological clock (or clocks), it has been found that all the cycles occurring within a single individual may not be in step with the same aspects of the environment. For example, intertidal crabs change color in time with local day-night cycles, and they also vary cyclically in their "running periods," the time of greatest activity. Interestingly, the light-dark period takes place over about a twenty-four-hour interval and corresponds with the appearance of the sun, but there are two running periods, one usually at night, the other during the day. The running periods arrive fifty minutes later each day and correspond with the rise and fall of the tide. (The crabs come out to forage on the beach at ebb tide.) So, we see that crabs are able to time both solar day and lunar day.

Figure 15.2 Badgers change their daily schedule as they grow older. Young badgers are active in the morning, but as they mature they become more active at night.

Biological rhythms are not always coordinated with the *onset* of some ecological event. In some cases, the adaptiveness of a particular rhythmic pattern is not apparent when the pattern begins. There are instances when a certain behavior must begin at such a time that it will *end* at a critical period. Here, the environmental event is "predicted." For example, a wide-ranging nocturnal animal must stop hunting and begin its trek home so that it will reach the protection of its burrow before daybreak. It may not be immediately apparent to an observer why an animal abandons a promising hunt to turn homeward.

## TIMING AND REPRODUCTION

Another example of "anticipatory" timing is seen in the onset of breeding in some species. In many birds and mammals, breeding usually occurs so that the young will be born at the time of maximum food abundance. The timing in such cases may be regulated through *photoperiod* effect. Photoperiod refers to the daylight hours in each twenty-four-hour period and, of course, it changes with the season over most of the earth. So, some animals begin breeding in response to changing daylength, long before their food appears. In order to adjust to such seasonal environmental changes, by the way, animals must normally be able to differentiate between ten-hour days (winter) and thirteen-hour days (spring).

For some fish species, increasing photoperiod signifies the coming of spring, with its warm water and burgeoning food, and their gonads rapidly grow in preparation for breeding. But fish, as well as other animals, may cue on seasonal variables other than light. Cichlids (*Tilapia macrocephala*) and sticklebacks respond physiologically to warmer water, while cyprinid fish become sexual in response to rain (Emlen 1973). Rainfall may stimulate many frogs and toads to migrate to breeding ponds and to begin calling and copulating (Aronson 1965).

A few animals, such as some deepwater marine animals and a few tropical species, such as the Bornean tree frog, seem to show no seasonality in their breeding behavior at all, but most animals do show distinct seasonal breeding periods (although the peak may be less pronounced in the tropics). Greater seasonality in breeding, of course, would be expected where the seasons are more pronounced.

In some birds, it seems that feedback mechanisms are operating so that once increased photoperiod stimulates the gonads to develop and to begin producing reproductive hormones, other factors such as weather, food, and the perception of nest sites may further increase gonadal development and hormone secretion, which then makes the animal even more responsive to these cues.

Interestingly, large animals may use different cues than smaller ones. Small exothermic animals (cold-blooded, those whose body temperatures vary with the environmental temperature) might be expected to respond primarily to temperature or moisture changes. They don't get around so well when it's cold, as you know if you have ever easily caught a lizard on a cool morning before it has had time to sun itself. Large endothermic, or warm-blooded animals (birds and mammals), are better protected against the whims of the weather because they usually maintain their temperatures within tight limits through metabolic expenditures. But, on the other hand, energy is required to maintain this steady condition and energy comes from food. Also, large amounts of food are needed for the production of eggs and the early care of the young, so birds and mammals might be more responsive to changes in food supply than cold-blooded species.

One reason some tropical animals show extended breeding seasons is because, in temperate zones, temperature is highly variable on a seasonal basis, and thus the numbers of plants and small animals that provide food for the endotherms also vary seasonally. In the tropics, however, temperature varies far less, and the seasons are marked, not so much by temperature change, but by differences in rainfall, probably a less critical environmental variable (in the sense that variation in rainfall probably places less stress on most species than does variation in temperature).

Again, let us distinguish ultimate and proximal factors. Rainfall or light are *proximate* factors in the timing of animals. The *ultimate* factors are the constancy of seasonal effects that have operated through the eons. Each season has traditionally brought its own gifts and burdens to the populations of the animals that fall under its influence. In time, the animals began to respond appropriately to the environmental cues that serve as harbingers of the seasons.

Now, let's have a look at some causative factors as they are viewed by Emlen (1973).

### Ultimate and Proximal Factors in Timing

Timing in reproduction is basically important in three ways: (1) in bringing the sexes together, (2) assuring the greatest number of healthy offspring, and (3) increasing the chances of the parents' survival to the next breeding period (in those animals that have more than one breeding period).

**Getting the Sexes Together** In some species, getting the sexes together is no problem. They live that way (Figure 15.3). In fact, in star worms (*Bonella viridis*), the male lives in the reproductive tract of the female! Other species, however, have a problem in getting their gametes together. The California grunion (*Leuresthes tenuis*) is a small fish that seasonally spawns three or four days after a new or full moon, a time corresponding to the highest tides of the month. One to three hours after the tides begin to recede, the females, hotly pursued by the males, ride the incoming waves and thrash their way up onto the beach, bury their tails in the sand, and deposit their eggs (Figure 15.4). The males writhe around the females and release their sperm; then they all catch the next wave out. The buried eggs are thus protected from the dangers of the sea until the next spring tide, a month later, brings high seas to carry the young out to sea. (I confess to some hesitancy in being too specific regarding the time of grunion "running," since I have spent long hours on California beaches in the dead of night waiting for their arrival. They never came. They always waited until I went home.)

The sexes of some species associate together throughout life. Some geese, as is well known, spend their lives together. Other birds may mate with the same partner year after year, the bond weakening in the nonbreeding season. In the case of migratory

Figure 15.3 With zebras, the sexes live together and so they do not have to coordinate their timing in order to bring reproductive individuals together at the proper time.

birds, it isn't known whether breeding partners stay together in the off-season, but established partners may rejoin each spring. In other species, the sexes may come together only briefly to copulate, as is the case in some cats. However, in all these cases, it is essential that the sexes respond to the same proximal signals that indicate that the time of mating has arrived.

**Surviving and Rearing Young** The timing of reproduction must be considered a compromise between factors of time, energy, and risk. The compromise should yield a pattern that is optimal for the production of the largest possible number of offspring. It is obvious that in some kinds of animals reproductive timing must also enhance the welfare of the parents. An animal that lives through three reproductive seasons is more likely to leave more offspring than an animal that survives only two. So, it is to the parents' advantage to have young at a time when, not only food is available for the young, but the parents (who are expending energy in reproduction) can find enough food for themselves as well.

As I will stress later, for the well-being of both parents and offspring, young should appear when food abundance is high and risk of predation is low. In yellow-headed blackbirds, the young hatch just as the local damselfly population emerges (Orians 1966; Willson 1966). The great horned owl (*Bubo virgin-*

Figure 15.4 The grunion is a small fish that must time its reproductive activities precisely. It spawns in beach sand three or four days after a full moon.

*ianus*) lays eggs in mid-December in Florida, but in mid-February in Pennsylvania and Iowa. Their timing means that the young owls will hatch just before the local appearance of the young rodents and at a time when cover for the rodents is poorest (Emlen 1973). So, here we see the ultimate influences on population differences. The owls do not respond to the availability of food, but to the expectation of the availability of food. The timing, it seems, is keyed to *ultimate* factors because I know of no way the owl could tell from its environment when the rodents were going to appear. If owls have some way of assessing this, however (and it is possible that they do), then they are responding to *proximate* cues. But for now we should probably assume that the prey cycles in each place have been constant enough, for long enough, that the owls have evolved the timing of their own reproductive efforts to coincide with the appearance of rodents.

Deer in Virginia normally do not bear young in the winter. Most temperate herbivores wait until the exuberant spring flush of green plants. But deer must gestate (be pregnant) for several months, so the breeding season must be dictated by other cues than food abundance, since they breed months before the fawns are delivered. The proximate cause here is probably *shortening* day length. If they breed in the fall, the young will appear in the spring.

Predators, such as wolves, also need food for their offspring and they tend to prey upon young animals, as we shall see, so they must ideally bear young at the same time the prey does. The same foliage, then, that provides the food for the prey also provides them with shelter or "screens" from visually hunting predators. So here is the second reason certain prey species bear their young at a time when plants flourish. This screening by plants, it should be mentioned, may have placed a great premium on the development of strong olfactory powers in some terrestrial predators. Thus, some of these predators have developed long noses, the better to search out prey with, while grassland predators may have tended toward visual acumen.

In some cases, breeding cues are associated directly with the availability of water. In dry areas, the usual strategy is to reproduce when the rains come. For fish, rains produce runoff that carries new nutrients into their water, signaling the production of food for their eggs and young. In Texas, the frog *Rana pipiens* has two reproductive periods corresponding to the two wet seasons. But the tree frog *Microhyla olivacea* reproduces any time between April and September directly following rainstorms. (Another frog, *Acris crepitans,* seems to respond more to temperature than to rain [Blair 1951].)

In Africa, water birds and swamp birds breed mostly in the rainy season (or early in the dry season, when insects are most abundant). Grassland birds follow the same strategy (Emlen 1973). Strictly insectivorous birds, however, seem to find cover more important than food, and tend to breed earlier in the rainy season when plants are profuse, but when insect populations may not have peaked. On the other hand, many large African ground birds and raptors nest during the dry season when cover is sparse. We don't know why the ground birds have chosen this period to breed, but the strategy of the raptors is apparent. They nest when there is an abundance of grassland young and when plant cover is minimal (Moreau 1950; Benson 1963).

In Central America, nectar drinkers such as hummingbirds nest in the dry season when the flowers are in bloom. Birds that eat grass seeds, not unexpectedly, nest when grass seeds are most abundant, early in the rainy season. Many Central American raptors nest in the drier parts of the year, as do the African species, except for the maneuverable kites, which can catch insects on the wing and so nest early in the rainy season when insects are about. In fact, apparently all insect eaters nest at about this time (Skutch 1950).

It should be pointed out that birds might not fall easily into trophic classifications such as "insect eater" or "seed eater." Some birds eat both insects and plant matter. Eugene Morton, a researcher at the National Zoo, and others have shown that parent birds may subsist largely on plant matter, while feeding highly proteinaceous insects to their growing young.

In desert mammals, water is necessary for milk production, so most births occur during whatever wet season there is. Some desert rodents, however, are physiologically specialized for water conservation (for example, by producing highly concentrated urine and subsisting largely on the water produced in metabolism). These rodents, including the kangaroo rats (*Dipodomys*) and the pocket mice (*Perognathus*), may breed year-round (Figure 15.5). Even some

Figure 15.5 Desert animals such as this Merriam's kangaroo rat are highly specialized for conserving water, and therefore their life cycles are minimally dependent on seasonal rainfall.

larger desert animals such as peccaries may breed all year, but their physiological adaptations are not well understood (Sowles 1966).

In many cases, there must be compromises between breeding at the optimal temperature, escaping from predators, and coordinating breeding with the food supply. For example, most arctic birds begin to breed in "anticipation" of a food supply. Since the productive season is so short in polar areas, species nesting there must take full advantage of whatever season there is. However, the eider duck (*Somateria mollisima*) nests on Arctic islands, but it waits until well after food is available to launch its reproductive effort. It seems that young eider ducks are preyed upon by foxes, so they wait until the ice that links their island to the mainland melts, thus destroying the foxes' bridges to their nests (Lack 1954). Predation is apparently a greater threat to nesting eiders than is the risk of reduced food abundance.

In cases where there are severe time limits on the reproductive period of a prey species, it might be to its advantage to reproduce quickly and synchronously (Darling 1938; Kruuk 1964). Predators would not only have difficulty in adjusting their own reproductive period to coincide with that of the prey, but there would suddenly be a glut on the food market that the predators would not be able to fully exploit. If a prey species nested asynchronously, it would provide its predators with a rather sustained food supply that could be exploited piecemeal, as it appeared. Sporadically flooding the predator with food, however, would result in many prey offspring escaping simply because the predator could never get around to them; and then suddenly the predator would find all the nests empty. The resulting food shortage throughout most of the year could then act to reduce predator numbers.

Where predation is low, young can be produced more leisurely. Slow development of the young means less chronic demand on the food supply. Thus hole-nesting birds, which lay eggs in protected places, tend to have longer developmental periods than their colleagues who nest in the open (Lack 1948; Nice 1957).

Predation may also affect the time of nesting in other ways. In the tropics, nest predation is very high, so rather than putting all their eggs in one basket and risk losing a large clutch to a predator, birds may raise several small clutches each raised in a shorter period of time. Small clutches also result in less intense activity around the nest, which might attract predators. Small clutches in the tropics might also be the result of tremendous competition for food, as we shall see later.

There has been some disagreement among ecologists regarding whether animals reproduce at the optimal time for the young or for the parents, particularly the female. Lack (1954) suggested that egg laying was timed for maximum availability of food for the young. But in 1966 he revised his views somewhat, noting that producing and laying eggs places a tremendous drain on the female and that reproduction must therefore be timed so that the female has adequate food before egg laying. The strain on the female may be pronounced. For example, the great tit (*Parus major*) is not a large bird, but the female lays nine to ten eggs in as many days. Thus, females with an average weight of 17.3 grams produce 17.5 grams of eggs in a little over a week. There is some indirect evidence that the timing of nesting in birds may correspond primarily with the female's need for food to give her energy and materials for the process ahead. For example, the pied flycatcher (*Ficedula hypoleuca*) lays its eggs so that the young hatch *after* the peak caterpillar period. Thus, the young are in the nest at a time of declining food supply, but food was abundant when the eggs were being formed (Emlen 1973).

I mention this elsewhere, but I will also bring it up here. Many kinds of prey animals, from insects to rabbits, are most active at dawn or dusk. My guinea pig is bold indeed at dusk, but stays close to his bowl (which is his "home") at other times. This periodicity may have evolved because dawn and dusk are brief periods (usually it is either "light" or "dark"), and predators are usually either diurnal or nocturnal. Thus, neither type of predator would be well adapted to hunting in dim light. And, another point: territorial birds tend to sing more during the early hours of the morning and in the late afternoon. Singing renders them conspicuous, not only to competitors, but to avian predators as well. However, most avian predators hunt by riding air currents. (This ability saves the energy of flapping flight.) Air currents appear as the earth is warmed by the sun and rising conduction currents are produced. So avian predators may have developed a means of hunting that expends less of their energy (soaring and diving), but they can only do it during the warmer parts of the day. Perhaps songbirds have timed their singing to correspond to the time when there are no rising air currents. These ideas remain to be tested.

We have noted the adaptiveness of precise timing, concentrating primarily on reproductive behavior to show the importance of timing in only one area. (See Emlen 1973 for further references.) Hopefully, we also have a better idea of proximal and ultimate factors in the lives of animals, although the two are not always easily distinguishable. Now, let's take a closer look at some biological rhythms and the search for their mechanisms, first noting some rather peculiar rhythms at the suborganismic level, where most of the research has been done.

## UNEXPECTED RHYTHMS

One of the intriguing findings of the clock studies is that even *isolated* parts of organisms are rhythmic. For example, cell-division-rhythms and contraction rhythms are found in tissue cultures (cells growing on an artificial medium [Goodwin 1963]). Mammalian hearts and adrenal glands maintain a daily rhythm for several days after they have been removed from an organism. Even a single neuron in an isolated ganglion, or groups of nerve cell bodies, continue to exhibit a daily rhythm (Jacklet 1969). So far, though, no circadian rhythms have been shown at any level below the intact cell.

The complexity of rhythms has been driven home by a number of unexpected findings. For example, some organisms are more sensitive to biological agents at certain times of the day than at other times (Holberg 1960). A given amount of the toxin produced by the intestinal bacteria *Escherichia coli* will kill about eighty percent of a group of mice when given in the early evening, but will have little effect when administered only eight hours later. Also, humans show a rhythmicity in their sensitivity to certain common pharmaceutical agents.

It is obvious that rhythmicity exists in a wide variety of animals, is expressed in many different ways, and may be coordinated with a number of different cyclical environmental events. It should be stressed at this point that we only know about the rhythmic behavior of a few species and we don't know much at all about individual variation. However, Menaker (1969, p. 683) reminds us, "For nearly all organisms at any level of organization, most behavioral, physiological, or biochemical variables which one might choose to measure can be expected to exhibit a daily rhythm."

## THE PHASE, PHASE SHIFTING, AND ENTRAINMENT

So, we see then that many of the natural biological rhythms are about twenty-four hours in length, a period that corresponds to the solar day. When rhythmic organisms are placed under constant conditions where, theoretically, no aspect of the environment is allowed to vary, their rhythms may drift slightly, say, to twenty-three or twenty-five hour periods. These periods of about a day are called circadian rhythms (*circa*, about; *diem*, day).

It is interesting that the direction of the circadian shift (to twenty-three or twenty-five hours) that organisms may exhibit after they are placed under constant conditions may vary according to the conditions in the controlled environment. For example, a constant high-intensity light may lengthen the cycle (the repeating unit), whereas a constant light of lower than

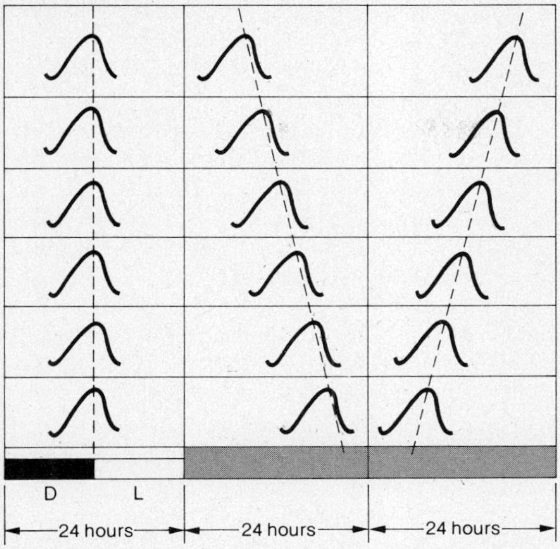


Figure 15.6 The circadian variation in a biological rhythm. The column on the left shows a hypothetical rhythm that appears in response to a normal light-dark cycle. The cycle repeats itself every twenty-four hours. If light and temperature are maintained at a constant level in the laboratory, the rhythm will shift, becoming longer or shorter. The cycles may lengthen to twenty-five hours as in the middle column or shorten to twenty-three hours as in the column on the right. Conditions can sometimes be experimentally manipulated to lengthen or shorten cycles.

normal intensity may shorten the cycle for some species. In other species, however, the light intensity may have the opposite effects (Figure 15.6). "Aschoff's rule" states that an increase in the intensity of a constant light lengthens the high-activity phase (free-running period, or FRP) for nocturnal organisms and shortens it for diurnal organisms (Hoffman 1965). Thus, with an increase in light intensity, the FRP of a mouse increases, while that of a sparrow decreases.

We see, then, that the time involved for the completion of an entire cycle may be altered by changing the environment. By lengthening the period of a cycle to, say, twenty-seven hours in a controlled environment chamber, the leaves of plants can be made to rise and fall every twenty-seven hours. Their "day" may also be shortened to eighteen hours (Figure 15.7). There is a limit, however, to how far the cycle can be pushed. For example, if very short "days" of about eight hours are offered, the organisms will revert to a circadian rhythm, but perhaps a short one.

## A Model

Biological clocks can probably best be visualized by considering a mechanical model, such as one described by Menaker (1969). This model assumes that daily rhythms behave in much the same way as do some physical oscillators (Figure 15.8). Here, imagine the behavior of a coiled spring, with a weight attached to one end, suspended from a rigid surface.

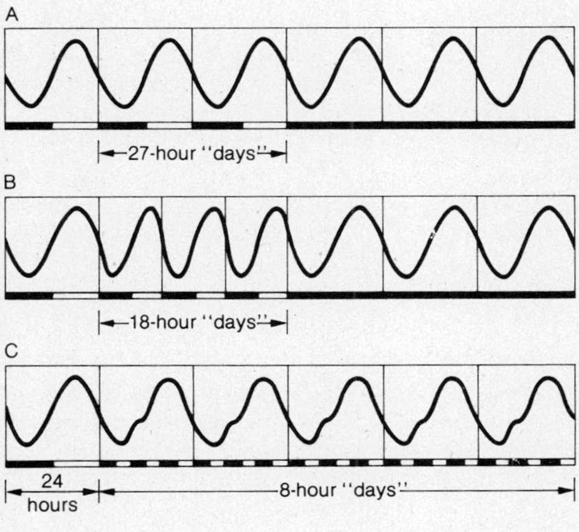


Figure 15.7 Effects of light-dark cycles on the period of a typical biological rhythm. In (A), after many days on a regular twenty-four-hour-day regime (the last of this regime is shown at the left of each row), the day is lengthened to twenty-seven hours for several days, and the organism adjusts accordingly. However, the effect is temporary. When the treatment is followed by constant darkness (darkness is indicated by the dark areas along the base lines), the organism reverts to its twenty-four-hour rhythm. In (B), the organism adjusts temporarily to an eighteen-hour day. In (C), however, when the organism is subjected to an eight-hour day, the extreme change is registered only in a distortion of the twenty-four-hour rhythm.

For purposes of discussion, we will assume there is no friction, so that once the oscillating (up and down) motion of the spring has started, it will continue without change unless disturbed by some external force. Various positions of the weight are shown as they would occur at different times (the black dots). Time is shown along the horizontal axis. We have already employed certain terms such as "rhythm" and "cycle" according to their common usage, but perhaps we should now tighten our definitions to comply with the clock model. The recurring oscillations comprise a *rhythm.* The repeating unit of the rhythm is referred to as a *cycle,* and the length of time necessary to complete one cycle is the *period.* In the diagram, the period of the rhythm is five seconds. *Phase* refers to a specific part of the cycle (or the position of the weight at a given point in the cycle). *Amplitude* describes the displacement of the mass, or the distance of the weight from the resting (coiled) position to the farthest point below the resting line.

We may now describe clock alterations in terms of the model. Figure 15.8B shows what happens if external force is applied at some point in the cycle, causing a *phase shift.* Phase shifts can occur in either direction or to any degree. If the external force increases the velocity of the oscillator, the phase is advanced by causing the "weight" to reach a point earlier than it normally would have. Conversely, retarding the oscillation, as is shown in the model, leads to a phase delay.

If one applies sufficient continuous external force with a certain period, the oscillator will take up the

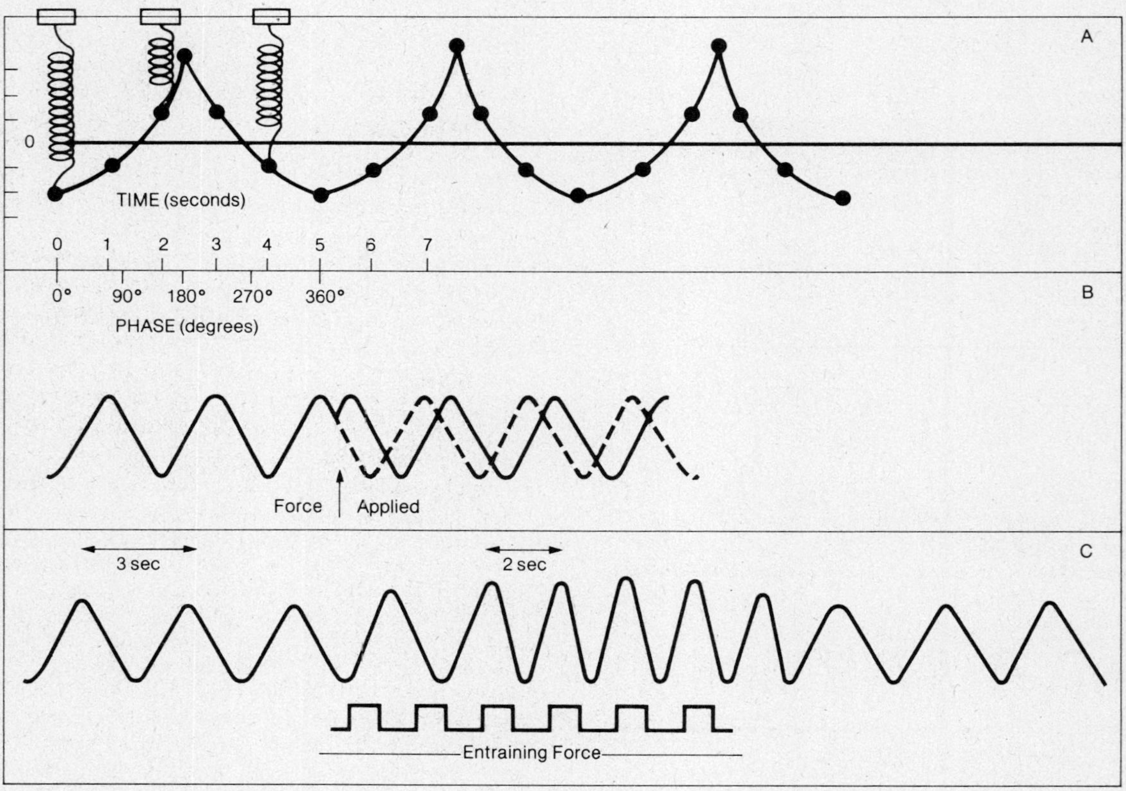


Figure 15.8 An oscillating model showing, in physical terms, the general characteristics of biological clocks. See the text for an explanation.

period of the applied force. This response is called *entrainment.* Notice that the entrainment does not occur immediately; instead, the new period is taken up gradually. Also, there is a fixed relationship between the phases of the oscillator and the external force. When the external force is no longer applied (recall the plant placed under constant conditions), the oscillator returns to its normal period.

Experiments have been done that demonstrate control of both the period and phase of cycles. A sparrow was kept under constant conditions of darkness and it showed a natural circadian rhythm of twenty-six hours. Then it was subjected to a cycle with three hours of light and twenty-one hours of darkness. Its period of activity (FRP) then began thirty-five minutes before the light went on each day (so the period changed from a twenty-six-hour to a twenty-four-hour basis through entrainment). When the light period was extended to eleven hours, with thirteen hours of darkness, the onset of activity began directly with the light, demonstrating a change in phase (phase shifting).

Phase shifting has also been demonstrated by reversing the light-dark cycle of a bean plant. The result is a reversal of the phase in the sleep-movement rhythm. So both the phase and the period of the rhythm are controllable. The limits within which entraining cycles can vary and still alter the normal period are different among species and even among individuals. So, oscillators are controllable, but only within certain limits, which are determined by their physical characteristics and the nature of the environmental input.

We are now beginning to see a basic pattern in the timing of living things. Plants and animals show rhythmic behavior even when deprived of the more obvious environmental cues. These cycles are modifiable, within limits, by known environmental factors. The flexibility of the cycles is ecologically imperative. As mentioned, the natural cycles are rarely of a twenty-four-hour period and so are somewhat out of phase with environmental events. Daily environmental cycles keep the organisms' clocks continually "set" so that behavioral phases occur at the appropriate time of day. Let's now consider a case from nature showing how rhythmicity may be influenced by environmental variables and how it is important to one species, the white-crowned sparrow. We will draw on the work of Farner and Lewis (1971).

Figure 15.9 The white-crowned sparrow is a long distance migrator that has been the object of considerable physiological research into the mechanisms of biological clocks.

## THE LIFE AND TIMES OF THE WHITE-CROWNED SPARROW

White-crowned sparrows (Figure 15.9) lead complex lives. Some populations spend their spring and summer in Canada and the northern United States, in fall heading south, going as far as Mexico. Then, in the spring, they return. Now, a bird spending the winter in Mexicali may well be ready to go north (or anywhere) by spring. But it has to know when to go. If it leaves too early, it may be caught by a cold snap. If it waits too long, the best nesting areas may be taken, or perhaps it will nest so late that it simply won't have time to rear its young before it is time to go south again. Nesting that far north is a tricky business; the timing must be precise because the productive season is short.

Now, not only must a bird be in the right place at the right time, but it must be in the proper physiological state. For example, if it is to reproduce, its gonads (testes or ovaries) must be developed before it starts to breed. But the white-crowned sparrow may reach its breeding ground with its gonads regressed (in an undeveloped state). It's better to migrate with small gonads simply because they are lighter, and with birds, every ounce counts. Some species shrink their gonads to about 1/200 to 1/300 their breeding size before they begin their long flights. The gonadal cycle is only one of several physiological changes that occur in a regular time relationship with the migratory pattern (Figure 15.10).

So the sparrow (or its body) must know precisely what time of year it is. And it does this by measuring the length of the day. But strangely enough, although the bird can see light, its physiological clock is not entrained by light entering the eyes. If the eyes are covered by a shield, or if the bird is blinded, its normal circadian pattern does not change; it still responds to changing day length. But how can a blinded bird be

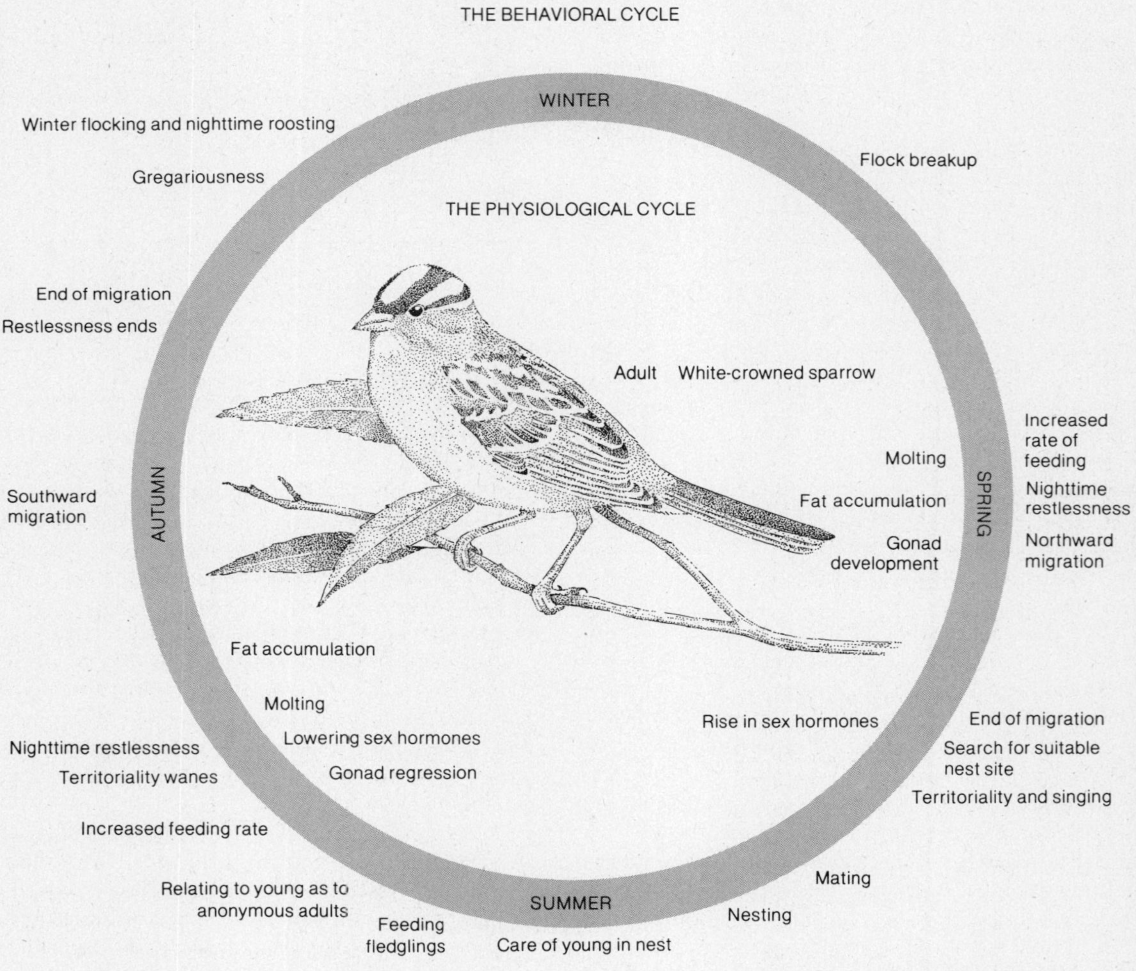


Figure 15.10 The behavioral and physiological cycle of the white-crowned sparrow. The behavioral cycle is shown outside the circle, and the physiological cycle inside.

sensitive to light? Apparently, the bird's light timer is located in the hypothalamus, and the photoreceptors that stimulate the hypothalamus seem to be located in the brain itself, perhaps in the pineal, according to Menaker and his group.

So how would a hypothalamic clock work? The "setting" of the clock (entrainment) may be possible only during certain "sensitive" periods (we don't know what brings on these periods). After the clock is set by the beginning of the light period, it is relatively insensitive to light changes for the next thirteen hours or so. Then it grows increasingly responsive to light change, reaching a peak about sixteen to twenty hours after the cycle started. Sensitivity then fades for a time, until the clock is set again, twenty-four hours from the previous onset of light.

Now, suppose the birds were placed under conditions where the light periods were twelve to thirteen hours long, with the nights spanning eleven to twelve hours. This would mean the photosensitive period would never be activated because there would be no light available during the light-sensitive parts of the cycle. But if the light period were increased to fourteen or fifteen hours, light would reach the brain during a sensitive period. The hypothalamus would thus be stimulated to release hormones that act on the anterior pituitary. This gland in turn would release various hormones, notably gonadotropins and prolactin. The gonadotropins stimulate the gonads to develop and exert their own physiological and behavioral effects.

Obviously, this scheme is testable. So Farner and his group performed the appropriate experiments (Figure 15.11). First, they set the clocks of a group of birds by exposing them to eight hours of light, followed by sixteen hours of darkness. The subjects' testes remained small since they were not receiving light during their sensitive period. Then a group of birds were put on an eight-light, twenty-eight-dark schedule. The clocks thus continued on a regular twenty-four-hour cycle, but they were receiving light during their sensitive period. The result is that, although the birds were receiving a lower ratio of light to dark, the testes began growing. The same result can be achieved if an eight-light, sixteen-dark schedule is broken by a light period seventeen to nineteen hours after the start of the day. Again, light would fall on the bird during its peak sensitive period.

So how does all this work in nature? For a bird wintering in Mexicali, after winter officially begins (December 22), days grow gradually longer. Finally, at some period in the spring light begins to fall on the bird during its daily sensitive phase, thus triggering the hypothalamus to activate the pituitary, which secretes gonadotropin. The gonads begin to grow, and the bird begins to eat voraciously, putting on fat stores it will need on its long trip north. The bird may actually increase its weight by thirty to forty percent. Later, it will become restless at night, hopping around nervously when, a few weeks earlier, it would have been sleeping. Finally, unable to bear it, it launches itself. Its directional mechanisms compel it to head north. When the males reach the breeding grounds, they set up territories, advertise them by singing, defend them against other males, and, hopefully, entice a female, build a nest, and raise their young.

Then, in early summer, the gonads dramatically and rapidly decrease in size. This marks the *photorefractory* period. During this time, the bird is completely insensitive to increasing day length. The time of breeding is over. There is no need to maintain the large gonads.

Some clues to the mechanisms of photosensitive periods have been provided by work with the white-throated sparrow, a close relative of the white-crowned (Meier 1973). It was found that there are regular daily changes in the blood levels of two hormones, corticosterone and prolactin. These changes do not occur simultaneously, however. The rise of prolactin lags behind that of corticosterone by about twelve hours in the spring, changing to four hours in the fall. The birds were kept under constant dim light and injected with corticosterone and prolactin at different time intervals throughout the day. It was discovered that one injection schedule could elicit the physiological and behavioral changes characteristic of fall, and another schedule those of spring or summer. The evidence indicated that the light receptors in the brain regulate the daily cycle of corticosterone and prolactin levels in the blood, and these hormone levels, in turn, exert their controls on the reproductive cycle of the white-throated sparrow.

Now, let's return to the question of how animal circadian rhythm is established. Is it controlled from within, a function of innate cellular changes? Or do the animals take their cues from the environment? Most researchers, you will see, subscribe to the former idea.

## ENDOGENOUS CONTROL THEORIES

Crickets chirp at dusk. The most obvious explanation is that crickets chirp in response to the sun's going down. However, there is some good evidence that we must look further for the real stimulus, far back into the history of the species. In other words, perhaps a cyclic nature within the cricket has appeared in response to regularly recurring light and dark cycles over the eons, and the physiological response to these cycles cannot be fundamentally altered by individual experience. Such a system of recurring behavior cannot be considered simply as "instinct," according to the ethological concept, since there is no evidence of releasers, IRMs, or other components of instinctive patterns. Learning or imprinting are also probably not involved, since even when several generations are reared under constant conditions, rhythms persist. The more prominent endogenous theories involve a rhythmic system within the cell based on a cyclic nature of its chemical constituents. For recent findings, see Schweiger and Hastings (1976).

Aschoff (1963) stated that, in the organisms that had been investigated thus far, rhythms are innate and inherited properties. The notion of the heritability of clocks is supported by such evidence as successful artificial selection in the laboratory for strains of insects with different rhythms. In other experiments, it was found that plants and animals that had been reared entirely under controlled, stable conditions suddenly became rhythmic on their own or turned out to be arhythmic (showing irregular fluctuations). In some cases, arhythmic organisms could be induced to become rhythmic in response to a single stimulus. For example, in nature, fruit flies

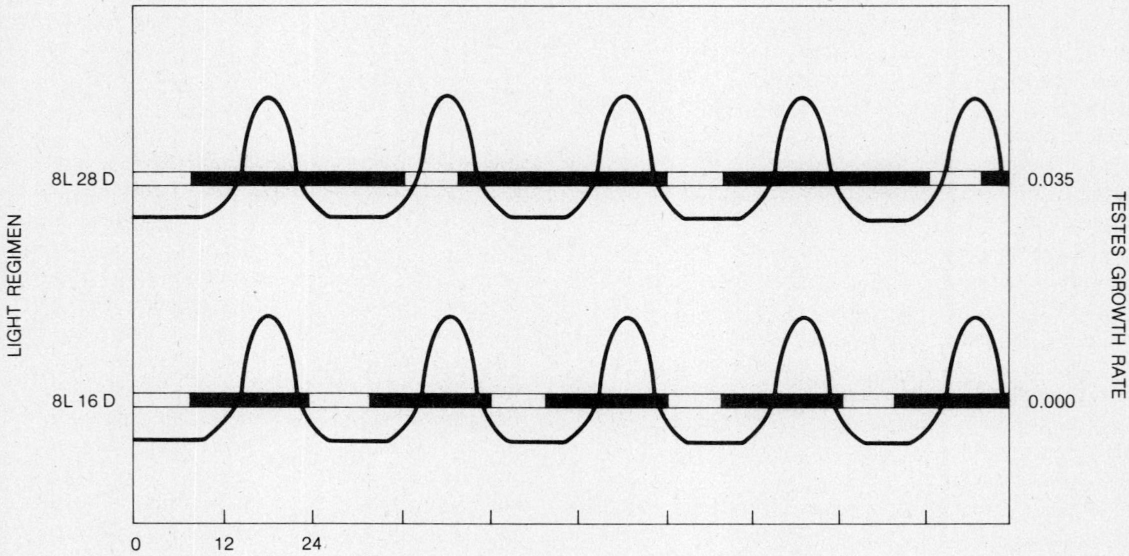


Figure 15.11 Control of testicular growth by changing photoperiod. On a regimen of eight hours of light (L) and twenty-eight dark (D), light reaches the birds during their light-sensitive periods, shown by the curves, and gonadal growth results. In 8L, 16D regimen, there is more light, but it does not strike the birds during their sensitive periods, and there is no gonadal growth.

emerge from their pupal cases only at dawn and thenceforth show the usual twenty-four-hour rhythm. However, if eggs are kept under constant laboratory conditions, they will hatch at all hours of the day. It was found that even in strains kept continuously under constant conditions for twenty-five generations, with each generation hatching randomly throughout the day, when the developing pupae of any generation were subjected to a single flash of light, those pupae all hatched twenty-four hours later (or at some multiple of twenty-four hours). Only the single stimulus was necessary to synchronize the clocks of all those insects for the rest of their lives.

The persistence of circadian rhythms under constant conditions is presented as evidence for an endogenous (or internal) control mechanism. A flying squirrel kept under constant conditions may show a period of twenty-three hours, thirty-six minutes, and since there is no known natural cycle of twenty-three hours, thirty-six minutes, the control is presumed to be internal. Also, other flying squirrels may show slightly longer or shorter periods, further evidence that the rhythms are not originating in the environment. These individual differences in circadian rhythm bolster the argument of the endogenous-control theorists. The lack of precision or synchrony, they feel, doesn't upset their theory at all, since evolutionary pressure could be expected to develop a clock that only *approximates* the twenty-four-hour day. Normal day-night cycles could then operate to make the small corrections necessary to adjust the system to local conditions. Day-night cycles, they say, just "set" the clocks. Since daylength changes with the season, a certain flexibility of rhythm, according to this interpretation, is essential.

The endogenous rhythm theory, then, does not rule out the possibility that timing is influenced by geophysical factors. It only states that the basic timing mechanism is not dependent upon any periodic changes in the environment, that the basic mechanism is innate. Menaker (1969) pointed out that the earth's environment cannot furnish adequate timing clues because of the variable nature of the environment. As examples of variable, and hence presumably undependable, geophysical factors, he cites variations in light intensity, temperature, barometric pressure, magnetic field, and ionization of the air.

As the endogenous theorists argue, then, any internal biological clock probably operates biochemically. The concept is basically simple. Biochemical reactions proceed at a certain rate. Thus, a complete biochemical reaction could correspond to a single rhythmic cycle. In fact, there are cyclic biochemical reactions known in which the end-product of a reaction triggers the initiation of a new reaction, thereby maintaining the cycle. By-products and intermediate products produced in the course of a reaction could themselves affect specific "targets" in a physiological system and account for the phases of a cycle.

As you might expect, objections have been raised to the notion of a strictly biochemical mechanism. For example, opponents point out that temperature greatly affects the rate of most chemical reactions. An increase in temperature of only 10°C doubles the rate of many reactions. If a clock could be made to run twice as fast by a 10°C increase in temperature, then it could be assumed that clocks could be used as peculiar thermometers, but they can't be. In fact, temperature seems to have very little effect on most biological rhythms. A shift in temperature of 10°C seldom even caused a change of twenty percent in the period (but changes as little as 3°C can cause phase shifts in some cases). Even these small effects might be the result of a change in the ease with which the rhythm is expressed in its milieu, rather than a change in the basic mechanism. It has been pointed out, however, that a feedback mechanism could account for the temperature independence of biochemical clocks (Figure 15.12) (Hastings 1970). It is very important, of course, that clocks are not greatly affected by temperature. If they were, cold-blooded animals, whose body temperature fluctuates with that of the environment, would be unable to develop reliable clocks. Interestingly, *alternating* temperature changes may influence the period (entrainment) more than temperature change per se. In some organisms, twelve-hour periods of 28°C alternating with 20°C can entrain a rhythm that will persist for a time when the organisms are shifted to constant environmental conditions (Palmer 1970).

Figure 15.13 shows a theoretical relationship of the basic *mechanism* and its observed *effects*. Determining their precise relationship is not unlike trying to determine how the mechanism of a clock works by

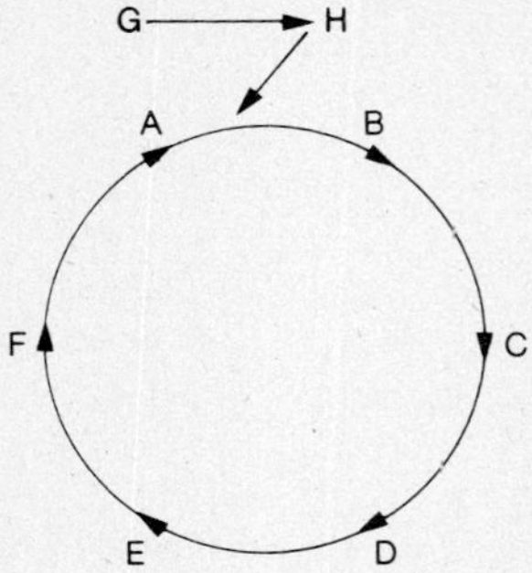


Figure 15.12 A possible mechanism for a chemical clock. *F* is the starting material. Temperature independence is achieved through a feedback mechanism ($G \rightarrow H$ and $A \rightarrow B$). The mechanism can work if $A \rightarrow B$ and $G \rightarrow H$ are temperature dependent as are most chemical reactions. This means that *B* and *H* would increase as the temperature is raised. However, if *H* is a specific inhibitor of the $A \rightarrow B$ reaction, the heat-induced production of *H* would slow down the production of *B*, thereby stabilizing the rate of the entire reaction.

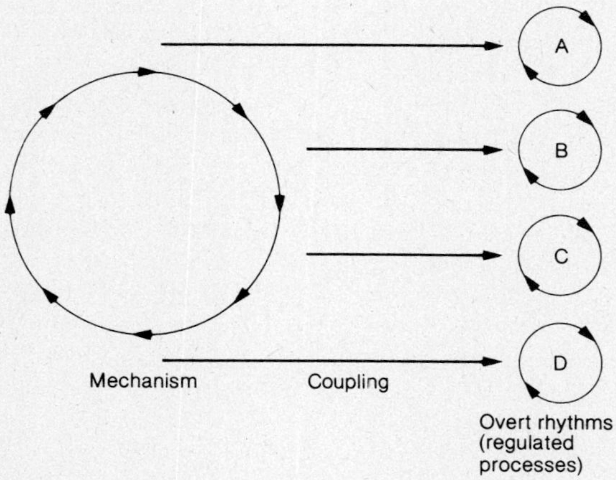


Figure 15.13 A model for a hypothetical mechanism of a circadian rhythm indicating a self-regulated mechanism that could be sensitive to environmental variables such as light and temperature. Such a mechanism would serve as the drive behind a number of observable cellular phenomena such as *A*, *B*, *C*, and *D*.

manipulating the hands. The problem is, the observable cyclic events (the hands) can be altered without affecting the basic mechanism. In keeping with the analogy, it may be said that there is an accumulating body of evidence regarding the behavior of the hands, but a dearth of information regarding the mechanism of the clock.

*Gonyaulax*, a dinoflagellate, provides an example of how the hands may be altered without affecting the mechanism. This organism has a glow-rhythm of bioluminescence, and another rhythm of photosynthesis (which corresponds to the presence of sunlight). Chemical inhibitors can block the rhythm of photosynthesis while not affecting the bioluminescence. But when the inhibitors are removed, the photosynthesis picks up right in phase (as shown by its relationship to the glow rhythm) showing that the basic mechanism was not affected.

The clock-and-hands analogy also shows how multiple rhythms with different periods could operate by the same mechanism. Recall the crab whose color period was in phase with the sun and whose running period was in phase with the tide. The two hands of a clock have different periods also, but they are driven by the same mechanism. The existence of a clock-and-hands scheme would mean that different mechanisms need not have evolved for every rhythmic system, and thus the endogenous theory becomes more plausible. This is not to rule out the existence of multiple systems, however, since it is possible that more than one mechanism may be involved in producing the numerous rhythms found in some organisms (Figure 15.14).

Manipulation of the hands of a clock can furnish some information about the nature of the mechanism, but such indirect information is, of course, of limited value. For this reason, there has been a great effort to "get behind" the observable patterns, to find the deep-seated source of the rhythm. The problems involved in the effort are enormous. In some studies, biochemical inhibitors have been used to try to alter the mechanism and thus to provide a glimpse into its workings. However, if an inhibitor is to be successful, its effects must first be reversible and, second, it must throw the clock out of phase by as many hours as it is allowed to remain in the system. (If the rhythm picks up again in phase after the inhibitor has been removed, then only the hands have been altered.) A true inhibitor of the mechanism should also be able to suspend the activities of the mechanism for extended periods of time (limited only by the effect of the suspension of the viability of the organism).

Inhibitory agents have, in spite of such limitations, provided some clues regarding the biochemical nature of clock mechanisms. Through such studies it has been found that clocks are generally insensitive to a great variety of chemical inhibitors, such as narcotizing agents and sublethal amounts of metabolic poisons. Some of the findings have been startling and very difficult to interpret. For example, poisons such as sodium cyanide may alter the *amplitude* of a rhythm, while the *period* remains essentially unchanged (Figure 15.15); thus, it seems, the clock itself has not been altered. Or has it?

It is clear that simple inhibition experiments are not likely to produce definitive answers to the question of biochemical mechanisms. Even simple metabolic pathways are often replete with complex systems of shunts and feedbacks so that blocking a specific pathway may have very little effect on the general cellular reactions. The administration of general or multiple inhibitory agents is likely to produce unanticipated, unseen, or severe effects that would render the agents less than useful as a research tool. In essence, general inhibitors are blunt instruments.

Some promising evidence is based on the effects of agents whose influence on the system is not really understood. Deuterium (heavy water, or $D_2O$), mentioned earlier as a rather unspecific agent that retards the clocks of homing in pigeons, has an unusually interesting but unexplained effect on clocks. It was found that the phototaxic rhythm of the flagellate *Euglena* was changed from twenty-three to twenty-seven hours when water in the medium was largely replaced by $D_2O$. Heavy water may also affect the rhythms of mammals. For example, deuterium in the drinking water of a deer mouse increased the period of activity as a direct function of the $D_2O$ concentration. In general, it is assumed that the deuterium affects the mechanism itself, since it can cause phase shifts (Hastings 1970). According to some researchers, deuterium inhibits the synthesis of enzymes by DNA.

The biochemical mechanisms of clocks remain clouded in mystery. We have been provided with only occasional tantalizing glimpses of what is really there. Educated guesses have often proven valuable in providing working hypotheses until more data are forthcoming. The guess of those who work with clocks is that the mechanism is based on the rate of protein synthesis. Such a system would involve the very regular building up and tearing down of those extremely complex molecular proteins called enzymes. Such a scheme, unfortunately, is riddled with unknowns. How can protein-based rhythmic systems be turned on and off? What is it that actually oscillates DNA activity, enzyme quantity, enzyme localization,

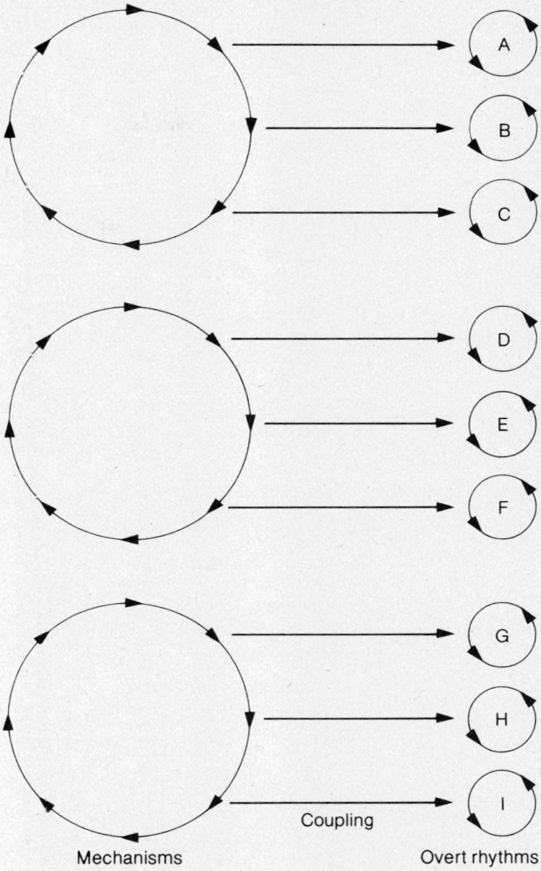


Figure 15.14 Although a single mechanism can drive several rhythms, multiple mechanisms may be present, each producing its own rhythms.

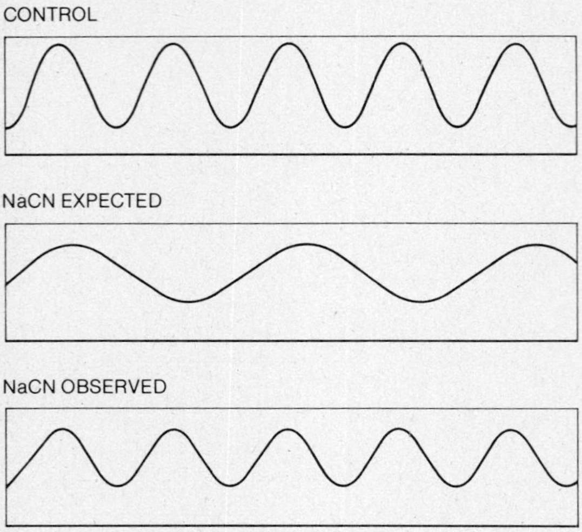


Figure 15.15 The effect of a sublethal dose of sodium cyanide, a metabolic inhibitor, on both the period and amplitude of a biological rhythm. It was expected that the period would be lengthened. However, the period remained the same but the amplitude was decreased.

or enzyme activity? Or are enzymes directly involved at all? Perhaps substrates, the substances that enzymes act upon, are the rhythmic component. After all, under some conditions, enzymes appear in response to the presence of their substrates.

The exogenous (environmental) control proponents accept the idea of a cellular clock mechanism, so the biochemical approach is of interest to both groups. So, now let's consider some of the arguments for exogenous control. The proponents of this theory are fewer in number than the endogenous control people, but arguments for the case have been concisely presented by Frank A. Brown (1970).

## EXOGENOUS CONTROL THEORIES

Some of the support for exogenous control stems from different interpretations of the same data cited by cellular control theorists. For example, two flying squirrels (*Glaucomys volans*) were kept in total darkness. One showed a twenty-three-hour, fifty-eight-minute period and the other a twenty-four-hour, twenty-one-minute period. Endogenous theorists argued that the difference indicated that each squirrel had its own *internal* clock that was independent of the environment. However, Brown (1970) interpreted the data another way. He argued that animals may generate periods other than twenty-four hours, while using a twenty-four-hour clock that is based on *environmental* cycles. For example, a man may get up later on weekends, but he is still using a standard twelve-hour clock (which is actually a twenty-four-hour clock as we acknowledge when we refer to A.M. and P.M. or "13:40" hours). Some people are "day-people," some are "night-people"; but they both use the same clock. Botanist R. J. Garrett claims that his heart doesn't even start until noon, while most of his friends seem to be bounding around at the crack of dawn. Using a twenty-four-hour clock to time a rhythm that doesn't have a twenty-four-hour period is called *autophasing*.

The exogenous theorists do not deemphasize the importance of biochemical endogenous clocks. They only argue that, ultimately, the "endogenous" timer is sensitive to environmental clues. While exogenous theorists maintain that the intracellular rhythm occurs in response to cyclic geophysical changes, there is no general agreement as to just what part of the environment is furnishing the clues.

Brown conceptualized the system as a two-layered ring (Figure 15.16). According to the model, there is a "core" of unchanging geophysical cycles existing under a changeable system that can be affected by such things as light, temperature, and food. The geophysical forces, then, can penetrate all experimental conditions so that what were thought to be constant conditions are not constant at all. According to the scheme, only the outer conditions can be experimentally controlled.

It is known that several diverse geophysical phenomena are cyclic in nature and therefore could be furnishing information to the inner ring. The most important, for purposes of this study, are those that fluctuate on a daily basis. The more obvious daily changes, of course, include temperature (discussed earlier) and light-dark cycles. But there are more subtle geophysical cycles. For example, the atmosphere is subjected to tides, just as are the oceans. (Atmospheric tides can be measured as barometric highs and lows.) Atmospheric tides, then, might have secondary effects on such geophysical entities as gravity, geomagnetism, electrostatic field, and background radiation. These tidal pressure changes are responsive to cyclic celestial events and they can therefore be predicted on a daily, or even an annual, basis.

There are also good indications of cues being provided by other phenomena associated with the rhythmic pulling of the sun and moon. Fiddler crabs, according to long-term records, show a diurnal metabolic change that fluctuates in a semi-monthly pattern. Careful monitoring over a number of years showed that their metabolic rates rose and fell in two-month cycles year after year. Then, in July, 1954, a strange thing happened. The graphs of the metabolic rhythms were mirror images of what they should have been. Although there was a wide search for possible causative factors, there was no explanation for the change. A year later, it was found that at the time of the sudden change there was a mysterious inversion of the daily rhythm of primary cosmic radiation, the particles from space that strike atmospheric molecules. This variation couldn't be the answer to the metabolic change, however, because the secondary cosmic rays, which are sent earthward by the impact of the primary particles and which strike the surface of the earth at such phenomenal speeds, are not related, in time, to the bombardment of primary particles in any reliable way. However, the primary radiation *is* directly related to fluctuations in the earth's geomagnetic field. The stronger the field at any time, the fewer the cosmic rays that strike the earth's atmosphere. The idea was exciting. Could organisms be reacting to daily fluctuations in the earth's geomagnetic fields? In order for such an explanation to be considered seriously, it would have to be shown that organisms are sensitive to exceedingly small changes in environmental magnetism.

Brown discovered that, in fact, some organisms are very sensitive to magnetic fields of the amplitude found in nature. He found that the pathway of the lowly and negatively phototaxic planarian would vary according to the compass direction in which the apparatus was pointed (Figure 15.17). Then he found that the path could be altered in a predictable way by placing a bar magnet across the top of the apparatus. The evidence indicated that the worms could read compass direction by magnetism.

Brown then found that the worms show a recurring monthly variation in the angle of their turns, and

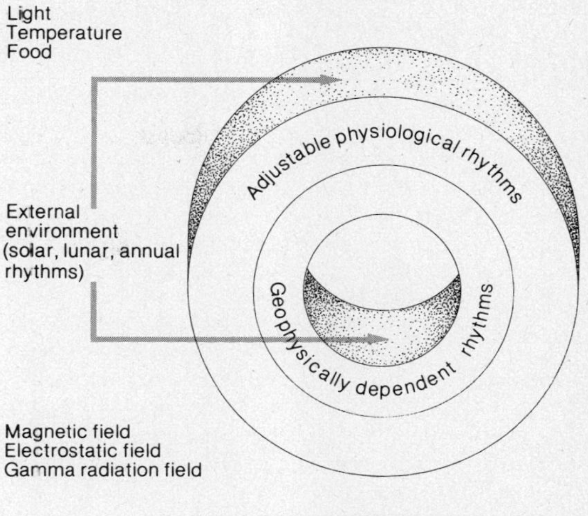


Figure 15.16 The exogenous theorist's representation of the twofold basis of biological rhythms. The outer ring is the source of patterns that more precisely fit the organism to its environment. It controls circadian, tidal, and annual rhythms. The inner ring, however, is the source of more primal patterns. These respond to the subtle rhythms of the geophysical environment.

that the direction of the turn can be altered by changing the direction of the light source *or* magnetic fields (Figure 15.18). An apparent turn of 180 degrees caused a phase shift of half a cycle. So, Brown believed, the monthly rhythms are apparently timed by subtle geophysical forces.

Brown's third finding in this set of experiments was that the degree of turning could be affected by previous trials. In other words, the worm is able to "remember" the directions of light stimuli in relation to geophysical clues (Figure 15.19). After establishing that the planarians possess these three abilities, he argued that it is just these characteristics that are essential for an organism to be able to set timed events by recurring directional events. The problem becomes not one of time, but of space, and the data are given as support for the idea that, ultimately,

Figure 15.17 Results of experiments demonstrating that flatworms can use magnetic fields to orient themselves. In (A) and (B) the apparatus is oriented north and east respectively. The worms moved in a twenty-five-degree direction in (A) and in about a twenty-eight-degree direction in (B), as is shown by the open circles in (C). In (A′) and (B′) a bar magnet augments the earth's natural field and the worms move in a direction of about twenty-three degrees. In (B′), however, the apparatus is not shifted, but the magnet is moved ninety degrees counterclockwise to simulate a rotation of the apparatus with respect to the earth's magnetic field. The worms move as if the apparatus had been shifted as shown by the black circles in (C). Notice the close correlation of the two pairs of points in (C). Standard errors are shown by the vertical lines intersecting the circles. The worms respond to magnetic changes irrespective of the direction of light from the flashlights.

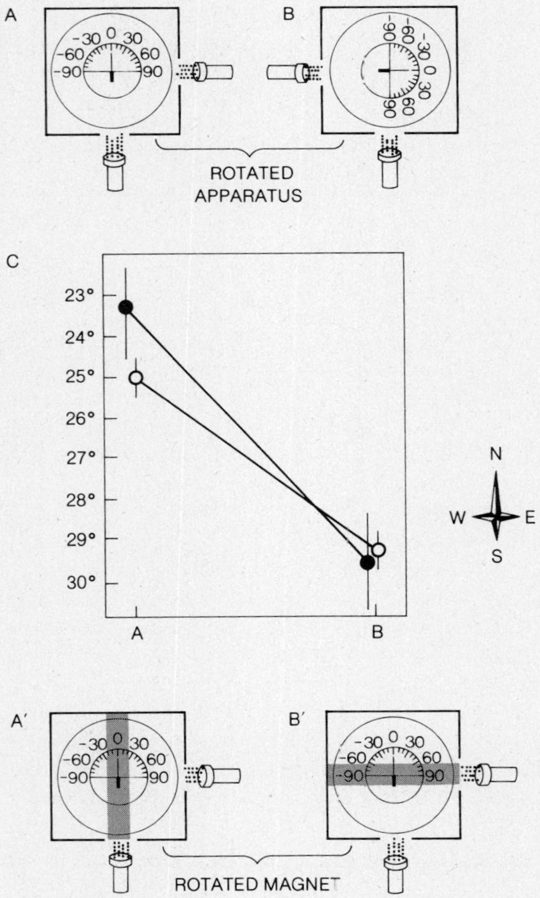


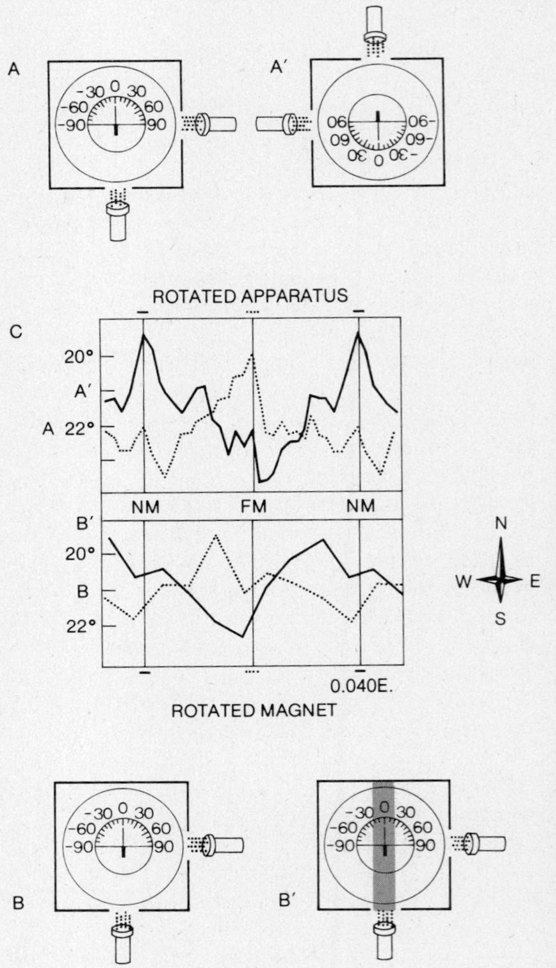


Figure 15.18 Experiments showing that the phase of the monthly variation in turning of the flatworm can be altered by changes in magnetic fields. In (A) and (A′) the changes in direction are reversed according to the compass-card in the apparatus. The results are shown in the top portion of (C) where the curves are 180 degrees out of phase. In (B) and (B′) the apparatus is not rotated, but the poles of the magnet are reversed. The worms behave in (B′) as if the entire apparatus was directed southward. The results of a comparison of (B) and (B′) are shown in the lower graph of (C). The turnings are 180 degrees out of phase. In (C), the circles indicate the average turning direction each day. NM designates the new moon and FM the full moon.

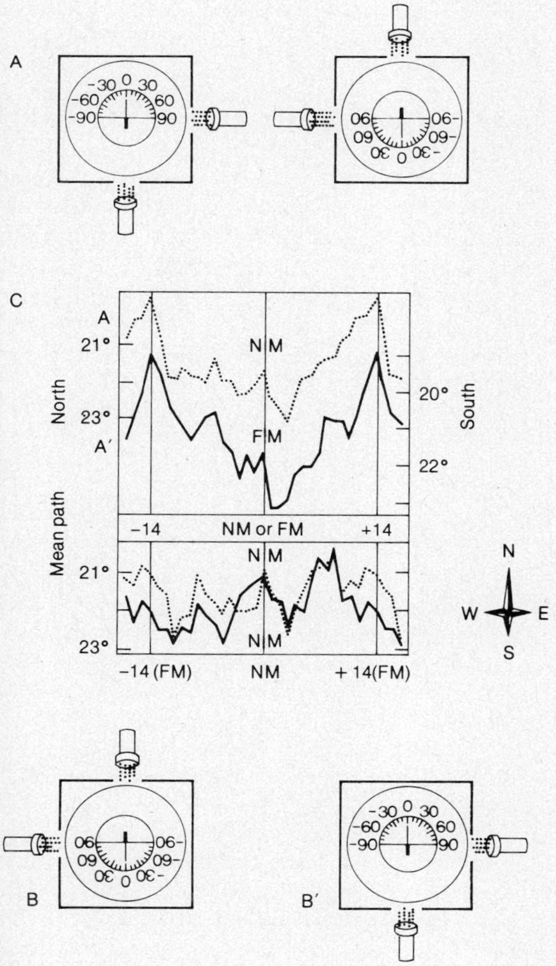


Figure 15.19 Evidence of a form of learning and memory in flatworms. (A) and (A′) are experimentally identical to (A) and (A′) in Figure 15.18, as is the upper pair of graphs in (C), except that one has been shifted by half a month. The result of the shift is that the curves may be superimposed showing that their shapes are similar. In (B) and (B′) the worms were first directed southward with respect to the lights. After they had established a pattern relative to this direction, they were quickly rotated so that they were oriented northward. The lower graph in (C) shows that the curves are different from (A) and (A′) but similar to each other. The worms continue to orient themselves as if the apparatus had not been rotated.

timing is by the perception of subtle rhythmic geophysical changes.

Part of the problem in arriving at an agreement regarding the basic source of timing is that much of the data can be interpreted as supporting either the endogenous or exogenous premise. The data collected by the endogenists can usually be used by the other group simply be inserting them at the variable control stage, the outside of Brown's two-layered ring. The exogenous school, by the way, was forced to abandon or modify many of their premises on the basis of experiments conducted in space tests where the experimental subject was presumably out of touch with earth's environmental influences.

As we can see, timing is accomplished by very complex mechanisms. The widespread occurrence of the phenomenon among living things indicates the high adaptiveness of the clocks and also suggests that several types of mechanisms could have developed through parallel evolution. This latter possibility serves as warning that the search for the mechanism may prove to be incredibly complex, and that when promising evidence does appear, extrapolation to other species may not be warranted. For further information regarding the nature of biological clocks, you might refer to Aschoff (1965a and b); Bunning (1967); Mayersbach (1967); Sweeney (1969); Palmer (1974); Schweiger and Hastings (1976).

# 16

# THE STRUCTURE OF COMMUNICATION

A little screech owl rocks on its branch and suddenly splits the deep quiet of the woodland night with a piercing shriek. In the cool African highlands, a big male gorilla sits quietly and stares directly at a smaller male. A colorful sage grouse assumes a bizarre posture and steps and whirls crazily on the American plain. Farther north, a spotted skunk walks on his front paws, tail high, toward a retreating lynx. In each case, an animal is communicating with other animals, although in different ways and to different ends. The owl's screech and the grouse's strutting have the same function: to gain a mate. The staring gorilla threatens another male, who quickly retreats. The skunk is also threatening, but here the communication is between animals of different species. Thus, communication is a diverse and widespread phenomenon.

In its broadest form, communication may be defined as any activity that alters the behavior of other organisms. Since it may be assumed that such ubiquitous behavior is highly adaptive, we are encouraged to search for the various benefits of communication. In our examples, we can see certain advantages to both the signaler and the recipient. The owl shrieks and gains a mate. The gorilla does not even need to rouse himself, much less fight, in order to cause another male to retreat. The skunk is not attacked. On the other hand, the recipients may also fare better. A female owl who hears the male gains a mate; the smaller gorilla retreats, and is not attacked; and the cat is saved from despair. In each case, the animals adjusted themselves with respect to other living things in the environment. The behavior of the recipient thus became more adaptive, based on an increased level of prediction. The perceptive cat could "predict" what would happen if he continued to approach the skunk.

In addition to signals that communicate intentions to act, there are those that permit a negative prediction. An animal, in other words, may signal that a certain behavior is *not* likely to occur in the near future. Darwin referred to this sort of behavior as the *principle of antithesis.* As an example, a cringing dog with tail tucked, ears back, and head down has the opposite appearance of a stiff-legged, bristling, belligerent dog with his head held high and his ears forward. We know the cringing dog is not likely to attack.

But communication has other advantages than enhancing prediction. It permits animals to lie. And the world is full of furry-faced liars, as we shall see.

## THE COMPLEXITY OF COMMUNICATION

So, communication functions *at least* in permitting a higher level of predictability. It helps animals to alter their position in their world so that they ultimately increase their success, measured in terms of their likelihood of leaving their kinds of genes in the gene pool. The business of lying is less well understood, but it obviously imparts *erroneous* information regarding the environment. This is certainly what we might expect in a competitive and predatory world. If a chimpanzee manages to look stronger than it really is by standing its hair on end, it may have access to more commodities than its intimidated fellows. If a moth can convince birds that it is dangerous, the birds may pass up an edible item. In other words, communication may not only help an animal adjust to its world, it may also hinder other animals from adjusting to theirs. Just keep in mind that communication is adaptively complex and has no single, easily summarized role.

The ways communication occurs, the environmental constraints on each type, the information contained in specific signals, and their level of adaptiveness make up an incredibly complex picture, and one that is only beginning to be unraveled. Scientists, of course, seek to explain their areas of investigation in as precise quantitative terms as possible. In the area of communication, however, we often find ourselves only able to make the most general of statements about the behavior. For example, trained researchers can make few statements beyond those of a dog owner regarding the meaning of the pet's facial expressions. In other instances, however, animal communication is being measured with a high degree of precision. To illustrate, some signal molecules used in chemical communication have been identified and even synthesized. Furthermore, molecular size and configuration have been related to specific signal functions. In another area, animal sounds are recorded and converted to visual images that yield themselves to incredibly precise description.

So, now let's investigate some of the various ways in which living things communicate, but we should first be aware of some general aspects of the phenomenon and how it developed.

## THE EVOLUTION OF COMMUNICATION

Tinbergen (1951) pointed out that communication evolves by at least two steps. Originally, a signal springs from some aspect of an adaptive pattern. If the hair rising on a chimpanzee makes him look larger and more formidable to competitors, it is adaptive. In time, hair raising alone may come to be recognized as a threat (see Darwin 1872). And then, in time, the pattern may become ritualized and appear in new contexts.

Basically, signals or displays could have arisen from three different sources: autonomic responses, intention movements, and displacement behavior. Signals springing from such sources need not interfere with the adaptive pattern from which they arose, or with other adaptive patterns that are being carried out simultaneously with the signal. To illustrate, it wouldn't do for spider monkeys to show anger by waving their arms. Angry monkeys would fall out of their trees. However, they are perfectly able to hang onto limbs while raising their hairs, an autonomic response. Raising hairs doesn't generally interfere with anything important that might be going on. Now, let's take a closer look at these types of signals.

First, sympathetic *autonomic responses* provide an ideal raw material from which natural selection can develop signals. For example, as a human prepares for "fight or flight," the skin may blanche to a lighter color as blood is shunted away from peripheral areas. One may go "pale with fury."

Second, *intention movements* may provide material for signals. An animal about to leap may bob its head so as to increase its momentum when it does leap. Head bobbing, then, may mean, "I am about to leap." One animal may stare at another as he sizes him up before an attack. The stare is a necessary part of the attack, and it may also be read as a threat. So signals derived from intention movements not only fail to interfere with adaptive patterns, they are originally an integral *part* of such patterns. Only later do they become ritualized and separable. (Most displays are considered to have undergone ritualization.) Ritualization may, in time, permit the intention movements to become highly exaggerated, stylized, and repetitious, so that their value as a signal is increased.

In the third case, displays may be derived from *displacement behavior.* A male herring gull confronting a strange female may be motivated to chase and court simultaneously and, hence, may do neither, but may preen vigorously. The displacement preening may be performed in a slightly different manner from normal preening, however, and thus because of the peculiarity of the pattern, or the context in which it is performed, it comes to be recognized as a signal.

## LEVELS OF COMMUNICATION

If we see a man slip on an icy sidewalk, we gain information regarding our environment. Also, in the absence of a misdirected sense of humor, we may choose to tell the next person about the ice. In the first instance, communication is unspecific, inadvertent; in the second, specific. In this example, it is clear that specific communication potentially carries much more information than inadvertent communication. The principle will become more apparent as we go along.

The lowest level of communication involves inadvertent, unspecific signaling. As an example, consider certain protozoa that produce chemical by-products of metabolism that then cause rather general reactions in other protozoa in the medium. As another example, a prowling leopard may break a twig and alert its prey.

The next level of communication is reached when the neural apparatus of an animal is programmed to respond to specific inadvertent signals of others. Schooling behavior of fish is possible because swimming movements of any fish can be influenced by certain highly specialized receptors that are tuned in to inadvertent vibrations produced by the movements of other fish.

Much more complex communication is found in those groups in which the message is produced by specific signaling devices. The potential variety in such systems is enormous, limited only by environment and anatomy. The potential complexity of such systems is increased by the possibility of discrimination within each signal type. In other words, a fish may not only react to the visual stimulus of a red belly on another fish, but his reaction may vary according to how red the belly is. A dog being threatened by another dog may respond in different ways according to the vigor of the threat.

The highest forms of communication are reached in mammals, culminating in the primates. It should be pointed out that even animals capable of complex forms of communication may also employ simpler types. As an example, Jane Goodall was, for a long time, unable to approach the wily chimpanzees of the Gombe Stream Preserve because they showed avoidance responses to her inadvertent signals. Chimpanzees, however, often use complex forms of communication, as we shall see.

It is among the mammals that true purposiveness in communication is most clearly demonstrated. A small dog hearing a noise at the door may bark loudly while glancing over his shoulder at his master to make sure his alarm is being received. He may have no intention of attacking an intruder, but he may wish to make the situation known to his master. In purposiveness, the action is performed with the response "in mind." Purposiveness in communication, it should be stressed, is by no means restricted to mammals, although probably, in general, it is limited to vertebrates (Figure 16.1). Among the vertebrates, the receptors may be no more sensitive than, say, those of an insect, and vertebrates may employ no more devices than invertebrates, but because of purposiveness and because they can read subtle clues and simultaneously analyze a number of them through gestalt perception, they have developed intergrading and flexible communication methods.

Figure 16.1 A goose begging at the Bronx Zoo. It assumes the peculiar posture and orientation in order to receive food from visitors, a purposive behavior.

## Four Aspects of Communication

The structure of communication is defined in four ways (Tinbergen 1951; Hinde 1970; Jolly 1972). They are signal, motivation, meaning, and function. The *signal* is the form the act takes—screaming, walking on forelegs, urinating, or whatever. It is the part we see, and some researchers (Jolly 1972) tend to equate "signals" with "displays" once ritualization has occurred. *Motivation* refers to the mental set of the animal doing the communicating. Here we must make assumptions—we're back to the question of trying to know the "mind" of the animal. But there are ways of getting around too broad an interpretation of motivation, as we learned earlier.

*Meaning* is determined by the reaction of the animal receiving the message. Thus, a stare may mean a threat among gorillas, but the animal being stared at is often able to make fine discriminations among threats. He can base these discriminations on the complete analysis of the sender's facial features, his age, stance, sex, behavior in the previous minutes, lifetime history of behavior, and his distance. *Function,* then, is the adaptiveness, the evolutionary advantage, of the signal.

Descriptions of signals may carry connotations relating to all four aspects. For example, the "alarm-bark" of baboons may mean that usually (1) the sender is alarmed, (2) the receiver becomes alarmed, (3) something threatens, and (4) whatever it is, it is not likely to get the sender just because he barked, and the bark may save other members of the troop.

Lemurs scream (*signal*) when seized, or when they see a flying hawk, or when another lemur screams (Figure 16.2). The *motivation* may be shock or sudden terror. It *means* "take cover" and it *functions* as an air-raid alarm. Langur males whoop when they see a rival troop. The *signal* is apparently *motivated* by surprise and excitement. The rival troop withdraws, and thus the whoop functions in spacing neighbors and allocating food. The *meaning,* then, is not the *functional* "overgrazing threatens our environment," but "get the hell out of here!" (Jolly 1972).

Figure 16.2 Lemurs, the fascinating social prosimians of Madagascar, scream when alarmed and thereby alert other members of the group.

## Signaling Devices

As would be expected in a world where animals come equipped with all sorts of sensory mechanisms, there are a number of signaling devices, some quite bizarre. The cooing of courting pigeons seems proper enough, but who would have expected an amorous male porcupine to walk toward the female on his hind legs with an erection and then unceremoniously drench her in urine?

Each signaling device has its special advantage, but these often exist at a price. The ideal communication system should be specific in the nature and emphasis of messges. It should be able to reach the receiver rapidly, to travel long distances, and to show a wide range of information content. It should also be able either to start and stop quickly or to persist in the environment for long periods, depending on the nature of its message. Obviously, no system can fulfill all these requirements, so the function of the signal, the anatomy and physiology of the animal, and the nature of the immediate and past environment must determine which qualities are emphasized.

Signaling devices are varied and may include electrical, tactile, visual, chemical, and acoustical means. Probably because of the nature of our own communication systems, we know more about acoustical and visual signals. Much of the information in this chapter is derived from studies of birds and primates, since their perceptive abilities are roughly similar to man's, and since researchers have focused a great deal of attention on them in order to learn something about the development of human communication. We will not restrict ourselves to these groups, however. The discussion of a wide range of animals should give us a clearer picture of the nature of communication, and we will come to realize that the signaling mode is dictated as much by environmental demands as phyletic relationship. For example, our own communication system may differ quite markedly from those of other mammals. They are, after all, more-or-less color-blind, relying heavily on olfaction. On the other hand, our vocal communication has many features in common with the sound signals of insects. In the following discussions, we will concentrate primarily on communication by sight, sound, and smell because these areas have been the subjects of the most intensive investigations.

## METHODS OF COMMUNICATION

We will begin by describing the physical properties and relative merits of each type of communication since such information is critical to our understanding of the environmental implications discussed later.

### Visual Communication

Visual signals are especially important in fish, lizards, and birds, and in diurnal (and a few nocturnal) insects, some poisonous amphibians, and many primates. Lorenz (1935) and Tinbergen (1948a) laid the groundwork for ethological investigations of visual signaling and, in the ensuing years, made continuing contributions to the study as new facets of the phenomenon came into focus. In spite of a growing army of researchers, the bulk of the data has come from work on only a few species. For information and bibliography, you are especially referred to Tinbergen's work on fish and birds (1951, 1953c, 1959, 1967) and for mammals, De Vore (1965) and Ewer (1968). For further discussions, you might see Lanyon and Tavolga (1960); Marler and Hamilton (1966); Smith (1977); Hailman (1977).

To the ethologist, visual signals are especially interesting because of the tremendous amount of information they may carry. This "information load" is possible because visual clues may be expressed in four basic ways: through color, posture or form, movement, and timing. We have seen that color is important in releasing fighting behavior in fish and birds and that it functions in mating and foraging patterns in some species of insects. Size, motion, and color are all involved in the male-female interaction of the fritillary butterfly (*Argynnis paphia* L.) (Magnus 1958). In other butterflies, males first react to females because of their color or their flight movements (Crane 1955; Stride 1958). The importance of timing can be seen in fireflies in which the sexes are attracted to each other on the basis of their flash intervals (Barber 1951).

One important advantage in such a potentially diverse system is the wide range of variation in expression that is possible. In some cases, an animal may convey only the presence or absence of a particular condition or motivation (Figure 16.3). In other cases, such as in aggressive behavior, it may be useful to express degrees of motivation (Figure 16.4). So, visual signals may be modified by changing their intensity and by mixing the four component factors. Of course, graded and mixed signals can only be important in those species that have developed receptors sensitive enough to register subtle differences and that are intelligent enough to interpret complex communication. The complexity of visual signaling also makes it possible for the same message to be conveyed by multiple means. Such redundancy

Figure 16.3 An example of an all-or-nothing display in the male cutthroat finch *Amadina fasciata.* At left is a low intensity courtship display. At right is a high intensity display of the same type. While the differences in the displays are slight, there are no intermediate states.

Figure 16.4 An example of graded display. At the left the bird, *Fringilla coelebs,* is showing a low intensity threat. The posture at center is a medium intensity threat and below is a high intensity threat.

may serve to emphasize and reiterate signals and thus to insure accuracy in the signal recognition. If a dog raises its hackles, bares its teeth, and rushes toward us, we are almost *sure* to get the message.

There are other advantages to visual signals as well. For example, they can be started or stopped immediately. If a displaying bird suddenly spots a hawk and freezes, its position will not be revealed by any lingering images. Also, when a sender wishes his position to be known, as when he is trying to attract a mate, there is never any doubt as to his exact position. The receiver can see him and therefore respond to him in terms of his precise location as well as his general presence.

A visual signal may be a permanent feature of the animal, such as in the elaborate coloration of the male pheasant and the facial markings of a male mandrill (Figure 16.5). It may also be a temporary or recurring signal, such as the reddening tail area of the female baboon in estrus and the annual growth of antlers in male deer. The signal may also be present or not, according to the dictates of the animal. An angry male baboon can expose his long canine teeth as a warning, but the weapons are covered at most other times. A threatening bird may display strikingly colored feathers that are kept discreetly covered at other times, such as when the bird is feeding or resting.

The dynamic properties of visual signals are often critical. Temporal patterning of the signal, for example, is important in a wide variety of species. The mysterious Caribbean fireworm (*Odontosyllis*) is about an inch long and spends its time burrowed in the bottoms of bays throughout the West Indies. However, on the fifth night after an August full moon, it begins a strange pageant. The worms crawl out of their burrows and swim to the surface of the warm tropical sea. They time their activities precisely, rising in that darkened hour between sunset and the time the moon rises, so there is nothing to interfere with what they are about to do. When they reach the surface, for the only time in their lives, they become phosphorescent. The water sparkles and shimmers with living, greenish, blinking lights. This is their time of reproduction. Females maintain a steady light for about ten seconds, while males, blinking once or twice a second, swim toward them. Suddenly, in a tiny phosphorescent explosion, clouds of eggs and sperm fill the water. Then the lights quietly fade and the worms sink slowly to the bottom, probably to die.

Visual signals have their disadvantages, of course. The most obvious of which is, quite simply, if the sender can't be seen, his signals are useless; and vision is easily blocked by all sorts of obstructions from mountains to fog. Also, visual signals are useless at night or in dark places (including the depth of the sea), except for light-producing species. Furthermore, visual powers weaken with distance, so visual clues are usually not employed for long-distance communication. As the distance increases, visual signals must become simpler and bolder and, of course, they can therefore carry less information.

Visual signals can only be maximized by nearness and contrast. It is not possible to increase such a signal by pumping more energy into it, except for light-producers. So visual communication is strongly controlled by the dictates of the immediate environment, in that its properties do not usually permit it to circumvent or overcome environmental obstacles, and it normally can't be "pumped up."

Visual signals stem from several evolutionary sources. For example, certain signals have apparently developed from intention movements. A dog with teeth bared, we know, is likely to bite. Other dogs and postmen now interpret the sight as hostile warning, but originally the dog was probably only moving his lips out of the way. Some signals may also have developed from involuntary events. In man, a reddening face may mean that he is embarrassed or angry. The reddening of the wattles in courting or fighting birds may also have occurred involuntarily, through autonomic action, and only later have come to be used as a signal. Displays could have also been derived from protection movements. The frown of some primates could have originally been a movement to protect the eyes before a violent encounter. Physiological functions such as respiration could have provided the raw material for the development of signals as well. For example, the air sacs of the male prairie chicken are brightly colored and boldly displayed in courting ceremonies.

Figure 16.5 A male mandrill has permanent markings that advertise his sex, but his signals of anger are temporary and dependent on his mood. It is probably often to his advantage to be related to as a male, but less frequently as an angry male.

Signals could also be derived from redirected or displacement behavior as was mentioned earlier. Preening in pigeons is now a part of their courtship behavior, but at one time it may have simply been the displacement activity of an excited and conflicted bird.

While visual communication involves other parts of the body, the head area seems to be particularly important, especially among birds and mammals. In one fascinating experiment, a monkey learned that when a light came on, he had to quickly press a bar or he would receive an electric shock. A second monkey in an adjoining cage could not see the signal light, but learned that he would be shocked unless he pressed his own bar at the right time. The only clue he had that the shock was coming was the expression on the first monkey's face. He quickly learned to read it and to press the bar in time (R. E. Miller 1967).

It is not surprising that facial expressions are more important in group-living animals such as wolves, than in solitary ones such as foxes (Fox 1970). Facial expressions are relatively subtle cues and would probably involve a high degree of learning. Such learning could probably only occur where the animals were exposed to each other for long periods of time. We will have more to say about this later, but the association of many signaling devices with the head region should not be unexpected since the anterior area is the location of the most complex neural centers.

**Habitat Influences** The structure of the habitat largely determines what sort of signals the animals living there will use. Because of a tendency for animals in a certain kind of habitat to communicate in similar ways, certain generalizations can be made regarding habitat-communication relationships.

In woodlands, birds that occupy the lower levels of the forest canopy can be expected to be dull in coloration in comparison to the birds that occupy the upper levels or live at the forest's edge. The lower canopy is often dark, and visual signaling would therefore not be efficient. Instead, animals that occupy dark areas (or those species that are in danger of predators) might be expected to rely more on sound or chemical signaling. By the same principle, long-distance visual signaling could only be effective in unobstructed environments, such as in grassland prairies or savannahs.

In general, birds that defend large territories herald their ownership by vocalization, while those defending smaller areas, such as nest sites, use visual displays. This may be because at long range, visibility may be more easily blocked by vegetation.

An animal's communication must also reflect an adjustment to the kinds of animals who share its habitat. Predators, of course, are an important factor in the development of any communication pattern of a prey species. Spatial distribution of nests or broods also may be influenced by predation (Chapter 23), and spacing or distance affects communication in important ways, as has been mentioned. Predation may also influence signaling in another way. If the risk of predation is great, prey species may be cryptically marked, thus reducing the opportunity to utilize visual signals except by movement or posturing.

Predators, as well as prey, may be cryptic and, therefore, subject to the same limitations in visual signaling. The conspicuous hood of the brown-headed gull may help it in signaling others of its kind, but it may be disadvantageous at times when the bird is searching for surface fish because the contrasting coloration may conceivably alert the prey. Here again we see a character seemingly preserved because of a *net* adaptive value. Its disadvantages are outweighed by its advantages.

One of the most critical environmental variables is the distance between the signaler and recipient, a factor largely determined by habitat structure, population density, and ethological influences. The greater the distance, the greater the probability of physical barriers interceding, of course, but distance presents other problems that must be met in specific ways. Long-distance visual communication, for example, requires simple, conspicuous, stereotyped patterns, whether in marking, posture, or movement. And for greatest visibility, these must contrast sharply with the background. A distant tiger in the grass would have little luck in trying to signal through posturing.

At this point, we don't know whether different markings can shift their overall significance with distance. In other words, a pattern may be conspicuous against a certain background at short range, but as distance increases, the pattern may become cryptic against the same background. For example, from close up, a zebra is much more conspicuous than a horse, and its markings may be useful in intraspecific communication. As distance increases, however, the horse becomes more conspicuous as the same bold markings of the zebra cause it to fade into the background.

## Acoustical Communication

Studies of sound communication have revealed new realms of animal interaction. The value of such studies stems primarily from the recent use of the sound spectrograph, a device that changes sound to visual markings, which can then be precisely measured and analyzed (Borror 1960).

Sound plays such an important part in human communication that it is surprising to discover that among other animals it is limited, for the most part, to arthropods (joint-legged creatures such as insects) and vertebrates.

Arthropods communicate acoustically in a variety of ways. In some cases inadvertently, as we see in insects that react to the flight sounds of others of the same species. In fact, the whine of a flying insect may be reproduced electronically to lure other members of that species. But in other cases, the sound signal is *not* a by-product of another pattern; sometimes it functions solely in communication. Arthropods, for example, utilize highly specialized friction devices involving almost every movable part of the body, including the wings, legs (and their evolutionary derivatives), and peculiar vibratory membranes.

Sound communication has apparently evolved independently hundreds of times in insects, and now occurs in tens of thousands of species. But in spite of this strong reliance on sound, the dimensions of arthropod signals are severely limited. In fact, the most complex arthropod sounds have fewer dimensions than most vertebrate sounds. For example, insects lack rhythmic fluctuations in frequency (melody) in their signals. Sounds by insects are characterized by differences in *time* distribution of the pulses, rather than by pitch or tone—qualities important in the sound signals of birds and mammals (Alexander 1960).

Vertebrates, as would be expected in such a diverse group of animals, show a wide variety of sound communication. In fish (Tavolga 1960, 1964), sounds are produced in a variety of ways, but usually by frictional devices or by manipulation of the air bladder. Since sound travels quickly in water, it might be expected that sound signals would be widespread among fish. However, such is not the case. Most fish are rather quiet and make good apartment pets. It seems surprising that fish rely on sound so little, but it may be due to the fact that sound sources are difficult to locate underwater.

Terrestrial vertebrates usually produce sound signals by vibrating special membranes as air is shifted in the respiratory system. However, they also produce nonvocal sounds. Rabbits thump the ground; grouse drum their wings; gorillas beat their chests; woodpeckers hammer hollow limbs (or drainpipes early on Sunday mornings).

Vocal communication is rather limited in lower vertebrates. The limbless caecilians (amphibians) have no ears and aparently receive sound through substrate vibration. They are thought to be incapable of producing sound. However, a few salamanders emit squeaks, and two genera apparently can whistle; one large California ambystomid can even *bark* (Bogert 1960)!

In frogs and toads, sound is important in mating and territoriality. (It is interesting that these anurans have "vocabularies" that are about the same size as those of the orthopteran insects—grasshoppers and crickets.) For a discussion of the acoustical behavior of amphibians, see Blair (1963).

Reptiles, although generally more specialized, have less well-developed sound systems than do the amphibians. Snakes apparently rely on substrate vibrations to detect sound, as do the caecilians. Snakes, crocodilians, larger lizards such as the Gila monster, and some turtles may hiss as a warning, but they are warning other species (since their own species could not hear such a sound). A few snakes, such as the coral snake (*Micruroides* spp.), have developed the rather gauche habit of drawing air into the cloaca (the common genital and excretory opening) and loudly expelling it when irritated. And, of course, some snakes in the genera *Sistrurus* and *Crotalus* have rattles, and these may be effective in repelling attacks (although the adaptiveness of the rattling has never been conclusively demonstrated). Certain geckos (Gekkonidae), however, have perhaps the most advanced vocal system among reptiles.

Reptiles may also communicate through startlingly loud sounds. Roaring and bellowing tortoises were described by Darwin after visiting the Galapagos Islands in 1839 at a time when the reptiles were breeding. Loud bellowing of enormous territorial alligators could often be heard in the deeper recesses of our great southern swamps (Beach 1944), until their skins became fashionable, lumbering interests took an interest in the trees, and "developers" drained the marshes of their life-sustaining water.

In birds and mammals, sound communication becomes much more complex and takes on new meanings and increasingly subtler nuances, and, as would be expected, the acoustical receptors in these groups are more complex than those of other vertebrates. The outer ears of many mammals are movable and can be shifted to focus on a sound source, but for most animals, the slight difference in the time it takes for sound vibrations to reach each ear is generally more important in locating the source. (If sound reaches the left ear first, its source is to the left.)

Sound may vary in frequency (the number of waves per unit time—usually expressed as high or low pitch), in volume, and in "quality." (Quality may be demonstrated by two people humming the same note at the same volume, whose voices are still distinguishable.) John MacKinnon (1974), who conducted an eleventh-hour study of the orangutan as their habitat was being decimated by loggers, was often able to recognize his subjects by the qualities of their vocalizations.

The possible variation permitted by such subtleties makes sound suitable for those vertebrates that have mental abilities that can take advantage of such a potentially complex system. But, at the same time, their ability to discriminate between subtle differences in sounds has posed special problems for the investigator who would like to be able to label each sound as having a specific meaning. He finds gradations and interpretations inconvenient.

It is true that primate vocalizations are less easily categorized than those of birds, but there are similarities in the types of sounds employed in similar circumstances by the two groups (Collias 1960). An example is the low, soft sounds by which both birds and mammals "speak" to their young. The significance of this convergence is underscored by the fact that the sounds of birds are produced by a syrinx and those of mammals by vocal cords. Convergent evolution indicates a high adaptive value of the common characteristic.

The relationship of the *characteristics* and *functions* of a signal can be demonstrated in studies of sound communication. For example, consider the difference in two types of responses of passerine (perching) birds to predatory birds. Passerines will usually mob a stationary owl or crow by swooping down on it, displaying, and calling loudly to attract other birds (Figure 16.6). Their sounds at such times are easy to locate; a typical sound spectrogram is shown in Figure 16.7. When a hawk is spotted overhead, however, the response is different. Here, the call is a high-pitched *tweee.* A bird in the open hearing such a sound heads for cover, crouches quietly, and sometimes takes up the plaintive, hard-to-locate call (Figure 16.8). Note that the high-pitched warning call is similar among the species (Marler 1961b), and probably several species are able to take advantage of the warning call of a single bird.

Figure 16.6 A mockingbird attacks a great-horned owl. The mockingbird's distinctive and easy-to-locate calls will attract other birds that will join in the attack.

Sounds that are easy to locate, such as mobbing calls, breeding calls, and signals between parent and young are usually repetitive, of brief duration, and of low frequency with the emphasis on the first part of the sound (Klopfer 1962). These sounds may be highly variable while maintaining critical qualities in common. The adaptiveness of signal variability is discussed later.

Sound signals, then, have a number of advantages. They may vary in several dimensions (frequency, volume, timing) and they may also be distinguishable at a very low level. In addition, sound may be raised above environmental noise by pumping energy into the system (making the sound louder). The transient nature of sound makes possible rapid exchange and immediate modification, but it does not permit the signal to linger without the sustained expenditure of energy. And finally, an animal does not necessarily have to suspend other activities in order to transmit or receive sound signals.

**Environmental Influences** There are a number of examples of the effects of environmental noise on communication. Naturalists tell us that the island-dwelling St. Kilda wren is much louder than the mainland species, which lives in a quieter habitat. If this is so, it may be an example of a species increasing the energy of a sound signal to overcome background noise rather than yielding to it and utilizing alternate means of expression, as we saw earlier in the Brandt's cormorant.

For burrowing and tunneling animals, such as prairie dogs (Figure 16.9) and pocket gophers, the texture of the soil may determine whether sound is used in signaling, although subterranean species probably do not rely much on sound communication. However, sound carried through the earth might reveal the approach of terrestrial predators, and finer, more densely packed soil particles would be a more effective transmitter.

Above ground, the type of vegetation may affect how well sound carries. For example, animals living on a forest floor can best signal over long distances if the wavelengths of the sound are longer than the average diameter of the trees (Konishi 1970). If an

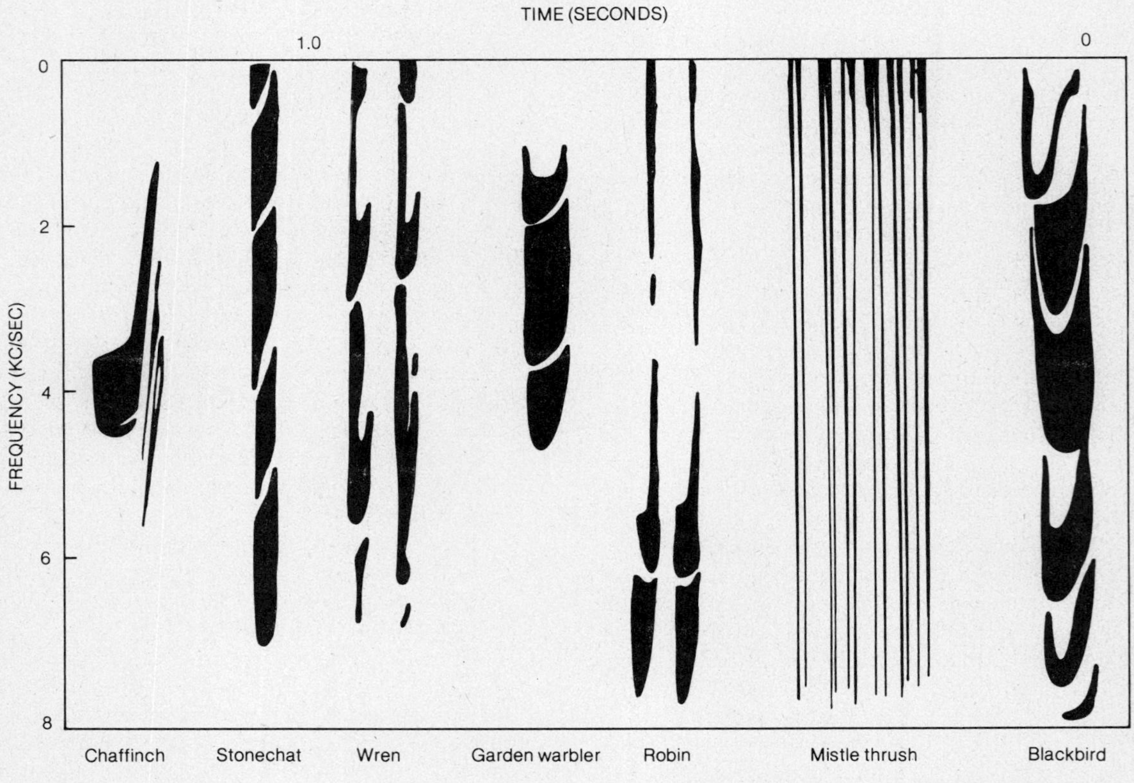


Figure 16.7 Sound spectrograms of mobbing calls of several species of British birds attacking an owl. Each sound has acoustical qualities that make it easy to locate.

animal living on the forest floor were to make sounds of a shorter wavelength, the energy of the sound would quickly be dissipated by reflection, diffraction, and absorption. Since higher pitched sounds have shorter wavelengths, denizens of the forest floor tend to produce lower sounds for longer-distance communication. An example of such a sound is the "cough" of a leopard (see also Morton 1970).

In the interest of not attracting predators, birds that build solitary or cryptic nests seldom advertise their nest areas by song. They either sing more softly around their nests or they sing someplace away from the nest area. This behavior makes sense, since the threat of nest predation may have dictated that they be solitary or cryptic nesters. Bird *colonies*, on the other hand, are usually built in protected sites where the likelihood of predation is reduced (Hoogland and Sherman 1976). And as we shall see, gregariousness itself offers some predator protection. Thus, colonial nesters often blatantly advertise the presence of their nests (see Crook 1969), a behavior that may repel nest site competitors.

We all know that birds sing more in the morning than at other times, but they may not be responding to the time of day, as such. Birds apparently sing in response to light and not the hour. It appears that birds increase their singing frequency in response to relatively rapid increase in the duration or the intensity of light in the habitat. To illustrate, birdsong is inhibited by overcast skies and even by solar eclipses. Singing may also be depressed by high wind, extreme heat or cold, and heavy rains. Logically enough, birds are most likely to sing on quiet, humid days when the sound carries well.

For some animals, the influence of time is even more specific. When at dusk the light intensity drops to about five footcandles, cicadas begin to sing so suddenly that the effect is startling; they fly in every direction for a short time and then, just as suddenly, they stop. The "dusk effect" is especially noticeable in tropical jungles when the quiet calm of early evening is suddenly shattered as if a great wind were sweeping through the trees. The following calm, if the chorus of another species doesn't swell, sometimes gives me the eerie feeling that something terrible has just passed. Such synchronization of intraspecific communication, of course, is highly adaptive for prey species that are thus exposed to predators for the briefest period of time. It may be that signaling and

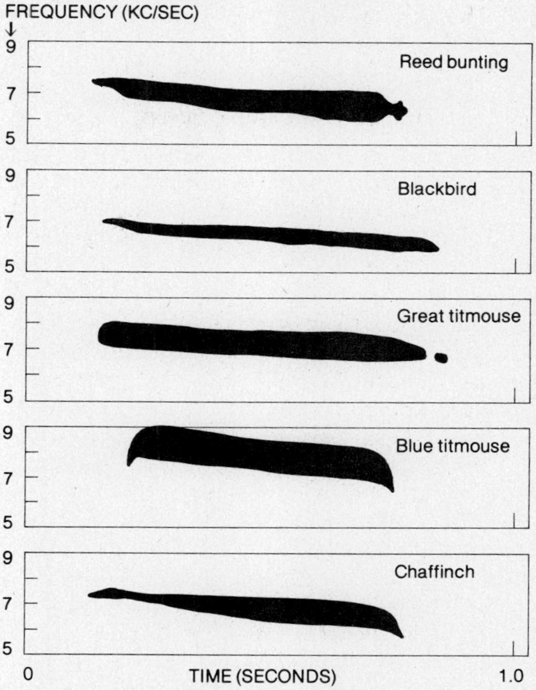


Figure 16.8 Sound spectrograms of calls of five species of British birds when a hawk flies over. These calls are usually high-pitched and extended, qualities that make them hard to locate.

general activity at dusk or dawn is adaptive in that it occurs in a period for which neither nocturnal nor diurnal predators are well suited, as I mentioned earlier.

As I also said, birds tend to sing more on humid days because humidity enhances the propagation of sound. The phenomenon can be testified to by anyone who has stood on a lakeshore and been startled by a whispered conversation between people out in a boat. Also, the humid air of the Arctic has been called "acoustically transparent" (Hunter 1965). Humidity may have greater impact on the lives of animals than has been believed. For example, it has been pointed out that perhaps one of the reasons desert animals have such large ears is to hear better in an acoustically poor environment. (Of course, heat dissipation through exaggerated appendages may be another reason.)

Long distance presents special problems in sound communication. Complexity of signals increases potential information content, and one way complexity in sound can be achieved is through the use of different frequencies. However, since high-frequency sounds attenuate faster than lower ones, frequency modulation or change of pitch cannot be used well in long-distance communication. Some frequencies would die out before others, and only parts of the message would get through. Variable frequencies, however, may be expected in many of those animals that gather to breed, or those that live in colonies where both high- and low-frequency sounds are likely to reach the recipient.

Harmonies are an interesting phenomenon. They are produced by uttering high and low notes simultaneously, and are frequently used by some social species. For example, colonial or semicolonial birds signal harmonically (Collias 1963; Hall 1962). Monkeys and apes that form social groups and do not mingle with other species show many harmonics and lack "purity of tone." On the other hand, those primate species that live in forests, move in loose groups, and mix with other species, thus increasing the likelihood of confusion in sound signaling, produce purer, even "birdlike," tones that are more easily distinguished (Marler and Hamilton 1966).

Figure 16.9 Burrowing animals, such as this prairie dog, probably do not greatly rely on sound communication, at least while underground. Certain soil types, however, transmit sound better than others and it is in these places that we should expect subterranean vocalizations.

## Chemical Communication

We have all seen ants running single file to sack our cupboards. We have also taken perturbingly slow walks with dogs that stop to urinate on every bush. Both kinds of animals are communicating by the use of chemicals. In fact, a wide variety of animal species, primarily insects, fish, amphibians, reptiles, and mammals employ chemical communication, usually by smell, but occasionally by taste, although the distinction between the two types of reception is not always clear, as we learned earlier.

In insects, chemicals produced by one individual may cause very specific behavioral changes in other members of the population. Such stereotyped response should not be surprising in the absence of great intelligence or insight in these animals. The immediate and rigid responses elicited by such chemicals were thought to be similar to those produced by hormones, so the chemicals were called *pheromones* (from the Greek *pherein,* to carry; *horman* to excite). Pheromones, then, are substances that an animal secretes into the environment and that cause a specific behavioral or developmental reaction in another animal of the same species. Interestingly, pheromones are not restricted to insects.

There are certain mammalian pheromones that are of interest to reproductive physiologists. Female mice, kept in a group free of males, will come into estrus (heat) infrequently and asynchronously. But if a male is introduced, the females will immediately become receptive all at once. They will also respond to male urine or the scent of an upwind male. However, if the olfactory bulbs of their brains are destroyed, the females will not respond (Bronson 1968). Such studies have revealed an interesting relationship between signals and physiology: pregnant female mice, when exposed to a strange male, will abort their fetuses and enter estrus again. Here, pheromones cause hormonal changes (Clulow and Clarke 1968).

In order to understand more fully the nature of chemical signals, let's briefly consider a case involving insect pheromones. The sexually receptive female silkworm moth (*Bombyx mori*) releases a sexual lure into the air. The substance is a complex alcohol called bombykol, and it is so potent that it arouses males that are presented with a glass rod that has been dipped in incredibly dilute solutions of the substance ($10^{-12}$ mg/ml, or one part in a trillion). In nature, the female stays put and releases her potent sex pheromones. Obviously, bombykol molecules dispersed by the wind would produce too weak a gradient for the male to follow, but once aroused, he could find the female by simply flying upwind in a tactic response (Case 1966).

Sex pheromones have been used effectively in the elimination of insect pests. For example, the Mediterranean fruit fly was exterminated in Florida by using traps baited with a sex-lure. In the northeastern United States, an artificial lure, disparlure, is being used to trap the particularly destructive gypsy moth (*Porthetria dispar*). Such lures are far preferable to pesticides, and more research attention should focus on such "biological" means of pest control.

Honeybees provide another interesting example of how one organism can affect another's behavior. The queen bee produces a "queen substance" that, when fed to developing females, inhibits their sexual development, causing them to become workers, replete with the complex array of innate patterns of the worker bee. When she dies, the inhibiting substance is no longer produced, and new queens can develop. The undeveloped ovaries of some workers also mature, and these eggs produce drones, or royal consorts, one of which will mate with the new queen in her nuptial flight. The behavior of each caste is very specific, and the caste of any individual is determined chemically by other members of the group.

**Physical and Environmental Influences** Wilson and Bossert (1963) analyzed the properties of various chemicals employed in signaling and described the characters of the types of chemicals according to their function (Figure 16.10). They found that the range of carbon atoms (the "backbone" of chemical chains) should number between five and twenty and that the molecular weight of the chain should be between eighty and three hundred. They reasoned

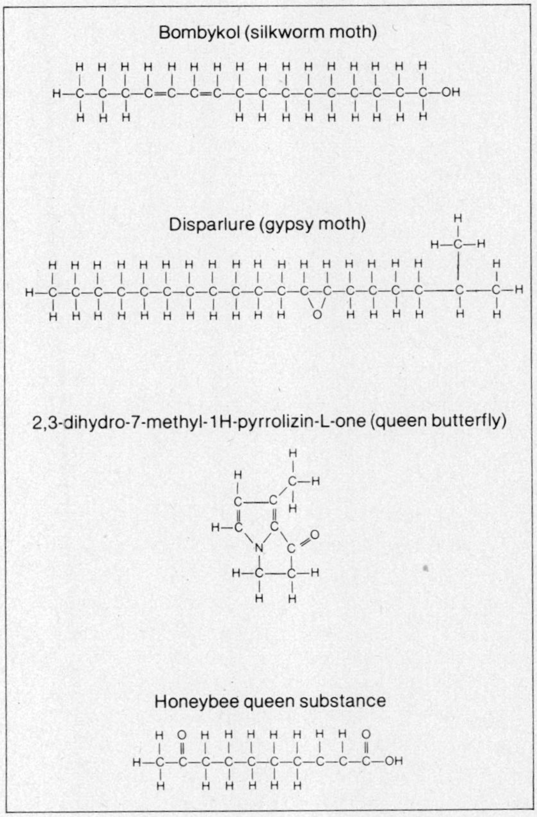


Figure 16.10 The specific molecular configurations of four insect pheromones. Our knowledge of their structure has enabled us in some cases to manufacture synthetic signals.

that at least five carbons are needed for diversity, otherwise the number of possible configurations necessary for specificity will be too small. The maximum size of the molecules, however, is determined by other factors. Since the molecules usually travel through the air, they must be small and light. Small molecules are also easier to make and store. As Wilson and Bossert predicted, sexual attractants are larger than the molecules used in other chemical signals, since large size increases the potential for complexity and complex signals must be used to ensure that only members of the same species are attracted. (For further information on insect pheromones, see Collins and Potts 1932; Dufay 1957; Schwink 1958; Barth 1961, 1962; Wilson 1968, 1970, 1971; Jacobson 1972.)

Animals could, and possibly do, increase the information content of chemical signals by mixing chemicals, but mixing has its problems. If one molecule is lighter than another, it may diffuse farther and confuse distant receivers. Instead of mixing, animals could release different chemicals sequentially, but the long life of lingering chemicals generally rules out such a system, since distant receivers might be likely to be confused by a number of chemicals.

It has been found that in spite of the complexity and specificity of long-chained chemicals, males of some species are stimulated by pheromones from females of other species (Schneider 1962). So, apparently, chemical distinctiveness does not completely ensure private lines of communication. Instead, species-specificity may be maintained in these cases by chronological factors such as differences in the time of day or the season that the signals are sent. In fact, a circadian rhythm has been found in the *responsiveness* of some male moths.

Some insects are capable of chemical discrimination within a much wider range of stimuli than is indicated in simple pheromone studies. For example, although bees use antennal receptors, they can distinguish up to forty-three odors (von Frisch 1950).

Apparently, in some insects, taste perception is much more limited than olfaction. Water beetles and honeybees discriminate only four tastes, comparable to sweet, salty, sour, and bitter, and they are unable to discriminate between various compounds within each class.

So, then, of the major types of communication, only chemical signals linger in the environment. Even the shortest-lived chemical persists longer than a visual, auditory, or electrical signal. Unlike any of the others, however, its effectiveness is strongly dependent upon movement in the surrounding medium. One result is that time and space take on special meanings; it is unnecessary for the sender and the receiver to be simultaneously coordinated in time and space as demanded in the former types of communication.

We need only be reminded that chemicals share the advantage with sound of being able to go around environmental barriers; but, unlike sound, chemical signals are difficult to modify on short notice.

## Tactile Communication

Many kinds of animals communicate by physically touching. For example, bees may touch a dancing worker with their antennae. Birds may preen each other, and the preening, whether it is displacement behavior or not, may strengthen social bonds and instill confidence in the one being preened.

Figure 16.11 Gelada baboons grooming each other. Mutual grooming is not only a way of keeping the fur clean, but it helps build reassuring bonds by continual touching.

Among dogs, a dominant individual may place a foreleg on the back of a subordinate. This is probably an intention movement to mount the subordinate. The more highly aggressive bitches, such as my German shepherd, will mount another spayed female in the "normal" sexual pattern, but without following the action with thrusting movements. It is apparently a signal of dominance, but it looks ridiculous.

Tactile communication is expressed in a number of ways among the primates. Primates, as an order, are contact animals. They touch a lot. One form of communication is in grooming behavior. Most prosimians, the most primitive group, scrape each other's fur with a "toothcomb," a device that consists of all the lower incisors and canines. Monkeys and apes, the groups more "advanced" than the lemurs and tarsiers, part the fur with their hands and dexterously remove fine particles with fingers and lips (Figure 16.11). Admittedly, such acts clean the fur, but from lemur to chimpanzee, it also provides a social function (Jolly 1972). In some species, grooming patterns may have developed from mother-infant relationships (as the infant moves from the teat, it continues to grasp and suck the fur [Anthoney 1968]), or in New World monkeys from precopulatory behavior (Moynihan 1967). Primates may also nuzzle, lick, kiss, hit, hug, bite, pat, and kick as they communicate.

Garden snails (*Helix pomatia*) are hermaphroditic (having both male and female reproductive organs) and so they present an interesting problem in mating. For them, the problem of sex recognition becomes one of species recognition, and the courtship is a bizarre one indeed. It begins with the snails slowly entwining around each other's slippery body. Their gentle touching gives no hint of what is about to happen because, suddenly each snail thrusts a calcarious arrow—a love dart—into the body of the other (Figure 16.12). They both recoil and, in fact, they can be grievously injured. Then they exchange sperm. Then, if they have not seriously wounded each other, they go their separate ways.

In some instances, to touch can mean *instant* death. The jumping (salticid) spiders (Figure 16.13), hunt by running and jumping upon their prey, and wolf spiders (lycosid) chase and pounce on theirs. Each have good eyes, as spiders go (Figure 16.14). The problem lies in their habit of attacking other arthropods roughly their own size. The male jumping spider, for example, is smaller than the female and is

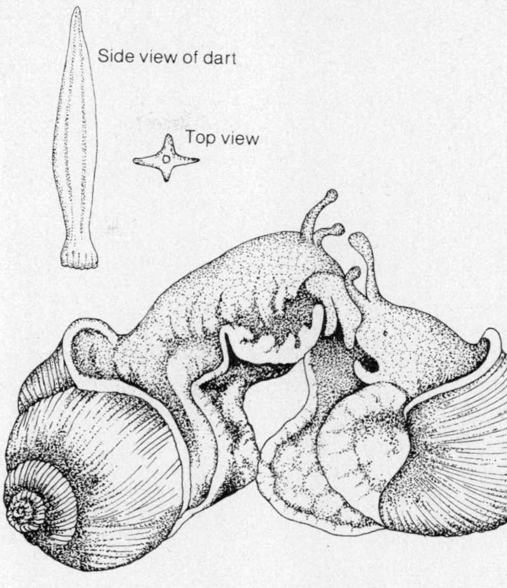


Figure 16.12 Garden snails in the act of mating. On the left is a "love dart."

Figure 16.13 A male jumping spider, seen here with its prey, leaps on its victim from some distance away. Thus as males approach females to mate, they are in danger of being mistakenly attacked.

Figure 16.14 The complex eyes of a wolf spider are very keen and specialized to detect movement.

therefore in great danger of being taken for prey when he approaches for mating. So he approaches her with extreme caution, signaling incessantly by waving his legs and dancing back and forth until she gives signs of receptivity (Bastock 1967). Finally, he may stroke her with his forelegs. In some species, his antics seems to put the female into a trancelike state, whereupon he deposits his sac of sperm with a specialized arm, and clears out. Males of other arthropod species are not so lucky. Among mantids, the approaching male is frequently seized by a voracious female. He is forced to deposit his sperm as she devours him. In fact, his mating movements often increase after she has completely eaten his head and inhibitory ganglia (which perhaps shows how much brains it takes to copulate).

Animals, then, communicate in a number of ways and with a host of adaptive effects. Any communication mode is a compromise because it will surely have its drawbacks. The modes available to any species will reflect its evolutionary history, since environmental influences over time will have selected for the kinds of signaling that will best enable an organism to be read *to its advantage* by those sharing its environment.

# 17

# THE ADAPTIVENESS OF COMMUNICATION

The world is full of living things, seemingly all attempting to communicate at once. You can perhaps sit right there and overhear conversations, smell perfume guaranteed to be seductive, hear dogs bark or mosquitoes whine, and you may be surrounded by books. You're probably reading one. The world is also full of signals that you are unable to pick up. You can't smell the bombykol that is driving a moth near you mad. Perhaps the white flower near you is an irresistible violet to a bee. Are the people behind you exchanging glances? Animals are communicating everywhere as each seeks an advantage in a rigorous world. Obviously, communication is important, or else it would not have evolved in so many different species. And natural selection would not have produced so many ways to ensure private lines and so many ways for those lines to be tapped. So, let's now see some of the ways communication is adaptive.

## RECOGNITION

Animal signals may roughly be categorized according to whether the message is between members of the same species or those of different species. Although the means of communication may be the same in the two instances, there may be distinct differences in the adaptive results. For example, intraspecific (within a species) communication usually functions in species and sex recognition, in identification of populations or individuals, in alarm, and in social coordination. Interspecific (between species) signals, on the other hand, may function in alarm, in threat, in intimidation, and in crypticity (the message, "I am not here"). In some cases, we see the same message conveyed between and within species, such as in the warning cries that alert members of the signaler's own species and other animals in the area as well.

Communication in the animal world has, of course, many roles. We shall explore only a few broad categories of these adaptations, beginning with recognition, and, as we progress, it will become increasingly apparent that there are few instances when a recognition signal will be useful in a single, narrow sense. Here, as in all biology, we are reminded that dogmatic categorizing is presumptuous.

### Species Recognition

It is, first of all, important for animals to know with whom they are communicating. Some species have little need to communicate with other species, so it is of utmost importance that an animal be able to establish its credentials within its own group. Just as important, an animal must be able to read the credentials of other animals: "Is he one of us?" In some cases there is no problem. Certainly, a fox is not going to confuse a horse with one of its own. However, in Chapter 11 we learned that birds may be imprinted to consider even a man as a conspecific (of the same species). Confusion may occur in other cases also. After all, a male moth can be fooled into approaching a female of another species if she emits chemicals similar to those of his own species. Some species of birds are strikingly identical and when they occur together, they may mate or attempt to mate.

Such mating mistakes are largely prevented by each species being born with distinctive morphological and behavioral traits that others of their species can recognize. It is interesting that some species are also born with certain sound repertoires. Nearly all insects that communicate largely by sound also overwinter in the egg stage. The result is that by the time they hatch, their parents are long dead, so no individual has contact with any member of preceding generations. But each year, the new generation produces the sounds of the species without ever having heard those sounds. In insect species with simpler songs, a deafened individual can produce the proper songs without even having heard *himself*. In other insects, it has been demonstrated that for some types of songs, a little practice is needed before the sound organs operate properly (Alexander 1960).

This is a good place to again remind ourselves that there is a close coordination between inherent and acquired characteristics to produce an adaptive response, but that the actual development in various species may involve quite different levels of input. Song development in birds, for example, shows clearly that the relative importance of innate and acquired patterns may vary from one species to the next. In some birds such as the roller canaries (*Serinus canarius*) and the European whitethroat (*Sylvia communis*), the normal song is produced even when the birds are reared in auditory isolation. The European blackbird (*Turdus merula*) will produce the proper song even when deafened while young. In these cases, genetic programming is important. On the other hand, the chaffinch (*Fringilla coelebs*) learns its song, but only in its first spring. After this, song learning is, for the most part, not possible. It has been suggested that certain critical parts of the chaffinches' song are learned through a rapid learning process similar to imprinting (Thorpe 1961). For a review of the importance of learning in song development, see Lanyon and Tavolga (1960); Thorpe (1961); Marler and Hamilton (1966). The development of song in the white-crowned sparrow has been mentioned and will be discussed in more detail in the next chapter.

Why is it so important that an animal not make species-identification mistakes? The answer is, in most cases, to ensure the proper performance of reproductive and aggressive behavior. Put simply, an animal must not waste time attempting to mate with another animal when the result of such a union, should it occur, is likely to be reproductive failure (Mayr 1942; Huxley 1942; Sibley 1957, 1961). Also, any animal's strongest competitor for available resources is likely to be a member of the same sex and species. So, these competitors must be recognized and, in many cases, either avoided or driven away.

But how is such recognition accomplished? How do animals know their own? One method is by the tendency of animals of a given species to congregate in the same places, thus one may behave toward those around him as if they were conspecifics (of the same species). Chemical stimuli are involved in some insect aggregations. Males of one species of beetle are attracted to places where other aggregations have recently occurred (Eisner and Kafatos 1962), and some species of butterflies aggregate in response to pheromones (Crane 1955). Of course, once individuals are drawn together, they may employ specific grouping and coordinating signals, mutual environmental attraction being only a means of crossing the first hurdle—achieving proximity.

In other cases, animals are drawn together by the *actions* of others of the species. Schooling in some fish is accomplished by each fish sensing the peculiar visual image and vibrations of other fish (Keenleyside 1955). Certain birds congregate by using specific calls to bring the group together (Stoddard 1931; Odum 1942; Hinde 1952). We have already mentioned sounds made by groups of migrating birds that may function in keeping the flock together when visibility is poor. "Grouping signals" are in most cases species specific, but there are notable exceptions, such as occur in mixed foraging flocks of tropical birds (discussed in Chapter 30).

Theoretically, proper species identification could be enhanced by a system of filters, whether peripheral or central. In such cases, signals lacking certain critical characteristics of the species would evoke no response and, in fact, might not even be perceived.

Species recognition may occur through any of several types of cues interacting in unexpected ways. W. E. Lanyon (1963), of the American Museum of Natural History, set up dummy flycatchers with two loudspeakers hidden nearby. The dummies were within sight of live flycatchers. He then played different songs over each loudspeaker. One of the songs was the normal flycatcher song. The other was of a different species. Only the dummy singing the "appropriate" song was attacked; the other was ignored. Apparently, the visual clues needed to be reinforced by acoustic ones. In another experiment, thrushes were presented with models of their own and four similar species and they attacked all the models indiscriminately. However, when the song of the attacking species was played, the attacks were restricted to the model of that species (Dilger 1956). Here, sound seemed to have a directing or restricting effect, since with sound, attacks on "incorrect" species were eliminated. (Some have criticized the lack of experiments to learn what sounds birds can respond to [Ramp 1965].)

In some cases the environment, or "context" of a signal, may function directly as a modifier of the message. The setting itself may help the parent of a young bird to know whether the offspring is crying in response to being lost, disturbed, or hungry. The honking of Canada geese may be related to defense of territory, defense of mate, competition for food, greeting of mate, a contact call, or an alarm call (Collias and Jahn 1959). Head tossing in gulls may be elicited in soliciting food, precopulatory display, or hostile encounters (Tinbergen 1952a; Moynihan 1962a). So the recipient of a conspecific's message must read both the signal and the context to know whether to feed, avoid, or copulate with the communicator.

In particular, animals may be strongly influenced by environmental dictates when sign stimuli are being utilized. As an example of the importance of context in releasing situations, recall the male robin that will not react to the red of another robin's breast outside of his own territory (Lorenz 1957). It must be pointed out that any tendency toward the same signal having a number of meanings, depending on its context, may not be a reversion to the simple; it might rather be a step toward the complex in that the meaning is dependent upon simultaneous recognition of the signal and the conditions under which it is given.

We need only be reminded that environmental effects on species recognition may be expressed in another way: some signaling is possible only at certain times of the year, such as during the breeding season (Marler 1956). As an example, the clucking of hens, which functions to gather their chicks, appears in response to a rising level of prolactin (a female hormone), which, in turn, is dependent upon daylength (Collias 1960). And we are aware that in many fish, members of the species, whether potential mate or competitor, are only recognized as such in the spring.

### Population Recognition

Once the species has been recognized, subtler distinctions may be made. Among some animals, there is a distinct predisposition to associate not only with others of the species but also with members of one's own population (or subgroup) within that species.

If we define a population as a normally interbreeding group of animals, we can see that it may, or may not, be the same as a species. You may recall that salmon of the same species may return to their own redds to spawn each year, thus effectively forming distinct reproductive units or populations. Differing populations are likely to form when subgroups of the species live in different kinds of habitats as each subgroup becomes more finely tuned to its kind of area. Fernando Nottebohm, an ornithologist, travelled through the mountains of South America and noticed slight changes from one place to the next in the songs sung by birds of the same species. Such changes are apparently related to the shifts in habitat types as one travels first at high altitude and then in valleys, passing from timbered areas, to grassland, and into the rugged timberline area. Each habitat-type influences the characters of its denizens by placing specific requirements upon each group, even if the animals are of the same species. It may have been advantageous for birds to sing songs of their subgroup because of the importance of mating with other birds having the same characteristics. For example, if the birds of the wetter lowlands are darker in color, resulting in their better matching their background, it would be to their reproductive advantage to mate with other birds that are also dark. If they mate with lighter birds, the young may be conspicuous and fall easy prey to predators. Darker birds may, of course, be genetically compatible with members of lighter-colored populations from drier areas, and some level of interbreeding between populations can be expected.

If the rate of breeding between populations is sufficiently low, it is possible that the populations may diverge, each along its own line, until such a time that they may become so different that successful interbreeding between them is no longer possible. When this happens, the populations are usually considered to have formed distinct species. At some point before this happens, however, the adaptiveness of population recognition probably increasingly shifts from an ecological basis toward a reproductive one. In our example, the advantage in choosing a mate from within the group was first ecological (to blend with the background); later, the advantage would have been reproductive (to ensure the production of viable offspring). So, there are species for which population recognition is important, the adaptive basis being ecological, reproductive, or both.

Population recognition has been demonstrated in a variety of species, so let's consider some specific examples of the phenomenon. For example, ants and honeybees recognize the smell of members of their own colonies, probably because of small differences in diet. The ethologist M. Konishi (1970) reported that cricket frogs may employ filters not only to select the sounds of their own species, but also of their own population. But the greatest amount of experimental evidence, not surprisingly, derives from studies of song "dialects" of bird populations (Sauer 1955; Gompertz 1961; Thielcke 1961, 1969; Marler and Tamura 1962; Grimes 1974). Conceptually, birdsong dialects are much like their human counterparts, many aspects held in common from one group to the next, with enough differences between them to make each identifiable. Sound spectrograph analysis (Figure 17.1), indicates that they often appear as a geographical mosaic.

As a more general aspect of dialect formation in birds, dominant song patterns may exist in all populations over a large area (Benson 1948; Thielcke 1961; Thorpe 1961). As an example, the songs of the

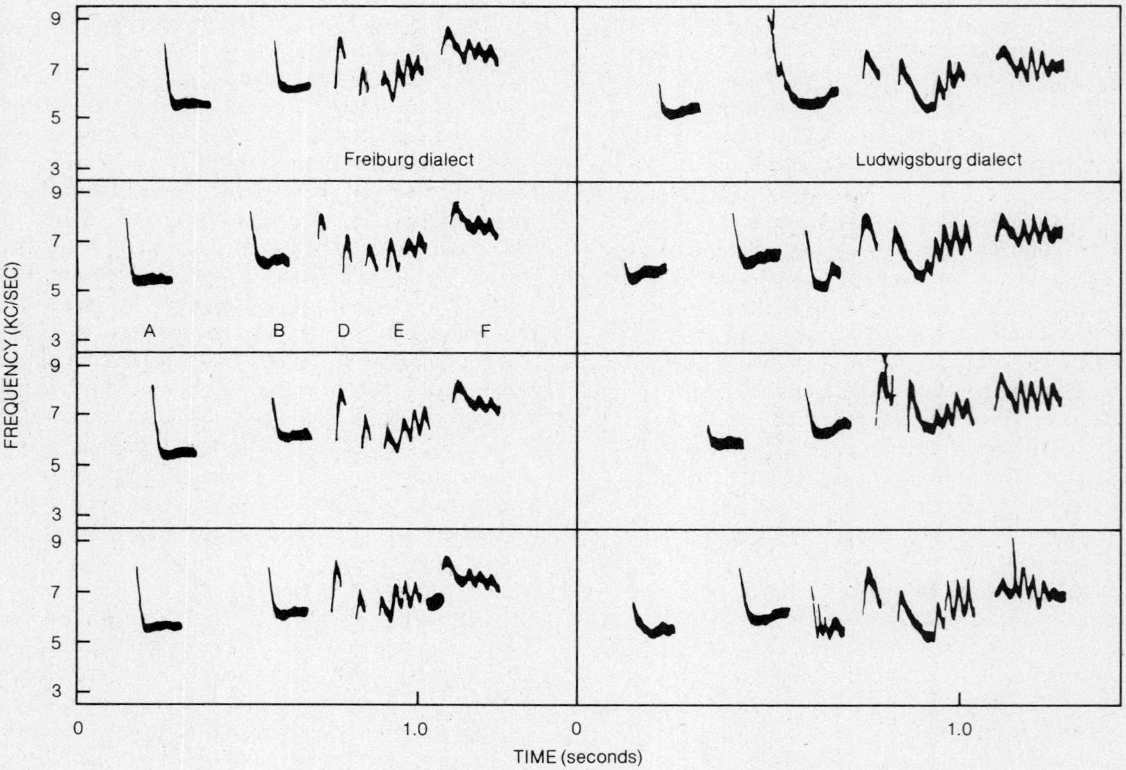


Figure 17.1 Sound spectrograms showing song dialects in the tree creeper (*Certhia brachydactyla*). The songs differ in birds from different geographical areas.

chipping sparrows (*Spizella passerina*) in the United States have shorter syllables, more syllables per song, and are uttered at a more rapid rate than those of the Mexican chipping sparrows (Borror 1959). The basis for the differences is not known. It may be due to such nonselective evolutionary processes as genetic drift or to adaptations to different "sound environments." (See the discussion of white-crowned sparrow dialects in Chapter 18.)

Interestingly, sometimes there are no apparent song differences in widely separated populations. Songs of the crested flycatcher *Myiarchus tyrranulus* vary little over extensive ranges in South America (Lanyon 1960a). There is also no discernible difference in widely separated populations of the house wren *Troglodytes aedon*. It seems unlikely that the song requirements for species are identical throughout their entire ranges, so it must be assumed either that there is pronounced genetic flow, little selective pressure on the songs, or that song patterns, for some reason, resist change (Marler and Hamilton 1966).

Sometimes there are even subgroups within a population. "Tribes" of birds may behave as an integral unit (Andrewartha 1961). In fact, sometimes individual birds may be evicted from a tribe, but still remain within the population. Families of mice (*Mus musculus*) are reproductively isolated from each other, even when they live a few feet from each other within the same barn (Selander 1970).

Social insects (ants, bees, some wasps, termites) present interesting social schemes. Here, the individual is not recognized as such (except the queen), nor is it merely recognized as a member of the population; instead, recognition is also according to caste. Workers clearly demonstrate caste recognition when they methodically kill bachelor drones.

There may be differences in the degree to which populations of a single species are able to utilize signals from other populations. In an experiment, the recorded distress calls of French crows provoked no response in Maine crows. However, when the French calls were played to Pennsylvania crows, the Pennsylvanians responded. Why the difference? It turns out that the crows in Maine migrate southward to Pennsylvania in winter and encounter no other crows because the Pennsylvania crows migrate even further south—to areas where other species of crows are found. The Pennsylvania crows, then, may have learned to react to the calls of other populations (Frings et al. 1958). In a sense, and to no New Englander's surprise, the Maine crows are more provincial.

## Individual Recognition

A person standing on the beach watching seabirds might be able to recognize several species. However, if she looks closely at, for example, the strongly marked Caspian terns (*Hydroprogne caspia*), she might note that they all appear completely identical in size, markings, and even vocalizations (Figure 17.2). She may be a highly experienced ornithologist, but she will have little or no success telling the birds apart. Actually, however, they are not identical. Each has its own characteristics that are immediately evident to other birds of the species. They apparently have no difficulty in recognizing each other at all.

The question arises, What is the advantage in birds being able to recognize each other individually? The answer is intuitively apparent when we consider the problems that might arise if we could not recognize our own neighbors, spouses, or children. It is just as important in other species for an individual to react to other members of the population according to "who" they are.

So members of a species or population must be similar enough, at least in some respects, to be recognized as belonging to the group; on the other hand, in many species, they must vary enough to permit individual identification. The question, then, is, How much can an individual vary from the rest of the population before it is no longer recognized as a conspecific? Is natural selection pushing toward difference and uniformity at the same time? On the basis of what we know about instinct and learning, we can formulate at least a partial answer. Those behavioral or morphological characters that serve as sign stimuli and are releasers of, for example, reproductive behavior, can be expected to vary only within very specific limits. On the other hand, some social interaction, such as mate identification, is based on individual recognition. In such cases, the identifying characters probably are learned, and are, therefore, more free to vary. It is possible, and even likely, that the ability to learn certain characters of

Figure 17.2 To the human eye, Caspian terns all look alike. However, they are individually distinguishable to each other.

others has been selected for, according to the Lorenzian scheme described in Chapter 12. For example, a gull might distinguish its mate on the basis of its voice or facial features, but would pay no attention to unusual qualities in its feet. Perhaps it is *unable* to learn much about feet. So, in essence, some characters are relatively free to vary, others are not, and, of course, the characters that fall into each category may differ even between closely related species.

Good examples of how much individual and species characters are free to vary are found in birdsong. The *song* (a sound normally used in advertisement) of birds is usually, but not always, restricted to males and is associated with territorial defense, establishment and maintenance of the pair, and control of the reproductive cycle. Some repertoires are extremely small, and, in fact, chipping sparrows have only one type of song (Borror 1959). Some finches have well-defined song vocabularies comprised of several patterns. Other birds, such as the brown towhee (*Pipilo fuscus*), may show so many individual differences that it is hard to find any common patterns at all (Marler and Hamilton 1966).

Also, it is hard to know what the function of extemporaneous vocalizing might be. Some birds seem to sing in response to no apparent environmental variable—for no "reason." Charles Hartshorne, ornithologist and philosopher, maintains that birds may sing for the joy of it, out of sheer exuberance (1958), in which case the individual songs could be expected to vary according to the mood of the bird. Such an interpretation is easy to accept for those who have listened to the chortling of small birds early on spring mornings.

The problem of investigating individual variation in song is complicated by the fact that individual variation is as common in birds with small repertoires as in those with large ones. And subtle variations in song may exist that are hard for humans to detect. On the basis of studies on a number of species, it was concluded that two individuals rarely sing identical songs (Borror 1961). Individual variation may thus express itself in subtle ways, even within restricted song types. It must be remembered that such variation should not be dismissed as a nonadaptive difference in the expression of personality or as an evolutionary side effect. The variation may be an important factor in the social organization of the species in permitting individuals to recognize each other.

Individual recognition, interestingly enough, is often based on facial features. Although one is not often taken aback by the variation in the faces of chickens, if a hen is removed from the flock and her comb affixed to the opposite side of her head, she will be treated as a stranger upon her return (Guhl and Ortman 1953). Facial features may also be important in individual recognition between primates. George Schaller (1963) was able to recognize individual mountain gorillas by their noses and surmised that perhaps the gorillas themselves use the same clues. And the Douglas-Hamiltons were able to recognize hundreds of elephants in the Manyara Preserve by their ears and tusks.

Finally, chemicals can also convey individual identity. Dogs can distinguish the trail of a particular dog or man, for example (although they have difficulties with identical twins) (Kalmus 1955). You undoubtedly recall Toby's assistance to Sherlock Holmes in *The Sign of Four.*

**Individuality and Social Systems** In the wild, individual recognition may be important to habituation and the formation of hierarchies. As mentioned in Chapter 11, habituation may occur between animals sharing common territorial boundaries. In an experiment, songs of the ovenbird (*Seiurus aurocapillus*) were played to territorial males in the field. It was found that the songs of neighboring males elicited far fewer responses than did those of strangers (Weeden and Falls 1959). So individual recognition may lower the levels of aggression in some species, as neighbors come to be recognized and ignored.

Also, individual recognition is the basis of *linear dominance* or the "peck order." In a peck order, a single individual generally dominates all the others. His immediate subordinate dominates all others except him, and so on. The last animal in the order is subordinate to all the other animals and may lead a frenetic life indeed. The peck order may exist within either sex, but rarely are both sexes included in the same linear order. However, the leadership "cliques" of certain primate species may involve both sexes. This has been the point of some debate. In some species, the female assumes the rank of her mate among other females. Lorenz reported that, among some birds, if a female's mate dies and she mates with a higher ranking male, she elevates her status in the population. So, not only are these female birds ranked according to their own attributes, but also according to whose mates they are.

It is interesting that the highest ranking males of polygamous species do not necessarily fertilize more females than do other males. In fact, they may be less successful in romance, presumably because they spend so much time defending their status. Such a disadvantage, however, would necessarily be restricted to only the highest ranking individuals. For most of the group, it could be assumed that higher status would usually enhance reproductive success, otherwise no one would take risks in order to increase status.

One's position in the peck order usually depends upon fighting ability, but factors other than pugilistic skill may also be a determinant. For example, as we all know, intimidation or "bluff" is an important weapon. Thus, many threatening animals tend to raise hair or feathers, stand tall, growl, and hiss in order to intimidate other animals. Interestingly, rank cannot always be predicted by size. Among wolves, the largest do not always dominate over the smaller animals; boldness, sheer chutzpah, seems to play a part. Thus, some animals may find themselves leaders of other animals who may have greater fighting prowess. The qualities of leadership are presently being explored by several research groups.

Interestingly, certain features associated with threat may determine one's position in a hierarchy. In chickens, if the comb and wattle (structures used in aggressive signaling) are removed from a hen, she will begin to lose more fights and may end up at the end of the pecking order (Marler and Hamilton 1966).

**Individual Recognition Between Species** David Barash (1974) of the University of Washington found that even in two "solitary" species, animals were able to recognize each other. He trapped male raccoons and red foxes and found that for each species, those trapped nearest each other were more likely to already have dominance hierarchies established than those trapped farther apart. Specifically, he found that neighbors initiated dominance-subordination behavior sooner than nonneighbors, when the animals were exposed to each other, and that their aggressive interactions were less intense.

The phenomenon of linear dominance, then, is ecologically and reproductively important for many species. It functions in permitting each animal to know its rank so that once the system is established the number and vigor of fights is reduced. Animals that spend less time fighting may then spend more of their time and energy in maintenance pursuits, such as grooming and foraging, and in mating and rearing young. The ecological benefits of linear hierarchies are discussed in more detail in Chapter 22.

There is an abundance of other data indicating that animals of different species can recognize each other individually. We all know of cats who associate with certain dogs while remaining wary of other dogs. Jane Goodall found that young baboons and chimpanzees can become "friends," at least for a certain period in their lives when their parent groups share the same area. It must be admitted, however, that we really know very little at present about the importance of interspecific individual recognition in the wild.

## COMMUNICATION AND REPRODUCTION

We shall now consider certain social interactions for which individual recognition may be important, at least at certain times. These interactions are involved in various ways with reproductive behavior.

### Communication Between Members of the Same Sex

We have already noted several instances of fighting and threatening between males. Such behavior occurs, less commonly, between females of some species. But any hostilities most frequently occur within a sex. Consider, for example, territorial birds. Males exclude other males of the species from the territory, and resident females usually only attack intruding females. Such intrasexual belligerence has its own adaptiveness that may, in some respects, be quite different from that of other types of hostile interactions. Intuitively, it would seem that intrasexual belligerence primarily must be related, in some way, to reproductive pursuits and, perhaps, secondarily to ecological competition. We will see if this is the case. But let's first learn something of the nature of intrasexual encounters.

Since intrasexual interactions are often of a hostile nature, it is important that effective communication systems reduce actual fighting and the likelihood of injury or death for the combatants. In fact, in aggressive contests, a winner is often decided on the basis of signaling rather than fighting. Male birds disputing a territorial border may never touch each other, but may stand apart, threatening and displaying, until a victor is decided, whereupon the loser flies away.

Even when real fighting occurs, the winner is almost always decided before serious injuries are inflicted. For example, males of the ensifera insects often spar with the forelegs, antennae, and hindlegs. It's noisy battle, and the sounds produced by each individual during such encounters reflect the dominance of one as the less noisy individual retreats (Alexander 1960).

We will discuss fighting in more detail later, but here we might note that when an animal does not wish to fight or finds he has confronted a superior animal, it is often to his interest to turn off the aggressive behavior of his opponent. The European jackdaw (*Corvus monedula*) averts attacks by turning its head and raising its silver tipped neck feathers (Lorenz 1952). This *appeasement gesture* induces the attacker to begin preening the intended victim instead. As Marler says, "Fights may end in an orgy of mutual preening" (1961a, p. 112). Here we see again Darwin's antithesis principle. The animal stopping the attack does just the opposite of what he would do if he were belligerent. He exposes the vulnerable back of the head, which, if hit, could result in his death, and at the same time he hides his weapon, the bill.

Intrasexual hostilities may also be averted if members avoid each other entirely. A singing bird may cause other male birds to avoid the area and a fight. In the field, two male crickets singing the same song will move away from each other (Burkhardt, Schleidt, and Altner 1967). Recordings of the male southern toad (*Bufo terrestris*) attract females, but repel males (Collias 1960). It must be pointed out, however, that such signals do not always cause conspecifics of the same sex to move away. In fact, in some cases a male's advertisement may actually attract other males. Such attraction would probably have a different selective basis than the aforementioned aggregation and mutual calling we find in some species and will be illustrated later when we consider lek species and males that set up adjacent territories for the *sake* of combat (the "dear enemy" syndrome).

In some species, signals by one animal may not only attract others of the same sex, but may arouse aggressive behavior in them. In the satinfin shiner (*Notropis analoctanus*), territorial males produce knocking sounds during courtship display and when chasing other males. Recorded playbacks of the knocks attract other males and elicit aggressive behavior in them (Stout 1963). The factors that determine whether an advertising male causes other males to approach or to withdraw are not entirely understood.

Obviously, not all communication between members of the same sex is hostile. Other types of interactions also occur, such as nonaggressive grouping signals and parent-offspring communication, but these will be discussed separately. We have only considered signals within sexes—signals that are related to reproduction—and it seems that the more important ones are aggressive.

### Communication Between Sexes

Darwin set forth the principle of intersexual communication when he stated that winning a mate is a highly competitive process and one in which displays are important. An individual must be able to persuade a member of the opposite sex to enter into reproductive activities with him or her—and *not* to take another mate instead. To this end, the highly motivated signals leading toward mating have become intense demonstrations of salesmanship.

Let us examine a few specific effects of individuals attracting the attention of the opposite sex. We know, for example, that a male chaffinch, singing loudly to nearby females, is letting them know several things: he is an unmated male, in reproductive condition, in possession of a territory, and nearby at that time (Marler 1961b). As conditions change, songs may also change, taking on new meanings. For example, the calling songs of crickets attract females in reproductive condition. If a calling male runs into another male, the sound will change to the shorter "rival's song" and may precede fighting. In grasshoppers, the rival's song is not a threat. The males are so promiscuous that they will mount anything even remotely resembling a female grasshopper—even other males. Here, the shorter call serves in sex identification—to keep males from ravaging each other in vain attempts to copulate. If the singing male grasshopper runs into a female, however, he changes to his courting song. He will sing this song for long periods with occasional attempted mountings as long as the female remains. The time she is willing to remain depends upon her own reproductive state (Burkhardt, Schleidt, and Altner 1967).

In some animals, all the signaling is not done by males. In some birds, pairs sing duets, and female grasshoppers may answer calling males (since it is hard to see in tall grass). Some female fish entice their males chemically (Tavolga 1956) just as in insects, and in many mammals, the vaginal odor of females in heat stimulates males. In some female primates, the "sexual skin" of the perineal region swells and turns red when they are in estrus. Estrous females also may behave differently than they would otherwise, for example, by standing still for the male's approach or assuming the "ready" position for copulation (Figure 17.3). Females may also fight, but they usually reserve their attacks for other females.

Interspecific matings may be avoided by signal complexity. As an example, consider the mating behavior of the cockroach (*Nauphoeta cinerea*) (Roth and Willis 1952). Cockroaches are attracted to the scent of their own species, so they are likely to congregate in the same places. Sexually mature males scurry about touching antennae of other cockroaches until they receive a specific chemical stimulus from a virgin female. When this happens, the male and female stand there stroking each other's antennae. Then the male turns his back, raises his wings, touches his abdomen to the ground, and maintains the pose for up to a minute. Secretory glands on his back are thus exposed, and the female comes forward and nibbles at the secretion. She them climbs onto his back. The male grasps her genitalia with his claspers and twists his body from under hers so that their bodies are in line but facing opposite directions, the normal copulatory position (for cockroaches). It is apparent that in such an elaborate mating procedure, mating mistakes will be

Figure 17.3 Females sometimes actively signal males previous to copulatory behavior. Female rats hold their head and rear high and entice males to mount them.

rare; other species just aren't inclined to go through those contortions (or don't think it's worth it).

In the cockroaches, the remorseless male leaves the female directly after copulation. The female develops the eggs, and once they are laid, she too is relieved of any responsibility for them. In other animals, however, the reproductive process may not be quite so carefree. For example, in birds, at least one parent must remain with the young to feed and protect them after they hatch. In some species, both parents remain with the offspring and share in the labor involved in rearing them. Differences in reproductive contributions of the sexes in bird species provide an unusually clear and fascinating insight into how natural selection may operate.

**Fitness and Sexual Selection** The *biological fitness* of an animal is determined by the proportion of genes it is able to contribute to the overall "gene pool" of the population. So, in one sense, fitness is measured by the number of offspring the animal leaves. The term "fitness" is somewhat misleading, since it really implies nothing about the animal's individual well-being. The meaning of the term becomes apparent, however, when the animal is considered in relation to the effort it makes in ecological and reproductive realms. Those animals that maximize their reproductive success will contribute most to the subsequent gene pool and therefore have the greatest effect on the species in the next generation. They are relatively fit. Those that emphasize any other aspect of their lives, instead, may benefit themselves more individually, but they will have less effect on the species because they will transmit fewer of their own kinds of genes. They are less fit. So natural selection is continuously favoring higher reproductive success. The population at any given time is composed primarily of the descendants of highly reproductive stock, which are therefore likely to be prolific reproducers themselves.

Fitness, however, is not simply a game of numbers. An intestinal parasite is not necessarily more fit than a hen because it lays thousands of eggs to the hen's one. Nor is a hen that lays many eggs necessarily more fit than another hen that lays fewer. If the hen that lays many eggs has specialized physiologically for rapid production at the expense of slower, normal development of the egg, none of her eggs may hatch, or the offspring may be sickly and die before they, themselves, reproduce. Or if a hen has a new brood before her previous ones are able to take care of themselves, her offspring will very likely die. In such cases, the hen would have contributed nothing to the gene pool and her fitness would have suffered.

If only a limited niche (Chapter 24) with its various commodities, such as food and nest sites, is available to a species, then a population may sooner or later come up against certain barriers to its continued increase. If the population size is limited by its environment, then specialization for leaving large numbers of offspring cannot define fitness. If the environment is limiting, fitness can only be increased by contributing one's genes to the gene pool at the *expense* of others, especially unrelated individuals, in the population. Therefore, fitness, for competing species, must be defined in terms of success in maximizing the *percentage* or *ratio* of one's genes in the overall gene pool. The principle of competition as discussed in Chapter 24, then, must be considered when developing any definition of biological fitness.

For now, we are concerned with how fitness may be affected by communication. We have seen the importance of signaling in mate selection. So, one means of increasing fitness would be to maximize the impact of one's signals (and therefore, presumably, one's attractiveness to the opposite sex). In fact, this evolutionary route has been taken by many species of animals. The peacock with his elaborate coloration, bold markings, and glaring presence must present a veritable barrage to the IRMs of the drab peahen (Figure 17.4). Since the female reacts to him on the basis of his appearance, it may well be that the bolder the markings, the longer the tail, and the more garish the demeanor, the more likely the female is to react to them. This seems even more likely in light of what we know about the possibility of boosting releasing effect, as demonstrated by the existence of supernormal releasers. It should be pointed out, however, that at present there is little hard data to show that fitness can be maximized in such a way in nature; the scheme, as presently understood, is based primarily on circumstantial evidence. If it exists, of course, it is an evolutionary trap. It forces males increasingly toward the rococo in order that they, rather than a competing male, might be chosen as a mate. However, conspicuousness is fraught with danger in the wild. If the splendid peacock attracts females, he must also attract tigers. It is apparent that such a male cannot be expected to live a long time, but for a while he may lead an active sex life indeed.

The payoff for the males in such systems is the attraction of a number of females in a brief span of time. Therefore, the males cannot be expected to assist in mundane family chores such as the rearing of the young. After copulation, they move on to the next conquest, leaving the female to raise the brood alone. In some birds such as the ruff, prairie chicken, argus pheasant, and cock-of-the-rock, the sexes meet only for copulation, and no pair-bond exists at all. The males are promiscuous and mate with as many females as possible; whereas the females, once mated, no longer respond to the overtures of any male.

The female, as is well known, is usually the drabber, dull member of these sexually dimorphic species. She is therefore much less likely to be subject to predation and may live through several more seasons than the male. She has taken a different evolutionary route, one which favors crypticity so that she may make her contribution to the next generation by seeing to it that her young reach adulthood. Her role is to tend to several broods over the years and to maximize the chances for each chick carrying her genes to reach maturity. Again, we lack an abundance of hard data to support such theorizing, but the available circumstantial data generally support the scheme.

The conspicuousness of males may not only be through color or markings, but through sheer size. Male Alaska fur seals (*Callorhinus alascanus*) may weigh six times as much as females (Amadon 1959). Male grackles may weigh twice as much as females. Robert Selander (1965) interpreted the larger size of male grackles as being adaptive in aggressive interactions arising as a result of competition with other males. It is important to realize that any showiness in males, whether in size or markings or sounds, may be functional both in attracting females and in intimidating other males, and that the relative importance of each effect may vary from one situation or species to the next. It may even be that, in some cases, sexual dimorphism in size is unrelated to reproductive success. Actually, there is only limited information regarding greater reproductive success in showier males (Kok 1972; Selander 1972; Maynard Smith 1966). In one experiment some years ago, a female jewelfish was allowed to choose between two males of

Figure 17.4 Peacocks increase their fitness by enticing peahens to copulate with them by a barrage of releasers.

different intensities of redness. The females chose to deposit their eggs in the end of the aquarium nearer the redder male. Few such laboratory experiments have been done, and there are even fewer data from the field, but our assumptions in such areas are important ones. We are probably guilty of jumping to generalization in this area because intuitively we can see how it could work.

In a few bird species, such as the button quail and certain phalaropes, the sexual roles are reversed (Figure 17.5). The females are larger and more striking than the males, and in these species, the females compete with each other for the attention of the males. In such cases, it is likely that the males have assumed a greater role in rearing the young. The "freed" females, on the other hand, may tend toward promiscuity, a condition referred to as *polyandry* (see Chapter 26). Unfortunately, specific data on the roles of the sexes in cases of reversed sexual dimorphism in vertebrates are meager.

The relative numbers of the sexes may be related to the mating system in some species. In birds, the females of Wagler's oropendola (*Zarhynchus wagleri*) outnumber the males six to one, and, not surprisingly, the males are promiscuous. In their mating, the female oropendola quarrels for a nest site and builds the nest alone, while the male ogles down at her with his blue eyes, raising his feathers, twitching his tail, spluttering, cackling, wheezing, and gurgling. When the nest is finished, she seems to notice him for the first time and accepts his amorous advances. After breeding, he leaves her to her nesting duties and flies away.

In other nonmonogamous bird species of more nearly equal sex ratios, reproductive behavior in the female may be harder to elicit, in one case released not by the presence of a garish male or even by a gaily decorated bower, but (in a more practical vein) by a male with a nest already built. The male crimson-crowned bishop bird courts one female at a time, builds a nest for her, mates, and then goes off to build another nest.

Just as the visible traits of the sexes may vary, some species show sexual dimorphism of the sound-

Figure 17.5 In some bird species, such as the red phalarope, the sexual dimorphism is reversed. That is, the females are larger and more striking. For example, there is one female in the group above. Which one?

producing apparatus. For example, in males of some ducks, the syrinx (sound producer) has a long bulla, lacking in females, which gives males a distinctive sound (Johnsgard 1967). Sound dimorphism would seem to be highly advantageous in prey species in that a female can recognize a male without his exposing himself, as it were.

In most bird species, sexes are almost indistinguishable in the field (and over ninety percent of bird species are monogamous [Lack 1968]). Even an expert is unable to distinguish the sexes of the flycatcher *Myiarchus* by its appearance, general behavior, or vocalizations. In other species, slight differences in morphology between sexes may be apparent. The sexes of the golden-fronted woodpecker (*Centurus aurifrons*), are almost identical except for subtle differences in the color of the head and a few tail feathers. In such species, mate selection is probably not due, in any important sense, to differences in physical appearance. Instead, sexual differences in *behavior* may be increasingly critical, and bonds may form relatively slowly.

**Sexual Dimorphism and the Evolution of Mating Systems** Sociobiologists have for years assumed that sexual dimorphism in size had a rather simple explanation. The assumption was based on Darwin's proposal that sexual dimorphism occurs in response to competition between members of one sex over members of the other sex. Males, according to this idea, are generally larger and more conspicuous in order to win more females. Robert Trivers (1972) at Harvard University pulled the prevailing ideas together and said that whichever sex invests the most in the offspring will tend to be in shortest supply and competed for. Among birds and mammals, females contribute more to the offspring and so males compete for them. Furthermore, he suggested, ecological factors such as food abundance and distribution affect the direction of parental investment and, as a result, the evolution of intrasexual competition. Predictions based on Triver's work were tested (Alexander 1974) and found to hold true for artiodactyls and pinnipeds. Furthermore, it was suggested that the larger size of human males compared to females is evidence of past mating systems in which males competed for females, success being measured by copulation counts.

The notion of sexual selection being at the root of sexual dimorphism has been challenged (Kolata 1977). It is noted, for example, that in three kinds of animals in which females are larger (golden hamsters, bats, and seals) the females do not compete for males, so their large size is due to something else. One wonders, then, could the large size of males be due to something other than competition?

## Communication Between Parents and Offspring

In many species, there is no need for the parents to be able to communicate with the young because there is virtually no interaction between the generations. In some cases, as we know, the entire parent generation dies before their eggs hatch the following spring. Even when parents and offspring interact, it may be anonymously. Insects usually leave their eggs and, when they hatch, the parents react to their offspring no differently than to any other young members of the species. Generally such species have "simpler" nervous systems so that the young are able to behave in an adaptive manner after a short period of development. Parental care, on the other hand, is believed to have developed in species that are behaviorally more complex (flexible) and for which more time is required for the young to develop advanced types of nervous systems. This parental care is demonstrated in such activities as feeding, cleaning, leading, protection, and alarm. Here we will consider only a few such patterns; other aspects of parental behavior are considered elsewhere.

In those species in which one or both parents care for the young, communication may be important in feeding behavior. We have discussed young gulls, which direct their pecks toward a spot on the parent's beak, eliciting food regurgitation. Similar signals operate in other species. For example, in the young of other bird species, such as the parrot finch shown in Figure 17.6, there are four large spots visible when the young birds gape to be fed. These spots act as releasers of feeding behavior for the parents. In some species, even the foraging of the parent is influenced by the behavior of the young. The more they cry, the more time and energy the parent puts into searching for food for them.

In most species, including birds, it is not likely that a parent with multiple young can recognize any of them individually as long as they are related to as "offspring," still without rank or sex. Such parents are, however, usually able to distinguish their own young from among those of other animals. (Young animals probably fall into one of two groups: "my young," or "not my young." The signals of one's own offspring may release parental behavior, whereas the same signals emanating from other young animals may not [Rheingold 1963].)

(It would be interesting to know what kinds of animals recognize their own young after they are grown. Chimpanzee mothers apparently do, and possibly African hunting dog mothers. Wolf mothers apparently do not. Notice we've said nothing about fathers. These are group-living animals, and paternity is not easy to determine.)

There are undoubtedly several basic ways a parent may come to recognize its own young. For example, ground-nesting gulls seem to recognize their young by their cries; cliff-nesting gulls by their location. Other animals may use other sensory clues. Nanny goats come to know the smell of their offspring by licking the newborn animals. If a newborn goat is removed from its mother before she has had a chance to lick it and is returned to her a few hours later, she will reject it. The ability of the mother to become associated with her young seems to be limited to the period directly after birth. A young goat apparently learns to recognize its mother by the process of elimination after being continually rejected by all the other mothers (Etkin 1967; Klopfer 1971).

In some species, on the other hand, specific signals may help the young recognize a parent. Lorenz found that the night heron (*Nycticorax nycticorax*) bows to its nestlings when it approaches, thus displaying its bluish cap and white plumes. It thereby suppresses any escape or defense behavior that the young normally show toward any large figure approaching them. Of course, there is no individual recognition demonstrated here—only the acceptance of a "friendly" large form.

In some cases, only restricted types of signals (sign stimuli) are used in the identification of young by the parents. Herring gulls (*Larus argentatus*) nest only a few feet from the nest of other members of the colony, and the parents vigorously defend their young. However, if one of their young should die, they may eat it. When they no longer hear its call and they see no movement, it loses all significance as offspring and becomes food (Tinbergen 1964). We see the opposite response in baboons. A mother will carry her dead offspring for days.

Although individuals in a group of offspring may not be distinguishable to a parent, differences in the *state* of the individuals may be. In birds, unfed chicks call more than fed ones (Muir 1954), and this probably has an effect on food apportionment by the parents. They feed the noisiest ones. Such a system would tend to ensure an equal amount of food for the offspring. In addition, it may have important ecological repercussions. Since the last of the young to hatch would be smaller than its siblings, its lack of vigorous pleadings might cause it to receive less food in time of scarcity, and so it would quickly die. The parent, then, would have a brood of a size compatible with the carrying capacity of the immediate environment and could, therefore, be expected to rear more young than if it had tried to feed them all. (See the discussion of *parental* manipulation in Wilson 1975b.)

**Imprinting in Precocial and Altricial Species** The question of imprinting takes on new facets when it is considered in reference to parent-young communication. A baby chicken, after hatching and drying, is up and about looking for food within an hour. Such birds that hatch at a late stage of development are called *precocial*. They still require some parental care, but they are independent in many respects. The bluejay (*Cyanocitta cristata*), on the other hand, is an example of an *altricial* species. These birds are born naked and helpless and they spend their early life sprawled or doddering about in

Figure 17.6 When the young parrot finch opens his beak wide, four large irridescent blue spots appear along its mandibles. These spots elicit feeding behavior in the parents.

their nest until a parent appears with food, when they feebly gape to be fed. The complete dependency of the young and the prolonged close relationship of parent and offspring in altricial species, could be expected to provide more time for the young to *learn* the characteristics of the species by observing their parents. The importance of imprinting in altricial species might therefore be reduced since time is not so critical a factor.

Sexually dimorphic species do not depend as much upon imprinting as monomorphic ones. The young are exposed to the female of such species more often than the male, as we have seen. Therefore both male and female offspring would tend to imprint on the mother and, of course, this would be disadvantageous in later interactions such as mate selection. So, as Klopfer (1962) pointed out, while precocial ducks might be expected to rely more importantly on imprinting than would altricial thrushes, selection for imprinting among ducks in sexually dimorphic species might be less than in monomorphic ones.

## ALARM SIGNALS WITHIN SPECIES

Alarm signals may function at several levels, for example, within populations, between parents and young, and even between species. The adaptiveness of warning systems, it seems, would be in preserving one's offspring or relatives. Of course, the animal giving the alarm is likely to be taking some risk, especially if it is warning of the presence of a predator. The acceptable level of the risk is largely determined by how closely related the average recipient of the alarm is to the animal giving the warning (as is discussed in the section on kinship selection). Any warning signal, of course, becomes part of the environment of any organism able to perceive it; so, members of unrelated species may react appropriately to the same warning signal.

Alarm signals may vary widely in their specificity. They may be very general warnings of an unspecified threat, as is demonstrated in barking dogs. ("What's that dog barking at?") They may also be of a more specific nature, as was discussed in regard to the high pitched *tweee* warnings of birds that mean "hawk overhead." But alarm signals may be even more specific. The California ground squirrel (*Citellus beecheyi*) has three different warning chirps; each announces the presence of either hawks, snakes, or mammals (Fitch 1948).

The information content of a warning may be increased by variations in the vigor with which it is expressed. The warning chirps of a merely anxious or suspicious bird may suddenly rise in intensity when it definitely *sees* the sinister birdwatcher (Nice 1943).

You will notice that warnings might be interpreted as communication regarding the nature of the environment—something that we stated earlier did not often occur in animals other than humans. However, the restricted classes of such communication, the paucity of species in which it occurs, and the rather stereotyped levels of information and responses indicate that the general rule holds. Nonhuman animals do not normally communicate information regarding their environment. When they do, it is of limited scope, as in warning signals or in signals attracting others to sources of food (whether inadvertently or purposefully, in both cases).

## INTERSPECIFIC COMMUNICATION

Since the biotic milieu of any animal includes other species, adaptations to their presence and activities may be expected. In some cases, such adaptations are critical, and so complex interspecific communication has resulted. It is not surprising that interspecific communication is especially important in birds and mammals in light of their complex life histories and relatively high mental capacities. However, interspecific signals, just as intraspecific ones, by no means function only through learning. There are even examples (discussed below) of communication of a sort between plants and animals, and in spite of recent contentions of certain writers regarding the perceptiveness of plants, plant–animal communication probably occurs in the absence of learning by plants. The point is, both learned and innate patterns may operate in interspecific communication.

### Defense Signals

Let's consider one of the most common and widespread types of communication between species—one whose operating mechanism we might "intuitively" understand—the maximizing of threat. We have all seen two unacquainted dogs approach each other stiff-legged, heads held high and hair bristling.

Such behavior makes each animal appear larger and perhaps more dangerous than it is. This is an example of how communication is often used to deceive, a usage that has not received enough attention.

The threatening quality of large size is obvious. In our own society, large, powerful men somehow initially gain our respect. Many animals may employ the size maximization principle when dealing with members of other species. A cat hunches its back in the presence of an aggressive dog. The dog's hackles are raised. In each case, the animal appears more formidable. Several displays of intimidation through an apparent increase in size are shown in Figure 17.7, but threat maximization is not always visual. The adult howler monkey (*Alouatta palliata*) roars his displeasure at human intruders in a resounding and awe-inspiring manner (Carpenter 1934).

It seems likely that such "lies" as hair raising may not work on conspecifics. Hair raising is probably now a ritualized pattern. Thus, a chimpanzee seeing another with its hair on end is not likely to think, "Here is a big chimpanzee," but "Here is an angry chimpanzee." However, a predator is not likely to be able to tap ritualized lines of communication and so is likely to believe he has encountered a large, formidable prey and may therefore leave it alone. The point is, the lie, "I am formidable," may only work on other species. One's own species may derive a different meaning from it. The idea remains to be tested.

Alarm calls were mentioned in reference to intraspecific communication, and we noted that such signals may function between species as well. If two species sharing a habitat are hunted by the same predators, there is an advantage in each being able to recognize the alarm of the other. We have noted the similar characteristics of the cryptic high-pitched alarm calls of different species of birds. It should be pointed out that interspecific communication may also occur that is not based on convergence of signals. Communicating species don't have to use the same lines—the signals can be one-way, although the advantages may be mutual. A well-known example is the relationship of the wary little tickbird and the rhinoceros. The tiny bird warns the nearsighted rhinoceros of intruders and thus protects his food supply—ectoparasites of the mammal.

Figure 17.7 Warning and intimidation displays in several species of animals. The displays serve to make the animal look larger and more formidable to its enemies. The animals shown here are (counterclockwise): frilled lizard, zorille, short-eared owl, and spotted skunk.

Prey animals may communicate with predators in a number of ways so as to reduce their efficiency. For example, a bluejay may attack a predatory owl that ventures near the jay's nest. Actually, the jay can do no serious harm, but its attacks are usually effective in driving the owl away. The owl, having no natural enemies, is just not equipped to handle attacks. Apparently, the jay has "short-circuited" the owl's normal hunting pattern by attacking instead of fleeing (Tinbergen 1965a).

A hunted animal may also confuse a predator by giving false clues. A moth may display "flash-colors" while it is in flight, but they disappear as soon as the moth lands. A pursuing hunter then continues to look for the flashing colors, and the moth goes unnoticed. A lizard may twitch its tail and thus divert a predator's attention away from its intended direction. Some species of fish have concealing stripes running through the eyes and often a false eyespot on the tail so that a predator is unsuccessful when it attempts to "head off" its prey (Figure 17.8).

The well-known white rump flashes (Figure 17.9) of several species of mammals (primarily ungulates) have for years been believed to function as alarm signals to other members of the population. It was assumed that the disadvantages of an individual attracting the attention of a predator to himself were more than balanced by warning others among which would be the signaler's offspring or kin. However, another theory of the advantages of such signals has also been advanced (Smythe 1970). Perhaps the signal is primarily intended for the predator. The system could work like this: there is a critical distance that a "chasing predator" must be from a "running prey" when he begins his chase, if he is to have any chance of success. When a prey animal, such as a deer, notices that he has been detected by a hunting predator, he bounds away, tail erect and rump patches flaring. He thus maximizes the impact of his presence upon the predator's sensory apparatus and induces the predator to begin the chase prematurely,

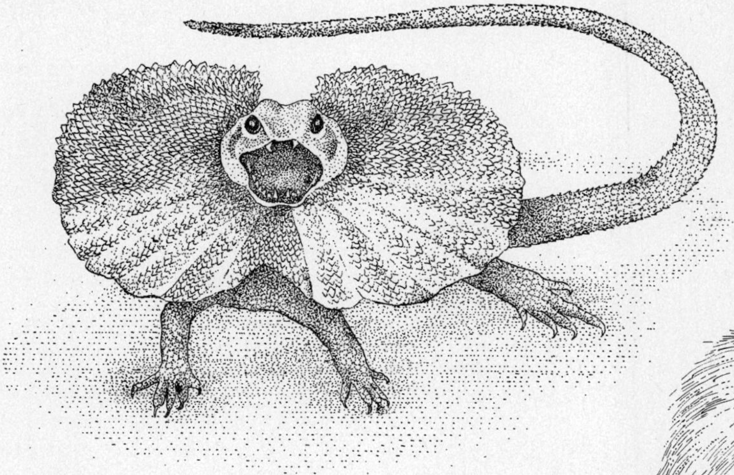

Figure 17.8 Eyes often receive inordinate attention from other animals. Some fish have capitalized on this by developing false eyespots near their tail and/or concealing their eyes, thus giving misleading information about their likely direction.

or while still outside the critical distance. According to the theory, when the prey had demonstrated his escape abilities, the discouraged predator would go elsewhere to search for food.

## I Am Not Here

"I am not here" is the message of cryptic species. Crypticity may range from simple markings that break the contour (Figure 17.10) to devices that render the animal almost invisible. There are many well-known examples of crypticity, including reed-nesting marsh birds that sit immobile with bills pointed upward, moths that seem to disappear on the bark of a tree, and strange toads that are almost invisible against the leaves of the forest floor (Figure 17.11).

Of course it wouldn't do for a green frog to crouch on a brown branch. Green frogs would do better clinging to a green leaf, and, in fact, that is where they are most likely to be found. There are two possible reasons for this. One is that a frog in the "wrong" place would attract the attention of a predatory snake and quickly disappear from the population. Another is that green frogs might show a specific attraction for a green substrate. There is some evidence for such specificity among moths. In an experiment, several kinds of moths were put into a box with each side painted a different shade of grey. They tended to settle on the side that best matched the reflectances of their forewings (Sargent 1966, 1968; Kettlewell 1956). The mechanisms for "color matching" among cryptic animals have gone largely unexplored.

Movement is also an important part of crypticity. Most cryptic animals remain immobile most of the time, or when they do move, they move so slowly that they attract as little attention as possible. The selective mechanism for cryptic species is similar to that for certain mimics, such as leaf mimics, discussed below, whose message is, "I am not of interest," and so we will consider them separately.

There are a number of species that do not spend much time in concealment, but that, at certain times, must hide. As an example, consider the large cats.

Some of these animals are so formidable that they can spend much of their time lounging in full view of other animals. It is only when they begin a hunt that they seek to "disappear" (Figure 17.12).

## No!

Certain animals, notably insects, communicate with their predators by saying, "No!" A wide variety of insect species can discharge noxious chemicals when they are captured. Some of these chemicals are powerful toxins or irritants and, in some species, they can be shot with some accuracy in several directions. In addition, in some species, two or more chemicals are mixed just before they are discharged, thus suddenly increasing the temperature of the mixture. This hot irritant has especially nasty properties. Thomas Eisner (1970) at Cornell has described in some detail the remarkable defense glands of insects (Figure 17.13). (Such signaling is by no means limited to arthropods, since the message of the skunk must be much the same.)

Figure 17.9 The white-tailed deer (*Odocoileus virginianus*) shows a conspicuous "flag" when alarmed. The tail is raised and the hairs of the underside are erected. This very visible pattern may be an alarm signal or it may act to entice predators to attack prematurely.

It is rather interesting that most insect poisons are not fatal to the predator. The adaptiveness of mildly poisonous defenses seems to be related to the intelligence or learning abilities of the predators. If the usual predators are vertebrates, with their high learning capacities, nonlethal dosages might be expected. It might be advantageous for a noxious prey species to allow vertebrate predators to live, since their diets may be influenced by tradition learning or by other modifiers of behavior; also, it is likely that such predators might then switch their diet to edible competitors of the noxious species and set the example for their colleagues to do so as well. However, in many cases, other arthropods, such as ants or spiders, may act as predators. It would appear that it would be more adaptive for a poisonous prey species to kill these predators outright unless the predators' foraging habits could be altered by bitter experience through learning. The concept of prey species "teaching" their predators has not received enough research attention at this point.

## Mimicry

One of the most interesting types of communication based on predator-prey interaction is *mimicry*. Mimicry may be defined in narrow terms, but we will consider the concept broadly. For our purpose it occurs when any organism takes on characteristics usually associated with another species in order to alter the reaction of yet other species, usually predators.

**Müllerian Mimicry** *Müllerian mimicry* is named after Fritz Müller, a nineteenth-century German zoologist. It refers to convergent evolution producing *common warning coloration* in poisonous or noxious species. The result is that predators have fewer signals to learn. Usually, warning coloration is bold and distinctive, such as we see in a number of bees and wasps. Poisonous animals that advertise their presence are not likely to be mistaken for palatable

Figure 17.10 The principle of disruptive margins. To see how they work, place the book in an upright position. Then back away. See which disappears first, the zebra or the pseudozebra. Note the disappearance of the figures at the bottom.

Figure 17.11 A toad, *Megophyrus nasuta,* of Malaysia, among dead leaves.

prey and they tend to have certain basic characteristics in common, which include: (1) bold, simple arrangements of colors, usually including black, white, yellow, orange, or red; (2) obvious movements in open view; (3) alarm postures that maximize the effect of the color and markings; and (4) gregarious social patterns or high density, both of which reduce the likelihood of capture by naive predators. (If a young predator catches one, he won't want to catch another one, but if the prey is dispersed, he may not catch them often enough to learn his lesson, which is not to attack animals that look like that. The longer it takes him to learn, the greater is the likelihood of any prey animal being attacked.)

**Batesian Mimicry** *Batesian mimicry* is named after the nineteenth-century English naturalist Henry Walter Bates and refers to a palatable species adopting the warning characteristics of a noxious or harmful species (Figure 17.14). The harmless species is called the *mimic* and the noxious one, the *model.* Actually, mimicry is rare among mammals, but let's consider one example from this group. The flesh of Bornean tree shrews (*Tupaia*) seems to have a terrible taste, thus most predators avoid them. On the same island, the palatable squirrels (*Sciurus*) resemble the tree shrews to such a degree that even most zoologists cannot distinguish them by the skins alone. The squirrels are undoubtedly protected from predators by their similarity to the shrews (Wickler 1968).

*In all cases* of Batesian mimicry, the relative numbers of the model and mimic are probably important. It seems that the mimic must maintain numbers low enough that it is exposed to predators less frequently than the model; otherwise attacks by predators would probably be reinforced often enough to cause the system to break down.

In Batesian mimicry, the mimic may take on the general appearance of the model, but in some cases only certain parts of the model are copied. For example, many predatory animals such as hawks, owls, cats, and wolves have eyes that focus directly ahead rather than being situated on either side of the head, as is the case in prey animals such as rabbits and songbirds (Figure 17.15). (Such predators are usually at the top of the food chain and, therefore, less concerned about what may be creeping up on them than what type of prey is ahead.) Animals that have frontal eyes may utilize them as a defense mechanism or a threat (Figure 17.16). Gaze directly at a strange dog (a small one) and notice his uneasy behavior. So, eyes have a considerable impact among animals. Several species of harmless animals have utilized the relationship between eyes and threat and have developed eyespots as a means of repelling predators (Figures 17.17 and 17.18). Such "eyes" usually appear to be large, wide-open, and staring straight ahead, although they may be sightless spots on the wings of a harmless insect.

Sounds may also serve as models in mimicry. Hissing and growling are sounds that accompany many truly dangerous reptiles, mammals, and even birds. Sound mimicry would probably be most successful in hole-nesting species that can normally be heard, but not seen. It is known that hole-nesting titmice, screech owls, and some nocturnal cockroaches hiss; and the coot (*Fulica americana*) growls when defending its nest against predators.

There is an interesting age element that may function in some types of mimicry. Young harmless animals may threaten in the manner of older, more dangerous ones. Young lions growl, cough, and spit in a most threatening way, and young platypuses may growl when they are disturbed (Burrell 1927). Such threats by young animals may be more than incipient behavior since, often, they are mimicking other animals. For example, baby snowshoe hares (*Lepus americanus*) and baby European hares (*Lepus Europaeus*), when threatened in their nest, will growl and lunge at a predator in the manner of a dangerous species (Severaid 1942; Hediger 1948). The prospect of being attacked by an enraged baby hare might give a predator pause just long enough for the hare to escape. (In what is perhaps the opposite sort of mimicry, the Goodmans (1976) have described snakes that may mimic young birds.)

**Sex-limited Mimicry** In some cases, selection in signaling devices must reach a balance between the need for crypticity to reduce predation, and the need to be seen by other members of the population. The result can be *sex-limited mimicry*. Among certain butterflies there are those species in which the male attracts the female by his visual characters, therefore the males maintain their mate-attracting properties, while the females are free to mimic noxious forms. In insect species that use olfactory clues in mate recognition, visual mimicry can occur in both sexes.

There is an unusual case of sex-limited mimicry among fireflies (Lloyd 1975) that deserves mention.

Figure 17.12 A hunting leopard. Its markings and coloration and behavior may render the animal almost invisible as it moves through its habitat.

Figure 17.13 The bombardier beetle is able to aim and squirt noxious chemicals at other animals.

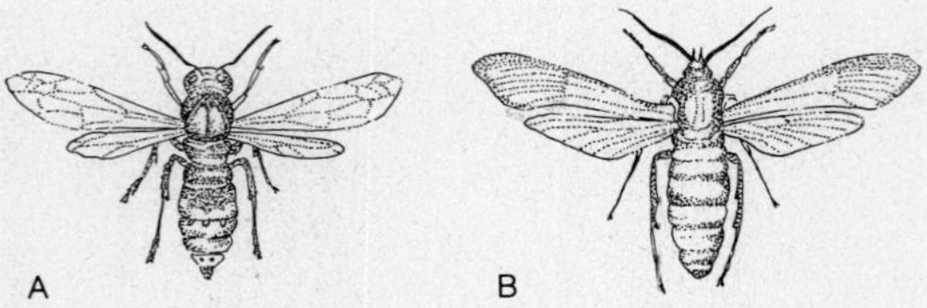


Figure 17.14 In Batesian mimicry, a harmless species mimics a harmful one. This innocuous hornet clearwing (B) is very similar to the hornet (A).

Figure 17.15 The eyes of the cat are located on the front of its head, those of the rabbit on the side. Two eyes, straight ahead, are often considered the mark of predatory species and many harmless animals mimic them.

Each species of these nocturnal beetles sends out short, rhythmic flashes that attract the opposite sex, and each species has its own flash duration and frequency—its own private lines of communication. But the females of the genus *Photuris* have capitalized on this system in a fascinating and deadly manner. These *femmes fatales* have developed the ability to mimic the flashes of at least three different species. When males draw near one of these females to copulate, she meets their advances by capturing and devouring them (Figure 17.19). Apparently, what stimulates a change in the flash pattern of the female *Photuris* is the sight of a particular flash pattern that indicates the presence of males of another species. Upon perceiving such a pattern, she begins the flash pattern of the female of that species. Strangely enough, the female pattern of any species may be quite different from that of the male.

Figure 17.16 The owl is trying to be inconspicuous by hiding his eyes, sleeking his feathers, and standing erect (A). Upon being discovered, it leaps into a stare, feathers abristle, in a most startling manner (B).

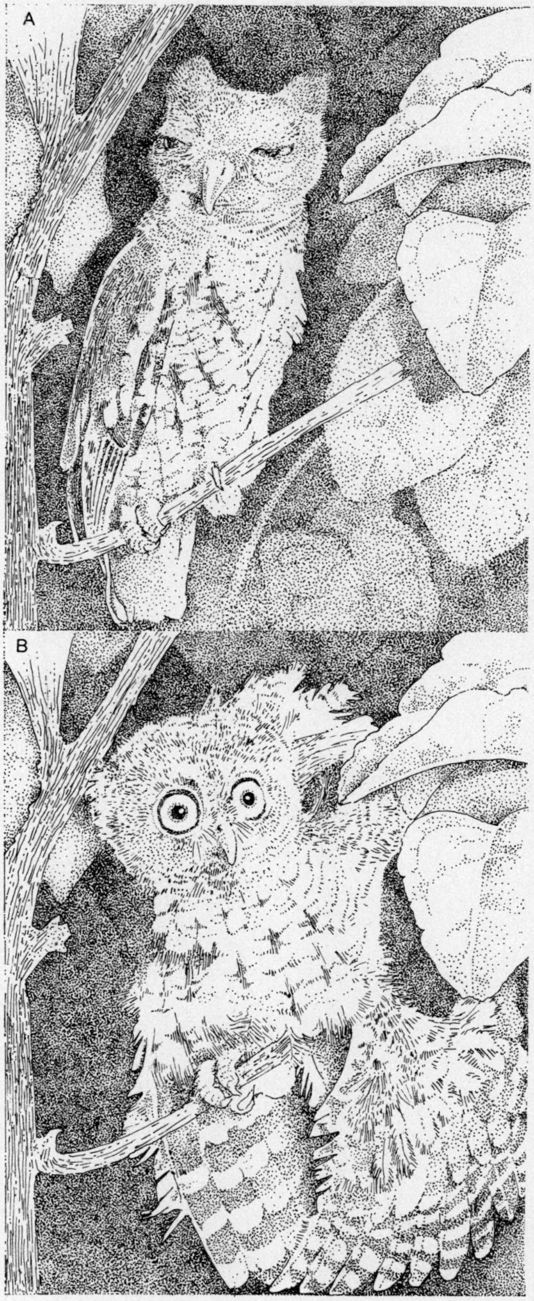


The firefly data bring to mind two questions. One involves the mechanism of the stimulus-response patterns. What sort of filtering systems are operating? It is probably too early to tell. The other question involves the phylogenetic development of the pattern. How did it arise? The *Photuris* females give off flashes that are different from their own mating flashes, as they walk or just before they take off or land. Perhaps these "nonmating flashes" are "intention movements." Since these kinds of flashes are not important in *Photuris* mating, they could have been free to change and therefore could have provided the raw material for the flash-mimicry.

**Automimicry** In Costa Rica, the milkweed (*Asclepias curassivica*) is avoided by cattle for good reason; if they eat it they may die. The plant contains glycosidic chemicals similar to the drug digitalis, but, fortunately, one side effect of ingesting these chemicals is severe vomiting. Since vomiting normally will be induced before enough of the plant is eaten to cause death, the usual result is severe temporary illness. But in the Danainae, the family of the monarch butterfly, the larvae may feed on the milkweed and incorporate its poisonous glycosides into their own bodies (Figure 17.20). Thus their own bodies become poisonous. Predatory birds made ill by catching one of these insects will avoid all monarchs, although other larvae in the population may have fed on other plants and are perfectly edible. Protection afforded an entire population by the noxious qualities of only some of its members operates through what is called *automimicry* (Brower 1969). In such a case, the model and mimic are of the same species. Since they have identical appearances, automimicry can be expected to be more efficient than Batesian mimicry.

The possibility exists that certain poisons (such as the alkaloids or glycosides of some insects) are tasteless to predators. However, these chemicals may normally occur with harmless, but flavored, chemicals. Thus, it may be that the predator uses the harmless chemicals as a clue to the presence of the

Figure 17.17 An edible hawkmoth caterpillar with eyespots on its back (the real head is visible at the extreme anterior) and a stinger at the posterior end.

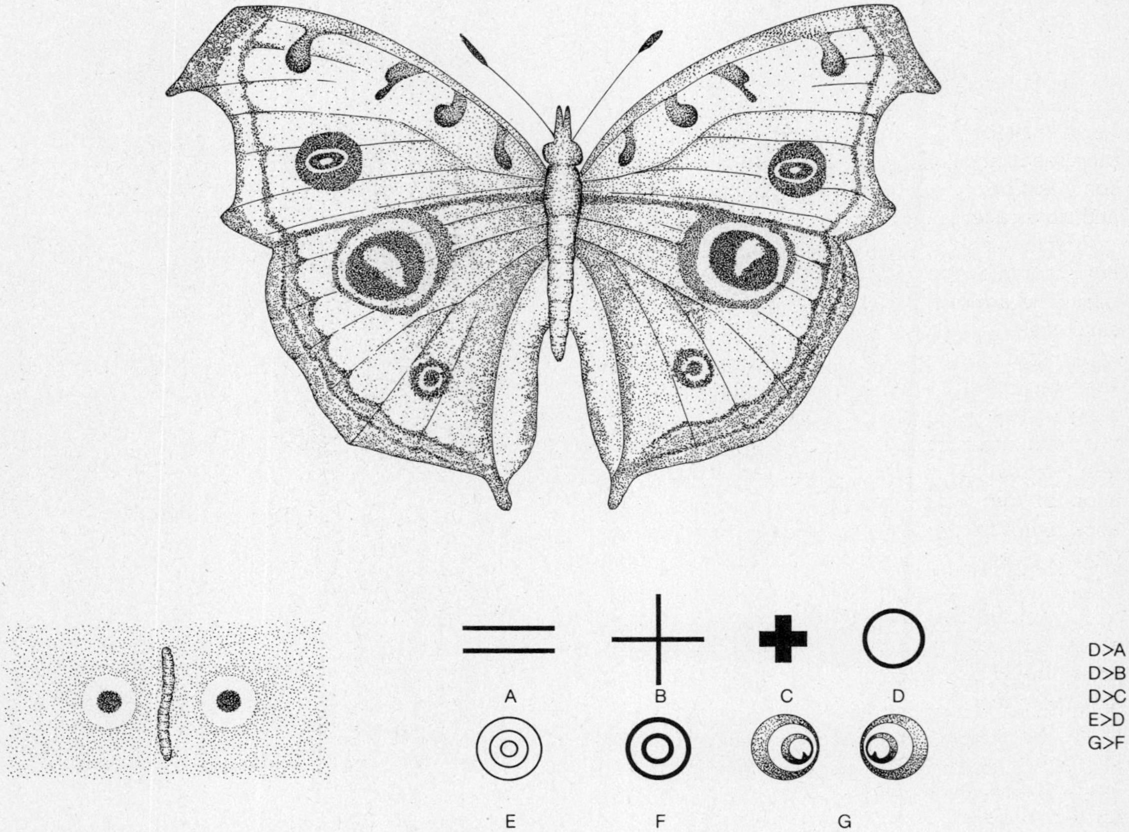


Figure 17.18 Top, the butterfly *Precis almana* uses eyespots as a defense. Left, in an experiment, the birds were trained to take food from a slide at a certain site. When a mealworm was placed on the slide, the birds normally took it. However, when various eyespot models were projected onto the slide, right, the birds would not approach it. The relative effectiveness of the various models is shown.

poison. If this happens, then certain edible species might develop the harmless *clue* substance in a form of chemical mimicry (Brower 1969).

Some species have something called weapon automimicry (Figure 17.21). Here, hair, ears, and facial markings accent the weapons of the animals, particularly in artiodactyls (Guthrie and Petocz 1970). In another kind of automimicry, some snakes raise their tails and move them as they would their heads. They may even "strike" with their tail. But if a predator should grab *this* head, it may find that the teeth are located at the other end (Pianka 1974).

**"I Am Something Else"** In another type of mimicry, perfectly palatable individuals may not hide, but may appear inedible. The twig caterpillar takes on the appearance of a broken twig on a limb. It doesn't hide; it appears to be something of little interest to a predator. Certain tropical katydids have leaflike wings that are true to their model, even to showing "veins" and little "holes" from insect attacks. One unabashed species of frog crouches on leaves in its imitation of a bird dropping. In these species, the animal may not move at all for long periods; in other cases, however, movement may enhance crypticity—for example, some leaf-hoppers that mimic leaves, sway gently to and fro as a leaf might be moved by a breeze.

Mimicry is primarily employed by prey animals, but in some cases predators may use it to draw prey near. The angler fish lies on the bottom and draws small fish to their fate by wriggling a wormlike appendage just above its cavernous mouth. When a small fish is attracted to the worm, it disappears in a lightning-fast slurp. By human standards of conduct, one of the most ignoble of the deceivers is a blenny (*Aspidontus*), a small fish that has evolved the coloration of the helpful cleaner wrasse. When a fish approaches for cleaning, the blenny rushes over and tears a piece of flesh from his unwary host (Figure 17.22).

Among the more fascinating examples of mimicry are those involving plants and animals. Such interactions should not be unexpected, since both great kingdoms are interdependent and subject to the laws of natural selection. Most of the signaling between plants and animals apparently is associated with insect pollination of plants. Insects are usually attracted to certain flowers for physiological reasons (Marler and Hamilton 1966), but in some cases, insects approach flowers for other reasons. In one

Figure 17.19 The *Photuris* female (left) attracts foreign males by mimicking the mating signals of their females. She has seized a male of another species here and is chewing on his thorax. When she is finished, only his eyes, feet, and wings will remain.

Figure 17.20 Larva of the monarch butterfly feeding on the leaves of the milkweed. The body of the butterfly thus becomes impregnated with powerful poisons that cause even edible members of the population to be avoided by predators.

Figure 17.21 Patterns of ear shape and facial markings that emphasize the visual effects of the horns in *Hippotragus equinus* (top) and *Oryx beisa* (bottom).

group of orchids, each species mimics the odor and appearance of a particular species of insect (Figure 17.23). The insect pollinates the flower while attempting to copulate with it (Baerends 1950). As another example of plant-animal mimicry, there are the fly-pollinated flowers that emit odors reminiscent of rotting meat (Johnsgard 1967).

**Reflexive Selection** Gairdner Moment (1962) described an interesting phenomenon in the animal world that operates on a principle opposite to that of automimicry. In some instances, members of a population look as *little* like each other as possible. In such cases, the probability of an individual having a certain appearance is inversely related to the number of other individuals in the population having that appearance. Selection thus works on a feedback system that Moment calls *reflexive selection.* The adaptive principle was specifically described by L. Tinbergen (1960) in his discussion of *search images.* (The idea of the search image was suggested by von Uexküll after he missed seeing a *flask* of water on the table because he was looking for a *pitcher.*) Tinbergen argued that a foraging bird will continue to look for food that has the appearance of the type he has recently found or "has in mind." The system seems to work through principles of stimulus generalization and reward expectancy. After a predator has taken a member of a prey population, it would be likely to continue searching for other morsels with that general appearance. Those individuals of a different appearance might then escape notice. If variation is thus selected for in prey types such as the brittle stars (Figure 17.24), predatory shorebirds may be unable to form a search image for the entire starfish population and so their foraging efficiency would be reduced. Obviously, reflexive selection could not work for animals that signal intraspecifically by visual methods. The ecological implications of such selection are important and should be studied in more detail in other highly variable or polymorphic species.

I should add that the formation of search-images may be ecologically important in other ways. It would seem that the development of search-images would be dependent upon the relative density of food types so that, as a prey species became scarcer through predation, the predator would lose that particular search image and would tend to drop that species from its menu. After the prey had recouped enough that it was in no danger of being decimated, it could again be hunted.

Figure 17.22 *Labroides dimidiatus* (A) and *Elancantinus oceanops* (B) are true cleaner fish. They are small fish that clean ectoparasites from the bodies of larger fish. Many of the larger fish even allow them to enter and clean the parasites from inside their mouths. Compare the color patterns on the two cleaner fish and that of *Aspidontes taeniatus* (C), which is a mimic. That is, it mimics the appearance of the cleaner fish. However, when it is allowed to approach the larger fish, it bites chunks of flesh from the would-be host.

Figure 17.23 Mimicry of female wasps by *Ophrys orchids* (bee, spider, and fly orchids). The result is that sexually oriented male wasps pollinate the flowers by moving over them.

**Mertensian Mimicry** Let's briefly consider the fascinating idea behind *Mertensian mimicry*, a concept developed, of course, by Professor Mertens. The coral snakes (*Micrurus*) are highly poisonous and most species bear a characteristic red, black, and yellow banding that makes the snakes highly conspicuous. These species sometimes occur with similarly colored nonpoisonous snakes (*Simophis*). It has often been assumed that the harmless species was the mimic, and the deadly one the model. It might be asked, though, if the models usually *killed* their attackers, how would the predators ever *learn* to avoid them? (There is no evidence that their major predators have any innate recognition system by which they would avoid coral snakes.) Mertens suggested that, actually, neither of these species is the model, but that both are mimicking a similarly marked, and *moderately* poisonous species, *Pseudoboa*, that inhabits the same areas. Survivors of attacks by *Pseudoboa* could learn to avoid snakes with the characteristic banding, and thus both the deadly and the harmless species could also profit by the lesson taught their common predators by a moderately poisonous cousin.

We have seen, then, that animals communicate via a bewildering array of techniques, sometimes with their own species, often with others. We have also found that communication is more than simply letting other animals know what you are about do do. Animals may tell each other things about the nature of the environment and they may lie to each other. In broad terms, however, the adaptive result is the same. Communication permits animals to adjust more closely to their environment and thereby to increase their likelihood of reproductive success. Because such success is often gained at another's expense, certain areas of communication are undoubtedly in continuous flux, shifting and changing, as animals do the things that tend to make the system work for them.

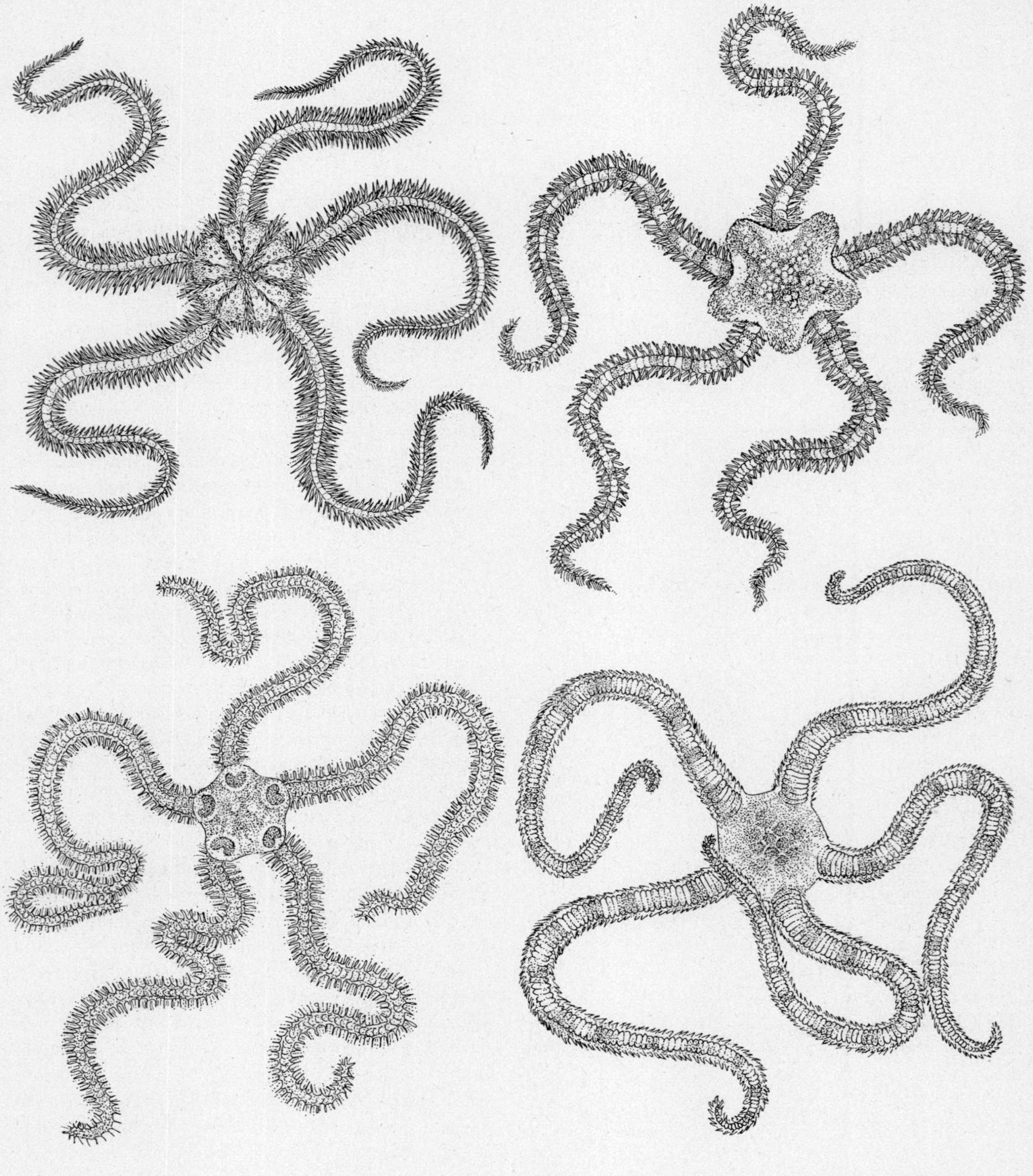

Figure 17.24 These four brittle stars (*Ophiopholis aculeata*) are all of the same species but have different appearances, thereby inhibiting the formation of search images in predators.

# 18

# THE DEVELOPMENT OF COMMUNICATION

A small bird who has never heard another bird begins to sing, and its song is recognizable as the song of its species. Another bird seems to be willing to duplicate any song it hears. Do birds learn their songs? A young rattlesnake recoils around its vibrating tail the first time it is threatened. Did it learn to rattle? What is the effect of the noise on other animals? Do they innately avoid such a sound or do they have to learn to associate the sound with being dead? How does communication develop?

Obviously, communication develops differently in different kinds of animals. So, in this chapter, we will concern ourselves with differences in the flexibility of communication. In other words, we will ask to what degree patterns of communication are innate. The other side of the coin is, to what degree can animal communication be influenced by learning? As we look at the development of communication, we will pay particular attention to three representative groups of animals.

## SONG DEVELOPMENT IN CRICKETS

One has to be very careful these days about saying, without careful qualification, that any particular behavior is inherited. So it is almost refreshing to hear that "the song pattern of each cricket species is stored in its genes," that the distinctive song patterns of the world's some three thousand cricket species are not learned, but are "encoded" in the insects' chromosomes (Bentley and Hoy 1974). As evidence, researchers found that when different cricket species are crossed, the offspring produce songs that are intermediate to those of the parents. Not only that, but in more complex breeding experiments, the song pattern could be shifted systematically according to the proportion of certain kinds of genes carried by the male cricket.

The commonest cricket song is the calling song, sung by males to lure sexually receptive females to their burrows. The songs of each species are highly distinctive, and the females have evolved the ability to pick out the songs of their own species from among a mixed chorus (Figure 18.1). No matter how females were reared (with silence, "wrong" songs, different population densities, diets, temperatures, light regimens, and so forth), when they matured, they always responded appropriately to males of their own species. Once the singing male has attracted a female, he switches to the courtship song that facilitates copulation. Then he sings the post-copulatory song (which the French, of course, call the triumphal song). This may help to keep the inseminated female around for a while so that her pregnancy is not interfered with by other males. Should one male invade the territory of another, a fierce fight ensues, which is accompanied by the aggressive song if the fight is particularly severe. The winner goes on singing this song after the loser has been booted out.

So, the communication system, in crickets, consists of males transmitting a variety of signals that are detectable by both males and females. It was found that no practice is needed for the insects to be able to sing the proper song and that acoustical feedback is unnecessary. In other words, the insects don't have to hear themselves sing in order to be able to produce the proper song. Apparently, all the necessary information for cricket singing is stored genetically and is "read out" in the normal translation of DNA, ultimately forming the appropriate neuronal and muscular devices necessary to produce and perceive the biologically appropriate songs.

Figure 18.1 A tethered female cricket "walks" on a Y-maze of styrofoam. Different male songs are played from the right and left sides, and her preference is shown as she continually comes to forks in her "road."

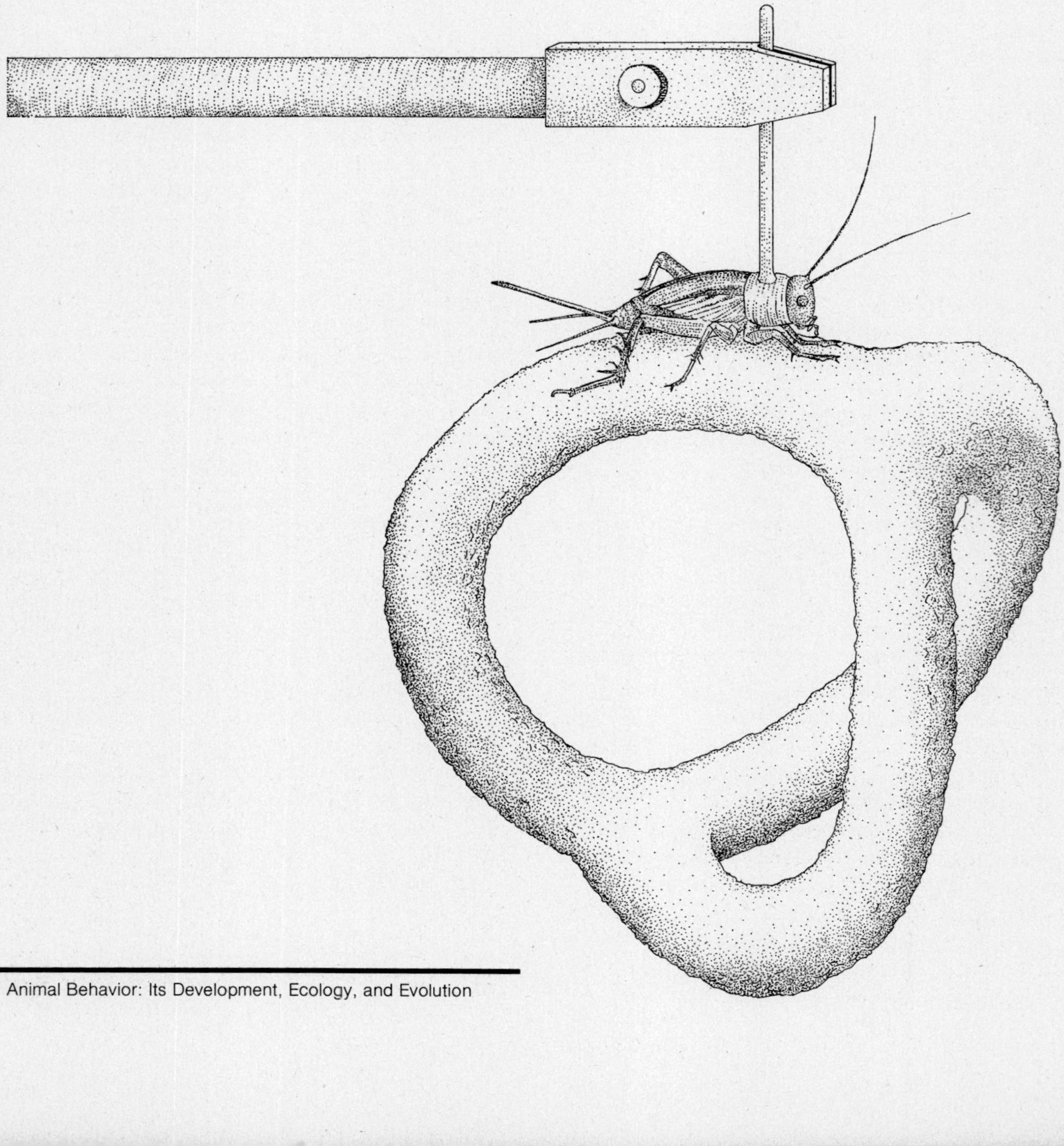

## SONG DEVELOPMENT IN THE WHITE-CROWNED SPARROW

Bird song has traditionally fascinated us. The songs may be melodious, or harsh and repetitious, but they remain, in part, as mysterious as they are fascinating. Of course, we do know certain things about them. For example, songs are almost always the same for any population of birds, and they usually carry certain specific sorts of information. A singing male may convey the messages: I am here; I am healthy; I am in reproductive condition; this is my territory; if you are a female, welcome; if you are a male, beware. Now, we are beginning to learn something about how a male comes to sing a particular song.

The basic question is, Since many birds have the physical ability to sing any number of songs, how do they come to sing the song of their own species, or even of their own specific population within the species? A fascinating body of evidence has emerged from several fronts (Marler 1959, 1970; Marler and Tamura 1962; Nottebohm 1970; Konishi 1965; Konishi and Nottebohm 1969; Thorpe 1973). Some of the results show surprisingly precise interweaving of learned and innate behavior to produce an adaptive pattern.

The white-crowned sparrow is a small finch that inhabits the Pacific coast of North America (see Figure 15.9 in Chapter 15). Its range is rather extensive, and different subpopulations across the range sing their own dialects. In some cases, the dialects are markedly different in populations living only a few miles apart (Figure 18.2). It turns out that if newly hatched males from different populations are put into soundproof isolation chambers, no matter which populations they come from, they all sing similar songs—simplified versions of their normal song. It seems, then, that they are born with the basics and later pick up the embellishments that distinguish local dialects, by listening to other males in their own populations. This song learning takes place primarily in the first three months of life (Marler and Tamura 1964). Strangely enough, during this time the young birds can't even sing. But they can listen. If males are taken from the wild in the autumn (after having hatched in late spring) and put in isolation chambers before they have ever uttered their first song, when they do begin to sing, it will be in the dialect of the population from which they were taken.

Birds put into isolation as hatchlings can be induced to sing any dialect by letting them hear tape recordings of that dialect—but they are receptive to such manipulation only during their first three months. After three months, the young birds won't learn anything about songs.

So, here we have a basic, simple. inherited song pattern (the one sung by isolates) that is subject to modification learning, resulting in locally specific embellishments. Such learning is possible only until a certain age and does not take place as a result of practice. Instead, the embellishments are "remembered" until such time as the birds begin to sing.

Once a bird starts singing, it may improve a bit with practice, perhaps as a result of learning to control its own muscles. Also, a young white-crowned sparrow may learn by singing "subsongs," softly uttered, garbled parts of real songs that they mutter during the nonbreeding season. These segments are similar to real songs, but likely to be out of normal sequence and a bit ragged.

Konishi (1965) deafened birds at different ages by removing the cochlea of the inner ear. He found that if a newly hatched fledgling is deafened or if a bird had been deafened after it had been trained, but before it had begun to sing, it will sing when its time comes, but only disconnected notes, quite unlike the songs of the isolates (Figure 18.3). The interpretation is that in order for these birds to produce their *inherited* song patterns, they must be able to hear themselves. So, it seems that proper singing is not simply the result of an innate ability to contract the muscles of the diaphragm and syrinx in a certain pattern, or not just the ability to produce a simple song, but rather some sort of neural template, which translates into a particular set of muscular contractions that produce the song. The template, then, serves as a *pattern* that the bird must match with its own song. And in order to make its own song fit the template, it must be able to hear itself and to match the song stored in its memory. But once birds have learned to sing the "correct" song via the normal sequence, if they are deafened, they will go on singing the correct song.

There is another facet to all this. White-crowned sparrows can only learn to sing the white-crowned sparrow song. If a male, isolated at hatching, is played the songs of another species, his song will be that of a bird kept in utter silence. It is as if for some reason he can't hear the song of the other species. But he can. It seems that the strange sounds simply do not "unlock" the IRM, if there *is* an IRM for birdsong.

It was suggested that perhaps the white-crowneds simply lacked the vocal apparatus for producing the sounds of other species. So they tested another species. The chaffinch is a bit more flexible and can sing a number of songs, but it behaves similarly to the white-crowneds in song development. (The syrinxes of both these species are almost identical to those of two closely related species that are excellent mimics

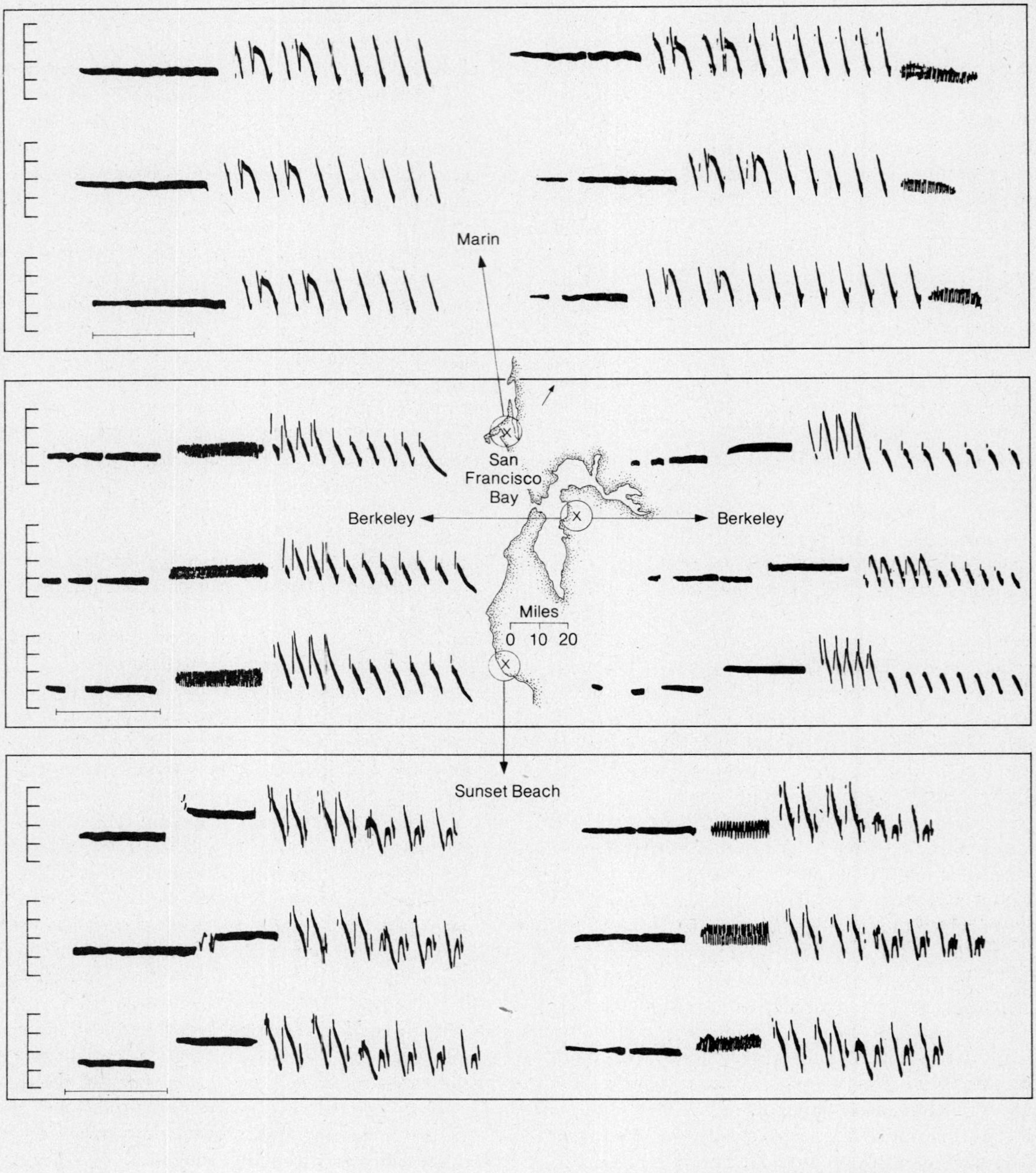


Figure 18.2 Sound spectrograms of the songs of eighteen white-crowned sparrows, recorded in three locations indicated around San Francisco Bay. The top three songs are from Marin County, the middle three from Berkeley, and the bottom three from Sunset Beach.

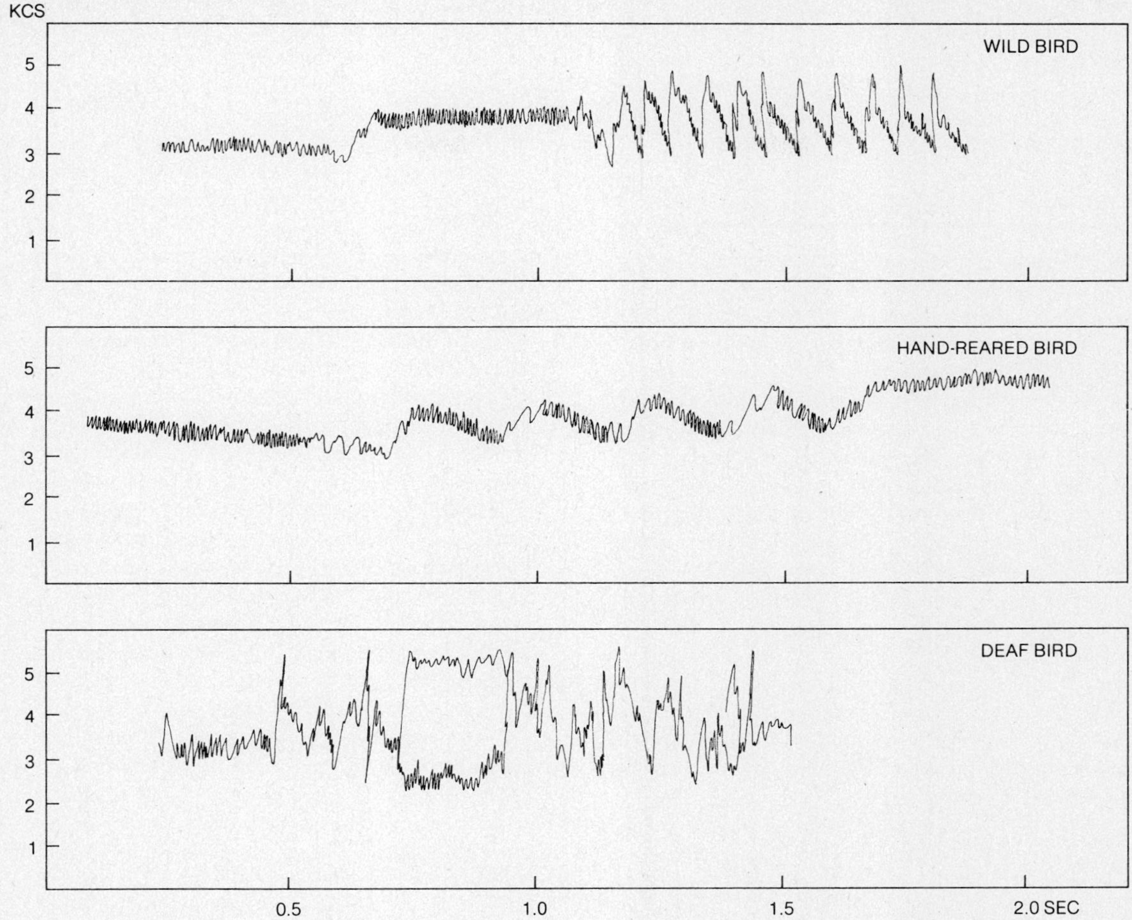


Figure 18.3 Sound spectrograms of songs produced by three white-crowned sparrows reared under different conditions. The lower bird was deafened before five months of age. It had learned its song, but was not able to hear itself sing it.

of other birds, the bullfinch and greenfinch [Thorpe 1961].) It turns out, though, that the chaffinch can imitate only one other species with any degree of success. This is the tree pipit. Although tones of these two species seem quite similar to the human ear, they differ greatly in their acoustical properties. So the chaffinch is *able* to mimic the sounds of another species, but it behaves as an isolate if it is reared in an environment void of the sounds of its own species. Here, again, we find inborn constraints on learning.

After its first spring, then, the song of the white-crowned sparrow is not subject to change through learning. It will keep pretty much its same song through the rest of its life. Such information should provide grist for the neurophysiologist's mill. However, the neurophysiologist must be careful in the selection of species because although the songs of zebra finches and chaffinches develop in much the same way as the white-crowned sparrow, Oregon juncos and indigo buntings may modify their songs for at least another year (Marler, Krieth, and Tamura 1964; Rice and Thompson 1968). Blackbirds often change parts of their songs throughout their lives (Hall-Craggs 1962), and mimes such as the mockingbird will learn to imitate almost anything at any time and have presented an enduring puzzle to evolutionists (Hinde 1969).

## COMMUNICATION AMONG PRIMATES

It should now be apparent what we are up to. As usual, egocentric humans are angling toward a discussion of themselves. We've just talked about crickets and sparrows, which one might think aren't like us at all (but beware), and now we're going to consider primates. Furthermore, a common cliché has it that chimpanzees are our closest relatives. (Chimpanzees are rather quiet about it.) So, now let's have a look at how other primates, particularly chimpanzees, communicate among themselves.

There have been two notable biases in studying the mechanisms of primate communication. First, almost all the work has been done on either baboons or macaques, two closely related genera of Old World monkeys, with only some recent work on chimpanzees. Secondly, because of the convenience of recording and the easy analysis of sound spectrograms, most of the work has been done on sound communication. Film can capture facial expressions, but it is notoriously difficult to analyze.

One of the greatest problems in studying communication systems in primates lies in our own emphasis on sound. We tend to hear an animal sound and then to ask, "What does it mean?" We *expect* sounds to carry "meaning" in primates, although we are willing to believe they may be less important in other animals. The bias is natural. For us, it is possible to completely express ourselves with sound. We can register anger over the telephone. We are able to write words from which someone years later can extract our precise message, writing simply being the visual expression of verbal ability. In nonhuman primates, however, sound may be of secondary importance in communication, used only for emphasis or redundancy. Facial expressions, gestures, and postures may be more important.

Humans also have another problem in dealing with sound communication. We tend to expect nonhuman sounds to be used as we use them, in a structured manner; we may even expect diction. Humans, of course, tend to place words in some sort of traditional order, as you know if you have ever tried to struggle through learning a foreign language. Also, we use names. In fact, it has been suggested that the comprehension of the development of human language rests on understanding the evolution of the system of naming (Lancaster 1968). It has long been assumed that nonhuman primates use no names, but recent work with chimpanzees, however, indicates that this may not be true.

Peter Marler analyzed higher primate communication (1965) and arrived at the conclusion that the systems are extraordinarily complex—much more so than those of, say, birds or rats. He also found that primates rely heavily on "multimodal signals." In other words, a gesture, sound, or facial expression standing alone might not normally constitute a complete signal. Signals must be, instead, considered in the context of a vast collection of cues, which includes gestures, sounds, facial expressions, postures, and the social milieu. By altering the expression of any of these, or changing their combinations or sequences, an incredible array of signals is possible.

Multimodal signaling is well suited for the mental abilities and the social systems of most higher primates. They are smart enough to interpret subtle changes, and they are usually highly social. This social contact not only means that they can react to each other as individuals, but proximity itself permits the use of different signal modes. They are close enough to see, hear, smell, and touch one another. As we know, long-range signals tend to be of a single mode and relatively simple in their components.

In addition to being multimodal, primate signaling is likely to be graded. This means that specific categories, without ambiguous intermediates, are difficult to detect, or perhaps they do not exist for some modes. Graded signaling, of course, offers increased opportunities for precision in expression. One may not only indicate, for example, that an animal is angry, but *how* angry. At the same time, it should be stressed that, for some modes, there do exist ungraded, or discrete, signaling patterns among higher primates (Lancaster 1968).

We will see in the following discussions that most primate communication is used in social signaling. And probably most signals function, directly or indirectly, in dominance-subordination relationships. Higher primates are particularly adept at expressing dominance, submission, anxiety, fear, and aggression. Of course, other forms of signaling are also important among primates, such as nonagonistic behavior between mother and infant, between mating individuals, and between peers.

### Chimpanzee Communication

First, we should review some aspects of chimpanzee natural history, as described by Jane Goodall (1968). Chimpanzees are nomadic in that they sleep in a different place each night. Their beds are made of bent branches and twigs and may be loosely lined with leaves. They tend to forage in the same general area, but the direction and distance of their wandering is unpredictable, varying with the location and abundance of food. On occasion, they may go far from their usual home ranges to take advantage of a particular fruit that is in season. Although adolescent and mature males, and more rarely, females, sometimes are found alone, chimpanzees tend to travel in loose-knit groups of from two to over thirty, in which they are usually able to recognize each other individually. Groups may consist of any combination of age or sex classes. Membership is usually temporary since individuals tend to wander off alone or to move to other groups. The most stable association is between a mother and her offspring, and dominance hierarchies with any group are fairly well defined. For example, when two chimpanzees meet on a narrow branch, if the dominance hierarchy is known, it is usually possible to predict which one will give way. Goodall reported to the National Geographic Society in 1978 that chimpanzees may be more viscious and violent than she had at first led us to believe. After eighteen years of closely watching the chimpanzees at Gombe Stream Preserve in Tanzania, she and her colleagues found that when a splinter group left the main group, males from the larger group would attack, maim, and kill single animals from the smaller group.

It is not an easy matter to cover the range of chimpanzee communication in any meaningful way. Earlier, I said that the olfactory powers of chimpanzees are about the same as that of humans, but the evidence is circumstantial. We really don't know that. In the development of their models, early ethologists had the advantage of not having paid much attention to primates, so their task was simplified. However, because they had looked mostly at birds and fish, they were confident that there should be no problem in recognizing displays or discrete communicative acts in primates. After all, one can easily tell when a gull is in the aggressive-upright position and when it is giving the long-call. The problem with primate displays is that they are not clear-cut; they tend to grade into one another. Marler (1969a,b), in fact, has stated that *every* chimpanzee call is part of a continuous series. Primate facial expressions are just as difficult to separate and identify (van Hooff 1967).

*Modal expressions* are those identifiable signals that are most often directed toward other members of the species. Crying, or the "cry face," is a human modal expression (Figure 18.4). Chimpanzees also make cry faces, but their cry face appears as a transitional form as they change from "whimper face" to their "bared-teeth scream-face" (Figure 18.5). Whimpering, then, is a modal form in chimpanzees, while in humans it is generally a transition form (Jolly 1972).

Thus, for humans, the problem of investigating chimpanzee communication is complicated. And, paradoxically, it is partly complicated by its apparent simplicity. Although many of the signals in chimpanzees have their obvious counterpart in human signals, we must be exceedingly careful in extrapolating from chimpanzees to humans. The expression may be the same—the meaning or function, quite different.

Another complicating factor in investigating chimpanzee signals is that their movements and calls are likely to be influenced by a number of variables such as the immediate environmental and behavioral state, the identity of the sender, and the state of other individuals in the group (Goodall 1968). Furthermore, primate signals tend to (or invariably) occur in clusters. In fact, Marler wrote:

*In most situations it is not a single signal that passes from one animal to another, but a whole complex of them, visual, auditory, tactile, and sometimes olfactory. There can be little doubt that the structure of individual signals is very much affected by this incorporation in a whole matrix of other signals.* [*1965, p. 583*]

Chimpanzees seem to rely little on olfactory communication. There is no evidence that they recognize individuals on the basis of odor. When estrous females are "presenting" to males, however, the male

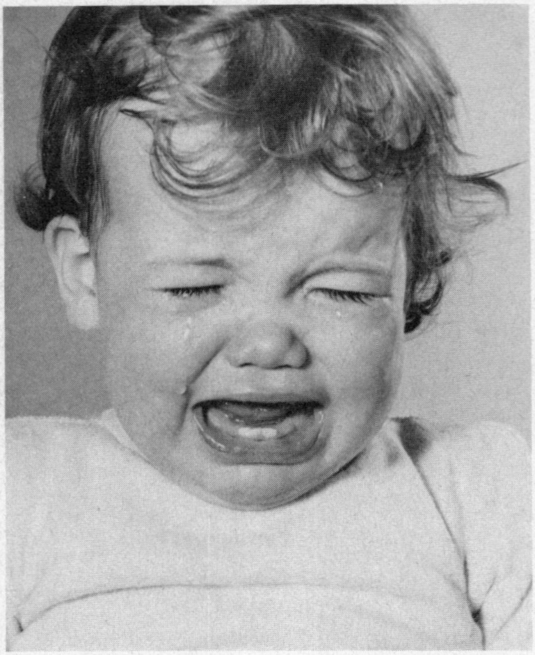

Figure 18.4 The human cry face. Other humans recognize the cry face, but other animals probably do not. The cry face in chimpanzees is a transitional signal leading from whimper face to bared-teeth scream face.

Figure 18.5 Whimpering is a discrete (modal) signal in chimpanzees, but is usually transitional in humans.

may touch her genital area and then indelicately smell his fingers. It probably doesn't require great olfactory ability on his part, however, to smell *something*, although the scent doesn't seem to be important in initiating reproductive behavior. As mentioned earlier, there is preliminary evidence that human females excrete odorous vaginal substances at the time they are most likely to become pregnant. but any pheromonal effects in humans remain to be substantiated. It has also been suggested that underarm hair may function in scent dispersal in humans (Goodhart 1960) (and sometimes, perhaps, in people dispersal) and that it evolved on just such an adaptive basis as a scent disseminator. So, it is possible that humans rely (or did rely) on olfactory communication even more than chimpanzees do.

The chimpanzee relies strongly on acoustical communication. Goodall (1968) said that not only can a human observor establish sex and age but even individuals on the basis of "panting hoots," a sound that apparently functions in long range communication, such as when one group approaches another from the distance. Such individuality in sound may also help to reunite two animals after a separation. Sound individuality may be generally important in nonanonymous groups (where individuals are recognized) that live in areas of low visibility, such as in heavy vegetation.

Harsh grunts over a wide frequency range are very important in sound communication among chimpanzees and other primates. Possibly such sounds permit a wider range of variation than is possible with clear, birdlike sounds. It has been suggested that human vocalization is a form of grunting that is shaped by the mouth cavity on its way out. Chimpanzees and baboons make the same "hominid grunt" and also alter the sound by changes in their facial muscles.

Chimpanzees are highly visual animals, as are humans, and many of the visual signals in chimpanzees are associated with the face. A subordinate animal in the presence of a dominant will constantly watch the latter's face, presumably looking for changes in expression that may reveal a change in mood (Goodall 1968). A young animal denied food may register its discontent by pouting with the lips. An attacked individual may scream at the top of its lungs with its lips drawn back from the teeth. Many primates grin when afraid (see Figure 3.4 in Chapter 3) or when confronted by a superior (Figure 18.6). Bodily posture is also important since confident or threatening animals hold themselves erect and may even swagger, while subordinates hunch over, crouch down, or lie down. Such submissive behavior, by the way, often conceals characters that are used in individual recognition—quite the contrary of a confident animal's behavior (Jolly 1972).

Chimpanzees may stand their hair on end when excited, as I mentioned earlier. They may do this when they arrive at a food source, meet another group, notice other chimpanzees in the distance, hear a strange sound, find themselves in an unfamiliar situation, threaten, attack, or just prior to copulation (Goodall 1968). It is apparent that chimpanzees may erect their hair in a variety of situations. Because of this, it seems that it might be due to a general arousal and that we do not need to assume that it has a specific meaning or function, or is a result of the activity of a specific neural center.

In feeding, greeting, and courtship, male chimpanzees may display erect penises. The vulva of the female may become enlarged in the same situations (as well as when her genitals are handled by a male), whether she is in estrus or not (Yerkes 1943). It is generally believed that in other primates, erect penises can serve as a threat to other males. In some cases, the penis and scrotal area are brightly colored in intimidating reds and blues (Eibl-Eibesfeldt 1970). The only time I saw such behavior among humans was in a New York subway and I must admit everyone cleared out fast enough.

Figure 18.6 Many primates grin as an appeasement gesture when confronted by a superior.

**Avoidance and Escape** A startled or frightened chimpanzee may show a variety of avoidance and escape patterns, in addition to its submissive behavior, according to Goodall (1968), upon whose work much of this discussion is based. If one chimpanzee is being chased by another, the chasee screams loudly, with the mouth open and lips retracted from the teeth and gums. The chase usually stops soon after it starts. However, if a chimpanzee is running from something other than another chimpanzee, it will usually remain silent. It seems, then, that the scream and fear grin are not simply unadaptive patterns that appear when the animal is threatened, but that they are part of an intraspecific communication system.

If a chimpanzee is startled by a swooping bird or a large snake, it will duck its head, as a man would, and fling one or both arms over its face. Or it may throw both arms straight up into the air. It will then calm down or hit toward the offensive object with the back of its hand. This "hitting away" motion has come to be incorporated as a threat gesture among chimpanzees.

Chimpanzees also tend to hide from threatening objects, such as humans. If not overly distressed, they may move behind a large tree trunk, continuing to eat and peering around every few minutes. They may also hide in old sleeping nests. Sometimes one chimpanzee may hide from another, especially if threatened. In other cases, they may try to stealthily sneak away, moving slowly while peering back, and fear-smiling at the aggressor if caught in the act.

**Conflict Behavior** Conflict behavior, sometimes called frustration behavior (Goodall 1968), consists of a number of patterns that appear when the animal is motivated by conflicting tendencies (Hinde and Tinbergen 1958). Such conflicted behavior may be expressed, for example, when (1) fear of another individual prevents the chimpanzee from achieving an objective, such as when a dominant individual takes food away from a subordinate, and the latter becomes angry and afraid; (2) when the chimpanzee is incapable of achieving a goal, such as when it can see a banana inside a box, but can't get at it; and (3) when another individual does not respond appropriately to a communicative signal, such as when an infant, unable to make a large jump to another branch, looks at its mother with a pout face and "hoo whimpers," but is ignored instead of being assisted. Most of the responses to such situations are in the form of displacement and redirected behavior.

The chimpanzee in a conflict situation may scratch itself, just as a man may scratch his head when he doesn't know what to do. Among chimpanzees, the greater the conflict, the more the scratching. Chimpanzees, as well as a number of other primates, may also yawn in such cases. Yawning in man occurs in mild conflict situations (Schaller 1963). In chimpanzees, it most often occurs in incompletely habituated individuals and, for example, in youngsters who can't take a banana in the presence of superiors (Goodall 1968).

When in a conflict situation, a chimpanzee may also make quick and ineffective grooming motions on another's fur or may rock back and forth, slowly at first, but with increasing vigor, although the stimulus strength remains the same. I have noticed that conflicted or psychotic humans may also rock back and forth, often with the arms crossed or clasped tightly over their chests. Goodall found that male chimpanzees may masturbate when frustrated, and one juvenile female was frequently observed making thrusting movements while rubbing her clitoris on the ground or on a branch, which may or may not have been conflict behavior. I don't recall having noticed precisely this reaction among humans.

Conflicted chimpanzees may also shake and sway branches while glaring at the stimulus, a behavior found in a number of other primates. In some cases, chimpanzees may rush away from the stimulus while slapping the ground, dragging branches, throwing sticks and stones, and pounding on trees or their chests. These patterns are most frequently performed by males. The branch-waving and branch-dragging displays are most likely to occur when adolescent or mature males approach food, when mature males join other chimpanzees, and at the onset of, or during, heavy rain (the "rain dance"). A youngster may also engage in temper tantrums—screaming, leaping into the air, hurling itself onto the ground, writhing about, and, in the midst of it all, casting glances at his mother (Yerkes 1943). Youngsters behaving in such a way are often ignored by adults.

All the aforementioned conflict patterns probably can fall under the heading of displacement behavior. None helped to solve the problem at hand, and if any conflict pattern has value, it is in communication, but we really don't know much about it. The communication question centers over how such actions alter the behavior of other animals. It would seem that, in such highly intelligent and ecologically labile animals as chimpanzees, displacement behavior might come to have a communicative effect without ever undergoing ritualization.

Also, as I mentioned, chimpanzees in conflict situations may also show redirected behavior. An animal that has been attacked by a dominant individual cannot retaliate directly so it may vent its rage by attacking the nearest subordinate. The action is appropriate to the situation; the startled target is not. Redirected behavior is common among primates. Actual fighting among chimpanzees is rare, however, partly due to the fact that (1) the animals usually know each other and so they behave within a hierarchical framework; and (2) there is a wide range of submissive gestures that "turn away wrath," as well as reassurance gestures, such as touching, hugging, patting, mounting, or kissing.

It seems, then, that chimpanzees learn a large part of their signaling repertoire. Their signals are modifiable because they are multimodal and graded, largely dependent upon context and influenced by individual recognition. The very complexity of chimpanzee communication argues for a strong learning component.

## THE DEVELOPMENT OF LANGUAGE IN HUMANS

Humans have a wide range of gestures and expressions, as well as nonverbal sounds (grunts, moans, laughter, and so forth). But in addition, we have developed a phenomenal ability to communicate verbally. In the following discussion, it should become apparent that the normal development of language in humans may not be particularly unusual in the animal kingdom.

### Language Development in Children

This book was written, in large part, in Spain where I tried to take advantage of the situation to learn Spanish—an "easy" language, but I found that it's *not* so easy. On the other hand, children all around me were speaking Spanish with no problem. They perhaps conjugated the future tense of *ir* incorrectly, but generally they were understood better than I was. We (the children and I) were both learning a language, but we were learning it through different methods. Children are able to learn a language with no systematic effort whatever. They just "pick it up." On the other hand, I know adults who have spent years in Spain without ever learning the language.

It might be said that children simply imitate what they hear. And this is undoubtedly true, but only up to a point. Consider this reconstruction of a conversation (Bellugi 1970):

*Child: My mommy buyed me some toys today, daddy.*
*Father: Did you say that mommy* bought *you some toys today?*
*Child: Yes*
*Father: What did you say she did?*
*Child: She buyed me some toys today.*

An adult learning the language would have tried to learn that the past tense of buy is an unusual case. A child is content to know that past tenses end in -ed. The child had never heard an adult say "buyed." The -ed rule was extracted from familiarity with other words. Rule extraction implies more than imitation, thus imitation is not the complete answer to how children learn language.

Perhaps another major language ability is found in children (Bellugi 1970). This is the capacity to put words in a "proper" sequence. At first the sequencing is crude, but it becomes more refined in stages. For example, a first-stage statement might be, "No put sock"; second stage, "He not bump that one"; third stage, "I don't want that one." No child ever heard its mother say, "No put sock" (at least I couldn't find evidence for it in the literature). But the child says it. It appears, then, that there is a development pathway for language learning in humans.

Children generally begin making two-word utterances about six months after they learn their first words. And even these two words are not randomized. Instead, they conform to the basic rules of grammar. These rules are neither direct copies, nor reductions of adult speech. Their regularity suggests that the organization of the two words is imposed by the child and *not* drawn from the adult system (Brown and Bellugi 1964).

It is often two years after the first two-word sentences appear that the child verbally expresses emotion. It seems that it would be just as easy to say "baby mad" as "baby blanket," but apparently it isn't. During these first years, it is evidently easier for the child to express emotion, not through words, but through something more similar to primate displays, such as pouting, crying, and tantrums. Even when adults express emotion verbally, they continue to rely strongly on display patterns to accompany the words. Pounding a desk, of course, is not much different from pounding a tree. And, then, a number of emotions, when elevated to high enough states, may leave adults speechless so that only displays can convey messages. The point is, then, that speech in the young child almost exclusively carries environmental information and only later can it express emotion. The first system is highly developed before the second appears.

Recently a number of investigators have been trying to get at how language-learning occurs by trying to teach chimpanzees human language patterns. The most famous of the "talking" chimps are Washoe, taught by the Gardners, and Sarah, taught by the Premacks. Washoe learned more than one hundred sign language words, while Sarah learned to manipulate plastic blocks as word symbols. The differences in language-learning in these chimpanzees and children were that an enormous amount of effort had to be expended to teach the chimpanzees what a human three-year-old picks up naturally. Also, the chimps never mastered the grammatical structuring of words. Current studies involve communication between humans and chimpanzees via computerlike devices. These experiments may clarify the differences in language ability in the two species.

In other experiments, monkeys were unable to make certain critical associations that humans make easily. For example, a monkey can easily be taught to choose a circle over a square when the circle is associated with reward and the square with pain (Ettlinger 1960). But if it is then presented with carved circles and squares that it can feel but not see, it will not be able to transfer what it learned visually to what it senses tactually. Adult humans have far less trouble, in general, forming certain kinds of associations received via different sensory modalities (Lancaster 1968). Perhaps this association ability permits man to learn *names*, a critical property in learning speech patterns for humans (Geschwind 1964). "Names," in this case, is not synonymous with "nouns." Naming, here, refers to special kinds of neural associations, thus "table," "running," "hard," "blue," "chatter," are names with which our minds associate certain things (nouns *and* verbs).

Between 1860 and 1875, two European anatomists concluded that there are two definable language centers in the brain, both in the same cerebral hemisphere. They examined the brains of people who had suffered from language loss (aphasia) at some time during their lives and found, in every case, lesions or damage in one or two areas of the cortex. The scientists were Paul Broca and Carl Wernicke, and these areas of the cortex were named after them (Figure 18.7). Damage to Broca's area effectively results in an inability to speak. Whatever is said is pronounced badly and haltingly, and surprisingly enough, rules of grammar are almost completely ignored. A person thus afflicted might say, "Horse . . . none." He also cannot write, although he can read and can understand speech normally. Broca's area, then, is probably important in word organization. Maybe it is here that the "rules of speech" lie.

If damage occurs in Wernicke's area, speech occurs in its normal rhythm, but the person afflicted usually only speaks nonsense sentences. ("This one here was like this but then that one didn't.") Both written and spoken language are completely incomprehensible to such people, although they can see and hear. It is believed that the auditory region of the temporal lobe (where sounds are perceived) relays the information to Wernicke's area for decoding and recognition. Also, this area may relay messages to Broca's area for organization prior to their expression as speech. So, it seems, then, that at least certain language abilities are centered in specific areas in the brain, and that the human brain is "programmed" for the rapid learning of language rules. What we're trying to do here, is to see whether human language is a function of inborn neural centers (or activities) that selectively incorporate environmental sounds into a learning program for speech. In a sense, I suppose, we're trying to see if there are parallels between language learning in humans and song learning in white-crowned sparrows. And it seems that there are (Marler 1970).

What are some of these parallels? Children undergoing language learning do not pick up the barks of dogs or the sounds of traffic, although they may have the physical abilities to mime such sounds well enough. Instead, in language learning they seem to filter out everything except human speech. Juvenile white-crowned sparrow males hear all sorts of bird songs, but they selectively pick out the adult white-crowned song to imitate. Young white-crowneds and young children produce relatively unstructured sounds early in life (subsongs in birds, babbling in children), which may be essential in learning to make the sounds of their species as they gradually compare their sounds with the sounds they are subconsciously trying to imitate. Both species, then, apparently rely on inborn neural centers to enable them to select the appropriate sounds from among the environmental noises for imitation and to rapidly learn the rules for language use.

It has been suggested that no such language center or innate ability exists in humans, since there are so many different human languages. But at the same time, it has been pointed out that, in spite of their differences, various languages have basic underlying similarities, and that these similarities indicate innate language-learning abilities in humans, such as exist in the white-crowned sparrow. In one case, two Danish children were reared in isolation by a deaf-mute aunt. They never heard language of any sort. But by the time they were teenagers, they were perfectly able to talk to each other; the language, however, was completely undecipherable to anyone who heard them once they entered "civilization" (Eibl-Eibesfeldt 1970). A recent news report described two grade-school twins, reared normally, who spoke only their own language. The girls would converse endlessly, but had to be formally taught to speak English, a language that they slowly learned.

## The Evolution of Language

It is likely that when those first, small-brained, early premen were fumbling around, trying to use pebbles as tools, their communication system was already highly developed. There is no reason to assume that their system of displays was any less developed than those of modern primates. And, in fact, perhaps a real language was beginning to develop at this time (Lancaster 1968). If the emerging language were only used in displays, or in the communication of emotion, however, it would have hardly been worth developing. They could already do these things well enough. But, on the other hand, if a rudimentary language imparted even a limited ability to *name* things, then the advantage of language takes on enormous implications, and the ability would undergo powerful selective pressures for its development and refinement.

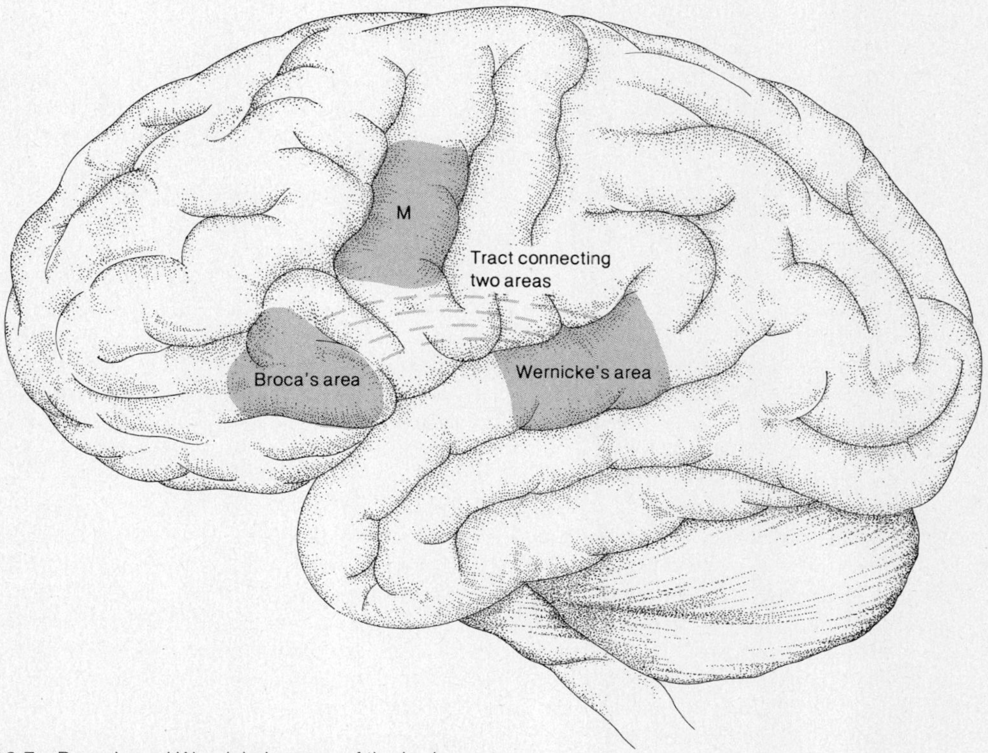


Figure 18.7 Broca's and Wernicke's areas of the brain. Notice that they are connected by a tract of neural fibers. The *M* region is the area responsible for motor control of the lips, jaws, tongue, and throat.

Now, language could serve a purpose not met in the primate display system.

The ability to name things, even on a very primitive level, would have strengthened or even initiated those factors that were causing this peculiar group of animals to assume a somewhat different evolutionary route than that of other animals. Even a very small vocabulary of names would have contributed greatly to the social or predatory actions that increasingly demanded coordination of effort and clarity in communication. Imagine the importance of a few key words suggesting major activities, such as hunting, and a few others indicating nearness in space and time, in combination with a few nouns. The impact on the behavior of the group would have been enormous (Lancaster 1968).

Consider an example of an ecological effect of language development. If the group has the ability to make environmental reference, it can alter its means of exploiting its environment. For example, if a social unit had always hunted and foraged together, they would have placed certain kinds of demands on themselves and their habitat. But if they could agree to split up into subunits, to forage separately and then rejoin later, their social unity would not be threatened; at the same time they could much more efficiently utilize the resources of a wide home range. Such ability may account for the fact that humans "need" far more space than any other primate. As we will see in later discussions, such an exploitive technique is particularly successful when food is rare, dispersed, or seasonal. It might also have great advantage if the food were "patchily" dispersed (abundant in small, isolated areas). Not only does dispersal increase the likelihood of more individuals in the group finding food, but agreeing to rejoin provides a means whereby successful individuals can share food with other members of the group. If no permanent meeting place is necessary, such as a home or den, then the foraging range could be vastly increased, taking the pressure off local food supplies and increasing the likelihood of more members of the group being fed. Thus, the emergence of language is related in important ways to changes in the ecology of the evolving species. Human language ability, then, even on its earliest and most rudimentary level, could have been a key specialization that sent our group off on its own very peculiar path among the animals of the earth (Klopfer 1976).

It is also interesting that, in a certain sense, our thoughts may be "prisoners" of our language. Peter Klopfer pointed out that language is usually structured in certain ways so that we have actors and actions, subjects and predicates, nouns and verbs (1973). In fact, all Indo-European languages demand such ordering. Thus we say, "The grass grows." That makes sense. But then we also say, "It thunders." We know what that means, but there's the problem. There's no logical reason that this sentence should make any sense whatever. *What* thunders? "It is thundering" makes even less sense on any logical basis. But our language demands that we think in terms of something-or-other thundering. Klopfer has pointed out that the impact of such language structuring on our perception of the universe has not been trivial. Even the effect on biology has been monumental. We think of organisms as discrete entities. You are not the same as a bush. If you say, "I jump," it has nothing to do with a bush. It doesn't jump. You do. But living things are *open systems*. The molecules that now comprise that bush may have once been a part of your body. You are simply an association of atoms that have been slowed down in their movements enough to give you whatever "substance" you have. The atoms will soon be gone, perhaps to join the bush, and other atoms will be slowed down enough to replace them. So, what jumps? Are you distinguishable from the bush? And what does "Subdue the earth" mean?

Some languages are free of these constraints (but they may contain their own kind of limitations). For example, in the Hopi language, one can speak of ten men, but not ten days. Ten days is a metaphorical concept, and statements of quantity, in Hopi, cannot be dealt with by metaphors. Also, a Hopi can speak of a drop of water, but he has no noun for water in general. Because of such language differences, then, one might assume that the Hopi views a somewhat different world than one who speaks English (Whorf 1956). It would be interesting to compare perceptive abilities in people who speak both Hopi and English with people who speak only one of the languages.

# 19

# POPULATION DYNAMICS

If you have spent much time in the field, you have probably noticed changes in the numbers of animals found there. The numbers of species change and so do the numbers within any species. In some cases, the changes are to be expected. For example, many insect populations regularly fall off each winter, but rise anew in the flush of spring. The numbers of birds in the North American forests also drop sharply each fall, but this is understandable since we know that some species migrate southward. However, some population changes cannot be predicted. In some years, bird numbers may dwindle so that hardly an individual of certain species can be found. Sometimes the populations recover, but sometimes they don't. Roger Tory Peterson, the noted ornithologist, noticed a decline in many species of songbirds around his home in Connecticut over the past several years. Have they simply moved to another habitat? Or are these species quietly slipping into oblivion? Into a very final extinction? On the other side of the coin, while some populations dwindle, others mysteriously swell. Occasionally, animals may become so abundant that they give rise to invasions or plagues.

Although the numbers of animals may rise and fall, one of the most impressive characteristics of many populations is their *stability*. One gets the impression that the rising and falling of numbers, whether regular or irregular, usually occurs within rather definite limits, and if well-established populations are observed over long periods of time, most will be found neither rising out of control nor dwindling into oblivion. If, somehow, such populations are maintained at certain levels through a regulatory mechanism—and some say they are not (Ehrlich and Birch 1967)—the question arises, How does the mechanism work? And does that mechanism function, say, for insects as well as vertebrates?

## THE EUROPEAN TIT STUDIES

Perhaps we can illustrate a few problems involved in the study of population control by reviewing one of the most comprehensive population studies ever done on birds. Hopefully, we cannot only illustrate some of the basic problems involved in such studies, but also gain insight into some of the mechanisms behind population changes, as described by Krebs (1971). We will return to many of these points later.

The continuing work on European titmice (Paridae) was begun in the Netherlands in 1912, supplemented by observations initiated in England by David Lack. Six species of the tit genus *Parus* breed in Europe, but we will consider only three; the great tit (*Parus major*), the blue tit (*Parus caeruleus*), and the coal tit (*Parus ater*) (Figure 19.1).

The tits proved rather easy to census in the breeding season, but incredibly difficult to observe at any other time because of their mobility. It was found, however, that tit populations fluctuate markedly from year to year, unlike some other bird species (such as the tawny owl, *Strix aluco*). A tit population not only fluctuates yearly, but it does it in tandem with other species of tits (Figure 19.2). This synchrony was most evident for the great tit and blue tit in a number of different habitat types across Holland (although it broke down in a few places). The synchrony suggested that climatic factors were exerting a control over the numbers of the birds. But, one wonders, what other factors could cause such synchrony?

Tits are hole-nesting birds, and the coal tits in these studies were provided with artificial nesting boxes. Could this have confused the result? Do too few nest sites limit tit numbers? It turns out that the boxes that were provided were not always occupied, and it was suggested that there were far more natural nesting holes than boxes in these woodlands (Perrins 1965). It seemed that insufficient nest sites were not limiting the population.

Was food abundance affecting the number of eggs laid? Reproductive rate in birds is largely determined by clutch size (the number of eggs in a nest) and the number of clutches laid each season. Several factors may influence clutch size in the great tit (Kluyver 1951; Perrins 1965). For example, the clutch size normally varies from five to twelve eggs, the smaller clutches being laid as the breeding season progresses (average clutch sizes were 10.3 in the first half of April and 6 in July). Clutch size also decreased as the population increased. In Marley Wood, near Oxford, in England, as pairs increased from thirty to sixty, the clutch fell from ten to eight. Birds breeding for the first time also lay fewer eggs. The implications seem to be that at high density, perhaps there is less food that can be converted to egg production by the female, and that young, inexperienced birds may have greater success in feeding a small brood the first time than in trying to sustain a large brood. Actually, in fact, food seems to generally be more important as a regulatory factor *after* the eggs are laid. In other words, food abundance is more important in determining how many young can be fed than in determining how many eggs are laid. (We will consider contradictory evidence later.) Food may also affect the survivorship of young indirectly, by operating differently on early and late broods. Caterpillars, a mainstay of the tits, are only abundant for two to three weeks in the spring (Gibb 1950), so broods arriving later may suffer from poor feeding conditions, and, as a consequence, late young may be lighter and less vigorous than their slightly older peers. Underweight young survive poorly after fledg-

Figure 19.1 Many of our rules governing population dynamics have been inferred from these European species, the great tit (top), blue tit (center), and coal tit (below).

ing (leaving the nest) and have virtually no chance at all once harsher conditions set in.

The habitat, itself, was found to influence clutch size. Clutches were lower in Oxford Gardens than in Marley Wood, and even within Marley Wood those birds living in areas with many large oaks had larger clutch sizes than did those in areas with fewer big trees (Perrins 1965). One might think that such habitat differences could exist because one area offered more food than the other, thus reducing competition. However, clutch sizes in the great tits of Marley Wood showed no relationship with the abundance of the bird's prime food source, caterpillars—at least *before* eggs were laid. Perhaps, high density depresses reproduction some other way, such as by causing more frequent agonistic encounters between parents, thus reducing the time and energy they would have to spend rearing young.

Populations inhabiting similar kinds of woodlands in different places may show different reproductive patterns. For example, there are marked differences in nesting behavior of English and Dutch populations of tits. Second broods in a single season are rare (less than five percent) in English oak woodland (Gibb 1950). But in similar habitats in Holland about thirty-five percent of the pairs renest. In pinewoods, twenty-eight percent renest in England; seventy-six percent in Holland. The reason for the difference is not understood. However, some interesting correlations emerge from these studies. The food supply is better later in the summer in pinewoods than in oak woodlands (Gibb and Betts 1963). In Holland it was found that second broods were most frequent: (1) among older experienced birds, (2) in years of low density, (3) in pinewoods (Kluyver 1951). Thus, food abundance would seem to be implicated as a factor influencing multiple (successive) nesting.

In the tit studies, predation exerted various levels of pressure from year to year and from place to place. Actually, in most years predation was low, and that which occurred was mostly due to squirrels (*Sciurus vulgaris* and *S. carolinensis*) and weasels (*Mustela nivalis*). Weasels occasionally took great numbers of whole broods. In Marley only between three and fifteen percent of nests of the three species of tits was destroyed from 1947 to 1956, but in 1957, fifty-seven percent of the nests was destroyed, mainly by weasels (Lack 1958). Predation on eggs seems to be a density-dependent factor, as we will see later (Krebs 1970).

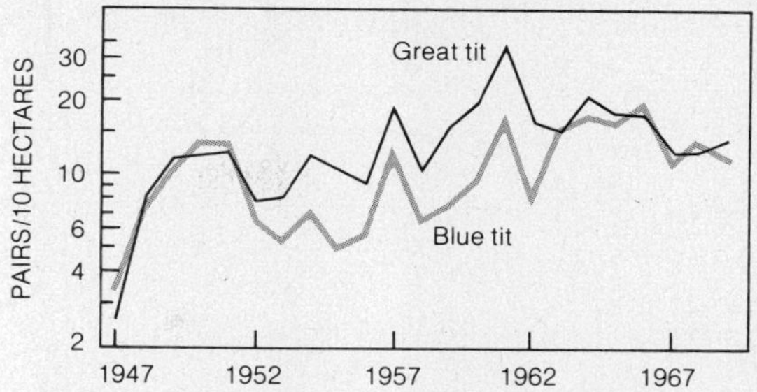


Figure 19.2 Changes in the densities of the breeding populations of the great tit and blue tit in Marley Wood, near Oxford, England.

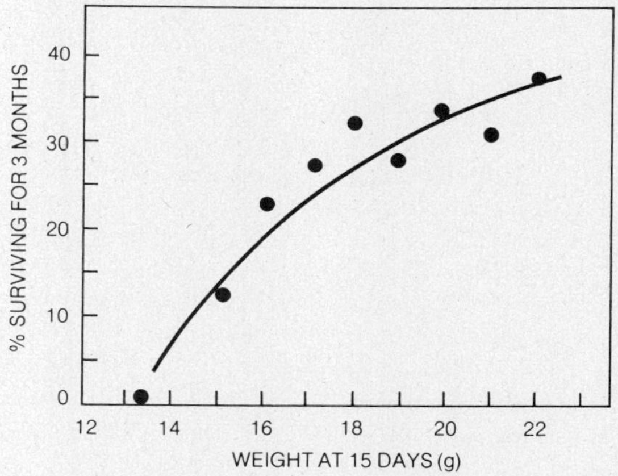


Figure 19.3 Survivorship of young great tits of different nestling weight after leaving the nest. The larger nestlings have greater success.

Outside of such losses almost no young great tits died in the nests. In fact, about ninety to ninety-five percent of those hatched left the nest successfully. But their troubles had just begun. Most young birds met their deaths sometime between leaving the nest and the start of winter, a time when lightweight birds were especially susceptible (Figure 19.3). Interestingly, the larger the brood, the lighter the birds will be by the time they are fledging (growing feathers).

Table 19.1 The relationship of brood size to survivorship of fledglings

| BROOD SIZE | BROODS | YOUNG | % RECOVERED IN AUTUMN | RECOVERED PER BROOD |
|---|---|---|---|---|
| *1–5* | *28* | *120* | *15.8* | *0.63* |
| *6–8* | *42* | *286* | *13.8* | *0.90* |
| *9–11* | *31* | *308* | *13.6* | *1.36* |
| *12–14* | *21* | *264* | *7.5* | *0.95* |

*Source: Krebs (1972).*

Apparently this is because food apportionment by the parents results in each offspring getting less.

In Marley Woods in 1959, broods of nine to ten had the greatest success, and as selection would have it, broods of these sizes were the most common (Table 19.1). It seems that the selective forces operating on tits had been in effect long enough for the brood size to begin to stabilize.

Density changes, it would seem, could probably be explained on one of two bases, differences in (1) how many young are produced, and (2) how many die or move out of the area. In Marley, the annual variations in the number of fledglings were not correlated with population changes (Lack 1964). The conclusion, then, is that changes in the population density of great tits in Marley in those days were produced by the mortality or movement of young birds. This conclusion is supported by estimates of the ratio of juveniles to adult birds in the fall. The mortality rate of older birds is rather constant and independent of age, so this sort of censusing indicates how many juvenile birds make it into their first fall. And it turns out that, in Marley Woods, at least, the ratio of juvenile to adult birds in the fall population is correlated with observed population changes (Perrins 1965).

Kluyver (1966) shed some light on the mechanism of regulation in great tit populations. He removed some eggs and young birds from nests on an isolated island in the North Sea. Of course, this could be expected to reduce the juvenile/adult ratio in the fall, and that fall there were, indeed, very few juveniles around. But the population did not decline! Although there was no immigration from the mainland, the population size remained high (Table 19.2). It seems that adults and juveniles both had a higher survival rate through the fall. This change in survivorship (discussed later) remains entirely unexplained. There was apparently some mortality or emigration factor operating in the summer or early autumn that altered both adult and juvenile losses according to the number of young birds entering the population.

Food shortage has been implicated in most tit mortality (Lack 1964; Perrins 1965; Chitty 1967) in the first few months after young birds have left the nest, beginning in June. This is the time that underweight birds would normally meet their fate. This finding was in disagreement with Kluyver (1951) who had suggested that the main factor responsible for fewer birds in fall is not the death of birds, but emigration. He believed that territorial aggressiveness in September forces a large number of birds to leave the area.

So, what we have to work with are three points: (1) breeding densities of the great tit fluctuate from year to year; (2) population changes are correlated with the number of juveniles in the woods between June and November; and (3) the numbers of juveniles in the population may be a function of their deaths or their being driven out by territorial adults. This last is the greatest point of debate.

Table 19.2 The effect of experimentally reducing clutch size on survival rate of birds

| | BREEDING BIRDS (AVERAGE) | FLEDGLINGS PRODUCED (AVERAGE) | ANNUAL SURVIVAL RATE | |
|---|---|---|---|---|
| | | | ADULTS | JUVENILES |
| *Control years (1956–1960)* | *52* | *274* | *0.26* | *0.11* |
| *Experimental "birth control" years (1960–1964)* | *55* | *110* | *0.54* | *0.20* |

*Source: Krebs (1972).*

Lack (1964) suggested that territorial behavior could not be limiting the density of birds in Marley because if it were, the yearly population fluctuations would indicate that territory size was highly unstable. Although he was aware of such instability, he assumed a certain minimal size of territories, which, alone, would result in the population never rising above a certain level. Perhaps territories of a certain size were necessary to reserve a minimal food supply. Interestingly, in the great tits in Marley Wood, territory size in any year does not seem to be related to that year's food abundance. What we need to know is, does territorial behavior ever keep potentially reproductive birds from breeding? If so, the population could change by territorial birds adjusting (or changing) their territory sizes from year to year.

When breeding birds were shot in their territories near Oxford, they were rapidly replaced by other birds (Krebs 1971). This seemed to demonstrate that some birds had been excluded from breeding by the territorial behavior of the residents. It was found that the replacement birds were moving into the woodlands from nearby hedgerows, where they may have had their own territories, but of an inferior quality. Hedgerows are marginal habitats for the great tit, and hedgerow birds raise fewer young than woodland birds. Thus, territoriality *was* apparently limiting breeding density in the optimal habitat.

But why do woodland birds hold larger territories in some years? Territory size is not related to food abundance and so the question remains unresolved. As Krebs (1972) asked, Should we even assume that the same factors control population changes in all parts of a species' distribution? He pointed out that at least five major problems remain unanswered even after years of intensive work on these titmice. (1) What causes some young to be underweight? (2) When does the survival rate of juveniles begin to rise in their first five months? (3) What role does immigration and emigration play in population changes? (4) What role does territoriality play? (5) How do females adjust their clutch size in relation to population density?

In the following discussions, when we consider other species, it may seem that we are answering some of these questions. And perhaps we are; but we can't be sure. *What controls one species may have little effect on another.* However, the titmice have provided us with an introduction to the problem of population regulation that should help to indicate the breadth of the problem.

We should keep in mind that some changes are *dynamic,* such as the annual rises and falls of the number of flies. Others are static, referring to changes in equilibrium and average numbers. Also, we will see that some changes are imposed on the population from the outside (extrinsic influences), and others from factors relating to the animal itself (intrinsic influences). As a general guide, we can separate the two as follows:

| Extrinsic influences | Intrinsic Influences |
|---|---|
| weather | physiological |
| predators | behavioral |
| parasites | |
| food | |
| disease | |
| shelter | |

And, as you might guess, the effects of such factors may be *multiple,* so that more than one operates at the same time; *variable,* so that they shift in relative importance; and *causative,* so that one brings on another.

## EVOLUTION OF REPRODUCTIVE AND MORTALITY RATES

Any species that suffers a high mortality rate must be blessed with a high reproductive rate if it is to continue to exist. And, conversely, highly reproductive populations can be expected to suffer heavy mortality if the population size is to maintain any semblance of equilibrium. Figure 19.4 shows the inverse correlation between survivorship and fecundity in lizard populations. It seems, then, that reproductive and death rates evolve together and tend to stay in some kind of balance, so that changes in one would tend to cause changes in the other (Pianka 1974).

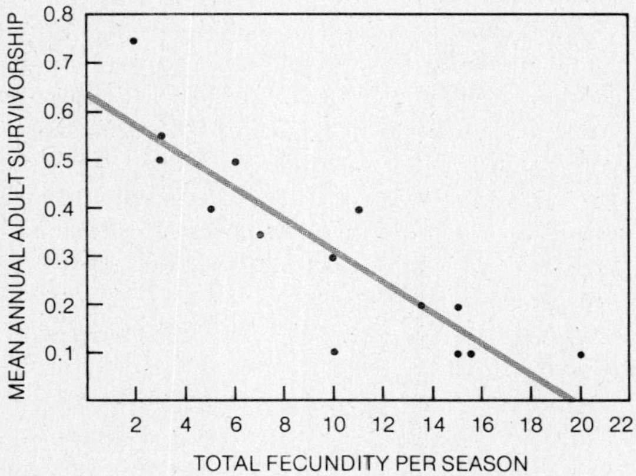


Figure 19.4 Fecundity as a correlate of the probability of surviving to the next reproductive period for fourteen lizard populations.

When the probability of producing viable offspring is low, natural selection can favor an immediate and all-out reproductive effort, even if it is at the expense of the organism's own survival. The organism could expend itself in myriad ways, such as by diverting some of the energy that might have gone for maintenance into reproductive efforts or by risking itself in attacking predators that threaten its young. Animals that have short life spans or those that have difficulty finding mates might also adopt such a strategy. In contrast, when the expectation of producing future offspring is high, organisms are likely to reserve a part of their energy for their own maintenance and to engage in a lower immediate reproductive effort. A wildebeest mother may drive away marauding hyenas for only so long; at some point she may desert her baby. But she will then possibly live through other reproductive seasons and, in the long run, produce more than if she had fought to the death for her previous offspring (Figure 19.5). It has been pointed out that natural selection should favor organisms that behave as though they were weighing their short-term expectations of reproduction against their long-term (Williams 1966a,b).

Generally, smaller organisms tend to maximize their reproduction at their own personal expense. The idea is that they have to get into the numbers game because of their generally shorter life spans. Figure 19.6 shows a positive correlation between body size and generation time. An advantage of small body size, then, is that more individuals reach reproductive states quickly. Shorter generation times mean greater opportunity for mutation to occur, and thus smaller animals can be expected to evolve faster to meet the demands of a changing environment. And then small animals need less food and can find more hiding places than larger ones. It might seem, then, that small size should therefore be generally favored.

But the benefits of increased body size must sometimes outweigh the loss in numbers of offspring, or else large body sizes would have never evolved. In fact, the fossil record frequently indicates a tendency for phyletic lines to increase in size through time (Newell 1949). Of course, some advantages of large size immediately come to mind. Larger organisms,

Figure 19.5 A wildebeest mother will protect her young only up to a point. After that, she increases her likelihood of reproducing if she flees and lives to mate later. This concept, by the way, suggests that older mothers should be more protective, since they will have less opportunity to reproduce again.

for example, are less likely to fall to predators. A moose is a formidable prey animal if the predator lacks a gun. Also large animals place a heavy demand on their environment and are therefore likely to be more spaced out (physically). Such dispersed animals would be less likely to transmit parasites or disease. (The relationship between body size and dispersion might not hold true for herding animals, at least at certain times of the year.) And then larger organisms have a smaller surface area in relation to their volume than do smaller animals, and would thus maintain a relatively smaller interface with the disruptive environment. Finally, the longevity associated with large body size permits large animals to gain the experience necessary to cope with some of their problems on an individual basis.

So, we see that there are two periods in animal life histories that could serve as the points of focus for population regulation. Animal numbers could be limited by (1) restricting the number of young produced, or (2) by affecting the mortality or longevity of individuals in a population. Such considerations are important to the study of animal behavior for a number of reasons. A short life span or an inability to compensate for eggs somehow lost could have important influences on how an animal is likely to behave toward predators, competitors, or potential mates. As we can see, the life strategies of animals, developed through natural selection, are a balanced and highly integrated mélange of interacting forces. In order to account for one, we must carefully consider the others. We will discuss such factors in more detail shortly.

## POPULATION CHANGES

Now, let's briefly consider how populations tend to change when they are unrestricted and how that change is altered when they run into environmental problems.

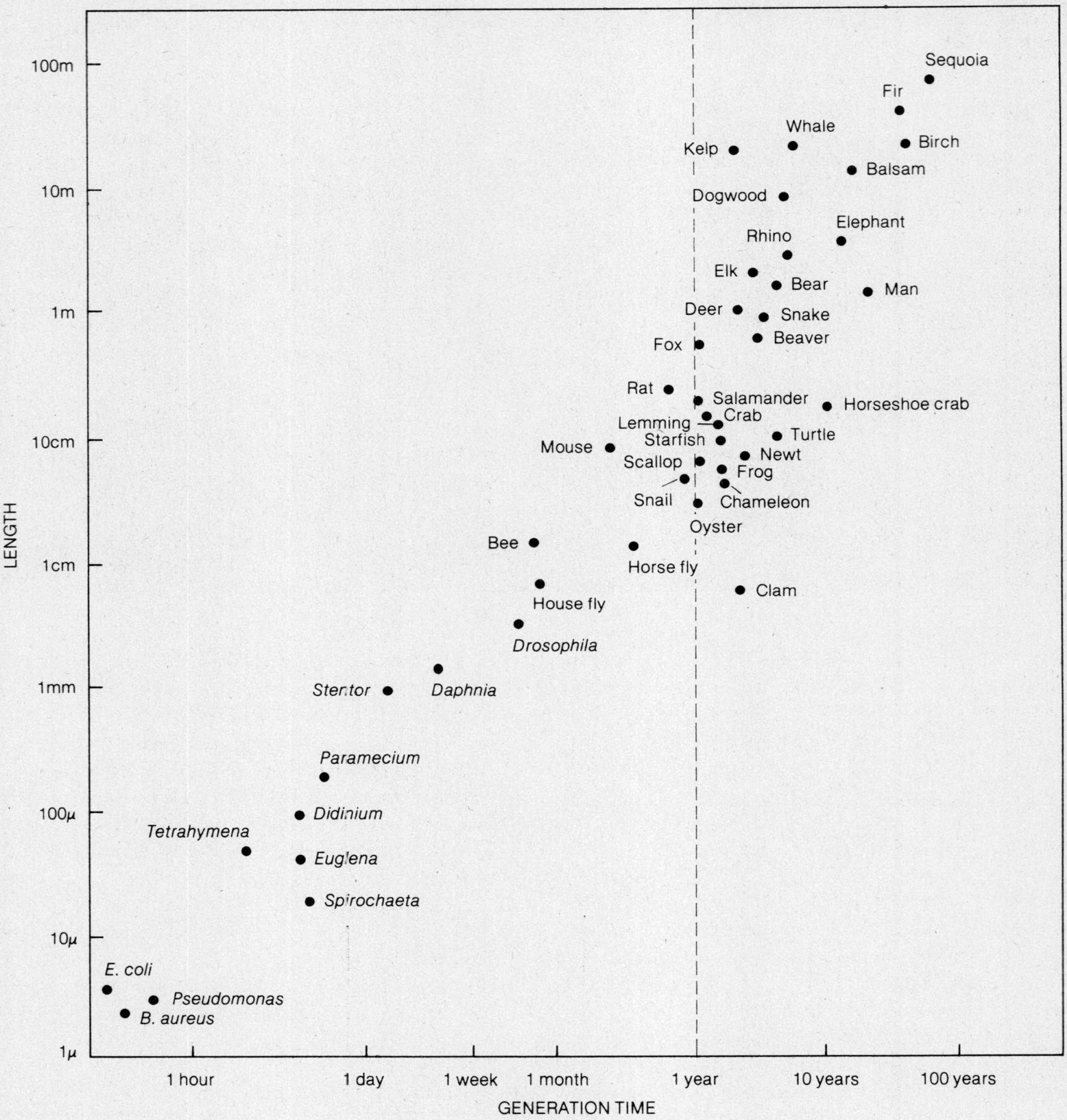


Figure 19.6 Size (as determined by length) plotted against generation time (a log-log plot) for a variety of plants and animals. Larger organisms tend to live longer.

## Population Growth Rate

If we were to place a few bacteria in a suitable medium, or a few sheep on an uninhabited but hospitable island, their population growth curves might be similar, at least at first. Since we are assuming that their new environment provides them with the necessities of life, their initial growth will be limited primarily by how fast they can reproduce, rather than by environmental factors. Figure 19.7 shows the growth for a theoretical population.

Note that in the beginning the *rate* of increase becomes greater with time because of the increasing numbers of individuals that are available to produce offspring. If each individual in the original group leaves two offspring, the population will remain fairly small for several generations (one bacteria divides to leave two, these leave four, these leave eight—all low numbers). But once the population increases a bit, the numbers begin to skyrocket. (What would be the result after only ten generations? Twenty generations?)

## Carrying Capacity, Environmental Resistance, and Biotic Potential

The growth rate of a population when it is unrestricted in any way is called its *biotic potential*. The full biotic potential of any species is not likely to be reached because of the inherent limitations imposed by the environment. Environmental resistance is encountered when a shortage in some commodity, such as food, reduces the population's rate of increase. Environmental resistance prohibits biotic potential from being reached and defines the *carrying capacity* of the environment. As a population approaches the carrying capacity of its environment, its numbers begin to level off (Figure 19.8).

In some cases, a population may overshoot the carrying capacity of its environment. When this happens, its growth rate may fall off abruptly, only to pick up again when the numbers fall below the carrying capacity. In more stable populations, the numbers may oscillate rather closely around the carrying capacity (Figure 19.9).

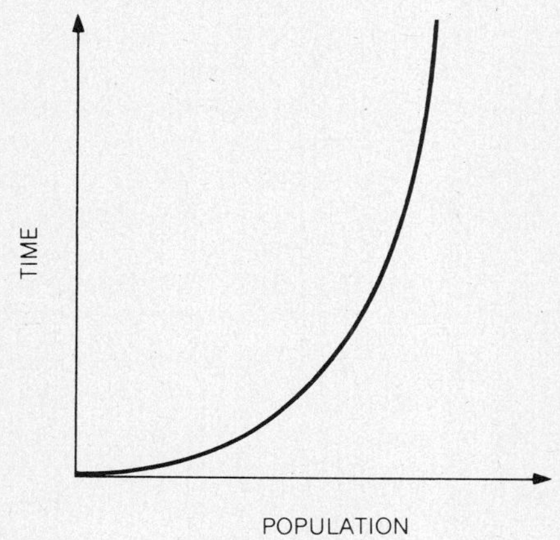


Figure 19.7 Theoretical growth of a population. Time is plotted against population size. In such a case, there is nothing operating to restrict growth, so the population is achieving its reproductive (or biotic) potential. The characteristic curve is produced because larger populations increase more rapidly than smaller ones.

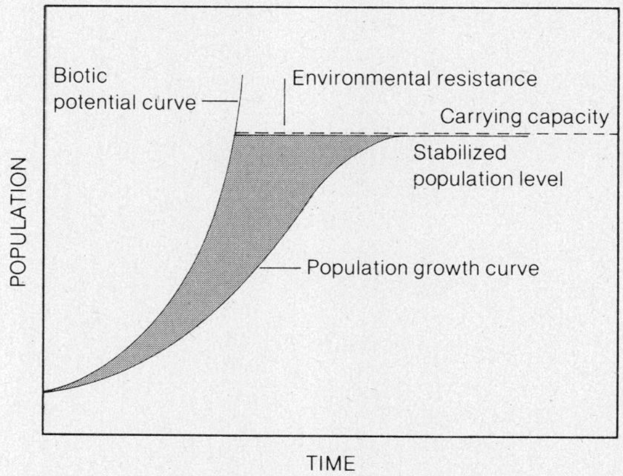


Figure 19.8 The theoretical relationship between biotic potential of a population is rarely, if ever, achieved due to the problems of living in a real world (the problems are collectively called environmental resistance). Theoretically, the population should stabilize around the carrying capacity of the environment.

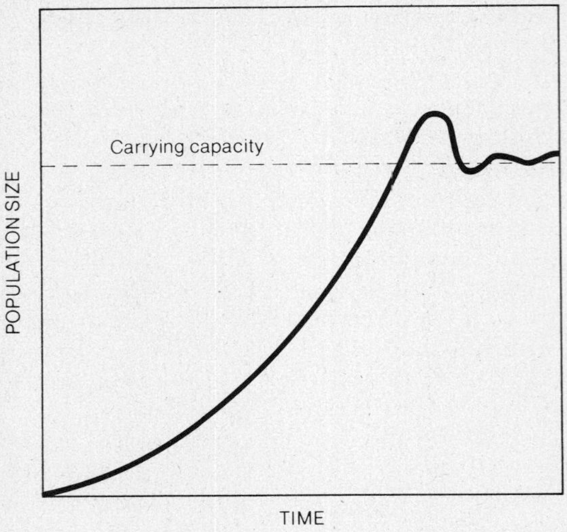


Figure 19.9 A theoretical population reaching the carrying capacity of the environment and then oscillating at that level.

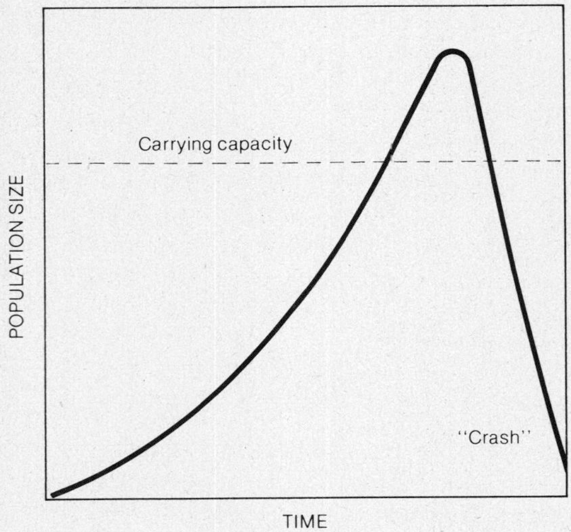


Figure 19.10 A population crash can result when numbers drastically overshoot the environmental carrying capacity. The resulting depression can be part of a cycle or a catastrophe.

If a population drastically overshoots the environment's carrying capacity, the result may be a severe depletion of numbers and is called a *crash* (Figure 19.10). In some species, crashes are a normal part of the population cycle. For example, the numbers of annual plants and of many insects crash each autumn, but their progeny appear each spring to reproduce madly the few months they have available.

Some crashes are more permanent. A bacterial population may rise rapidly, but if the population lives in a petri dish on a laboratory shelf, it must eventually run out of food even as it befouls its environment with its own waste products. When it crashes, it never recovers. As another example, if environmental resistance on any group of animals—that is, resistance in the form of predators (including hunters), disease, or food and water limitations—is somehow reduced, those animals may increase to the point that they eventually devastate their habitat. Oversized herds of elephants may eat all the acacia plants in their habitat, which if exploited normally, would have continued to provide food for generations (Douglas-Hamilton 1975). Such devastation lowers the carrying capacity of the habitat for long periods afterward. In fact, it may permanently lower the habitat's carrying capacity, even in nature. For example, if stripping an area of its foliage should increase erosion, the vegetation may never recover.

Now to return to factors influencing reproductive rates, it seems that any such factor must theoretically operate in two ways: by affecting (1) the number of offspring that are produced, and (2) the longevity or mortality of the organisms in the populations.

## HOW MANY OFFSPRING ARE PRODUCED?

The reproductive potential of most species, even the slowest breeding ones, is incredibly high. Charles Darwin estimated that a single pair of elephants could leave over nineteen million descendants in only 750 years. The fact that the entire world was not teeming with elephants indicated to him that some elephants were not reproducing and that the reproducers were somehow "selected" by the environment.

While elephants are very slow breeders, other species are more prolific. For example, the housefly's reproductive potential, for a single year, is shown in Table 19.3.

But what determines the rate at which an organism reproduces? It should be apparent that simple

Table 19.3 Biotic potential of the housefly[a]

| GENERATION | IF ALL SURVIVE BUT 1 GENERATION | IF ALL SURVIVE 1 YEAR BUT PRODUCE ONLY ONCE | IF ALL SURVIVE 1 YEAR AND ALL FEMALES PRODUCE EACH GENERATION |
|---|---|---|---|
| 1 | 120 | 120 | 120 |
| 2 | 7,200 | 7,320 | 7,320 |
| 3 | 432,000 | 439,320 | 446,520 |
| 4 | 25,920,000 | 26,359,320 | 27,237,720 |
| 5 | 1,555,200,000 | 1,581,559,320 | 1,661,500,920 |
| 6 | 93,312,000,000 | 94,893,559,320 | 101,351,520,120 |
| 7 | 5,598,720,000,000 | 5,693,613,559,320 | 6,182,442,727,320 |

*Source: Adapted from E. V. Kormondy,* Concepts of Ecology *(New Jersey: Prentice-Hall, Inc., 1969), p. 63.*

[a]*The numbers are based on each female laying 120 eggs per generation and half of these developing into females.*

laws of energy would not allow an elephant to become pregnant several times a year. There is just no way that a female could find enough food to produce that many offspring. And what would happen to baby elephants as younger brothers and sisters appeared on the scene, each demanding the mother's attention?

Since some animals must limit their reproduction, then, it may seem paradoxical that, according to some researchers, each animal reproduces to its biological limit. Can it be true that some animals leave as may offspring as possible, while at the same time actually slowing their birth rate? If they do, then we are left with a whole new set of questions about how animals maximize their reproductive success.

## Natural Selection of Birth Rates

It might seem that in a competitive situation an animal could exclude its competitors by simply overwhelming them with the sheer numbers of its progeny. Is this the most common tactic? Under what conditions can we expect other strategies? Let's examine the principle in the animal world. In other words, let's see how natural selection may have determined birth rates.

Some animals, placed into new habitats, have demonstrated their reproductive potential most graphically. For example, when the ring-necked pheasant (*Phasianus colchicus*) was introduced to a small island, its numbers increased from 8 to 1,898 in five years (Einarsen 1945). And then you are probably aware of what happened when the rabbit was introduced to Australia. Australia was overrun. It is important to realize, however, that in both cases the animals were taken from their normal habitat and possibly relieved of those environmental pressures that might have held their numbers down if they had stayed in their home area.

Some of the most interesting and best-documented work on how selection can operate on birth rates comes from studies on clutch size, such as those we just considered that involve European tits. According to Lack (1954, 1966, 1968), most birds rear as many young as possible. This is not to say, however, that they lay as many eggs as possible. Rather, the number of eggs per clutch has evolved as the maximum number the parent can successfully *feed* (at least in altricial species, those in which the young are fed by the parents). Larger clutch sizes do not necessarily result in more young (Lack 1966). In fact, in the European swift (*Apus apus*), the clutch size has not become fixed by evolution and it varies from one to three eggs. It was found that the average number of young that survived to the fledgling state for two-egg clutches was 1.9; for three-egg clutches, 2.3; and for artificially increased four-egg clutches, 1.4. So, those females producing two eggs were better off than those laying four. Why should this be? Obviously, more mouths to feed result in less food for each offspring. The principle is demonstrated in Figure 19.11.

We've redefined fitness as a measure of an animal's reproductive success, and in birds it's determined by the number of hatchlings times their probability of survival divided by two (parents), and can be plotted against egg number (Emlen 1973) (Table 19.4). In the case of the European swift, fitness is increased by anything that shifts clutch size towards three. But that ideal number could change as the environment changes. For instance, the number of young the swifts were able to rear can vary with the weather (Lack 1966). A great proportion of the larger broods (of three) survived in good summers when there was an abundance of insects. But in poorer years, clutch sizes of two proved more successful.

It should be pointed out that many birds are physiologically capable of laying more eggs than they do. If an egg is removed from a nest, the female can often replace it. In other words, under unusual conditions she *can* lay eggs that would normally never have seen the light of day. A yellow-shafted flicker (*Colaptes auratus*), in fact, was thus stimulated to lay seventy-one eggs in seventy-three days (Bent 1939)! The evolutionary principle demonstrated here has come to be called "Lack's hypothesis." Simply stated, animals tend to produce the number of eggs or offspring that result in the maximum number surviving to reproduce.

In some cases, selection may have even allowed for mortality of young (Cullen 1957a). Gulls, species that suffer high predation, may normally only be able to feed two young successfully, but they often lay three eggs, since at least one of the young is likely to fall prey. In the kittiwake, a gull that nests on cliffs and does not suffer from predation, the usual clutch size is two. Clutch size may be adjusted to predatory pressures in other ways as well. For some species of preyed-upon gulls, smaller clutch sizes may make

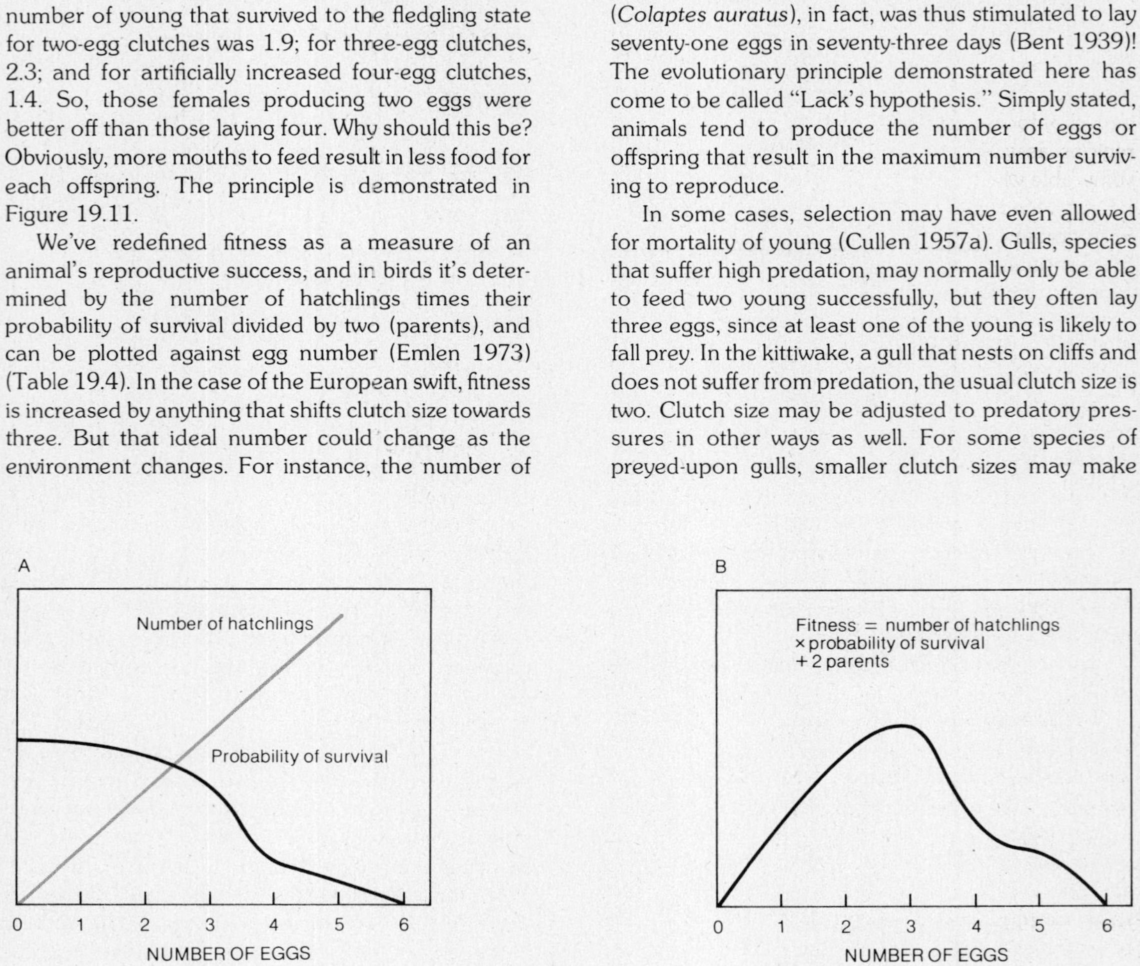


Figure 19.11 Hypothetical curves demonstrating that (A) as the number of hatchlings increases, the probability of their survival decreases, and (B) fitness is increased by anything that moves the clutch size toward three.

broods less conspicuous to hunting predators. The gull *Furcatus*, which often loses its young to predators, lays only one egg, but compensates *behaviorally* by nesting in cycles of less than twelve months (Hailman 1965).

It is obviously more advantageous in terms of biological fitness to rear large broods, so aside from responding to predation, why do so many bird species rear small broods? Part of the answer may be related to the *time* involved in raising broods of various sizes. *The principle is:* the same quantity of food can raise a small brood quickly or a large brood slowly. So, when would an animal want to raise its offspring quickly? Rapid development of young is imperative in areas where the nesting season (or period of food abundance) is short—such as in polar regions, as we learned earlier. Also, where young are vulnerable while in the nest or den, this stage should be shortened, and one way is to rear fewer young. High predation, of course, would reduce the number of individuals utilizing the food supply; so, with the resulting increase in available food, slightly larger broods might be expected where clutch size is dictated by predation rather than by a short season. But this notion remains to be tested.

In bird species that have relatively precocial offspring, clutch size probably is largely a function of the food available to the laying female. More food means more nutrient material that can go into egg production. However, it should be noted that egg laying stops while food is still abundant, so it is not entirely dependent upon the *immediate* presence of food. This is not to say, of course, that egg laying could not be *initiated* by high food levels and that the hen might fail to respond to the same stimulus later in the season.

The size of the egg is probably of great importance to precocial species, since a large egg means more food for the embryo, permitting hatching to occur at a relatively advanced stage of development. This advantage must be weighed against the increased strain on the female in developing large eggs. A very large egg, with a very exhausted mother, is at no advantage; so where eggs are larger, fewer are likely to be laid, all other things being equal.

Since the number of offspring attempted may be affected by food abundance (but you recall that the correlation was not shown in the tit studies), there must be different strategies for species that have an abundant and relatively constant supply, for those that are subject to a regularly fluctuating (seasonal) supply, and for those with an irregular food supply. As an example of an adaptation to the latter, consider the great frigate bird (*Fregata minor*), which breeds in the Galapagos Islands. It is believed that a peculiar breeding pattern has resulted from erratic and nonseasonal food supplies. Frigate birds have long incubation periods; the young develop slowly and have the ability to fast if necessary; they leave the nest at a relatively advanced stage of development and are fed by the parents for extended periods even after they can fly (which may be an adaptation to give the young a chance to learn the fine art of piracy—robbing other birds of their catches). Finally, the birds

**Table 19.4 Optimal and observed brood sizes in Swiss starlings**

| BROOD SIZE | BROODS OBSERVED | YOUNG PER NEST RECOVERED 3 MONTHS AFTER FLEDGING |
|---|---|---|
| *1* | *65* | *0* |
| *2* | *164* | *0.04* |
| *3* | *426* | *0.06* |
| *4* | *989* | *0.08* |
| *5* | *1235* | *0.10+* |
| *6* | *526* | *0.10* |
| *7* | *93* | *0.10+* (sizes 7 and 8) |
| *8* | *15* | |
| *9* | *2* | — |
| *10* | *1* | — |

*Source: Lack (1948).*

breed only every other year, and so the drain on the parents and on the food supply is diminished (Nelson 1967).

In a competitive system, the number of young appearing on the scene each year can be expected to affect the mortality rate, and the arguments of the precise relationship of birthrate and mortality rates can become rather esoteric. Lack, for example, believed that natural selection has resulted in animals attempting to produce as many young as possible each season, and that the mortality rate is adjusted to the reproductive rate. On the other hand, Alexander Skutch, a noted field ornithologist, suggested that in the tropics one reason more young are not attempted is because of low adult mortality. It has been pointed out that the two theories are not entirely incompatible (Fretwell 1969).

The occurrence of multiple broods in times of high food abundance indicates a certain relationship between available food and the number of offspring attempted. Lack (1954, 1966) provided references to such correlations. (Lisa Salmon, the "Bird Lady" of Jamaica, told me that she provided food for a pair of Jamaican woodpeckers while they raised three successive broods.) (Also, food abundance may result in more eggs being laid by lizards [Hahn and Tinkle 1965].)

Another consideration was raised by Professor Wynne-Edwards (1962) who suggested that birds may not lay as many eggs as possible, but that, instead, they practice a form of "birth control" at high density. The concept entails a form of "group selection" where selective mechanisms move to enhance the welfare of the group at the expense of the individual. In other words, under some conditions an individual might diminish its own reproductive effort in order to benefit the group as a whole, perhaps in this case sacrificing its own reproductive output to keep the population numbers within viable limits. The implications are contrary to our idea of fitness where each individual contributes as many genes as possible to the population at the expense of other individuals. Wynne-Edwards has been roundly criticized (cf. Lack 1966), but so far the question, by its very nature and the current limitations of our research techniques, has firmly resisted final resolution. However, this is not to say there are not those willing to take up a position and argue as if the data were in.

There is another interesting way that reproductive rate can be depressed—one that has interesting implications for humans. High population density may, itself, depress reproductive success. At high densities, for example, mammals may show lowered fecundity, at least in the laboratory (Christian 1963). The effect is probably rare in birds, but, as we know, Lack (1958, 1966) found reduced clutch sizes at higher breeding densities in great, blue, and coal tits, and others have reported similar findings for other species. In another study on tits, fewer females attempted second broods in high-density years (Kluyver 1963). The mechanism by which such a control could operate is not clear, but physiological explanations have been offered. For example, the effect of social stimuli upon gonad development has been shown for several species of birds (cf. Phillips 1964). The effects of density on population numbers will be discussed in more detail in Chapter 20.

## POPULATION CONTROL THROUGH MORTALITY

As was mentioned earlier, and it's an obvious point, populations can be regulated by the deaths of its members. Of course, probably no single factor, or type of factor, exclusively regulates any natural population. In living systems, complexity is the rule, and undoubtedly population regulation occurs through a variety of mechanisms. These mechanisms may be *abiotic* (nonliving), such as weather, or *biotic,* such as predation or competition. We will consider various mechanisms separately, but first, let's discuss the larger question of the evolution of death and the implications of death for human populations.

### The Evolution of Death

If, as some would have it, death is the "natural extension" of life, we might ask, Is it logical that death would be one of the constants among virtually all the myriad living forms? Why do living things age and die? There is very little concrete evidence, so again we find ourselves immersed in speculation.

In the beginning, when certain droplets in the primordial seas began to be a little more organized than the rest, they may have been able to grow and, when reaching a certain critical size, they fragmented on a rather mechanical basis—the way coacervate droplets behave in a laboratory today. If, as some have suggested, life sprang from such "organized" droplets, only later developing the ability to reproduce through means other than fragmentation, we can imagine how death could have occurred for these early forms. They were undoubtedly imperfect structures. Natural selection simply hadn't had time to adapt them to their environment very well, and we assume that they all eventually perished. Imperfection, then, may have accounted for the phenomenon of death among early life forms. But what about later? Surely selection could have developed immortal, or exceedingly long-lived, forms over the billions of years that life has existed on the planet. Why hasn't this occurred? Is it possible that there has been selection for death?

We could begin to consider the selective "logic" of death by noting two things. First, the environment changes, and second, commodities (the things necessary to sustain life) are limited.

Many of the earth's changes are historic and obvious. Not only have continents moved, islands formed, landmasses submerged, ice ages come and gone, and weather patterns shifted, but man has, in a very final sense, altered the earth's surface in the past few thousand years. Many changes are going on before our eyes, and some are perplexing. The grasses of our plains are slowly but surely being replaced by woody plants (Krebs 1972), and we don't know why.

As we will see in following discussions, populations can be limited by shortages in all sorts of commodities, from trace minerals, to food, nest sites, water, and space. In such cases, competition must exist, to one degree or another, at least at times.

Now, as a variable population is confronted with changing conditions, some members of that population can be expected to be able to cope with the new conditions better than others. So, if a population is going to survive in a changing habitat, its best bet lies in its members producing variable offspring. The more changeable the habitat, the greater premium there is in producing offspring that are somewhat different from each other. As new generations appear, their myriad genetic variations may provide material on which new kinds of environments can exert their molding effect through natural selection. Many in the new generations will fall by the wayside, to be sure, their kinds of genes lost from the population. Individuals with other kinds and combinations of genes, though, may fit the new environment well. These, then, will continue and reproduce, at least until conditions change again.

But, consider these successfully reproducing individuals in the context of a competitive situation. After an organism finishes its period of reproduction, unless it can somehow assist in rearing its descendants, its very presence will be detrimental to those descendants. It will leave fewer genes than an organism in a different part of the habitat that reproduced and then died, so as not to interfere with the success of its offspring. This is especially true when old and young individuals utilize the same commodities.

If new generations help insure success in a changing world by being genetically different than their forebears, and if a parent that remains in the population after its reproductive period may be a detriment to its offspring, then the phenomenon of death can be associated with the end of the reproductive period. When reproduction is over, it is best to die.

Why then, one might ask, isn't there selection for longevity *with* an extended reproductive phase? The answer would seem to be because a single individual cannot change fast enough, or to such degree, to adapt to new environmental conditions. Thus, any long-lived individual would likely sooner or later encounter a situation with which it couldn't cope and it might then be lost from the population. If it had put its energy, not into longevity but into producing offspring, some of them might have survived the new conditions quite nicely, and that organism's genes would have continued in the population. But genes that promote extreme longevity would be lost.

Also, an animal's immediate offspring would be more similar to it than those several generations removed. So, variability can be directly related to the number of generations produced. Variation through generations would be especially great in sexually reproducing species, where the genes from different parents are randomly assorted at meiosis as their offspring enter reproduction.

If the biological importance of death can be granted, one might ask, why do females live longer than males? In the United States, women live several years longer on the average. But, as usual, one can find primitive tribes where men outlive women. Perhaps we should find out what the *usual* condition is.

If the female reproductive period ends at about age forty-five, why do women go on to live out their "three score and ten"? It will be recalled that humans require parental care and that a woman who has had a child late in her reproductive period, say, at forty-five, will likely live long enough to see to it that the child bearing her genes successfully enters its own reproductive period. She may even be able to contribute to the rearing of her grandchildren. The longevity of females, then, may be based on their traditionally being directly involved in nourishing and rearing children. On the other hand, a male dies nearer the time he is no longer capable of inseminating or attracting females with any notable success. It should be pointed out, at the same time, that human males are capable of inseminating women past the time that they would individually be able to actively defend or provide for any progeny. Older men are not often chosen as mates by reproductive women, however, possibly because senescence diminishes their secondary sex characteristics and they lose their appeal as sex objects. Their continued longevity may be related to their value as leaders or repositories of information. This role, of course, would be most valuable in a species that was verbal, intelligent, and essentially political.

It is interesting, also, that in May–December marriages, men are almost always on the Christmas end. Almost no young men marry old women. One reason for the limited successes of old men may be that the power they lose physically is regained politically and that women are attracted to powerful men, however it is measured.

Senescence, in either sex, may be a mechanism that reduces sexual attractiveness in an individual that is likely to be unable to successfully perform the role of a parent. But senescence, on this adaptive basis, may not be due to group selection, as it might seem at first. Instead, if an individual is related to other members of the group, it may be to his or her advantage to fail to reproduce in old age so that reproductive efforts of the group will be restricted to more vigorous (and related) individuals.

It has been noticed that people who live under extremely harsh conditions seem to "age" quickly. A person of thirty or forty may look very old indeed in some cultures (Figure 19.12). Premature aging may be an adaptation to a rigorous life in which many will not live to successfully care for their offspring until the offspring would have normally been able to go it alone. Perhaps the early loss of sexual attractiveness is an adaptation to a short, dangerous life and ensures that individuals are not attracted to others who probably won't live long enough to successfully rear a child.

Figure 19.12 The bushmen of the Kalahari Desert live under harsh conditions and seem to age quickly. Perhaps rapid aging is an adaptive trait in short-lived populations in that the senescence signals the probable early end of the reproductive period. This woman, making ostrich eggshell beads, is about 40, but shows signs of more advanced age.

Death is undoubtedly more important on an evolutionary basis for some species than others. As examples, certain plants and animals pinch off "buds" from their bodies that then disperse to begin an existence elsewhere, but they contain a genetic complement identical to that of the parent, so genetically they *are* the parent. Planarians divide by fission (and a popular interpretation is that this means they need never die), and there is no change in the genetic component from one generation to the next, except by mutation. Many intestinal parasites self-fertilize and thus lose certain benefits of sexual reproduction. When we discuss *r*- and *K*-selection, you might like to consider the two selective schemes in the context of death. For other ways of looking at the evolution of old age and senility, see Medawar (1957); Williams (1957); Ricklefs (1973); Klopfer (1962).

## Death and the Size of Human Populations

It has been suggested that since humans bestow such intense care upon their offspring, selection might push for an early death in weakened or malformed human infants. In fact, note in Figure 19.13 that infant mortality is high in the first year. Hamilton (1966) suggested that modifier genes exist that shift susceptibility to disease, and the expressions of malformations, to the earliest possible age. Thus, if the child has a high probability of dying, an early death will give the parents a better opportunity to replace it. It has also been noted that fetal mortality is greatest in the first part of pregnancy. Ideally, if an embryo is going to die through some malfunction, it should do so when it will be resorbed by the mother's body and thus negate the risk of the mother's losing the baby through miscarriage at a later time.

The question arises in such cases, Where is the selection operating? If it is acting on the baby, causing it to be removed from the population (if it would have died anyway), we can see support for the group selection argument because the organism sacrificed itself for the reproductive good of the group (or the parents). As mentioned, the notion of group selection is in disfavor with most biologists. The system, however could also operate through kin selection (as discussed in Chapter 32), wherein the death of a defective organism increases the likelihood of its parents reproducing again and producing children with genes in common with itself. In addition, the selection could be acting on the parents, not the offspring, causing them to produce genotypes that self-destruct if they happen to be found in offspring that are unpromising. This question is probably much more important than it seems on the surface. If

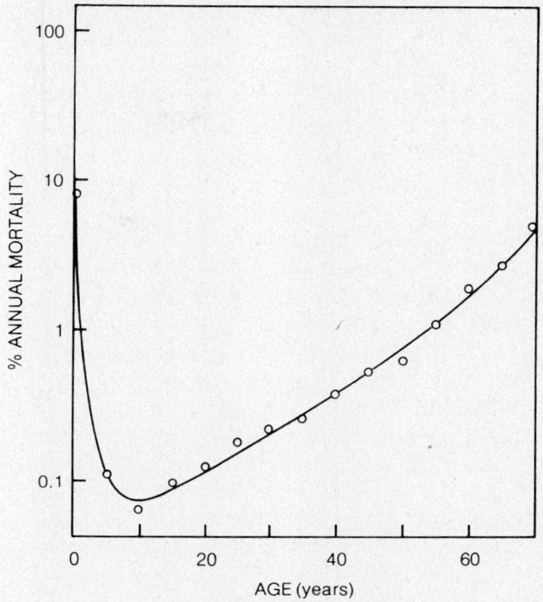


Figure 19.13 Relationship between age and death in humans (here Costa Rican females).

selection can operate on an individual by acting on its offspring, at what point will it begin to act solely for the good of the offspring? Are you someone's reproductive effort? Are there conditions under which it would be better for your parent's reproductive effort if you were to be removed from the population? At what point did you lose your expendability?

**The Size of Human Families** At a time in our history when fertility can be artificially increased to the extent that quintuplets are no longer a rarity, and at a time when almost any child born will live through its reproductive years, it is important to consider how many children are "normal" for us. In other words, if the number of offspring among animals is arrived at through natural selection, what is "our" number? And, perhaps more importantly, what are the consequences of exceeding that number, now that we are able?

The human female normally has the ability to produce one child a year for over thirty years, but hardly any couple has thirty children (one woman had sixty-nine children—the record). In fact, are humans physically or psychologically prepared to deal with thirty children in their lifetime? Or were we never "intended" (by the abilities granted us through natural selection) to have so many children? Could it be that our reproductive ability is an evolutionary adjustment to a historically rigorous life in which most offspring were not likely to survive? In other words, could the reproductive potential of the human female have evolved in response to a traditionally high death rate for children?

Children, as is the case with the young of other animals, are particularly susceptible to a number of parasites and diseases that will have much less effect on them later in life. Also, in spite of the parents' best efforts to watch over them, children are likely to suffer accidents and, in some cases, to fall to predators because of their slow period of development and extended reliance upon older members of the group. Even today, in stressful situations such as the drought that swept across a wide band of Africa, the children are among the first to die, along with the very old and infirm.

Now, normally, if a child should die in its first year (a risky time for children), the mother might quickly become pregnant again. If the child should continue to live, however, she might not become pregnant again for some time. This could be partly due to her paying more attention to the child than to her mate. But there is another factor at work, too. In women, suckling babies somehow inhibit the secretion of the follicle-stimulating hormone (FSH) and the luteinizing hormone (LH) from the pituitary gland. Thus eggs are not released from the ovaries. The result is that pregnancy is not likely to occur while she is nursing. Nursing can continue for as long as three years, thus lowering the probability of the mother becoming pregnant again for that period. It seems, though, that the hormonal inhibition only occurs in marginally nourished women. And, perhaps, this has been the historical condition of mankind.

The high reproductive ability of humans might also be an evolutionary response to bleak prospects of an adult surviving through the entire reproductive period. There is evidence that in some earlier civilizations, the life expectancy was only about thirty years (Table 19.5). Thus, selection might move toward maximizing reproductive success in the few adult years traditionally available to humans.

Table 19.5 Estimated life span in human populations[a]

| POPULATION | YEARS |
|---|---|
| *Neanderthal* | *29.4* |
| *Upper Paleolithic* | *32.4* |
| *Mesolithic* | *31.5* |
| *Neolithic Anatolia* | *38.2* |
| *Austrian Bronze Age* | *38* |
| *Classic Greece* | *35* |
| *Classic Rome* | *32* |
| *United States, 1900–1902* | *48* |
| *United States, 1950* | *70* |
| *United States, 1975* | *72* |

*Source: After E. S. Deevey, "The Probability of Death." Copyright © 1950 by Scientific American, Inc. All rights reserved.*

*[a]It should be pointed out that these are average life expectancies. Many people died young and many made it to old age in all these cultures.*

But what about now? As Table 19.6 shows, a child born in a developed country these days is likely to live through its first critical years. Mortality decreases after the first year, is low through the teens, and then gradually increases to about the age of sixty, when it rises rather rapidly. This profile is especially true for developed countries, such as the United States, but is less representative of underdeveloped countries. The reasons for such longevity, of course, involve developments in health care, industrialization, and higher "standard of living." So, now we find ourselves able to successfully rear almost every child we produce, but our reproductive *abilities* hark back to a time when most children died before they, themselves, reproduced.

The problems stemming from this change are not to be minimized. At this point, we have no real idea of their extent, however. Are humans, who evolved with the physical and psychological properties necessary to rear few children, equipped to deal with larger numbers of children? Has the much-discussed "battered child syndrome" anything to do with association with too many children? How do children with

Table 19.6 Child mortality rates 1–4 years (for latest available year)

| CONTINENT AND COUNTRY | RATE PER 1,000 CHILDREN (AGES 1–4) PER YEAR |
|---|---|
| AFRICA | |
| *Mauritius, 1971* | 4.5 |
| *Réunion, 1967* | 5.5 |
| *United Arab Republic, 1960–62*[a] | 37.9 |
| NORTH AND CENTRAL AMERICA | |
| *Barbados, 1969* | 1.3 |
| *Canada, 1973* | 0.8 |
| *Costa Rica, 1973* | 3.0 |
| *Dominican Republic, 1960–62*[a] | 10.8 |
| *Guatemala, 1970* | 39.6[b] |
| *Mexico, 1973* | 19.9[b] |
| *Puerto Rico, 1970* | 1.0 |
| *Trinidad and Tobago, 1972* | 6.3[b] |
| *United States, 1974* | 0.7 |
| SOUTH AMERICA | |
| *Argentina, 1970* | 15.7[b] |
| *Chile, 1971* | 3.0 |
| *Colombia, 1960–62*[a] | 17.4 |
| *Equador, 1960–62*[a] | 22.1 |
| *Peru, 1960–62*[a] | 17.4 |
| *Venezuela, 1960–62*[a] | 5.9 |
| ASIA | |
| *Ceylon, 1960–62*[a] | 8.8 |
| *China (Taiwan), 1960–62*[a] | 7.2 |
| *Hong Kong, 1974* | 4.4 |
| *Israel, 1974* | 5.8[b] |
| *Japan, 1974* | 0.9 |
| *Pakistan, 1968* | 17.4 |
| *Philippines, 1960–62*[a] | 8.4 |
| *Singapore, 1974* | 4.1 |
| *Turkey, 1966–67*[a] | 14.7 |
| EUROPE | |
| *Austria, 1974* | 0.8 |
| *Belgium, 1973* | 0.6 |
| *Bulgaria, 1974* | 6.4[b] |
| *Czechoslovakia, 1973* | 0.8 |
| *Denmark, 1973* | 0.6 |
| *Finland, 1973* | 1.0[c] |
| *France, 1972* | 0.8 |
| *Germany, East, 1974* | 1.2[c] |
| *Germany, West, 1973* | 0.8 |
| *Greece, 1974* | 0.8 |
| *Hungary, 1974* | 0.8 |
| *Ireland, 1973* | 4.3[b] |
| *Italy, 1972* | 0.8 |
| *Netherlands, 1974* | 0.7 |
| *Norway, 1974* | 0.7 |
| *Poland, 1974* | 0.8 |
| *Portugal, 1974* | 2.2 |
| *Spain, 1970* | 0.9 |
| *Sweden, 1974* | 0.4 |
| *Switzerland, 1974* | 0.8 |
| *United Kingdom—England and Wales, 1973* | 0.7 |
| *Yugoslavia, 1973* | 1.7 |
| OCEANIA | |
| *Australia, 1973* | 3.9[b] |
| *Fiji Islands, 1960–62*[a] | 3.7 |
| *New Zealand, 1971* | 0.8 |

*Source:* Demographic Yearbook, *United Nations (1975).*

[a] *Average annual; latest reliable data.*

[b] *Includes deaths of children under 1 year of age.*

[c] *Includes deaths of children under 1 year of age through 14 years.*

large numbers of siblings fare in comparison to their peers from smaller families? In general, what are the effects on parents and children of the successful production of increased numbers of offspring? Of course, the ultimate problems lie in the effects of overpopulation on a hungry planet.

We have seen, then, that the consideration of what determines population size is complex, beyond the pale of what one might first imagine. Among birds, for instance, such factors as weather, number of nest sites, habitat type, age of breeders, food abundance and cyclicity, predation and little-understood tragedies that befall fledglings, all have their molding effects on populations. Other behavioral factors such as emigration and territoriality also mold populations. In addition, we have seen that reproductive and mortality rates are the products of natural selection and are dependent upon the "life strategies" of the animals in question. Humans are not immune from such selection, but our selection may have molded us to fit into a world quite different from that which we have recently made for ourselves.

# 20

# DENSITY-INDEPENDENT AND -DEPENDENT POPULATION CONTROL

Populations can be changed by a variety of factors, as we know. However, such changes can generally be classified according to whether those factors are influenced by the density of the population itself. So, let's take a closer look at these two basic kinds of regulation. We will begin with nonliving factors and then consider the more complex living ones.

## DENSITY-INDEPENDENT CONTROL

We know that certain populations drastically fall off each year. Annual plants, for example, usually die in the fall, leaving their seeds to represent them in the next season. Some annual animal species suffer the same fate (Kormondy 1969). You may have noticed sluggish and dying grasshoppers on your lawn as the nights grow colder and autumn draws on. In winter, their numbers are reduced to virtually zero as their hope lies in overwintering eggs. The following spring will initiate a period of rapid population growth, and over the years the grasshoppers will produce a growth curve of rather regular seasonal peaks and depressions.

Just as seasonal weather change can alter population numbers, so can irregular or unusual weather patterns. Extreme drought may decimate populations of both plants and animals. Untold millions of birds have perished because they started their northward migration in spring only to be caught by a late cold snap.

It is important to realize that these kinds of population changes are not affected, to any great extent, by population density. The parching sun can destroy a whole field of corn as easily as a single plant. Hence, it provides an example of *density-independent control.*

Notice that I refer to population density, not numbers. The difference between population density and population numbers is apparent. Numbers are ecologically not very interesting, but the numbers within any given area are. A population of thirty thousand elephants becomes more interesting to the ecologist when it is made clear that they all live within the boundaries of one park.

Probably the best-known study that shows density-independent control is a fourteen-year program involving a tiny insect pest, *Thrips imaginis,* which infests roses and other plants in southern Australia (Davidson and Andrewartha 1948a,b). The study revealed that the population density of *Thrips* was related to the weather and that populations swelled in good weather and crashed when it became too hot or dry (Figure 20.1). However, the study has been severely criticized (Solomon 1957; Nicholson 1958; Smith 1961).

### Multiplicity of Factors

The noted geneticist Theodosius Dobzhansky (1950) proposed an interesting hypothesis regarding geographic influences on population control. He suggested that in temperate zones, mortality is relatively independent of the genetic makeup, the behavior, or the morphology of the animals and, furthermore, that mortality is not greatly affected by population density. He reasoned that temperate conditions are periodically so severe that deaths are more often caused by weather than by that other much-suspected factor—competition. The adaptive response to harsh

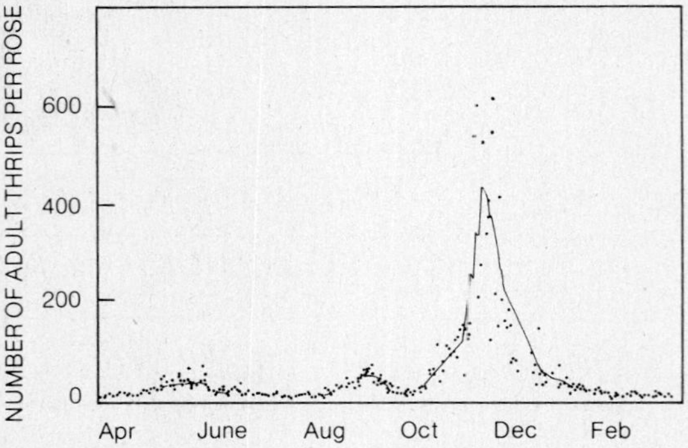


Figure 20.1 Numbers of *Thrips imaginis* per rose, from April 1932 to April 1933 near Adelaide, Australia. The points show daily counts. The curve is a moving average for fifteen days.

weather, according to Dobzhansky, should therefore be to increase the reproductive rate at the possible expense of rearing "quality" young; in other words, in the face of density-independent factors, leave as many offspring as possible (this strategy is now called *r*-selection). In the more hospitable tropics, he said, a greater percentage of the young could be expected to survive, so most mortality would be due to other factors, such as competition. Therefore, in the tropics, natural selection should constantly be moving toward making the individual better adapted for competitive interaction (this is called *K*-selection). The evolutionary accent, then, under more constant conditions, is on the quality of the individual, not on mass reproduction. Ecologist Eric Pianka (1970) noted that in the tropics birds lay fewer eggs and development is slower; therefore, it seems that parents are increasing their fitness by prolonged care of the young, making them better able to compete (*r*- and *K*-selection are discussed in more detail later in this chapter).

There is another interesting geographic phenomenon associated with reproductive effort. Many birds and some mammals produce larger clutches or litters at higher latitudes than at lower latitudes (Figure 20.2), even within the same species. The following hypotheses, using data from birds, have been proposed to account for the phenomenon (Pianka 1974). Keep in mind that the explanations are not mutually exclusive. You will also note that while latitude may be the primary factor influencing clutch size, it may operate through biotic mechanisms. This interrelationship should clearly illustrate the multiplicity of effects operating in nature and point out that our efforts to blithely categorize them are purely artificial, done for the sake of convenience. The hypotheses, then, are:

(1) *The daylength hypothesis.* In late spring and summer, days are longer at higher latitudes, and thus diurnal animals have more time to gather food and feed larger broods.

(2) *The spring bloom hypothesis.* Many temperate birds migrate, but few tropical species do. In temperate latitudes, during the spring, a profusion of plants and insects flourish, so food is presumably abundant. But because resident birds have suffered a long winter, their numbers are down. And the rigors of the journey have reduced the numbers of migrators making it back to the temperate climates. Thus, spring populations of birds in temperate areas may be rather small. This means they are subjected to comparatively little competition for the commodities. They are, therefore, able to gather more food and produce and feed more young than their tropical counterparts. Tropical birds would suffer no periodic reduction in numbers and competition would remain high.

(3) *The predator hypothesis.* There seem to be more numbers and species of predators in tropical than temperate zones. Nests, for example, suffer very high levels of predation in the tropics. It has been suggested that predators may actually locate bird nests by following the activities of the attentive parents. Since birds must make more trips to nests with large broods than small ones, the greater level of activity around the nests harboring large broods is likely to attract more predators. This hasn't yet been proven in the field, but in support of the hypothesis, it has been shown that clutch sizes in hole-nesting birds, which are relatively free of nest predators, do not vary as much with latitude as do clutch sizes in birds that have more exposed nests. As further evidence,

tropical islands support fewer predators than do adjacent mainland areas, and tropical island birds tend to have slightly larger clutches (MacArthur and Wilson 1967).

It should also be pointed out that clutch size often increases with altitude, and none of the above explanations covers this phenomenon, since altitude does not necessarily affect daylength, competition, or predation. But the climatic uncertainty of high altitudes (due to instability or unpredictability) could result in reduced competition at higher elevations (Cody 1966).

So, density-independent or abiotic population control is not as distinct from what we will term biotic control as one might have thought. It should be apparent that population density has no meaning unless it is considered in relationship to other characteristics of the habitat, including the abiotic ones (Andrewartha 1961). Consider a population that declines in density near the boundary of its habitat. The reason may be because the peripheral areas are not entirely satisfactory, due, for example, to some character of the habitat such as an increasingly rocky terrain. Usually, the decline of one population along its range is accompanied by an increase in numbers of some other species, better adapted to that type of habitat; and so interspecific competition may increase in such areas and keep numbers of both species low. The pressure on the population in the marginal area, then, is due to both abiotic and biotic factors, and these may not be independent.

The varying relationships of factors may be illustrated in another example. The red-bellied woodpecker ranges from the East Coast to central Texas. Its western boundary is partly determined by a change in vegetation-type and partly by the appearance of a competitor, the golden-fronted woodpecker, in Texas; but competition does not satisfactorily explain the red-bellied's absence off the Carolina coast. Density, then, is only meaningful in the context of environmental variables.

I should add that living organisms can, by their presence, alter the environment so that abiotic control increases its influence. The concept will not be discussed in detail here, but you might like to refer to a paper by A. J. Nicholson (1957b).

In discussing abiotic-control mechanisms, we cannot neglect the influences of man's ability to alter his environment far beyond his natural biological impact. It has already been demonstrated that so many dangerous chemicals have been added to the air, especially over large cities, that increased mortality rates in these areas have resulted—affecting primarily the very young, the very old, and the ill. The Ehrlichs (1970) describe a 1968 report by UNESCO, which states that within about twenty years this planet may be uninhabitable for man because of the quality of

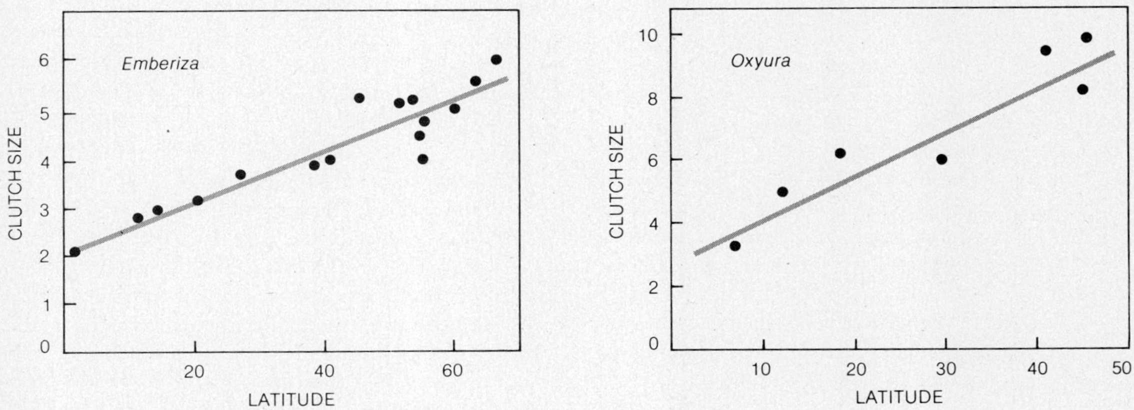


Figure 20.2 The relationship of latitude and clutch size for two genera of birds, *Emberiza* and *Oxyura*, showing a trend toward smaller clutches nearer the equator.

the air. A host of other startling reports indicate that we have already deposited so much mercury into our waters that by the time it has all been altered biologically, its effect may render most of the earth's water dangerous to human consumers. There is evidence that our effluent has even altered weather patterns. But for the most part, the warnings by scientists have resulted in little more than some families changing to white toilet tissue. Nevertheless, man's phenomenal powers to alter the environment may soon bring the real meaning of density-independent mortality into sharp focus.

## BIOTIC CONTROL

In considering the control of animal populations, one must keep clearly in mind the type of animal being investigated. For instance, Lack (1954) stated that the numbers of birds, carnivorous mammals, certain rodents, large fish (where not fished), and a few insects are limited by food. On the other hand, the numbers of gallinaceous birds (which "scratch" for food), deer, and plant-eating insects (a group that comprises most of the world's species) are limited by predators or parasites. Actually, Lack probably should have given more weight, in this early book, to the population-limiting effects of territoriality and dominance hierarchies, for reasons discussed below.

One should also be aware of the interactions of control mechanisms. Some researchers have pointed out that there can be no single model to describe population control, but that the reason is the controls may be essentially different depending on what trophic level is being considered (whether decomposer, plant, herbivore, predator, and so forth). It was noted that since the world is covered with green plants, herbivores obviously do not consume all the food that is available to them. Also, as we shall examine more critically in this chapter, when predators are removed, prey seem to increase in numbers. Thus, herbivore populations must be limited by predators (Hairston, Smith, and Slobodkin 1960). Ecologist William Murdoch (1966) was the first to point out several critical flaws in this reasoning. Among his arguments he noted that the presence of herbivore "food" does not mean food doesn't limit herbivore populations, especially when the food is plants. Other researchers have indicated that mature, leafy vegetation probably contains too little protein and too much cellulose to make it a suitable food, except in enormous quantities, which would be difficult to handle. Certain vegetation may also contain toxins (Whittaker and Feeny 1971).

Lack (1970) presented four main points in his argument for food being the limiting factor in populations of birds:

(*1*) Apparently few adults die of predation or disease.
(*2*) Birds are usually more numerous where food is more abundant (Figure 20.3). Seabirds are scarce in the tropics, for example, except over the plankton-rich Humboldt Current. If warmer waters shift into the region, the plankton die, and so do the fish and birds.
(*3*) Birds living in the same habitat depend on different food sources. When a food is superabundant, different species may tend to converge in their foraging habitat.
(*4*) Fighting for food occurs more often in winter, a time of low food supply.

Lack also admitted that although food probably limits population density in most birds and a few other types of animals, there are species that are held in check by other mechanisms, such as predation, and that therefore rarely reach the food barrier.

Lack was undoubtedly correct in stressing the importance of food as a population regulator. Birds are so sensitive to food levels that it may be possible to assess the abundance of food by observing the birds' behavior (Stenger 1958; Kear 1961; Kress 1967; Kilham 1958a; Wallace 1969; Ward 1965). While there are those who argue that the available food is probably never depleted or is sufficient throughout *most* of the year, and that food, therefore, cannot be a significant factor in population control, others (cf. Willis 1966a) point out that bird populations are undoubtedly regulated by the *lowest* amount of food present during the course of the year. Lack agreed that food probably does not exert its population-limiting effect on many species throughout the year to the same degree it does in winter. Interestingly enough, if there is a lag between food increase and its utilization, as described in Chapter 23, there probably *is* a food surplus through most of

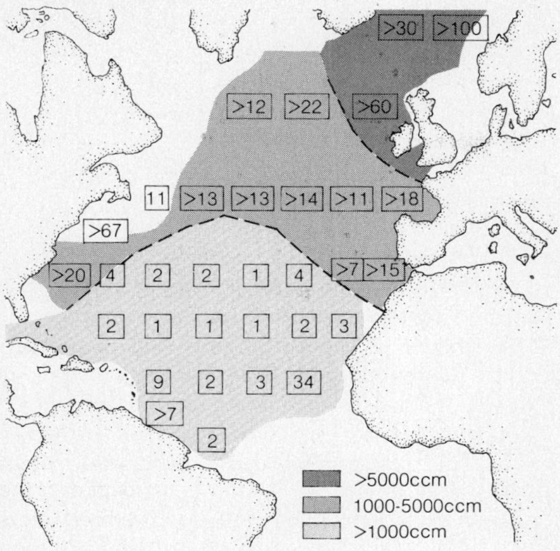


Figure 20.3 Seabird numbers as related to macroplankton density in the Atlantic Ocean. The shading indicates the densities of macroplankton. The numbers in the boxes indicate the number of seabirds observed in that area each day.

the year. The role of food competition as a population-limiting factor is discussed by Hutchinson (1959) and Watson and Moss (1970), evidence for both arguments being presented.

Too often, discussions center on whether one factor or another is operating in population regulation. Such arguments, again, sometimes seem to imply that an either/or situation exists in nature. Such reasoning is patently unrealistic. As an example of the complexities of the question, J. H. Crook (1970b) stressed that at one period the key factor could be extrinsic, such as food, while at another time the critical influence could be an intrinsic social one, such as territorial behavior. There is also evidence that population regulation may occur through a number of factors operating at the same time, although one may be more important than another (Watson and Moss 1970).

Where populations are regulated through mortality rates, the determinants can interact in a compensatory manner so that if the effects of one increase, another may decrease. When guppies (*Lebistes reticulatus*) were artificially culled from a population, cannibalism—the guppy answer to overpopulation—decreased, and vice versa (Watson and Moss 1970).

## DENSITY-DEPENDENT CONTROL

A. J. Nicholson (1933, 1954) suggested that the stability of natural populations could only be accounted for by living things acting on each other in a responding and sensitive system. He envisioned a form of *density-dependent control,* the theory being that an increasing population will grow indefinitely unless, as it increases, stronger controls are brought to bear on it (Figure 20.4). Conversely, a decreasing population will become extinct unless any depressing influence reduces its pressure as the population lowers. Obviously, the most likely type of control would be one that was sensitive to population density, such as we find in a living entity, such as a predator. Abiotic control could not be expected to operate in such a fashion. After all, in times of drought, as a population began to die off, the sun could not be expected to respond by reducing its effect. (But again, we must guard against oversimplifying. It has been suggested that all species, if they increase beyond certain limits, will deteriorate in quality, or become less viable through stressful interaction, so that such effects as the weather will have inordinate influence (Chitty 1960). The result, then, is that weather, under some conditions, can be a density-dependent regulator.)

What has come to be the conventional view of density-dependent mortality involves the influences of food supply, territories, hierarchies, disease, and predation. It is based on the work of Nicholson (1933) and Smith (1935) and has been reviewed by Lack (1954, 1966, 1970).

The basic concept is not a difficult one. Let's consider an example and see how a prey species, in times of high population density, could be controlled through density-dependent mechanisms. First, it seems obvious that as the prey species increased, so would its predators, rising in response to an increased food supply. Second, even as the number of predators increased, the prey species might put

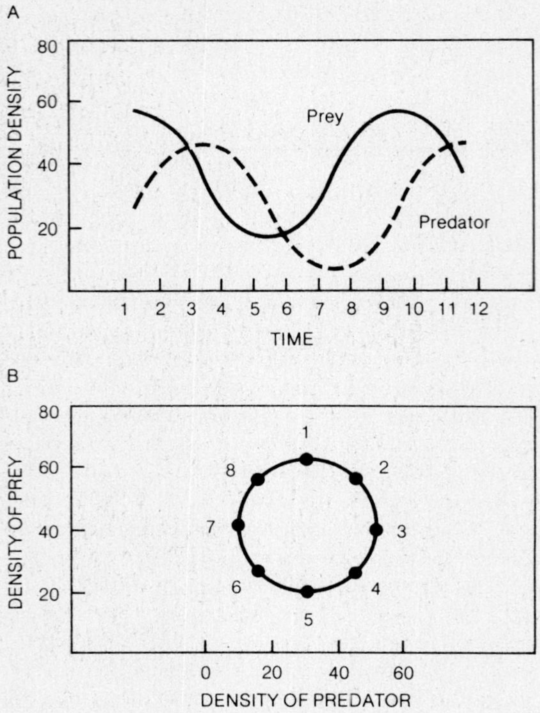


Figure 20.4 The theoretical relationships of predator and prey numbers according to the classical Lotka-Volterra oscillation. As prey numbers fall off, the predators starve and begin to die, relieving pressure on the prey and allowing their numbers to recover. (A) shows the numbers plotted against time, while (B) shows them plotted against each other.

increased pressure on its own food supply, thus tending to reduce its own numbers. Third, at high-population density the prey species would increase the probability of parasites or disease spreading from one individual to another. The net effect of such pressures (predation, competition, parasites, disease) would be to lower the numbers of the prey. As their numbers decreased, their predators might begin to starve or switch to different food sources. At the same time, the level of the prey's food would then begin a comeback in response to lower demands placed on it. Parasites and disease could not be transferred as easily with the prey's numbers low, so, at some point, the prey's numbers would begin to rise again.

It seems, on an intuitive basis, that the stability of animal numbers could most easily be accounted for by a system involving density-dependence. But because of complexities such as the interaction of abiotic and biotic influences, density-dependent control has been notoriously difficult to demonstrate in the field. The difficulty is compounded by the fact that such influences are not always immediately reflected in the population, and the "lag" thus impedes correlations.

Some data on density-dependent regulation of populations are available, however. One type of evidence comes from field studies such as a fifteen-year study of bobwhite quail (*Colinus virginianus*) in Wisconsin (Errington 1945). The summer population was assumed to be ten times that existing in April, based on the number of first-year young found. These results (Figure 20.5) show that the higher the summer population, the higher the loss before November. In other words, the loss was density-dependent (in this case due to changes in death rate, not birth rate). (Lack [1968, p. 117] demonstrated a different sort of density-dependent effect—one that is not apparent until considered on a long-term basis.)

In a study of the feeding ecology of tits (Gibb 1954), it was found that food was especially short in midwinter so that possibly mortality was more strongly related to density-dependent factors at one season than another. High mortality could be expected in winter due to the density-independent effects of weather. But if the severe weather reduced the food supply as well, so that a portion of the population starved as a result of losing the competition for food, then the weather could heighten its effects by exerting density-dependent regulation. Here, we have not only an interaction of regulating factors, but the possibility of a shift in importance from one to the other according to the season.

And then there are spatial effects on population regulation. On the basis of a study of two species of barnacles occupying adjacent intertidal strata along a rocky shore, it was determined that the lower limits of the species occupying the higher zone, *Chthamalus stellatus,* was largely set by the action of biotic factors, including competition for space with another species, *Balanus balanoides,* which occupied the adjacent lower zone. *Balanus* was able to crowd out *Chthamalus* by virtue of its higher population density and faster growth. In some cases *Balanus* more directly influenced *Chthamalus* by covering, smothering, or undercutting it. Whereas *Chthamalus* was limited in the lower part of its range by biotic factors, in the upper reaches it was limited by abiotic factors such as heat and desiccation at low tide (Connell 1961). So, here is a clear example of a species being limited in different parts of its range by entirely different factors.

It must be admitted that density-dependent regulation can fail under certain conditions. Our own species has provided us with one of the clearest examples of such failure in our relentless hunting of the great whales. They have become so reduced in number over the past few decades that, at present, their numbers are perilously small and some species have been placed on the endangered list. The demand for whale products has not diminished, however, especially in Japan, and the whaling fleets are only now cutting back their efforts, partly in response to a general outrage at their behavior, magnified by the efforts of the Greenpeace Foundation, and partly because they can't find enough of certain kinds of whales to make the slaughter profitable (a density-dependent effect).

A clearer example of the premises of density-dependent regulation failing before the human appetite is provided by the fishing industry. In spite of increasing worldwide efforts to catch fish, some of the catches have begun diminishing. There is not likely to be any relaxation in fishing because of the needs of the world's burgeoning human population, and so fishing efforts are actually being increased. Perhaps man provides poor examples of animal behavior because of his unique position, but he sometimes beautifully illustrates certain points by behaving in ways no other animal will.

And lastly, it should be pointed out that density-dependent factors may be viewed as positive or negative. For example, if *death rate* increases with increasing density, this is an example of positive density-dependence. On the other hand, if *birth rate* drops off as the population density increases, we have a case of negative density-dependence (Figure 20.6).

Ecologists, at this point, are strongly divided over the relative importance of density-dependent and density-independent factors in the control of populations. To consider the various arguments, you might refer to Andrewartha and Birch (1954); Lack (1954, 1966); Nicholson (1957b); Orians (1962); Horn (1968b).

### *r*- and *K*-Selection

We have seen that density-dependent and density-independent population control are often associated with very different sorts of ecological conditions. Theoretically, in highly variable or unpredictable environments, catastrophic mass mortality may occur. In some cases, this may have little to do with the qualities of the organisms or with the size of the

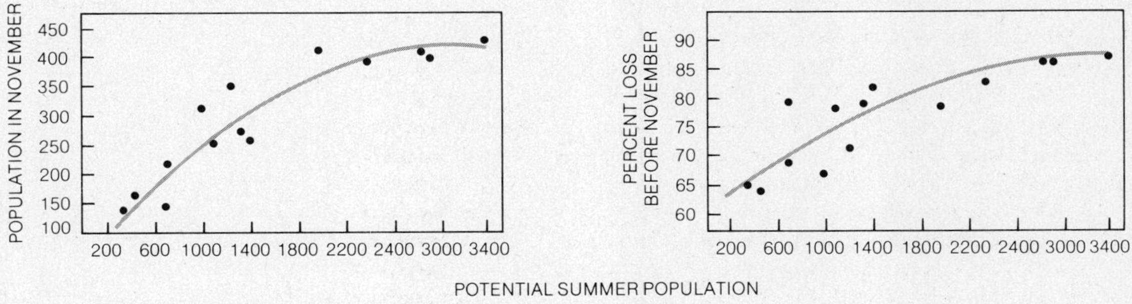


Figure 20.5 The population of the bobwhite quail and its losses by autumn.

Table 20.1 Some of the correlates of *r*- and *K*-selection[a]

| | *r*-SELECTION | *K*-SELECTION |
|---|---|---|
| *Climate* | *Variable and/or unpredictable; uncertain* | *Fairly constant and/or predictable; more certain* |
| *Mortality* | *Often catastrophic, nondirected, density independent* | *More directed, density dependent* |
| *Survivorship* | *Often Type III* | *Usually Types I and II* |
| *Population size* | *Variable in time, nonequilibrium; usually well below carrying capacity of environment; unsaturated communities or portions thereof; ecologic vacuums; recolonization each year* | *Fairly constant in time, equilibrium; at or near carrying capacity of the environment; saturated communities; no recolonization necessary* |
| *Intra- and interspecific competition* | *Variable, often lax* | *Usually keen* |
| *Selection favors* | *1 Rapid development*<br>*2 High maximal rate of increase, $r_{max}$*<br>*3 Early reproduction*<br>*4 Small body size*<br>*5 Single reproduction* | *1 Slower development*<br>*2 Greater competitive ability*<br>*3 Delayed reproduction*<br>*4 Larger body size*<br>*5 Repeated reproductions* |
| *Length of life* | *Short, usually less than 1 year* | *Longer, usually more than one year* |
| *Leads to* | *Productivity* | *Efficiency* |

*Source: Pianka (1970).*

[a]*Type I survivorship shows little mortality until some age and a steep increase thereafter (humans); Type II shows constant death rates over a period of time (some lizards); Type III has very high mortality rates when very young and high survivorship afterwards (fish and insects).*

population of which they are a part. These organisms rarely reach population densities great enough so that they fall into strong competition. On the other hand, populations that live under more stable and predictable environments may reach high densities where individuals fall into strong competition.

In a "competitive vacuum" produced by the first set of conditions, the best reproductive strategy is to produce as many offspring as possible, as soon as possible. In the absence of strong competition, these offspring stand some chance of surviving, even if they aren't particularly good specimens. Their survival is largely a matter of probability—and the probability of any of them reaching adulthood is low. So, rather than carefully produce a few superior young (all of which are *almost* certainly doomed), a reproducer has a better chance if it produces multitudes of young in the "hope" that a few will make it.

On the other hand, a more stable, less risky, environment may harbor species that are blessed with longer lives and a greater probability of reaching adulthood. Competition is likely to be high in such places because so many populations tend to maintain rather stable and high levels. In any such "saturated" environment, density-dependent effects may be pronounced. Therefore, the best reproductive bet is to produce offspring that can compete successfully. Since this may require larger offspring, fewer can be produced. MacArthur and Wilson (1967) call these two different forces *r*-selection and *K*-selection, respectively, after symbols in a logistic equation. I mentioned these earlier with respect to communication.

Now, no individual is going to be completely *r*-selected (but we will point out later that intestinal parasites come close) or *K*-selected (but chimpanzees are disproportionately *K*-selected, as we shall see). Actually, the two values make sense only when comparing one kind of organism with another. So, generally, we think of an *r–K continuum* and an organism's position somewhere along it—in a particular environment and at a given point in time (Pianka 1970, 1972, 1974). The above qualification implies that the organism's position along the continuum may change, depending on the circumstances. We will see some evidence of this in following discussions. Table 20.1 lists some of the correlates of *r*- and *K*-selection.

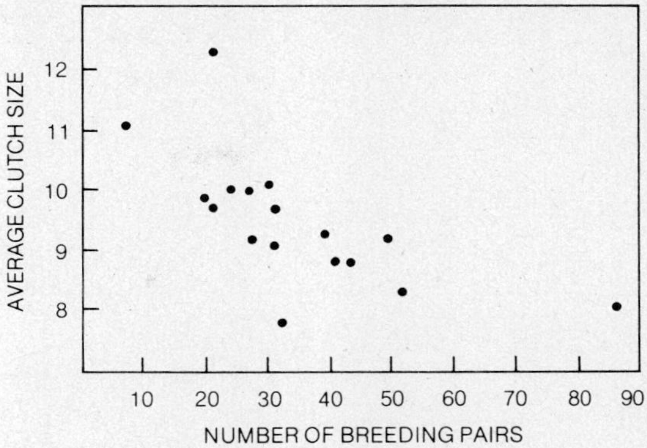


Figure 20.6 An example of negative density-dependence. Average clutch size is plotted against density of breeding pairs in an English woods over a seventeen-year period.

Populations, then, may be regulated by a host of effects. Some of these may operate independently of the density of the organisms and these are often purely physical factors. Others are sensitive to the density of the population and may operate according to a negative feedback system.

# 21

# TERRITORIALITY

The first inklings that animals possess anything like territorial behavior appeared in writings hundreds of years ago. Then the topic was more or less neglected until 1920, when Eliot Howard wrote his book, *Territory in Bird Life.* After watching birds chase each other through the woods, and noting what seemed to be a proprietary interest in certain areas, Howard decided that the behavior functioned: (1) as a way to form and maintain mating pairs, and (2) as a way to regulate population density and decrease competition on the food supply needed for the young. The vigorous and continuing debate over the function, and even the definition, of territoriality is testimony to the depth of the problem (or to our ineptitude in understanding such matters). The ubiquity of territorial behavior among so many diverse animal species indicates a high general adaptiveness of the phenomenon and suggests that there are no easy answers.

Although territorial behavior appears in all sorts of animals, it has mainly been studied in laboratory fish and in wild birds. We will therefore focus on these groups, primarily birds, but for discussions of other species, you might refer to papers on territoriality among insects (Lindauer 1961; Lin 1963; Shields 1967; Bush 1969; Spieth 1966, 1968; Alexander 1961; Wilson 1973); marine invertebrates (Johnson 1959; Dingle and Caldwell 1968; Connell 1963); fish (Gerking 1959; Chapman 1966; Magnuson 1962); frogs (Wiewandt 1969; Emlen 1968; Test 1956); salamanders (Grant 1955; Hutchinson 1959); and diurnal lizards (Milstead 1967; Hunsaker and Burrage 1969). Reviews of mammal territoriality are provided by Ewer (1968) and Eisenberg (1966).

Generally, territoriality is most likely to be found among complex and mobile animals, therefore we can expect it among birds as well as fish and mammals. We can look for it in certain groups among the other animals. Among reptiles it is more likely to be found among the mobile lizards than the more sedentary turtles and snakes. Frogs and toads probably boast the highest incidence of territoriality among amphibians, and among arthropods (the joint-legged animals), crabs are often territorial as are ants, wasps, and butterflies (Brown and Orians 1970).

## WHAT IS A TERRITORY?

We will define a territory in the usual broad sense by calling it any defended area, without attributing to it any special properties that might prejudice our discussion of its effects. (We should keep in mind, however, that its role may be quite different in different species.) Territoriality has also been defined as "space-associated intolerance" (Eibl-Eibesfeldt 1970), and it will become apparent that the intolerance is directed most strongly toward competitors. We are able to talk about territoriality in rather general terms because of its high degree of similarity in vastly different species. The surprising similarity encourages searching across species lines for general underlying adaptive themes. However, the study is, at the same time, made more difficult by the appearance of very different territorial behavior in some very similar animals. For example, the male Norway rat (*Rattus norvegicus*) chases strange rats only on special paths that he has marked. On the other hand, the black rat (*Rattus rattus*) defends the entire area that is crossed by his paths (Telle 1966).

Territories are rather specific areas and do not, as a rule, extend to all the places where an animal may wander. The area in which an animal might be found at any given time is called its *home range*. Home ranges may overlap and animals meeting within home ranges (but outside territories) do not normally fight. The hamster (*Cricetus cricetus*) defends its den and the area around it, but will retreat from other hamsters it may encounter out in the fields; baboons defend their "sleeping trees" and feeding grounds, but they meet peacefully with other troops at watering holes. Interestingly, for some species the territory is not a stable geographically defined area. Some animals shift their territories according to where the food is and others carry their territories with them—"individual distances" that other animals must not violate.

Territoriality may be expressed in a number of ways. In species that form pairs and live apart from other pairs, territories are usually defended by the male against other males, sometimes with the help of the female. In many species, she shows a special intolerance for other females (toward whom the males may appear more amenable). At certain times such as in winter, the sexes of some species may hold separate territories, their belligerence dissipating in spring, enabling them to come together for mating. In some bird species, a number of pairs may join and, in a community effort, defend a mutual territory against members of other groups; in other species, pairs may live close together, each pair defending an area around its nest from other pairs; and other bird species come together during the breeding season to hold small, temporary, individual territories on a communal display ground. So, we see that territoriality includes a wide variety of behavior.

In spite of the diversity of territorial behavior, the selective mechanisms may be basically the same for most kinds of animals. It has been argued that territorial aggression will be employed to defend commodities when they are dependable (available and accessible) and within certain limits (in terms of time and energy) (Brown 1964). So the underlying theme, then, would be one of economics. Natural selection would probably not favor territoriality when key commodities are not readily accessible. For instance, when a food supply of a species cannot be defended bcause of its mobility or its transient nature, no territorial system is evolved to defend it, and any territory that develops in that species is likely to be restricted to the nest site. The general lack of territorial behavior of certain birds such as colonial seabirds, nomadic and social-feeding songbirds, and aerial feeders seems to lend credence to the generalization. In short, defendability of commodities is a requisite for the development of territorial systems. Defendability is defined in terms of the time and energy that would have to be spent in defense and the physical description of the habitat.

Before going into the evidence for the adaptiveness of territories, we should first consider a few general hypotheses that have been proposed to account for the behavior, including those:

(*A*) Associated with reproduction
- (*1*) securing a nest site
- (*2*) aiding in the formation of the pair
- (*3*) reducing the likelihood of disturbances from rivals
- (*4*) permitting females to select the most vigorous males

(*B*) Associated with habitat
- (*1*) permitting increased inbreeding within a population as a means of more finely adapting to the local habitat

(*C*) Associated with food
- (*1*) permitting animals to become familiar with the precise location of food and water
- (*2*) permitting more time and energy to be spent in foraging (or reproduction) once territories are set up
- (*3*) securing adequate food for the pair and young
- (*4*) dividing resources between dominant and subordinate individuals with consequent reduction in fighting and stress

(*D*) Associated with predation, disease, and parasitism
- (*1*) permitting knowledge of precise location of cover
- (*2*) dispersing, which results in a lowered

probability of being detected by predators
(*3*) increasing the distance between animals, thus lowering the likelihood of the transmission of diseases or parasites

(*E*) Associated with limitation of numbers
(*1*) preventing overpopulation and overexploitation of resources through means of natural selection

(See Lack 1954; Hinde 1956b; Tinbergen 1956, 1957; Schoener 1968; Krebs 1971; Klomp 1972.)

Of course, we must remember that just because territoriality originally developed on one basis is no reason to exclude the addition of other advantages, or to ignore the possibility that any adaptive quality may change. We must also keep in mind that the advantage territoriality bestows on one species may not be equally adaptive for all species, so, in effect, territoriality should be perused with an unprejudiced eye for each species. Apparently, territoriality may offer similar advantages for many species, but that similarity cannot be assumed. Also, the appearance of any adaptive activity within a territory does not mean it occurs as a function of territoriality (Klopfer 1969a).

## THE LABILITY OF TERRITORIES

It must be pointed out that territories are rather flexible entities and their shape and size may be influenced by a variety of ecological and evolutionary factors. For example, the physical features of the habitat may determine territorial boundaries through what might be called *landmark influence.* For example, a pair of nesting Puerto Rican woodpeckers I observed claimed an entire wooded area that was surrounded by cultivated fields.

In some cases, as landmarks change, so may territories. A newly planted row of elodea may be accepted by a cichlid fish as its territorial boundary even if it results in a decrease in the size of his previous territory (Eibl-Eibesfeldt 1970). A population of brook trout was almost doubled by adding physical structures to the habitat—structures that apparently provided extra territorial stations (Saunders and Smith 1962). Also, partridges have smaller territories in hillocky areas with many "landmarks" and larger ones in open flat ground (Watson 1964).

Territories are apparently rather easily changed or lost altogether from populations by shifts in selective pressure. As evidence, recall the differences in the territorial behavior of the closely related brown and Norway rats; or consider the scrub jay, which is aggressively territorial, and the closely related Mexican jay, which maintains weakly defended territories (Brown 1963).

*Site influence*, in its simplest sense, refers to the appearance of territorial behavior only at certain places, such as immediately around the nests of some birds. The adaptiveness, here, is obvious: they are defending their nests. But site influence operates in more subtle ways as well. For example, some large wading birds, such as herons and egrets, have both nesting and feeding territories and their belligerence may change depending on which site they are defending. The nests are located near each other, but just out of the effective pecking range of the next pair. Here, then, tolerance is rather high. But when the birds move to their feeding areas along streams, their aggression increases and the space between them widens (Festetics 1959). Site influences are made graphically clear in those species that show that their defensive interests are focused on the site and not on its contents. For example, in some seal species, females are attracted, in part, by the territory a male has aquired. The male does not defend the female or hinder her from leaving, but he zealously guards his territory and she joins him on it. In some gulls, mates, young, and nests can be changed around without causing undue distress to the parent. Their attachment and interest seems, rather, to be directed toward territory that contains the nest. Of course, such behavior would be totally unadaptive were it not for the fact that, under normal conditions, the odds are that the territory will hold commodities, mates, or nests, or aid in acquiring them.

Territory size may be subject to *competitive influence*. Apparently because of differences in competitive pressures, when the population is dense, territories are smaller; in low-density years, territories may be larger. An animal may even decrease the size of a territory it has already staked out because of the appearance of a particularly aggressive newcomer. As further evidence of competitive influence, an animal may increase its territory size upon the death of a neighbor. Competitive influences actually encompass several phenomena and because of its breadth, its conceptual utility is diminished. What it really refers to is the relationship between an animal's territory size and the number of its competitors. The number of competitors at any time, of course, is related to the food supply, present and past, variables that we will discuss shortly.

Time also affects territory size. As we know, species that are territorial at one season may be gregarious at other times. Pairs of Puerto Rican woodpeckers maintain strongly defined territories in spring, but in winter they move in flocks through the forested mountains. *Time influence* is undoubtedly critical in the concept of territory. Any area may be valueless at certain periods, and the appearance of territoriality must coincide with the area's increase in worth. The investigation of this correlation, however, is made more difficult because some species may defend winter territories, not for their immediate value, but to ensure ownership in the breeding season (Lack 1968).

Leyhausen (1956) observed that cats (*Felis domesticus*, sometimes referred to as *Housus cattus*) maintain temporally defined territories. A number of male cats can use the same area, but at different, well-defined hours of the day. In other words, there seems to be a sequential ownership. Thus, the owner at the time is avoided by other cats, and only those areas with old markings are considered safe to enter.

Short-term territories are unusual in the animal world, but we will refer to one more example of an interesting time factor. In Costa Rica, three species of hummingbirds were observed coming to a certain species of tree every morning and moving about freely, showing no territoriality. As the day wore on, however, territorial behavior began to emerge, with each species maintaining a certain dominance-subordination relationship with the others, until about midafternoon, when all species would abandon their territories and leave (Stiles and Wolf 1970).

## TERRITORIAL ADVERTISEMENT

Of course, if territoriality is to function with any degree of efficiency, the defended areas must be marked somehow so that other individuals will be aware of the ownership. As we saw in Chapter 16, territories may be labelled in a number of ways, such as by using vision, sound, and chemicals. Let's review a few ways communication is used directly in territory advertisement.

Visual advertisement may utilize posturing, movement, and, in flyers, aerial displays. This type of display may include an element of threat since, if the territory holder and intruder can see each other, the very sight of the stranger may have evoked the territorial display.

We are aware that territories can be advertised by sound because we are familiar with the territorial songs of birds. Other territorial sounds include the growl of the toadfish, the roar of the bull alligator, and the chirp of the male cricket (Moulton 1956; McIlhenney 1959; Alexander 1957; Haskell 1957). Some animals make their territorial sounds without ever seeing a competitor. Many male birds, for example, sing without sighting another male. The reason could be because they are likely to be hidden deep in their habitat and cannot rely on being easily detected by other animals; nor can they always see an intruder. So they sing just in case. Birds are also mobile animals, and a transient male might inadvertently violate a territory and initiate an encounter that neither bird wants unless the territory is continually advertised.

Several species—mostly mammals, with their well-developed sense of smell—mark territories with special scents (Burkhardt, Schleidt, and Altner 1967). The badger, marten, and mongoose deposit secretions from glands at the base of their tails; some ungulates utilize hoof glands, and some rabbits have both chin and anal glands. The giant galago, a primate, urinates on the palms of its hands, rubs it into the soles of its feet, and walks about on its territory. The brown bear urinates near its boundary, rolls in the urine, and then rubs against trees. The little Madagascan tanrec hedgehog sits on the place it wants to mark, wets the place, scratches itself on the rump, and rubs its scent on the wet place (which may be on its chagrined keeper). Territorial dogs may register their size by urinating as high as they can on some object. If a strange dog can't urinate as high as the previous dog, it has been said that he may leave the area. (If only man could solve his disputes with even *this* degree of elegance.)

Some mammals may even mark their mates as their property. The male tree porcupine (Erithizontidae) rushes toward his female on his hind legs and unceremoniously drenches her in short spurts of urine. Short-headed flying phalangers (*Petaurus breviceps*) mark their territories with urine, but graciously use special glands to mark their mates. The sexes of phalanger clans or tribes may mark each other and thereby develop an odor-spectrum specific to the group—one in which the odor of strong males predominates.

Scent-marking may act also as a means of making the animal feel at home, reassuring the owner, and thus imparting a certain home-related confidence. A frightened badger in a strange environment can be calmed by letting it sniff a previously marked object (Eibl-Eibesfeldt 1970). The smell of a strange mark elicits hostile behavior in males of several species of mammals and probably has a repelling effect on intruders.

It should be pointed out that, in spite of the above examples, most mammals are nonterritorial. It makes sense. Mammals are endothermic (which means they normally require a lot of energy in the form of food) and many of them are large. So they tend to exploit large home ranges. The larger the area, the more costly it is, in terms of time and energy, to defend. So, territoriality just isn't worth it for most mammals. As I said earlier, it should be expected mainly in those species exploiting concentrated, easily defensible food sources. It has been argued that territoriality has evolved in red squirrels (*Tamiasciurus*), but not in other squirrels, because the food of red squirrels (cones of conifers) is produced in great abundance over a short time and is therefore easily stored and defended (C. C. Smith 1968). Among primates, the fruit-eating arboreal species exploit a temporally and spatially unpredictable food supply in an area with poor visibility (forests) and, therefore, they are nonterritorial. But the leaf and bud-eating monkeys (for example, *Alouatta* in the New World, *Colobus* in Africa, and *Presbytis* in Asia) exploit a more abundant (though not particularly localized) and predictable food supply and they are strongly territorial (Marler 1969a; Carpenter 1958).

Territorial signals may be closely associated with mating behavior. In some birds, courtship is often interspersed with attacks as the male oversteps the fine line separating the two patterns. Males that overstep too often, of course, drive the female away and so do not contribute to the next generation. It should be made clear, though, that territorial advertisement and mating are not always served by the same signaling mechanisms. In some cases, the signals *must* be different. For example, closely related species living in the same area may utilize *common* territorial signals since they are strong competitors, but *different* mating signals since hybridizing is usually unadaptive. The red-bellied woodpecker (*Centurus carolinus*) and the golden-fronted woodpecker (*C. aurifrons*) are very similar in appearance (Figure 21.1) and vocalization, and where they come together along narrow bands in the Southwest, they are able to recognize each other's territorial signals and so they maintain mutually exclusive territories, but they use different signals in identifying potential mates, so they do not interbreed. Thus, two types of signals are obviously separate and distinguishable.

## THE EFFECTS OF TERRITORY ON NUMBERS

Now that we have some idea of the nature of territoriality, let's consider its role in the regulation of population numbers.

Territoriality, as mentioned, obviously has a high adaptiveness because it occurs so widely and because it persists in populations in spite of some of its serious drawbacks. And, in fact, the drawbacks are enormous. It probably results in higher risk of predation for the territory holder because he must generally advertise himself and his territory to competitors. Territories make demands on time that could be used in other pursuits such as courting, grooming, and resting. Also, occasionally a territorial animal must fight to preserve its territory, and fighting is dangerous. In sum, the cost of territoriality places the individual at certain disadvantages and so it can be expected to lower the population numbers to some degree. But its main role as a population regulator likely lies in its spacing effects, as we shall see.

The argument over the role of territoriality in population density, by the way, is far from concluded. Most such studies have, thus far, been conducted on bird populations. For reports that purport to show such a role among birds, see Jenkins, Watson, and

Figure 21.1 The red-bellied (left) and golden-fronted (right) woodpeckers overlap in a narrow zone in Texas. Their similarities result in interspecific territoriality, but they do not interbreed, apparently because of differences in mating signals.

Miller (1963) on grouse and ptarmigan; Delius (1965) on skylarks; Tompa (1964) on song sparrows; Patterson (1965); Coulson (1968) on colonial seabirds; see Klomp (1972) for other references. Lack (1954, 1966, 1970) seemed to be increasingly willing to accept the role of territory in population regulation; on the other hand, Brown (1969) critically assessed the available data and concluded that there is no justification for some of the sweeping generalizations that have been made regarding the population-regulating effect of territoriality. Watson and Moss (1970) argued that territoriality is important in population limitation, but they stressed that the evidence is, so far, circumstantial, indirect, and inconclusive. Krebs (1970, 1971) compared the effects of territoriality with other factors operating on populations of the great tit and concluded that territoriality has, at most, only a weak density-dependent effect that operates only a little more strongly in high-density years than in low-density years.

Obviously, then, the issues are complex. So, let's take it one step at a time. One of the first questions to be asked is whether territorial behavior acts to *disperse* populations, thereby lowering population density. It is true that once territories have been established, there is usually no room for additional animals. After a certain number of territories were established by sticklebacks in a tank, no other males experimentally added could obtain territories (Van den Assem 1967). At high densities, some male trout and salmon were unable to obtain territories and were constantly chased by territory holders (Kalleberg 1958).

Studies of several other species (Crook 1970b) indicate that territoriality may cause nonbreeders (those unable to get a territory) to move to less optimal areas. The result of such movement is to lower population density as individuals move out of the better areas and, concomitantly, to lower the population size, the absolute numbers. The latter may occur for two reasons. First, animals may not be able to breed as successfully in the poorer areas. Second, the suboptimal areas into which animals are chased by territory holders can be expected to hold fewer commodities such as food and nest sites than the areas they left. Thus, less successful individuals may find less food, or suffer higher predation because of a lack of hiding places. It has been shown that, for a number of species, populations may be regulated by dispersal or by dispersal and mortality combined (Watson and Moss 1970).

## Floaters

In some species, the unsuccessful bidders for territory do not move into suboptimal areas, but remain in the area as "floaters," to surreptitiously compete with the landowners. They are also in a position to be able to move into a territory should a vacancy occur. Their presence has been cited as evidence that territoriality limits population by keeping some individuals from breeding. While this is seemingly the case, in a species that has floaters territoriality might have a reduced effect on population regulation, since the optimal area is shared by the owner and the furtive floater. In these cases, unsuccessful members are not banished to inferior habitats. On the other hand, floaters might help to regulate populations by causing the territory owners to find less food in their own territories due to the clandestine competition with these nonbreeders. The resulting food reduction might cause the territorial owners to attempt fewer offspring.

But how good is the evidence that floaters exert a density-regulating effect in the places where they live? Actually, not very. It has been shown (Stewart and Aldrich 1951; Hensley and Cope 1951) that in several species of birds, including tits (Krebs 1971), when territory holders were shot, new males (presumably floaters) immediately moved in, took over the territories, and began to breed. They were previously unable to breed, although food in the area was abundant (Carl 1971).

## Problems of Synthesis

So we see three possible ways territoriality may limit populations. Populations may be limited first, by high death rates because some individuals are forced into a suboptimal habitat; second, because a certain part of the population is prohibited from breeding; and, third, because of a reduced fecundity of territory holders due to time and energy spent in territorial pursuits or due to competition with floaters. But the picture is not yet entirely clear because there is little hard evidence. Much of the earlier data purporting to show such regulation have been imprecise and misinterpreted. For example, in the experiments in which floaters moved in when territory holders were shot, females tended not to be replaced (Brown 1969). The implication is that all of the females that could breed did breed. This was found to be the case in a study of blue grouse (Bendell and Elliott 1967) in which it was discovered that younger males were excluded from breeding by older dominant males, but that all the females bred. The situation may be different in other species, however. For example, in another study, five males and two females were shot from a territorial population of twelve males and seven females of ptarmigan (*Lagopus mutus*) in Scotland, in the spring (March 15). Two weeks later, the area was occupied by thirteen males and seven females, all territorial.

Fretwell and Lucas (1969) came up with a novel idea regarding the role of territory in the limitation of bird numbers. They constructed a model showing how birds could effectively distribute themselves over the available habitat. They suggested that, in heterogeneous habitats where some areas are better than others, if no bird is expelled from any habitat, there is an "ideal free distribution." If, on the other hand, settlement is obstructed by territorial behavior of the residents, at least in the most suitable habitat, then there is an "ideal dominance distribution."

In the first case, the birds start settling in the best habitat (that is, where they can produce the most young). As the number of birds there increases, the habitat becomes less ideal, until it is not even as good as what was second best. So, new settlers now shift to this previously neglected area, until its rising population density decreases its value and causes new birds to move to the third best area, and so on. Eventually, all the birds will have arranged themselves so as to equalize the *average* success rates of residents in the different habitats. It seems that territoriality in such a system would not limit population numbers.

In the case of the ideal dominance distribution, territorial birds exclude settlers from the best habitat, but are forced to accept a number of settlers at the expense of the compression of their own territories. But some excluded birds, rather than spending time and energy in fighting (when they would probably lose anyway, since established birds are rarely defeated on their own grounds), simply move into a less suitable habitat. But, there, they have lower reproductive success than birds on better territories. Here, territoriality would obviously limit the size of the population the following year. It would therefore be to a bird's advantage to try to squeeze into the optimal habitat. If field studies show that birds living in habitats that harbor high densities (preferred areas) have consistently higher reproductive success rates, this would indicate that territorial behavior functions in limiting population density.

There has been some discussion of the genetic basis of population regulation (Chitty 1967), but there is, as yet, little real data to support the idea. It has been suggested that population declines are caused by genetic selection for aggressive individuals that are at an advantage in crowded populations, but are generally less viable in other ways. At low densities, selection for less aggressive, but more fecund, individuals takes over, and the population rises.

**The Buffer Effect** Three levels of territoriality in birds have been described based on their relative population densities (Brown 1969). The idea is that territory sizes at low densities are not limited by competition, and no birds are prevented from nesting in good habitats. At middle densities, some birds are excluded from preferred habitats, but may utilize poorer habitats. And it is here that we may see a *buffer effect.* Kluyver and Tinbergen (1953) first described the effect in titmice that inhabited mixed woods where food is abundant. The most successful birds held territories of rather stable sizes year after year and drove other titmice into adjacent pine woods where food was less abundant. In contrast to the territory stability in the mixed woods, in the pine woods the territory sizes varied widely from year to year as a function of the population numbers. When population numbers were high, territories in the pine woods were necessarily small. As more birds were crowded in, an excess population was absorbed by the "piney sponge." While the mixed woods were buffered, then, pines were considered unbuffered. It must be pointed out that the buffer principle has not yet been widely demonstrated, but models have been divised to show how it could work (Brown 1969).

It is apparently only at high densities that some birds are prevented from breeding anywhere. They often set up a breeding surplus or reserve that exists in and around territories or on separate ground. Only under such conditions is the elimination of territorial birds followed by the immediate reoccupancy of the territory, as was described earlier. Studies of species with high population densities, then, support the idea that territoriality limits the size of the breeding population, since it is apparently under these conditions that we find nonbreeding prospective territory holders lurking in the vicinity. Or so goes the theory.

## Territories and Social Structure

The degree to which the breeding failure of some males affects population numbers depends on the social structure of the species. For example, among the great reed warblers (*Acrocephalus arundinaceus*), males driven out of optimal areas may set up territories in suboptimal regions where it is impossible for any females they might attract to build nests. Rather than attempt to breed with these unsuccessful males, then, the females return to the optimal areas and mate with an already mated male who has a suitable territory (Kluyver 1951). So we see that, in this case, the key to population regulation lies in how many *females* are excluded from breeding. Females cannot always choose an already mated male, however. In many species, established females eject newcomers (see Bustard 1970 for a description of an "intermediate" condition).

The individuals, which for some reason are excluded from breeding, can be considered to comprise a "surplus" population, or a breeding reserve. These individuals, usually males, may be able to take their place in the breeding population at some later time and their addition to the reproductive community is termed *recruitment.* Recruitment, then, is simply the addition of new breeding adults into the population. Recruitment is partly a function of the number of young that survive to the breeding season, coupled with social effects such as hierarchy or territoriality, which may determine those that will be allowed to breed.

In some species, the breeding surplus is hard to find. We have mentioned the rapid replacement of songbirds in their territories by males that were not apparent until the original owners had been shot. In other species, the surplus is more easily defined. For example, when the nesting ground of some colonial seabirds is limited in size, those of lower rank may be conspicuously relegated to nearby areas where they are not able to breed.

In some species of seals, such as elephant seals and those seals slaughtered in the Pribilof Islands of Alaska each year for fashion's sake, nonterritorial males group together in bachelor herds. These are usually young males that have not yet engaged in much territorial fighting and are therefore relatively unscarred until they die beneath the club. A young male that survives the yearly kill may be able to join the breeding population by displacing an aging bull through combat (Figure 21.2). The old bull, himself, may join the surplus population where he is likely to remain until his death.

Let's consider another kind of social implication of surplus populations. Gelada baboons (*Therophithecus gelada*) sleep on the bare cliffs of Ethiopian canyons. During the day, in the bountiful rainy season, they move slowly in great herds over the rich land. But when the harsh dry season approaches, they break up into harems (one-male reproductive groups) or all-male troops. The all-male troops tend to wander away from the canyons and, in so doing, they apparently reduce competition for food with the harems. Since it is likely that the canyons provide the best source of food and the safest resting places, the all-male troops suffer higher mortality during this period than the reproducing groups. John H. Crook (1970b) believed such a system would permit a reproductively optimal allocation of resources by channeling food to reproductive females. If all the male groups stayed in the cliffs, they would probably be excluded from many of the commodities by the dominant breeding group and would suffer higher mortality than they would in the less optimal areas where, at least, any existing food is available to them. (See also Jarman 1968.)

A large breeding surplus increases the likelihood that any individual will end up as an "extra," failing to breed. Thus, the long-term reproductive success of breeders may be reduced, since their offspring may be relegated to the surplus. Thus, certain adaptations to a persistent breeding surplus can be expected. For example, delayed maturation may be an adaptation, if not to the surplus itself, then to the conditions that brought it about. Selander (1965) explained that a large breeding surplus in birds decreases the probability that an inexperienced first-year male can successfully breed; his chances are so low that it is advantageous for him to not make the attempt until a later time. An older male, conversely, is likely to be more successful due to his greater strength and size and through the increased benefits of learning. Of course, adapting to a surplus through delayed matu-

Figure 21.2 Male elephant seals fight for mating rights. Breeding is limited almost exclusively for those who hold territories in the breeding area, and competition for this real estate is intense.

ration is more likely to be found in species having a high rate of survival (Brown 1969). In those with a high mortality rate, the emphasis could be expected to be toward an all-out attempt to breed as soon as possible.

It has been suggested (Brown 1969) that surplus birds have been found under only five conditions: (1) in polygamous species and species with promiscuous males; (2) in monogamous species with an unbalanced sex ratio of adults; (3) in species with shortages of nest sites; (4) in species living in areas where man has artificially raised the maximum population density; and (5) in monogamous species living in natural habitats relatively free of human interference. Surplus birds can easily be found in the first four categories, but we need clear data from the fifth if we are to know whether territoriality, in fact, limits population density.

### Inheritance of Territories in Birds

In some resident species of birds, the first-year males may remain in their parents' territories and help in nest building, defense, and care of young (Skutch 1961). The system could have arisen as follows: as the density of competitors increases, the likelihood of a young bird being able to acquire a territory decreases. At the same time, a territorial male must spend an inordinate amount of time and energy in maintaining the territory in the face of so many competitors. Offspring, which are subordinate to their parents and which would normally be driven away when they matured, might be allowed to stay under these conditions and participate in territorial defense and all family affairs—except breeding (Brown 1969). Both the parents and the helpers benefit. The offspring of territorial birds, or nonrelated subordinates willing to play the role, might, by remaining, increase their chances of taking over the territory when the dominant bird dies (Rowley 1965). In a study of the superb blue wren (*Malurus cyaneus*), it was found that groups with "helper" males produced 1.9 fledglings annually as opposed to 1.2 for those without helpers.

## TERRITORY AND FOOD

If territory is important in regulating population density, then we may ask upon what basis the resulting dispersal is adaptive. Intuitively, we can see the advantage of expelling other animals that are after the same commodities. So, let's consider the most obvious commodity: food.

It has been shown that some types of territories include a food supply at least during part of the year (Holmes 1970). Many observers believe that territorial behavior appeared through evolution primarily in response to a need to defend food supply, but the evidence, at present, is not conclusive by any means (Brown 1964; Murray 1971); much of it is deductive, negative, or full of exceptions. Even the best evidence has serious shortcomings. Lack (1966) noted the rarity of species that actually feed within their territories. He also mentioned that of those species that do feed within their territories, many do not protect the food there. The European robin, for example, prevents other robins from singing and breeding in its territory, but does not interfere with their feeding there. Klopfer (1969a) and others have conversely pointed out that some bird species allow intruders to feed within their territories but not to display there (Jenkins 1963; Brown 1964; Keenleyside and Yamamoto 1962). What is needed is a careful comparison of foraging success in animals with territories and those without. The apparent simplicity of such a study is deceptive. However, again, in the absence of any conclusive information we forge ahead.

### The Evolution of Territories

It's great fun to hypothesize, secure in the knowledge that no concrete contradictory evidence is ever likely to turn up. So let's indulge ourselves in speculation as to how territorial behavior may have developed, with but one premise—one that places even fewer strictures on our theorizing. If we acknowledge that any present adaptiveness of territorial behavior does not necessarily reflect the original advantage of the pattern, then we can ignore the myriad variations in territorial behavior that we find about us. We can begin by noting that there are indications that territories may hold food and that successfully territorial males are often "preferred" by females.

We might first ask how territory holders came to be preferred as mates by females unless the territories held food for the offspring or unless the ability to acquire a territory indicated the general vigor of the male. If the territory were only considered a yardstick of male vigor, then we might ask how this particular measure evolved. Why did males begin to defend territories in the first place? They may have begun by fighting over food, the largest and strongest males winning. Thus, size and strength would be selected for. Males, of course, might be evolutionarily freer than females to increase in size due to their not being required to meet the physiological demands of gestation. Smaller females would be unable to compete with them in fighting, so territories would generally be held by larger or more aggressive males. Successfully fighting for food has much more immediate advantage than fighting for other commodities, such as nest sites, and could, therefore, through its strong reinforcing effect, come to be established in the behavioral repertoire rather quickly. (Fighting for females might be less effective in general because of its seasonality, or periodic advantage, and fighting for space per se is unimaginable.) Once males learned to associate an *area* with the food it contained, a general site-related belligerence might be expected. Females, smaller than males, could conceivably become dependent to a degree on the food that the male had secured. The male might initially allow females into his defended area for purposes of copulation or mating. But then he would also increase his own reproductive output by allowing his reproductive partner to avail herself of his food. Aggressiveness in males could then be selected for through the generations until they even become intolerant of other males anywhere in their proximity, at least at certain periods, such as when breeding or in times of low food supply. Once the system became well established, patterns could become ritualized or altered as competitive conditions changed (Figure 21.3). The limited value of so general a hypothesis is apparent, but in any such widespread and diverse phenomenon as territoriality, its development must be explained in as broad terms as possible.

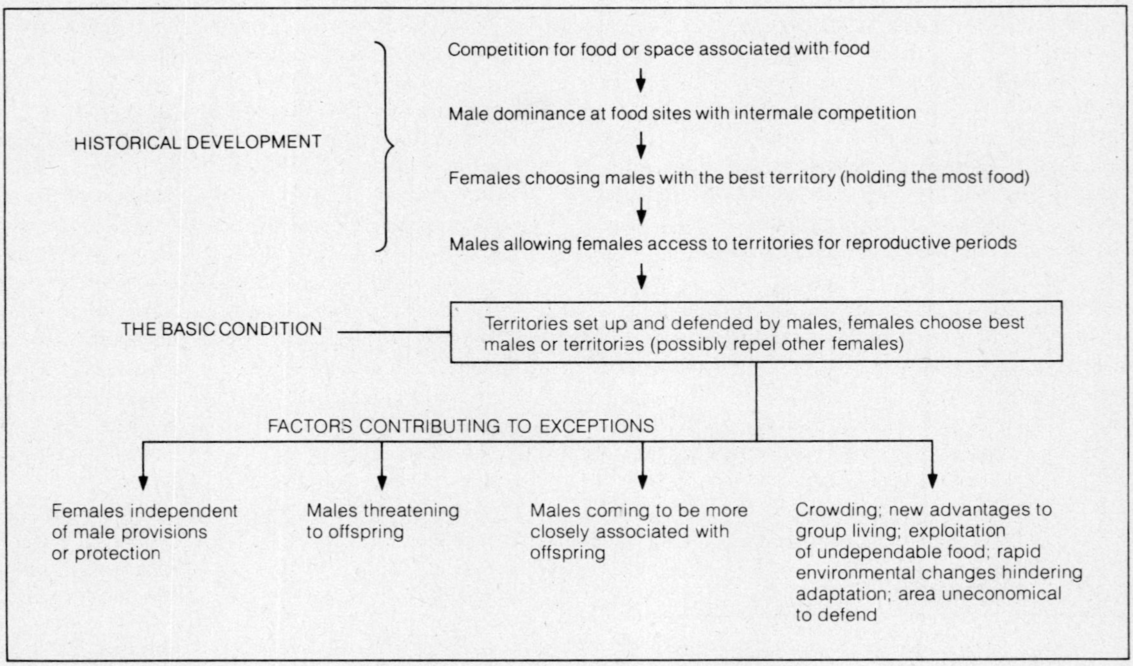


Figure 21.3 Theoretical evolutionary factors leading to territoriality and some causes of its variation.

## Territory and Food Allocation

Ecologists continue to argue over whether spacing behavior (such as territoriality) limits breeding populations in vertebrates, or whether breeding populations are limited some other way (such as through starvation) *before* the animals arrange themselves spatially.

There are indications that territories are adaptive in ensuring a food supply. The evidence, if circumstantial, is compelling. For example, generally, both territorial and nonterritorial species have denser populations and smaller feeding areas where the food supply is better (Brown 1964; Kluyver 1951; Stenger 1958). But the evidence that territory size is directly influenced by food abundance is weak and generally based on the inverse relationship that usually appears between them (Schoener 1968; Holmes 1970).

Also, we must remember that the factors that influence territorial size are not necessarily those that compel the animal to take a territory in the first place. Undoubtedly, the establishment of territory is generally an innate pattern, subject to internal and external influences. On the other hand, territorial *size,* as indicated by its very flexibility, may be generally much more subject to environmental influences. Of course, there are limits to how much territorial size may vary, as we have seen. And, as I pointed out, some species show a tendency toward maintaining their usual territory size no matter what the environmental conditions, while other species simply hold territories of a minimum size. However, if territories are related to food stores, territory size should generally reflect food abundance. What evidence is there of such relationships and how could it work?

Food-related physiological conditions could influence the general level of aggressiveness in animals and thus help to determine the limits of their territory size. It has been suggested that the animal might adjust its territory by estimating the available food, either visually or through assessment after a brief period of foraging (Watson and Moss 1970). If such assessment indeed occurs, the investigation of the neural chain of events would be particularly interesting. It is unlikely that the animal thinks, "Since I see a lot of food and have had no difficulty getting it, I will take only a small territory and save myself a lot of time and trouble." It is more likely that good nutrition tends to physiologically reduce aggression, or that an animal tends to remain in an area in which he has found abundant food, unaware that the level may be high elsewhere as well, and that this reduced roaming results in a smaller territory.

It has been found that very large territories occur only in certain kinds of species. Probably the largest territories are held by animals that utilize them only for foraging. A solitary carnivore, such as a lynx or a hawk, has very little chance of finding a prey animal within any given area of land that it happens to be surveying and, too, the presence of other individuals would disturb it and reduce its success; it therefore needs a large territory. Large territories may also be advantageous for other reasons. European wrens feed inconspicuously and alone in large territories that apparently permit them to avoid attracting predators.

In general, predators hold larger territories than herbivores or omnivores, presumably because of their more sparsely distributed food. Also, large predators hold disproportionately larger territories than smaller predators, probably because there is less food available per unit area for large predators than for small ones (Schoener 1968). Even among herbivores, home ranges are larger for those that must search for food, compared to those who stand knee-deep in it, and larger herbivores need larger home ranges (Figure 21.4).

If the function of territory is to secure at least minimal food resources, then one might expect fewer animals of any single adaptive type (that is, competitors) where the habitat is less productive. A few cases illustrate that the principle is valid in at least some instances. For example, plant productivity in desert habitats (Figure 21.5) is less than 0.5 grams of dry organic matter per square meter per day (Odum 1971). It improves in prairie, shrub, and mountain forest habitats, varying from 0.5 to 3.0 grams, but moist lowland forests (Figure 21.6) produced from 3 to 10 grams. Interestingly enough, as primary plant productivity increased from desert to grassland, pine forest and deciduous forest, so did the number of birds utilizing that habitat (Table 21.1).

Further evidence that territory size may be related to food abundance is indicated by the changes in how a territory is utilized as different demands are made

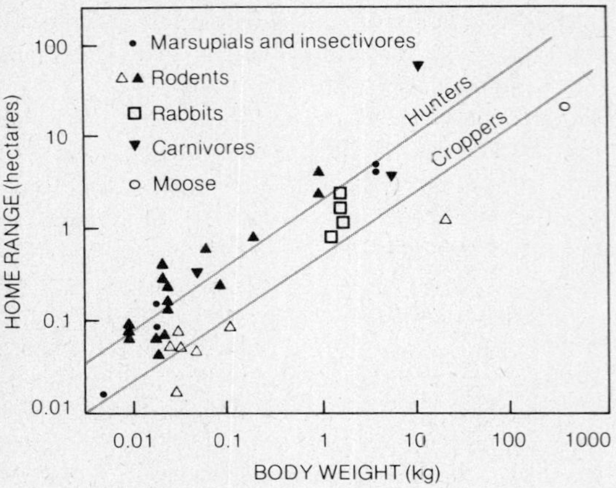


Figure 21.4 Relationship of body weight to home range size in mammals. Note that home ranges of hunters (solid symbols) are about four times larger than the home ranges of croppers (open symbols). Hunters, here, are herbivores that search for fruits and seeds. Croppers graze and browse on leafy vegetation.

on it. In the ovenbird (*Seiurus aurocapillus*), the portion of the territory actually used by the adults is largest when the young are being fed and food is being gathered at the greatest rate (Figure 21.7).

Territory can also be related to food in another way. If food in the habitat is unpredictable, either spatially or temporally, there may be no territoriality. For example, African weaverbirds live in a wide range of habitats including grassland, savannah, and deep forest (Crook 1964). Most of the forest species feed on a predictable and well-dispersed supply of insects and they nest in solitary pairs on large, well-defended territories. On the other hand, most of the savannah and grassland species are seed eaters. They must rely on the irregular appearance of seeds in certain localized areas. None of the seed eaters defend feeding territories. In fact, the savannah birds nest in huge colonies in the few trees that are scattered about their habitat, defending only small areas around their nests. Grassland birds exploit transient and localized food supplies, so they do not defend feeding areas, but "neighborhoods" of birds do defend small nesting areas.

## Territory Size and Parental Food

There is some interesting evidence that territory size for any generation may be partly dependent upon the food that was available to its parents. For example, rats fed on protein-poor diets produced offspring with a reduced exploratory drive (Cowley 1970). It had been demonstrated earlier that exploratory drive was associated with aggression and spacing behavior in some rodents (Calhoun 1963). So, during perennial periods of low food abundance, the tendency in these species might be toward smaller territories. If such is the case, it seems that *preservation* of food might not always account for the territorial behavior of these animals. They are, of course, social species, so possibly in times of low food abundance, intragroup fighting, with its time and energy expenses, might be reduced.

It is well known that red grouse (*Lagopus scoticus*) population density can vary widely. Territorial behavior limited population density in red grouse breeding grounds in Scotland, but it was found that breeding density varied widely (from fourteen to thirty-four birds per hectare) in different years (Jenkins 1963). Apparently, the territorial behavior was being influenced by some environmental factor, such as food abundance. But the environment could have been exerting its effect on any population by acting on the parents. The quality of food eaten by the wild red grouse affects the aggression and, therefore, the territory size of its offspring. It seems that with good food, the eggs will be more likely to produce healthy cocks that, when grown, will take smaller territories (Watson and Moss 1970). The advantage of smaller territory sizes when food is abundant should be

Figure 21.5 Deserts are poor producers of plant material and can sustain only a limited animal population (opposite page, top).

Figure 21.6 Deciduous forests are good producers of plant material and can therefore harbor large and complex animal systems (opposite page, bottom).

Table 21.1 Relationship of bird density and plant productivity

| HABITAT | LOCALITY | TOTAL BREEDING PAIRS PER 100 ACRES | PAIRS OF 3 MOST COMMON SPECIES PER 100 ACRES |
|---|---|---|---|
| *Desert* | *Mexico* | *22* | *4, 3, 3* |
| *Grassland* | *Saskatchewan* | *92* | *28, 18, 11* |
| *Ponderosa pine forest* | *Colorado* | *290* | *45, 35, 35* |
| *Deciduous flood-plain forest* | *Maryland* | *581* | *83, 69, 59* |

*Source: From Audubon Field Notes, after Ricklefs (1973).*

obvious. If smaller territories harbor adequate supplies, the territory holder is relieved of the risks of trying to maintain a large territory with its longer borders to defend.

The time and energy spent in territorial defense during the nesting season can have variable effects. For example, it may reduce the attention that can be shown to the young through what is termed "aggressive neglect." It has been suggested that interspecific aggression may cause the dominant species to reduce its brood size as an adaptation to the reduction in time available for nesting duties (Ripley 1961).

## Conceptual Problems

Of course, the biological world can always be counted on to provide us with exceptions, and so there are cases where the determinants of territory size are not known. Certain species of mustelid weasels apparently hold a stable number of territories over their range despite fivefold increases in the numbers of their main food, a rodent (Lockie 1966). It is suggested that they and a few other species, such as tawny owls and golden eagles, hold territories as large as they can so that food is always available except under the most extreme conditions. It may be that the food supply of certain predatory species that seek elusive prey is difficult to assess before territorial size is established, or that certain prey species fluctuate and therefore cannot be predicted by the predator. In such cases, one would expect selection for territories as large as possible. Stability of territory size, then, is not necessarily an indication that the size was not ultimately related to food supply.

One of the most serious weaknesses of ecological studies is clearly demonstrated in the preceding discussions. We are still firmly at the "second-guess" or post-hoc stage—hypothesizing after the fact; our predictive abilities are almost negligible. If rats have small territories when food is scarce, we say, "This is to conserve energy that would have been wasted in squabbling." If rats, on the other hand, had shown larger territories when food is low, we would probably say that it was because larger territories would ensure a large food supply. Put simply, we are simply not at the point where we can synthesize unitary rules or make predictions with any real success. Our predictive abilities would probably be greatly increased if we had more accurate means of measuring environmental variables, but we are really not very good at such things at present. It is extremely difficult to assess food abundance for any species, and even then our findings are likely to be applicable in very limited ways. This, then, is the reason that any discussion of the role of territory must involve convoluted discussions full of exceptions.

## Interspecific Territoriality

One of the arguments against the food-value theory of territory is that the areas are not usually defended against other species with similar food requirements (Hinde 1956b). However, there are many examples of interspecific territoriality and these may lend clues in the search for the adaptiveness of territoriality.

In some cases, species may be so similar in morphology and behavior that where they occur together, they may behave shabbily toward each other simply as a result of mistaken identity. As was mentioned earlier, the red-bellied woodpecker (*Centurus carolinus*) and the golden-fronted woodpecker (*C. aurifrons*) are so similar that they are difficult to distinguish where their ranges overlap, at least for the casual observer (see Figure 21.1). These species exclude each other from their territories, but they are much more tolerant toward the distinctly different red-headed woodpecker (*Melanerpes erythrocephalus*), except perhaps around nest holes.

It should be pointed out that interspecific territoriality is often much less vitriolic than intraspecific territoriality. Animals are generally more tolerent of other species. For example, fiddler crabs of the genus *Uca* live along the shores of the Indo-Pacific and America. A number of species may live on the same beach, and while any male tends to chase any other male, males of different species build burrows closer together than those of the same species.

It may be difficult to separate "tolerance" from the effects of an imbalanced time-and-energy-expense factor. Red-winged blackbirds (*Agelaius phoeniceus*) exclude tricolored blackbirds (*A. tricolor*), which appear on their territories in late spring, attempting to set up breeding colonies in the red-winged's marshy domain. At first the red-wingeds are highly aggressive and chase away all intruders, but their task becomes insurmountable as more and more tricoloreds appear on the scene. As one tricolored is being chased away, another moves in from behind. Finally, the red-wingeds are forced to give up trying to hold territories if they stay in the area. So much time and energy must be spent in trying to repel intruders that it is simply no longer economical to attack them (Orians and Collier 1963). This is not to say the red-wingeds become less aggressive, but the effects are the same.

Orians and Willson (1964) considered what sorts of factors contribute to interspecific territoriality in birds. They suggested that such studies can yield more precise information on the *ecological* ramifications of territoriality than intraspecific studies, since interspecific interactions are less likely to have other complicating (for example, reproductive) implicatons (Johnson 1963). They pointed out that when two species that have previously been isolated come together, several outcomes are possible:

(*1*) They may not be reproductively isolated, so they may hybridize and "swamp" their differences.
(*2*) One species may outcompete and thus eliminate the other.
(*3*) They may get along together with no problems.
(*4*) They may occupy different parts of the habitat because of differences in habitat selection.
(*5*) They may become interspecifically territorial, excluding each other from parts of the area.

The latter is most likely to be expected when two very similar, perhaps related, species first contact each other. In time, however, it might be expected that the energetically expensive territorial behavior would be replaced by some less demanding system,

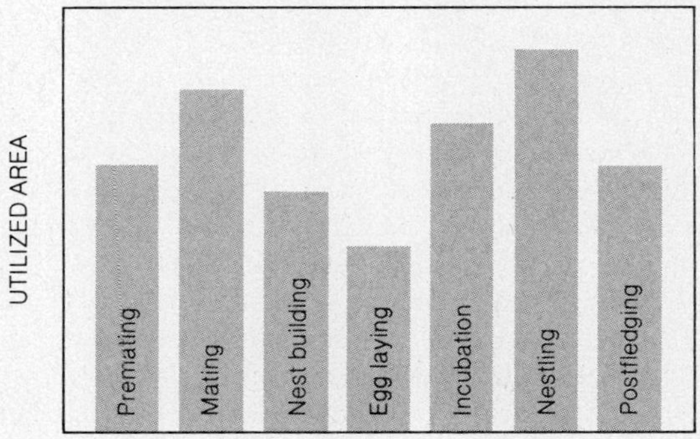


Figure 21.7 Territory utilization by ovenbirds at different stages of their nesting cycle. The utilization is expressed as a fraction of the total area (T) used by the pair during the entire nesting cycle.

such as habitat partitioning. If the interspecific territoriality continues, it would probably do so under only certain conditions. It would occur if there were a number of other competing species in the area—species that would not allow the two species to separate ecologically by competing with them. In other words, species *A*, which is changing its habits to avoid the competition with species *B*, might then begin to fall into stronger competiton with species *C* or *D*. Thus, species *A* is constrained from making that kind of shift. It cannot, then, shift its specialty, and it cannot become more of a "generalist," taking a wider variety of food types or foraging in more kinds of places, because that food and those places are already being exploited by other species. Also, in "simple" habitats, such as grassland, there may not be room for ecological expansion, since simplicity of habitat is generally correlated with a paucity of species, including food items. And then, if species *A* restricts itself to utilizing only a part of its former resources while *B* exploits the rest of the resource spectrum, it may not find enough food. There simply may not be enough for either if they split the resources. Resource splitting, however, is possible if both *A* and *B* suffer a reduction in numbers, and such occurrences may be widespread in nature, although there is little direct evidence for it.

Interspecific territoriality may be maintained by gene flow emanating from the centers of two competing populations toward the zone of contact (Selander and Giller 1963). Any tendency toward ecological or behavioral shifts to accommodate the interspecific competitive situation in the contact area would be swamped by new genes arriving at the "front line" bearing the old "no compromise" traits that permeate the bulk of each population.

Orians and Willson (1964) concluded from their study of interspecific territoriality between red-winged blackbirds and yellow-headed blackbirds (*X. xanthocephalus*) that there is a marked advantage in living in "better" habitats (those habitats to which a species is best adapted) and that they are worth fighting for. Therefore, they figured, probably all cases of stable (long-term) interspecific territoriality involve species with very similar preferences in habitat. (See Lanyon 1957 for supporting evidence from meadowlarks.)

There are those who suggest that no useful purpose can be served by studying interspecific behavior in the search for the primary adaptiveness of territoriality. Perhaps, they argue, interspecific territoriality is only a kind of misdirected intraspecific aggression and is, in fact, unadaptive, persisting only under certain conditions (such as where two populations narrowly overlap). Most authors disagree with this assessment but Wynne-Edwards (1962) and W. D. Hamilton (1964a,b) argue without data that mutual interspecific competition contradicts the hallowed competitive-exclusion principle (discussed in Chapter 24).

We have learned that habitat complexity may influence interspecific territoriality. Ornithologist Lester Short has data showing that in three species of woodpeckers (highly specialized feeders) of Baja California, interspecific territoriality occurs in simpler habitats (such as those with only a few species of trees, relatively little variation in tree height, and with limited altitudinal variation). He found the species coexisting peacefully in complex and diverse mountain areas where they partitioned the habitat, each species utilizing certain parts of the vegetation. Thus, in complex areas, competitors may overlap geographically by occupying different parts of the same habitat. Theoretically, it is only when competitors are forced to utilize the same parts of the habitat (or the same "niches") that they become territorial toward each other. The concept is discussed more fully in Chapter 24.

Finally, interspecific territoriality may be related to the evolutionary appearance of the competitors in the ecosystem. An example will illustrate. In Europe, as many as six species of titmice may be found in the same area. They forage in mixed parties outside the breeding season, and when nesting, their territories may overlap. In North America, on the other hand, no more than two species can be found together. These, however, exist with heavy competition from warblers and vireos, and the two titmice are mutually territorial. Apparently tits evolved in Europe and were coadapted, but when they reached North America, the competitive interaction with established species was so severe that they were forced into interspecific territoriality to maintain themselves (Orians and Willson 1964).

Generally, then, interspecific territoriality would be expected when three conditions are met: (1) the habitat is simple (thus impeding physical partitioning), (2) the species are food specialists, and (3) other competing species are present (Orians and Willson 1964).

Territoriality, then, may be rather simply defined, but the simplicity of the definition is misleading. Part of the problem of developing clear rules that are both simple and accurate is due to the many conditions under which territoriality appears and the ease with which it can sometimes be changed. At this point, we can identify certain environmental traits that seem to be associated with territoriality (such as defendability), but the phenomenon is so incredibly diverse and suffused with subtle permutations that we find ourselves still trying to induce its nature, thousands of years after it was first observed.

# 22

# HIERARCHIES AND DISPERSION

When one considers the spectrum of animal behavior, it seems apparent that territoriality is but one end of a social continuum. At the other end is the hierarchy, in which individuals are recognized and related to according to their rank. The relationship between the systems is suggested because some behavioral patterns seem to fall into either category, and because in some cases, territories and hierarchies achieve the same adaptive end. For example, once established, they both decrease the likelihood of hostile interaction. And we are reminded that sometimes territorial aggressiveness can change according to whom the territorial animal is confronting. So, let's explore the evidence that indicates that hierarchies and territoriality are evolutionarily related.

It often seems that territoriality is so variable and its appearance so unpredictable among the species that it is not possible to define it. Although such diversity has stimulated great interest in the phenomenon, our inability to form hard and fast rules has been perturbing to those of us with orderly minds. House mice (*Mus musculus*), for example, give us problems because they may or may not be territorial. Norway rats *(Rattus norvegicus)* also show striking irregularity in territorial behavior. Some populations defend ony the territory around the opening to a burrow, while others show dominance hierarchies but no territorial defense. The only regularly occurring territoriality in rats is nest defense by the female rats (Scott 1966). So, territoriality in rodents is not only variable, but it becomes indistinct as it grades into hierarchy.

Whether populations of house mice show territoriality or hierarchies is at least partly a function of population density (Southwick 1970). It appears that they are territorial at low population density, but as conditions become crowded, the system breaks down and is replaced by a hierarchical system in which all the animals share the area. (The same phenomenon has been noted in salmon.)

Just as increasing density can cause territorial populations to become hierarchical within a single generation, there is some evidence that high density may also set the stage for hierarchies on an evolutionary basis. The Hispaniolan woodpecker (*Centurus striatus*), undoubtedly descended from territorial birds of North America, is unusual among woodpeckers in that it nests colonially. Each pair defends a small area around a nest hole, but several pairs may occupy a single tree. The birds have extremely high population densities in large parts of the island, and it has been suggested (Ashmole 1967) that, as the population density of the island-colonizing birds increased, it finally reached a point where a territory holder was at a disadvantage because of the disproportionate amount of time spent in defending the area from so many potential interlopers. As a result, aggressiveness was reduced and territories were given up. The birds then were able to come together to nest colonially (advantages are discussed in Chapter 29). There is as yet no direct evidence of hierarchies in this species, but such a system could be expected, at least at feeding stations. (Notice, again, the influence of environment upon social patterns.)

Interestingly, in some animals, territoriality itself may be the *basis* for congregation. Male Allen hummingbirds (*Selasphorus sasin*) are highly aggressive and solitary by nature and apparently their great spirit of belligerency causes them to take territories next to other males just for the sake of combat (Armstrong 1965). In such a case, hierarchies can be expected, but they haven't been found yet. (Keep in mind that a spatial effect operates in territorial systems. Animals generally become more imposing the deeper they are into their own territories. Perhaps true hierarchies, then, should be defined as "space independent.")

Actually, territories and hierarchies may exist simultaneously. Prairie dogs (*Cynomys ludovicianus*) inhabit territories that are divided into family holdings. Each family excludes members of other groups, but within each family the membership is arranged hierarchically (King 1955). In such complex systems, we may ask how the social patterns arose. What, for example, is the relative importance of instinct and learning? It is believed that in prairie dogs the young learn the limits of the family territory, either through communication of the information by the elders (Klopfer 1962), or through being thrashed upon violating the territory of another family. And in some cases, the integrity of family borders is well maintained, indeed. In house mice inhabiting barns, interfamilial borders are so strongly demarcated through territoriality that gene frequencies may vary between families only a few feet apart (Selander 1970).

And then, as another example of the replacement of territories by hierarchies, we find group hierarchies, reminiscent of human gangs. In groups of Welder turkeys (*Meleagris gallopavo*) in Texas, hierarchical rank is established through fighting by first-year males. After males of a family or "sibling group" establish their intrafamilial rank, they then fight with other groups to establish a "group hierarchy" as well. The larger groups usually win. Then the dominant males of the dominant groups do almost all the breeding. The dominant males of a subordinate group only breed when a dominant group is not around (Watts and Stokes 1971).

In turkeys, the polygynous mating system can be of the *lek* type (where the males come together to display on a common arena and females choose from among them), such as that shown by Welder turkeys, or of the *harem* type (where males fight for and "own" females), as we find in the Eastern wild turkeys. And the system may be dictated by the environment. For example, in Texas the climate is dry and severe, food appearing for brief periods when it rains. Females, therefore, must be brought to a breeding state quickly, and groups of displaying males on a lek may serve as a powerful releaser of reproductive behavior (more so, anyway, than a single despot operating in a system with a better food supply).

John H. Crook, in his weaverbird studies, found that those populations living in woods form small social units, thus setting the stage for hostile or territorial interactions between groups, one group usually displacing another. In woodlands, the food supply is widely dispersed, dependable, and available year-round and could thus easily be exploited by little groups of belligerent birds. However, in grasslands, with their seasonal and localized food supplies, the birds are under greater pressure to find food and cannot rely on any area holding food at any given time. Thus, they form large flocks. In the latter case, there would be little advantage in territoriality, but some hierarchical or "role" system might be expected. So, again, we see the strong influence of the environment on social structure.

## HIERARCHIES AS POPULATION REGULATORS

It is difficult to know the importance of hierarchical systems in the regulation of populations because hierarchies demand a high and sustained level of complex social interaction, a requisite that not many animals meet. We have considered the population-regulating effect of territoriality, but the potential role of hierarchies in the regulation of populations should also be stressed. For excellent reviews, see Crook (1970a,c).

An interesting social system is found in bullhead catfish (*Ameiurus*) (Todd 1971). At high densities, they form dense communities where no territories are held and all members are free to move peacefully through the area. At lower densities, however, a "territorial hierarchy" develops. In this system, the dominant fish takes the largest and most protected territory, and those of lower status take smaller and more exposed areas. Dominance is established by fighting, and status is apparently reflected chemically. In an experiment, a dominant bullhead was removed from his territory and subjected to a thrashing by a "hatchet man"—a larger and more formidable fish. When he was returned to the tank, his subordinates no longer fled. Apparently, he was emanating different kinds or concentrations of chemicals. (The importance of chemical communication was demonstrated when territorial fish ceased their aggression after water from a communal aggregation was pumped in.)

Among bullheads, the relationship between dominant and subordinate is more than a system of partitioning the territories. The boss bullhead may act in some ways as a defender of the group. (Some dominant mammals play a similar role.) If a strange fish approaches the group, the subordinates rush into the territory of the dominant male. The boss then engages the stranger in combat. After the fight, the boss allows the subordinates to remain in his territory for a time before driving them back to their own areas.

There are certain ramifications for population regulation in such fighting. The loser of a bullhead fight often survives, but, for some reason, fails to grow normally thereafter even if placed alone under optimal conditions. Size is, therefore, not entirely a result of access to commodities. Fighting to establish dominance could thus be a means of *magnifying* the differences between the stronger and frailer individuals. In times of stress, then, the population might quickly be reduced to viable limits without placing undue stress on all its members. Some physical correlates of rank have also been noted in mammals (Figure 22.1).

The higher ranking individuals of peck orders (as hierarchies are sometimes called because of early studies with chickens) probably have an increased likelihood of survival due to easier access to commodities, and subordinates likewise benefit by not wasting time in fights that would have been lost anyway (Lack 1968). Measurements of foods eaten and foods available for some species, such as the coal tit and the wood pigeon (*Columba palumbus*), showed that the number of individuals surviving each winter was proportional to the remaining food supplies. The correlation, it has been argued, was most likely due to competition for food and the existence of a hierarchy functioning through individual recognition.

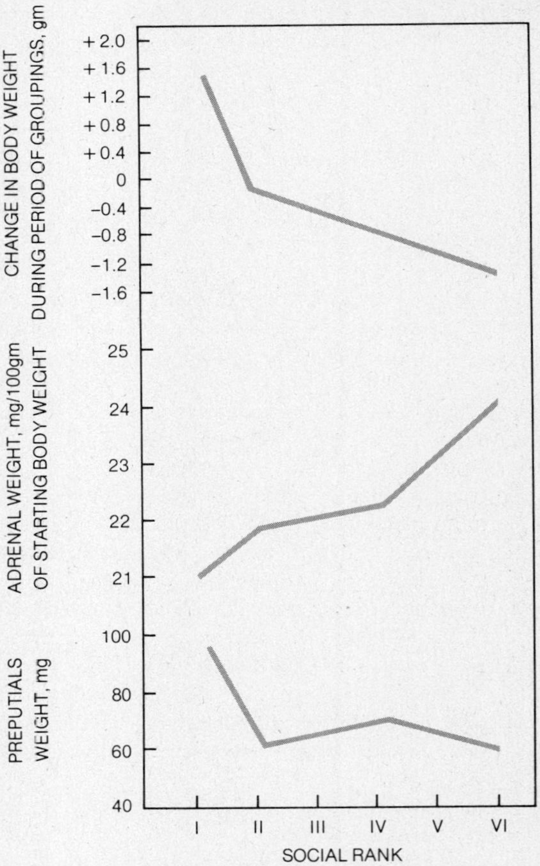


Figure 22.1 Changes in the social rank are correlated with changes in adrenal, body, and preputial weights for wild stock house mice. I is the highest rank, VI the lowest. The study is based on fourteen populations of six mice each.

We have already noted an effect of stress operating through hierarchical systems in bullheads. Stress effects of subordinance in the wild have not yet been clearly shown, but Wynne-Edwards (1962) and Lack (1966), in rare agreement, suggest that hierarchies may act as "social guillotines" when food becomes scarce by causing the quick elimination of certain, less vigorous, segments of the population and ensuring a larger food supply for those at the top of the order.

Such a selective system of population control could only be expected to work when commodities such as food became gradually scarce and thus whittled away at the population. There are times, however, when the environment can exert a sudden and full effect, and the hierarchical system can fail in its role of protector of the powerful. In flocks of wood pigeon populations, those forced to the front of the group got less food by being pushed and hustled by those behind and, as a result, they suffered the greatest mortality in times of food shortage (Murton, Isaacson, and Westwood 1964). One year, however, heavy snowfalls put all the pigeons into strenuous conflict for the few *Brassica* plants that were available as food, and the whole population suffered a drastic weight reduction at the same time. Perhaps the effectiveness of social factors in limiting mortality in populations must depend on certain critical environmental variables such as weather, habitat type, and the dispersion of the group (Crook 1970b).

It should be clearly understood that neither territoriality nor hierarchy depresses populations by *ensuring* the deaths of certain individuals or segments of the population. If subordinate individuals were always unable to secure territories, or if those at the lowest levels of hierarchies were assured of death or of never reproducing because of having lost an encounter, it would be to their advantage to fight to the death to achieve higher status. However, fights to the death are rare. It can be assumed, then, that subordinate or unsuccessfully territorial individuals generally increase their likelihood of surviving and reproducing by temporarily assuming the consequences of their low status. Displaced animals or those of low status are at a disadvantage, but their chances are better than if they had continued to fight. By maximizing the probability of their *survival* they may be able to change their *status* at a later time.

## Hierarchies and Roles

The ethological approach to social behavior has traditionally been confined to descriptions of pair interactions. Such descriptions, of course, virtually define the ethogram, and this is about as far as we have gotten toward integrating the separate social components. In mammals, such a description is woefully inadequate (Crook 1970a). The approach is about as effective as discovering the nature of the game of football by analyzing a list of all the observed moves of each individual player. Part of the problem in studying social behavior lies in the difficulty of determining dominance. The usual methods of assessing rank are especially inadequate when dealing with higher animals. This is not surprising, since social structure can be expected to be highly complex in gregarious "learning" species that are able to utilize subtle and graded communication methods. (For a discussion of the effects of learning on baboon social systems, see Rowell 1966.)

Dominance concepts have now largely been replaced by the concept of *roles* (Bernstein and Sharpe 1966; Rowell 1966; Gartlan 1968; Crook 1970b). Roles are defined by the relative frequencies with which individuals perform certain sequences. When the "behavior set" of an animal is determined, it is said to reflect a role. For example, the "control animal" in a group of capuchin monkeys (*Cebus capucinus*), a species that shows no definite hierarchical system, is the one *most likely* to confront external dangers and to end intragroup frictions. Such an animal may not be the "leader" of the group, however, in terms of determining group movement (Bernstein 1966). In primate groups, it is one thing to say an animal *has* a role and quite another to *define* that role. To look at the matter another way, since each individual shows certain responses in relation to older animals, dominant animals, subordinates, peers, and so forth, its behavior can be described in terms of the sum of its reactions and this, then, would be its role (Crook 1970a).

Consider roles in macaque monkeys (*Macaca*). There is the control male, the central subgroup males (competitive with others in that group, but dominant to peripheral males), peripheral males, isolate males, and central and peripheral females. Such roles are not fixed and immutable, and changes of position in the role hierarchy are usually correlated with changes in general behavior. It is interesting that males may shift their roles many times in a lifetime, while females are socially far more stable. Their stability thus gives them a certain "baseline" importance in determining the roles of males of the group (Crook 1970a).

In baboons and macaques, two conflicting social pressures contribute to role changing in males: (1) the ever-increasing freedom sought by the dominants, which constrains subordinates; and (2) the efforts of the subordinates to keep from being constrained. The resulting tensions may cause shifts in role hierarchy. Role hierarchies, then, reflect relative status in a complex system of possible relationships. And roles can change, each change resulting in a number of new kinds of relationships and hence a change in levels of constraint or access to commodities.

Certain roles, as would be expected, are characteristic of some species and not of others. For example, the harem "overlord" exists in hamadryas baboons (*Papio h. hamadryas*), but not in gelada baboons (*Theropithecus gelada*) or in macaques.

Interestingly enough, rank may be "bestowed" by the subordinate (Rowell 1966). After all, a leader cannot exist if no one agrees to submit. In some higher primates, at least, subordinates almost always choose to leave any commodity to a dominant who wishes to compete for it. This is expected, since the subordinate would probably lose the commodity anyway, so, in order to avoid such situations, he simply avoids dominant animals. Such avoidance means that the subordinate will often not even attempt activities that he probably could have accomplished, and so the dominant gains even more freedom by the subordinate's self-imposed constraints.

The behavior of the subordinate may have an effect on population numbers if its acquiescence results in its death. In competitive situations, the subordinates are likely to be deprived of commodities and hence weakened. Also, in social interactions the stress produced by defeat may cause behavioral peculiarities that result in a reduced chance of breeding. A furtive, nervous individual may not make out well at all. Either of these results is likely to have a depressing effect on the numbers of individuals.

Among macaque monkeys, rank may be influenced by kinship (Koyama 1967; Sade 1967; Missakian 1972). Although there are exceptions even among macaques, and the "rules" cannot be applied wholesale to other species, it does appear that the trend is quite evident in stable monkey troops that show low levels of immigration. In general, it has been found that:

*(1)* Ranks of females of the same age are equal to those of their mothers.

*(2)* A mother is dominant over her female descendants.

*(3)* Daughters of dominant females tend to be dominant, unless the daughter of a subordinate female is significantly older.

*(4)* Among sisters, the younger are dominant, possibly because the protective mother always resolves disputes in their favor when they are very young.

*(5)* Males follow the same rules as the above for about two years—or when they begin to differ in size and behavior.

*(6)* After the age of three, older brothers dominate younger ones, probably as a function of their larger size.

*(7)* Males generally assume the rank of their mothers.

(It should also be pointed out that although we usually assume individuals within all sorts of primate groups to be related (Buettner-Janusch 1963), generally, we cannot be sure of their geneology.)

Although subordinate male macaques have problems in gaining access to commodities, they are left with few options. They can become solitary and later take over a small branch group; they can affiliate with another male and the two can rise together in rank, and, lastly, they can act as "uncles" to infants of high-status adults and so become affiliated with their superiors. Adult male macaques attach themselves to one-year-old animals when their mothers are entering a new birth season and the previous young are being ignored. These males sit with the infants and hug and groom them, accompanying them when the troop is on the move and offering them protection. They are usually middle-class males that show low levels of hostilities and are oriented toward high-caste animals (Itani 1959). By caring for high-caste children, subordinate males are tolerated by the parents and some may gradually rise in rank by their affiliation. Male care has not been found in all troops, so it may be that the phenomenon is a localized "cultural" one. Aside from its influence on roles, "uncling" may have an additional adaptiveness that operates through kinship selection since, in a troop, the offspring are likely to be related to their protector.

## SPACING

The dispersal pattern of organisms over the habitat is an exceedingly critical factor in their natural history. It would obviously influence mating behavior. (Are potential mates numerous or rare?) It might also influence food finding and protection, not to mention competition. In fact, a number of aspects of the ecology and evolution of a species' behavior are related to how it is distributed over its habitat.

Basically, animals may be found in three distribution patterns: random, regular, or clumped (Figure 22.2). Random dispersal in animals is probably rare, but has been reported in clams (Jackson 1968) and a few other species. Regular spatial arrangements may be seen in colonially nesting seabirds when each nest lies just out of reach of the nearest neighbor. Clumping arrangements may be found in animals that form clans or families, such as lions or mice, or in reproductive groups, such as deer herds, in which a number of females are kept together by the activities of a male. So let's explore some of the ways animals manage to arrange themselves with respect to each other.

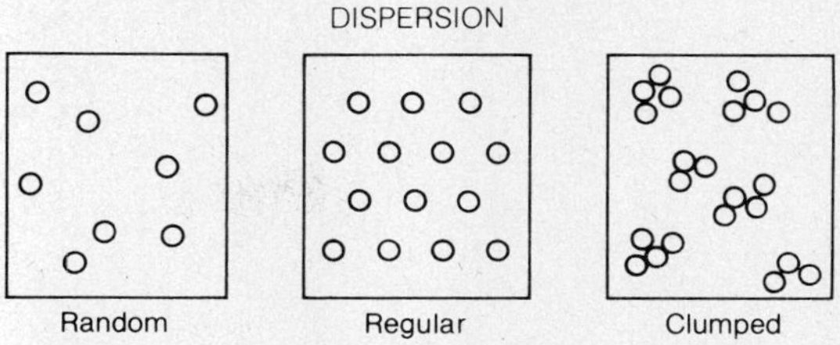


Figure 22.2 Three basic distribution patterns of animals. Random distribution is rare among well-studied species. Regular patterns are found in territorial animals in which individuals arrange themselves with respect to each other. Clumping is a trait of family groups.

### Individual Distance

Some animals deliberately maintain a certain distance between themselves and others of the same species. Under a given set of environmental and physiological conditions, that distance may be rather specific and inflexible. In foraging flocks of chickadees and kinglets, for example, one individual usually will not tolerate another within a distance of two to four feet (Morse 1970). Of course, there are circumstances under which the distance will not be maintained, depending on the prevailing environmental or social conditions or the physiological state of the animal. Even the most solitary of mammals must at times permit a member of the opposite sex to draw near for purposes of mating.

*Individual distance*, or individual space, is usually defined as a defended area that moves with the individual (Hediger 1955). The highly social starling (*Sturnus vulgaris*) will sit peacefully with its comrades on a telephone line as long as no other bird infringes on its individual distance, which is a matter of inches (Figure 22.3). Such an infraction usually results in an immediate threat by the established bird. Thus, we see the birds perched at quite regular intervals. In other instances, individual distance may be quite large. Birds seeking territories will often avoid those areas where they hear another of their species singing. (So both territories and individual distances may be maintained by two behavioral phenomena: aggressive expulsion of an intruder and avoidance behavior on the part of the newcomer.) But territories are essentially different from individual distances, as becomes obvious when we consider the sorts of things that can change the two. In cold weather, for example, some animals change their individual distances to zero as they huddle together to conserve heat. Any parallels drawn between the two phenomena should probably go no further than to note they are both spacing mechanisms.

In mobile species, individual distances are notoriously difficult to measure, so field researchers have resorted to quantifying aggressive encounters as a measure of tolerance (Recher and Recher 1969). One study involved migrant shorebirds that stopped to feed along the mudflats of beaches. Their interactions were scored according to intensity. Threats received a value of 1, supplanting of one bird by another was given a 2, active pursuit a 3, and actual fighting, 4. Among western sandpipers, then, intraspecific encounters had an average score of 2.5. Interspecific encounters with other shorebirds were less frequent and less intense. Of the 138 interspecific interactions recorded, 88 were between species

Figure 22.3 Starlings sit peacefully next to each other on a telephone line. They are regularly spaced according to their tolerance of each other's proximity.

so closely related that they were placed in the same genus. Thus, tolerance was lowest between the strongest competitors.

Social groups may show something akin to individual distances. For example, coveys of partridge (*P. perdix*) may have home ranges that greatly overlap, but the coveys remain separate and do not mix (Blank and Ash 1956). Wild sandhill cranes (*Grus canadensis*) space themselves rather evenly, both as individuals and as flocks (which are probably families) (Miller and Stephan 1966). Hence, there is probably also a movable group space corresponding to individual space in some species (Brown and Orians 1970).

An interesting phenomenon that is associated with individual distance, but which is probably less important ecologically, is *flight distance*. Flight distance is that distance maintained by an animal that, when encroached upon, results in the withdrawal of the animal. Such distances are not rigidly species specific, but may be strongly influenced by learning. You may have noticed that you cannot approach English sparrows too closely on the street. But the same species may be seen around zoos hopping in and out of cages filching food from caged animals and, here, they even seem to ignore zoo visitors. Flight distance can only be correctly determined in those animals that have no particular attachment for an area. (A bird on a nest might be much more reluctant to fly than it would when feeding.) Flight distances may also vary according to the other protective devices of the animal. For example, small animals often flee sooner. Larger, cryptic, or dangerous animals have a shorter flight distance and some do not flee at all.

It is noteworthy that many animals on oceanic islands show no fear of man, a relative newcomer to their environment. They may be approached closely and even touched (Figure 22.4). Collectors about to shoot tiny island birds have found the little birds fluttering over to sit trustingly on their rifle barrels. Of course, such trust in humans has resulted in the slaughter and even the total extermination of many species.

## Dispersion

Some species are highly sensitive to crowding and at high densities they tend to reduce or stop their breeding activities. In other cases, they may breed when crowded, but unsuccessfully. For example, high density drastically reduces the reproductive output of field mice (*Peromyscus*). On the other hand, brown and collared lemmings fall at the other end of the rodent "sensitivity spectrum." They continue to breed at any density (Eisenberg 1967). It is in these latter two species, then, that we can probably expect the greatest levels of dispersal—spreading out from the established geographic range as a result of overcrowding. Dispersion at this level may be called *geographic dispersion* to distinguish it from spatial arrangements on a smaller scale; or those instances in which a population oscillates between one area and another.

Overcrowding may occur in some species when the young begin to mature and to circulate as adults or, at least, at a higher level of competition with adults. Then, mass emigrations or irruptions may occur. Depending on the species, the dispersing animals may be young or old, but in any case they leave, and thus the species is constantly invading new areas, sometimes successfully, sometimes not. The dispersing animals may continue to move until they have removed themselves from strong competition with the parent population, driven by an urge to maintain a certain distance from others. We have considered the mass movements of such animals as lemmings, squirrels, and antelope. And we have also seen that such movements are not restricted to vertebrates. Swarming bees leaving an overcrowded hive, you recall, will reject any new site within a couple of hundred yards of the old hive.

The dispersion of some animals is apparently brought about by the increased level of contact between crowded animals. It may simply be triggered by excessive, albeit benign, interaction. In other cases, the interaction may not be benign at all. Some animals leave, it seems, to escape increasingly hostile behavior within the group. And then, in some species, dispersal is brought about, not so much by the presence, as by what they do, even when they are neither hostile nor benign. Tinbergen pointed out that in certain species of ducks, a male that has established a territory and already has a nesting female will greet any new female with alarmingly rough attempts at mating, showing almost no normal courtship behavior. The new female, reacting to his crass behavior, withdraws. The male's behavior is probably adaptive in keeping the new female from disrupting the nesting behavior of his mate, and the newcomer's withdrawal results in an increased likelihood of her finding an available mate, but the immediate result is a reduced population density.

Figure 22.4 Many island species, safe from predators on their distant homes, have lost their fear of other animals, including man. Such trust has often resulted in wanton slaughter, in other cases it has provided remarkable photographs.

**Some Advantages of Spacing Out** The importance of dispersion has been illuminated by studies of the selection of nest sites of birds. Lack (1970) listed some species that defend territories around their nests, but that forage in groups, even at the height of the nesting season. It is believed that in such cases the birds maintain territories as a device to space themselves out, thus avoiding attracting predators by the high level of activity that occurs around groups of nests. Nesting dispersion could, then, be expected in species that suffer high risks of predation.

Nest dispersion, it should be added, may be rather labile. Tinbergen (1952b) found that one group of undisturbed herring gulls usually nested at a certain density, but after the nesting grounds were raided by a prowling dog or fox, the birds spread out more. On Bear Island, nonterritorial eider ducks (*Somateria*) nest close together; but on the mainland, where foxes can reach the nests, they nest solitarily (Lack 1970).

It is increasingly apparent, then, that predation may influence nest dispersion in birds. As further evidence, nests of solitary or cryptic species of weaverbirds are not advertised by song, presumably so as not to attract predators. However, colonial weaverbirds usually nest in protected sites, and here nests are blatantly advertised over long distances (Crook 1969).

To investigate the effectiveness of dispersion as a predator defense, Tinbergen and his group laid out artificially camouflaged hen's eggs in plots of different densities and attracted wild carrion crows (*Corvus corone*) by placing a conspicuous egg in the area. The crows, after eating the conspicuous egg, continued to look around and searched longer among the more dispersed eggs, but found fewer.

The spacing of concealed nests in birds is paralleled in other species of animals that have not, as yet, been as extensively observed. It is known for example that females of the poplar hawkmoth (*Loathoe populi*) and the eyed hawkmoth (*Smerinthus ocellata*) disperse their eggs rather widely, presumably reducing the chances of predation (Lack 1970). In spite of the emphasis here given predation as a determinant of nest dispersion, other less well-documented factors may also operate to produce the same effects. For example, nesting patterns may be influenced by food abundance or distribution as is discussed in Chapters 23 and 25.

### Who Goes and Why?

Young animals are more likely to leave the group than older ones in a variety of species. Many invertebrates, and even some vertebrates, produce larval stages that travel especially well. For example, the free-floating medusae stages of some coelenterates drift aimlessly with the ocean currents, eventually settling in new areas. In this case, the young are specially adapted dispersal phases of a usually sessile species.

Among vertebrates, young animals tend to be the ones to go. In some cases, they are simply forced out by territoriality, competition, and sometimes direct parental pressures. Among birds, the young are often driven out of the nesting area at about the time they become sexual competitors with their elders. The same physical qualities that would help a young bird attract a mate, undoubtedly also attract the hostile attentions of older, stronger, and more experienced birds. (As long as young birds retain their juvenile plumage, they are usually tolerated.) The same principle operates among some mammals. In the beaver (*Castor canadensis*), only one family occupies a given pond (Figure 22.5). As soon as the young of any year mature, the offspring of the previous year are chased out and forced to find a new pond (Bradt 1938). Some of these excluded animals move into areas where competitive pressures are low enough that they can survive better than in the home area. In general, then, regularly occurring dispersion ensures that the boundaries of the main reproductive group are likely to be continually pushing outward, their genes testing new areas.

Most of these wanderers must eventually settle down. Continued wandering reduces the advantages of latent learning, subjects the animal to increased predation, and periodically causes the animal to find itself in completely unsatisfactory circumstances. Thus, they tend to establish themselves once they reach an even marginally suitable area, or so it would seem in the absence of hard evidence.

Figure 22.5 Among beavers, each year's young are chased away by the parents as the next brood appears. The result is that young animals are forced to invade new, unproven areas.

### Genetic Emissaries

But why are dispersers usually young? Why are young birds, for example, such notorious wanderers? First-year herring gulls that were captured, marked, and released on Kent Island, off New Brunswick, were captured one year later, an average of 1,380 kilometers (about 800 miles) away. Second-year birds averaged 695 kilometers, while third-year birds and older moved only an average of about 465 kilometers (Gross 1940). (I am not completely convinced, however, that young herring gulls are driven away by older birds, since I have spent hours watching herring gull interactions around a submerged bait box at Santa Barbara, California. Often a first- or second-year bird would take over the box and successfully defend it against all comers. Possibly, though, the aggression of the adults exceeds that of the young only during reproductive periods, since their open ocean food supplies are normally neither defended nor defensible.)

Perhaps the greater dispersal distance of the young of some species is not due to their being chased, but is simply due to their being young, unattached, and explorative. It is known that in other vertebrates, such as some monkeys, innovations or novel behavior almost always springs from the young. Perhaps it is not coincidental that the greatest dispersers into potentially new or novel situations are the most behaviorally flexible and inventive elements of the group.

It is conceivable that the exploratory behavior of young animals, which results in their sometimes encountering hostile conditions, is a parental device to aid in spreading their own genes. It might be to a young animal's advantage to remain with the group and to behave conservatively, but it is perhaps compelled to disperse, explore, and experiment by a nervous system bequeathed it by its parents. If parents produced young that were inclined to take chances, to go to new places and try new things, the parental genes would effectively be invading or attempting to invade new areas without personal risk to the parent. In a sense, the young would be *genetic emissaries* of the parents.

One would expect the parents to reduce any risk taken by young that were energetically expensive and to be willing to freely gamble young in which they have little investment. Thus, we find dispersion and exploration rather low in chimpanzees, limited mainly to handling new objects in the environment (under the mother's watchful eye), looking into new parts of the habitat, while remaining close to the group and trying new foods. Birds have less invested in their offspring and the young are great colonizers, travelling broadly. In some cases, first-year birds do not navigate very well and are likely to end up in new places, bearing their parents' genes. Finally, tapeworms have little invested in their fertile eggs and they scatter them to the four winds, the passive dispersal being effectively the same as purposeful exploration by birds and mammals. The notion of genetic emissaries is a form of parental manipulation, an idea that will be discussed later and one that raises intriguing questions about precisely where natural selection operates. (When is an animal able to cease being an expendable emissary of its parent and to begin seeking to maximize its own success, or at what point do offspring become masters of their own fate?)

### Dominance and Dispersion in Humans

Individual distance has been shown in humans (Hall 1969; Felipe and Sommer 1966). In an experiment, David Barash (1973) recorded the responses of college students (who may or may not reflect society at large) to intrusions upon their individual distance, sometimes called "personal space." His study area was the reference room of a library at State University College at Oneonta, New York. The room contained nine rectangular tables (224 by 108 centimeters), which accommodated five students along each side. When he spotted a student alone at a table, reading or writing, Barash would approach and sit directly next to the student with the chairs separated by their usual distance (30 to 56 centimeters) or pull his chair to within 8 to 15 centimeters of the student's chair. He then recorded the students' responses, particularly how long it took them to get up and leave. His controls were other students sitting alone whom he did not approach. Barash performed the experiments dressed in a suit and tie (obviously a faculty member and, hence, a "dominant male" among students) and also dressed in jeans and a casual shirt (apparently a "peer"). The students left sooner when he appeared to be a "dominant male" faculty member.

Barash's findings should not have been surprising. Ethologists have observed that dominant animals usually possess greater personal space than subordinates (McBride 1964). Thus, in keeping with theory (a situation accompanied by some relief and often a measure of surprise), the students retreated more quickly from a "dominant" than a "peer." The students also showed responses other than leaving, such as slightly turning away or arranging books or clothing so as to construct barriers between themselves and the experimenter in the ten-minute interval during which the trials were conducted. They objected vocally only seven times to the "faculty intruder," but eleven times to the "student intruder," showing somewhat greater inhibition in the presence of a dominant. (See Leibman 1970 for a review of the literature.)

## DENSITY AND ABERRANT BEHAVIOR

We have learned that there is evidence that overcrowding may result in behavioral abnormalities that can then reduce the population size and that one result of high density may be a reduction in reproductive success. The phenomenon has been described for a wide range of animals. For example, at high densities, mammals show reproductive depression (Christian 1963). Although the effect is probably rare in birds, reduced clutch sizes are correlated with higher breeding densities in the great, blue, and coal tits (Lack 1958, 1966), and fewer females of tits attempt second broods in high-density years (Kluyver 1963). There are similar findings for other species. It is believed that the reduced reproductive success may be partly due to density-related stress reactions.

### Density and Stress Reactions

It may be argued that fewer reproductive attempts at high density simply reflect an adaptation to a reduced food supply brought about by the severe competition that must exist under crowded conditions. However, appearing with these changes in reproductive success (and, perhaps, in some cases occurring with no change in reproductive patterns) are certain interesting physiological changes. For example, population "crashes" in voles (*Microtus agrestis*) have apparently not been related to food, predators, or any other component of the environment, but, instead, to the number of voles themselves (Chitty 1952, 1957, 1960). Voles from large or declining populations reared fewer offspring than normal, and the young suffered an exceptionally high mortality from the usual hazards encountered by voles. At high population densities, the voles fought more frequently and were found to have enlarged adrenal glands and spleens (characteristics often associated with hemolytic anemia). Other physiological effects of crowding mice into small cages have been clinically described as increased hematopoiesis, reduced growth, thymic involution, elevated plasma corticosterone, delayed or prevented sexual maturity, delay of the onset of sperm formation, reduced weight of accessory sex organs, reduced rates of ovulation and implantation, reduced milk supply of mothers, increased spontaneous abortions, and total reproductive failure. Aside from that, they did fine.

Strangely enough, when vole populations were undergoing such a crisis, even *isolated* individuals showed increased size of the spleen. It was then suggested that perhaps the crashes are due to periodic changes in the behavior of genes and are regulated by a sort of biological clock rather than environmental conditions. The idea has not yet been adequately tested, but if it is true, then the timing mechanism undoubtedly arose as a response to historical waxing and waning of vole numbers that were, in turn, reflections of some environmental factor. A genetically controlled periodic increase in fierceness could act to ensure commodities during a simultaneous increase in crowding. The physiological concomitants, such as the enlarged spleen, would theoretically be less detrimental than attempting to be amiable when population numbers are too high.

It is known that some animals become more sensitive to their own numbers at certain times of the year. For example, muskrats (*Ondatra zibethica*) seem to be more conscious of the presence of others during the breeding season, and when a certain level of crowding is reached, the "fiercer or more determined" individuals drive out or kill the others. Those who are driven out suffer high mortality from predators (Errington 1943).

While another widely held hypothesis assumes that overcrowding causes increases in strife and aggression, it suggests that the major physiological changes occur, not in the overadrenalized aggressors, but in the *losers* of conflicts. In confined animals, it is known that pathological changes may occur in the endocrine systems of subordinate individuals and their young (Christian 1968), but little data are available on wild populations. The effects of social stimuli, specifically upon gonad development, have been shown for several species of birds (Phillips 1964). For example, in chickens it was found that subordinate status depressed growth of the gonads (Flickinger 1966). In spite of such findings, it must be stressed that high density does not always depress reproduction and that much of the above evidence is only circumstantial.

Crowding can cause a host of mysterious effects that have no easy explanation, either physiologically or evolutionarily. For example, crowded animals may be subjected to such stress that they behave in ways detrimental to themselves. Toads (*Bufo bufo*) survive for years in captivity if left alone, but suffer high mortality if crowded. At such times they may starve to death in the presence of food (Elkin 1960). According to our earlier definition of fitness, such behavior must be considered as unadaptive and not as some sort of self-sacrifice for the good of the population.

It should also be pointed out that the endocrine effect of high density is not ubiquitous, even among closely related species. Crowded populations of *Mus* mice showed higher death rates and lower birth rates, but no increase in adrenal weight (Strecker and Emlen 1953). The vole (*Microtus pennsylvanicus*) was subjected to artificial population rises in large outdoor pens enclosing their usual habitat, but researchers found no correlation between population size and adrenal, thymus, or spleen weights, and no secondarily induced susceptibility to other stress (Clough 1965). However, in both these cases the populations may have been limited before stress, which would have affected the endocrine system, had time to develop. We have very little information about the population levels necessary to produce endocrine reactions and how often these levels are reached in the wild, but see Hamilton (1937); Chitty (1952); Kalela (1957); DeLong (1967); and Hoffman (1958) for information on wild rodent populations. For general reviews, see Christian, Lloyd, and Davis (1965); Myers (1966); Thiessen (1964); Brain (1971).

A preliminary study on rats (Calhoun 1963) has particularly ominous overtones. In one experiment, eight mice were placed in a situation where food and water was abundant and where they were kept healthy and free from disease. They were allowed to multiply to over 2,200 in two years. But as they reached their peak number, their social structure began to fall apart. They were unable to complete fighting or mating behavior. They continued to ramble about, eating and drinking, but unable to perform social acts. There are those who see implications here that might provide insight into the sometimes horrifying antisocial behavior of people in large cities.

## Density and Habitat Heterogeneity

It may be that density per se should not be singled out as the primary factor in bringing about the sort of stress I have described. Such stress may not even be due to the increase in competition that arises from crowding. Rather the crucial factor may be simply the increase in the level of *interaction* with other individuals. If this is the case, then the physical structure of the environment becomes increasingly important. A complex and heterogeneous environment should afford increased opportunity for an animal to seclude itself, and so, in complex habitats with abundant hiding places, animals would be likely to interact less often and therefore be subject to less stress. In rats, where food was unlimited, but not much space and cover were available, there was high litter mortality because of excessive fighting among adults (Brown 1953; Southwick 1955). In the brown rat, when nests were small and cover was poor, fighting increased and reproduction decreased although food was abundant (Calhoun 1963). In another experiment, a population of brook trout was almost doubled by adding physical structures to their habitat (Saunders and Smith 1962). And it has been discovered that the home ranges of partridges are generally larger in fields with poor ground cover (Jenkins 1961). So, in some cases, population density might be increased without detrimental effect by reducing the opportunity for social interaction. One wonders if perhaps the unwillingness of crowded people to become involved with others is the result of an adaptive mechanism that permits them to exist under crowded conditions; in order to avoid developing a severe antisocial reaction, the presence of so many others is simply not acknowledged.

Hierarchies, then, are simply at one end of a behavioral spectrum that includes territoriality. The two phenomenon can be shown to be closely related, at least developmentally. Hierarchies can act as social guillotines that, under harsh conditions, deprive the weak of essential commodities and hence regulate populations. Young animals usually occupy the lower end of a hierarchy and their status may cause them to emigrate, the result being that new areas are continually being invaded by the most adaptable and innovative members of the population. Emigration results in a lowered population density and possibly reduces stress by reducing the level of interaction between members of the group.

# 23

# PREDATION AND PARASITISM

In spite of very strong family ties, occasionally a young elephant may become separated from the group. The remains of the unfortunate infant might well be found the next day baking in the hot afternoon sun, a pride of potbellied lions basking nearby. Lions do kill young elephants and so it might be assumed that they affect elephant populations. But they don't—at least not to any important degree.

All sorts of animals are preyed upon by all sorts of predators, but the central question to the ethologist is, Does predator *X* effectively reduce the numbers of prey *Y*? What about parasites? What sort of effect do they have on populations? As we shall see, the answers are far more complex than they at first might appear. So, in this chapter, we will begin by considering the effects of predation, giving special attention to the wolf, but first reviewing a few classical experiments.

## SOME EXPERIMENTS IN PREDATION

The Atlantic salmon (*Salmo salar*) is an important commercial and sport fish along the east coast of Canada, but naturalists there have been worried about its declining numbers. It was known that predatory birds, particularly kingfishers and mergansers, were preying on the young salmon stock, so it was decided that the birds should be shot (Elson 1962). An average of 54 mergansers and 164 kingfishers were shot each year for six years along a ten-mile stretch of the Pollett River in New Brunswick. The results are shown in Table 23.1. Not only were the percentages of salmon that survived for two years increased, but other species of fish doubled their numbers. The conclusion seemed obvious: predators reduce the numbers of their prey. But, of course, the situation is not always that simple.

In his classical predator-prey experiment, the Russian ecologist G. F. Gause (1934) devised a simple predator-prey system using two microscopic protozoans, *Paramecium caudatum* as the prey and *Didinium nasutum* as the predator. When both were placed in a medium that contained bacteria (as food for the paramecium), *Didinium* overate its prey and then starved to death (Figure 23.1A); thus, both predator and prey were depleted. When the experiment was repeated with sediment added, the paramecia had a hiding place, and again, *Didinium* went extinct, but this time the paramecia recovered (Figure 23.1B). Then, when Gause allowed "immigration" by introducing new individuals of both species at regular intervals, predator-prey cycles resulted (Figure 23.1C). It was believed that the cyclic condition best represented what usually goes on in nature.

There have been attempts to examine predation and competition simultaneously. In one experiment, a beetle (*Callosobruchus chinensis*) was used as the prey and two species of predatory wasps as competitors (*Neocatolaccus mamezophagus* and *Heterospilus prosopidis*) (Utida 1957). The prey was provided with a constant food supply. The wasps had similar life histories and they were both dependent on the beetle for food. It turned out that the population densities of all three fluctuated wildly (Figure 23.2), but four years and some seventy generations later, all three were still coexisting. Furthermore, close examination indicated that the two wasps tended to fluctuate out of phase with each other. It seems that

*Heterospilus* was more efficient at exploiting the beetles at low prey densities, while *Neocatolaccus* was more efficient at high prey densities. Thus the competitive advantage of each wasp shifted with changes in the numbers of beetles. It was suggested that the beetle population fluctuated in response to changes in predatory pressures, and by density-dependent changes in the beetle's own reproduction rate. Thus we see that "stable," but fluctuating, systems are possible, yet another example of the complex interactions of living things.

## A COMPLEX PROBLEM

One more experiment will help to illustrate the complexity of predator-prey systems and their relationship to other biotic aspects of the environment. A predator was removed from one area to see what effect it had on its prey, while allowing it to remain in an adjacent area as a control (Paine 1966). A major predator along the rocky intertidal seacoast of the Olympic Peninsula of Washington State is the starfish (*Pisaster ochraceus*). It was assumed that if it were removed, the numbers of other species would increase. But when the starfish was removed, there was a drastic *decrease* in the number of species remaining. Marine invertebrate species in the area dropped from fifteen to eight. But how could the removal of a major predator cause a *decrease* in species number? It seems that the intertidal zone was a "space limited" system. Thus, in the absence of predation, the more efficient "land grabbers" dominated the area and forced other species out. A mussel (*Mytilus californianus*), it turns out, was the most successful space occupier. Presumably, predation on the mussel by the starfish left other areas open for less efficient colonizers. So, here, a predator was permitting the coexistence of otherwise incompatible species. If, as is suspected, species diversity contributes to community stability, predators may have far-reaching ecological effects indeed.

In another unsettling instance, we find that a predator actually increases the numbers of individuals in its prey population. When fishermen (acting biologically as predators) take larger fish, the result is an increase in the number of young fish that are able to mature. Then breeding begins earlier and more eggs are laid. The result is an increase in numbers of fish (when overfishing occurs, of course, numbers are reduced) (Lack 1968). Such an effect could only be expected in species that are limited by food and that are so abundant that moderate predation does not bring the population below the level where food can exert its limiting effect. In other words, in such a case, predation must not be the prime regulator of the population.

In other cases, predators seem to have no effect on prey numbers when they might have been ex-

Table 23.1 The results of removing birds that prey on salmon, beginning in 1947.

| | YEAR | SALMON FRY PLANTED | SURVIVING TO SMOLTS (2 YRS.) (PERCENT) |
|---|---|---|---|
| *No bird control* | *1942* | *16,000* | *12* |
| | *1943* | *16,000* | *6* |
| | *1945* | *249,000* | *2* |
| *Bird control program* | *1947* | *273,000* | *8* |
| | *1948* | *235,000* | *6* |
| | *1949* | *243,000* | *8* |
| | *1950* | *246,000* | *10* |

*Source: Krebs (1972).*

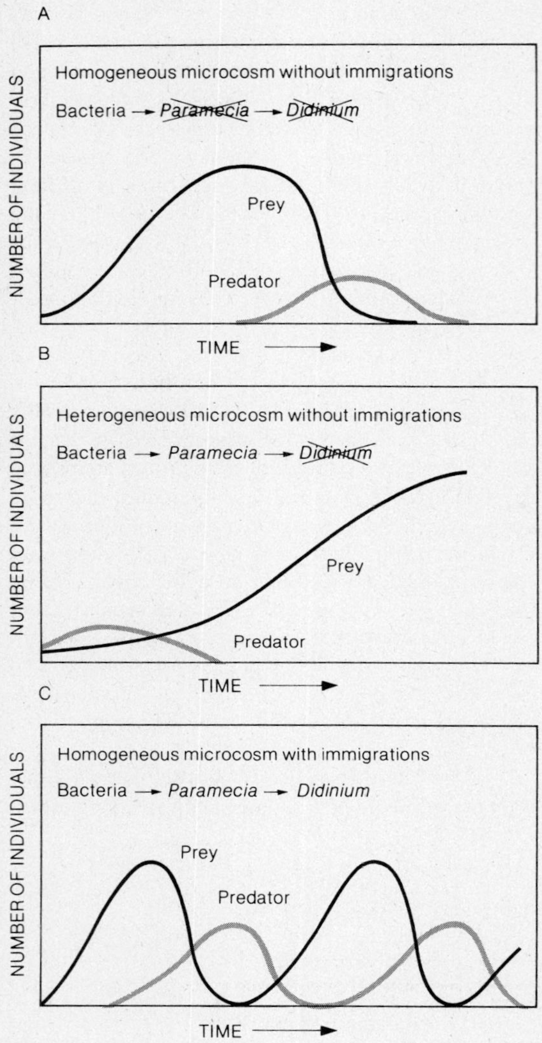


Figure 23.1 The three classical experiments by Gause. In (A), a simple system, the predator quickly decimates its prey and then starves. In (B), the system is more complex, and the predator goes extinct before it can exterminate its prey. In (C), if prey and predators are periodically added, the numbers of both will oscillate.

pected to. For example, populations of ruffed grouse (*Bonasa umbellus*) fluctuate greatly, so, in one area, kindly hunters, who wanted to kill the birds themselves, had a series of predator-control experiments carried out. A total of 557 predatory birds and mammals were removed from about 2,000 acres in New York, while an adjacent area was left untouched as a control. The results are shown in Table 23.2 (Edminster 1939).

Strangely enough, it turned out that predator removal greatly improved nesting success in the grouse, but for some reason, this did not lead to higher population densities in the fall. Something else was killing the birds. The populations were allowed to recover for a year, then the experimental and control areas were reversed. The results were the same. Predator control reduced nest losses, but still the chicks died. The conclusion was that predator control could not increase grouse densities.

Skeptics groused about the experimental technique, arguing that the birds and the predators were free to move on and off the experimental areas. So, the experiment was repeated on Valcour Island (1,050 acres) in Lake Champlain, New York. The results were the same and they have never been satisfactorily explained.

The complexity of predator-prey interaction can also be seen in a hypothetical example from birds. It has been suggested that birds, such as tits, probably take small percentages of any prey in spring when many prey-species are available, but that they exert severe pressures on their prey when only a few prey-species are present, as in winter. Any food-effect on the *bird* population, then, is exerted in winter. The numbers of the *prey*, such as spruce beetles, on the other hand, are limited by the bird's foraging success in winter. So we see strong seasonal effects. As certain food species disappear in late fall, pressure is put on the birds and they, in turn, put pressure on the beetles in winter. Apparently, then, while depressions in numbers may coincide in the two species where they coexist, the populations may be regulated simultaneously by entirely different density-dependent mechanisms (here, food competition and predation). Such information should discourage generalizing regarding the population-regulatory effects on different species simply because their population curves are correlated in time.

I have mentioned that a decrease in prey numbers can result in predators starving or switching to

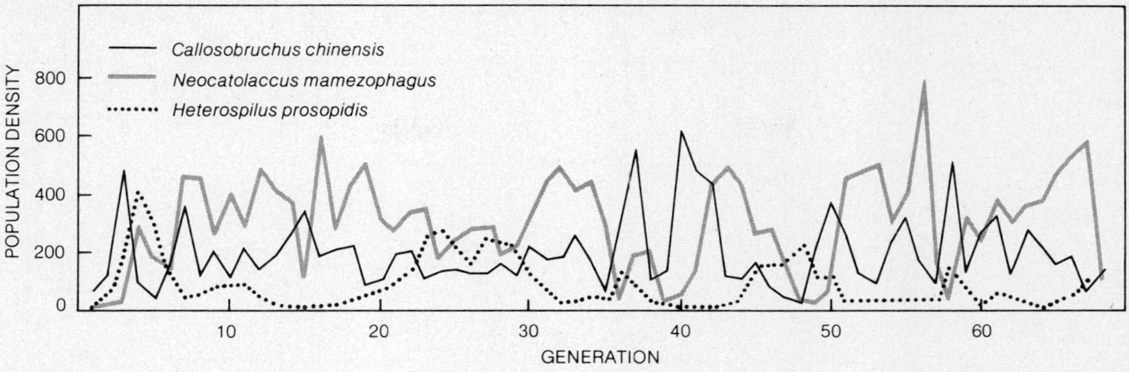


Figure 23.2 An experiment testing the effects of both competition and predation. A beetle, *C. chinensis*, and its two competing predatory wasps were watched for some seventy generations. Although their numbers fluctuated wildly, they continued to coexist, apparently because of differences in the success of the predators at different prey densities.

different food sources, and therein lies the essence of another important law of nature: a predator does not normally eliminate its prey. Of course, there are stories of introduced species wiping out their prey. For example, the introduction of the mongoose on Puerto Rico is locally given credit for the absence of poisonous snakes (which weren't there to begin with), and the release of scores of wild pigs is said to be responsible for the annihilation of reptiles on Belle Isle in the Detroit River. Such accounts of a species being wiped out are almost always anecdotal and probably untrue unless a defenseless, exposed, and available prey was attacked by predators efficient enough to decimate it in a very short period of time. Unfortunately, these criteria have undoubtedly sometimes been met.

There are certain specific characteristics of animals that dictate whether they are candidates for limitation by predators. Some prey animals are more dangerous, cryptic, elusive, or aware than others, and any of these traits may discourage predators from relying too heavily on them. However, such animals may constitute a meager, but relatively *constant* part of a predator's diet, and may therefore figure importantly in maintaining predator numbers. A wolf may attack a porcupine only as a last resort, but if the wolf is hungry enough, it will brave the quills. Porcupines are present year-round and easy to run down, so they can serve as an emergency reservoir. On the other hand, more palatable or available prey may be quickly removed as an important part of the predator's diet.

The degree of reduction in a predator's numbers as its prey dwindles depends, in part, on the characteristics of the predator. For example, a predator that forages in a rigid or stereotyped manner is more likely to be influenced by its primary prey's numbers than one that is more flexible. A flexible predator is more likely to be able to hunt for its usual prey in new ways or new places, or to switch to new kinds of food.

Certain types of predators tend to exert more pressure on their prey than do others, simply because of their hunting techniques. A predator can be characterized, for example, according to whether it hunts by olfaction or vision, and visual hunters might be more successful than olfactory hunters (trailers) when the prey reaches low numbers (Holling 1959). To show the importance of predator characteristics on prey limitation, you might consider the entirely different results predicted by two opposing theories that assume differences in the appetite and prey-finding abilities of predators (Thompson 1939; Nicholson 1933)

It should be clear, then, that highly complex relationships exist between predator and prey. Generalizations cannot be expected to come easily, and,

Table 23.2 Predator-control experiments on ruffed grouse.

| | 1931 | | 1932 | |
|---|---|---|---|---|
| | PREDATORS REMOVED | NOT REMOVED | PREDATORS REMOVED | NOT REMOVED |
| *Nest lost (percent)* | *24* | *51* | *39* | *72* |
| *Chick mortality (percent)* | *57* | *67* | *54* | *55* |
| *Adult loss (percent)* | *11* | *15* | *32* | *21* |
| *Grouse population density in fall (birds per 100 acres)* | *13.0* | *9.8* | *18.7* | *18.0* |

*Source: Edminster (1939).*

in what is apparently the very early stages of our predator-prey investigations, we should extrapolate from one species interaction to the next only with great care.

### Compensatory and Noncompensatory Predation

L. B. Slobodkin, in his "How To Be a Predator" (1968), suggested that, in an optimal system, a predator would primarily take those animals that were about to die anyway so as to alter the "natural" pattern of mortality as little as possible. (That is, it would tend to take those animals, or those numbers of animals, that would probably have succumbed to some other danger anyway.) Thus, the optimal prey would be old ones—or young ones that have not yet reproduced. Young should only be taken, of course, if they are normally subject to high mortality and would probably have died in any case. The idea is supported by Errington's (1946) theory that predators take the prey number that is in excess of the carrying capacity of the environment.

Errington described two entirely different types of predation: *compensatory* and *noncompensatory*. In the former case, predators tend to take a heavy toll of certain segments of a population when the prey density exceeds a certain level. Compensatory predation could function in those cases in which only a limited number of hiding places or nest holes is available so that, when the population of prey rose above a certain threshold level, some individuals would be forced into more exposed areas where they would be more likely to be caught. In such a situation, if one animal falls prey, it alters the probability of another animal also being taken. For example, whenever an animal is captured, the result is less competition for hiding places. Another effect is, if most exposed animals are removed, there will likely be fewer predators attracted to that area. Compensatory regulation also implies that if one regulatory factor, such as predation, decreases, another (for example, competition for food) may increase its effect. The concept of compensatory predation is important to any discussion of the regulatory function of territoriality. General compensatory influences are considered again in our discussion of the wolf.

Noncompensatory predation operates when no part of the prey population is more available than any other part. While in compensatory predation, an animal being taken reduces the likelihood of another one also falling, in noncompensatory predation, the death of one individual has little bearing on the survival of any other individual in the population. Such behavior may be visualized in the interaction of coyotes and rabbits. When rabbit numbers are low, it does not mean the population is safe from coyotes. Instead, the coyotes apparently search out the remaining rabbits in an intelligent, selective manner. It would seem that one could expect differences in the learning abilities of predators that operate according to the principles of compensatory and noncompensatory predation.

## Delayed Density-Dependent Effect

We learned earlier that predator populations do not rise and fall in synchrony with their major prey populations. According to the classical theories (Lotka 1925; Volterra 1926; Gause 1934; Nicholson and Bailey 1935), prey numbers in undisturbed conditions rise steadily for several generations and then fall, while those of predators do the same, but are a little behind the prey in their timing (see Figure 20.4 in Chapter 20). The generally accepted underlying principle seems sound, and there are some indications that such a pattern actually occurs in nature although it has never been clearly demonstrated. The "lag" phenomenon may appear because the death rate of the prey at any time varies with predator numbers and the latter are partly a function of the number of prey in the past (Varley 1947, 1953). In other words, as prey numbers decrease, some time elapses before the predators begin to starve. Then, as prey numbers swell, it takes time for the predator numbers to rise in response to the new food supply. The phenomenon is called a *delayed density-dependent effect.* The length of the predator's delay, of course, is dependent to a degree upon its breeding characteristics. Fecund ferrets would be expected to lag less than the slower breeding wolf.

## Dampening

It may be that population regulation, by whatever means, functions by "dampening" the oscillations rather than by maintaining numbers within any particular range. "Dampening" is different from "maintaining." The effect may be similar, but the mechanisms are not. Whereas maintenance may be due to mechanisms that keep a population within some limit, dampening simply smooths out the peaks and valleys at *any* population level. Dampening might conceivably occur in a number of ways. For example, the switching of predators to a different source of food when prey numbers are low has a dampening effect on fluctuating prey numbers (at least according to some models [Murdoch, Avery, and Smith, 1975]). Thus, when a variety of prey types is available, the number of each is likely to be more stable than if any were the sole food of a single predator. Apparently, when an alternate prey is available, the numbers of the preferred prey will continue to fluctuate, but not so widely. They do not reach the highs and lows that they would if they were a primary food supply of a predator, forced to hope for dwindling predator numbers to reduce pressure on themselves.

Obviously, predators that are able to switch food supplies would also tend to maintain more stable populations since their food might never become severely reduced. The tawny owl populations near Oxford, England, have been remarkably stable in spite of great fluctuations in some of their prey (Figure 23.3). Apparently the owls readily switch foods if one prey species falls into short supply. As another example, drastic fluctuations have been described for populations of insects in cultivated woods that were dominated by a single species of planted trees (Voûte 1946) (see the implications of simple habitats in Chapter 25). In mixed woods, where there was a greater range of food for insectivores, insect populations were more stable. Instabil-

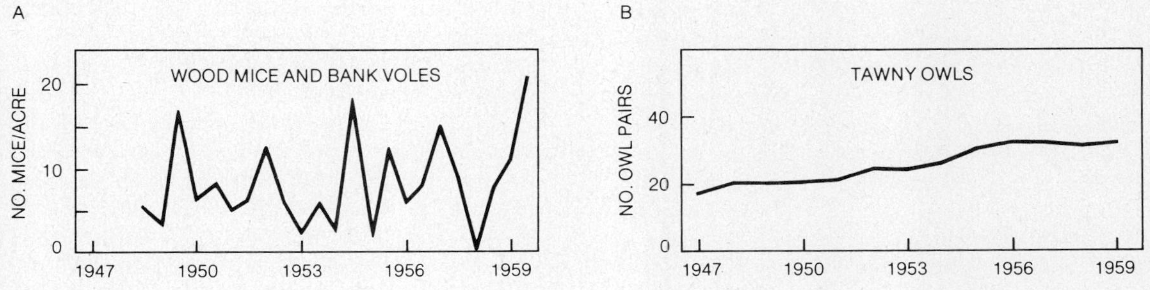


Figure 23.3 In spite of drastic fluctuations in prey, the tawny owl near Oxford, England, maintains rather stable numbers, apparently through prey-switching.

ity in many kinds of populations is probably important, since if numbers fluctuate wildly, there is a danger that they might fall so low that the population might be unable to recover.

### Tactics and Population Stability

By use of simulation models, four major tactics for population stability of both prey and predator have been developed (Holling 1959). First, a "system stabilizing" predator would be one that could alter its physiology so that in low prey densities its own reproduction does not fall off. In an experiment, one predatory terrestrial bug (a pentatomid) lowered its metabolism when moved to a low diet, therefore more of its energies could be devoted to producing offspring.

Secondly, some predators are wasters, killing all the prey at a single location. (A mink may destroy all the hens in a henhouse, but devour only one.) Such animals normally hunt dispersed prey and have no innate inhibition against wasting. Other predators, which normally hunt aggregated animals, only kill one and often return to the carcass until it is consumed (lions that hunt herds of ungulates are an example). It is apparent that stability of the system will result only if each predator is primarily exposed only to its usual prey type.

The third concept is an interesting one and intuitively apparent when brought to light. If a primary predator (for example, a ferret) is, in turn, attacked by a secondary predator (for example, a hawk) the oscillations of both the primary predator and its prey may become dampened by the effect of the secondary on the primary predator.

And lastly, increases in the numbers of predators that do *not* attack members of the prey population randomly, but that tend to choose a certain type of prey from among the population, could result in multiple attacks on a single prey and thus help to stabilize prey oscillations. Random attack refers to an equal probability that any individual in the prey population will be attacked. If the probability is greater for some than others, then the predator (or parasite for that matter) will have less overall effect on prey numbers. As an example, African hunting dogs will rush a group of ungulates, each chasing its own prey. But if one should start gaining on an animal that may be very young, old, or weakened, the others will break off their chases and assist in that kill.

### Predation and the Distribution of Animals

Surprisingly enough, the local distribution of some species seems to be limited by predation (Krebs 1972). For example, the common mussel (*Mytilus edulis*) is widespread on rocky coasts throughout the world. But within the sheltered waters of the Lough Ine of Ireland, the mussel is rare or absent (Figure 23.4). Its absence cannot be explained by its failure to reach the Lough. When pieces of rocks with mussels attached were placed in these waters (in other protected coves and certain rapids and along the open ocean coastline), only the latter group survived (Figure 23.5). The reason was soon discovered. It turned out that three species of crabs and one starfish were preying upon the mussels of the Lough, but these species were absent along the open ocean coastline, and it was here, of course, where mussels were abundant. Large mussels could survive in those areas where the local crabs could not open large shells (Kitching and Ebling 1967). (At this point, we really don't know how important predation is as a limiter of prey distribution for very many animals.)

In other cases, food may limit the distribution of the predator. An example of a species that is limited in its distribution by its food species (this time a plant) is a rare fruit fly called *Drosophila pachea*, which breeds only on the stems of senita cactus (*Lophocerceus schottii*) throughout the Sonoran desert. Apparently there is something in the plant that is necessary to the fly's development. The substance is probably a unique sterol—a kind of alcohol. Strangely enough, the same plant also produces something, probably an alkaloid, that poisons other species of Drosophila.

The red tree mouse (*Phenacomys longicaudus*) is an aboreal mouse, as you may have guessed, and it has the most restricted diet of all North American mammals. It feeds entirely on the needles and twigs of the Douglas fir. It ranges from north of San Francisco up to the Columbia River. From here, the tree continues northward, but the mouse stops. Apparently the Douglas firs simply delimit the mouse's *potential* range. Other factors may determine which parts of the forests it actually utilizes. Evidently it does not disperse very well, since it is even

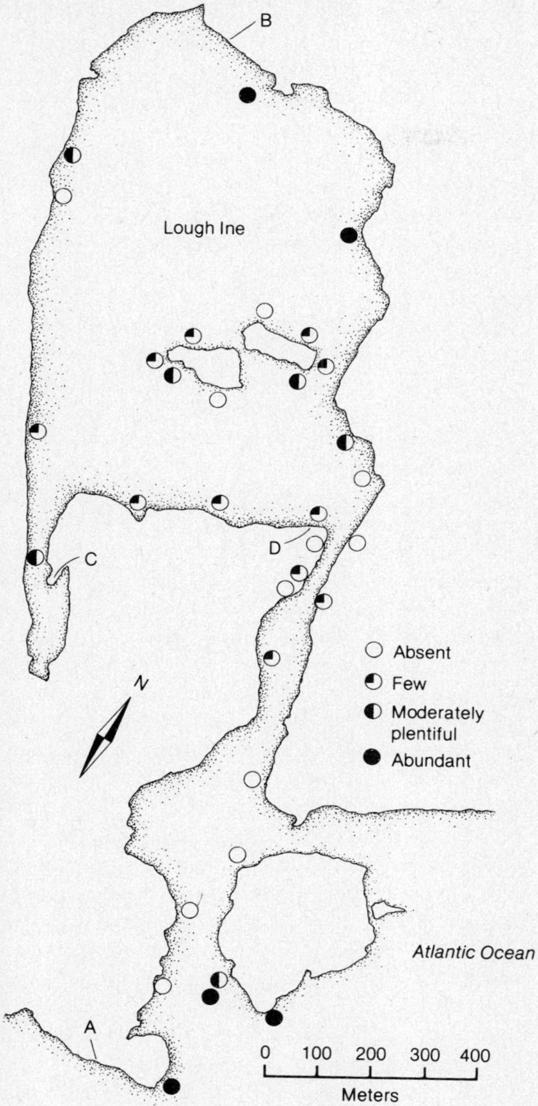


Figure 23.4 Distribution of the mussel *Mytilus edulis* in Lough Ine and the nearby Atlantic coast of Scotland. Points A through D are described in Figure 23.5.

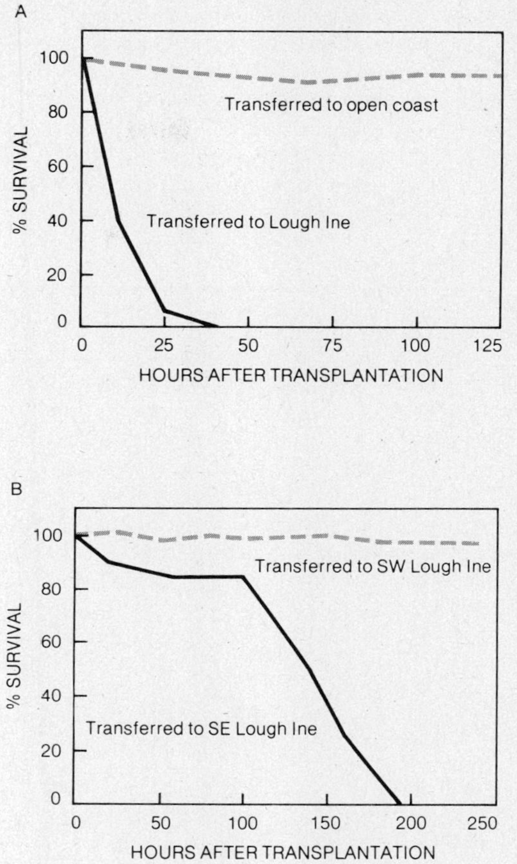


Figure 23.5 In (A), small mussels transplanted to open coastal areas (A in Figure 23.4) survive, but disappear if placed in the sheltered waters of Lough Ine (B through D in Figure 23.4). In (B), large mussels transplanted to the southwest (SW) Lough area (C in Figure 23.4), where they occur normally, survive, but disappear if transplanted to other parts. Large mussels could survive only in those areas where the local crabs were unable to open large shells.

absent from isolated patches of Douglas fir well within its range (Hamilton 1962).

By considering plants in a discussion of predation, I don't mean to imply that herbivores interact with their food items in the same way carnivores do. It is true that in some cases, they undoubtedly do behave in a similar manner, but in other cases, such as the red tree mouse, the herbivore may be closer to a parasite than to a predator in its association with its food. The point is that the range of some animals may be limited by the presence of potential predators and that the distribution of animals, both herbivores and carnivores, may be influenced by their food items.

## THE WOLF AS A PREDATOR

*One day I watched a long line of wolves heading along the frozen shore line of Isle Royale in Lake Superior. Suddenly they stopped and faced upwind toward a large moose. After a few seconds the wolves assembled closely, wagged their tails, and touched noses. Then they started upwind single file toward the moose.* [L. David Mech, *The Wolf*]

There has been quite a bit of attention in very recent years to large predators, their life histories, how they kill, and the effects of their predation. L. David Mech (1970) has provided us with well-researched insight into the nature of the wolf. His story is based on years of watching these magnificent predators, especially on Isle Royale (Figure 23.6), and he has provided answers to some long-standing questions. Is the wolf an effective killer? Yes. Does it ever kill more than it can eat? Sometimes. Does it attack humans? No. Is it always successful in its hunts? No. In fact, for every 100 attacks on moose on Isle Royale, only 7.8 were successful.

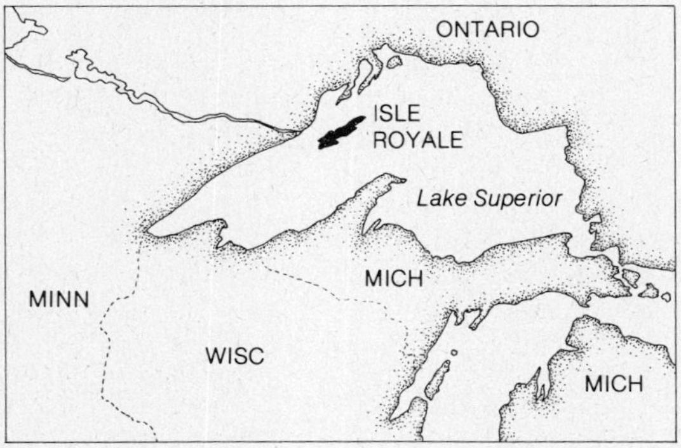


Figure 23.6 Isle Royale, an island near the Canadian border, where intensive studies of the wolf under natural conditions have been conducted in recent years.

And we have been provided with other very specific information. For example, at prey densities in excess of 25,000 pounds of prey per wolf, wolf predation apparently does not limit the population of that prey. But where prey densities range from 7,400 to 24,000 pounds per wolf, the predation does control the prey population.

One of the first things that should be mentioned about wolves is that they are continually on the hunt. They may attack a prey animal only minutes or hours after a successful kill and when they are on the move, they are hunting. Their stomachs can hold up to twenty pounds of food, and it is apparently rapidly (or incompletely) digested.

They probably locate prey most often by scent (although they may also stumble across it). Mech found in fifty-one hunts in which he could determine that the wolves were trailing their prey (moose) through the snow, forty-two cases involved direct scenting. (They once detected a cow and her twins $1\frac{1}{2}$ miles away.) When the lead animals catch wind of the prey, all pack members stop and stand alert, concentrating intently on the moose. Then they may carry out a peculiar ceremony, standing nose to nose and wagging their tails. If they are in deep snow, they just pile up behind the leader who then sets out straight for the prey. As they near their quarry, they quicken the pace and seem anxious to lunge ahead. But they hold themselves in check, alert and tails wagging. The restrained approach may permit the wolves to draw near to their prey without sending it into flight. Of course, if they are scent tracking, the wind would not alert the animal.

When the prey realizes its situation, it can do one of three things: go after the wolves, stand its ground, or flee. Only once did Mech observe a moose head straight for a pack of wolves that was stalking it, and the wolves fled. The second best strategy seems to be to stand ground. As soon as wolves realize their quarry has spotted them but is not running, they stop stalking (Figure 23.7). If they proceed, they do so cautiously. There may be an advantage in this wariness. First of all, a moose is a strong and dangerous animal, easily capable of killing a wolf. If the animal doesn't run, then, it may be out of a spirit of

Figure 23.7 On Isle Royale, the moose is a primary prey of the wolf packs that inhabit the island. The wolves normally will not attack an animal that appears unafraid and stands its ground. This moose stood its ground for five minutes, after which the pack gave up and left.

confidence in its own prowess. Perhaps it has successfully dominated other moose and has not developed the habit of retreat. Also, if it doesn't run, it doesn't waste energy and can therefore probably put up a better fight. In fact, on Isle Royale, any moose was safe from attack as long as it stood its ground. But if it ran, it was chased and attacked. (Smaller or weaker animals, such as deer, do not have the option of standing; they could certainly not fend off a wolf pack.)

But another factor may enter in here. A wolf may need the sight of a running animal in order to stimulate its closing rush. Mech believes the sight of a nonmoving animal inhibits the rush response. They may still approach, but slowly. (You may have noticed that your dog will chase anything that runs from it. On my afternoon runs on Ibiza I often tried to outrun two short dogs who invariably chased me. When they were close enough to bite, I stopped and lunged toward them, sending them off in a cloud of dust. One of John Fentress' hand-reared wolves would run from approaching sheep and chase retreating ones. A five-month-old wolf pup he owned always ignored horses until he saw one roll in the dust. The sight of a downed animal was just too much and he had to be restrained.)

So, if the prey runs, the wolves give chase. If the quarry is fast enough in its initial run, it may get away. Wolves are very perceptive about such matters, and if the chase looks hopeless, they stop. But no one knows what kind of cues they use to determine whether to call off the chase. It may go on for miles, but usually covers only a very short distance. Of forty-one observed chases, thirty-two were called off within half a mile. Mech saw one that lasted three miles, and Lois Crisler, who also studies wolves, saw a caribou chased for five miles. Apparently, there is a bit of myth regarding the long-distance chase of wolves. Usually a stalk and a quick chase does it.

Interestingly enough, the prey runs no farther than necessary. When the wolves give up, the prey (whether moose, deer, caribou, or Dall sheep) usually stops and turns around to watch the wolves. Thus it doesn't waste energy in running needlessly.

Wolves are widespread (though vanishing) animals, and they must confront a wide range of prey types. Since they are largely carnivorous, their best energy bet lies in bringing down large ungulates. While one wolf may be able to do this, in rare instances the wolf social system permits the strength of numbers. This means they can hunt various large and formidable animals. Some of their prey are herding animals, some are solitary, some live in forests, others on tundra, and others on jagged, rocky slopes. Some depend on strength and some on speed. Probably no group of wolves specializes in more than one or two of these types of animals. It would be interesting to see how wolves that rely primarily on deer would go about attacking a moose.

Wolves attack different animals in different ways. For example, moose are usually bitten on the rump area with one wolf sometimes clinging to the large, rubbery nose. If the kill is not at first successful, the wounded moose may stiffen and weaken and be brought down some days later. On the other hand, caribou are usually attacked about the shoulder and neck.

Jane Goodall and her group, studying various African predators, found that different kinds of predators tend to kill in specific ways. Hyenas and African hunting dogs may disembowel their prey and begin devouring it while the unfortunate animal is still on its feet (Figure 23.8). Cheetahs have small mouths and kill by neck bites or by strangling. Lions may strangle prey or suffocate them by holding the prey animal's nose in their huge jaws. And, apparently, some animals just die of shock or heart attacks when they are attacked.

Figure 23.8 Wild dogs of the Serengeti killing a hartebeest. These animals come under severe condemnation by moralizing humans because of their habit of disemboweling and eating living prey.

We have concentrated on the behavior of wolves as they hunt moose and other large ungulates. But wolves are opportunists. They will also eat mice, insects, fish, rabbits, birds, or just about anything they can catch. Their food range is enormous. They will eat certain kinds of berries and will even attack powerful bison.

Wolves have also changed their diet to conform to the presence of man. For example, they raid garbage dumps. But more to their detriment, where man has replaced wild game with domestic animals, the wolf's diet has changed accordingly. While wolves almost always dispose of all the remains of wild game, they sometimes eat only parts of domestic animals such as cattle and sheep, possibly because human activity drives them from such kills. This behavior has brought down the wrath of government agencies at the behest of lobbying stockmen. Wolves have traditionally brought bounties and their appetites make them easy to poison. A most disagreeable habit of wolves from a modern human viewpoint is that they will kill and eat dogs and cats. And they may be cannibalistic under some circumstances—although there is no evidence of wolves actually killing other wolves for food (they may eat the carrion) or of members devouring even the corpse of a member of their *own* group.

So, what kinds of animals do wolves take from among their prey populations? After all, their success rate is surprisingly low. Of 131 moose detected by one pack on Isle Royale, only 6 were killed. And of the 131, 54 escaped without even being stalked by the wolves. Since so many attempts are made and since the success rate is so low, it seems that primarily disadvantaged prey would be taken. The prey could be newborn, inexperienced, malformed, sick, old, wounded, parasitized, starving, crippled, or just plain stupid. Of course, the removal of most of these types would have a "sanitizing" effect on the group. A sick or parasitized animal would not be able to spread its condition to others. Malformed or stupid individuals would not be able to pass along their kinds of genes. Old, wounded, or crippled animals would not be able to interfere with group movement, in the case of herding animals, and would not compete for food or other commodities. Several studies have shown that wolves primarily kill animals less than 1 year old, or those who have lived at least half the usual life span for that species in the wild. In a sample of ninety-three deer killed by wolves in Minnesota, fifty-nine percent of the adults were at least 4½ years old. Among four hundred deer killed by hunters, only twenty percent were in this age class. It is hard to see how killing young animals could benefit the herd other than by simply reducing the herd's density. But we must keep in mind that wolves are not in the business of raising ungulate herds, so perhaps their strategies are not optimal to the herd. On the other hand, perhaps the young that they take are not well defended by the parent, due to some inferior physical or behavioral quality of the mother, or perhaps the young animal is not particularly vigorous and cannot escape with the mother, so the sanitizing effect of predation could even operate on young animals, but undoubtedly with less precision than it works on older ones.

But what is the evidence? There is an accumulating body of data, much of it circumstantial, that wolves do kill a disproportionate number of diseased or starving animals. Mech (1970) states, "Because most of the figures were based merely on the examination of the bones or partly eaten carcasses of wolf kills, it is remarkable that such high rates of maladies were found. Such a situation strongly suggests that, if large numbers of intact carcasses could be examined, almost every animal killed by wolves could be classified as young, old, or otherwise inferior." Wolves would undoubtedly take any animal that they could. The problem is, they have no choice. Strong, healthy animals provide too great a risk.

Do wolves control the populations of their prey? For a number of reasons, it is difficult to make an unequivocal statement, but consideration of the problem should provide further insight into the complexities of predator-prey interactions. First of all, it should be pointed out that, in many places where such studies have been made, ungulates may have been particularly abundant. The law (and perhaps the "rites of manhood" among human males) has been harder on the wolves than on the ungulates. Wolf (and coyote and lynx and lion) pelts are proudly displayed, and wolf collars on cheap coats sold in department stores are very popular with some sets. While wolves have been hunted, ungulates (just as ducks) have been "managed." The results are apparent in the overwhelming number of deer in New Jersey. And mallards are going through an unprecedented boom in their numbers due to strict hunting laws. Possibly, at one time when ungulate numbers were lower and the numbers of wolves higher, wolf activity may have been a major factor in the control of ungulates (Pimlott 1976). The point is, because of such artificial controls on wolves and their prey, it is difficult to know just what sort of population control is imposed by the predators.

Now, at first glance, it would seem that if a wolf killed a moose, it would be reducing the moose population. But it is possible that the wolf-factor could be removed from a prey population (by killing all the wolves) and that the numbers of prey would *not* increase. This might be due to Errington's "principle of compensation" mentioned earlier: one mortality factor increasing as another decreases. If, for example, a deer herd is being ravaged by disease, few deer will die from other factors. But if the disease is wiped out, more may starve or be caught by predators. This illustrates another aspect of population control: different mortality factors may come into play at different population levels. Disease and starvation, for example, are usually not important until the herd has built up to an unusually high level.

In discussing the relationship of predators to prey, one must consider three major variables: (1) the density and reproductive potential of the prey species; (2) the variation in predator-prey ratios for different prey species and also for the same prey in different areas; and (3) the possible difference in the amount the predator does eat and what it might eat if food were readily available.

Wolves might have different effects on populations with different densities and reproductive potential. For example, deer may reach a density of thirty per square mile, while moose are much rarer. Therefore, more deer would have to be killed than moose in order for wolves to control deer populations. Deer may give birth when only one year old, and they often produce twins or triplets. But bison may not breed until three or four years old, and then they only produce one calf per year. Removing a bison, then, has a greater effect on the population than removing a deer.

Also, since a bison is four or five times as large as a deer, wolves would have to kill more deer than bison in order to feed themselves. A ratio of thirty moose per wolf, is, in fact, roughly similar to 150 deer per wolf. This factor is partly compensated for by the fact that larger animals are usually sparser.

There are cases where wolves have controlled prey numbers. For example, on Isle Royale there were about thirty moose (averaging eight hundred pounds each) per wolf (from 1959 through 1966). So, there were about twenty-four thousand pounds of moose per wolf. The 210-square-mile island supports about 600 moose in late winter, and about 225 calves are born each spring, but only about 85 survive to be recruited into the herd. There are about twenty-three wolves on the island and these kill about 140 calves and 83 adults each year for a total of 223 animals. Thus, the annual production of the moose herd is balanced by the average number of wolf kills. The implication is that the wolves are taking enough moose to control the herd.

As further evidence of the regulatory impact of the wolves, the moose have lived on the island since about the first of the century, but the wolves did not arrive until 1949. In the early 1930s, the moose herd was estimated at one thousand to three thousand animals (Murie 1934). They apparently overbrowsed their area and their numbers fell off drastically because of starvation and disease only a few years later. The herd recovered, but began to starve again in the late 1940s (Krefting 1951). However, since the wolves arrived, moose numbers have been lower and more stable than ever before, and the vegetation is beginning to recover.

In other places, wolves have failed to control their prey. In a study conducted in the national parks of Canada, one area contained about one wolf for every ten square miles. Three more areas had almost no wolves. There was no significant difference in survival in the big game of the two areas (Cowan 1947). The predator-prey ratio was three wolves for every four hundred head of game (including bighorn sheep, moose, elk, deer, mountain goat, and caribou). If the prey weighed an average of 300 pounds each, the predator-prey ratio would be 90,000 to 120,000 pounds of prey per wolf. Mech (1970) summarized a series of such studies where prey-weights were calculated. They show that wolves definitely control prey at predator-prey ratios at 7,400 pounds per wolf, at 15,000 to 22,500 pounds, and at 24,000 pounds. At 23,000 pounds, there is apparently two-thirds control by wolves, and no control is reported at ratios of 25,000 to 50,000, 52,500 and 90,000 to 120,000 pounds of prey per wolf. Such figures cannot be taken too literally because of censusing and weight-estimation problems, but it does seem likely that when the prey does not far outstrip the wolves in sheer weight, the predators will exert a major population controlling influence. Care must also be taken in extrapolating from these figures to any other predator. For example, African lions are much larger than wolves (and larger animals normally need to consume less food per unit weight than do smaller animals). Also, lions live where it is warm and the energy provided by food is not as important in maintaining body temperature as it is for the wolves of colder areas. In addition, cats lack the endurance of canids and stalk their prey very carefully, saving energy for a short rush—energy that need not be replaced by kills. Differences in social systems may also make extrapolation difficult. Male lions may expend almost no energy in the hunt, but merely walk over to the kill and take charge after the females have brought down the prey, thus they may require fewer calories than if they actively hunted. In any case, however, perhaps this glimpse into the role of wolves as predators will give us an idea of how one system works and an appreciation of some of the complexities involved. You may wish to consult Mech (1970) for details and extensive references.

## FORAGING STRATEGIES AND PREDATION

There have been several valiant attempts to synthesize unifying principles out of the mass of information we have regarding foraging behavior of animals. Some have suggested that one way of looking at it is as a matter of simple economics. What sort of effort will yield the greatest return per unit of energy expended? We all solve such problems every day. We may walk, rather than drive, to a grocery store that is just around the corner. And while we are there, we may pick up bread as well as milk so that we don't have to make two trips. Simple. So how does an animal (or how does natural selection) solve its "shopping" problems?

Matter and energy constitute the profits gained from foraging and those profits are spent in growth, maintenance, and reproduction. But foraging has its costs—some high, some low. A bird foraging on open ground is vulnerable to hawks. A grazing bull could be spending that time copulating. So, an optimal strategy maximizes profits and minimizes costs or risks (which, over the long run, may be converted to costs). Natural selection, then, is acting as an efficiency expert. It "fires" those genes that do not maximize their production.

Let's take an example from a predator-prey relationship. Prey come in different sizes and catchabilities. Since a prey of a given size will yield a certain amount of energy, how much energy should one expend in trying to catch it? If the prey is theoretically economical on the basis of size, how should the predator go about catching it? An optimal predator should not waste time catching fast, small animals that contain so little energy that more energy is expended in the chase than can be extracted from their flesh. An optimal forager would also avoid searching for prey at inappropriate times and in inappropriate places. But what is optimal in one kind of habitat is rarely optimal in another. If a cheetah chases a baboon from an open grassland into a wooded area, the strategy of giving chase at seventy miles per hour is no longer optimal. Speed loses its advantage if the baboon runs up a tree. Cheetahs don't climb very well and baboons do, and the baboon isn't about to come down while the cheetah is there. The cheetah will be more successful if it returns to the grassland to resume its hunt. The baboon's environment, then, is heterogeneous. It may contain grassy savannahs and wood. Let's take a closer look at heterogeneous environments and then see how they affect foraging strategy.

### Patchiness and Predation

Environments may change gradually from one place to another, but often the change is rather abrupt (Emlen 1973). Usually, environments take the form of a mosaic and are called "patchy." As one flies in an airplane over the countryside, it is easy to see how man has succeeded in making the terrain below patchy. A crazy-quilt pattern of fields, forests, and towns, dissected with the sharp lines of long roads attest to our diligence. This sort of patchiness is easy to see. But patchiness may take other forms, too, forms that emerge if the size of the habitat is scaled down. From the air, it may appear that an entire field is one large patch. And to a large animal grazing in the field, the field may also appear to be relatively homogeneous. As a buffalo grazes in the field it is not aware that one foot is on a rock and another on a clump of grass. So, although the buffalo hasn't noticed it, the field is actually broken up into smaller, dissimilar parts. Patchiness at that level is not of interest to the grazing buffalo, but it is more important to the lizard under the rock that was searching for insects in the grass until the @* @!! buffalo came.

For the lizard, then, this environment is patchy. It may search for prey in one kind of area and then in another. And some areas are more productive, as well as more dangerous, than others. Selection will tend to make the lizard relatively efficient in several, if not all, the patch types in its habitat. But at a cost. A lizard that blends with a rock is likely to be conspicuous among grass, and vice versa. Therefore, since the lizard is likely to be conspicuous much of the time, selection may opt for a very wary lizard that, because of its "spookiness," is likely to leave a good hunting area when there is really no danger. Or selection could take an intermediate route and make the lizard only *somewhat* inconspicuous on either grass *or* rocks.

Natural selection could also produce a lizard that blended well with grass and that only lived in grass and avoided rocks. The lizard would have a lower probability of being seen by a predator, but a whole range of rock-living prey would be unavailable to it. The lizard could then increase its efficiency (specialize) in hunting denizens of grass. So, because of its small size and terrestrial habits, then, a lizard may be highly responsive to a patchiness that would be hardly discernible to a larger animal.

In the case of the buffalo, the differences in the substrate beneath its feet would have very little selective influence. Natural selection would adapt the animal to the sum (or average) condition. This same environment, then, could be said to be *coarse grained* for the lizard, but *fine grained* for the buffalo. There may be entire areas nearby, however, that are unsuitable for buffalo, such as a sudden, steep, rocky hillside. In this case, both the grass-adapted lizard and the buffalo would be stopped by the same environmental "patch." Since lizards climb more easily than buffalos, a rock-adapted lizard might scarcely notice the incline, and so it might ignore this kind of a patch. The point is that patchiness is not solely the function of the environment, but also of the kinds of animals that live there.

The concept of patchiness must remain a rather broad and flexible one because of the ways patches can be defined. For example, patches can move. If one is a horsefly in a pasture, one must follow the group of patches that comprise the herd. If one is an African horsefly in the Serengeti Plains of Africa, when the great herds seasonally gather to begin their long treks across the plains, one's food source is much more fine-grained than when the ungulates are dispersed at other times of the year. As a general rule, less time is spent "searching" for food in fine-grained environments, especially if the grains repeat themselves at any predictable frequency.

Patchiness can also vary due to the behavior of the animal. In other words, the way an animal *utilizes* its resources can, itself, be coarse or fine grained. When, in the course of its daily activities, an organism encounters and uses its resources in the same proportions in which they actually occur (that is, does not select from among them), the animal can be said to use its environment in a fine-grained manner. Animals that spend disproportionate amounts of time in certain kinds of patches (selecting patches) use the environment in a coarse-grained manner. Fine-grained utilization may be forced upon the animal if it is large as compared to the patch size—such as a buffalo that doesn't notice stones or individual tufts of grass in its habitat.

Now, consider the ways insectivorous birds might forage in a mixed forest that harbors several kinds of trees and an abundance of insect species (MacArthur and MacArthur 1961). We will assume that no single bird species can exploit everything in the forest, therefore they must pick and choose, employing coarse-grained utilization to some degree. A bird could solve the problem by foraging in only one kind of tree as it flew over all other kinds, but a lot of time and energy would be spent in flying around. On the other hand, a bird could eat only one category of insect food (for example, those species within a certain size range), searching among all species of trees. Or, a bird could compromise. It could specialize regarding the part of the tree in which it foraged and the manner in which it foraged, once there. So, a bird might select a layer in a forest canopy and from this take a broad range of prey, utilizing a number of tree species and prey species. The compromise solution seems to be the best way to exploit a variable environment—better than specializing in places or prey, and in fact, this is how birds usually divide the resources in mixed forests (Figure 23.9). Only where food is extremely concentrated would it pay to specialize on a food type. For example, many parrots are highly specialized for eating certain types of fruit. Where these foods are found they are usually superabundant and very dense, and the parrots are "picking in high cotton."

There are basically three options open to animals that live in coarse-grained environments (Emlen 1973). First, species that tend to be "viscous," those that don't move around much, are likely to produce offspring that will continue to inhabit the parental patch type. Thus the species can, in time, become very well adapted to that patch. Pikas (*Ochotona princeps*), the little animals that entertain resting backpackers in the Sierras, live in rocky patches high in the mountains, but although they resemble rabbits, they almost never venture into subalpine forests or meadows.

Second, there are "vagile" species that can move around among patches, but prefer to rear their young in specific sorts of places. When some factor, such as high population density, lowers the desirability of one kind of patch, or makes it unavailable, they will move into the next best area. Recall the tits that were forced

Figure 23.9 The feeding zones of five species of the North American warblers in spruce trees. The darker areas indicate where each species spends at least half its feeding time. By exploiting different parts of the tree, the species reduce their competition for food and thus they can occupy the same habitat.

from mixed woodland into pinelands by high densities of territorial birds.

The third response to coarse-grained environments can be seen in viscous species in which local populations randomly disperse their gametes so that the offspring are forced to live wherever they happen to settle. This means those unfortunate enough to settle on undesirable patches are likely to die out. Probably clear-cut cases of this are rare among animals. Even current-dispersed barnacle larvae show some "choice" in where they live, and wind-dispersed insects can simply walk out of an area not to their liking.

Most environments are neither completely fine-grained nor coarse-grained, but fall somewhere in between. And, as we have seen, patchiness is a function of the characteristics of the animal that lives in the habitat as well as the physical characteristics of the habitat itself. A wolf pack may have to cross both forest and tundra as it chases a moose. A woodpecker moves about easily and may forage on the ground, flycatch, or search the bark of trees for insects. Also, although most environments are patchy, in a sense, the patches blend. As you walk to the edge of a field, and begin to encounter saplings, at what point can you state unequivocally that you are in a forest? It should be apparent that patchiness and the quality of the patches affect what food is taken and thus the nature of both predator and prey that inhabit any area. With these variables in mind, then, how can an animal maximize its foraging success?

## The Economics of Consumer Choice

MacArthur (1972) reviewed for us several aspects of optimal foraging theory. The problem is enormously complex, so he began with several preliminary assumptions in order to make some order out of the chaos. His assumptions were:

*(A)* The environment is essentially repeatable (patchy), with a certain given probability of finding

a commodity (for example, suitable microhabitat, nest site, or food item).

*(B)* There is a continuous and "normal" curve for the spectrum of food items, intermediate types, therefore, being the most common (an assumption that is violated by certain insects that feed only on one food, and, perhaps generally by herbivores, which are hindered from exploiting some of the most common potential foods by plant defenses).

*(C)* Similar animals forage in much the same way, thus any animal is best able to exploit food sources that fall between those that are optimal for two "neighboring" species. This means an insectivore with an intermediate jaw size will forage on insects appropriate to that size and will forage slightly less efficiently on insects that are somewhat larger or smaller. These latter insects would best be taken by slightly larger or smaller insectivores.

*(D)* No single phenotype can be maximally efficient on all prey types. Increasing efficiency on one prey type reduces efficiency in foraging on other prey types.

*(E)* A species tends to maximize its total intake of food.

After setting forth these assumptions, MacArthur broke down foraging into four phases: (1) deciding where to search, (2) searching for palatable food items, (3) deciding whether to pursue a located food item, and (4) pursuing it. Search and pursuit efficiencies are determined by the animal's morphology (assumption C) and the environmental repeatability (assumption A). An efficient animal wouldn't search for something too big to eat, and whether it pursued a particular food item would depend on the probability of finding another one in a repeating environment. So assumptions (A) and (C) are closely associated with phases 2 and 4. Since phases 2 and 4 are essentially covered in the assumptions, MacArthur considered the essential variables to be phases 1 and 3, that is, where to forage and what to pursue.

Obviously, an optimal consumer should forage where the greatest yield would be expected. The decision whether or not to pursue should also be easy. A predator either pursues a prey item it has discovered or finds another and pursues it instead. Both decisions ultimately result in the forager beginning a new search, so the best decision is the one that returns the greatest yield for the time expended. Therefore, an optimal consumer should pursue an item only when it cannot expect to locate and catch a better item during the time it would take to catch and eat the first one.

So, how does all this apply to the real world? Consider some examples. Many animals such as birds, which pick up insects from among the foliage of trees, spend much of their foraging time *searching* for prey, but relatively little time chasing, capturing, and eating the little insects. The insects just sit there or move slowly about, so the chase and capture is easy, and most are swallowed with a single gulp. So in "searchers," the mean search time per item eaten is large compared to the mean pursuit time, which is negligible. Hence, the optimal strategy is to eat every palatable insect encountered. This means that searchers can usually be classified as food generalists.

"Pursuers," which spend little time (and energy) in finding their prey, but a great deal of effort in capturing it, should be very selective in what they pursue. Lions live among their prey, but their prey is likely to be fast. Many of them can outrun lions, so lions have to be careful about which ones they are willing to put their energy (and time) into chasing. Thus, pursuers should generally be more specialized foragers than searchers.

Hyenas may have to search far and wide for a wildebeest to bring down on dark African nights. But when the great herds gather to move across the grasslands, they may find themselves in a food-dense environment. Thus their strategy should change. We really don't know at this point if it does.

Other animals may live among a high density food supply year-round. Food-dense environments result in a lower mean search time, compared to that expended in a food-sparse area. Thus, an optimal consumer should restrict its diet to the better types of food items when food is abundant. Actually, the balance is not always easy to reach. Young or newborn prey animals, for example, require less time for a predator to kill and eat, and more of their parts are digestible than those of full-grown animals, but they tend to have a lower protein and a higher water content. So, here, pursuers shift toward searching (and generalizing, since many prey types may be available—whether distinctions are made according to the condition of the animals or by species). But, at the same time, some discretion regarding food can be expected. The degree of discretion exercised, of course, would be partly a function of the range of palatability of the available food and how easy unpalatable items are to identify.

The principles of economics, then, are being used by ecologists, and increasingly by ethologists, to predict strategies of foraging and predation. The strengths and weaknesses of such theorizing obviously lie in the assumptions. And, it may be argued that maybe our generalizing at this point is premature. Perhaps natural historians haven't yet told us enough about animals. It has been only very recently, for example, that we found that chimpanzees use tools, that African hunting dogs have leaders, and we still aren't really sure how wolf packs are formed. We need people in armchairs developing their theories. And we need people out there seeing what animals are really up to. Occasionally someone does both. MacArthur did.

### Foraging and Feeding Efficiencies

We have seen that it is important to know how an organism allocates its time and energy budgets. It should also be apparent that conflicting demands are made on the matter/energy obtained through foraging. The ratios of the matter/energy that go into growth, maintenance, or reproduction tell a lot about the animal and, in a sense, they help to define its niche. But such budgets are flexible. In fact, shifting the time and energy budget is a very effective means of coping with environmental change. I recently watched a line of large black ants stealing seeds that I had placed out for birds. The large-headed "soldiers" were carrying seeds in their powerful jaws just as were the smaller "workers." When the group was disturbed slightly, however, it was the soldiers that appeared to be the most aggressive, and the one that made it onto my hand was most effective indeed. The ants that had all been behaving identically had shifted their behavioral patterns (energy budgets) to cope with a new environmental situation.

The shifts may be even more pronounced and unpredictable in vertebrates. The birds outside my house at present are establishing breeding territories, and the woods are full of song, and chasing and fleeing birds are everywhere. I expect all this to end shortly, though, as the birds begin to shift their energy into mating and raising young.

As another example of a budget shift, consider sexual differences in animals that show parental care. The energy drain may be quite distinct for the sexes. Adult female red squirrels that are lactating, consume an average of 323 kilocalories of food per day whereas a male eats only about 117 kilocalories (Smith 1968). At this point, we lack critical information on the difference in demands placed on the sexes of most animals.

We have compared two strategies of foragers, searchers versus pursuers, to introduce us to the idea of the economics of ecology. But there are other foraging strategies. For example, carnivorous animals can sit and wait (the old sit-and-wait strategy). In this case, the predator ambushes its prey. Web-building spiders, barnacles, boas, pythons, Cooper's hawks, and goshawks are sit-and-wait predators. At the other extreme, a carnivore can actively hunt for its prey as do most fish, as well as jumping spiders and wolves (Pianka 1966b; Schoener 1969a,b). Hunting obviously requires more energy than ambushing, but in order for the first to work, certain conditions must be met. There must be a fairly high prey density, high prey mobility, and/or low energy requirements of the predator. In the hunting tactic, the most important factors are the distribution of the prey and the searching abilities of the predator (Pianka 1974). Actually, the situation in nature is rarely an either/or condition. Ambushing and hunting are but two ends of the spectrum, and although most animals probably do substantially more of one than the other, probably most animals do some of both. In addition, some ambushing animals expend some time and energy to lure their elusive prey (Figure 23.10).

A number of factors can influence a predator's time and energy budget. For example, rock pipits (*Anthus spinoletta*) feed along the English coastline. They were observed for two consecutive winters, one mild and one harsh (Gibb 1956). In the mild winter, with total daylight exceeding 9 hours, they spent a daily average of 6½ hours feeding, 1¾ hours resting, and 45 minutes in territorial fighting. The next winter was harsher and food was much scarcer, and they

Figure 23.10 An angler fish is a rather sluggish creature that feeds on some small and elusive fish that it lures to their deaths by wriggling a wormlike appendage over its cavernous mouth. When a fish is within range, the prey soon disappears with a quick slurp. The angler fish, then, is an ambusher that attracts prey to its vicinity.

spent $8\frac{1}{4}$ hours feeding, 39 minutes resting, and only 7 minutes in defending territories. Evidently, the combination of reduced food levels and extreme cold demanded that over ninety percent of the daylight hours be spent in securing food.

Energy budgets are critical to the predator. It must be efficient. When prey items are too scarce, the predator simply dies. And scarcity is a relative term. An animal can starve in the midst of what is apparently abundant food. In some places, during the daylight hours in winter, English tits must find a food item about every $2\frac{1}{2}$ seconds (Gibb 1960)!

The question also arises, What is the optimal prey size for any type of predator? Holling (1964) used simple geometry to calculate the prey size that mantids *should* best be able to handle and then he looked to see what prey size the animals actually preferred. He first estimated the prey size a mantid (Figure 23.11) should handle easiest (Figure 23.12). He then offered five hungry mantids prey animals of different sizes to see which size they would attack most readily. The results were just as every good biologist would hope for (Figure 23.13). The mantids were reluctant to attack prey that were noticeably larger or smaller than the theoretical optimal size. The correlation may illustrate natural selection operating to produce effective prey-catching devices at the same time it produces a behavioral propensity to attack prey that can be handled by that machinery.

Note that the mantids were more favorably disposed toward prey items larger than optimal than toward those at the smaller end of the spectrum. It may well be better strategy for an animal to ignore prey that is smaller than some set limit and to spend that time in searching out larger prey. Larger prey, of course, yield more energy per capture, even considering the added energy expenditure in handling and killing larger animals.

Smaller prey, on the other hand, are usually much more abundant than larger prey, so most animals encounter and eat more smaller prey items than one might expect on the basis of energetics. In addition, since small animals that eat small prey encounter their food much more frequently than larger animals that need larger prey, larger animals tend to eat a wider range of prey sizes. The reason is, although they can handle larger prey, the scarcity of these animals causes them to attack smaller prey, also. (I have no proof, but I would suspect that the curve produced by Holling's mantids in Figure 23.13 would be skewed in the opposite direction for wolves, especially when several populations are lumped together. It seems that wolves would take a few large ungulate species, but that there wouldn't be much difference in the sizes of the largest animals they could bring down—for example, moose, elk, caribou. On the other hand, the range of prey smaller than optimal would be much wider and might include many species ranging down to hares, birds, and mice.)

Because of the heavy demand made on the food supply and the wide range of food sizes taken by larger animals, predatory species must generally be of markedly different sizes in order to inhabit the same area. The larger the predators, the greater the size differences between different species must be (MacArthur 1972).

The social system of the predator also plays a role in the range of prey sizes it takes. While a single wolf *can* bring down a mouse, wolves tend, instead, to group together to pull down moose. Coyotes, on the other hand, are much less social, and where they occasionally may take a larger animal such as a sheep, they are terrific mousers and rely primarily on small animals caught by individual effort.

One can also apply the principles of predator economics to herbivores. Herbivores don't have to chase their food and they may live surrounded by it. They are likely to be constrained mainly by plant defenses such as thorns or toxins. Plants are high in cellulose, however, and cellulose is difficult to digest, so the herbivores must expend a lot of energy and time in extracting the nutrients from their food. To cope with this problem, herbivores have long intestines through which their food passes slowly and they harbor intestinal fauna that can break down cellulose; they also must eat an enormous amount of this "low quality" food and they may ruminate or chew their cud. The nutrients in animal food are much more available and of a higher quality (closer to the specific requirements of the predators). Therefore, a predator can *afford* to expend more energy in acquiring its high quality food than can a herbivore.

Figure 23.11 The praying mantis is a predator that takes food within a specific size range, based on the size of its foreleg. This mantid is eating a young katydid.

## PARASITISM

Because of the often more delicate relationship of parasite and host compared to that of predator and prey, we find an incredible array of specializations by which parasites maximize their benefits while minimizing their effects on the host. Not only does the host sometimes serve as the ecological niche, as well as food supply of the parasite, but in some cases the relationship has developed social aspects that require highly specific communication between the host and the parasite (such as between ants and some of their insect parasites, as we shall see).

A parasite, of course, must have certain characteristics to qualify for the title. Basically, a parasite is a species that interacts with a host species and derives benefit at the expense of its host. In other words, damage is done. Since it is true that, usually, the host must remain alive in order to be of use to the parasite, it might seem that parasites would have little effect on population numbers. And, in some cases, that is the case. Some small worms living in the intestines of mammals take nourishment from their host, but their needs are so small that the host is at no great disadvantage. In other instances, however, the parasite may directly or indirectly bring about the death of the host or the host's offspring. Parasites may also weaken their host so that its reproductive output is decreased (resulting in fewer potential hosts for the next generation of parasites); or the weakened host may fall to predators (Cowan 1946). In either of the

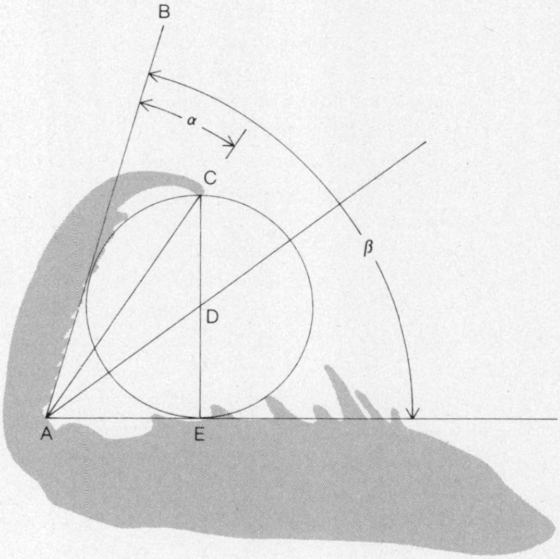


Figure 23.12 The dimensions measured to learn what the theoretical optimal prey size should be, based on how easy a mantid could be expected to handle spheres of various sizes with this type of foreleg.

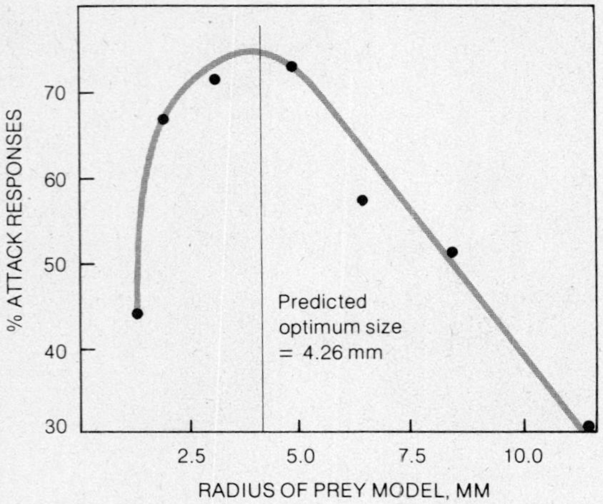


Figure 23.13 The percentage of prey of various sizes actually attacked by hungry mantids. The most frequently attacked prey were those that were predicted to be of optimal handling size.

latter two cases, the parasite loses as well as the host. In situations in which the reproduction or vigor of the host is impaired, it may be assumed that the parasite-host relationship probably developed relatively recently and that the host has not yet built up effective defenses against the parasite, nor has the parasite had time to ameliorate its effects.

An example will point up the disastrous results that can occur when parasite and host are not well adapted to each other. In 1904, the sac fungus (*Endothia parasitica*) was accidentally introduced into North America. In China, from whence it came, it is held in check by a variety of control mechanisms that do not exist in North America. Sadly, the majestic American chestnut proved defenseless to the fungus, and by the late 1940s, the tree that had dominated our Appalachian forests was virtually extinct (Kormondy 1969). Americans are now witnessing the same sort of "devastation rather than regulation" as our elms are falling to the newly introduced Dutch elm disease.

Actually, in some systems, the coadaptation of parasite and host may occur remarkably quickly. In one laboratory study, a population system of the house fly (*Musca domestica*) and a parasitic wasp (*Nasonia vitripennis*) was maintained for twenty generations. At the end of that time, it was found that both populations showed significant evolutionary changes (Pimentel, et al. 1963). The host had become more resistant to the parasite, and the parasite had become less virulent to the host (Table 23.3). Selection had apparently reduced the intensity of interaction between the host and parasite in that very brief period of time.

When prey animals are suddenly subjected to a new predator, they are less likely to be wiped out than if they are faced with a new parasite, although the predator's attacks are geared to result in death of the prey. This is because of the stronger operation of density-dependent effects when the attacker doesn't live within, or on, the attacked.

There are certain other differences in the effects of predators and parasites. It is true that both are seeking out another animal and are, therefore, subject to some of the same influences of prey distribution and number. However, the securing of food for the predator often immediately ensures primarily its *own* success, while a greater percentage of food for an established parasite goes into providing food for its *progeny*, usually in the formation of eggs. Also, the life history of the parasite probably must be more strongly coordinated with the habits of its host, since a parasite is normally more species-specific in its target and less "switching" is possible.

Parasites such as trematodes often have a number of stages in their life cycles and, interestingly, the regulation of their population numbers may differ at each stage (Figure 23.14). For example, small fish may prey on the aquatic stages of some parasites, subjecting them to density-dependent regulation. But at another stage in our example, the egg that passes out with the bird feces may be subject to density-independent drying by the hot sun. As a rule, the more the stages in the history of any organism differ, the more opportunity there is for different kinds of regulation.

## Host Defenses

Host species, of course, are continually developing defenses against their parasites through natural selection (Osche 1962, 1966). As examples, grass frogs, toads, and Alpine salamanders free themselves of leeches by sitting in the sun, subjecting any leeches to intolerable heat (Eibl-Eibesfeldt 1970). The amphibians are probably not too fond of the sun either, but those individuals that fail to sunbathe remain parasite-infested and probably sicken or die sooner than their fellows, and therefore their nonsunbathing genes are less likely to be represented in the next generation.

In leaf-cutting ants (*Atta cephalotes*), the larger maxima castes cut circular sections from the blades of leaves and carry them back to the nest. While the worker is involved in its task, it is vulnerable to attack from small parasitic flies of the Phoridae group. The risk has been lessened because members of the smaller minima case now "ride shotgun" on the leaf, jaws open, snapping at any flies that draw too near (Figure 23.15).

Other species allow themselves to be cleaned of parasites by "cleaners." Probably the best-known example is that of the rhinoceros and its tickbirds. In addition, there are the fascinating cleaner wrasses, small fish that maintain cleaner stations or "barber shops" in the sea. Species that normally do not tolerate each other gather at these stations and peacefully wait their turn to be cleaned of parasites.

In some cases, the answer to parasitism might be the development of more specific recognition signals. An example of parasitic behavior that is permitted by unspecific communication is found in the slave-making ants. The fertilized queen of the red ant *Formica rufa* may enter the nest of the related *F. fusca* if she can't find a nest of her own species. The *fusca* ants are unable to distinguish her from one of their own species, and if no other queen is present, she will be adopted and cared for as their queen. All her eggs are, of course, her own species, so the nest is soon entirely taken over by *F. rufa* (Eibl-Eibesfeldt 1970).

The fertilized queen of *F. sanguinea* has a more bizarre method of parasitizing her cousins. She is given to entering the nest of *F. fusca* and kidnapping a few of the larvae. She defends the larvae against recapture by their parent population and raises them herself. When they mature, they care for her as she lays eggs of her own species. For some reason, the

**Table 23.3 Evolutionary changes in the host–parasite system of the house fly (*M. domestica*) and a parasitic wasp (*N. vitripennis*) after twenty generations of interaction in the laboratory.**

| | REPRODUCTIVE RATE (AVERAGE NO. PROGENY PER FEMALE WASP) | PARASITISM RATE (PERCENTAGE OF FLY PUPAE PARASITIZED) | LONGEVITY OF FEMALE WASPS (DAYS) | LONGEVITY OF MALE WASPS (DAYS) |
|---|---|---|---|---|
| *Control wasps on control flies* | 140 | 51.7 | 7.0 | 1.6 |
| *Experimental wasps on experimental flies*[a] | 46 | 39.6 | 4.6 | 1.4 |
| *Control wasps on experimental flies*[b] | 68 | 46.0 | 5.2 | 1.4 |
| *Experimental wasps on control flies*[c] | 123 | 52.6 | 6.6 | 1.7 |

*Source: Pimentel et al. (1963).*
[a]*Measures evolutionary changes in both hosts and parasites.*
[b]*Measures evolutionary changes in host resistance.*
[c]*Measures evolutionary changes in parasite virulence.*

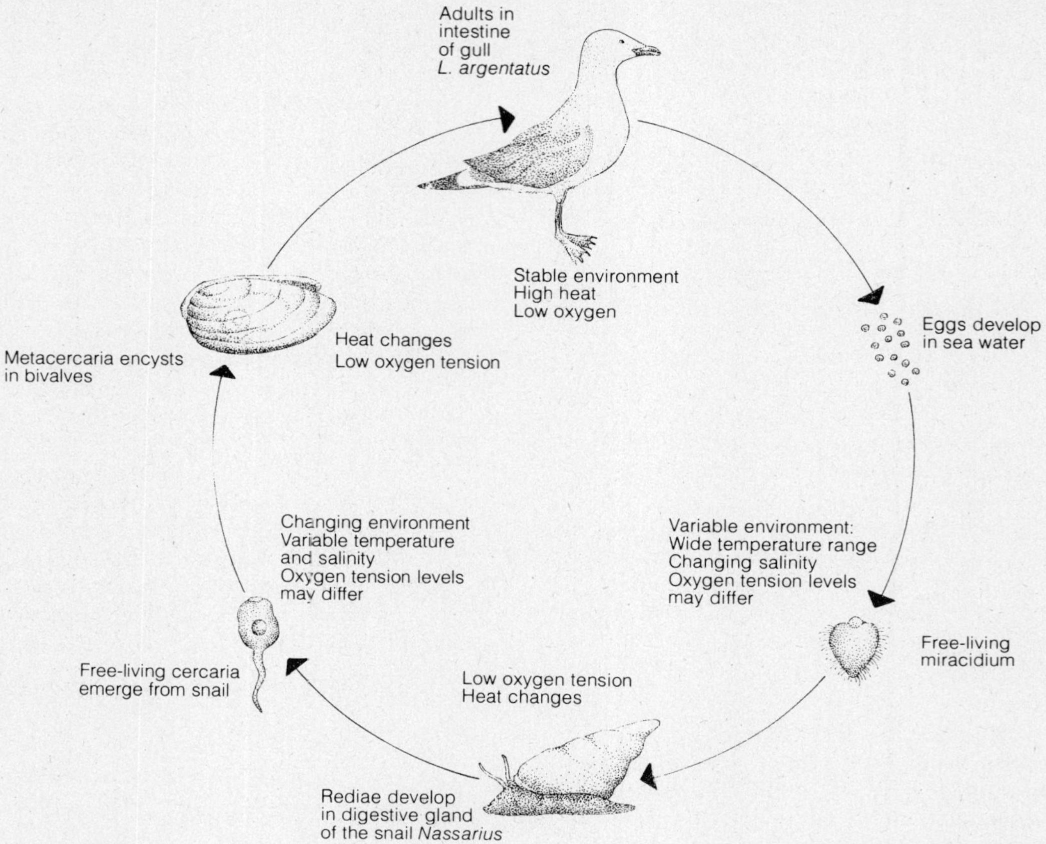


Figure 23.14 The complex life cycle of the trematode *Himasthla quissentensis*. Such complexity is not rare in the life histories of parasites. It is apparent that each phase in the cycle has its own particular risks and that very few individuals reach maturity. The reproductive solution to the high mortality rate in such species is increased fecundity.

queen is later killed by her own offspring, but in any case, eventually all the ants in the nest are of one species—*sanquinea*. The close evolutionary relationship of the ant hosts and parasites probably has resulted in their having similar signaling devices which accounts for the fact that the parasitic queens are not attacked immediately when they enter foreign nests. The story of "social parasitism" among ants is a fascinating one described in detail by Sudd (1967) and Wilson (1971a).

## Brood Parasitism

Finally, to illustrate the variety of population-altering interactions that can occur between parasites and hosts, we should consider the *brood parasite*. Brood parasites are species of birds that lay their eggs in the nests of other species, usually leaving them to be cared for entirely by the hosts. There are about eighty species of brood parasites and each has its own special methods of operation (Figure 23.16).

The parasitic cuckoos show remarkable and diverse specializations in their reproductive behavior (Lack 1968). But the basic strategy is simple. When a female finds a nest at the proper stage of development, she quickly lays one egg in the nest and flies away. (The cuckoo is a relatively large bird and one cuckoo chick is about all the host parents can raise.) But around this simple act, a syndrome of specializations have arisen. For example, cuckoos have a loud song, an adaptation to their relative scarcity. Also, the sound is simple because the young cannot learn it from their own parents (innate vocal patterns may lack embellishment). The song is also distinctive for each species, permitting easy recognition where more than one species of cuckoo may be present.

At least one cuckoo species will destroy any nest that is too advanced for it to lay its eggs in. The result is that the host parents may begin a new nest. The cuckoo can then watch and wait until the nest is at the proper stage. Then, when both parents are gone, the lurking cuckoo moves in. The females are able to lay eggs very quickly and so they are likely to be able to leave before the parents return. Even hole-nesting birds can be parasitized because the female cuckoo has an extensible cloaca and can therefore lay her heavy-shelled eggs into openings that she cannot enter.

Of course, some of the hosts have developed defenses. For example, most cuckoo hosts will leave a nest in which an egg appears that is different from their own, or they may throw out the strange egg. Stephen Rothstein (1970) described several behavioral, morphological, and ecological factors correlated with such rejection behavior. But, as a response to the host's rejection, some cuckoos have evolved mimetic eggs. Even a single cuckoo species may mimic several different types of eggs. The eggs of an individual cuckoo, of course, probably all look alike, but different birds may lay eggs of different appearances. It is difficult to understand how egg mimicry could have evolved unless females imprint on a nest type, and return to such nests or at least to that general area with some degree of persistence, so that distinct evolutionary lines can occur. Lack believed such a system could only be maintained by the male having no influence on the egg's appearance, since, if he did, he might carry genes for mimicking a different type of egg, causing the system to break down. No matter what the appearance of the egg, though, it has a short incubation period and therefore usually hatches first, before the host's eggs. The young cuckoo is therefore older and stronger than its nest mates and is likely to attract more of its parent's attention by its very vigor. In certain species, such as the honeyguide, the brood parasite may also kill the host nestlings or push them out of the nest (Figure 23.17).

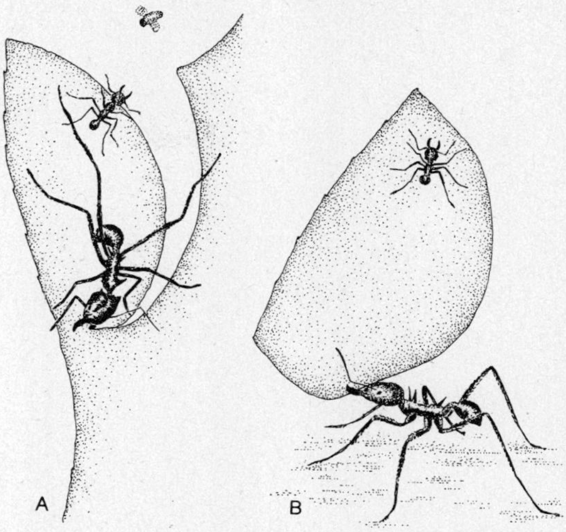

Figure 23.15 Defense against a parasitic fly (Phoridae) by a member of the caste of minima workers in the leafcutting ant (*Atta cephalotes*). In (A), as a larger caste member cuts the leaf, the minima chases away a fly. As the carrier transports the leaf (B), the minima acts as a guard.

It is interesting and not entirely unexpected, that young cuckoos often bear a striking resemblance to the young of their hosts (Figure 23.18), even to exhibiting special feeding releasers.

Rothstein (1970) discussed the evolutionary implications of nest parasitism. As one experiment, he placed both real and artificial (plaster) cowbird eggs in the nests of a number of species, and in most cases the birds immediately rejected the egg (he called these species *rejectors*) by burying it in the nest with new twigs, pecking it, or simply tossing it out of the nest. Other species (*acceptors*) accept the egg, hatch it, and rear the offspring as their own. As a result, they suffer reduced reproductivity. Rejectors have the following features: (1) eggs unlike those of the

Figure 23.16 Two brood parasites (the cuckoo and honeyguide) and their hosts.

Figure 23.17 A newly hatched and blind honeyguide (*Indicator indicator*) kills a barbet (*Lybius bidentatus*), the host nestling, with special mandibular hooks on its bill.

cowbird, (2) large bills with which they can pick up cowbird eggs, (3) long association with the cowbird through sharing the same habitat, (4) large, conspicuous nests, and (5) large population sizes. Features 3, 4, and 5 would contribute to rejection behavior by increasing the likelihood that the nests would be found by cowbirds. Thus, cowbirds could place strong selection pressures on the species to develop rejection behavior; (1) would permit the host to recognize the cowbird egg, and (2) would give it the means to do something about it.

The cowbird egg is usually rejected by tossing it out of the nest, undoubtedly the most efficient means of rejection. Egg burial and nest desertion are clearly less adaptive and may result in the host's losing its own brood for that season. Also, birds may be preadapted to carrying away strange eggs, since they carry away the droppings of their own young as a means of nest sanitation. This behavior would simply be a new use for an old pattern, and thus could be expected to develop rather rapidly.

If a cowbird's egg appears in a nest late in the host's nesting period, it probably won't hatch anyway, since the host will be off finding food for its hatched young while the cowbird egg still requires incubation. Thus, cedar waxwings (*Bombycilla cedrorum*) are more likely to accept cowbird eggs that appear in their nest late in the incubation period. Even if the cowbird hatches, it may soon die because the cedar waxwing is a fruit and seed eater and may not be able to furnish the protein necessary to sustain a cowbird (Rothstein 1976a,b,).

How do rejectors tell their own eggs from the eggs of cowbirds? The robin (*Turdus migratorius*) seems to respond by a system of additive stimuli (Rothstein 1970). (Recall the law of heterogeneous summation.) Cowbird and robin eggs differ in three major ways: in color, markings, and size. Eggs differing from robin eggs in only one respect are accepted. Eggs differing in at least two respects are rejected. Other experiments (Rothstein 1975) have shown that relative numbers are not important, at least in some species. Recognition, in these cases, is not by noting the "odd" egg. Rejector species apparently can recognize their own eggs, no matter what the ratio of "real" to "other" eggs is.

The additive stimuli effect in egg recognition implies that at least some birds recognize their own eggs innately. But some species may have to learn what their eggs look like. Rothstein replaced the first egg laid in the nests of eleven gray catbirds (*Dumetella carolinensis*) with artificial cowbird eggs. At six nests, the cowbird egg was rejected either before or shortly after the second catbird egg was laid. Later in the breeding cycle, the catbirds showed even less tolerance for cowbird eggs. Thus, it seems that the catbirds recognize their own eggs at least partly through learning processes. It may be that catbirds learn the appearance of their eggs in the first breeding season and that the six rejectors were experienced birds. There are no data to support this suggestion at present, however. Other experiments indicate that catbirds learn the appearance of their own eggs and that, in catbirds, the rejection of "novel" eggs is an innate behavior (Rothstein 1974).

It was hypothesized that learned egg-recognition permits the host the development of a wider range of antiparasite devices (Rothstein 1974). A long-standing argument states that nest parasites are continually evolving eggs that increasingly tend to mimic the appearance of their hosts' eggs, so that finally the parasite's egg will be totally indistinguishable from that of the host. This argument neglects the prospect of the host's egg continually diverging from the appearance of the improved parasite's egg. So, there may be an evolutionary, and unending, "chase" in the appearance of eggs. However, if egg recognition on the part of the host is innate, that host may tend to reject its own eggs if they are different from the established type, unless there is another genetic change (or a pleiotropic effect) that permits recognition of a new kind of egg. Birds that learn the appearance of their own eggs can more easily cope with changes, and their eggs could take two different evolutionary pathways. They could become polymorphic, so that the population produced multiple, but

Figure 23.18 The fledglings of the brood parasite *Eudynamis scolopacea* have plumage that resembles that of *Corvus splendors* (center), the species in whose nest they are hatched.

discrete, egg types, thus rendering mimicry difficult, or they could undergo directional selection and "run from" the mimicking egg (Rothstein 1974).

One may wonder, How did brood parasitism arise to begin with? It has been suggested that it could have begun with chance "dumping" of eggs in other nests if a bird's own nest had been destroyed before it had laid its eggs (Hamilton and Orians 1965). Interestingly, the phenomenon is not unusual among birds. If such dumping resulted in high reproductive output, the stage would have been set for the evolution of parasitism.

It was suggested that brood parasitism was most likely to develop under the following conditions: (1) the dumper is not ejected from the territory; (2) the young from dumped eggs hatch before the other eggs, resulting in increased attention by the hosts; (3) the protoparasite should tend not to imprint; (4) the protoparasite should be omnivorous; (5) the host should be altricial (so that its own young remain in the nest); and (6) brood parasites are least likely to decimate their hosts, and hence themselves, if they are rare as compared to their hosts. Rarity will also retard the development of antiparasite devices by the hosts.

We see, then, that both predation and parasitism act in complex ways to influence population size. Probably the effects of predation as a tool of natural selection are manifest in more ways than those of parasitism. The experiments so far have provided us with new insight, while unveiling the depth of our ignorance about how predators really behave and how far-reaching their effects are. At this point, we are in great need of careful long-range field experiments such as those that have been conducted on the wolves of Isle Royal.

# 24

# COMPETITION

"Nature, red in tooth and claw" has too often been the image of competition in the wild. Remembrances of fictional violence between leopards and pythons simply confuse the picture of what competition really means. After all, there is nothing for a python or a leopard to gain in such a fight—unless one intends to eat the other. But then, neither is competition completely portrayed by two wolves fighting over a prey carcass, although the advantages of winning in this case are more obvious. So, how, then, can competition best be portrayed? Perhaps by two plants of slightly different types growing side by side on a sun-warmed slope of an alpine meadow. One plant is slightly taller than the other. As it happens, the taller plant tends to shade its neighbor just a bit. Thus, the tall plant is able to capture more sunlight and increase its food supply over that of the shorter plant. The taller plant, as a result, leaves a few more seeds, each giving rise to a new, taller plant. Time goes by and the seasons come and go until, one day, the taller plants are in abundance, and scarcely a single plant of the shorter type is to be found. Interestingly enough, the presence of the taller plant seems to have had little effect on the bushes, trees, or grasses in the area, none of which it resembles much at all.

Let us return momentarily to our leopard-python fight. If the leopard attacked the python with the intention of eating it, then we are no longer talking about competition. Predator-prey relationships are quite different from competitive ones. For example, it is usually to an animal's advantage to eliminate its competitors, but certainly not its prey. When the vigorous and intelligent dingo was introduced to Australia, it did not kill off its primitive marsupial prey, but its hunting prowess was so superior to its competitors, the Tasmanian devil and the Tasmanian wolf, that these animals disappeared from the Australian continent and, today, are found only on islands to which the dingo has not been introduced.

In this chapter, we will concern ourselves with how animals compete, what they compete for, and some of the results of that competition. It will become apparent that there is some disagreement about these matters and, in fact, about whether competition is an important influence in nature at all. The answers to many of the questions will seem to be intuitively apparent, and one might wonder, since the answer is so obvious, why the question arose at all. According to George Bernard Shaw, the inspired guess comes first and the data that are supposed to verify it are only a dodge to knock it into the thick heads of uninspired numbskulls. The scientific world may be full of uninspired numbskulls with reasonably thick heads, but, nonetheless, time and again, cold hard data have burst many a theoretical balloon, even those that seemed eminently logical—except that they just didn't have much to do with the real world. Even when intuition was upheld by experiment, the results often proved more fruitful than anticipated (see Lack's 1968, p. 309, discussion of one of Tinbergen's "obvious" experiments).

We will see that for such a complex and encompassing subject as competition, no well-defined and succinct description of its operation can be gleaned even from the massive amount of information available. The discrepancies only serve to point out that we haven't yet entirely discovered the nature of the phenomenon. However, in spite of some contrary arguments given below, it should become apparent that competition exerts powerful influences on living things. It can be argued on an intuitive basis that perhaps every other sort of interaction operates ultimately to increase the likelihood of success in competition.

## PROBLEMS IN DEFINING COMPETITION

An ancient oriental saying states, "Two tigers cannot live on the same hill." It's hard to know precisely the point of some old sayings, but this one could mean that some large carnivores cannot live in close proximity because of the problems of *competition.* But what *is* competition? We will adopt Birch's (1957) definition: Competition occurs where animals utilize common resources in short supply or when animals harm each other while seeking resources that may not be limited. The latter instances may be much rarer in nature. Conflicts over commodities *not* in short supply should, for now, be considered as (1) evolutionary hangovers from times when those commodities *were* in short supply, or (2) the result of a mechanism to keep general aggression at a high level in case commodities should fall into short supply at another time—for example, if they are seasonal.

Competition is often used as an explanation for an observation when the term itself *explains* nothing (Andrewartha 1961). But rather than toss out the phrase, we will simply remind ourselves that it is a broad term that encompasses a variety of observed phenomena in the natural world. It will become apparent that the evidence that competition is a factor in natural selection is, when carefully examined, inconclusive. It may seem that the concept of competition is derived largely from circumstantial evidence, a measure of faith, and a lack of a better explanation. For a brief historical review of the idea of competition and resource partitioning in communities and suggestions for new approaches, you might see Schoener (1974) and Diamond (1978).

## THE EFFECTS OF COMPETITION ON EVOLUTION

It cannot be denied that the concept of competition is a central theme in much of modern ecological theory. But because its role in natural selection remains so ill-defined, the argument over its influence goes on. Many ecologists consider competition to be among the most critical of ecological and evolutionary phenomena, while others say the concept doesn't really help us to understand nature at all. Three basic reasons for the different viewpoints have been suggested (Pianka 1974). First, competition is a surprisingly subtle and elusive phenomenon in nature, difficult to precisely describe and quantify. Second, evolutionary ecologists, interested in ultimate causation, may see competition as more important than those ecologists with a physiological (proximal) bent (Orians 1962). And, finally, there may be a natural and very real distinction between relatively *r*-selected and *K*-selected organisms (Pianka 1970), and thus the investigation of competition can become very complex, especially if organisms or populations can shift along the *r*-*K* continuum. So, competition is a confusing issue. Perhaps in the following discussion you can absorb some of that same spirit of confusion.

## THE NICHE

All living things have their place in nature. Their surroundings—the place where they are found—is referred to as their *habitat.* The habitat can be defined several ways. For example, an animal may live in a desert habitat or specifically in a briny pool in that desert. Furthermore, it may live in a certain part of the pool, its microhabitat. No matter where an organism lives, however, it is forced to interact with its surroundings, including other forms of life that it may contact directly or indirectly. The sum of all such reactions along with the organism's requirements is the *niche.* Again we encounter an operational term that is mainly useful only as a general concept and that furnishes much raw material for scientific haggling. For material relating to more detailed descriptions and operational models of the niche, see Hutchinson (1959); Levins (1968); Pianka (1974); Whittaker, Levin, and Root (1973).

### Niche Theory

In his famous "Concluding Remarks" at the Cold Spring Harbor Symposium, G. E. Hutchinson (1957) probably set forth the most influential modern definition of the niche. He broadly defined it as the total range of conditions under which the individual (or population) lives and replaces itself. Unfortunately, Hutchinson's niche dimensions did not include behavioral factors, and thus his niche is similar to what some ecologists would call the microhabitat, or the "habitat niche," as opposed to the "food niche" or some other variable. In fact, the modern concept of the niche has become so broad that some (Weatherby 1963) have suggested that it be restricted to food alone. Others have said that perhaps we should *specify* the dimension by talking about "food niche," "place niche," "habitat niche," or whatever. In any case, in spite of some problems with Hutchinson's definition, it has proven to be a useful one.

Hutchinson's niche theory had important implications for behavioral ecologists. For example, it distinguished between the *fundamental niche* and the *realized niche*. This distinction carried with it the notion that an animal's potential niche is seldom fully utilized at any point in time or space. In other words, an animal usually *can* interact with its environment in more ways than it *does*. What it *can* do is its fundamental niche; what it *does* do is its realized niche.

Ecologist Eugene Odum (1971) included behavioral coordinates among the niche's dimensions. He said the niche is "the position or status of an organism within its community and ecosystem resulting from the organism's structural adaptations, physiological responses, and specific behavior (inherited and/or learned)." He emphasized that the ecological niche depends not only on where an animal lives, but what it does. In one quotable definition, he said that the habitat is the animal's address, whereas the niche is its profession.

**Deterioration of the Environment** Earlier we discussed how an environment chooses its animals. The implication is that only certain kinds of animals can live in certain kinds of environments. As an animal becomes more in accord with its environment, we can speak of *adaptation*. Living things can adapt in innumerable ways, as we know. *Populations* can adapt genetically (a means that includes changes in instinctive behavior), or *individuals* can adapt through physiological and developmental means (which include learning).

As we saw earlier, adaptation is a massive set of compromises, for the organism must adjust to countless, sometimes opposing, environmental demands. If predation forces an animal to be skittish, it may be a less efficient forager. Sometimes, however, different adaptations complement each other. For example, a northern carnivore can bring down larger prey if it increases in size, and by growing larger, it can also conserve more body heat by reducing its surface area in relation to its volume.

But even as animals change to adapt to their environment, the environment may be changing. Environmental changes that reduce overall adaptation are collectively termed *environmental deterioration,* an unfortunate term, since it is none the worse off, and it may only be "deteriorating" with respect to a single species. Such deterioration, though, means that animals must continually change to conform to it. So the chase is on; the environment changing and the animal adapting right behind it.

**Fisher's Model and Specialists Versus Generalists** A simple model that involved both adaptation and environmental deterioration was developed by Fisher (1930). He argued that no organisms are perfectly adapted to their environments. They all fail, to differing degrees, to conform to the demands placed on them. If, however, we were to imagine a perfectly adapted organism, then other animals could be compared to the theoretical ideal. Fisher's mathematical model accounted for an infinite number of variables to which an organism must adapt, but we will only consider three, following Pianka (1974).

If an ideally adapted organism lies at point *A* in the three dimensional continuum (Figure 24.1), an actual organism may fall at, say, point *B*, some distance (*d*) away from *A*. Changes in *A* that move it away from *B* are considered as environmental deterioration from *B*'s point of view. Fisher noted that small undirected (or random) changes in either the organism or the environment have a fifty-fifty chance of being to the organism's advantage (that is unless the movement describes an arc of the radius *d*). Anything that reduces *d* is advantageous to the organism. If *B* is shifted toward *A*, however, it should not be shifted farther than the distance *d* or it will overshoot its mark, and so great changes have a higher probability of being detrimental than small changes. As point *A* shifts on a daily, seasonally, or geologically timed basis, *B* must track it by individual change or population change (mutations).

Now, would environmental deterioration due to a change in *A* be more detrimental to a "specialist" or a

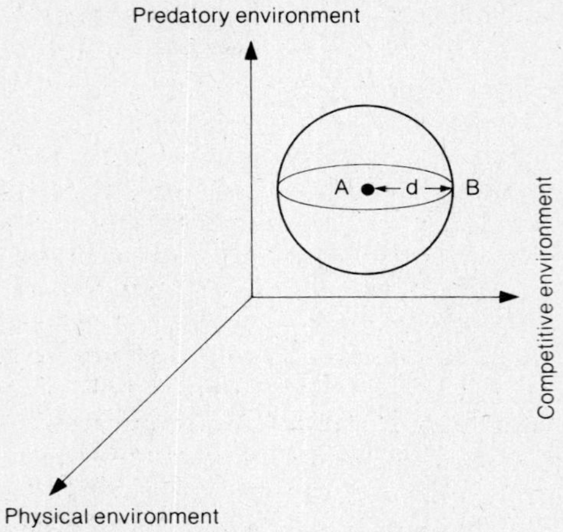


Figure 24.1 Fisher's model of adaptation and environmental deterioration. If *A* represents a perfectly adapted organism, another organism, *B*, can be compared to it. It is to *B*'s advantage to reduce the distance *d*. As *A* changes, *B* should tend to track it. As the distance *d* increases, the environment is said to deteriorate with respect to *B*.

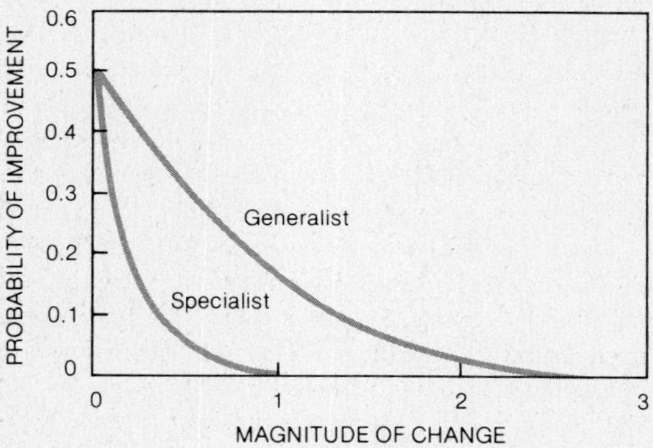


Figure 24.2 The probability of large and small shifts in the environment improving the lot of the specialist and the generalist. A random change of a given magnitude is more likely to improve the level of adaptation of a broadly tolerant generalist than a specialist with its narrow tolerance.

"generalist"? Obviously, specialists, with their narrow tolerance limits, generally suffer more by environmental deterioration than do generalized organisms with more flexible requirements (Figure 24.2). More specialized organisms are likely to have less well-developed homeostatic mechanisms than generalists. (These mechanisms act to keep the body's processes within more-or-less narrow, tolerable limits. For example, when the oxygen level in the blood is low, a homeostatic mechanism triggers faster breathing and increased heart rate until the proper oxygen level is restored.)

There is another point here. Fisher's model is applicable only to nondirected changes in *A* or *B* that alter *d*. These would include unpredictable climatic changes on *A*'s part or random mutations on the part of *B*. However, there *are* such things as directed changes, such as those caused by predators or competitors. Predators tend to reduce an organism's compatability with its environment (note that predation is one of our three axes that must be adapted to). The directed changes caused by competitors, on the other hand, can increase or decrease an organism's level of adaptation, depending upon whether those changes involve avoidance of competition (to avoid competitors, an animal might increase its risk of predation) or the improvement of competitive abilities. The latter would bring an animal into closer harmony with its environment. It should be apparent, then, that the ecological niche is the sum total of the adaptations of an organism, and that we can speak of the niche of an individual, a population, or a species.

One might wonder, then, if generalists do better at "tracking" the optimal niche, and optimality is always changing, why have specialists developed at all? When would they ever have an advantage? To begin with, these two terms, specialist and generalist, do not accurately describe most animals. It is more realistic to try to determine where an animal falls in the specialist-generalist continuum *for a specific resource*. The woodpeckers of Hispaniola will eat anything from a lizard to a larvae to a lemon and are thus rather generalized. They are more specific about *where* they search, however, almost exclusively utilizing the surfaces of trees. And they are definitely specialists when it comes to nesting, building L-shaped (or 7-shaped would be more precise) holes at heights between about fifteen and one hundred feet in trunks or vertical branches over about eight inches in diameter. So, is a woodpecker a generalist or a specialist? But for the moment, let's just consider specialization in foraging. Here, we can discuss specialization or generalization within either an individual or a population. As mentioned, an *individual* could adapt to (or track) the optimal niche *A*, if the individual is behaviorally flexible and *A* is moving rapidly. But if the optimal niche moves so slowly that it is only apparent over geological time, then any adaptation must be accomplished genetically by *populations.* So, depending on the conditions set by time, either consideration is appropriate. In actuality, the adaptive responses of an individual or a population may not be very different. All tapeworms probably behave about the same and have for untold generations. On the other hand, chimpanzees are flexible and "unpredictable," whether one is considering an individual chimpanzee or the history of a group.

Generalists, then, are jacks-of-all-trades (so you can predict what the disadvantage will be). They can capitalize on a number of food items, and when one runs low, switch to another. They can take advantage of what is available and are, therefore, opportunists. If the optimal niche is changeable, they can change with it by simply adjusting their behavior.

But, while generalists are the masters of no trade, specialists are jacks-of-one-trade. They are beautifully adapted to a particular way of life. They are attuned to locating, capturing, eating, and digesting a few food types and they do it with an enviable efficiency. If their food source should play out, though, they have few alternatives. Their perceptual, behavioral, and physiological systems simply may not permit them to shift to another food source. Their "strategy" did not account for such changes. And why not?

We can conceptualize the difference in the two stategies by considering the movements of point *A*, the optimal type, in our diagram. If *A* remains stable, changes very slowly, or changes cyclically and with a predictable regularity, stereotyped individuals or populations will be selected for. Koala bears evolved with a constant supply of eucalyptus leaves, their sole source of food, so their digestive tract, feet, and food preferences have evolved to take advantage of these aromatic leaves, which many other herbivores avoid. They have no contingency plans for the time when eucalyptus groves will be decimated to make way for pasture; pastures are a new thing in the history of koalas. But as long as eucalyptus trees stand, koalas are in business; their *A* has not shifted.

On the other hand, if *A* is unstable, unpredictable, or associated with high heterogeneity, then generalists will evolve. If the weather is unpredictable, or food populations come and go, or if many kinds of foods regularly appear at about the same time, but in such ratios that none, alone, can sustain a forager, a generalist will do best. Food generalists, then, can be expected in spatially or temporally discontinuous and/or unpredictable situations.

It might be added that probably generality per se cannot be selected for. Instead, the generalist is probably a failed specialist. As he attempts to adapt to current and local conditions, those conditions change, and he begins a new adaptive patten, still retaining some of his former adaptive ability, perhaps encoded in genes. When *A* shifts again, he is off in a new direction, but now with *two* residual abilities that may be periodically reinforced as *A* shifts back to a previous position. As this continual shifting and adapting takes place through time, the generalist will have developed a broad range of tolerances and abilities that can be called upon when *A* shifts yet again. The only thing that keeps a generalist from becoming a specialist is the failure of *A* to remain stable long enough. Of course, if the optimal condition *A* should settle at some locus that *B* could have tolerated for a short time, but could not tolerate for a long time (long enough to adapt to), then *B* may perish. So, in some cases, it may be better for *B* that *A* continues to shift so that its *mean* position is at least suitable for *B*.

These, then, are some of the theoretical problems involved in niche parameters and niche utilization and the conditions under which two different foraging strategies may have developed.

**Competition and Niche Size** An animal's competitors can, but do not always, strongly influence its ecological niche. For example, the realized niches of relatively *r*-selected organisms are determined primarily by their physical environments. On the other hand, realized niches of more *K*-selected organisms are probably more strongly influenced by their competitors. And the positions along the *r*-*K* spectrum can change with time. In temperate zones, in early spring when life suddenly flushes anew and species seem to literally appear from nowhere, annual plants may be relatively *r*-selected. Their energies go into making seeds. Species that bloom later in the season, however, when the landscape is already covered with competitors, may be more *K*-selected (Gadgil and Solbrig 1972). Even within a given species, populations in different parts of the geographical range may be more *r*-selected than others (Pianka 1974). (And see Wilbur, Tinkle, and Collins, 1974.)

Niche size may be affected in various, and sometimes unexpected, ways by interspecific competition (discussed later in the chapter). A prey species that forces out competitors (which may also be prey species) in one area but not another, increases patchiness in the predator's food supply. Therefore, predators (which also compete among themselves) should ideally restrict themselves to the patches with the highest yield, thereby decreasing the breadth of their "place niche" and possibly their "food niche." In addition, a predator can alter the niches of other predators in rather unexpected ways. For example, a predator that puts pressure on its food supply more or less equally across its range, can force a competitor (another predator) to expand the range of its food types (the breadth of its "food niche") and thus to treat the environment as more fine grained. The reason is, in a food-sparse environment, an optimal forager can't afford to pass up potential prey items. Thus he must become more of a generalist as he takes mere suboptimal food items (MacArthur 1972; Pianka 1974). So, we can begin to appreciate the complex effects species behavior may have on niche size. We will have more to say about how niche sizes change later.

## The Principle of Competitive Exclusion

The concept of niche division is an important one, since classical ecological theory, as we have been discussing it, says that no niche can be occupied by two species simultaneously. The *principle of competitive exclusion* (Hardin 1960)—also called Gause's law, the Volterra-Gause principle, or Grinnell's axiom—states that such a situation simply cannot last. No two species interact with their environment in identical ways, therefore one will have a net advantage, no matter how slight. The idea is that, in time, the species with the advantage will replace the other species at least in their zones of overlap (Figure 24.3).

## Niches and Competition

It seems, on the surface, that the niches of many competing species have overlapped for a long time. In fact, many very similar species may be found in the same habitat, apparently interacting with it in very similar ways. After all, some forests are composed of many kinds of trees, all needing the same things: sunlight, soil, water, and minerals. In "stable" forests, are not trees perpetually competing for the same things? In Lake Baikal, an ancient lake in southwestern Siberia, there are no fewer than 239 species of gammarid crustacea, tiny animals related to brine shrimp (Kozhov 1963). Over two hundred species of cichlid fish live in Lake Malawi in Africa (Fryer 1959). The animals comprising these groups are so similar in their ecological requirements that one is intuitively led to believe that at least some of them must be occupying the same niche. There is an impressive body of evidence, however, that indicates that, if they have been associating for long periods, these similar species are probably reacting with their environment in different ways that, granted, may be subtle, but nevertheless critical.

**Laboratory Evidence** In the 1920s and 1930s, there began to be a growing interest in how competing species interact, and several researchers set about to solve the problem in the laboratory. Usually what they did was grow two species together in a test tube and watch what happened. What usually happened was, after scores of generations, one species died out (R. S. Miller 1969). In 1934, G. F. Gause, a Russian biologist, grew laboratory cultures of two closely related species of the protozoa *Paramecium,* together, in the same medium. In most of these experiments, one species eventually perished. When he grew two species of yeast together, he found complications. He discovered that species 1 inhibited species 2 by production of alcohol, but that species 2 affected species 1 to a much lesser degree. So, he changed the growth medium frequently to prevent the build-up of waste and found that one species would still eventually replace the other, but that the "winner" could change from one experiment to the next. Replacement was obviously now due to competition. Recently, similar experiments were done using two species of the flour beetle *Tribolium.* The results

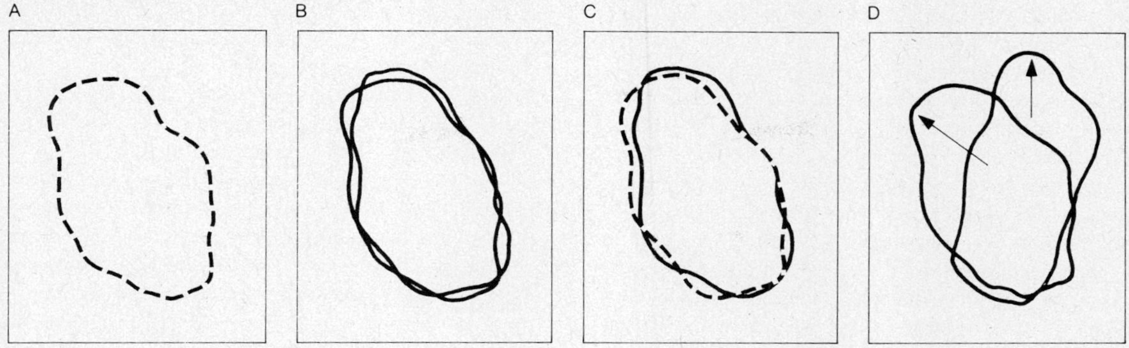


Figure 24.3 Theoretical results of competition for an ecological niche. The background represents all the possible environmental factors such as temperature, other species, oxygen, cover, food, and so forth as projected on a plane. An ecological niche, then, would be any unique set of these factors such as in theoretical niche A.

In (B), two niches are similar but not identical. They are each occupied by a different species. These species hold most, but not all, of the niche in common with the other species. Because each animal's niche includes so many factors that are identical to those held by the other species, it is logical to assume that they largely occupy the same habitat. The information here, however, says nothing about how closely the animals are related.

In (C), it becomes apparent that the two ecological niches held too many traits in common in order for each to remain filled. Gause's law states that two species cannot maintain the same niche indefinitely, but that one species will come to replace the other through competitive interaction.

In (D), a different situation arises. In this case, the unshared parts of each niche are sufficient to maintain each species for a longer time. If the species are not constrained (by other competitors, environmental barriers, and so forth) from expanding into new areas, the species will reduce competition between them by exploiting still other facets of the environment. The result is a divergence of life styles.

were the same. When two species were grown in the same jar of flour, one always died out. But, again, *which* species depended on the conditions of the experiment. For example, *T. castaneum* usually won under warm, moist conditions, and *T. confusum,* under cooler and drier conditions. When grown separately, however, both grew better under warm, moist conditions.

**Field Evidence** Cormorants are beautiful, large birds that suffer horribly during the periodic spills from offshore oil-drilling rigs. They are particularly interesting to watch as they feed. They are diving birds, and often two very similar species of cormorants can be seen surface-diving for fish off the same rocky coast, seemingly both feeding on the same thing. But if you were to dive with them, you would see one species stop its descent and begin to chase fish at intermediate depths, while the other continued to swim down to capture the fish on the very bottom. Since different kinds of fish live in different places, the birds were obviously utilizing different food sources, as was borne out in stomach analysis of the birds (Lack 1945). Most of the stomach contents (over eighty percent) of the great cormorant consisted of sand eels and herringlike fish that swim above the bottom, while the stomach of the shag cormorant was filled with bottom-living forms such as flatfish and shrimp.

Another example can be gleaned from bee studies. Four species of honeybees live in the English countryside. They are all about the same size, but their tongues are of different lengths. One species has a long tongue, the other three have shorter ones.

It turns out that the lengths of the corolla tubes of the flowers they visit also differ, and each bee species specializes on certain flowers according to the length of its tongue. The long-tongued bee (*Bombus hortorum*) visits flowers with long corolla tubes. The three short-tongued bees visit a variety of flowers, but this competition is reduced because only one of these three, *B. lucerum,* visits flowers in the open field. The other two forage primarily in woodlands. One might think, then, that at least these last two use the same resources, but *B. pratorum* is active earlier in the season than *B. agrorum*. Thus their niches are separated in *time*.

It should be mentioned that some people have argued that direct competitors *can* coexist (Ayala 1972; Harger 1972).

Ecologist J. H. Connell (1973) theorized that, for many species, populations rarely reach densities high enough to permit competition for resources. He argued that, instead, physical conditions or predation reduces the numbers of animals, particularly while they are in their young stages, so that their numbers remain so low that they rarely fall into competition. In fact, he suggests that the "niche shifts" that often occur on islands (discussed in chapter 25) may be due to predation differences rather than to the absence of competitors on islands. His arguments imply that the realized niche may not be so much a function of competitive restraints as is usually assumed—that competition may not be such an important evolutionary or ecological factor after all. His suggestions run counter to the mainstream of current ecological theory, which tends to place great importance on competitive influences. Nevertheless, his ideas at least provide a framework for further testing and they suggest a reanalysis of old data, data that he himself uses to substantiate his argument.

But in the best-known field demonstration of all, MacArthur (1959) showed statistically that a group of closely related warblers, living in the same woods and apparently sharing the same niche, had actually subdivided the habitat and therefore were occupying different niches. It turned out that each species foraged in different parts of the same trees (see Figure 23.9 in chapter 23). The birds, then, were actually occupying different microhabitats, each taking food from a restricted area, the location of which was undoubtedly partly dictated by the presence of the other species. It should be clear that such separation through specialization can only be accomplished in a heterogeneous environment that offers the opportunity for separation.

Just as animals may separate themselves spatially and behaviorally, they may also separate themselves in time. The same area may be occupied by different species as seasons change. The northern areas that the robin abandons each fall come to be occupied by more nothern species retreating from yet more severe winters. The competitive situation in different areas, then, must change at least partly on a seasonal basis. The robin's absence from South America each spring must alter the complexion of the competitive situation there.

Animals of the same species may also occupy different niches in the same habitat by dividing commodities by age or sex. For example, there are significant differences between the diets of young and old birds in the black-faced dioch (*Quelea quelea*) in Africa. Also, at certain times, males take different food than do females (Ward 1965). I will have more to say about the ecological separation of sexes later in this chapter.

### Adding New Species

In spite of our wish to discover firm natural laws, we find that niches *do* overlap. In some cases they broadly overlap and in other cases, barely. The degree to which they overlap varies from place to place and from one type of competitive situation to the next. The amount of overlap is very directly related to competition and may be reflected in the number of species of similar adaptive types occupying any one area. So, what really determines niche overlap? One way to approach the problem is to see how species can be added to a community.

Suppose we consider the distribution of several species of insectivorous birds along a single resource continuum, say, the size of insects in a forest. In Figure 24.4, we want to add species to community *A*. Obviously, this could be done by increasing the insect abundance while the bird variety remained constant. This way, the variety of insect species would not change, only the numbers of individuals in each species. Thus, each bird species could specialize on a narrower range of insects and, as they restricted their prey types, we could add more bird species as in *B*.

Let's take another approach. If the number of insects in each group remained constant, but more *kinds* of insects were introduced, condition *C* could result. The overall amount of food is increased, therefore more species can be added. Each bird species would go on exploiting insects just as it had, but the range of exploitable insects would increase and so could the kinds of exploiters. Under some conditions, the amount of ecological overlap, and hence competition, could be increased as in *D*, or the birds could restrict the range of sizes of insects that they ate, thereby making room for more bird species, as in *E*. So *B* and *C* require increased productivity. *D* and *E* do not. Of course, no species is going to move over to make room for another species. They will only move over or restrict themselves when it is to their own advantage to do so. It is hard to imagine cases where *D* or *E* could occur unless competition for insects was very low in *A* to begin with.

It is important for us to have information of this sort if we are to discover the ways the environment can influence the behavior between species. We are not primarily interested in foraging behavior here, but in how environmental elements, particularly food, can influence social patterns by altering such things as population density, species density, and predator-prey relationships.

## DISPERSAL AND ADAPTIVE RADIATION

Adaptive radiation occurs when a subpopulation becomes isolated from genetic exchange with the parent stock. This could happen in a number of ways, but generally the process is visualized as a subgroup becoming geographically isolated from a larger population and finding itself under a new set of environmental conditions. If, subsequently, the frequencies or kinds of genes in the isolated group change, new

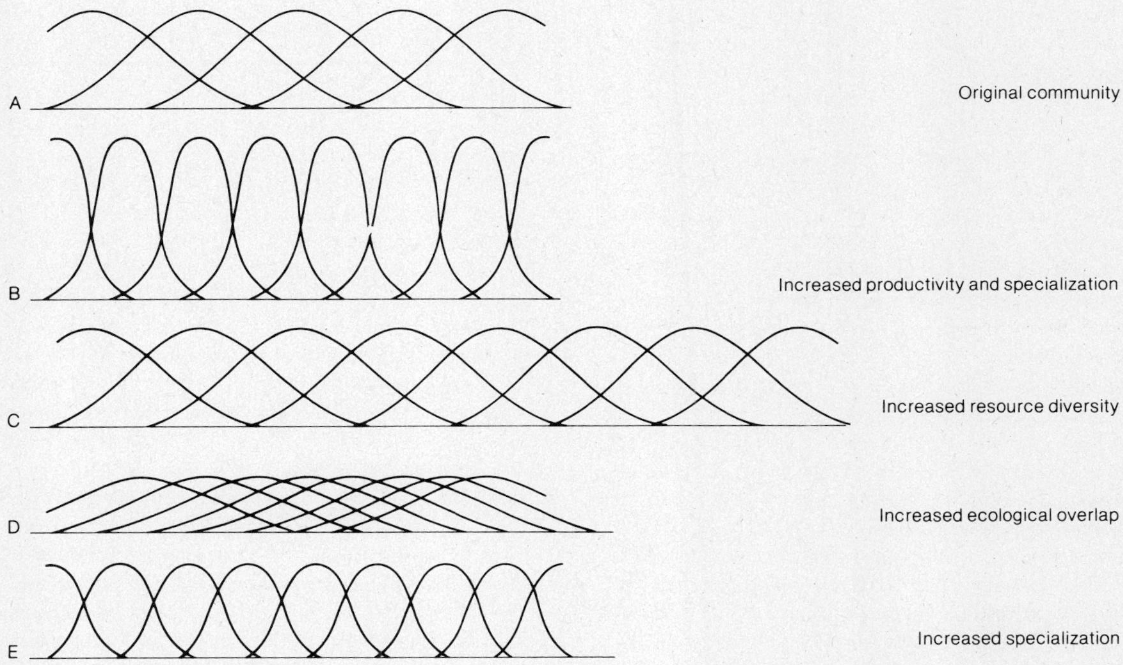


Figure 24.4 How resource utilization along a continuum can theoretically be changed to accommodate more species. Alternatives *B* and *C* require increased productivity of the community, but *D* and *E* do not.

adaptive types may be formed through a process called *adaptive radiation*. Normally the term implies that the subgroup changes enough that it can no longer breed with the parent stock. In this case, *speciation*, or the formation of a new species, occurs.

The classical example of adaptive radiation is in a group of birds usually referred to as "Darwin's finches," which inhabit the Galapagos Islands. The birds attracted Darwin's attention when he explored the islands during the voyage of H.M.S. *Beagle*, and what he saw served as an early impetus to his development of the theory of evolution. These nondescript little birds are four to eight inches in length and both sexes are drably colored in browns and grays. Species are easily distinguished on the basis of their bill structure. In one species, the bill is heavy and finchlike for seed eating; in another it is long and curved downward for gathering nectar from cactus flowers. Others have parrotlike, chickadeelike, or warblerlike bills and each bill type permits a specialized type of foraging (Figure 24.5). One species, the woodpecker finch (*Cactospiza pallida*), has a bill that is specialized for excavating wood. So it tends to forage like a woodpecker, but the bird lacks the long bill and protrusible tongue of the woodpecker. It compensates for this with a behavioral pattern. It breaks off a cactus spine or twig and holds it in its beak, probing insects from cracks and crevices (Figure 24.6). Whereas woodpeckers often impale their prey with their long bills or barbed tongues, the woodpecker finch generally uses its probe to flush insects from their hiding places. Then it drops the probe and grabs the insect. It isn't a very efficient means of foraging, but the bird has a food supply virtually all to itself, and so the technique is productive. Another Galapagos finch, *C. heliobates*, has also been seen using the technique on occasion. It is interesting that *C. pallida* not only forages in the same way as woodpeckers, but it has shifted its foraging location to more closely parallel that of woodpeckers. It now often looks for food on the vertical surfaces of the trunks of trees. Similar "stick-probing" behavior has been observed in the brown-headed nuthatches in the southern United States (Morse 1968), and in the island-dwelling New Caledonian crow, *Corvus moneduloides* (Orenstein 1972).

Apparently, the island species are descended from a common ancestor, a finchlike bird that colonized the Galapagos from South America. As the species became established on one island, some members of the population made it to the other islands. On each island, the birds, more or less isolated from populations on other islands, followed their own lines of evolution; each species adapting to the specific requirements of its own island. Finally, each group reached a point where its appearance, feeding patterns, and mating system were distinct from those of all the others. Some of these new species were rejoined later and their differences increased, perhaps reducing competition. Today there are fourteen species in all (Lack 1947). Each of the principal islands has at least three species and some have as many as ten.

The adaptive radiation of Darwin's finches, then, illustrates that the behavior and ecology of organisms may shift as a response to environmental opportunity, a shift that may be encouraged by competitive interaction.

## ECOSPECIES

So, animals may resemble each other morphologically and behaviorally because they are related—having sprung from the same stock through adaptive radiation. On the other hand, they may look and behave similarly, even though they are unrelated, if they have been subjected to similar evolutionary pressures. We can expect the animals inhabiting similar habitats to have certain features in common although they may be quite unrelated. These animals are, in essence, ecological equivalents or *ecospecies*.

The pronghorn (*Antilocapra americana*) of North America and the smaller antelope of Africa exploit their grassland environments in similar ways and may be considered ecospecies. So may the American bison (*Bison bison*) and the African buffalo (*Syncerus caffer*) (Figure 24.7). In fact, their plant food species, in many cases, may also be classified as ecospecies (Emlen 1973).The wolves (*Canis lupus*) of the northern hemisphere (note that we do not distinguish between American and European wolves) are ecologically equivalent to the African hunting dogs (*Lycaon*). Both species hunt large ungulates and bring them down through group efforts.

It should be apparent by now that the term ecological *equivalent* may be a bit strong. After all,

the wolves frequently hunt in pine forests, but the hunting dog has never seen a pine tree. Wolves often lose deer in deep snow (Mech 1970), but hunting dogs don't have that problem when chasing their long-legged prey in African grasslands. Also, hunting dogs have a different social system; sometimes the dominant individual of the group is a female—an unusual condition among social mammals. So, while different species may not be ecological equivalents in a strict sense, the term has a nicer ring to it than "ecological approximate."

Figure 24.5 These three Darwin finches have different bill shapes, yet they all come from Indefatigable Island in the Galapagos. The small ground finch *Geospiza fuliginosa* (left) and the medium ground finch *Geospiza fortis* (right) feed mainly on seeds. Note the shape of their bills. The cactus ground finch *Cactospiza scandens* (center) feeds on both the cactus and its fruit. For this, it needs a longer and sharper bill than the small or medium ground finches. All three share the same habitat.

## Convergent or Parallel Evolution

Ecospecies arise from *parallel* or from *convergent evolution*, their similarities being due to the same sorts of selective pressures (Figure 24.8). The physical similarity produced in unrelated forms can be startling. You may have caught small catfish from a river in Arkansas and then fished in the Gulf of Mexico from a pier off Corpus Christi, Texas, and caught a fish that looks much the same—undoubtedly a catfish. And the small bottom-cleaners in your aquarium may be any of a number of distantly related species, but they are almost always called a catfish of some sort. And all these fish have certain traits in common. They eat mostly from the bottom, they all have spines on their back (since any danger is likely to be from above), they all have sensitive barbels, or whiskers (since it is often dark on the bottom, chemical and tactile receptors are important), their mouths are located underneath, and they tend to escape by quick dashes followed by immobility. They're all "catfish," but the products of entirely different lines of evolutionary descent. So, if you're a

Figure 24.6 The woodpecker finch (*Cactospiza pallida*) uses a tool to help it feed. The bird uses cactus spines in much the same way continental woodpeckers use their long bills, to probe deep into crevices to extract prey from their burrows.

Figure 24.7 The American bison (above) and the African buffalo (below) are ecologically equivalent species because, although they are not closely related, they interact with their environments in similar ways.

fish and you live on the bottom, the chances are very good that you will have a set of characteristics that will cause people to call you a catfish. And you may be unrelated to a host of other animals bearing the same name.

As other remarkable examples of convergent evolution, the African longclaw (*Macronyx croceus*) is a pipit that is ecologically very similar to the eastern meadowlark (*Sturnella magna*). Both birds live in grasslands and forage by probing and prodding the ground with their bills. They are about the same size and have similar bill sizes and general body plans, as one might expect. But who would have ever guessed that they would both possess a yellow throat and belly, a brown back, white outer tail feathers, and, especially, a white line over the eye and a black "V" on the chest. What do an eye-line and a black "V" have to do with grass foraging? We haven't a clue, but the occurrence of convergent evolution in so many, seemingly unimportant, respects leaves us with a whole series of unanswered questions. Is the "V" adaptive for ground foragers? If so, how? Is it an incidental result of gene linkage, or of pleiotropy? If so, why should these species have converged genetically as well as phenotypically? Or are they more closely related than we think? Note the similarities in adaptive, as well as apparently unadaptive, traits in the species shown in Figure 24.9.

It should be made clear that convergence can produce strong potential competitors among unrelated species. The virtual extinction of the Tasmanian wolf, a marsupial from Australia, after the introduction of the dingo, a "placental" wild dog, attests to the competition between these two kinds of animals.

## COMPETITION BETWEEN SPECIES

Competition normally occurs between individuals that utilize the same things, whether those things be food, shelter, or members of the opposite sex. Therefore the strongest competition should appear between members of the same species. In some instances, however, members of different species may fall into strong competition, producing some rather interesting results.

### Balancing Intraspecific and Interspecific Competition

Intraspecific and interspecific competition may have opposite effects on the variation within a population. To illustrate, consider a hypothetical case where there is no interspecific competition. The population will distribute itself along a wide resource continuum (say, a food resource) so that all individuals are subjected to about the same level of competition from other members of the population. Since there will be an optimum food resource, most individuals can be expected to utilize that resource, or something

CONVERGENT AND PARALLEL EVOLUTION

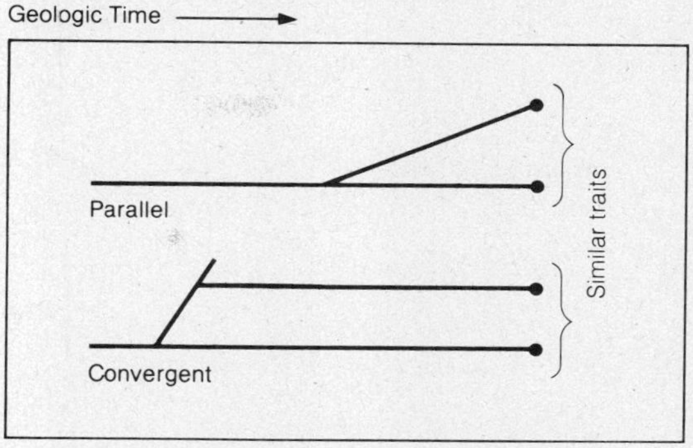


Figure 24.8 Parallel or convergent evolution may produce similar species that interact with their environments in much the same way, although they may not be closely related. In parallel evolution, the resultant species separated from the ancestral stock more recently than in convergent evolution.

close to it. But there may eventually be so many individuals competing for the optimal resource that it becomes advantageous for increasing numbers to begin to utilize suboptimal resources that may be subjected to less exploitation. Thus, they reduce the level of competition for the optimal resources. Since the suboptimal resources can sustain fewer individuals, high levels of competition may rather quickly be reached there, so some members of the population may take yet other resources, even further removed from the optimal type (Figure 24.10). The result is that individuals end up behaving in such a way as to equalize the demand along the resource continuum. Thus, all members of the population are subjected to about the same level of intraspecific competition. MacArthur (1972) called this the "principle of equal opportunity," and it means that intraspecific competition forces a broadening in resource utilization.

Interspecific competition, on the other hand, would tend to *restrict* the range of resources a population uses. The organisms at the "tails" of the curves in Figure 24.10 would fall into competition with species of similar adaptive types, but ones that were better adapted to use those resources. A population subject to interspecific competition (that is, most populations) are, in effect, boxed in by their competitors. Pianka (1966) hypothesized that, at equilibrium, the sum of intraspecific competition should be balanced by the sum of all the interspecific influences operating on a population. This, he admits, cannot be *quite* true because inherent behavioral and physiological limitations must also be operating to restrict the utilization of resources by any organism. Thus, probably any organism is more restricted in its range of resource utilization than competition would dictate. (For a dissenting view, see Cody's 1973 description of "character displacement crossovers," where island species may drastically alter their position along a resource continuum, some species even switching relative positions. The principle is discussed in our section on character displacement. Also see Crowell 1961.)

## Field Evidence of Interspecific Competition

In spite of the compelling logic behind our assumptions regarding competition, we must admit that much of our evidence is circumstantial. For example, we have noted that red-winged blackbirds attempt to exclude tricolored blackbirds from their nesting area. Such aggression is demanding and would have been "selected out" of the red-winged behavioral repertoire were it not beneficial, at least at some time in their histories. The advantage apparently lies in excluding similar types that might offer severe competition. Of course, the red-winged does not size up the tricolored and then attack what he fears may be a competitor. Rather, the tricolored, which looks remarkably like the red-winged, probably merely presents a certain configuration or pattern that releases aggressive behavior in the red-wingeds. The sight of a wading bird, adapted to an entirely different ecological niche, would release no hostilities. Neural mechanisms that permit instances of mistaken identity (aggressive behavior toward other species) in red-wingeds are very likely *maintained* through selection and are not simply unadaptive and incidental patterns. Strong hostility of the red-wingeds toward tricoloreds undoubtedly reduces the numbers and competitive influences of the latter to some degree where the species overlap. This all seems logical enough and may well be correct. But we don't *know*

Figure 24.9 Examples of convergent evolution. (A) *Rasbora* (above), a tropical Asian cyprinid fish, and an African characin, *Neolobias*. (B) An Australian agamid lizard, *Moloch horridus* (above) and the North American iguanid *Phrynosoma coronatum*. (C) The African civet (above) and an American weasel. (D) The eastern meadowlark of America (above) and an African motacillid, the yellow-throated longclaw.

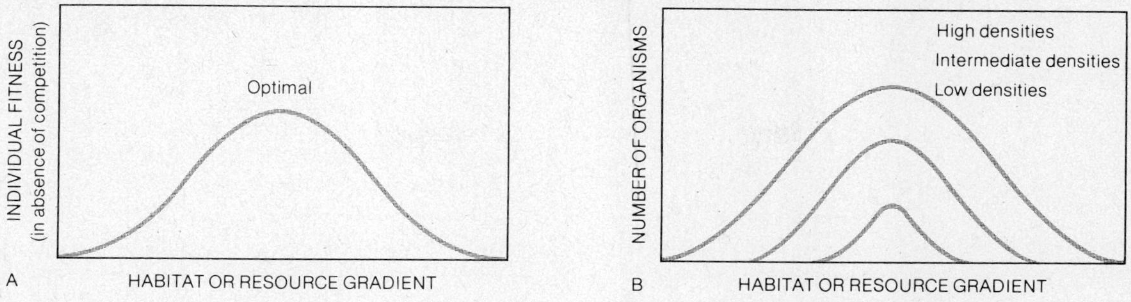


Figure 24.10 An expanding population, in the absence of competition, might disperse itself along a resource continuum as in (A). In (B), as the population density increases, competition intensifies at the optimal and forces more individuals to exploit suboptimal resources.

that the birds compete at all until we have demonstrated that one deprives the other of something essential. It is very difficult to prove such a thing, so we gather our circumstantial data and bind it with a matrix of assumptions.

But let's go on to other specific examples showing how species may presumably reduce competition between themselves. Consider four types of honeycreepers, *Diglossa lafresnayer lafresnayer, D. carbonaria atterina, D. cyanea cyanea,* and *Conirostrum cinereum fraseri,* which inhabit the same areas and the same parts of the habitat and exploit identical ranges of foods (Moynihan 1962b). They differ in the *degree* to which they utilize each food type, however. Also, although they are found in the same habitats, each species seems to prefer a certain area more than the others do. Individuals tend to avoid members of other species, so they don't fight much; however, individuals frequently quarrel with members of their own species. So each avoids other species, since there is no advantage in socializing with them anyway, but they react to intraspecific competition by distributing resources through quarreling. There are obvious reproductive advantages in maintaining social contact with one's own species, even if it, at times, leads to dissension.

The argument for the existence of strong competition between similar species is buttressed by a study on titmice (Hartley 1953). It was found that, although five sympatric species of titmice were separated in foraging height, preference for tree species, and parts of trees, the degree of separation changed with the season. In fact, at times they didn't separate at all. Such times, it was found, coincided with periods of superabundance of food—when competition would be lowest.

By a combination of differences in foraging levels and habitat preferences, certain vireos have been able to occupy the same geographic area (Hamilton 1962). It was suggested that the species-specific differences that reduce competition in vireos also inhibit attempts at interspecific matings and therefore function in reproductive as well as ecological isolation. Dobzhansky (1951) proposed the term "isolating mechanisms" for any of several processes or situations that act as barriers to gene interchange (Lack 1971). Those interested in genetics will wish to review revisions of the concept by Mayr (1942) and Dobzhansky (1951).

We have primarily considered competition between species of similar types. However, in some cases competition exists between very different kinds of animals (see the discussion of allelochemics for examples from plants). The tiny mouselike Australian phalanger (*Tarsipes spenserae*) climbs among flowering bushes and, with its long protrusible tongue, sucks up nectar from the flowers. The basic plant–animal relationship, then, is much like the one occurring between flowers and bees (von Frisch 1964). Possibly, some competition exists between these mammals and insects, but we lack hard data. And then, lake birds such as grebes (*Colymbas*), and large fish such as pike (*Esox*) both feed on small perch; but their foraging patterns are so vastly different that any competition that exists between them must be of an indirect sort and, perhaps, buffered in a variety of ways so that no true rivalry is likely to ever exist (Wynne-Edwards 1962). Such indirect competition is probably common in nature.

A very direct sort of competition exists between some species of seabirds. In the Galapagos Islands, the frigate birds (*Fregata magnificens*) have specialized in taking food away from other predatory seabirds though piracy. A frigate circles high until it sees a booby or other seabird capture a fish. Then it swoops down and pecks at the successful hunter until it regurgitates its prey—which the frigate bird then skillfully catches in midair. In at least one case, a tropic bird was killed in such an attack (Eibl-Eibesfeldt 1970). Galapagos seabirds show many peculiarities that are believed to be a result of their interaction with the marauding frigate birds. For example, the swallow-tailed gull (*Creagrus furcatus*) hunts only at night, and the dusky gull (*Larus fuliginosus) has* cryptic gray feathers. Perhaps since the frigate birds and gulls are after the same prey, this is a most direct sort of competition (Eibl-Eibesfeldt 1970). Wynne-Edwards (1962) argued that the attackers in such cases must be regarded as semipredators, or parasites, since they are dependent upon the "victim's" continued presence. Intense competition between the attacker and the victim might be the attacker's undoing if it were strongly dependent upon its ill-gotten gains, since it might depress the number of its victims.

## ALLOPATRY AND SYMPATRY

Competition is most effectively reduced, of course, when the geographic ranges of the species do not overlap, a condition known as *allopatry*. One might ask, then, why allopatric species should even be considered in a discussion of competition. To begin with, there is always the possibility that allopatric species may somehow be joined. Such union has occurred many times in nature, as well as by the hand of man. Thus allopatric populations are potential competitors.

As we know, *sympatric* populations are those that coinhabit a certain geographical area. Each species, however, is not necessarily sympatric with the same groups of species throughout its range, so the competitive situation for a species may change from one part of its range to another. Except in very simple and/or strongly demarcated ecological areas, any species would probably be free of any given competitor at least in some part of its range.

The condition of occupying adjoining areas with no significant overlap is known as *parapatry*, but the situation is probably rare in nature. Parapatric populations probably only exist where two ecotypes are sharply demarcated. Powerful influences would be needed to maintain discrete but adjoining populations, so any ecological separation probably would be reinforced by behavioral activity through mutual exclusion (Figure 24.11).

It should be apparent that the ecological niche of a species will be reflected in its morphology. Even an untrained observer viewing a heron and owl side by side could tell which one was a wading bird. (I hesitate to be overly confident, however, since after once leaving a group of pre-med students to identify the skeleton of a large pig, I returned to find that after careful consideration, they had concluded it was a leopard!) But in other cases, the phenotypic differences may not be so obvious. For example, long-distance migratory birds have more oxidative enzymes in their muscle tissue than nonmigrators or migrators that move shorter distances (Wallace 1966a). Thus, animals may adapt in ways so subtle that an observer would have difficulty in detecting differences that might permit differential niche utilization.

Given the strong influence of the niche on the morphology of its inhabitants, it should also be apparent that critical differences will exist between sympatric species as they utilize the same habitat in different ways. For example, we might expect differences in the morphologies of MacArthur's sympatric species of warblers. The differences, of course, would be related to their specific foraging modes. This means we can expect physical traits to be affected more than others. Thus, on the basis of their different foraging habits, we might expect the greatest differences between competitors to be in their *trophic*, or foraging, appendages (Hespenheide 1971, 1973), while body size and wing length might be less affected. We are reminded that differences in forag-

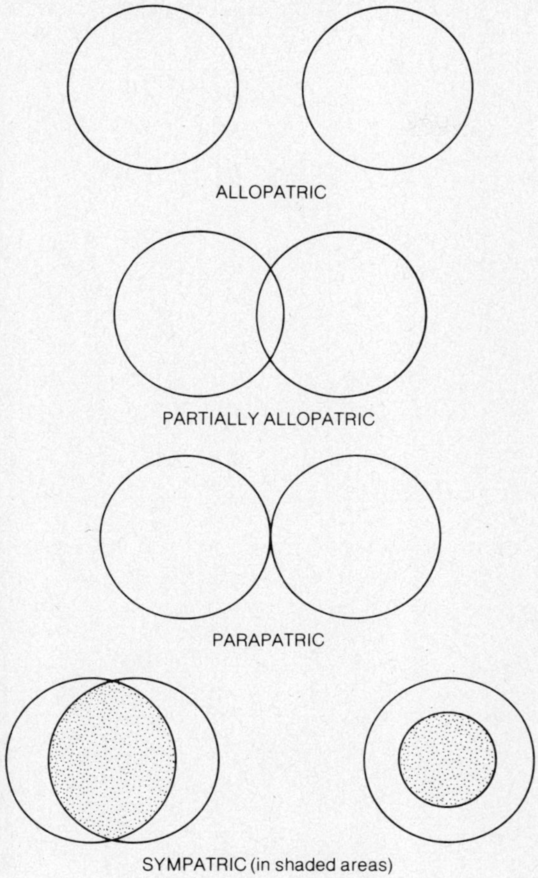


Figure 24.11 Geographic relationships of populations. The distinctions are somewhat arbitrary, since one cannot say, for example, when a species is sympatric as opposed to partially allopatric.

ing behavior, habitat choice, and bill shape between sympatric and closely related species of birds are believed to be adaptations for reducing competition for food (cf. Lack 1954, p.144; Mayr 1963; Selander 1966; Wallace 1974a).

## CHARACTER DISPLACEMENT

One of the primary forces that causes a species to occupy a certain niche and to assume a particular morphology appropriate to that type of niche is, as we have seen, the presence of competing species. The effect of competition in such cases is most clearly shown when the competing species overlap in only part of their ranges. In such cases, the strongest morphological and behavioral differences appear in the zones of overlap. In Figure 25.12, species 1 and 2 overlap in only part of their ranges. Note that they diverge in bill size to a much greater degree where they are sympatric (B), than when they are allopatric (A). The selective mechanism is apparently one that eliminates those members of either sympatric population that tend toward similarity with the other species. Any such similarity would, of course, increase the level of interspecific competitive interaction and place these individuals at a disadvantage.

The phenomenon is called *character displacement* (Bonner 1965). The general theory of character displacement was advanced by Brown and Wilson (1956), and since that time many such cases have been described. Unfortunately, the concept has been invoked rather uncritically in cases where morphological differences betweeen species may be due to something else. The result has been a growing question regarding whether character displacement is an important influence in nature, or whether it exists at all. At this point, however, it is probably safe to assume that it exists under certain circumstances.

Character displacement has been described on islands where similar sympatric species diverged in bill length more than those same species on continents (Grant 1968). Such a trend would not be unexpected in light of what we know about differences in continental and island niches. In the Galapagos Islands, those islands with more than one species of finches of the genus *Geospiza* usually exhibited character displacement (Lack 1947). On Bindloe and Abingdon Islands, there are three species of ground finches and their bill sizes do not overlap (Figure 24.13). On Charles and Chatham Islands, where only two of these species, *G. fortis* and *G. fulginosa,* occur, they still do not overlap in bill size. But *G. fortis* does not share Daphne Island with *G. fulginosa* and is intermediate in bill size there. Where *G. fulginosa* occurs without *G. fortis* on Crossman Island, *its* bill is also of intermediate size. The habitats on these islands are roughly similar, so it is supposed

Figure 24.12 Theoretical model for character displacement. Where species 1 and 2 overlap, they may diverge in critical ways as a means of reducing competition. Here, the bill size of 1 becomes smaller as that of 2 becomes larger where the species coexist, but are more similar where the populations are allopatric.

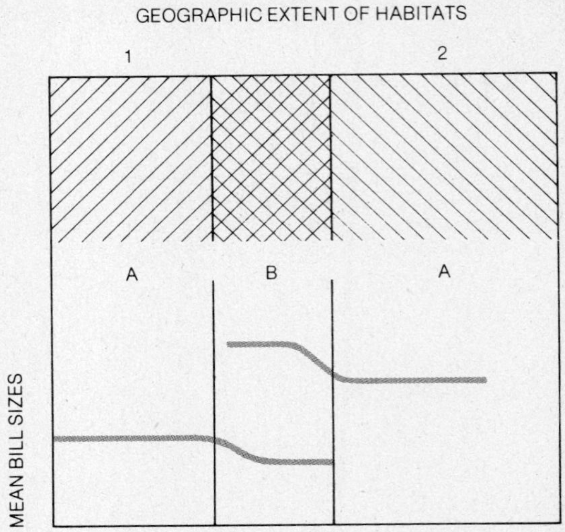


that these bill-size distributions are strongly influenced by competition, and that character displacement is one result of the competition. Character displacement was found in other traits as well, such as in nest sites, host specificity in brood parasites, habitat preference, and foraging behavior (Lack 1947).

For an argument regarding the validity of the most commonly cited examples of character displacement, see Hespenheide (1973).

## CHARACTER CONVERGENCE

Sympatric competitors may reduce their competition in ways other than by increasing the *differences* between them. They may become more *similar*. For example, there may be increased voice similarity between species of *Parus* where they overlap (Dixon 1961). They may also hold mutually exclusive territories in such areas. Thus, perhaps if competing species became more *alike* in certain characters in areas where they overlapped, they might begin to treat each other as conspecifics in the partitioning of resources. They might not, then, tend to force each other out. The phenomenon of increasing similarity in sympatric competitors is called *character convergence* (Cody 1969, 1973).

How could such a system work? And what characters would be affected? Would these species be expected to converge in all characters? It has been suggested that when two species compete for some commodity, perhaps selection for differences useful in mating signals will be held in check by a need for those signals that are used in competitive encounters. So, in situations involving interspecific competition, it would be advantageous for competing species to be able to recognize the other's *threat* signals, as do

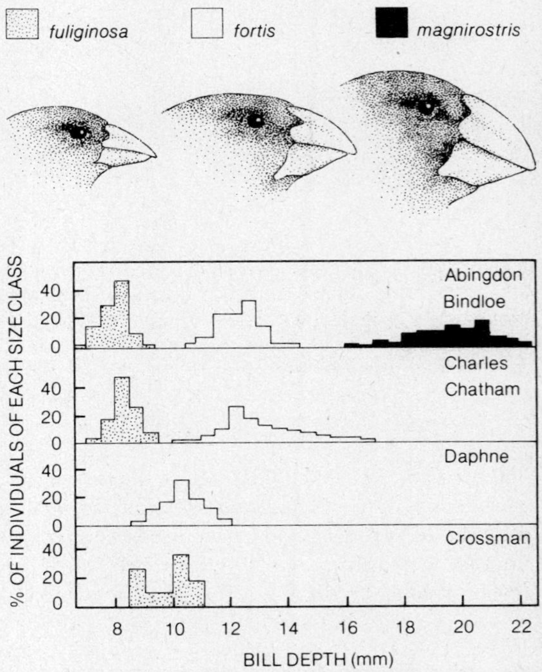


Figure 24.13 Frequency distributions of bill depths in ground finches of the genus *Geospiza* on the Galapagos Islands.

some species of tits (Stokes 1962a,b). Among these species, if an aggressive bird raised its wings, its interspecific opponent would flee. If it fluffed its feathers, its opponent often attacked. Of course, any such closely related species could be expected to use some of the same signaling devices. We are not concerned, however, with similarities due to "vestigial" patterns; we are interested in any similarities that are being maintained or increased through natural selection. So far, examples of character convergence have only been described for species that show interspecific territoriality. If two sympatric species are morphologically and behaviorally similar (suggesting strong competition), then mutual exclusion from territories would be adaptive.

So, we have seen that competition is a complex phenomenon. It obviously exists and it may operate in intricate and subtle ways in some cases, while being glaringly obvious in others. The central question does not focus on its existence or even on its methods, but on whether it is, in fact, an important tool of natural selection as it operates on wild species.

# 25

# HABITATS

One of the most intriguing questions in biology is a deceptively simple one: Why do animals live where they do? The scrub jay of California looks much like the bluejay of North Carolina. And they behave very similarly. So, why doesn't the bluejay live in California? Why do some blackbirds live only along water near forests, while others live in the forest itself? When you see animals in their natural habitat, why are *those* animals there? Why not others?

## HOW HABITATS SELECT THEIR ANIMALS

As we have seen, certain kinds of habitats tend to favor animals of a certain appearance and behavior. So, in a sense, the environment dictates the kinds of animals that can utilize it. For example, cold climates favor larger or stockier bodies. "Bergmann's rule" states that the higher the latitude, the larger will be the animals that live there. This is even true for ectothermic (cold-blooded) animals. To illustrate, there is a steady increase in the size of fish from tropical to temperate to polar waters (Lindsey 1966). Larger body size means that there is more mass per unit surface area, and it is through the surface areas (the skin) that metabolic heat is lost. The physical law states that, for a sphere, as its surface area is squared, its volume is cubed. Bergmann's rule may also apply to humans. For example, Eskimos are much more stocky and rounded and would tend to dissipate heat more slowly than African tribesman, who tend to be tall and slender, with a greater surface area for their body weight (Figure 25.1). Another generalization called "Allen's rule" states that extremities are likely to be smaller in colder regions (Figure 25.2), a trend also exemplified in humans. Thus, a race of people who have evolved in equatorial areas might have trouble invading polar regions, since their body types would be disadvantageous in that area.

Forests offer poor visibility and select for forms that rely strongly on hearing or smell, but in open country, vision is at a premium. However, forest monkeys would probably not fare well at all if they tried to invade the kind of grassland habitat where the sharp-eyed baboon lives. Their lack of success here could be due to the shortcomings of forest-adapted physiognomies in a grassland environment or because they would encounter strong competition from the baboons. Chimpanzees, however, live in both kinds of places. (It would be interesting to see if individuals from the two habitats show a difference in their sensory abilities. One might also wonder if chimpanzees with impaired vision or hearing, when offered a choice, would tend to spend more time in the place that maximized the usage of their best traits.)

Since denizens of the forest are less dependent on sight, they are more likely to be nocturnal than are prairie species. But, if one is short enough, prairies can be forests. Thus, small grassland mammals are more likely to be nocturnal than are larger mammals that can see over the grass. Nocturnality has a number of benefits for smaller animals. It protects them from the devastating effects of the sun (smaller endothermic animals have extensive surface areas for their mass and tend to lose water quickly through the rapid breathing that is necessary for them to maintain a constant body temperature). And, since most predatory birds and many mammals are visual hunters, nocturnality would reduce the likelihood of small animals falling to such predators.

Since visual communication is potentially much more specific, localizable, and detailed than olfactory or sound communication, most highly social species, such as baboons, are usually found in environments that promote visual signaling, such as grassy savannahs. Forest mammals are much more likely to be solitary, as are orangutans.

We cannot carry such generalities too far because you can probably sit right there and think of numerous exceptions to any of these rules. We only wish to reiterate the point that the nature of the environment strongly influences what kinds of species can live there. Actually, it might be said that an environment selects only the kind of organisms that *can* live there. The organisms themselves sometimes make the decision about whether they *do* occupy a hospitable habitat.

## HOW ANIMALS SELECT THEIR HABITATS

So, not only does the habitat determine the character of its denizens, but animals may come to *choose* various habitats or parts of habitats. The animal may make its choice on the basis of individual experience, as a dog probably comes to associate shade with coolness, or because of some innately based tendency, such as that which causes yellow-headed blackbirds to forage closer to the water's edge than red-winged blackbirds. Again, the two behavioral bases are not separable. If the yellow-headed began to associate the water's edge with food, any innate tendency might come to be strengthened through individual reward.

As was mentioned earlier, there is a general tendency in many kinds of organisms for the young to launch out and find new places to live. Their movement may be as simple as seed dispersion, as we see in coconuts washed to sea, which end up in a new and possibly different kind of habitat than their parent

Figure 25.1 Some human populations have adapted to specific sorts of environmental conditions. The Eskimo is short and rounded with short appendages and thus tends to conserve body heat by minimizing surface area with respect to mass. The Watusi lives where it is warmer and his height and exaggerated appendages help to dissipate heat.

Figure 25.2 The Arctic fox has reduced appendages, such as short ears, while the desert fox can afford longer ears without worrying about them freezing. Larger ears also help the desert fox to hear better in dry air, which does not conduct sound very well.

tree. A thistle can be carried long distances in the fur of a fox. Animals may also be carried along and deposited passively, having no choice in the matter of where they end up. The eggs of bottom-dwelling sea animals may be randomly dispersed by ocean currents silently holding sway beneath the great seas. The odds, however, are against them; most such passively dispersed organisms end up in unsuitable habitats and perish (Wolfenberger 1946; Wecker 1963). Dispersion is not always passive, however. A young animal may have a clear choice in the matter of where it will live. That choice will be an important one, since the habitat will certainly influence its reproductive success. Young tree squirrels that choose grasslands don't leave many of their kinds of genes in the next population.

Since habitat selection is a critical process, animals with special habitat requirements must possess efficient means of recognizing a good habitat. But not only must they be able to distinguish optimal habitats, they must also be able to tell the difference between the second and third best, in case the optimal area is not available (Orians 1971).

We are only now beginning to understand a few of the ways animals choose their habitats. For example, some animals may be sensitive to harmless "clue substances" that indicate the presence of dangerous or inappropriate physical or chemical conditions (Andrewartha and Birch 1954; Elton 1927). These are usually simple sorts of animals and their responses are stereotyped, similar to taxic responses. For example, estuarine animals, moving down a river toward the sea, may react negatively to slight rises in the salt concentration of the water they encounter and reverse themselves, following the gradient back into somewhat fresher water.

We find an apparently more complex form of stereotyped behavior in the choice of egg-laying sites by some insects. Among certain beetles, butterflies, and wasps, the female unerringly selects certain kinds of plants and animals on which to lay her eggs. These hosts are of no use to her at all, but will be essential to the larvae when her eggs hatch (Andrewartha and Birch 1954). In such cases, her behavior is rather rigidly stereotyped, as she responds to certain clue conditions (releasers) in the environment. For highly stereotyped animals, then, many areas or commodities that are perfectly suitable will go unutilized because they lack the proper clue substances. Even among birds, the choice of habitat is frequently based on certain conspicuous features of the environment that may have little to do with the actual physiological requirements of the species (Lack 1933). There seems to be a "psychological factor" that a habitat must possess in order to be chosen by birds, but which may be independent of the utilizable qualities of the habitat. However, this psychological factor could probably be explained on the basis of physiological mechanisms if more precise data were available (Wecker 1963).

What qualities of the habitat might birds key on? As usual, the answer is complex. Different things are important to different birds. In shrub- and forest-dwelling birds, habitat selection (or occupation) has been associated with height, spacing, the type of vegetation, and the presence of song perches (Lack 1933; Kendeigh 1945; Guillion 1960). The distribution of birds in certain kinds of communities is influenced more by the *structure* of the dominant vegetation than by the *kinds* of trees in the area (Pitelka 1941). For example, hole-nesting birds are likely to prefer areas where holes exist or can be made. It may not be important what species of tree the hole is in. Chipping sparrows showed a marked preference for areas with trees bearing thin pine needles rather than broad leaves. On the other hand, we know that certain tits prefer broad-leaved woodlands and move into pine forests only when forced to do so by their territorial colleagues. When kittiwake males return in the spring to the area where they were born, they do not seek out a nest site on the flat ground. For them to do so, with their behavioral repertoire, would mean almost certain death. Instead, they find a strip of rock a couple of feet long and about six inches wide, and it is here that they set up their tiny territories. Conversely, a herring gull had *better* pick a nest site on flat ground, in the same kind of area where herring gulls have been successfully nesting for untold generations.

One of the primary questions in studying habitat selection is, How much leeway does the animal have? Is it born with innate preferences, or does it learn the qualities of good and not-so-good habitats? We can make certain immediate judgments about some animals, such as planaria. They undoubtedly react innately to physiochemical stimuli. But what about differences in preferences between forest-dwelling baboons and those that live in savannahs? Let's consider two specific cases of habitat choice, one from insects, the other from mammals.

**How Honeybees Select Habitats** Martin Lindauer (1961) found that he could "read" the dance of honeybees and predict where new hives would settle. This was rather surprising. How and why do honeybees communicate their preference regarding a new home? And how had Lindauer tapped their lines of communication? Perhaps it would be helpful to trace the events leading to the formation of new hives. It all begins in the late spring when worker bees prepare several unusually large cells. The larvae that are hatched in these cells are fed a special "royal jelly" and they develop into fertile queens, instead of sterile workers like larvae in other cells. When these queen larvae hatch, they will fight to the death to determine which will be the new queen, or the first one to hatch may kill the others before they emerge. In any case, though, the old queen leaves before *any* of them hatch, and about half the bees in the hive go with her. She and her entourage fly to a place near the old site and hang together in a huge cluster on a branch or part of a fence or whatever. At this point, they still haven't determined where they will settle down and begin a new hive. From this mass of insects, however, scout bees fly out to reconnoiter the area. When they have found a crevice or hollow log that might serve as a new home, they fly back and perform their dance as described in Chapter 6. Lindauer was able to determine where the suggested sites were from these dances.

But since a number of bees may go out at one time, several scout bees may be dancing simultaneously, each having discovered what it considers to be a suitable place for a hive. The quality of the location, it turns out, is signaled by the length and vigor of the dance. If the dancer is convincing enough, the other bees fly out to that area to inspect it, and when they return, they too may dance to verify the original statement. Then more bees go there. There are usually a number of groups dancing simultaneously, each describing a particular site. Finally, one site wins out as most of the bees begin to dance in favor of it. Those that had inspected inferior sites cease to dance or switch their vote. Once general "agreement" has been reached, the bees all fly to that spot and begin to build a new hive.

But what determines a good site? And how is it distinguished from a better site? It turns out that honeybees are astonishingly sensitive to a number of factors. Since bees overwinter (as is described in Chapter 13), one of the most important qualities of a good site is its protective qualities. In southern Germany, where it can become quite cold in winter, bees prefer holes in the ground. Their second choice is hollowed trees, with man-made straw-basket hives coming in a distinct third. Size is particularly important since, if it is too small, not all the bees can get in. If it's too large, it will be drafty and cold. Shelter from wind is also important. In one case, the bees were equally divided between two sites in an open field. It took them six days to make a decision, and it was only made when the wind began to blow. That did it. They took the more sheltered site.

Another factor also operates here. The group that follows the old queen out will tend to choose a site that is farther from the old hive, all other things being equal. One must keep in mind the sacrifice a queen bee must make to cover long distances of, say one thousand feet. After all, she is adapted for laying eggs. Her massive and cumbersome body hinders her mobility; she is a terrible flier. But Lindauer watched queens struggle over long distances in order to reach sites that would result in lessened competitive interaction with members of the old hive, now serving their new queen. She has an interest in the welfare of that new queen, which is her daughter. So, her best reproductive bet lies in minimizing competition (or foraging overlap) between the hives.

**How Prairie Deer Mice Select Habitats** We now have an interesting story about how deer mice (*Peromyscus*) select their habitats (Wecker 1963). About fifty-five species of deer mice occur throughout North and Central America (Blair 1950). The deer mouse is extremely variable over its range, about sixty-six "subspecies" having been identified. These, however, can be divided into two broad groups. One, the prairie deer mouse (*P. maniculatus bairdi*), has a short tail, short ears, and lives in grassy fields. The other, *P. maniculatus*, called simply the deer mouse, has a longer tail and ears and it lives in forests, places that the prairie deer mouse avoids.

A number of studies were conducted to try to detect any differences in the physiological tolerances of the two species, but none were found. In fact, the prairie deer mouse was perfectly able to tolerate simulated forest conditions when forced to do so in the laboratory.

When woodland mice and prairie mice were given a choice between artificial "woods" and "field," it was found that each group preferred the type of area that they normally inhabited (Harris 1952). In order to determine the relative effect of ultimate and proximate causation, an experiment was devised to see whether the preference was primarily inherited or acquired. A one hundred-by-sixteen-foot enclosure was constructed with a nest box at either end of the enclosure, subdividing each half into a series of compartments (Figure 25.3). Mice were free to pass from one compartment to another through "runways" and, when introduced into the enclosure, could move either way, toward woods or grass. This was possible because the long axis of the enclosure fell across a natural wood-field interface, thus dividing the enclosure into two clearly distinguishable halves, woods and grass. Five of the runways were electronically equipped to record the passage of a mouse and the hour that it passed.

It turned out that wild-caught prairie mice spent most of their time in the grassy end of the enclosure. Also, the offspring of wild-caught mice that had been reared in either the laboratory *or* the experimental enclosure also showed a strong affinity for the grassland end of the enclosure. First-generation laboratory animals, of course, had never seen a field. On the other hand, while the offspring of animals that had been reared in the laboratory for many generations showed no preference at all, when they were reared under field conditions, they showed a marked preference for the field half of the enclosure. They could not be made to prefer woodland by being forced to live in woodland when they were young (Wecker 1963).

Wecker drew the following conclusions:

*(1)* The choice of the field environment by the prairie deer mouse is normally predetermined by heredity.
*(2)* This preference can be reinforced by early field experience, but this experience is not a necessary prerequisite for subsequent habitat selection.
*(3)* Early experience in other places (such as woods or laboratory) is not sufficient to reverse the normal affinity for the field habitat.
*(4)* Raising the grassland stock in the laboratory for twelve to twenty generations could reduce the hereditary control over habitat preference, thus increasing variability in responses when the group was tested.
*(5)* Laboratory mice retain the ability to utilize early field experience in learning to select the field conditions, but early experience in the woods or laboratory did not increase the likelihood of their choosing an area other than the field.

The principle of "habitat imprinting" (Thorpe 1945) was invoked to explain the influence of early experience.

We can see from these examples that animals may be acutely sensitive to a variety of cues when selecting a habitat. Such sensitivity has obviously been developed over eons of animals associating certain cues with certain kinds of habitats. Under some conditions, such preferences may be inherited; under other conditions, learning (perhaps in the form of imprinting) may play an important role.

## TEMPERATE AND TROPICAL HABITATS

At this point, we might pick up some valuable insight into aspects of behavioral ecology by considering differences in how species behave in two different kinds of zones: temperate and tropical.

It has been noticed that there are pronounced differences in the biological components of temperate and tropical areas. One of the most appparent differences is in seasonal change. In the temperate spring and summer, life flourishes and food abounds; but in winter, life is harsh. The winter landscape is one of bleakness and, at such times, most species are likely to find little food.

The seasonal fluctuation, however, cannot be explained entirely on the basis of food abundance (Figure 25.4). Winter also has its density-independent effects, such as cold temperature, that reinforce such density-dependent mechanisms as food abundance. Seasonality may also cause shifts in the relative importance of density-dependent and independent mechanisms. In temperate spring and summer, for example, populations may become more controlled by density-dependent effects than in winter.

While seasonal change in the tropics is much less pronounced than in temperate climates, it is also true that the seasonal dry spells of the tropics can drastically change the complexion of the wilderness there. Still, changes in rainfall produce nowhere near the ecological effects of changes in daylength and

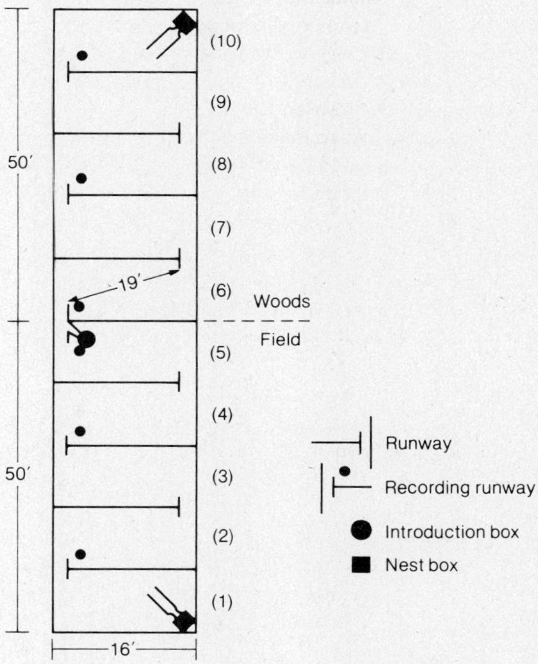


Figure 25.3 The experimental setup of Stanley Wecker, used to determine innate habitat preferences of two kinds of mice.

temperature (except in such places as dry, tropical deciduous forests). Thus, tropical niches are relatively stable and, in general, so are the numbers of animals in the tropics.

The stability of tropical weather is responsible for another ecological phenomenon. The reproductive, or fruit-bearing, period of each type of plant may be seasonal, but since there are usually no intolerable seasons, different types of plants can reach maximum productivity at different times of the year, producing a year-round food supply for many tropical species. This increased stability of the food supply may be partly responsible for the increased stability of animal populations in the tropics.

### Latitude and Species Diversity

It has been found that, as a general rule, there are more species at lower latitudes—which usually means in tropical areas—than there are in higher latitudes (toward the poles). For example, consider the latitudinal clines of ants, snakes, and birds (Tables 25.1, 25.2, and 25.3). Similar gradients have been found for a number of marine invertebrates and other animals (Emlen 1973).

The correlation between latitude and species diversity even appears in the fossil record. For example, more Cretaceous planktonic foraminiferans (tiny, shelled, marine animals) are found at lower latitudes. This is somewhat surprising, since during the Cretaceous period, climatic belts were less well defined and there was generally a milder worldwide climate than exists today (Stehli, Douglas, and Newton, 1969). There are also more families of Permian brachiopods (marine animals with bivalve shells) from lower latitudes in the northern hemisphere. Among fossorial bivalves and mammals, the ones in cold areas also extended into warmer latitudes, while many species were endemic to the warmer, lower latitudes. Also, it is assumed that the more cosmopolitan, or widespread, families are older. The evidence, then, suggests that species tend to arise in the tropics, some spinning off into temperate regions. If rate of speciation can be taken as a measure of evolutionary rate, it would therefore seem that evolution is more rapid in the tropics.

But why should evolution proceed faster in warmer areas? The answer is not related to body temperature, since both ectotherms and endotherms (and even plants) are affected. It is true that more solar energy strikes the lower latitudes, but why would this result in speciation? The answer is not at all clear, as Emlen (1973) pointed out.

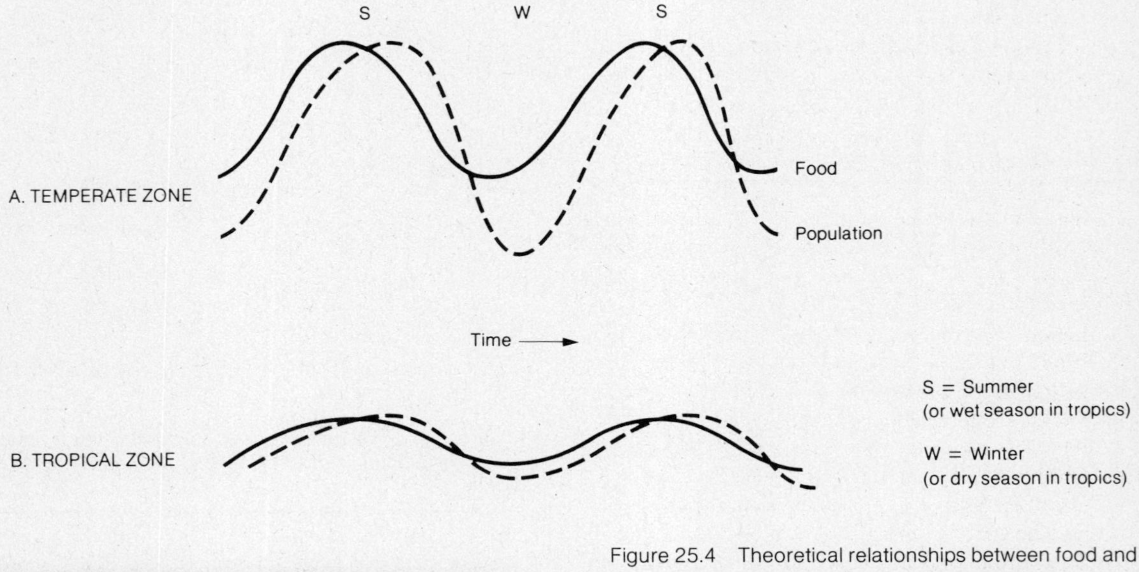


Figure 25.4 Theoretical relationships between food and the population utilizing it in temperate and tropical zones.

There are two prevailing explanations. The *non-equilibrium hypothesis* states that the tropical species are older and that temperate and arctic species are on their way to a maximal species density that, when reached, will remain at about equilibrium. They simply haven't arrived at that point. On the other side of the coin is the *equilibrium hypothesis* that states that, although temperate and arctic species exist below the species densities of the tropics, they have about all the species they can handle. If speciation were to increase in the temperate zone, extinction rates would rise to match it. So, in effect, all areas are "saturated." According to this hypothesis, regional variation in species diversity would reflect the local saturation level.

Ecologist Eric Pianka (1974) summarized the major theories that seek to explain the higher number of species in the tropics. He noted that any factor can be classified as primary, secondary, or tertiary, depending on whether it acts through the physical environment alone, through both the physical and biotic environments, or through the biological environment alone. His points are, in essence, as follows:

(1) *Evolutionary time.* This theory assumes that diversity increases with the age of the community and that the tropics are older. The notion is open to question, but it assumes that temperate zones have not been occupied (especially since the species-depressing effects of the ice ages) long enough for their species to have become well adapted to them.

(2) *Ecological time.* Here, the idea is similar but the time involved is shorter; the problem is not that there has been insufficient time for speciation, but for dispersal. Newly available areas have just not yet been fully colonized. The major weakness of this idea is that it ignores the phenomenal dispersal ability of most organisms.

(3) *Climatic stability.* Low seasonal change means that organisms do not have to develop broad tolerances for a wide range of environmental conditions. Thus, in the tropics, specialists arise, each utilizing rather narrow niches. The finer the habitat is divided, the more species that can be fitted in.

(4) *Climatic predictability.* Even when the climate is variable, it may be cyclic and predictable. Thus organisms can adapt to a certain cycle of conditions and *patterns* of resource availability.

**Table 25.1** Number of ant species at various latitudes.

| | AREA | NUMBER OF ANT SPECIES |
|---|---|---|
| *(North)* | *Alaska (Arctic)* | *3* |
| | *Alaska (all)* | *7* |
| | *Iowa* | *73* |
| | *Utah* | *63* |
| | *Cuba* | *101* |
| | *Trinidad* | *134* |
| *(Equator)* | *Sao Paulo* | *222* |
| | *Missiones* | *191* |
| | *Tucuman* | *139* |
| | *Buenos Aires* | *103* |
| | *Patagonia (all)* | *59* |
| | *Patagonia (western side)* | *19* |
| *(South)* | *Tierra del Fuego* | *2* |

*Source: Kusnezov (1957).*

**Table 25.2** Number of snake species at various latitudes.

| | AREA (PROVINCE IN ARGENTINA) | NUMBER OF SNAKE SPECIES |
|---|---|---|
| *(North, equatorial)* | *Missiones* | *55* |
| | *Corrientes* | *51* |
| | *Entre Ríos* | *34* |
| | *La Pampa* | *15* |
| | *Buenos Aires* | *22* |
| | *Rio Negro* | *5* |
| | *Chub* | *5* |
| | *Santa Cruz* | *1* |
| *(South)* | *Tierra del Fuego* | *0* |

*Source: Serie (1936).*

Table 25.3 Number of breeding bird species at various latitudes.

| | AREA | NUMBER OF BREEDING BIRD SPECIES | (AUTHORITY: CITED IN DOBZHANSKY, 1950) |
|---|---|---|---|
| *(North)* | *Greenland* | *56* | *Salomonsen* |
| | *Labrador* | *81* | *Peters* |
| | *Newfoundland* | *118* | *Peters* |
| | *New York* | *195* | *Parkes* |
| | *Florida* | *143* | *Grimes* |
| | *Guatemala* | *469* | *Griscom* |
| | *Panama* | *1100* | *Griscom* |
| | *Colombia* | *1395* | *de Schauensee* |
| *(South, equatorial)* | | | |

*Source: Dobzhansky (1950).*

Temperate climate is much less predictable than tropical weather and animals have to be physiologically prepared for a variety of conditions. This preparation, in effect, is similar to having a wide theoretical niche in that the organism may have to respond to the environment in a number of ways at any given season.

*(5) Spatial heterogeneity.* Structurally complex habitats offer a greater number of microhabitats than simpler areas. Complex habitats, such as exist in the tropics with its heavy and diverse vegetation, offer more places for animal species to live than do simpler and biotically impoverished areas.

*(6) Productivity.* In habitats with little food, foraging animals cannot afford to bypass potential prey items. In more productive habitats, they can be more selective. More productive habitats, then, encourage narrow diets. Because each species uses less of the total *range* of available foods, the same spectrum of food types will support more species (Figure 25.5). Also, productive habitats may support more species if resources, which are too sparse to be of much use in a nonproductive area, become abundant enough (MacArthur 1965).

*(7) Stability of primary production.* The maintenance of high numbers of animal species is strongly dependent on the stability of plant species. This is actually a secondary or tertiary mechanism because the plants themselves are continually adapting to changes in climatic conditions. Hence, they offer a buffer between primary (climatic) changes and the animals of the habitat. Since there are more plants in the tropics, there is a more pronounced buffering effect and, hence, greater stability. Environmental stability, as we know, increases the number of species. It should also be pointed out that plants could add to instability, for example, by blooming erratically. Pianka suggests that in this list, numbers 3, 4 and 7 could be included under "temporal heterogeneity" to parallel "spatial heterogeneity," number 5.

*(8) Competition.* Diverse communities, such as tropical rain forests, are thought to harbor populations that are near their maximum densities with the result that both intraspecific and interspecific competition are likely to be keen. *K*-selection probably predominates under such conditions and specialization is probably the rule. The resulting small niches make high diversity possible. Populations in temperate and polar regions, on the other hand, are believed to be less stable and well below their potential maximum levels. In these areas, organisms must primarily adapt to the "physical" world rather than the biotic one and

*r*-selection prevails. Competition is lower and these populations typically have broad tolerance limits and rather large niches.

*(9) Rarefaction.* Rarefaction refers to repeated density-independent removal of organisms from a community. For example, severe winter cold snaps may wipe out large segments of some populations. We recently nearly lost the Eastern bluebird in this way. Through rarefaction, competition is reduced *without competitive exclusion.* Rarefaction probably occurs relatively less frequently in the tropics.

*(10) Predation.* Predators can act as rarefying agents by reducing the level of competition among their prey. As we saw earlier, predation can actually increase the number of species in an area if it reduces the extreme competitive effects of its prey on other animals. Because predators tend to prey on the more abundant prey types, predation is density-dependent and thus promotes prey diversity. Predation, we know, is higher in the tropics.

It should be obvious that several of these factors could have operated or may be operating now to increase the numbers of species in the tropics. It should also be apparent that the differences in species numbers in tropical and temperate areas have wide implications for the evolution of animal behavior.

Orians (1969), in his investigations of the number of bird species in tropical forests, noted that the increased number of species in these places is due to the presence of ecological types that are not represented in the temperate forests. This suggested to him that the increased diversity was permitted by the greater range of resources permanently above "threshold values" in tropical forests. On the basis of preliminary studies of tropical communities at different altitudes, he tentatively minimized the role of temperate-zone instability as a factor reducing the number of bird species there. Noting the complexity of the problem, however, he stated, "It is sufficient for the present to note that it is dangerous to generalize too much about 'the tropics'."

Perhaps we can state, though, that in the tropics, population numbers are stable and high and that populations maintain levels near the environment's carrying capacity. One result of stability is steady competition, which means that although food is generally present year-round, because of high demands on it, it may actually be in short supply for most of that time (see Figure 25.4). On the other

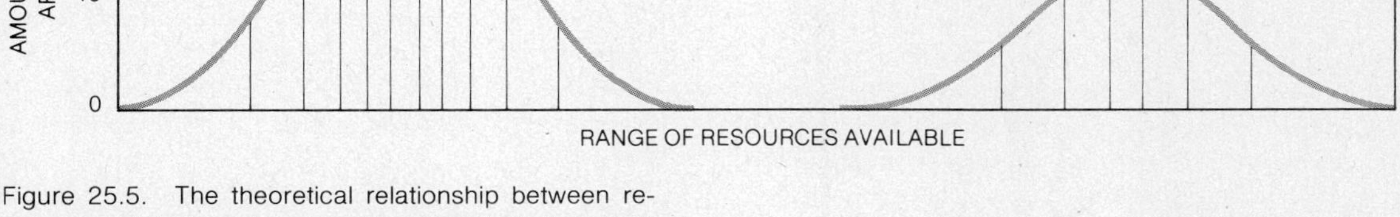


Figure 25.5. The theoretical relationship between resource availability and species diversity. The left curve is twice the height of the right one and indicates a greater level of resources. Also note that the bars under the curves are almost identical in area. The system at left is able to support more species because of its higher level of resources (standing crop).

hand, in temperate zones, populations tend to be unstable, animals and their food fluctuating out of phase, so that animals may periodically find themselves with a superabundant food supply.

In spite of periodic abundance, in the temperate zones animals must be geared for the seasons of lowest food supply when competition would be at its maximum. They can adjust to these periods in a number of ways, such as by altering their behavior as winter approaches, changing their diets, or entering a resting phase. However, in some cases, such flexibility could be expected to place a great behavioral "load" on them, especially on relatively stereotyped species that would require genetically mediated responses to correspond to each potential situation. An alternative to such a load is increased behavioral flexibility associated with learning. The problems here are those attendant with the development of greater learning ability.

There's another, rather disconcerting, possibility here. Could tropical species diversity be maintained through means that are essentially opposite to all the premises we have been discussing? For example, what if temperate species are more narrow and inflexible in their behavior than those in tropical areas? It could be that temperate species maintain strong ecological separation, perhaps with some degree of flexibility, even in periods of food abundance. The advantage of strong niche separation, even in good times, would be to establish and maintain behavioral patterns that would see these animals through the recurring lean periods. The constant separation would minimize behavioral adjustments when winter finally came. In fact, some researchers (Klopfer 1969a; Klopfer and MacArthur 1961), in noting the great number of bird species in the tropics, have theorized that the phenomenon is probably due to increased tolerances for niche *overlap,* as well as the ability to occupy marginal habitats. Many biologists, however, believe that the constancy of conditions in the tropics assures the relative stability of food and other commodities and promotes increased behavioral stereotypy in foraging and other activities—the stereotypy stemming from reliance on the same food and cover throughout the year—and the result is decreased overlap. Or so the story goes.

The large number of species in the tropics is itself believed to contribute to the stability of the area. A basic, if unproven, premise in ecological theory is that diversity increases stability in an ecosystem. If the premise is true, stability may be enhanced by behavioral *stereotypy* in animals. For example, the large number of trees found in lowland tropical forests may be partly the result of highly specialized predation of animals on seeds and seedlings, which results in no species of tree becoming dominant (Janzen 1971b). The impact of this statement is reduced when one realizes that behavioral *flexibility* may also contribute to stability through mechanisms of compensation. If one species should become extinct or otherwise reduce its impact in an area, its ecological role could be assumed by other species as they expand the range of their behavior.

Increased species diversity in an area indicates that each species is demonstrating increased *habitat selectivity.* But this does not mean that tropical diversity is high *because* tropical species exhibit an innate tendency toward narrow and stereotyped niches, and that this, in turn, permits their crowding together. Instead, the species are probably *forced* into their narrow niches by strong competition—competition that is not regularly reduced by seasonal hardship. There is some evidence for this. For example, the bananaquit, a small tropical bird, is restricted to a narrow niche, but in the absence of competitors it is capable of immediately broadening its niche and occupying new areas (Klopfer 1969a). We'll say more about this shortly, especially with regard to Crowell's opposing evidence.

In ecological terms, certain problems arise if tropical and temperate zones are defined simply by their proximity to the equator. In reality, strong differences exist between the ecosystems at comparable latitudes north and south of the equator. Charles Darwin was struck by the appearance of ibises, flamingos, caracaras, parrots, hummingbirds, and other "tropical" species amid the frozen sleet of Tierra del Fuego at the tip of South America. Because of the distance from the equator, this was considered a temperate area, but there it was, full of tropical species. It has been suggested that *constancy of environmental conditions* should be a primary criterion in the definition of "tropics" (Moynihan 1971). Thus, the constancy of what has been called the south temperate zone would permit the area to be reclassified as tropical. The north temperate zone remains biologically distinct from both equatorial and southerly latitudes due to the unique qualities and quantities of its life. For a discussion of differences in the biota of the tropics and temperate zones, see Kiester (1971).

## COMMUNITIES AND SUCCESSION—A FEW DEFINITIONS

Ecologists know very little about communities or ecosystems. The terms themselves are abstractions. In simplistic terms, an *ecosystem* is a theoretical entity that is essentially self-sustaining. The energy entering and leaving it is theoretically measurable. It is considered to be independent of other ecosystems (accordingly, a pond or a field might be an ecosystem). The organisms that live within each ecosystem comprise a *community* (for a brief introduction to the relationships, see Wallace 1978a). Communities are seldom clear-cut and distinct. They almost always grade into each other. If, then, we consider communities as "open" instead of "closed," we can visualize a continual interchange of materials, energy, and organisms between communities, and thus the concept becomes more useful and realistic (Pianka 1974). Where relatively distinct communities abut, the overlap or intergrade may be called an *ecotone.* These areas may be thin, geographical lines, but they can be rich in species since they contain representatives from adjacent communities. Where the change from one community to another is more gradual, we find *ecoclines,* and the animals that inhabit such areas may show clinal changes also. Ecoclines may be spatial or temporal. A temporal ecocline is marked by changes in the relative abundances of component populations as some species become extinct and new ones invade the area. Such a change is called a *succession. Primary succession* refers to the development of communities from bare, lifeless rock. *Secondary succession* refers to changes that take place after the life of an area has been destroyed. It is the latter that will briefly concern us here as we try to set the stage for our discussion of colonization.

In the farming areas of the southeastern United States, fields are often cleared of all standing vegetation in preparation for planting. Some of these fields are later left idle. Thus we have repeatedly witnessed community succession (Figure 25.6). First, annual and fast-growing weeds move in, then broomsedge, then small perennials, followed by shrubby perennials. Pine trees move in behind these, finally to be replaced by hardwood oak and hickory trees. The plant structure is, not unexpectedly, critical to the kinds of birds and other species that live there. So, as the plants change, so do the animals.

Presumably, the plants and animals of each stage alter the environment in such a way that species that follow find it suitable. The process is not very fast as humans measure time, and the sequence described here may take over a hundred years. When the oak-hickory stage is reached, it continually replaces itself, and thus the community is considered "stable." This is the *climax* community. In the course of succession, production usually exceeds respiration. In other words, not everything that is produced is used up. Thus, leaves and other organic materials may accumulate, rot, and form soil. The added soil contributes to new growth and new kinds of growth. At the climax stage, production roughly equals respiration so that organic materials cease to accumulate (Pianka 1974; Whittaker 1970).

Actually, there are marked geographical differences in successional events. What happens in Georgia is not necessarily what happens in Arizona, politics aside. In deserts, for example, the first invaders are usually the climax species. Oak forests never appear. Any succession is short. There is virtually no soil and the sparsely distributed plants do not produce their own "microclimates," to the degree that plants in heavily vegetated areas do. The earth's biomes are generally extensive plant communities; but disruptions caused by human activities, lightning, droughts, floods, and the like, keep large parts of the earth constantly undergoing succession. I might add that ecologist J. M. Connell is presently giving our definition of successsion, and the role we attribute to it, the skeptic's fish-eye. Revisions may be in order.

## WHAT KINDS OF AREAS ARE THE EASIEST TO COLONIZE?

Climax communities are, by definition, ecologically rather stable places. They have been *K*-selected for a long time, and their residents have nudged, molded, expelled, and encouraged each other in myriad ways until what is left must be a finely attuned and interdependent community. It can be assumed that

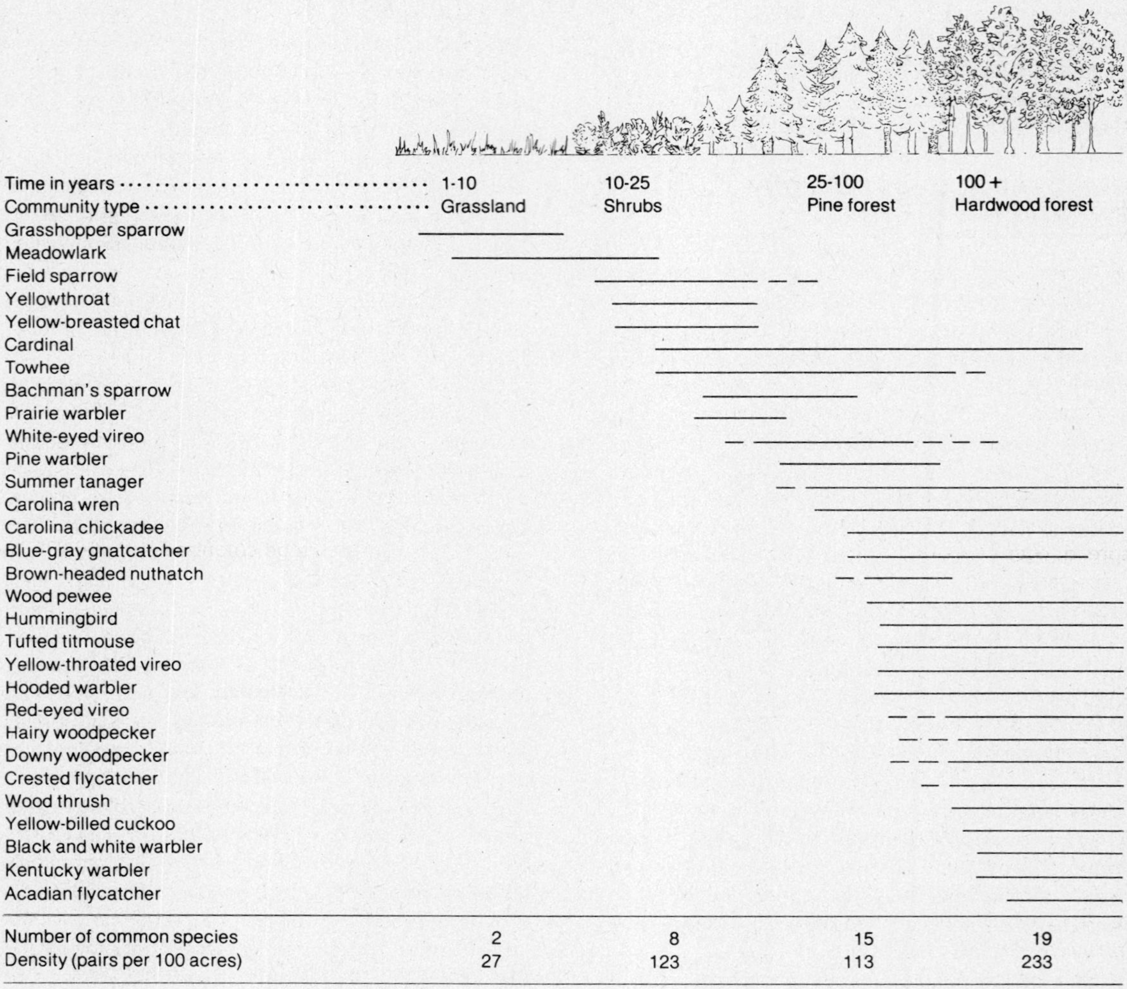


Figure 25.6 Typical secondary succession of vegetation and bird life in an abandoned field in the southeastern United States. Climax communities are of low productivity, using about as much energy as they produce. Clearing land for farming is simply a way of initiating succession and reaping the great productivity of early successional stages.

there is no uncontested resource, nothing useful, that some species is not already using. A new species is therefore not likely to be able to work its way into such a community. It probably could not replace any of the species in the system—not on their home ground. A climax community, then, is about the worst place imaginable to try to colonize.

Communities in their early successional stages are much more vulnerable. An invader of an early community might effectively disrupt the normal competitive pattern. It is most likely to be successful if it is highly *r*-selected. It might, at least for a time, be able to displace less fecund, more *K*-selected species.

Not only are early successional areas vulnerable to colonizers, but "impoverished" areas are also susceptible to invasion. These communities have resources that are not being utilized and can thus admit additional species with a minimum of niche shuffling and competitive rearrangement. Impoverished areas may include islands, peninsulas, and geologically newer areas.

Whether an area is impoverished or not cannot normally be determined until after the fact—after it has been successfully colonized by some invading species. If it is colonized, it was impoverished. We know now that with respect to amphipods (marine, shelled animals), Britain has been impoverished in the recent past. The freshwater shrimp (*Eucrangonyx gracilis*) and the brackish water amphipod (*Gammarus fasiatus*), both from North America, have spread over large areas of England and Ireland, respectively. The telling feature of their having invaded an impoverished area is that there have been no *detectable* disturbances in the biota of the area (Elton 1958; Emlen 1973).

## WHAT KINDS OF ANIMALS MAKE THE BEST COLONIZERS?

The answer to that is animals that don't try to crowd into climax communities. The best colonizers, then, will probably be any animal invading early successional communities, peninsulas, or islands. But the real question here is, What qualities do good colonizers have in common?

To begin with, they should be *r*-selected for the reasons mentioned earlier. Emlen (1973) had the feeling that successfully introduced species have had high *r*-values. He reminded us, for example, that of all plant-eating insects (an *r*-selected group) imported into the United States, about twenty-five percent have become established (De Bach 1965), and for weedy, annual plants (another *r*-selected group), the figure may be even higher. Also, larger species generally have lower *r*-values, and larger animals usually make poorer colonizers.

A good colonizer should also be able to quickly shift its realized niche (or its point *B* in Figure 24.1 in Chapter 24) to avoid competition and to capitalize on any new opportunities. Thus, the colonist should be a generalist.

Since areas must often be invaded and reinvaded before a species is finally able to take hold, a good colonizer should also be a good traveller. Birds are good travellers, snakes and mammals often aren't. Also, it does no good if only one mammal reaches an island, unless that animal is already pregnant. If a group of animals of a species invade an area, however, the likelihood of successful mating is increased. Thus, flocking birds tend to be good colonizers.

Mayr (1965) in his long studies of island birds, finds successful island colonizers are likely to have a number of traits in common: herbivory, social tendencies (among birds he lists starlings, sparrows, and social thrushes as good colonizers), travelling ability, behavioral flexibility, and a tendency toward what is, in effect, exploratory (or latent) learning (see also MacArthur and Wilson 1967). So, certain kinds of areas are particularly vulnerable to colonization and they are most likely to be colonized by animals with certain behavioral traits.

### Supertramps

Recently, bird populations on several islands near New Guinea were investigated, including the fabled Krakatau, which was once rent by volcanic activity, and several other volcanic islands that had been devastated in the same fashion and then recolonized over the past couple of hundred years. Also, several smaller islands whose bird populations had been wiped out by tidal waves were surveyed. It was found that the colonizing birds employed two major "strategies." One group was highly *r*-selected and dispersable; these were called "supertramps." As soon as an island became habitable, the supertramps moved in, reproduced madly and spread like wildfire. Once the slower-dispersing *K*-selected species arrived, however, the supertramps gave way before the better competitors, eventually to be replaced by them. But by that time, the supertramps were busy invading other areas. Whenever they found islets that were too small to support stable populations, or a recently devastated island, they bred on nearly a year-round basis, filled the island, and dispersed new emigrants ever outward. By the time *K*-selected birds arrived at the hospitable islands, the tramps had already invaded new temporarily empty islands.

It was surprising to find that where *r*- and *K*-selected species temporarily coexisted, the *K*-selected species were still in the minority long after they should have begun to severely depress the *r*-selected species. It was theorized, however, that these *K*-selected species were actually too efficient at exploiting resources and that their competition reduced the support-system of the island below the level it would have reached if it were only being exploited by supertramps. For example, *K*-selected insectivores may be so efficient at catching insects that they reduce the insect populations far below the levels that the supertramps would have. *K*-selected fruit-eaters may eat fruits that are not yet developed or fully ripened and thus reduce the potential food value of the fruit crop. Thus, on older islands, inhabited by *K*-selected birds, the bird population may be much lower than on islands that the supertramps alone have colonized, since the inefficiency of the supertramps allows the food diversity and abundance to remain high (Diamond 1974).

## ISLAND BIOGEOGRAPHY

Islands have long fascinated biogeographers and are now attracting the attention of other sorts of ecologists as well (Wallace 1978b,c). One reason for this attention is the long-standing argument over why islands have fewer species than mainlands. Another reason is that islands present "natural experiments" for biologists—peculiar ecological situations that may differ from the mainland and from one island to another—yielding clues as to the role of various factors in natural systems. For example, because islands usually support fewer predators, we can get some idea of the effects of predation on populations by comparing otherwise similar mainland and island situations.

### Habitat Variation on Islands

Whereas all continents boast a wide variety of habitat types (with the possible exception of Antarctica), greater variation occurs among islands. Some islands are nothing more than flat stretches of sand barely rising above sea level and, at times, they may even be submerged, such as the islands of the Chagos Archipelago of the Indian Ocean. Others are great bodies with tall mountains and lush valleys, as we find in the southeast Alaskan chain. It is therefore difficult to generalize regarding island habitats or niches; instead, we will focus our attention on general rules that may help in predicting the biota of any island.

Area and distance from the nearest mainland largely determine the types of niches islands offer. Lack (1969) pointed out that smaller, more distant islands are usually low lying, and therefore have fewer vertical transition zones. Vertical transition zones, a function of altitude, are directly correlated with the number of species an island can support. This means taller islands have more species. And taller islands are likely to be closer to the mainland. Actually, size alone would probably generally affect the complexity of islands by simply increasing the opportunity for variance in the habitat. And larger islands are usually nearer the mainland. So the nearest islands tend to be tall and large and to boast the greatest number of species.

The distance between islands and continents affects niche dimensions in several ways. One way is by determining what other kinds of animals any colonizer is likely to have to deal with. The first problem involved in colonizing an island is getting there, and it is not equally probable that any two species will reach an island. Some species get around easier than others. No one is surprised to find birds on islands. Birds are generally good travellers. They may discover islands while in migration, or mainland birds may be blown out to sea by storms. So any animal reaching an island is likely to confront birds. That animal does not necessarily have to put up with snakes, however. Reptiles are poorer travellers and even if they were to drift about on log rafts, they would confront special physiological problems in being continually wetted with seawater. A walk through a forested island of the West Indies is enhanced by the presence of many species of colorful birds and the knowledge that no poisonous snakes are about.

The effects of distance on a colonizer are inversely related to the travelling ability of the species. In other words, both good and poor travellers may colonize offshore islands, but more distant islands, especially the more recent ones, will be inhabited primarily by good travellers. Of course, older islands have an increased probability of being reached, even by unlikely colonizers.

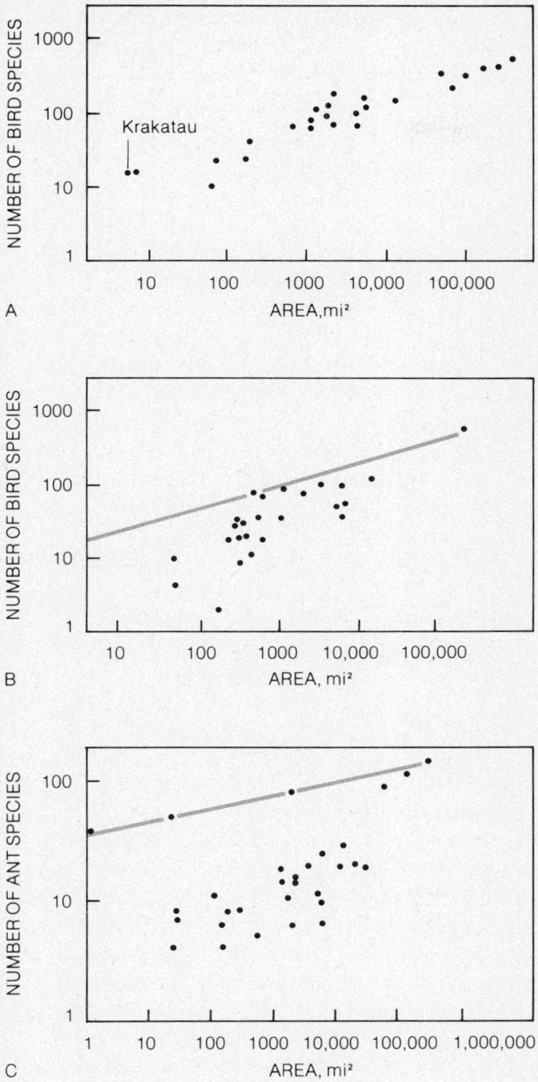


Figure 25.7 Some species–area relationships. (A) the relationship of land and freshwater bird species on the Sunda island group of southeast Asia; (B) the relationship on more remote islands in the South Pacific (the line is drawn through the two islands closest to their mainland source to show how drastically species density departs on the more remote islands; (C) the relationship of ant species to island size in various Molluccan and Melanesian islands. The line shows the number of species in increasingly large areas of New Guinea. The points show smaller islands.

The condition of having fewer species on islands than mainlands should not be unexpected, since some species have particular problems in reaching islands. But besides problems of distance, the potential colonizer might not find a suitable habitat once it reaches an island. One reason is related to the difference in the sizes of the island and the colonizer's mainland home; all the habitats found on a continent are not likely to be represented on an island because there isn't room. But even considering their small size, islands still have fewer species than would be expected. The point is this: fewer species are found *per available habitat* on islands than on continents.

Figure 25.7 shows some of the empirical evidence regarding the relationship between island size and the number of species of various types found there. One can also see the depauperizing effect of distance in Figure 25.7B.

Not only do islands support fewer species than mainlands, but they have far fewer at the higher trophic levels. Thus, there are fewer island predators. One reason may be because islands are isolated, by definition. When you're considering a twenty-square-mile island, you can (theoretically) be rather precise about the number and kinds of species that are likely to be living there, or are likely to be living there within a certain time span. If you compare that island with a twenty-square-mile area of mainland, however, you're comparing it with only a *sample* of the mainland. Rare species may occur in the mainland sample at any time if they are "just passing through." Also, wide-ranging animals may be moving in and out of it at all times. Such species might not be represented on islands. In addition, on the mainland even areas that are no more hospitable than islands may boast a great variety of animal life because of continual, exploratory immigrations from adjoining population centers. Islands cannot be so easily invaded. Thus, the species density and diversity is likely to be higher in a mainland sample. Predators need a lot of space, as we have seen, but because of the paucity of species on islands, here they need even more space. Twenty square miles of land may support a mountain lion on a mainland, but that same cat may need thirty or forty square miles on an island (Pianka 1974).

## Immigration and Extinction

One reason for the paucity of island species is their high rate of extinction. Islands are likely to offer simple habitats, as we have seen, and in the biological world simpler systems are apparently more unstable; thus, any biological component of a simple system has an increased probability of becoming extinct.

But there are also other reasons for the reduced number of species on islands. Any species adapted for continental living is not likely to be concomitantly well adapted for islands, and so colonizers may find it rough going until they adapt to their new conditions. However, some species are able to fit into the island situation in which they find themselves, at least well enough to survive until they can better adapt to their new habitat. But it should be pointed out that such preadaptation does not necessarily ensure success. To use an example suggested by Klopfer (1969a), a continental grassland species may require half an acre per bird. Upon reaching an island, the birds may find that only a quarter of an acre is available. So, some of the birds may move into adjacent woodlands (assuming that this habitat is vacant). The grassland species may then, in time, begin to adapt to its new habitat. However, if a woodland species should appear on the island, the grassland species occupying the woods may become extinct, unable to compete successfully with the newcomer. On the other hand, the new woodland species may encounter a strong competitor in the well-adapted descendants of the grassland species, and may itself become extinct. (There generally seems to be an advantage in "being there first"—but see Morse [1971] for an exception.) No matter which species becomes extinct, however, the result is a simpler biological situation than that which exists on the continent from whence all the species came.

MacArthur and Wilson (1967) summed up the relationship of colonization and extinction to island distance and size as follows:

*(1)* Islands farther from the mainland have lower colonization rates.

*(2)* Smaller islands have higher extinction rates than larger islands.

*(3)* Increased size of distant islands has a greater effect on increasing the number of species than does increased size of islands nearer to mainlands.

(In other words, large size is increasingly correlated with successful colonization as the distance from the mainland increases.)

Since animals must rely on other organisms for their sustenance, we may ask, "But what about food?" Whether an animal is plant eating or flesh eating, if it exists on an island, it is dependent upon food species that have their own problems of colonization. It was argued that the fewer species of birds on smaller outer islands is not due to the dispersability of the birds, but to lack of food (Lack 1969). It follows from what we have learned that any island-colonizing animal will have fewer food sources at its disposal and will be likely to be forced to change its diet quantitatively if not qualitatively.

A change in diet does not necessarily mean a change in foraging behavior. For example, it has been found that certain species of birds on Bermuda take food from a wide spectrum of habitat types that, on the American continent, are occupied by other species. The Bermuda birds, however, have not changed their foraging patterns either quantitatively or qualitatively (Crowell 1962).

## An Equilibrium Theory

We have been discussing islands as "impoverished," but this may be a biased and imprecise term. I had an uncle some people might have tried to pin that label on, but he argued that whereas he may have had less money than others, in actuality he just couldn't *use* any more. And so it is with islands, at least according to MacArthur and Wilson's equilibrium theory (1963, 1967).

Because islands typically hold fewer species than a comparable area of mainland habitat, it had long been assumed that they were below their saturation level, that they *could* hold more species. MacArthur and Wilson suggested that perhaps islands are supporting as many species as possible. If this is true, they reasoned, the rate of immigration of *new* species to an island should decrease as the number of species on the island increases. In other words, immigration (or colonization, which might be considered successful immigration) and extinction are inversely related (Figure 25.8). This means that the immigration rate must drop to zero if the species

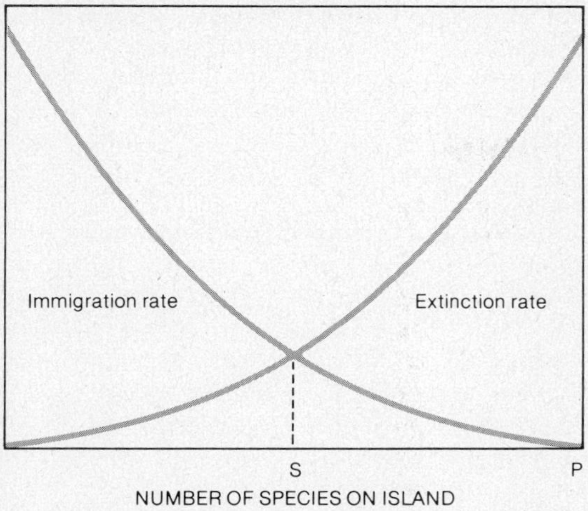


Figure 25.8 Theoretical shape of immigration and extinction rate curves in the equilibrium model of species diversity on islands. At first the immigration rate drops rapidly as the best colonists become established on the island. The extinction rate increases more rapidly at high species number because of greater competition between species.

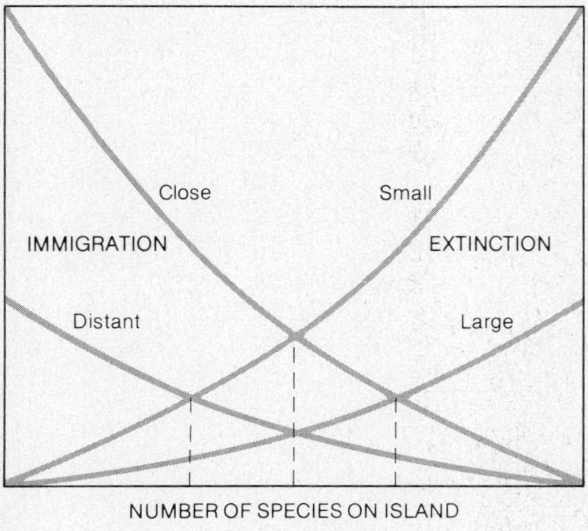


Figure 25.9 Relative number of species on small, distant islands and large, close islands predicted by the MacArthur–Wilson equilibrium model. The number of species on large, distant islands and small, close islands is intermediate.

density of the island were to reach the species density of the mainland "source area." Furthermore, the rate of extinction of species already present on the island should *increase* as the number of species on the island increases. This seems logical enough since, as more species utilize an island's resources, the average population size for any species should fall off as the increasing competition results in some competitive exclusion.

A state of *dynamic equilibrium* is reached when the rate of immigration equals the rate of extinction. In such a case, the species *number* stays constant, although the species *composition* may change.

The size of islands and their distance from the mainland may affect the species density in predictable ways, according to the equilibrium model (Figure 25.9). Smaller (and therefore, usually, topographically simpler) islands suffer higher extinction rates than larger (and presumably more complex) islands. Smaller islands also have a slightly lower immigration rate since they present smaller "targets" to colonizers. Islands that are more distant should also have lower immigration rates. Extinction should be largely unaffected by the distance of islands from their source areas, per se, but if immigration is low because of that distance, extinction should be correspondingly low at equilibrium.

**Support for the Equilibrium Theory** When Krakatau erupted in 1883, it continued to belch its molten bile for over three months, and strange and spectacular skies were seen over large areas of the world as sunlight filtered through the airborne debris. The island became completely covered with molten lava, and for months afterwards, rain falling on the island rebounded skyward with a crackling hiss. It can be safely assumed that nothing on the island survived.

Repopulation of the island by colonizers from Java and Sumatra proceeded rapidly, and by 1921 the number of bird species on Krakatau was about the same as that on a nearby island of similar size. Between 1921 and 1933, the kinds of birds on Krakatau changed, but the species number remained about the same. Birds are a mobile species though and probably reached maximum density rather quickly. The behavior of the bird populations is just what would have been predicted by the equilibrium model that states that at some point, immigration would equal extinction.

Although bird species were at equilibrium, slower-dispersing plants were still being added to Krakatau in 1933, the last time a count was made. There are two points to be gleaned from this. First, the occurrence of equilibrium for one kind of species does not mean other species have also reached that point. Second, birds rely, directly or indirectly, on plants, so as the number of plants increased, the theoretical equilibrium point for bird species could have risen accordingly.

Later, an experimental test of the equilibrium model was devised. Simberloff and Wilson (1969) carefully censused several tiny mangrove islands in the Florida Keys. Then they fumigated each of them, killing all the arthropods. Over the next two years, they carefully monitored the process of recolonization. Actually, within two hundred days, the numbers of species on the islands had stabilized (Figure 25.10). Turnover rates remained high, but the numbers of species remained relatively constant over the two-year period.

## Colonization and Interspecific Relationships

It has been found that where several species of warblers are sympatric, the dominant species attain high density but are less flexible in their behavior (Morse 1971). Possibly, the more *subordinate* species among a group of competitors would show a greater behavioral flexibility as an adaptation to being pushed into suboptimal habitats. Thus, subordinate species, with their greater plasticity, might be better candidates for the colonization of small islands where the sort of ecological conditions that they confronted on the mainland do not exist. There are no supporting data at this time, however.

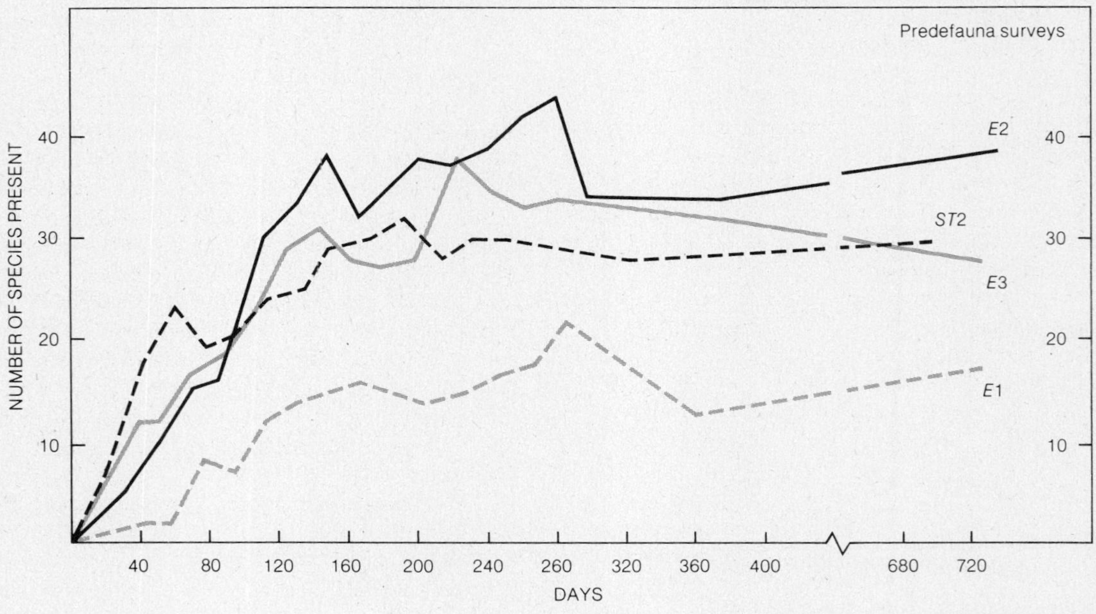


Figure 25.10 Colonization of four tiny mangrove islands in the Florida keys by arthropods after the islands had been depopulated by fumigation. The populations rose quickly, but soon levelled off to near their previous levels.

## Niche Parameters on Islands

We have referred to niche "dimension" because niche parameters on islands and continents *are* different, although the differences may be hard to visualize. Consider a model utilizing a two-dimensional concept of depth and width (Figure 25.11). If we assume depth to be the *amount* of food of a certain type that is available, such as the weight or mass of all tree-dwelling ants, width, then, could describe the *range* of available food types, for example, if referring to size, from small to large ants. We can see, then, that continental species theoretically have deep and narrow food niches as opposed to the shallow and wide food niches on islands. The model might help to explain why island species move so rapidly into unoccupied niches: they quickly begin to compete among themselves for a shallow food supply and are forced to expand outward.

We should also understand that such theories are intended to account for places other than land surrounded by water. The world is full of islands. Any relatively isolated species may be considered insular. Alpine species that can only exist at certain high altitudes may be restricted in their movement from one mountaintop to another, since they are unable to pass through the inhospitable valleys. Strong isolation of populations may permit them to diverge in their evolutionary directions, and one beneficial result, of course, might be a finer adaptation to each specific habitat. Caves can be islands. So can ponds. We will consider other ecological effects of insularity in our discussion of dimorphism. For now, let's just be aware that island niches may be described in theoretical terms that may apply to a variety of places, but the general utility of such theories remains to be demonstrated.

For further discussions regarding island colonization, you might refer to MacArthur and Wilson (1963); Carlquist (1965); Baker and Stebbins (1965); Grant (1968); Schoener (1970); Lewontin (1965); Emlen (1973); Pianka (1974); Lack (1976); Wallace (1978b).

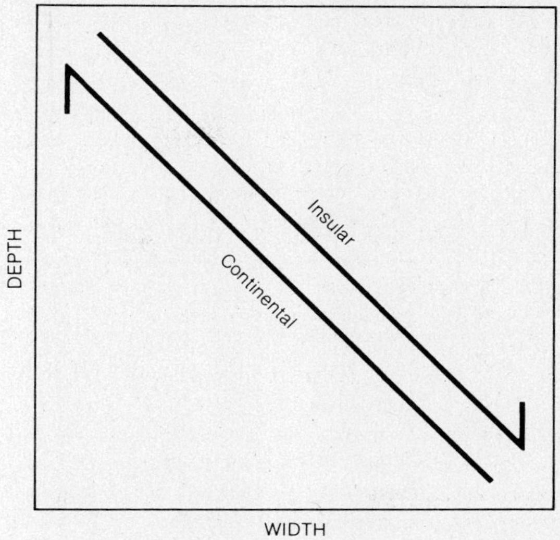


Figure 25.11 Niche parameters in continental and insular situations. See the text for an explanation.

Also, it should now be apparent that habitats shape both morphology and behavior. Their relationship is such that no student of ecology should attempt to interpret either without considering the other. For example, after noting high variability in trophic appendages of any species, one should be able to make tentative assumptions about the food and foraging patterns of that species and also to formulate some idea of what competitive pressures may be operating in its habitat. Such assumptions could then be tested, and in the biological world, providing material for testing is the noblest role of assumptions.

# 26

# THE SIGNIFICANCE OF SEX

A quick survey of an appreciative world reveals that sex is prevalent. Its ubiquity indicates that it is of great adaptive value, but its importance in the evolution of species is not entirely clear (Williams 1974), aside from its role in mixing genes and producing variation in populations. How did it evolve? In what ways is it useful? Why are there almost always two sexes, not more? Let's explore the question of sex, first learning about the differences in the ecological roles of males and females.

## SEXUAL DIMORPHISM

When the sexes of a species are distinguishable, the phenomenon is called *sexual dimorphism.* Obviously if sexes exist, so does sexual dimorphism, so we are really concerned about the *degree* of sexual differences. Most of us have long been aware of certain differences in the sexes. Boys are not usually hard to tell from girls, and male dogs are distinguishable from bitches. In other species, sexual differences are even more pronounced (Figure 26.1). However, we may have trouble determining the sex of our kitten, and the sexes of hyenas and elephants are very difficult to distinguish. Only experts can sex baby chicks, and almost no one can sex black ducks on the basis of external appearance. Obviously, then, levels of sexual dimorphism may vary widely.

### Sexual Dimorphism and Sexual Selection

Darwin, in 1871, set forth the idea that sexual differences are due to a particular type of evolutionary pressure called *sexual selection,* which he distinguished from natural selection. In his words, "Sexual selection depends on the success of certain individuals over those of the same sex, in relation to the propagation of the species; while natural selection depends on the success of both sexes, at all ages, in relation to the general conditions of life." This means the cryptic color of the female pheasant is the result of natural selection, whereas the garish aspects of the male must have developed through sexual selection as a mechanism to attract females. It can be argued that sexual selection is, in fact, just a form of natural selection. A careful reading of Darwin shows there is probably less argument between him and his recent critics (for example, Wynne-Edwards 1965) than it would at first appear.

We might note at this point that the adornments and conspicuous behavior of males have two rather distinct, but not mutually exclusive, functions. First, they might aid in attracting females. Second, in some species, the special male qualities may help in competitive conflicts with other males. Among fur seals, the huge males lumber ashore and establish territories. The smaller females that come ashore later are apparently attracted to the territories themselves and it doesn't seem to make much difference to them which males have succeeded in obtaining them. Once a female enters a harem, however, the male does not permit her to leave (Bartholomew 1970). It seems, then, that the large size and belligerent behavior of the males are not so much to attract females as to repel other males. It should again be stressed that in some species the character that is used in competition and fighting may also be the very one that attracts females. For example, the red shoulder epaulets of male red-winged blackbirds repel other males and attract females.

## How Do the Sexes Differ?

Sexes may differ in a number of ways. First, we can assume that they will differ in their primary sex characters, or "plumbing," but they may also differ in size. When the sexes of vertebrates are dimorphic in size, it is generally, but not always, true that the males will be larger than the females. The difference is sometimes negligible as in some birds, but often pronounced. For example, male fur seal males may be ten times larger than females.

Second, sexual dimorphism may be expressed in fighting accoutrements. In some species, males possess special fighting weapons, lacking in females. For example, consider the antlers of bucks or the enlarged canine teeth of male baboons. Birds do not usually show sexual dimorphism of weapons, but there are exceptions, such as in the long spurs of roosters.

Third, sexual dimorphism may appear in display apparatus, such as in coloration, markings, or appendages. We have already considered the development of the bold markings and coloration of the male peacock and pheasant.

Finally, sexes may differ behaviorally. In some species of birds, the sexes are similar or identical in appearance, but are dimorphic in behavior. The males of some "monomorphic" passerine species of birds sing loudly or make themselves conspicuous in other ways, while females are much quieter and furtive. The appearance of sexual differences in behavior indicates a possible dimorphism in the nervous system. It is sometimes argued that the nervous systems of the sexes are actually identical in these cases, and that behavioral differences appear because of differences in the kinds of hormones that are circulating in the blood and that activate only particular parts of the nervous system. Such may be the case for some species, but a case can also be made for the "dimorphic nervous systems" theory. For example, fighting behavior in the males and females of wild rats and mice is vastly different, the greater aggressiveness of males presumably being due to the level of male hormones in the blood. However, injections of male hormones into females and into castrated males produced two different results—an indication that the nervous system is

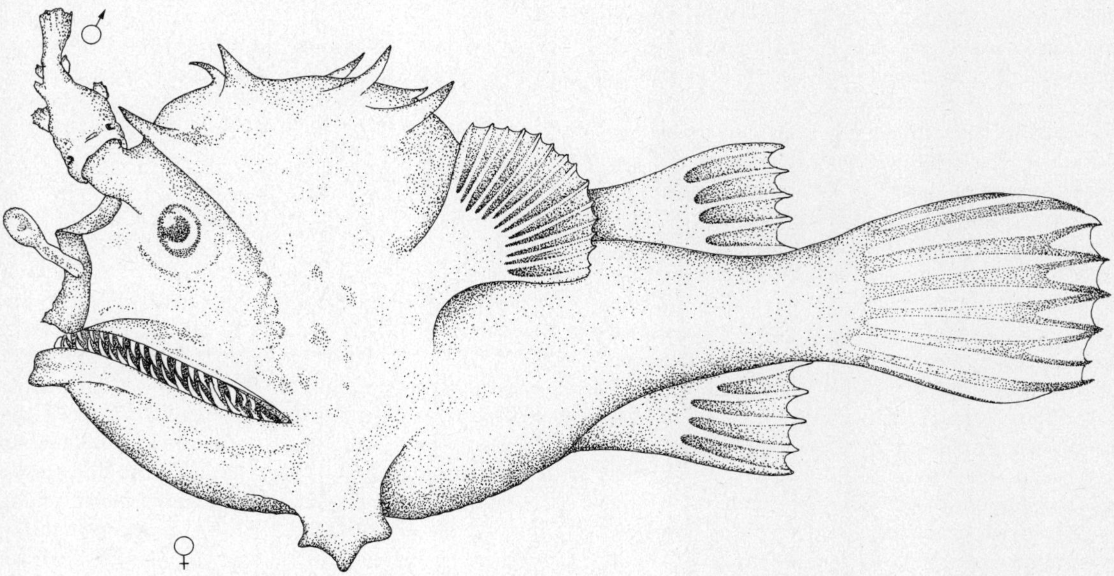


Figure 26.1 Female deep-sea fish with a tiny male fused to her head. This is one of the more extreme examples of sexual dimorphism among animals.

essentially different in the two sexes. The levels of aggression rose in both, but they fought differently. Females cannot be made to fight like males; they continue to use the female's patterns of movement (Scott 1966). We also noted earlier that embryonic hormones can affect the development of the animal's behavioral patterns in later life. Thus, hormones themselves may bring about fundamental differences in the nervous systems of the sexes.

## UNUSUAL SEXUAL SYSTEMS

We usually think of animals being either male or female, but there are a number of other sexual systems, some of which pose rather interesting problems. For example, there are hermaphroditic species in which individuals bear the organs of both sexes at the same time. Such individuals arise rarely in humans, as genetic abnormalities, but the condition is the normal one for many species including some parasitic worms and some fish such as *Rivulus marmoratus* (Harrington 1961).

In other cases, an individual may be first a female, then a male (*protogynous hermaphroditism*). For example, in the cleaner fish (*Labroides dimidiatus*), a male may dominate a harem of three to six females who, in turn, maintain a linear dominance hierarchy. If the male dies or disappears, the most dominant female begins a transition into a male. The entire process may take only a few hours. In about two weeks, the new male is able to release sperm! The advantage here, of course, is that no more males are present than are needed (Robertson 1972).

Probably the most bizarre sexual system, however, occurs in what are called unisexual species, such as some fish (Schultz 1961) and salamanders (Wilbur 1971). Here, the males are apparently not even of the same species as the females. Males of other species are duped into copulating with these females but their sperm simply initiate development in eggs that have not halved their chromosome number through meiosis. The sperm contribute nothing genetically. Obviously, a number of interesting questions arise here. How did such a system get started? Why do the males participate?

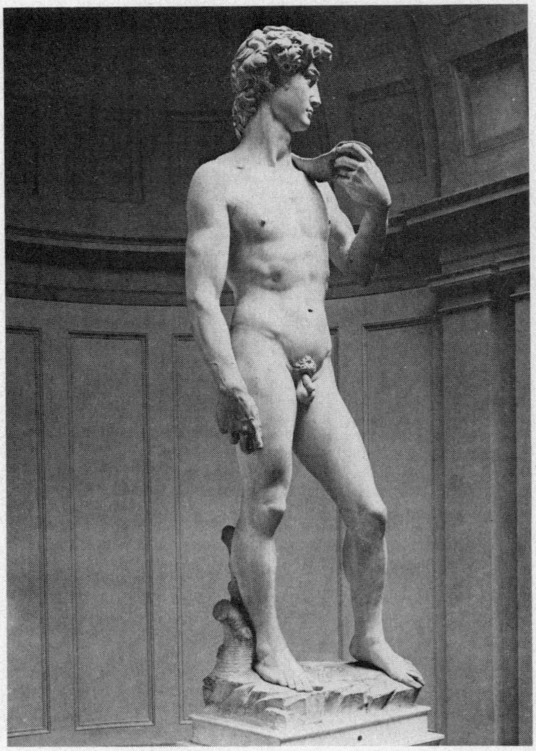

Figure 26.2 The heroic male figure is muscular, broad-shouldered and narrow-hipped, as seen here in Michelangelo's *David*. Men with the opposite traits are generally not considered sexually attractive, even in today's world.

## THE ACCENT OF SEXUAL DIMORPHISM IN HUMANS

Humans are sexually dimorphic in spite of some currently fashionable denials. In fact, historically, the differences have been encouraged and exaggerated. The heroic figures of Michelangelo are, even today, somehow more appealing than less masculine figures would be (Figure 26.2). Those qualities that say "male" have undoubtedly been important in the male's genetic contribution to succeeding generations, hence it has been adaptive to advertise those qualities. After all, muscular men have traditionally been of great value to the family and tribal unit. The shoulder area has been of particular importance, since the upper body and arms would have been involved in defensive fighting as well as in hurling weapons to bring down game. A woman was, at one time, well advised to choose a broad-shouldered man. It is probably on such a basis that men's clothing has traditionally accented the shoulders (Figure 26.3).

Height has also been advantageous in the masculine role. In fact, height itself is often intimidating. In the meetings between the Americans and the North Koreans at the 38th parallel, the Koreans insisted on higher chairs to negate the psychological height advantages of the Americans. The psychological effect has also not gone unnoticed in the design of police and military clothes (Figure 26.4).

In other cases, maleness has been more blatantly advertised. Many primates display by showing their penises (sometimes erect) and scrotal areas (Figure 26.5). Humans have also used the penis as a display (Figure 26.6). One of Hemingway's friends reported that when the *great aficionado* of bullfighting once tried on a matador's "suit of lights," he was chagrined to notice that he didn't bulge to the degree that the bullfighters did. He only found out later that their *machismo* was surreptitiously accented by as many as four handkerchiefs.

Female characters have also been enhanced and exaggerated through fashion. Female clothing has traditionally been designed to accent the width of the hips (important in childbearing) and the development of the breasts (important in nursing). In fact, it is probably not coincidental that the very things that often attract males are those things that are most important to successful childbearing (Figure 26.7). Sexual dimorphism in humans, then, has been important in the attraction of the opposite sex and, especially for males, in the intimidation of members of the same sex.

## THE ECOLOGY OF SEXUAL DIFFERENCES

It is a mistake to try to explain the presence of a single morphological or behavioral factor without considering the overall ecology of the organism. In order to satisfactorily account for any observation, several factors may have to be considered at once. To see the complex interweavings of morphology, behavior, and ecology, consider an example from baboons: the males are twice as large and heavy as the females and

Figure 26.3 Males tend to accent shoulder size although shoulders are no longer as important as they once were in acquiring commodities for the family unit.

Figure 26.4 Height is often emphasized as a gesture of intimidation.

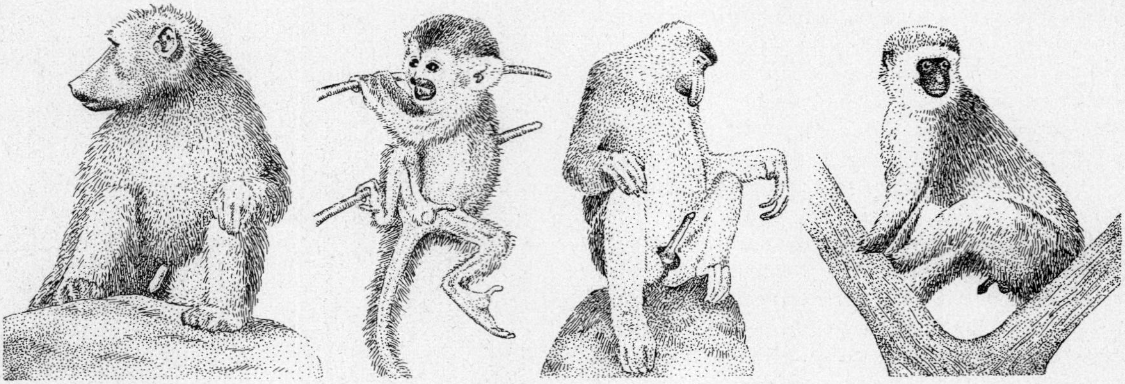

Figure 26.5 Male primates often display their genital areas, sometimes with erect penises, as sexual advertisement.

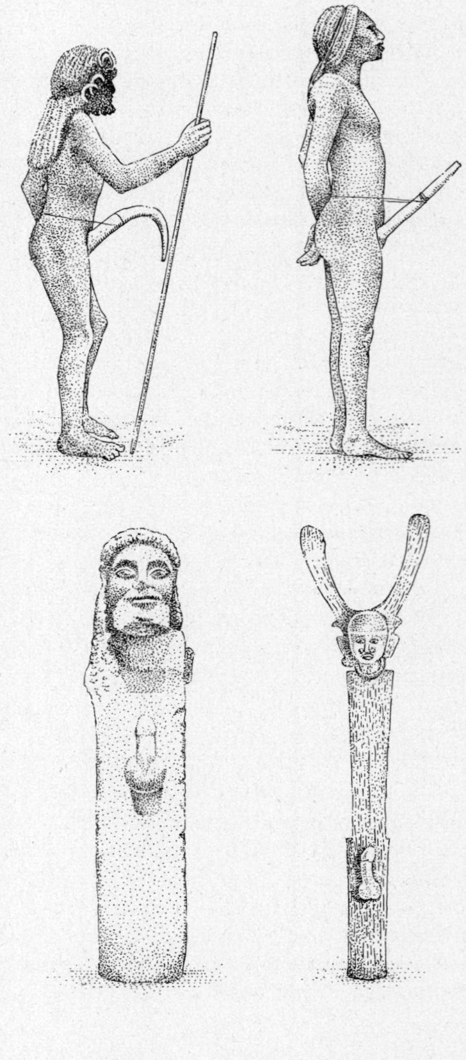

Figure 26.6 Some human groups use genital displays, including erect penises, and may advertise other signs of maleness as well, such as figures with beards (Herme of Siphnos, 490 B.C.) or head ornaments (house guardian statue from the island of Nias).

have well-developed canine teeth. They are also highly aggressive and defend the young from predators. They are so formidable that they can even repel cheetahs and, in a group, they have been seen teasing lions. In patas, another primate, sexual dimorphism is also pronounced, but in this group the males are lean and very fast. Usually one male in a group is distinctively conspicuous in size and coloring. He apparently functions in drawing the attention of predators to himself. He doesn't often fight; instead, he uses his remarkable speed to escape any predator he attracts. His bold markings and distinctive behavior may also enable the troop to see him well enough to follow him to new foraging places through tall savannah grass. He makes himself conspicuous to the group by moving ahead and climbing into trees (Hall

1964). So, in both cases there is strong sexual dimorphism, but to different adaptive ends, based on the overall ecology of the species.

Dean Amadon (1959) at the American Museum of Natural History postulated that in most birds in which the male is larger, the female is probably of optimal size in terms of the ecological demands of the niche. The male, he figures, has exceeded this size in response to evolutionary pressures of intrasexual competition (sexual selection). In other words, the female best reflects the ecological dimensions of the niche. If this is true, the large size of males places them at some disadvantage. Large size may enable them to win conflicts over commodities and mates, but in some species, including humans, a net disadvantage is reflected in their shorter life span (Figure 26.8). Of course, there is a limit to which males can diverge from optimum size in their efforts to win conflicts and attract mates. At some point, their overall fitness must decrease due to the increased hazards of conspicuousness and other problems attendant with large size. Certainly, any increase in size beyond that appropriate to the successful exploitation of the available food would be selected against. Thus, the upper limits of male size are more or less established.

We have seen that there are notable examples of "reversed" sexual dimorphism, the females being larger than the males. In the red phalarope, the male is smaller and duller than the female (see Figure 17.5 in Chapter 17). The brightly colored female courts the male with all elaborate displays seen in males of many species. But, here, things are different. He builds the nest and, after the female lays the eggs, he assumes all incubating responsibilities. Keep in mind that reversed sexual dimorphism among vertebrates is the exception rather than the rule. While such exceptions are interesting in themselves and cannot be ignored, the *usual* condition must be considered of prime importance when discussing evolutionary phenomena, and when the sexes of any vertebrate are of different sizes, the male is usually larger.

## Sexual Dimorphism and Foraging Patterns

There is no question that in many species sexual dimorphism promotes mating success, as was postulated by Darwin. Lack (1968) affirmed Darwin's position by noting that large size in males is probably associated with sexual selection because it is so often found in promiscuous or polygynous species (polygyny is discussed later in the chapter). However, others believe that sexual dimorphism need not necessarily arise on such an adaptive basis, but, instead, that size difference may have ultimately functioned in reducing competition for food between the sexes (Selander 1966).

Figure 26.7 In women, breasts and hips are traditionally accented, since these are structures associated with child bearing and child rearing. Sexiness, then, may actually be based in reproductive ability, or the illusion of it.

Sexual dimorphism in size may indeed have important effects on sexual differences in foraging. For example, consider the birds of prey in which females are larger (Storer 1952), the sexes taking food of different sizes (Schoener 1965), or types (Mueller and Berger 1970). As in birds of prey, most sexual dimorphism involves body size. However, although overall size dimorphism may reflect differences in foraging behavior, most of the studies of the relationship of sexual dimorphism to foraging behavior in birds have involved measurements of bill size alone (Amadon 1950; Selander 1966; Wallace 1978b).

In the context of ecological theory, it is apparent that strongly sexually dimorphic species exist only under special competitive situations. We know—rather, we believe we know—that sexes are unlikely to diverge if the species is occupying a narrow niche because the parameters of that niche are a reflection of restrictive competitive pressures. It is suggested, then, that since sexes of certain species *have* diverged, those species are likely to be occupying a rather wide niche—a situation permitted by low levels of interspecific competition. Among owls, the insectivorous (insect-eating) species show much less sexual dimorphism than those that prey on vertebrates.

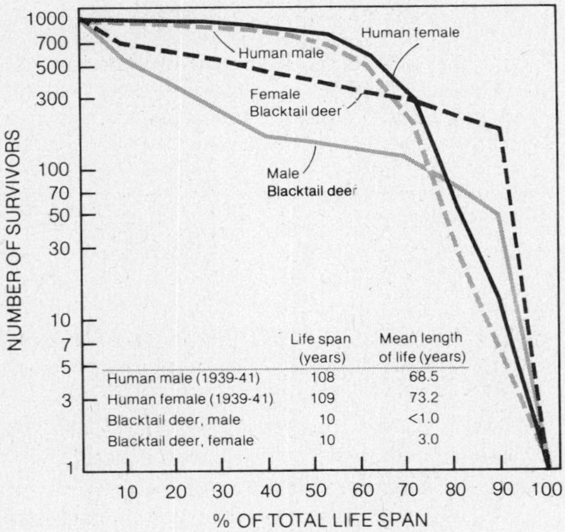


Figure 26.8 Different survival rates between the male and the female in humans and in the blacktail deer. In most mammalian species, the male has a higher mortality rate.

One explanation is that insects fall within a narrow size range and vertebrates do not (Earhart and Johnson 1970). This means a wide food niche exists for vertebrate-eaters according to our previous definition of niche parameters. These predators also probably occupy a shallow niche because of the paucity of vertebrates in comparison to the abundance of insects over most areas; thus the criteria for the kind of a niche that would allow divergence of the sexes (a broad niche) is met for owls that eat vertebrates.

Although we may expect sexual dimorphism in species that hunt variable and relatively sparse prey under conditions of reduced competition, in some cases sexual dimorphism may be related to insularity, which, in turn, affects the level of competition. Two species of *Dendrocopus* woodpeckers provide evidence to this effect. *Dendrocopus arizonae* is found in rather insular situations among the western mountains. The sexes of these insular populations differ in bill size and foraging patterns. The males probe and hammer more, and the females are more likely to peel bark. On the other hand, the sexes of the relatively monomorphic and wider ranging *D. borealis*, a close relative, diverge much less in their foraging behavior (Ligon 1968).

Considering the niche dimensions that are associated with sexual dimorphism, we might expect to find the most dimorphic species on islands, and in fact, the best examples of sexual dimorphism in vertebrate trophic appendages occur in such places (cf. Amadon 1950; Selander 1966; Schoener 1970). Let's again look at our friend the woodpecker, this time to provide us with an example of the relationship of sexual dimorphism to competition and social behavior.

### Lessons from West Indian Woodpeckers

The larger islands of the West Indies (and a few of the smaller ones) have been colonized by woodpeckers that apparently came from the North American continent (Figure 26.9). Hispaniola (comprised of the Dominican Republic and Haiti) provided highly suitable conditions for its colonizers, and today the birds are found in abundance there. The Hispaniolan woodpecker (*Centurus striatus*) may be seen darting about in the coconut plantations of the lowlands as well as in what remains of the montane hardwood forests. They are noisy, not particularly shy, and easily observed.

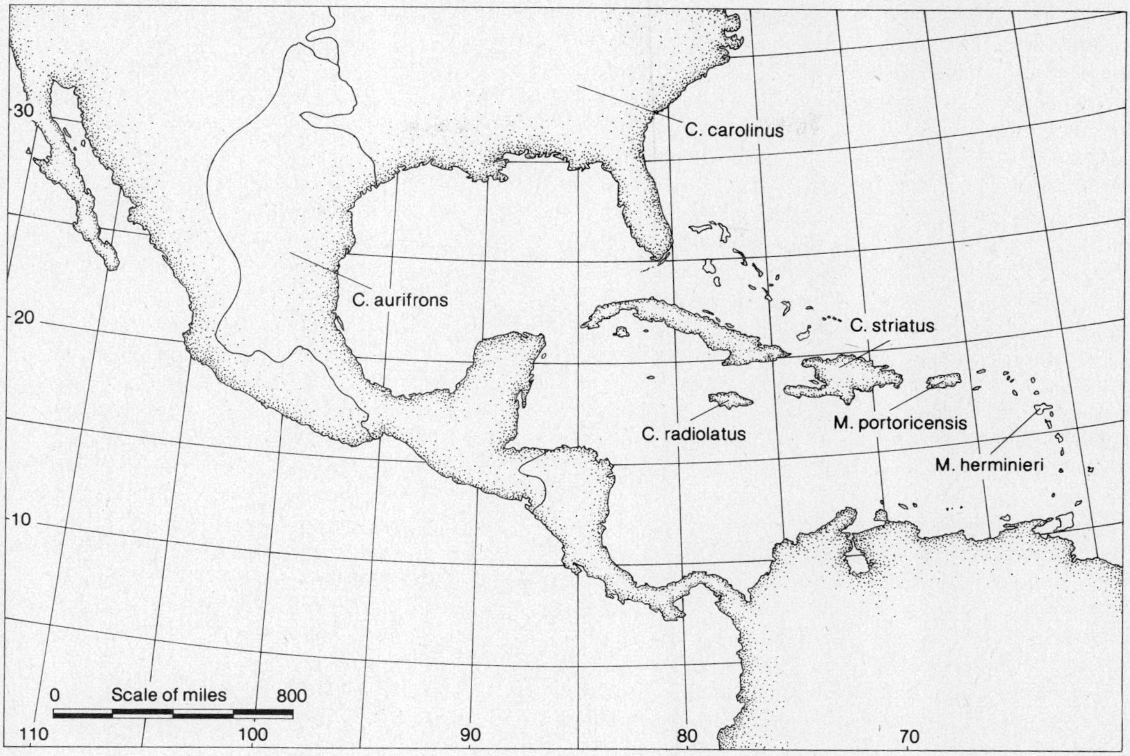


Figure 26.9 Ranges of various species of woodpeckers of the genera *Melanerpes* and *Centurus*. It is believed that the Antillean islands were probably invaded from North America.

This species is sexually dimorphic in bill size (Selander 1966; Wallace 1974a) to such an extent that an observer can "sex" the birds by silhouette alone (young birds of both sexes present a problem, however, since their bill sizes are similar to those of adult females). There are also sexual differences in the number of barb groups on the tongue (Selander 1966) and the supporting bones of the tongue, and the jaw muscles (Wallace 1969).

It is interesting that in bill length the sexes of the Hispaniolan woodpecker have diverged from the presumed continental progenitor in different directions. In other words, the males of the Hispaniolan woodpecker have larger bills than males of either the golden-fronted or red-bellied, their nearest continental relatives, while the insular female has a smaller bill than its continental cousins. Furthermore the insular female's bill is displaced more than the male's. In other words, the bill of the Hispaniolan male is not very much larger than those of males of the continental species, while the bill of the insular female is quite a bit smaller than the continental female's (Figure 26.10).

The greater displacement of the bill of the female may have occurred through a dominance relationship with the male. Such a relationship can be demonstrated in numerous ways. A female quietly feeding on a tree limb may suddenly be "supplanted" by a male. The more aggressive male simply swoops directly at the female and she must move to avoid being jostled. In another woodpecker species, the males tended to drive the females farther down the tree by their attacks (Kilham 1970). The male takes pretty much whatever place he wants on the tree and the female is forced into other areas. Thus, the greater morphological change in the females of insular species is not surprising, since they were probably the ones forced to do most of the adapting. Such dominance interactions between the sexes of birds is not restricted to woodpeckers, I should add. For example, in weaverbirds, females may lose weight in periods when the food supply wanes, a time when males actually increase in number. The changes may be due to dominant males pushing the females into less productive areas (Ward 1965). The point is, perhaps dominance relationships between sexes have greater influence on foraging ecology than has yet been realized.

Field studies of the Hispaniolan woodpecker show sexual differences in foraging behavior (Selander 1966; Wallace 1974a). Both sexes take food primarily by hammering to excavate burrowed food, probing into crevices with the bill and tongue, and gleaning food from the surfaces of limbs. However, the sexes differ in the degree to which they employ each technique. Males hammer and probe more with their larger and stronger bills. Females tend to glean more often.

Sympatric species of similar adaptive types differ in the size of trophic appendages by a factor of 1.2 to 1.4 (taken as the ratio of the larger to the smaller). The average of such differences has been found to be about 1.28 (Hutchinson 1959). This figure has been used as an indication of the level of difference in size of trophic appendages that must exist between species in order for them to be able to overlap in habitat (but see Willson's 1971 argument). In the Hispaniolan woodpecker, however, the *sexes* diverged even more than this (Wallace 1969). It seems, then, that the *sexes* of the Hispaniolan woodpecker may have separated to at least the extent as many sympatric *species* on continents.

It was found that several environmental factors greatly influence the foraging behavior of the Hispaniolan woodpecker. For example, the Dominican population at one end of the island foraged differently than the Haitian group at the other end, especially during the dry, winter months when food was scarce. In the bountiful spring months, the differences between the two populations disappeared. Apparently, in periods of low food supply, the foraging pattern is more strongly influenced by local environmental conditions.

Figure 26.10 Hispaniolan woodpeckers (top) and golden-fronted woodpeckers (bottom). The male is shown at the top of each pair.

There were also seasonal differences in the degree of separation of the sexes. In spring, the sexes overlapped much more in foraging technique than during the lean winter months when each sex was forced to use its particular morphology to maximum advantage.

It turned out that the degree of sexual separation could be correlated with habitat complexity. One group of birds foraged both in coconut groves and in nearby wooded areas in the Dominican Republic. Among the coconut trees, most of the foraging was done on the long trunks. In the wooded areas, the birds were able to take food not only from the trunks but from the limbs, twigs, and leaves of the "three-dimensional" canopy. When the foraging data from the two areas were compared, it was found that the sexes foraged more differently in the "simpler" coconut area.

We must keep in mind that the original colonizers of the islands were probably very similar in bill size, much as the continental species are today, and that their bill size shifted because of peculiar demands made on the animals. Although recent colonizers may at first suffer little intraspecific competition because of low numbers, later they can be expected to increase that competition as their numbers grow (Lewontin 1965). This intraspecific competition is apparently an important pressure behind the niche expansion of the species. Although interspecific competition forces a species toward its optimal habitats, that is, the "adaptive peak" of the species, strong intraspecific competition may reverse the trend and cause the expansion of the species into suboptimal habitats (Svärdson 1949) to which it may adapt, thus effectively widening its niche. We considered the principle earlier in our discussion of island habitats.

Since the sexes are able to separate and thus utilize different parts of their habitat, it may be assumed that they have increased the commodities available to the species. Thus, the Hispaniolan woodpecker undoubtedly increased its food supply as the sexes foraged in different ways, presumably taking food of different types. The result is an increased food supply, so we should expect to find an increase in population, and in fact, the Hispaniolan woodpecker is nothing if not abundant. By their sheer increase in numbers, they have intensified their levels of social interaction and competition, and the results have illustrated important relationships in behavior and ecology.

One of the peculiarities of the social system is that this species nests colonially. Most species of woodpeckers are belligerent and territorial, but I have found as many as twenty-six pairs of the Hispaniolan woodpecker nesting in a single tree. They, by no means, always nest colonially, however. In many cases, a pair will nest entirely out of sight of other pairs, and in coconut groves, only one pair usually nests under the fronds of each tree. Such trees are often closely spaced, but the birds inhabiting them do not show territorial behavior (except around nest holes). Those trees harboring more than one pair are usually large dicotledenous ones such as those used for shade in the cacao plantations. The factors triggering coloniality are not known; nonetheless, the gregarious tendency is strong in the species.

Various theories have been proposed to account for the high level of sociality in the Hispaniolan woodpecker. It has been suggested that there is an interaction of sociality and population density (Ashmole 1967). Perhaps the colonizers from the continent were territorial when they reached the island. Then, once the species became established, sexual separation in foraging patterns permitted an increase in population through an expanded food supply (Van Valen 1965). As population density increased, so did social interactions, including territorial encounters. Finally, the population reached such numbers that those individuals trying to maintain territories were at a disadvantage because they spent so much of their time confronting the multitude of usurpers. Territoriality, then, became selected against on the basis of its distortion of the time-energy budget. Less aggressive and nonterritorial birds were then at a selective advantage. Once territoriality had been abandoned, the species was free to move toward the advantages of gregariousness and coloniality.

**The Relationship of Bill Size and Social Behavior in Birds** Perhaps sexually dimorphic species of woodpeckers taking different foods results in their foraging closer together, thus facilitating social behavior (Kilham 1965). There is preliminary evidence that this may, indeed, occur in the Hispaniolan woodpecker. In Figure 26.11A, *A* and *B* represent the bill sizes of two species that utilize a food continuum. Birds in this complex of species that have bill sizes at either end of the distribution for the species are more rare because of interspecific competitive pressures (shown by the arrows 1, 2, 3). In Figure 26.11B, however, *A′* and *B′* represent sexes, and the distribution of bill sizes has changed. Arrows 1′ and 3′ still represent interspecific competitive pressures that reduce the numbers of birds with bills of those sizes. But 2′ represents intraspecific pressures and is stronger than 1′ and 3′. Thus, the curves for the sexes of the dimorphic population in Figure 26.11B are distorted from the normal, bell-shaped, distribution. The reversed, or mirror-image, skewness is evidenced by the fact that, in sexually dimorphic species, the median measurements in bill size for the sexes fall farther apart than the mean measurement. In the more sexually monomorphic species considered here, this is not the case.

There are three factors that might contribute to the stronger pressures at 2′. First, there will be intersexual competition for food and, if we may assume that bill dimensions reflect the food of the birds, those birds with bills of intermediate sizes can be expected to suffer from strong competition from the other sex. Second, selection against intermediates could be reinforced by sexual selection, especially since, as we know, visual animals such as birds are particularly attentive to features associated with the head. If males or females were identified, even partly, by bill length, the sexual dimorphism of the bills in the population would be strengthened.

Finally, selection against individuals of both sexes with intermediate bill sizes may operate on an ecological basis to facilitate social interaction. To illustrate, if the bill sizes of the sexes are different, then the sexes may forage together in the same parts of trees, without competing too strongly, by taking food of different types or sizes. Thus the sexes may, through increased proximity, maintain stronger pair bonds. The test, of course, is whether sexually dimorphic species, in fact, do forage closer together than monomorphic species. I am pleased to report that in a comparison of four species of woodpeckers of differing morphologies and foraging modes, I found that such a relationship does exist (Figure 26.12). The woodpeckers I looked at were *Centurus radiolatus* (Cr) of Jamaica, *C. carolinus* (Cc) of central Texas, *C. striatus* (Cs) of Hispaniola, and *Melanerpes portoricensis* (Mp) of Puerto Rico. The degree to which the sexes differ in bill size is related to their foraging proximity as determined by the height at which they forage in trees.

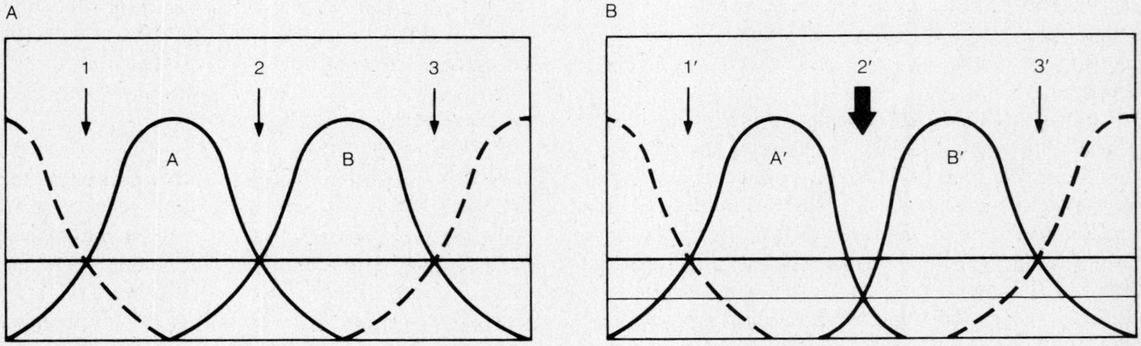


Figure 26.11 Theoretically, species *A* and *B* are limited by interspecific competition, shown at 1, 2, and 3. The result should be normal distributions of the species for any character. However, if *A′* and *B′* are sexes of the same species, there should be an increased depression of numbers where they overlap at 2′, for reasons discussed in the text.

So, we see that sexual dimorphism may have special implications for population density and social behavior, and that the difference between the sexes in size of trophic appendages may even be stronger than that between species because of a need for social interaction.

Now, let's consider for a moment the peculiar case of the Jamaican woodpecker. Almost all the woodpeckers on the West Indian islands are strongly sexually dimorphic in bill size like the Puerto Rican and Hispaniolan species. This includes the little-known Guadeloupe woodpecker (*Melanerpes herminieri*). (On Cuba, a large, complex island near the mainland, there are five species of woodpeckers with varying degrees of sexual dimorphism in bill size.) But the Jamaican woodpecker is remarkably sexually monomorphic in bill size, even more so than the red-bellied woodpecker of Texas. Why? If the trend in these neotropical islands is toward sexual dimorphism, why should the Jamaican woodpecker have taken a different tack? Field studies have shown that, in at least one area of the species' range, the sexes feed in almost identical ways, depending largely on gleaning insects from the surfaces of trees (Wallace 1974). There is another unusual quality about the Jamaican woodpecker. Both sexes are larger than the other woodpeckers, both in body size and bill length. The Guadeloupe woodpecker is as large, but it is sexually dimorphic in bill size, the bills of females being much smaller than the males. In the Jamaican woodpecker, both sexes have large bills. So, the question is, Why hasn't the Jamaican woodpecker become sexually dimorphic like West Indian woodpeckers are "supposed to"?

It may be because they developed large bills *instead* of sexual dimorphism. In the sexually dimorphic species, the niche was divided as each sex exploited the habitat in somewhat different ways; and thus they could forage close together without strongly competing for food. I should add that bill size in woodpeckers may be particularly indicative of their diet since they rely strongly on insects, and insects come in specific sizes. Seeds and berries comprise a large part of the diet also, and these come in discrete sizes as well. One may assume that in the sexually dimorphic species, the smaller-billed females take smaller food items; the larger-billed males take larger food items. But in the Jamaican woodpecker, both sexes are able to take a wide range of food types because of the large bills of both sexes. So, they both exploit the same kinds of food, reducing competition by increasing the range of food sizes available. Since they forage primarily by "gleaning" (at least in some areas), the larger bills probably would not interfere with their taking smaller food items, such as ants, at the same time allowing them to take larger insects, seeds, and berries. Since these items are usually of discrete sizes, this means they are actually taking more *species* of food items. Now this is probably an advantage *only* in neotropical or tropical areas where harsh, density-independent factors (such as frigid winters) do not act to reduce a large part of the food supply all at once. In more tropical areas, the breeding season of many species is extended, and the climate permits various food items to enter their productive stages at various times throughout the year. Thus at almost any time, *something* is coming

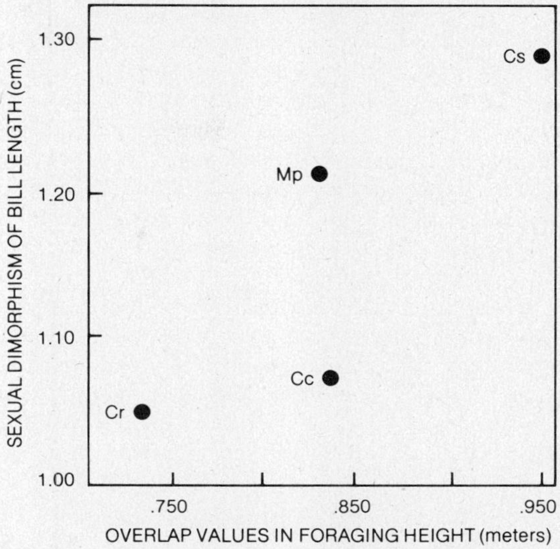


Figure 26.12 In a study of four melanerpine woodpeckers (see text), it was found that sexual dimorphism in bill size is correlated with foraging proximity as indicated by foraging height. Sexual dimorphism, then, may be a means of reducing competition for food so that the sexes may remain in closer social contact.

into production. The Jamaican woodpeckers, with their large bills, are thus able to sustain themselves at those times when relatively large food items are abundant, as well as when small food items predominate. Therefore, they are assured of a more constant food supply. Instead of the sexes more finely dividing the niche, then, each becoming a specialist in food of particular sizes, both have increased the range of food sizes available to them and have become generalists.

But why? Why have most West Indian woodpeckers become sexually dimorphic, while this one increased in size? Two possibilities come to mind. First, the sexes of the Jamaican woodpecker forage farther apart than the dimorphic species (Wallace 1974), a fact that would not be unexpected if they are taking the same kinds of foods. They probably take the same food in different places. It may be that the ultimate causation here was some factor such as predation (operating on either the Jamaican population or their continental forebearers). Two birds foraging close together would have attracted the attention of predators to the activity in a restricted area. On the other hand, birds foraging singly might attract less attention. So perhaps there were particular predation pressures that forced the sexes apart and permitted the development of large, similar bills.

Second, the differences in the two strategies may be couched in terms relating to preadaptation, although such notions are impossible to prove, since we know almost nothing about where the West Indian birds came from or what behavioral patterns their forebearers may have had. *But*, suppose the narrowness of the realized niche of the continental progenitors of the Jamaican woodpecker was due, at least in part, to a tendency to *avoid* members of competing species. This aversion would have incidentally pushed the sexes together. Once on their island, then, they would have been free to separate, thus paving the way for identical, generalized foraging habits for which increased size would have been advantageous. The island birds that became sexually dimorphic and foraged in close association, on the other hand, could have sprung from stock in which the sexes associated closely in their continental niche because of a strong mutual attraction of the sexes. Once on their islands, then, they could have been released to become sexually dimorphic so that even closer association was permitted (Figure 26.13).

Obviously, we are dealing largely in speculation here. It seems, however, that for some reason, many island species are larger than their mainland counterparts (Carlquist 1965). For the Jamaican woodpecker, large size may be an adaptation for being a foraging generalist where food of various and discrete sizes comes into production throughout the year. The selection for large size may have different bases for other "island giants" (Wallace 1978c). As an example, on the Spanish island of Majorca there are populations of giant snails, primarily species of *Iberellus*, *Archelix*, and *Helix*, that are apparently large forms of smaller populations that share the island. No one knows why the gigantism arose in certain populations, but all the "giants" are found at elevations above one thousand meters. The zones of these island giants (which may be twice as large as the smaller forms) are clearly demarcated from other populations. Interestingly, the zone of demarcation roughly corresponds with the summer cloudbase level. Thus the large snails live where it is perpetually damp. The lowland snails are subjected to summer droughts, when they go into periods of dormancy, so the highland snails have a longer growing season each year (Parrack 1973). The point of all this is that size and social patterns may be inextricably interrelated and either character may be modified through any number of complex and subtle factors operating on the population.

## MATING SYSTEMS

Mating systems vary widely and may be described according to the kind of bond that forms between the sexes. These bonds may be short-term or long-term, and an animal may or may not maintain more than one bond at the same time. Sexual dimorphism, as we learned earlier, is related to the duration of pair bond in a general way in that monomorphic vertebrate species tend to form longer-lasting bonds. Brief pair bonds are those that form only long enough for fertilization or copulation to take place. Sustained pair bonds may form well before fertilization occurs and may continue for varying periods of time. For example, in some birds that have sustained pair bonds, the bond is dissolved early after the nest is completed, while in others it may continue for a season or for a lifetime.

Figure 26.13 Perhaps the continental forebears of the sexually monomorphic Jamaican woodpecker (left) tended to remain in close association, partly out of an aversion to other species or members of other pairs. Once they reached the island, they would have borne the same trophic structures and foraging patterns, but would have been free to separate and to grow larger as they became stronger generalists. Sexually monomorphic island colonizers (right) on the other hand could have reached their islands with a strong affinity for each other, which could have resulted in their tending to become sexually dimorphic as they sought to remain in close association. Since these are questions of preadaptation, there is no real evidence for either argument.

An individual may form pair bonds with one or several mates. Those individuals that form bonds with several members of the opposite sex may do so almost simultaneously, so that bonds exist with several mates at one time, or sequentially, by dissolving one pair bond before initiating the formation of the next. It may be argued that no pair bond at all forms in those species that meet only briefly for copulation, but at such times the intense mutual interest shown between partners and the aggression toward mating competitors indicate that the relationship does hold important characteristics in common with sustained bonds.

## The Evolution of Mating Systems

At this point, let's define the major mating systems more precisely. When members of a pair stay together at least for a season and cooperate in rearing the offspring, the pairing is defined as *monogamy*. *Polygamy* refers to having more than one mate. If a male mates with more than one female in a season, the system is called *polygyny*. And *polyandry* refers to females taking multiple mates in a single season. Monogamy is generally considered to be the most primitive condition (and hence the most common). Polyandry is exceedingly rare.

But why would a species tend toward one system and not another? As usual, there is no simple answer. Fortunately, Gordon Orians (1969) at the University of Washington has been able to work through the massive data available and to make some sense of it—to synthesize some general rules (see also Selander 1965, 1972; Verner and Willson 1966; Verner 1964; Bartholomew 1970).

To begin with, we must consider some points made earlier regarding sexual selection. First, there is a difference in the kinds of *parental investments* made by males and females. Parental investment is any activity of an animal that increases the probability of having offspring that live to reproduce, but that extracts a cost in terms of the parent's ability to have additional offspring (Trivers 1972). As an example of qualitative differences in investment, an *r*-selected individual would tend to drain its own reserves to generate even more offspring, while a *K*-selected individual might risk its life in defending its existing offspring from a predator.

**Evolution of Sex** We are primarily interested in *differences* in the parental investments of the sexes. One primary difference is that females tend to produce gametes (eggs) with enough food surrounding them to sustain any resulting embryo for a time. Sperm, the male gametes, are generally much smaller, carrying almost no food. But many more sperm are produced than eggs.

Possibly the sexes evolved from ancient forerunners that were virtually identical. Sexual differences would have arisen with the advent of sexual behavior. And sexual behavior is actually only a means of getting the genetic material of two individuals together, your personal experience notwithstanding. Somewhere along the line, one kind of individual would have tended to contribute its genes to the genes of the other. (Meiosis would have arisen later.) The "adder" would have been the forerunner of males, the "addee," of females. The genetic mixing would have provided increased variation in the offspring, and after all, variation provides the raw material for natural selection. So "gamete fusion" would have had a high adaptive value indeed.

So the question was, Who gets to be the male, and who the female? Here we have to clear our throats and mumble another "assumption" about how perhaps in the early populations some individuals produced lighter or more mobile gametes, and the animals that produced these, then, became males. The larger gametes that waited passively to be fertilized would have been produced by animals that were primal females.

Through the long processes of natural selection, the sexes would have become more distinguishable, the sperm more mobile, and the eggs more food-laden. There would have also then developed the vast array of morphological and behavioral traits that bring males and females together and permit them to do those things necessary to get the sperm to the egg, no matter how indelicate.

But then there would have arisen competition for the privileges of having one's gametes participating in fertilization. And, the system being structured the way it was, this competition may have been more rigorous for one sex than the other. So, let's leave the surmisings of prehistory and see what a closer look at modern species can tell us. (See Trivers 1972.)

**The Roles of Males and Females in Reproduction** It is apparent today that males and females make unequal investments in the production of gametes. For example, among birds a single egg may weigh fifteen to thirty percent of the female's weight, and an entire clutch may equal her weight (Lack 1968). A male bird may ejaculate billions of sperm at once, but the energy thus expended can probably be made up in a few moments of foraging. Bird eggs are so large because of the massive amount of yolk they contain. That yolk is all the food the developing embryo will have until it escapes from its protective prison.

In placental mammals, the egg only needs enough stored food to last until the zygote (fertilized egg) can eat its way into the uterine wall where it can take its food and oxygen from the blood of the pregnant female. Thus the mammalian egg is small. Although a male's ejaculation is much more spectacular and energetically expensive than the silent and inconspicuous release of an egg from an ovary, should that egg become fertilized the female's contribution will dwarf anything that the male has done—reproductively speaking. She must first endure pregnancy, a time when she is so slowed and swollen that she may not be able to forage efficiently or evade predators. Much of the food she does find goes into the maintenance of the developing zygote (or zygotes). Finally, she literally risks her life at the moment of birth. If she survives this, then, she must be closely affiliated with the offspring for some time afterward since she is the sole source of its milk. In higher primates, including humans, the mother-offspring association may last for years. And where is the male all this time? If we're talking about peacocks or mountain lions, the male has disappeared into the sunset. If we're talking about woodpeckers or wolves, the males are there, catching food and feeding and defending the young.

Now, in polygynous systems, if a male impregnates an inferior female, there is little loss for him because he is continually looking for new conquests, and maybe the next female will be better. In monogamous species, however, a male that bonds with an inferior mate may waste an entire reproductive season. But even in these species, he has less invested in the offspring than she does, even if he is a bird, but particularly if he is a mammal. Thus, and this is an important point, the female must be exceedingly careful about whom she mates with—more careful than the male. If she mates with an inferior male, her considerable reproductive effort may be wasted, since his genes may produce inferior offspring that may die or be unable to mate successfully. If she is to maximize her reproductive success, she *must* mate with the best male she can find. Since the reproductive expenditure of the male is less, he is likely to be less choosy, thus mate selection becomes the "lady's choice" for many species.

But what does she look for? What is it that tells her she has found a good male? External cues may be highly significant for some species. The length of the plumes of a male bird or the size of the antlers of an elk may provide such clues. After all, they indicate that that male has been able to find the food to grow such accoutrements and that he has been wily enough to avoid predators in the process. If the female is making her selection based on such factors, then it is to the male's advantage to appear as "extravagant" as possible, to demonstrate his "machismo," to make every effort to be chosen over other males. The outcome of such a system is increasing male promiscuity and higher selectivity among females. Male promiscuity may even increase because if his garishness causes him to fall to predators more frequently, he would have to intensify his reproductive efforts in order to leave as many of his genes as possible in the time available.

It should be clear, then, that in these cases, while the male's contribution to a particular set of offspring is small (perhaps only the contribution of his genes), his *overall* reproductive contribution is maximal. After all, he is putting energy into those plumes and antlers—energy that could be used in foraging or the growth of other body parts. And, as he makes himself as conspicuous as possible to females, he also makes himself conspicuous to predators. So his life expectancy may be shortened. His "philosophy" may be that expressed in an old song once popular in Arkansas, "Live fast, love hard, die young, and leave a beautiful memory."

The female may be attracted to males on the basis of factors other than their physical appearances, of course. We learned earlier that in some species, females will not mate with a male unless he already has a territory. And, in some cases, she will choose an already mated male with a good territory rather than an unmated male with a poorer territory. In other species, a female may be attracted to a male on the basis of his behavior. Now, let's consider some specific points regarding the three primary mating systems.

**Monogamy**

To illustrate monogamy, let's consider a few examples from birds, since the great majority of bird species are monogamous. We should first remind ourselves that in species that must feed their young, the clutch size probably has been evolved according to the number of young that can be fed. Since more food can be brought by two parents than by one, an adult will tend to leave more offspring *per union* if it has only one mate, provided the sex ratio of breeding adults is about at equality. If a bird stays with its mate, both can help to feed the brood. If a bird tends to take more than one mate in sequence, then, generally, only one parent is available to feed the young, since the other is likely to be off seeking another partner.

Monogamy can also result if the numbers of the sexes are unequal. We have considered the weaverbird in which males force females into suboptimal foraging areas in times of food shortage and thus, in many cases, cause their death. The result is a chronic shortage of females. Therefore, the males are monogamous, since there would be little advantage in leaving one female to struggle with a few eggs while seeking out another scarce mate. Here, presumably, monogamy has risen as a result of socially induced mortality.

Probably, many of the examples of bird species that immediately come to mind for you are sexually dimorphic in coloration or morphology, so you may have been surprised to find that in by far the greater number of species, the sexes are monomorphic (Lack 1968). Under certain conditions, such as that of high predator risk, the sexes may both be cryptic; but in species that are able to flee, defend themselves, and are in little danger from predators, both sexes may be conspicuous. In either group of animals, however, an individual is likely to have only one mate at a time, since monomorphic species tend toward monogamy.

Orians (1969) predicted that while monogamy should be the predominate pattern among birds, it should be relatively rare among mammals. And it is. The difference is apparently due to the differences in male roles in mammals and birds. Since female mammals nurse the young, the male may play no role at all in bringing them up. This will be especially true if the female is able to provide food for herself. A female cat is perfectly capable of bringing home the bacon, or mouse, and when she has kittens, she is likely to drive males off her territory. (Males would compete for food and they may be cannibalistic.) In the case of birds, the male can do everything for the young that a female can do, so it may be advantageous to have him around to help out. It is therefore also much more disadvantageous for a female bird to mate with an already mated male than it is for a female mammal. Why should a female cat care if the amorous tomcat facing her has also inseminated another cat? She may even appreciate his experience. If he has a superior genotype, she should use him. A female bird, on the other hand, mating with an already mated male would only have half the help she would otherwise have in rearing her young. And a male who spreads himself too thin might contribute so little to each brood that all would die and he would leave none of his genes in the next generation.

Terrestrial carnivores *are* often monogamous. While herbivores often stand knee deep in their food, so that the mother is able to get all the food she needs for the production of milk, wolves must act together to bring down large game. Therefore, gestating females may benefit from males bringing food to them and their nursing young. It may be instructive to review in your mind the reasons wolves or coyotes would tend more toward monogamy than mountain lions or house cats.

## Polygyny

*Polygyny* is the condition in which a pair bond exists between one male and several females, either simultaneously or sequentially (Lack 1968; Lanyon 1963). In these species, the males are often garish, conspicuous, and relatively short-lived. In altricial birds, the condition usually can only arise if food is abundant enough for a female working alone to be able to gather sufficient amounts for herself and a small brood. Lack pointed out that among closely related species, grassland seed eaters, which usually suffer no food shortage, are more likely to be polygynous than are forest-dwelling insect eaters that seek comparatively sparse food. Polygyny, for some species, then, can be considered an adaptation to the food surplus when the sex ratio is about equal. Polygyny may also arise in populations where some males are forced into suboptimal habitats through territorial behavior of dominant males.

We know that in some species a female can leave more offspring by mating with an already mated male in a good territory than with an unmated male in a poor territory. Shortage of nest space, strong differences in the suitability of territories for breeding, and "patchiness" in food distribution all select for polygyny and hence tend to increase the numbers of surplus males. In addition, it has been suggested that polygyny would evolve when there was a marked seasonal abundance in the food supply (Armstrong 1955; Crook 1964).

The relationship of food levels and breeding systems does not hold for all species. Lack (1968) reported that in some species that live amid abundant food, the males are monogamous and, by helping the females, they raise even larger broods than the females could alone. (For a discussion of how polygyny could have evolved in seals and sea lions, see Bartholomew 1970.)

Polygyny, as Orians predicted, is more likely to be found in birds with precocial young that can run around, than in those with helpless altricial young. If a young chick is able to walk around with its mother and feed itself, what can the father contribute? His best reproductive bet lies in impregnating another female. Altricial birds, on the other hand, need a lot of help, and it may take both parents to feed them. Young titmice, for example, may eat over one thousand food items a day (Royama 1970). There are exceptions to these rules, of course. For example, gull chicks are precocious, but for unknown reasons, the parents are monogamous. It may be because food is hard to find in the oceans, or because the young must attain a large size rather quickly and, hence, need a lot of food in a short span of time.

The red-winged blackbird is another exception. It has altricial young, but it is polygynous. The reasons, happily enough, can be explained on an ecological basis. First of all, the birds breed in marshes (Verner and Willson 1966), and marshes are peculiar areas. They may be very high in insect productivity, and the food may be patchily distributed. During the red-winged's breeding season, food is so abundant that a territory that can be easily held by a single male may hold enough food to support several females. But since some areas are so much superior to others, it is again to a female's advantage to mate with an already mated male with a superior piece of real estate than with an unmated male with an inferior territory (Orians 1969).

**Polyandry**

*Polyandry* presents special problems to behavioral ecologists. In such a system, the female plays the dominant role in courtship and she may take several mates each season. She may be morphologically or behaviorally more conspicuous than the male, and after laying the eggs, she leaves the male to tend them by himself. Because of his fatherly responsibilities, he may only be able to mate once each season. Polyandry is rare in animals and there is, as yet, no ecological clue as to what causes the condition to develop.

Wife-sharing exists in the Tasmanian hen (Maynard-Smith and Ridpath 1972), and it was found that mating groups of one female and two males (usually brothers) were able to rear more offspring than simple pairs. In these cases, polyandry was simultaneous rather than sequential, and the basic scheme of rearing offspring was much the same in triangles as in pairs.

According to theory, on an ecological basis polyandry should be rare among animals (Orians 1969). Although a female would obviously benefit from having more than one male protect her and bring food to her offspring, it would not be to a male's advantage to care for another male's offspring rather than seeking to insert his own genes into the population.

In any case, in spite of its unlikelihood, polyandry does occur in a few bird species (Hays 1972; Oring and Knudson 1972). For example, in one species of tinamou, a small quaillike bird, the female produces several clutches of eggs each season, each time with a different male. She then leaves him with the eggs and goes off to find another male. The males assume all the duties of incubation and child rearing.

Perhaps the reversal of sex roles could have sprung from a time when the birds were monogamous and the male took the first turn incubating while the female went off to feed awhile and to recuperate from egg laying. To make a long story short, she never came back (Orians 1969).

Tinamou chicks are able to run around and feed themselves shortly after hatching, so the male could successfully tend to a brood alone. Such behavior is probably highly adaptive for a tinamou female since, this way, she can spread her eggs around. (Tinamou nests suffer high predation.) Strangely enough, even under such conditions the male manages to be promiscuous. He will mate with a number of females in turn as each adds her eggs to his clutch. It is to his advantage to get all his eggs in the basket as soon as possible, since a shorter incubation time reduces the chances of egg predation.

So, we have seen that once sex evolved, sexual dimorphism was inevitable. Sexual dimorphism has been expressed in a number of ways throughout the vast animal kingdom and its adaptiveness has often been ecological as well as reproductive. It is incorrect, however, to attribute sexual dimorphism or a particular sexual behavior to one adaptive cause or another. The morphology, behavior, and ecology of sex are highly interrelated in complex feedback systems, and their effects can rarely, if ever, be isolated.

# 27

# THE DEVELOPMENT OF AGGRESSION

Animal aggression has long captured our interest, perhaps because in recent years we have frequently been concerned by our own aggressiveness. But we have perhaps been too often caught up in the drama of the thing, branding it as evil and fascinating, but forgetting its adaptive effects. After all, when an aggressive animal moves against another, the result is often a redistribution of resources, and so aggression may spring from a spirit of competition. It should be made clear at the start, however, that by no means is all aggression the result of competition, nor does all competition involve aggression. Furthermore, aggressiveness (an aggressive spirit) does not always lead to aggression (the act). Theoretically, an aggressive animal may be benign in its behavior, and aggression could occur in normally peaceful animals. In our discussion, we will consider aggression to be an action rather than a motivational state (but see Moyer 1969, 1971).

In spite of the fact that aggressiveness, and one of its expressions, fighting, has captured our interest for centuries, it is only recently that we have tried to critically analyze the phenomenon in an effort to discover its ecological importance. Our recent efforts probably began with a remarkable paper by Collias (1944a), a report that remains a landmark in that it predates what are often regarded as current findings.

The levels of aggression vary widely across the animal kingdom. Some species are highly aggressive, or prone to aggression, and others are not aggressive at all. Even closely related species can be very different in their aggressiveness. Patas, for example, show almost no real aggression. These primates live in woodland-savannah habitats of Africa in groups of about fifteen individuals. The groups are strictly territorial and, while males of different groups may threaten each other, they apparently never fight, and within groups, no aggressive behavior has ever been seen (Hall 1965a). Baboons live in a similar type of habitat, but they may engage in fierce fighting, especially in areas of high population density (Devore and Hall 1965b).

It is perhaps more than coincidental that some of the world's most successful animals are characteristically both aggressive and adaptable (Southwick 1970). In a sense, then, they may be described as both behaviorally and ecologically aggressive. "Ecological aggressiveness" refers to the ability to invade and colonize new areas, whereas behavioral aggressiveness is undoubtedly important in competitive situations. If the word "aggressive" is to be used in an ecological context, however, it should probably be applied to the more adaptable species showing high levels of exploratory behavior. In any case, we are primarily concerned here with what is called "behavioral aggressiveness."

Why is one species aggressive and another not? Why are members of a species more aggressive at one time than another? Why do normally placid individuals suddenly and violently explode? Why are some individuals in a population much more aggressive than others? The answers may be synthesized through a consideration of social structure, environment, and evolution, and regrettably, our level of information, and our synthetic abilities are not sufficient to yield very many unqualified answers at this time.

## INTRASPECIFIC AGGRESSION

Aggression, in most cases, can be expected to be directed toward competitors—the higher the level of competition between two animals, the greater the likelihood of aggressive interaction. Obviously, then, aggression can be expected to occur predominantly between members of the same species. We have already discussed the ecological implications of aggression-related territoriality, hierarchies, and individual distance, all of which may result in a division of commodities. Such phenomena, of course, are most often parts of intraspecific social systems.

The relationship of aggression and competition has been indicated in a few simple experiments. In tests with rhesus monkeys, it was found that when food was in short supply, a strange animal introduced to the group was most often attacked by those with whom it would compete most directly (Southwick 1970). Juveniles attacked introduced juveniles and females attacked females, and so forth, seemingly in response to a kind of social template. In the flicker, the male has a "moustache" at the corner of the mouth, which is lacking in the female (Figure 27.1). If the female of a pair is captured and a moustache is painted on her, when she is released she will be attacked by her own mate who apparently views her as a competitor for his mate, wherever she is.

Adults of many species are generally inhibited from attacking subadults. Subadults, of course, may be the offspring of any adult, and they are likely to offer little competition in many areas, and so, it is to the advantage of both adults and young that the young be recognized as such. Therefore, the young of some species of birds do not grow the colorful adult plumage until they are sexually mature (Figure 27.2). Thus, the drab young do not trigger the adults' hostility while they are still dependent on them and before they can defend themselves or flee from attacks. They begin to look like adults at the time when they begin to compete directly with the older birds. In many respects, it would be to their advantage to keep the juvenile appearance. They wouldn't be attacked, but then they would not be able to mate either. The same appearance that attracts hostility from competitors also attracts mates.

The relationship between aggression and competition is not always clear. Consider the interactions around the nest sites of weaverbirds (Collias and Collias 1962, 1964). The nests are arranged communally, several in one tree. The male builds the outer shell and attracts females to it by hanging at the entrance and singing. A female accepts him by entering the nest (or by copulating), and then she finishes the nest. When a female in a tree indicates that she is ready to enter a nest, competition for her attention becomes fast and furious, and neighboring males may swoop in and knock the female away from another male's nest entrance. The female suffers the buffeting, but the competitor is another male.

Female weaverbirds also compete among themselves for nests, although they are much less aggressive than males. If one female hesitates at a nest entrance, another may rush past and enter the nest herself. Younger females take longer to choose a nest, vacillate more, and are likely to pause before entering. So they are likely to lose, jostled aside as older females dart past them. The older females are much more decisive, and if their behavior can be considered aggressive, it may then be assumed that learning has probably influenced their level of aggression.

Aggression can, upon occasion, be intersexual. Sometimes a male weaverbird continues to display at the entrance after a female has accepted the nest. In

Figure 27.1 In the flickers of North America, the sexes are recognized on the basis of a "moustache" that is present on males. If a moustache is painted on the female, she will be treated as a competing male by her mate.

one case, the female inside reached out and grabbed the displaying male by his head feathers, holding the rascal tightly while he flapped about, trying to free himself. When he got away, he shook himself and straightened his feathers, but he stopped displaying at her doorstep.

In one case, I noted an instance of what could have been signaling to avert misdirected aggression between two highly agitated birds of opposite sexes. While participating in the International Indian Ocean Expedition a few years ago, I was able to investigate a number of rarely visited and uninhabited islands in the Maldive group. Once, I found a pair of nesting black-naped terns on a rocky beach and as I approached they took to the air and wheeled overhead, diving at me and driving me away from their single egg (you can occasionally get an impressive whack by disturbing nesting seabirds). When I moved a safe distance away, they both returned and stood over their egg, still apparently agitated, but "facing away" from each other and thus not risking eliciting the aggression of a mate that was assisting in the defense. As soon as I drew near again, however, here they came! When I backed away and they settled down, they again faced away (Wallace 1966b). This was apparently an instance of an appeasement gesture being used not only as a means to reduce aggression between animals, but possibly as a mechanism to permit more effective and coordinated aggression. On other occasions, I have witnessed birds redirecting their aggressive behavior toward a mate in the excitement of driving an intruder away from the nest.

Of course, unchecked intraspecific aggression would result in havoc. In the maintenance of aggression as a response to competition, it is important that the aggression occur only at certain levels and within certain contexts so that the delicate balances of social systems are not upset.

Figure 27.2 From bottom to top: first year, second year, and adult herring gulls. The adult plumage does not appear until the third year. Subadult birds are not likely to breed successfully and they are also not attacked by adults in the breeding season.

## INTERSPECIFIC AGGRESSION

Aggressive interactions may, on occasion, occur between animals of different species. On the basis of what we know about the advantages of aggression, we can assume that, in most cases, interspecific aggression is directly related to competition for commodities. For example, the pugnacious starling (*Sturnus vulgaris*) may eject woodpeckers from a hole after the woodpecker pair has laboriously excavated the nest site. Neither the tree nor the woodpeckers hold any special interest for the starlings until the hole is made; then they become aggressive toward the woodpeckers and actual fighting may occur, the woodpeckers usually losing.

Normally, the behavior of a predator toward its prey probably does not involve aggression. As Lorenz has so often pointed out, the lion about to spring for its prey shows no more aggressive signs than a man does toward his dinner. However, the prey's reaction toward its predator may involve real aggression. A terrified wild pig, cornered by a young lion that has not been able to kill it on the first attempt, may charge the lion with all the signs of true hostile aggression—tail high, ears back, squealing loudly. Obviously the lion is confronting an angry animal. At this point, the young lion may show an integral part of the aggressive pattern, but at the opposite end of the spectrum—fear.

It is important to recognize that fear is a component of aggressive behavior. Thus, aggressiveness in animals should not be considered simply as hostility (Tinbergen 1968), because hostility may be tempered by, and in some cases may change to, fear. The two components can be distinguished in territorial encounters between two birds holding adjacent territories. The territory owner shows aggression in chasing an intruding neighbor back into his own territory; but once in the intruder's territory, hostility diminishes and fear mounts in the first bird as he finds fear being replaced by belligerence in the bird he had been chasing. Belligerence and fear, then, should probably be considered as parts of the same behavioral continuum.

## CROWDING AND AGGRESSION

The relationship between overcrowding and aggression can be demonstrated quite simply by a visit to New York City or a morning drive on a Los Angeles freeway, but such behavioral reactions to crowding are not exclusively human. We have already considered some of the lengthy series of experiments on the effects of overcrowding in animals begun by John Calhoun in 1947. As a part of these experiments, he placed five pregnant Norway rats into a quarter-acre outdoor pen. The rats had unlimited amounts of food and water and did not suffer from predation. They were allowed to interact freely and, over the next twenty-eight months, the population never exceeded 200, finally stabilizing at 150. Considering the biotic potential of rats, the population could have reached 5,000 in that span of time. Calhoun estimated that 5,000 rats can be accommodated in ten thousand square feet of space and remain healthy if they are kept in small cages eight inches square. So, why did the population level off at 150 in the "wild" state? The answer seems to be that, at high density, aggressive behavior increased to the point where normal behavior patterns, including reproduction, were disrupted. Dominant rats garnered territories around the food supply, and other rats, forced to the periphery, had a lower survival rate, and a high percentage of their offspring died.

Then, domesticated albino rats were placed into a ten-by-fourteen-foot room for sixteen months (Calhoun 1962a, 1962b). The room was partitioned into four sections with litter mates confined to each section (Figure 27.3). At forty-five days, the mothers were removed and ramps were installed so that the rats had access to all four sections. Food and water was available in every quarter and the space could have supported forty-eight rats comfortably. But the population reached eighty, at which point it was artificially stabilized by removing animals. So how did the crowded rats behave? Surprisingly, they did not distribute themselves equally throughout the four quarters, but tended to gather in one area, which Calhoun called a "behavioral sink." Two of the quarters had only one entrance and these quarters were monopolized by strong males, each of whom admitted no other rats save those in his harem. Most of the other rats fed together in the sink. Some of the most interesting results centered over the behavior of the subdominant rats. For example, in many cases pregnant females did not come to term, or their nesting behavior was so disrupted that the young could not be cared for properly. While the aggressively dominant males continued to exhibit normal behavior, the subordinates either (1) avoided both fighting and sex; (2) became sexually hyperactive, constantly following females, but breaking the normal sequence of events and showing abnormal copulatory behavior; (3) became pansexual, mounting anything male or female, young or old, receptive and nonreceptive; or (4) withdrew from all social activities and roamed about chiefly when other rats slept. Male

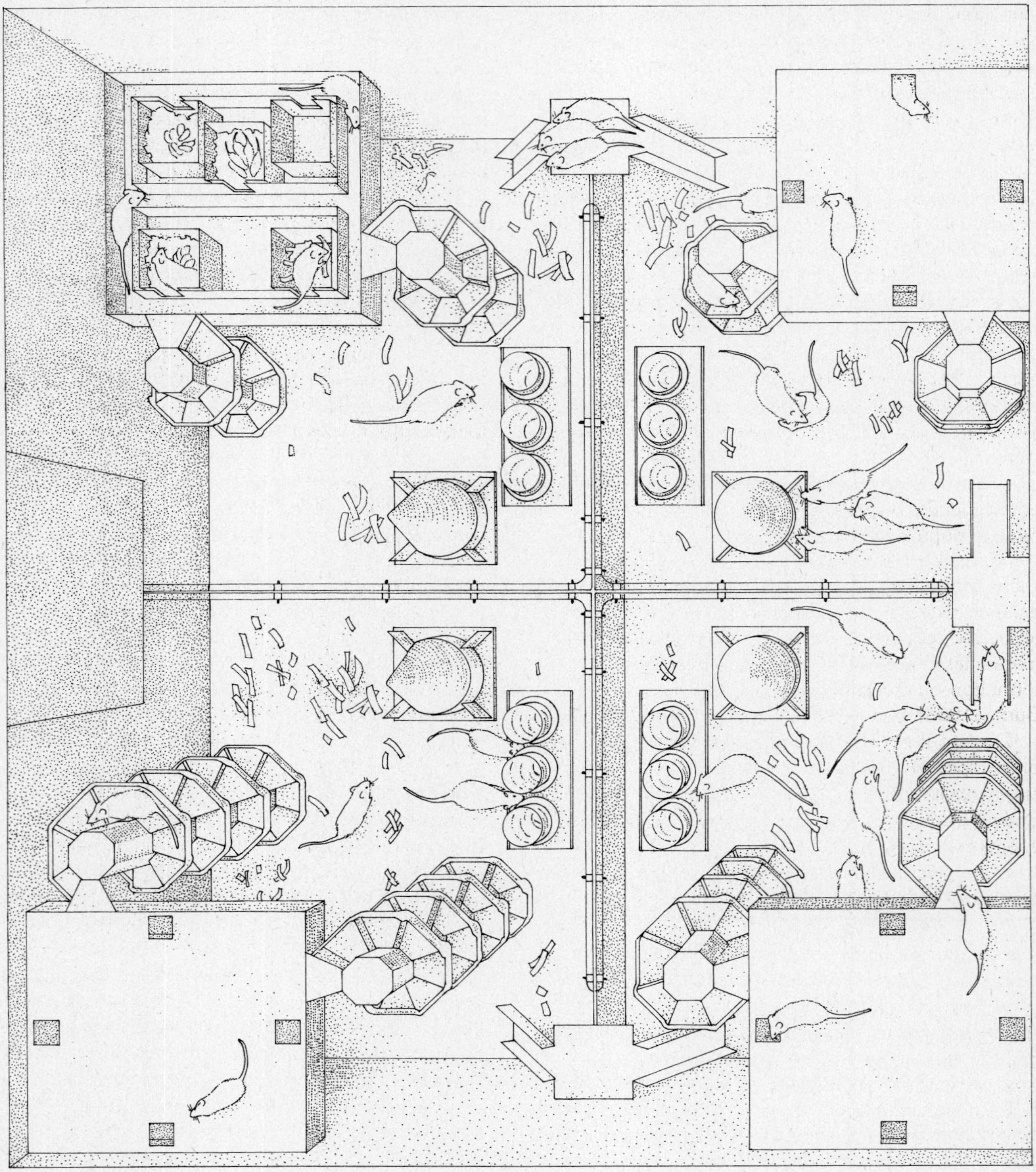

rats often engaged in unprovoked fighting. Those hardest hit by the high level of aggression were females and young, and their numbers were reduced disproportionately. Female mortality was 3.5 times that of males, partly because of the rigors of giving birth.

Calhoun's work has been criticized on the basis that he used a disorganized population to begin with. In other experiments where rats and mice already had established a dominance hierarchy in nature before they were crowded together, aggression and social aberrance were not so pronounced. It is difficult to say at present whether, in nature, crowding in wild rodent populations increases aggressiveness, but normally crowding leads to emigration, and emigration is often a response to increased aggression as we know. (See Archer 1970 for a review and critique.)

Charles Southwick (1970) described experiments he conducted at the University of Wisconsin with mouse populations. He found that as crowding increased, social disintegration appeared. The mice spent increasing amounts of time and energy in fighting. They became wounded, their health deteriorated, parasitic infections increased, and they frequently cannibalized their young. For further information on crowding behavior, see Myers (1966); Burns (1968); Stokes (1969).

Density also increases aggression in a number of other species. Baboons fight more frequently in larger groups and in areas of high population density. Rhesus monkeys in India normally show a high level of aggression, but hostile encounters are much more frequent in cities, towns, and temple grounds where they are crowded and can easily see each other, than in the forest areas where population density is lower and trees interrupt their vision (Southwick 1970). (For research showing the effects of crowding on endocrine function, see Thiessen and Rodgers 1961.)

The implications of such findings for man are alarming when we realize that Manhattan harbors about seventy-five thousand people per square mile (Ehrlich and Ehrlich 1968), and the entire greater New York area population may well be thirty million by the mid-1980s. Also, as we have seen, it seems that in spite of small depressions in rate of population increase here and there, the world's population is expected to double in thirty to thirty-five years and then again every few years after that. It may be argued that our population will be brought under control by "natural means" before such a situation occurs. That may be true, and the prospect is terrifying.

Figure 27.3 The experimental enclosure devised by John Calhoun to test the effects of overcrowding in rats. The four pens are divided by an electric fence, each pen being a complete dwelling unit containing conical food hoppers and three watering dishes. The elevated nests are reached by winding staircases. Ramps connect all but two adjoining pens (numbering clockwise from top left, 1, with the nests exposed, and 4). In pens 1 and 4, dominant males keep harems and repel intruders. The three rats peering into pen 1 are called "probers." They are deviant types produced by the high population density. Most rats congregated in pen 2, a "behavioral sink."

## IS AGGRESSION INSTINCTIVE?

Although aggressive behavior is common among animals, including man, from what source does it swell? There are those who believe that it is basically innate and subject to all the laws of true instinctive patterns. Some maintain that aggression is learned, and yet others believe that both learning and instinct are necessary for the successful development and expression of aggression.

The general acceptance of the innateness of aggression can probably be traced to Lorenz's *On Aggression* (1963). The idea is that aggression is based on an inherited spontaneous tendency whose neural properties may be the same as those of any other instinct. On the other side, it is argued that, even if aggression does have an inherited basis, its expression is not inevitable. Those taking this position have argued that humans are born with the *ability*, more or less, to pole-vault, but that few of us do. They argue that aggression depends as much upon experience as on any innate factor.

In fact, it has been demonstrated in a variety of species that aggressive behavior can be strongly affected by experience (Scott 1960; Kahn 1951; Kuo 1960, 1961; Ulrich 1938; Montagu 1968; Barnett 1964). However, it should be apparent that because some factor, such as learning, can influence the expression of aggression, does not mean that factor is the underlying *basis* of the phenomenon. Those who propose that aggression is based solely on individual experience offer a number of possible explanations for the formation of the trait. It has been suggested, for example, that aggressive behavior is gradually learned early in life through pain as a result of competing for food and play-fighting (Scott 1960). Other researchers (Dollard et al. 1939; Berkowitz 1962; Montagu 1962) believed that aggression is the result of frustration (any impediment to reaching a goal). Some proponents of the experience argument maintain that if aggression were instinctive, there would be appetitive behavior for fighting and that it has never been demonstrated (Craig 1928). (Although there is contrary evidence that indicates an appetitive phase in some animals, as will be pointed out.)

Marler (1957b) partly resolved the problem conceptually when he argued that aggression must be a reaction to environmental influences, while he stressed the importance of endogenous influences, such as hormones, which render an animal susceptible to specific stimuli.

The proponents of the theory of instinctive aggression include two prominent ethologists: Konrad Lorenz and his student, Irenäus Eibl-Eibesfeldt. Let's examine some of the evidence indicating that aggression is largely innate, which, to me, is the stronger case. I hasten to add, however, that I do not mean the condition is "instinctive," a blind response to some sort of releaser for aggression, and an action that builds its own specific energy. The issue is a complex one. For example, it can be argued that there may be a lowering of the threshold for aggression, but that this is not the same as a buildup in an "energy" for aggression that demands release.

There is an unusually clear demonstration in one species that aggression is innately controlled (Lorenz 1963; Rasa 1969). In the highly aggressive cichlid fish *Etroplus maculatus* and *Geophagus brasiliensis,* the males *must* fight before they can successfully pair with females. If no other male is around, they will attack the female and may even kill her. Then they will search for a female to breed with. The behavior seems to indicate a damming up of a "drive" (action-specific energy) that must be released before the next pattern (involving mating) in the sequence can be carried out. The fighting urge is quickly fatigued, and aggression stops well before the male is physically exhausted. We can see here certain important similarities with instinctive patterns. It is certainly true that many aggressive patterns can be shown to operate through systems involving fixed action patterns, IRMs, and releasers. Recall the sticklebacks, which react in precise ways to simple decoys with red undersides. Such behavior alone stands as evidence of a close relationship between aggression and instinct.

As further evidence indicating an instinctive center of aggression, it has been found that rats will learn mazes in order to kill mice, and thus, it is presumed, to perform consummatory aggressive behavior. (It may be argued that this is simply predation, since rats sometimes eat mice.) Squirrel monkeys, after receiving a painful shock, will learn a task in order to be able to attack a ball (Azrin, Hutchinson, and McLaughlin, 1965). It can be argued that aggression in this case is *learned* as a result of painful experience; nevertheless, the response does show the classical symptoms of the presence of an appetitive behavior for aggression.

### Are There Appetitive and Consummatory Phases of Aggression?

Do animals go around looking for fights so that they can release their aggressive energies? Will a well-fed dog actually get up and seek something to attack? It is true that territorial birds "patrol" their territories, driving away intruders, and other animals may do the same. But does this mean they're looking for opponents for the sake of starting a fight? Does the level of aggression in a bird making its rounds rise when it spies a competitor in its territory, or is it in a belligerent mood to begin with? The only way to answer such questions would be to test the bird's aggressive threshold at these times, and, so far, this has proven impossible.

If we can show that the performance of aggressive behavior can act as a reward, would this mean the act is consummatory? Such reinforcement was mentioned earlier. It has been shown that both fighting cocks and fighting fish will learn to perform an act in order to make an attack (Thompson 1963, 1964). But, of course, it can be argued that these are artificially selected species, developed for their belligerence. Nonetheless, the reinforcing aspect of aggression under any circumstances invites its comparisons with other unquestionably instinctive tendencies, such as eating or drinking.

In some experiments, rats and mice have been reared in isolation so that they had no opportunity to learn aggression. When conspecifics were introduced to their cages, however, the isolated animal attacked, showing all the normal fighting and threat patterns (Eibl-Eibesfeldt 1963). Jungle fowl (the type used in cock fights) have been reared in isolation and found to be *more* aggressive than "socialized" animals. If left in isolation, they would even attack their own tails. (We are reminded of the lowering of the threshold for courting behavior in caged doves.)

It should be pointed out that the existence of species-specific patterns of fighting in isolates does not clearly indicate the innateness of aggression, since fighting or threat behavior may be *derived* from other types of patterns—such as feeding—which are instinctive.

Lorenz argued that if fighting is instinctive, it will tend to go off "in vacuo" through a buildup of its action-specific energy. The implications for humans are readily apparent. However, the view has been challenged by the eminent Robert Hinde who maintains that fighting is derived principally from the situation, and he adds there is no reason to postulate causes that are purely internal to the aggressor.

Tinbergen emphasized that the disagreement between the gentlemen and their exponents is purely semantic, since the ethological instinct theory fully accounts for the interaction of internal and external variables. Lorenz's reference to the "spontaneity of aggression" must be predicated upon the existence of a certain set of environmental conditions. After all, the male stickleback does not always react to an intruder the same way. His response may vary widely, from none at all to a full attack. The specific response is, as we know, related to his internal state and that, in turn, is strongly dependent upon the environment.

Lorenz has also maintained that a redirected attack exhausts the aggressive urge, a finding that would strongly support the instinctive argument. Tinbergen (1968), however, did not entirely agree. He stated that from observations of animals (and even soccer matches), we can infer that aggressive behavior has two simultaneous and opposite effects: a waning effect and one of self-inflammation. The latter may have the effect, when full-blown, of contributing to mass hysteria in human crowds (Figure 27.4).

It must be admitted that the appearance of a behavioral pattern is not evidence of the existence of a neural "center" from which it emanates. If aggression is instinctive, however, such a center (an IRM) theoretically must exist. There is some experimental evidence that indicates the existence of such centers in some animals. For example, the electrical stimulation of specific areas of the central nervous system can elicit aggression in a number of species (Kaada 1967). However, it has been argued that the existence of an internal physiological mechanism, which needs only to be stimulated to induce fighting, does not prove the existence of a "fighting instinct."

The issue is far too complex to be considered in depth here. But at least we have been able to see the lines of reasoning of some of the arguments. Some obviously hold more water than others, and a host of new questions is raised. For example, if aggression is instinctive, and if fear is qualitatively the same as aggression, does fear raise the threshold for aggression? When someone startles you, why do you feel a sudden, peevish aggression toward them that usually quickly wanes? Are you as easy to startle again in the seconds that follow?

## ARE HUMANS INSTINCTIVELY AGGRESSIVE?

An argument that may be as critical as any that science has approached centers over whether man possesses an instinct for aggression. Any successful social action designed to inhibit the initiation of war must ultimately incorporate the answer, since systems of control based on the two prevailing theories would be vastly different. Also, the theories are important in respect to such everyday matters as safety in the streets, sports, and television programming. For example, does television violence *teach* or *release* aggression?

The argument over whether man harbors an aggressive spirit that is deeply seated in instinct has been raging for years. Several popular books by Ardrey, Lorenz, and Morris have defended the evolutionary position, and the resulting publicity has caused the argument to be treated as a new one. The flurry of attention to an old idea has not been welcomed in all quarters. Traditionally, psychologists,

Figure 27.4 Violent behavior in humans may be self-inflammatory so that aggression feeds on itself. The phenomenon does not support the notion of instinctive aggression in which a consummatory act raises the threshold for that act.

sociologists, and anthropologists have been obsessed with the cultural determinants of human behavior. The result has been their tacit, and sometimes vehement, denial of man's biological heritage. It seems that the prevailing notion is that man is basically a sensitive and agreeable creature, and the insinuation that he could harbor a darker side is simply not popular. It is understandably more pleasant to believe that humans are imbued with a gentle and loving spirit and are driven by sinister outside forces to commit acts of aggression. The result of such thinking has been an effort to manipulate the environment, institutionally or culturally, in an attempt to decrease our contact with the things that would teach us aggressiveness. In spite of the money and research attention given this approach, has it worked? If it has, why are our prisons full? Is it because of better law enforcement? Or is it that we simply haven't developed the proper cultural-modeling techniques? And why war?

It isn't at all clear when man incorporated warfare into his normal pattern of behavior. It could have been in defense of hunting rights, or as a result of overcrowding, ritualism, or religion, or, perhaps, when stored surpluses became worth stealing and defending. But nevertheless, the pattern is firmly established among our own species. In fact, its virtual universality among human populations suggests that it is probably even a "pleasurable" activity for us. A foreign diplomat once lamented that it was unfortunate that the United States had been cheated out of the "glories" of war in Southeast Asia.

Killing, itself, is certainly pleasurable. People spend a lot of money and travel long distances for the pleasure of killing animals. On Santa Cruz Island, off California, hunters from Los Angeles pay dearly for the privilege of being flown in with their high-powered rifles to kill the hapless descendants of the domestic sheep that were once raised on the island. The sheep are literally tame enough to be hit with rocks. A friend of mine caught one with his hands. The killing cannot be for the stringy meat of these animals; it is obviously done for the pleasure of killing (with the head taken as a memento). In Washburn and Lancaster's (1968) words, "Men enjoy hunting and killing" (Figure 27.5).

The idea that man enjoys many kinds of violence is apparent by the popularity of boxing. The more violent a fight is, the "better" it is. "Full contact" karate is growing in popularity at the time of this writing. And why do people who can't skate enjoy watching hockey? In a televised auto race in which a man was killed, the incident was replayed no fewer than six times as viewers sat spellbound watching his death. There is indeed ample support for the notion that people generally enjoy violence and killing. Open today's paper and notice the themes of movies—keeping in mind that Hollywood is *highly* attuned to our whims and preferences.

Undeniably, in so brainy and social a creature as the human animal, there is a strong culturally based component of behavior. And to assume the existence of innate patterns or tendencies in man is not to deny that they can be modified culturally. We rarely see couples copulating wildly in romantic little restaurants no matter what their mood. Nor does the occasional discovery of a peaceful tribe in some mountain refuge refute the existence of an aggressive instinct in the human species in general. Who knows to what degree the members of this tribe have learned to control themselves? And there is another point to be made here. Even if it could be shown that this tribe, in fact, harbored absolutely no aggressive urges, its condition might be explained on the basis of strong selection for pacifism in an isolated group. But we are not particularly interested, here, in the behavioral *extremes* of our species, but in the *prevailing* condition for the group in general. How do *most* people behave?

As an aside, it is interesting that many human aggressive gestures are very similar to the gestures of other aggressive primates (Figure 27.6).

In one rather interesting experiment, it was demonstrated that violence could be provoked in children, which when discharged, caused a measure of relief (Hokanson and Shetler 1961). An experimenter first angered a group of students (a simple task), with the result that they showed an increase in blood pressure. The group was then divided, and one group was given a chance to observe this experimenter in a series of tasks and to administer electrical shocks whenever he made an error. The other group could only inform him of an error by flashing a light. Those who believed they were administering painful shocks showed a rapid decrease in blood pressure, while those who were not thus "released" did not. In other experiments, it was found that being allowed to hurl verbal insults also provided a measure of relief. The release of the tension was of short duration, as is the case in other types of instinctive patterns.

Figure 27.5 There are countless examples of the indisputable human affinity for killing. Every autumn, for example, thousands of doves, ducks, and geese fall from the sky under the guns of novices. People also hunt inedible animals, taking a part of the corpse as evidence that they are successful killers.

If one wishes to debate whether aggression in man is innate, then he must also ponder the existence of *inhibitors.* Recall the jackdaw that abandoned his attack and began preening his intended victim when the latter turned his head, exposing his silver nape feathers? Can aggression in man also be blocked or inhibited by signals on the part of the attacked? Certainly if they exist they are very weak, as is demonstrated by the frequent occurrence of what is termed "unnecessary" violence in robberies—often against helpless or pleading victims. While it has been suggested that familiarity acts as an inhibitor (Eibl-Eibesfeldt 1970), and one is less likely to attack someone he knows, criminologists indicate that most murders are carried out by persons who knew the victim—but this may be a function of an increased level of contact.

It is apparent that humans *can* suppress aggressive behavior in others by adopting certain nonbelligerent gestures. "Smile when you say that," and "A soft answer turneth away wrath," are not without their instructive elements. But it is not known what *specific* behavior stops aggression or how effective it is in general. Smiles or grins are generally important among primates in turning off aggressive behavior (Figure 27.7).

Early investigators of Australian aborigines found that adults often brought children with them when meeting strangers in the belief that no one would harm a child. We know that among dogs, puppies are rarely attacked or bitten, and young birds do not elicit aggression from older birds until they develop adult plumage. Among humans, childkilling is often considered outrageous—even by "civilized" man at war. Does the sight of a child inhibit aggression in man? We know that soldiers have been able to deliberately

Figure 27.6 Humans share many gestures with other primates including aggressive ones. Here we see the open-mouth threat in the red uakari and a human.

kill children at close range, but the action is rarely applauded and, in fact, is often condemned by the citizenry. Still, there is no real evidence that humans are especially inhibited from killing children.

If man does harbor any system of inhibitors, to *any* degree, our technology has brought us past the point where it could operate effectively anyway. The long-range gunner or bombardier is certainly unable to see the pleading faces of the people, or their babies, so there can be little opportunity for inhibitors to work. In fact, those soldiers may only be filling breeches or pushing buttons with no feeling of aggression at all. The aggression decision may have been made miles from the carnage by men whose greatest danger is sticking themselves with the pins they are moving on maps. Perhaps no one feels aggressive, except, perhaps, those unfortunate people who are angered at being shot at. The point is, because of the complexities in dealing with human motivation, it is hard to know precisely which of our behaviors should be labelled as aggressive. Swinging at someone with a "tire iron" probably goes in the aggressive category. Killing people miles away with computer-directed artillery may not.

I remember a discussion with a very properly accoutered co-ed at a large university a few years ago. She informed me in the course of the conversation that she was a regular churchgoer, loved children, had small brothers and sisters, loved her parents very much, and that they were proud of her. Our conversation began because of a sign she was carrying that said, "Bomb the Hell out of Hanoi." Perhaps it is true that some forms of aggression could be inhibited if the victim could somehow be personalized.

I also recall reading articles, during the Vietnam conflict, of B52 pilots talking about their missions. The bombardiers said they sat in their small, comfortable compartments and never even saw the ground. They were well out of range of enemy artillery, and one told of eating a sandwich while he monitored the computers that opened the bomb bay doors and released the deadly cargo. He felt no aggression, or sense of having been to battle, whatever. He may have gone to the officer's club that night, had a drink, traded a few jokes, retired to his BOQ to some music and a little reading before bed. But somewhere back there, people had died. Perhaps some people who were only four years old. We all know that almost no officer would have been able to enter a Vietnamese hut and kill a four-year-old with

Figure 27.7 Grinning among primates may inhibit aggression.

his bare hands. He would likely have vigorously defended it against anyone who tried. But one day he killed it from the air. He didn't see it die.

Perhaps herein lies the problem that will ultimately lead to the earth being inherited by (or experimenting with) a new dominant species—ours being only one of an endless succession of animals that had its chance but finally failed. That problem is technology. It is easy, very easy, to kill. I have seen "squab shoots" in Spain, where pigeons are released from boxes (with the "macho" sportsman standing with his shotgun already shouldered and aimed at the box). The birds are blown from the air a couple of feet from the ground before they can even get their bearings. Children are hired to slam wounded birds into the ground until they are dead. And the admiring women sit on the hoods of parked cars and look on. But all this is beside the point. We were talking about killing our own species. (There are those who would argue that, actually, this is *not* beside the point—that it is important to realize that humans *enjoy* killing, and that other humans are not exempt as subjects.)

Anyway, killing other humans would probably be less common if it were necessary to do it with one's bare hands. Perhaps most of us are able to respond to submissive or pleading gestures of other people. Perhaps our system of inhibitors simply hasn't been able to evolve as fast as our weapon systems. It is easier to stop choking someone than it is to recall a bullet.

Man has developed certain signals that may not serve as inhibitors, but that act to deter aggression by giving the potential aggressor certain information. For example, open-handed greetings, including handshakes, show that no weapon is being carried. Bowing places a man at a distinct disadvantage and therefore indicates peacefulness as well as a measure of trust. These, of course, are learned patterns, but they are nevertheless generally effective in dissuading aggression. It is interesting that the most exaggerated and intricate forms of "polite" behavior have developed in countries that have the highest population densities, such as in Japan, where social protocol is of great importance.

So who, then, is aggressive? We're all willing to label a mugger on the streets of Washington, D.C., as aggressive. Violence, after all, is a component of the most blatant forms of aggression. Getting hit on the head by someone—a person deliberately out to rob you—hurts. But what about the transactions going on in the clean, well-lit buildings whose shadows shield the mugger? Is it aggression when a well-scrubbed and highly respected politician allows some company to continue to pollute, thereby increasing its profits, in exchange for support from the polluters? Pollution hurts, too, and the politician is, quite simply, after the same thing that the mugger was after. Is it aggression when a politician trades a vote, puts in a word, or sells himself to a heavy campaign contributor, with the result that mass transit construction is stifled, or that the price of some commodity is allowed to rise unnecessarily, or that prices are kept from being lowered? Automobile fumes and high prices hurt too. But how can such quiet, mannerly "business" dealings realistically be called aggressive? How can they *not* be?

Figure 27.8 Mahatma Gandhi, a nonviolent, but perhaps extremely aggressive, man.

Perhaps Mahatma Gandhi was one of the most aggressive men who ever lived, if one would wish to measure aggressiveness by motivation. This great chronicler of nonviolence aggressed against a social structure he believed was wrong and he won (Figure 27.8). Could a less aggressive, less assertive person have accomplished what he did? Yet his life is generally believed to be the antithesis of aggressiveness. The point is, aggression, especially among humans, may be hard to define, and thus the problem of investigating its motivation and historical origins becomes enormously complicated.

## THE MATURATION OF AGGRESSION IN NONHUMAN PRIMATES

In a study of the maturation of instinct in nonhuman primates, it was argued that aggression is innate in certain primates and that it is probably a solid component of the biological heritage of man (Deets and Harlow 1971). In their experiments, the researchers deprived rhesus monkeys (*Macacus indicus*) of the opportunity to learn social behavior by isolating them at various periods of their lives.

In one set of experiments, infants were placed in isolation only a few hours after birth. They had no opportunity to learn aggression, therefore if aggression were a learned trait, it should have never appeared. However, at the period in their lives when they would normally have begun to show aggression, they began to react in new ways to certain stimuli. When a glove was drawn across the cage of one of the isolates, it began to respond by threatening its keeper. Just as often it would attack. However, there was never an object upon which the aggression could be released, with the result that the monkeys turned to attacking their own bodies. They bit themselves so badly that in some cases serious wounds resulted, and one individual even blinded itself!

Just as these animals "suddenly" developed aggression at some point in their lives, they also, rather spontaneously, began to show the desire to affiliate with members of their group (affection) and to fear certain things. It turns out that in normally reared animals, these behavioral characteristics (affection, fear, aggression) appear at rather predictable times in the individual's history and in a certain sequence.

During the course of their development, monkeys first show a tendency to affiliate with members of their own species. During their early weeks, they do not fear the others nor do they show any aggression toward them. After a few months, however, they begin to show fear reactions, not only toward other animals, but also toward any unusual inanimate objects. Interestingly, they develop these fear reactions whether or not they have ever been exposed to threatening situations. The isolates, however, tend to develop fear behavior a little later than their normally reared counterparts.

In another set of experiments (Sackett 1966), infants were separated from their mothers almost immediately after birth. The infants were placed in total isolation, with any environmental sounds masked by "white noise." There was no physical or visual contact with other monkeys. The isolates were shown slides of other monkeys engaged in various activities, including threatening ones. In early weeks of life, the infants showed no reaction to the pictures, but at about three months there was an abrupt change in their responses: when the threat pictures were shown, the infants became fearful. It seemed that some innate mechanism, one that produces fear under threatening conditions, had suddenly matured.

It is not until the end of the first year that aggression becomes part of the behavior repertoire. At about this time, the young primates engage in vigorous play, which includes biting, threatening, and hair-pulling. Occasionally, the roughhousing gets out of hand and breaks into brief episodes of barking, threatening, and physical abuse. When the victim cries out in pain, the attack stops. As such aggression emerges, a dominance hierarchy is formed that will render actual fighting unnecessary in later life. It seems that aggression, just as fear, must mature within the individual.

In several other experiments, monkeys were isolated during critical periods of their lives when one or another of the behavioral responses was maturing (Deets and Harlow 1971). It was found that if a monkey is isolated from other monkeys during the time that the innate tendency to affiliate and the fear of unfamiliar stimuli are maturing, the monkey becomes rather permanently fearful of social contact. The infant in such a case shows no affection for other monkeys and does not learn to control his fear of them. Later on, as aggression matures, the isolate's fears may temporarily suppress any expression of his aggression. But not for long. The isolate proves to be hyperaggressive when allowed to associate with other monkeys later on.

If a monkey is not isolated until the fear response has matured, but before aggression has reached fruition, a different social pattern emerges. The developing infant will have learned to control his fear of other monkeys, but will not have learned to control his aggression toward them. Such isolates showed little fear of their peers, but they were very aggressive and came to dominate the control monkeys. These isolate infants differed from the first group in their reduced level of fear and in their ability to show some degree of affiliation.

In monkeys that were not isolated until maturation of fear and aggression were established, there was little behavior abnormality after eight months of solitary confinement (except for a brief period of heightened social activity when first returned to the group).

If the patterns of affection, fear, and aggression are indeed innately based, then the importance of learning becomes apparent. Affection, fear, and aggression seem to swell within each animal at a certain time, no matter what its experience. However, in order for these innate patterns to be of benefit, they must be tempered and moderated through the molding effects of experience. Primates are "learning animals" with extended parental care, so there exists mental facility and the time for experience to have its effect.

The importance of the timing of the onset of fear, aggression, and affection should now be apparent. For example, if fear developed before affection, the individual might have difficulty forming bonds. However, affiliation appears first. A monkey growing up in the wild stays close to the group and they protect him from danger. He may be reckless and foolhardy in the early weeks of his life, but later on, fear emerges, and the infant begins to avoid many dangers on his own. Interestingly, in those first few months, the mother's protective responses wane as the fear pattern develops in the infant (Hansen 1966). Also, the individual usually does not fear those animals that it grew up with.

While the advantages of fear and affection are intuitively apparent, we may have more trouble seeing how aggression can be useful in social animals. It must be remembered that many nonhuman primates live in groups in which it seems that aggression would be disruptive. However, the importance of dominance hierarchies has already been discussed, and aggression would function in the development of the group structure. In fact, group living might not be possible in the absence of aggression.

One might then wonder why aggression would not disappear within the individual once his rank was established. However, the question neglects the complexity of social groups. After all, there may well be other advantages in aggression. For instance, dominance hierarchies change as individuals seek the blessings of higher rank, and such changes can take place through aggression.

Aggressiveness might also be maintained in order that it could be directed to out-group members. Primates are not usually territorial animals in the strict sense, but groups do usually avoid each other where their ranges may overlap. Among some species, when groups encounter each other, there may be vigorous threats or even fighting. In many primates, it is extremely difficult for a strange individual to be accepted into an existing group.

Out-group aggression may have important ecological effects if it functions in dispersing groups so that the resources in an area may be most efficiently exploited.

Such aggression may be based in kin-selection if we are willing to assume that cooperation within the group can increase foraging efficiency and the level of protection. Since groups are cohesive units, there is a strong likelihood that the individuals are related to one degree or another. Under such conditions, there is a reproductive advantage in cooperative behavior. Any altruism may result in the likelihood of the altruist's kinds of genes being propagated through relatives. In order to increase the advantage of cooperative behavior, then, it is best if a related group excludes outsiders.

The idea of "getting at" the roots of human behavior by examining other animals is rife with dissension. It should be made clear that there is no intention here to say that because monkeys show a certain developmental scheme, humans do too. On the other hand, *if* we are going to extrapolate from the behavior of other animals to ourselves, it is probably better to choose those species that are most similar to us (Washburn and Hamburg 1968). The controlled study of other animals, which is not often possible with man, may give us certain insight into our behavior, perhaps providing us with testable hypotheses.

In fact, there are indications that the same maturational sequence of affection, fear, and aggression occurs in humans as in primates (Deets and Harlow 1971). A human baby accepts any adult with equal enthusiasm at first. For example, in the second month, the infant smiles at any face. Later, it becomes more discriminating and smiles only at familiar persons. By the sixth or seventh month, attachment to specific individuals is apparent. The baby is distressed when removed from them and generally cannot be quieted by strangers. In another month, the infant begins to show overt fear of strangers. So, affiliation and attachments normally appear in the human before fear reactions develop. Aggression, according to the primate data, should appear in the first four or five years of human life, but there is as yet no verifying information. Until very recently, psychologists have concentrated on learning processes and correcting behavior in older children and have almost ignored the innate development of human behavioral patterns.

In one of the few studies conducted, it was found that aggression is established early and is relatively stable, particularly in boys. The level of aggressiveness in the preschool years appears to indicate how aggressive the person will be through early adolescence in females and through adulthood in males (Kagan and Moss 1962).

There is also dismaying evidence that children who have spent their early years in socially sterile institutions show an impaired ability to form social relationships. In addition, they show difficulty in controlling aggression (Yarrow 1964). Even in the face of such evidence, however, psychologists have generally continued to ignore the ontogeny of behavior (Blurton-Jones 1972). It is reasonable to predict that in ensuing years there will be increasing attention to the biological heritage of man in the search for solutions to his social problems.

# 28

# THE PHYSIOLOGY AND ADAPTIVENESS OF AGGRESSION

What causes aggression? What happens when a placid animal becomes aggressive? What changes occur in the millions of impulses dancing through the central nervous system? Do they take new directions? Are new areas of the brain activated? Do neural centers begin to interact in new ways? Are there fundamental differences in individuals that we would call aggressive and those we wouldn't? Do aggressive people have different chemicals circulating in their blood than their more benign comrades? Are their central nervous systems different?

Some of these questions can be answered by considering anomalies—when aggression goes wrong. But the important questions for the biologist center over what happens when aggression goes right. How is it adaptive? Is aggression beneficial? Can fighting be good? Let's now consider the physiological roots of aggression and why it is so common among animals.

## THE PHYSIOLOGY OF AGGRESSION

A brief discussion of the physiological and anatomical mechanisms associated with aggression might hopefully give us better insight into this most critical of behaviors and help us to understand more about motivation.

### Chemical Influences on Aggression

I once had a friend in college who was actually quite formidable physically. But in an aggressive encounter, he would lose his usual articulate manner, his voice would rise to a squeak, and his jaws would flap uselessly as he tried to make some telling point. At these times, he appeared most unintimidating indeed. The root of his problem was probably the epinephrine (or adrenaline) liberated into his bloodstream by his adrenal glands, which generally raised his level of physiological arousal and excitement (in his case, too much). It is known that adrenaline levels and general excitement rise in humans who feel aggressive, as was shown by monitoring the blood of a group of college students who had deliberately been insulted and provoked (Hokanson and Shetler 1961; Hokanson and Burgess 1962).

Blood chemistry changes, then, can be correlated with aggressive behavior. Strangely, if an animal has been *defeated* in aggressive conflict, marked changes may occur in its blood chemistry. Frank Bronson and B. E. Eleftheriou trained a group of a single strain of mice to fight as "gladiators." The gladiator mice were then matched with untrained mice and, of course, quickly defeated them. Instead of giving the old "thumbs down" to the losers, though, the researchers separated the mice and sampled their blood for chemical differences. It was found that differences existed and that the effects were long-term. Even mice that had previously (but not recently) been defeated had more than three times as much of the chemical indicator of adrenal activity (a hormone, unbound corticosterone) in their blood than mice that had never been defeated. Thus, it seems that defeat caused a relatively permanent change in the physiology of the animals. It has been found that lower-ranking mammals can be distinguished from dominants by white blood cell counts, fat deposits around the adrenal glands, blood cholesterol levels, secretion of hydrocortisone, ascorbic acid levels in the adrenals, and overall adrenal weight. It would be interesting to know if there are permanent physiological effects of continually suffering defeat in young humans. Could the *noblesse oblige* behavior of the wealthy be partly correlated with blood chemistry? Do the "new rich" behave differently from the "old rich" (families with a history of wealth as opposed to those who recently made it the hard way)? Are there differences in the blood chemistry of the two groups? What about blood differences in all human social strata? For details of further biochemical effects on behavior, see Garattini and Sigg (1969).

Chemical studies have also shed light on the neural mechanisms that control killing behavior. Tiny tubes were implanted in the brains of two groups of anesthetized rats that had proven themselves to be either spontaneous mouse killers or nonmouse killers. The tubes were positioned in both hemispheres and aimed directly at the lateral hypothalamus. Thus, those specific parts of the brain could be bathed in solutions injected through the tubes. It was found that carbachol, a drug that acts like acetylcholine and therefore *increased* neural activity, would induce peaceful rats to kill mice. Another drug (atropine methyl nitrate) blocks neural activity and when it was administered to killer rats, it suppressed mouse killing (Smith, King, and Hoebel 1970). Other neural chemicals had no effect either way, and if the tubes were moved even one millimeter from the lateral hypothalamus, even the appropriate drugs had no effect. Thus, both chemical and anatomical specificity were demonstrated.

## The Limbic System and Aggression

As we learned earlier, the limbic system is composed of a series of structures in the upper brain stem and lower cerebrum and is intimately involved in motivation. It is sometimes called the "old brain," you recall, because it arises earlier in both embryonic and evolutionary development. The "new brain," or neocortex, is added secondarily. We have gone through most of these names before, but perhaps giving you their quaint old origins will take away some of the mystery of the terminology. Thus, the limbic system includes the thalamus (which means bridal chamber), the epithalamus (roof of the chamber), hypothalamus (floor of the chamber), hippocampus (seahorse—because of its curved shape), amygdala (almond—because of its shape), cingulum, or cingulate gyrus (belt), and septum (fence) (Figure 28.1).

The limbic system has long been implicated in aggressive behavior. Almost a century ago, researchers found that if the amygdala or hippocampus were cut, monkeys became unusually tame. In 1937, in a more sophisticated operation (Kluver and Bucy 1937), incisions were made in both temporal lobes of the brain of wild monkeys, at the same time cutting the amygdala and hippocampus. After recovery, the monkeys were gentle, playful, hypersexual, and had lost many of their other inhibitions (for example, they might put any strange object in their mouths). Temporal lobe lesions in squirrel monkeys may cause them to lose their rank in the dominance hierarchy (Plotnik 1968). Such lesions in free-ranging monkeys impair various social patterns including threats and submissive gestures (Johnson 1972).

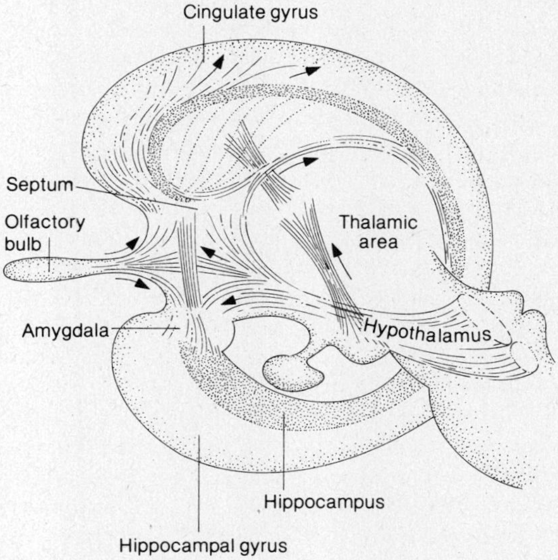


Figure 28.1 The limbic cortex and associated areas. Pathways from the hypothalamus lead through the amygdala and the septum into the limbic cortex (the darkened area). The limbic system is believed to coordinate and elaborate emotion and aggression.

Interestingly, in some cases, when the amygdala was cut in cats, they showed increased irritability and rage, while in other cases, they became extremely docile. Even vicious wildcats have been tamed by severing the amygdala, and wild Norway rats, subjected to the same treatment, lose all aggressiveness, as well as their fear of novel objects. Such findings are not necessarily discrepant, because the amygdala contains many nuclei, some of which have opposing functions (Johnson 1972). The point is that there are certain brain centers that are intimately associated with aggressive behavior, but they cannot be pinpointed in all cases. And, as we will see later, there is no single "seat" of aggression in the brain.

## The Physiological Relationship of Aggression and Predation

Electrode-implant experiments have drastically altered the behavior of animals, no matter what their natural tendencies or early experience might have been. In one experiment, electrodes were implanted in the lateral hypothalamus of cats that had always gotten along well with rats. A few millionths of a volt of electricity could be administered at the will of the experimenter. It was found that when the postoperative cat was placed in a cage with a rat, they continued to get along. Low-intensity stimulation would cause the cat to get up and prowl around the cage, still ignoring the rat. At a slightly higher stimulus, however (presumably above some threshold), the cat would leap on the rat, biting it behind the neck in the fashion of all "rat killers," usually killing it. Such attacks are strangely "unemotional," lacking in any sign of aggression or "rage," and have therefore been called "quiet-biting" (Flynn 1967). The cats are highly selective about what they bite at such times. They prefer real rats to dummy rats, anesthesized rats, toy dogs, or rubber blocks. They almost never attack "irrelevant" (nonprey) objects, such as rubber blocks or the experimenter. The lack of aggression, of course, would be expected if some sort of "prey-attack" center were being stimulated.

However, the experimenter should take heed: if he stimulates the more medial hypothalamus instead of the lateral area, the cat will then become a veritable Frankenstein, lashing out at anything (Roberts and Kiess 1964). The pupils dilate, the back arches, the hair stands on end, the claws are extended, and the cat growls, hisses, and urinates. Here we have a most disagreeable animal. A cat in this state will attack a rat, another cat, or anything it can reach. It is said to be "enraged" by those it *has* reached.

Quiet-biting, by the way, has also been elicited in the rats themselves. In rats, when the posterior hypothalamus is stimulated, natural mouse killers will kill faster, but in order to cause nonkillers to quiet-bite, the anterior and medial hypothalamus must be stimulated (DeSisto 1970), thus the "centers" for the behavior are different in cats and rats. Actually, quiet-biting attacks have been elicited from fifty-three points in the ventrolateral hypothalamus (Woodworth 1971). Also, quiet-biting rats will bite live or dead mice (rats sometimes prey on mice as was mentioned earlier), but never other adult rats. In such cases, the victim seems to be chosen on the basis of its size. Stimulated rats will attack mice and pups of their own species, but will ignore adult rats or guinea pigs.

It should be apparent by now that whereas ethologists do not normally consider predatory behavior to be aggressive, there may be physiological elements common to the two types of patterns, perhaps even stemming from the same parts of the brain. In spite of any close relationship suggested by such data, though, we must remember that neural areas only a very small distance apart may produce entirely different patterns when activated, and that the quiet-biting (probably related to predation) is markedly different from the attacks of an enraged animal (probably related to true aggressive attacks).

Quiet-biting seems to be a species-specific consummatory behavior, neither completely rigid nor plastic (Johnson 1972). For example, quiet-biting may only be elicited if there is an appropriate victim available. But, here, previous social experience may influence the strength and direction of an attack. For example, cats with more social experience will attack more vigorously, persistently, and selectively than cats raised in isolation. But it is important to realize that attacks can be elicited in animals isolated from birth. Thus, social experience is not necessary for coordinated killing directed against specific sorts of victims (Roberts and Berquist 1968). It should be clear, then, that just because a pattern is not performed, it does not mean that it lacks the neuromotor apparatus for its performance. After all, we now know that, at least in some species, proper stimulation of the hypothalamus can produce female sexual behavior in normal males and male behavior in females.

Three main points can be gleaned from the studies of elicited aggression. First, the mammalian brain appears to possess the organization that gives every individual the potential for attack no matter what his previous experience. Second, there are different *kinds* of aggressive behavior. There is no single stereotyped pattern that *is* aggression. One person may wish to recognize only attack, defense, and flight. Another may choose predation, inter-male conflict, fear, irritability, territoriality, and maternal defense. Finally, it is apparent that there is no "aggression center." Destructive attacks and killing can be elicited from widely separated parts of the mammalian brain. These undoubtedly include parts other than just the hypothalamus. In fact, the amygdala may be more important, overall, in the normal performance of aggression. You may wish to consult Johnson (1972) for further references.

## SEX AND AGGRESSION

It is probably unwise to assume that adjacent areas of the brain necessarily deal with related kinds of behavior, but such a relationship exists in some cases. For example, as shown in Figure 28.2, the face control centers are located adjacent to each other and near the tongue. The center associated with the elbow is between the wrist and shoulder. Thus, anatomical areas that lie in proximity to each other are often associated with brain parts that are adjacent.

The principle may or may not hold true for motivation centers, but physiological studies show that sex and aggression, which are apparently interrelated emotionally, are also closely associated in the neural centers. Mapping experiments of the monkey brain (MacLean and Ploog 1962) show that neural structures that, when electrically stimulated, immediately cause erection of the penis are located within a millimeter of certain rage centers. Also, brain damage in humans has been known to affect both sexual and

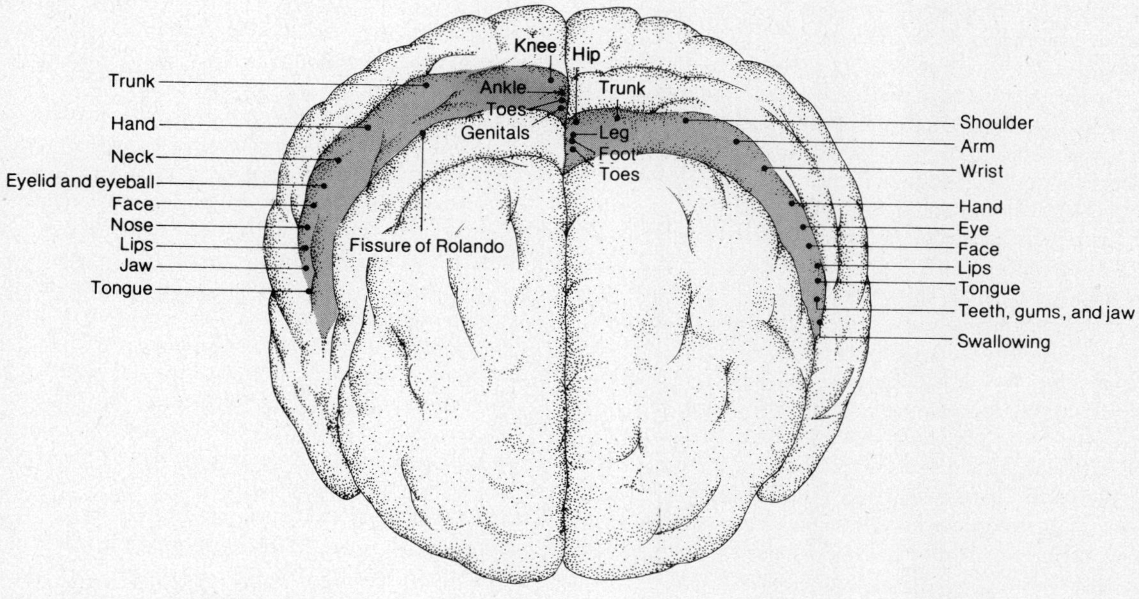


Figure 28.2 Function maps of the cerebral cortex (top view) showing sensory areas (left) and motor areas (right).

aggressive behavior. In one case, an extremely aggressive woman assaulted a number of people with razor blades and broken bottles. She also masturbated eighteen to twenty times each day and frequently participated in group sex and lesbian activities. Neurological examination revealed that portions of the right frontal part of the temporal region were malfunctioning. An antiseizure medicine was administered to help control the abnormal neural activity, and both her aggressiveness and sexual appetites fell. Apparently, one could not be controlled without altering the other. All that this sort of evidence indicates, however, is that aggressive and sexual centers may lie near each other. It does not *necessarily* mean that these centers interact in their expression.

It is also interesting that a familiar two-word expletive in which one person suggests that another engage in sexual intercourse is actually an aggressive phrase.

### Sex Hormones and Aggression

It has long been known that a stallion is more aggressive and dangerous than a gelding. And savage, bad-tempered bulls can be reduced to plodding oxen by a simple stroke of the knife. A "fixed" dog is less aggressive and an ex-tomcat stays around the house more and gets in far fewer fights. In earlier times, boys were castrated so that their bell-like voices wouldn't change and they could continue to sing for royalty. Eunuchs could also be trusted in the harem. In the United States, prison convicts were first legally castrated in Indiana (Le Maire 1956) (this has since been repealed). Germany, Switzerland, and the Scandinavian countries, especially Denmark, have also legalized the operation for criminals. However, the punishment is rarely used and then mostly, but not always, reserved for sex offenders. The result of castration in human offenders is a general pacificism and a low incidence of recidivism. In any case, it seems to be generally recognized that maleness, or more specifically, the male hormones (produced by the testes) are strongly implicated in aggressiveness.

Aggressiveness in women may also be related to hormones. The premenstrual period of women is often accompanied by heightened irritability and aggressiveness. It seems to be associated with a drop in the level of progesterone in the blood. In fact, most crimes by women are committed during this premenstrual period (Moyer 1971). But changes in the level of aggression caused by changes in female hormones cannot compare to the changes that occur with varying levels of male hormones in the blood.

It is interesting that hormones specific to the sex begin permeating the embryonic body early in its development and that an individual beginning the developmental pathway toward becoming a male can be shunted into the female pathway by the injection of female hormones at the appropriate stage. The result is a "normal" female. The reverse experiment has also been performed (Goy 1970). Thus, the sexual identity of an individual is not necessarily fixed at fertilization as was previously thought.

Although there are a number of different male and female sex hormones, the one most important to aggression is testosterone. It is produced in the gonad and it initiates the physical changes of puberty as its level in the blood rises at this time. But puberty often also marks an increased aggressiveness. For example, rats don't fight much until puberty, at which time the aggressive incidence picks up dramatically. If young rats are injected with testosterone, they will begin fighting earlier (Levy and King 1953). Testosterone is also present in females, but only in minute quantities.

If testosterone is injected into hens or castrated chickens, they will develop the comb and plumage of a rooster and their aggressive behavior will intensify (Figure 28.3). In fact, such injections can change the animal's rank in the peck order (Table 28.1). Hens injected with testosterone become more aggressive and dominant and fight their way up the peck order, sometimes going from the bottom to the top. Roosters that are either castrated or treated with female sex hormones become more timid, and their status in the peck order drops. Testosterone may also have an effect on territorial behavior. If ring doves are injected with the hormone, they tend to enlarge their territories (Bennett 1940).

Female mice are usually much more docile than males, but if they are given male hormone treatments right after birth (when such treatments are highly effective—the effect is reduced later in life), they breed only with difficulty as adults. In fact, they may attack and seriously cripple a courting male (Bronson and Desjardins 1968).

There's another peculiar thing here. Certain hormone manipulations that affect aggressiveness in males may fail to do so in females. Castration, for example, lowers aggressiveness in males, but the removal of the ovaries has no effect on female aggressiveness. Female hamsters are larger than males, and if unreceptive, a female may kill her suitor and she may kill him with or without her ovaries. If estrogen is administered, she may grow even more aggressive, but higher levels of progesterone (another female sex hormone) have no effect. For some reason, though, if both estrogen *and* progesterone are administered, she may become *less* aggressive. (But see Payne and Swanson 1971.)

Such results can be accounted for as we make a major point. That is, the nervous systems of the sexes of adult mammals may be structurally different. It has been known for some time that prenatally administered hormones may cause a permanent structural change in the part of the nervous system that affects adult sexual characteristics (Phoenix et al. 1959). Later, researchers found that when lesions were made in the hypothalamus of male and female guinea pigs, entirely different sorts of changes in sexual behavior occurred (Goy and Phoenix 1963). For example, preoptic lesions completely eliminated male sexual behavior, while not altering female sexual behavior at all, and that anterior hypothalamic lesions, which caused the cessation of female sexual behavior, only partly interfered with male sexual behavior (Singer 1968).

Other results have shown that an area called the paraformical medial forebrain is vital to the performance of male sexual behavior, but has nothing to do with female sexual patterns. In addition, the effects of septal lesions in rats depend partly on whether the animal is male or female. If androgen or estrogen are given to female mice on their day of birth, they both facilitate the developmental differentiation of androgen-sensitive areas for the rest of the animal's life. These, in turn, affect adult aggressive behavior. The evidence stems from experiments that show that when female rats are treated with testosterone as adults, aggression rises somewhat in normally reared females, but the ones that received androgen or estrogen at birth fight as much as normal males. All three fight more than females given a placebo at birth (Johnson 1972). Also, if testosterone is injected into pregnant monkeys, the female offspring will be masculinized, although these tendencies will not appear until later in life. I'm citing these experiments to indicate that perhaps one reason for the difference in the levels of aggressiveness in male and female mammals is that the *brains themselves* are sexually dimorphic. This should not be too surprising, since it is only reasonable to assume that if the sexes are rather distinctly different in size, strength, and speed

Figure 28.3 Testosterone injected into baby chicks will cause premature development of masculine traits. These are two eighteen-day-old chicks, the one at top having received such an injection.

Table 28.1 Changes in peck order among hens after testosterone injections.[a]

| SOCIAL ORDER | | | |
|---|---|---|---|
| HEN NUMBER | BEFORE INJECTIONS | DOSAGE OF TESTOSTERONE | AFTER INJECTIONS |
| | *(Number of other hens pecked)* | | |
| *RG* | *7* | | *6* |
| *RW* | *7* | | *6* |
| *YY* | *6* | | *4* |
| *BR* | *6* | | *4* |
| *GG* | *4* | | *3* |
| *RR* | *3* | | *1* |
| *BG* | *2* | *0.75 mg* | *1* |
| *RY* | *1* | *1.00 mg* | *7* |
| *BY* | *0* | *1.25 mg* | *7* |

*Source: Allee, Collias, and Lutherman (1939).*

*[a]Hen* BY *rose from the bottom to second from the top after testosterone injections. The top hen,* RY, *took her position only after testosterone injections.* BG *fell ill during the experiments, and the injections had no effect.*

(whether considering extremes or means), the neural apparatus that controls the two types of bodies could also be expected to be different.

Not only do the sexes of some monkeys differ in their reproductive behavior, but also in emotionality, aggression, play, and food preferences (Goldman et al. 1974). In addition, brain maturation proceeds at different rates in males and females, and this apparently affects learning ability. The researchers removed the orbital prefontal area of the cortex in eight animals of both sexes (using nine more as controls). When all were ten weeks old, they were tested in habit-reversal experiments. Learning ability was reduced in the males that had been operated on, but not in the females. In a wide range of other learning experiments, they demonstrated that, up to about a year of age, cortical damage impaired learning in males, but not in females. However, after the age of fifteen to eighteen months, the brain damage affected both sexes to about the same degree. It was concluded that the functions of the orbital area develop earlier in males than in females.

## BRAIN DAMAGE AND HUMAN VIOLENCE

I arrived to begin my doctoral work at the University of Texas only a few short weeks after a student there had climbed to the top of the university's observation deck and in the next ninety minutes shot thirty-eight people, killing fourteen. I came to know some of the people who were shot at and some survivors of victims. The human tragedy was beyond comprehension. The student's name was Charles Whitman, and to all outward appearances he was a normal, clean-cut, hardworking young man. But he started to become filled with a nameless rage that came and went. He spoke to a campus psychiatrist about his uncontrollable violent impulses. He received no real help. Perhaps his own words can best describe what happened. On the evening of July 31 he wrote:

*I don't quite understand what it is that compels me to type this letter. Perhaps it is to leave some vague reason for the actions I have recently performed. [At this point, Whitman had harmed no one; his wife and mother were elsewhere in the city, still alive.] I don't really understand myself these days. I am supposed to be an average, reasonable and intelligent young man. However, lately (I can't recall when it started) I have been a victim of many unusual and irrational thoughts. These thoughts constantly recur, and it requires a tremendous mental effort to concentrate on useful and progressive tasks. In March when my parents made a physical break I noticed a great deal of stress. I consulted a Dr. Cochrum at the University Health Center and asked him to recommend someone that I could consult with about some psychiatric disorders I felt I had. I talked with a doctor once for about two hours and tried to convey to him my fears that I felt overcome by overwhelming violent impulses. After one session I never saw the doctor again, and since then I have been fighting my mental turmoil alone, and seemingly to no avail. After my death I wish that an autopsy would be performed on me to see if there is any visible physical disorder. I have had some tremendous headaches in the past and have consumed two large bottles of Excedrin in the past three months. It was after much thought that I decided to kill my wife, Kathy, tonight after I pick her up from work. . . . I love her dearly, and she has been as fine a wife to me as any man could ever hope to have. I cannot rationally pinpoint any specific reason for doing this. I don't know whether it is selfishness, or if I don't want her to have to face the embarrassment my actions would surely cause her. At this time, though, the prominent reason in my mind is that I truly do not consider this world worth living in, and am prepared to die, and I do not want to leave her to suffer alone in it. I intend to kill her as painlessly as possible. . . .*

Later in the night he did kill both his mother and wife, and then wrote:

*I imagine it appears that I brutally killed both of my loved ones. I was only trying to do a good thorough job. If my life insurance policy is valid please see that all the worthless checks I wrote this weekend are made good. Please pay off all my debts. I am 25 years old and have never been financially independent. Donate the rest anonymously to a mental health foundation. Maybe research can prevent further tragedies of this type.*

The next morning he began shooting from the top of the tower. Luckily, his timing was off and he had missed the class changes by about ten minutes. Finally, a policeman worked his way to the top and out onto the observation deck from where Whitman was shooting and killed him. A postmortem examination revealed a tumor (of a highly malignant type called glioblastoma multiforme) about the size of a walnut, imbedded in the area of the amygdaloid nucleus (Sweet, Ervin, and Mark 1969).

It has been estimated (Mark and Ervin 1970) that ten million Americans have obvious brain damage, while another five million may be less seriously affected. Such damage can be caused by a number of factors. Tumors may be the cause, and damage can occur at pregnancy or during delivery, by accidents, by prolonged reduction in blood sugar levels, Rh incompatability of the parents, drug use, oxygen deprivation, or a number of other factors. Unfortunately, the behavioral results of such damage may be masked for years. Changes in aggression, by the way, are most likely to occur when the limbic system itself has been damaged.

As another example, the dreaded disease rabies is a killer that has been survived by only one human (a boy, in 1971, who was bitten by a rabid bat). Rabies literally means "rage" and it is caused by a virus that attacks the medial hippocampus of the limbic system. When contracted by a human, even the most gentle and agreeable of people are transformed into terrifying individuals filled with rage and violence.

Often the organic roots of violent behavior among humans are not investigated. The person is simply punished or subjected to some sort of social rehabilitation program. A case history will illustrate: Jennie was a teen-age girl reported to be a model child. But, once, when criticized for playing records too loudly, she suddenly flew into a violent tantrum. She smashed everything in her room, and the police had to be summoned to restrain her. Subsequently, her moods varied from "angelic" to "devilish." Finally, she was institutionalized after smothering to death a baby whose crying had annoyed her. Her brother had a history of epilepsy, and this suggested to doctors that her violence might be caused by organic brain disease. She was referred to the Massachusetts General Hospital, where careful tests of her brain activity revealed abnormal electrical patterns in the

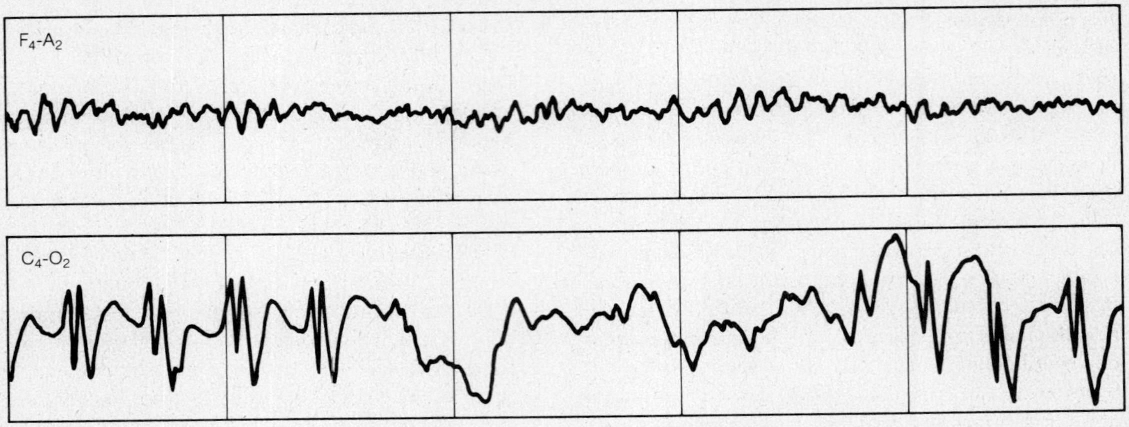


Figure 28.4 Normal (top) and abnormal (bottom) brain-wave patterns in specific parts of the brain as seen on the electroencephalograph.

hippocampus. Because she had attacked several other babies, vigorous remedial efforts were called for. An attempt was made to localize the diseased brain tissue by probing with electrodes. Electrical recordings (Figure 28.4) of brain activity were made from these depth electrodes when the patient was (a) relatively relaxed, (b) angry, and (c) when she heard a baby crying. The EEG (electroencephalograph) patterns proved useful in localizing the diseased brain tissue, since abnormal EEGs in certain areas could be evoked by the sounds of a crying baby. Unfortunately, she was not able to receive further medical attention because she was sent to a state institution.

## FIGHTING

Fighting is one expression of the aggressive mood. It is potentially the most dangerous expression and, in some cases, it has therefore become one of the most highly stylized forms of interaction among animals. The widespread occurrence of fighting in the animal world, and the existence of ritualized or "tournament" fighting, paradoxically indicates a strong selection for a dangerous behavior. Obviously, since fighting is dangerous, it would have been selected out long ago if it were not highly adaptive. Instead, it is maintained in a wide variety of animals, although in some cases, incredibly complex rituals have developed to keep its risk at a minimum.

Fighting is highly variable across the spectrum of animal species. Some do not fight at all, and others may fight to the death. Let's consider a few examples from fighting animals. One of the more bizarre cases occurs in certain sharks. Their eggs hatch in the uterus, and the shark pups are thus born alive. However, the first young shark to hatch within the uterus turns within his mother's body and cannibalizes his brothers and sisters as they emerge from their eggs (Kenney 1968). Such behavior may have elements seated in predation as well as aggression, but it is a most effective means of reducing the numbers of potential competitors.

Elephants are normally peaceful enough with their own, but occasionally serious fights do occur. Carrington (1958) described such conflicts:

*They fight head to head, either attempting to push one another backwards with their foreheads, or intertwining their trunks and engaging in a titanic tug of war, so that one animal is drawn into the other's tusks. The struggle may continue until one of the contestants is killed. This usually happens through the weaker elephant breaking away, thereby exposing his unprotected flank to the tusks of his opponent. Then thus caught at a disadvantage an elephant can be disembowelled with a quick lunge of the head. The bodies of elephants are sometimes found with their heads battered in, their abdomens ripped open, and great holes torn out of their sides, bearing witness to the ferocity of the struggle.*

Such fighting would appear to be unadaptive. But another kind of elephant behavior that also is difficult to explain is that of "rogues"—solitary and ill-tempered bulls. Some rogues have been famous in their time, such as the rogue of Kankankote, a village fifty miles from Mysore in South India. In the 1870s, this rogue terrorized travellers on a road leading to the village, viciously attacking people, stomping them to death, and tearing them apart with his trunk and legs. Carrington (1958) states, "It seems that the typical rogue is either mad, like a homicidal maniac in our own species, or is berserk," although the distinctions are not made clear. The "madness" could be brought on by the pain of a toothache, overstrained tusks, disease, or a gunshot wound. If Carrington's analysis is correct, the behavior is obviously aberrant and unadaptive—an evolutionary dead end that must arise anew each time it occurs (Southwick 1970).

But we are primarily concerned here with the adaptive aspects of the fight. What good is it? Where does it fit into the big picture?

## The Adaptiveness of Fighting

One of the most obvious explanations of fighting is that it is a behavioral result of competitive interaction, and therefore winning bestows the same advantages as successful competition. Put another way, there is no need for an animal to try to "outeat" his competitors if it can forcibly exclude its competitors from the food.

As we have seen, fighting may actually function in reducing aggressive interaction. It need not occur every time there is a dispute over commodities or mates. Instead, its usefulness may lie in its role as a mechanism for establishing a hierarchy and thus avoiding future conflict. Once the hierarchical system is established, fighting might appear rarely within a group; probably occurring mainly when an animal attempts to change rank, such as when a young animal grows up and challenges his superiors. Actually, any animal of unestablished rank may be forced into frequent conflict for a time. For some species, the introduction of a strange animal into the hierarchy results in a round of fights until the position of the newcomer is established.

Fighting also plays an important part in the establishment of territories, as has been mentioned. Once territories are established, fighting is greatly reduced. Such reduction could occur through a number of influences, such as (1) the formation of nonbreeding groups that do not violate territories; (2) "floaters" that do not challenge territory holders; (3) habituation by territory holders to owners of nearby territories; and (4) the high level of aggressiveness in territory holders within their own areas that could lead to the quick expulsion of an intruder.

It is obviously important that fighting occurs as infrequently as possible because of the dangers involved. The danger is not only that posed by the immediate threat of bodily harm per se, but by its aftermath as well. An injured animal is a weaker animal, and the results can be far-reaching. In one case, a dominant squirrel monkey fell to the lowest status after suffering injuries from a "successful" hierarchy battle (Rosenbaum and Cooper 1968). Injured individuals might also suffer higher predation through being incapacitated, or injury could lead to interrupted or altered feeding patterns, thereby placing the animal at a competitive disadvantage.

### Interspecific Fighting

Since fighting is usually a result of competition, we can expect to find it occurring most often between members of the same species. When fighting occurs between members of different species, those species are usually in direct competition, and very similar in appearance (Figure 28.5), or are prey animals that have initiated attacks on their predators. In some cases, however, highly unlikely and unadaptive fighting may arise (Figure 28.6).

There are cases in which fighting is not a direct result of competition. For example, there are at least thirty-five species of ants that depend to some degree on slave labor by other species (Wilson 1975). The techniques by which they acquire their slaves are among the most sophisticated behavioral patterns found anywhere in the insect world. In fact, some species are so specialized as raiders that if they are deprived of slaves, they are unable to care for themselves and they die. Basically, the slave-making ants swarm directly to a colony of slave-types and kill

Figure 28.5 A quarrel between females of two different, but highly similar, species: a purple finch and a song sparrow. Such interspecific quarrels may develop when the appearance of an unrelated animal elicits aggressive responses that are adaptive on an intraspecific level. These species probably arrived at their similarity in markings through convergent selection for crypticity. In similar sympatric species, aggression may be minimized by differences in utilization of the habitat.

any defenders that put up any resistance at all. They then carry off the pupae of the slaves. When the pupae hatch, they treat the slave-makers as sisters (they're all females—see Chapter 29). Thus the slave-makers receive the benefits that would have been bestowed upon the sisters of the slaves. The slaves go right to work. They are invariably highly organized and social species, much more so than their "masters." Since the slaves are invariably sterile, the slave-makers must replenish their stock by yet new raids. When not raiding, some slave-makers do nothing but groom themselves and beg from their hard-working slaves in a kind of "social parasitism" (Wilson 1975).

## Reducing the Dangers of Fighting

When fighting occurs between members of the same species, there are distinct advantages to keeping physical harm to a minimum. In keeping with our concept of fitness, it would be disadvantageous for an animal to harm a prospective mate, watchful members of the group, a sibling, or even other individuals that may share a significant number of gene types through kinship.

In fact, in most species, fighting rarely leads to real injury. All-out fighting, when it occurs, is usually between members of a relatively harmless species. The dangerous ones are usually those that can also run fast and get away from each other and those that have some system of inhibiting aggression (Eibl-Eibesfeldt 1970).

Among dangerous animals, when one realizes it is beaten in a fight, it usually stops the attack by signaling its submission. Hamsters (*Cricetus*) have no inhibitions against fighting to the death, but they rarely do so because after a few bites are exchanged one animal usually flees. Apparently, the ability of members of a species to retreat quickly can compensate for the lack of killing inhibition, but the system is not infallible. Males from a pride of East African lions (*Felis leo*) will sometimes kill a strange male that wanders into their area (Schenkel 1966). It is not known whether fleetness is a very effective protection against the willingness of lions to kill members of other prides.

We are already aware that fighting may be rendered less dangerous by the use of ritual or ceremonial fighting. Such "tournament" fighting may take the form of vocalizations, displays, or "pushing" contests in which the stronger animal is decided without injuries being inflicted on either combatant.

Tournament fighting, of course, is most important among dangerous animals. Rattlesnakes can kill each other with their poison, but they never bite. Instead, they move so that they lie side by side, then they raise their heads and try to push each other to the ground. When one is pinned, he is allowed to retreat (Figure 28.7).

In some cases, tournament fighting is abandoned for real fighting. Lava lizards fight each other through ritualized thrashing with their tails. In one fight, however, a tailless lizard didn't seem to notice his disadvantage and flailed away with the stub. After continually losing territorial conflicts, day after day, he seemed to become fed up, turned around, and bit an opponent on the tail, thus winning the fight. Such dirty pool indicates at least some degree of flexibility in this species.

In the marine iguanas (*Amblyrhynchus cristatus*) of the Galapagos Islands, males defend territories by head-to-head pushing contests (Figure 28.8). When one drops to the ground, the fight is over. Rarely, one may bite the other on the back of the neck. Females, on the other hand, may engage in damaging fights as they compete for scarce egg-laying sites. Females of some species seem to lack well-developed inhibitions against inflicting injury, but in the case of this iguana the more serious fighting in females may be the result of an increased level of competition.

Figure 28.6 A highly "unlikely" fight. Here a hare attacks a raccoon. The raccoon had approached the hare in the darkness. The hare, apparently unable to get away, responded by thumping the earth. As the raccoon approached, the hare leaped over it, giving it a drubbing on the head with its hindleg after which the raccoon retreated.

## Morphological Specializations for Fighting

In some species, there are special morphological adaptations for tournament fighting. For example, in fiddler crabs, which fight by pushing each other with slightly opened claws, small protuberances on the front of the claws increase purchase so that relative strength can be measured more easily. When the fighting becomes more intense, the crabs may hold on with their claws, but here again, special structures ensure that the animals grasp each other just so, and no serious damage usually results (Crane 1966).

**Aggression in Horned Mammals** Horned mammals have provided important information on how damage may be avoided in fights. First of all, horn size may give direct information regarding fighting ability

Figure 28.7 Fighting male rattlesnakes (*Crotalus ruben*). Each male tries to push the other to the ground, but the deadly poisonous snakes never bite each other.

(Geist 1966a), and the fight may therefore be avoided entirely as the smaller-horned individual chooses not to fight. Strangely enough, when intraspecific fighting does occur, the horns are rarely used to inflict injury. The Beisa oryx (*Oryx gazella beisa*) uses its long rapierlike horns to impale attacking lions, but in intraspecific fighting these antelopes only spar with the tips. The shafts of the horns also provide purchase for head-to-head pushing contests (Figure 28.9).

A belligerent oryx will not attack the vulnerable flank of an opponent even when he has the chance. Instead, he will hold his attack until the opponent is facing him (Walther 1958). Keep in mind that the attacker is not dissuaded from injuring his opponent through any spirit of benevolence: he is *unable* to pierce the flank of his opponent. When the system works, the male oryx could no more gore an opponent than fly! But remember, this same oryx would wreak havoc on an attacking lion who was unfortunate enough to expose a flank. It is important to note that fighting a competitor and fighting an attacking predator have been shown to be controlled by different parts of the central nervous system (Kaada 1967).

Systems of inhibition in fighting, of course, are better developed in some species than others. Some horned mammals, such as the bushbuck (*Tragelaphus scriptus*), may attack in an unrestricted and unoriented manner. This smallest-horned member of its genus is not inhibited from goring an opponent at all. It might be noted that both the fighting structure *and* the fighting behavior are rather primitive in this species (Ewer 1968).

By the way, Robert Ardrey, in his popular book *African Genesis* (1962), failed to recognize that there is no necessary connection between the aggressive nature of man and his predatory past. Not only are predation and aggression probably qualitatively different, but plant eaters are no more peaceful among themselves than are carnivores or hunters (Eibl-Eibesfeldt 1970).

It is interesting that hornless female antelope show no compunction against butting each other in the flank (Figure 28.10), but, of course, such attacks are usually harmless. In those species in which both sexes are horned, usually neither sex attacks vulnerable areas.

In some species of horned animals, the horns or antlers play little part in interspecific fighting. For example, probably the most effective weapons in deer are their slashing hooves. These weapons are so effective that deer have even been able to repel attacking mountain lions, as depicted in David Wolper Production's moving film, *Say Goodbye*.

A quick overview of African ungulates reveals a wide variety of horn types, although the animals often have common predators. If the horns were subject to selection on the basis of predator defense, one would

Figure 28.8 Like many other animals, the male marine iguana (*Amblyrhynchus cristatus*) of the Galapagos Islands will defend its territory against intruding males. When the intruder approaches, the occupant of the territory nods his head and struts. He then lunges at the intruder and they push against each other head-on (top). When one iguana feels he cannot win, he drops to his belly (bottom).

expect to find a tendency toward similarity due to convergence. The strong differences between species tend to support the idea that perhaps horns were developed for intraspecific fighting rather than for predator defense (Eibl-Eibesfeldt 1970), with some animals secondarily adding an antipredator role.

So, if horns are not primarily used to inflict injury—at least in some species, it might be expected that they might assume forms that reduce their effectiveness in battle. Such may be the case in bighorn sheep. The strongly curved horns are not used in repelling predators; instead, a mountain lion faces dazzling and powerful thrusts of the hooves. But when the males engage in intraspecific fighting, the large horns provide strong surfaces for spectacular but relatively harmless butting and pushing contests.

Another explanation for the development of elaborate head weapons in ungulates involves sexual selection. The bizarre and exaggerated horns could function in releasing mating behavior in females. No matter what the original function of the horns, if predation is low enough and if females came to identify males on the basis of their horns, the situation could easily get out of hand (Eibl-Eibesfeldt 1970). Larger and more elaborate horns could develop, which might be almost useless in predator defense, but which could increase fitness in the males by more effectively attracting females. (Remember, the effect of a releaser can be boosted as we saw in our discussion of supernormal releasers.)

Even in ritualized fighting, such as "pushing contests," however, fighting cannot always be considered a safe activity. There is always that dangerous moment when the weaker animal turns to escape. In the elephant, as we know, and also in the hippopotamus, eland, and pig (Ewer 1968), the victor may seize this instant to deliver a lethal blow. Thus, fighting in these species cannot be considered completely stylized. Even in the more harmless fighting, the loser may retreat unhurt only because of his haste, not because the victor is constrained from inflicting serious injury.

It shouldn't be surprising that even stylized fighting sometimes yields injuries. After all, it must test not only strength but determination and pugnacity, if it is to fulfill its biological function. The fight *cannot* be a good-natured, gentlemanly affair—a sporting event.

Figure 28.9 Two fighting bulls (*Oryx gazella beisa*) use their horns as weapons, but only the upper third portion. Later in the fight, head-to-head pushing may occur, but in such fights the horns are not directly employed.

Figure 28.10 The hornless females of the Nilgai antelope fight by butting against their opponent's flanks fifty percent of the time. However, horn-bearing males or females of other species rarely attack in such a manner.

The combatants are probably highly incensed and blind with fury in spite of the limiting motor patterns they are compelled to follow.

In general, then, those species in which fighting is highly stylized possess lethal weapons. But not all species possessing lethal weapons fight in a stylized fashion. This may be because their livelihood depends on their efficiency as killers (Ewer 1968). Thus they cannot develop any tendencies that would temper this efficiency. The fighting technique is, to a large extent, dependent on the nature of the weapons, and if the weapons are molded through natural selection, to be employed in killing prey, harmless fighting may not be possible. In these animals, threat and intimidation may become increasingly important.

### Social Influences on Fighting Behavior

Social influences on fighting behavior have been demonstrated in the laboratory, using mice and rats, although these species are not representative of most social animals. For example, they show little inhibition against serious biting. In fact, in almost every fight, someone gets bitten. Also, rats and mice are unusual in that there is often only one dominant and uninjured animal in any group and no status among the rest.

If a strange mouse is introduced to an established group, it may be inspected for hours, but eventually it will be attacked and killed. Rats and mice both seem to lack any inborn system of aggressive inhibitors or appeasement gestures. Scott (1970) maintained they also lack threat signals and tournament fighting, but these patterns have been observed by other researchers. Eibl-Eibesfeldt (1963) described Norway rats that threaten by approaching their opponents broadside, following this with boxing and wrestling in true tournament style until one animal falls over. If that animal doesn't retreat, serious biting may then result. Actually, the switch from tournament fighting to real fighting is not unusual among animals.

Laboratory experiments on social animals indicate that socialization may be important in fighting behavior. Generally, exposure to other animals decreases the likelihood of fighting. Animals that have been isolated after weaning tend to attack more often and retreat less often than "socialized" animals (Kahn 1951). It may be that socialized animals fight less often from the operation of *active inhibition* (reduced probability of fighting after being bitten in a fight) (Scott 1966). The phenomenon can also be explained by *passive inhibition* (not fighting as a habit that forms by simply not fighting).

Fighting within a species may arise not only out of conflict between individuals but between groups as well. We have learned of ant colonies that attack strangers on the basis of their scent. Rats and mice may show the same sort of social intolerance. Groups of white-footed mice are peaceable as long as they have grown up together, but they kill members of other groups (Southwick 1970). One group may also engage another in combat. Rhesus monkeys that come abruptly upon another group may engage in fierce intergroup fighting. Troops of baboons may also attack other troops. Howler monkeys show intergroup hostility, but avoid injury by merely howling and shrieking in an intimidating manner.

## ALLELOCHEMICS

Whittaker and Feeny (1971) reviewed the concept of *allelochemics* in a delightful paper. Let's consider a few of their findings that pertain to competition.

Attack and defense may involve not only physical force but chemical agents as well. Earlier, we noted the effective chemical defense of the bombardier beetle. Chemical agents may operate between animals, between plants, or between plants and animals. Allelochemistry involves the study of the chemical interactions by which one species affects the growth, health, behavior, or population biology of another species.

The growth and distribution of a number of higher plants can be altered by chemicals released from other higher plants, a phenomenon called *allelopathy*. One type of shrub, the hard chaparral, was studied by C. H. Muller in the hills around Santa Barbara, California. The rain that falls on the plant washes its poisons into the surrounding soil. These poisons, mostly phenolic compounds, inhibit the germination and growth of herb seeds present in the soil. The effectiveness of the inhibitors can be seen after a fire because the destruction of the chaparral releases the herb seeds from inhibition. The ground then begins to bloom with a profusion of annual herbs for a few years until the regrowth of the shrubs subjects the herbs to new inhibition.

The similarities between allelochemics and fighting behavior in animals should be apparent. The shrubs do not compete with herbs by simply trying to catch more sunlight or acquire more water, but by directly excluding them. Of course, aggressiveness per se is not involved, unless one wishes to stretch the definition. However, the development of such a competition-reducing mechanism may have occurred on the same adaptive basis as exclusion-through-fighting in animals.

Just as fighting in animals has its undesirable consequences, so does the formation of inhibitory chemicals by plants. (The term *dangerous* implies only a potential—as Gandalf described his own dangerousness in Tolkien's *Lord of the Rings*.) The chemicals, after all, are poisons and thus can harm their makers. They may be especially poisonous to certain "successional" producers (Odum 1971). The chemicals are produced, though, because successional species are vagabond populations that dominate a community for only a short time at any given place, and then they are gone anyway; thus, permeating an area with a poison to which competitors might have lower tolerance than the producer could have immediate, short-term advantages.

Allelopathic compounds may also be rendered more harmless to the producer by being chemically bound in the producer-plant so that their effect is diminished until they are released into the environment. The maintenance of such compounds and the development of safety precautions place a heavy metabolic load on that producer. Substances, then, that the producer could otherwise use for growth and repair are diverted into the manufacture of the poison systems. The development of the system in the face of such strong disadvantages is testimony to its high adaptability.

## PLANT DEFENSE AGAINST ANIMALS

The curious naturalist may wonder why green plants continue to dominate the earth's surface in the face of so many animal consumers. One reason, as we have seen, may be because herbivore numbers are not limited by their food, but by predators, and, therefore, they do not increase until their food is consumed. Another reason may be because adult herbage is simply uneconomical for most herbivores to consume, as Gordon Orians and his group at the University of Washington have said. And then some plants are able to repel herbivorous animals mechanically, as with thorns, or chemically, through allelochemicals.

I had the opportunity to look at several different plant habitats in Spain. On Ibiza, a small island in the Mediterranean, there were two areas between my house and the ocean, two hundred meters away. One was grassy and the other sparsely inhabited by plants, mostly low-lying shrubs. It seems the plants in the two areas protected themselves from herbivores in two different ways. The grass was highly synchronous in its reproductive efforts. Seeds were produced and dispersed in a very brief period of time. One week the seeds were there, the next week they were virtually all gone. A seedeater would probably not have been able to successfully exploit a crop that appeared in such abundance all at once. Two weeks later, a new crop of grass seeds would appear that would only last for a brief period.

In the shrubby area, I made two transects, and of fourteen species of plants growing there, eleven were aromatic (and hence inedible to most herbivores). Thus, the shrubs went untouched by grazing sheep. Later, high in the Sierra Nevadas of southern Spain, I made three transects and found two species of grasses (plus a few low-lying conifers and some mosses). The area was grazed by domestic sheep. Here the plants might have had a different "strategy." The grasses were clumped randomly several feet from each other, thus producing a patchy environment to any grazer. As a sheep moved from one clump to the nearest clump, it would have wended an erratic path, so more and more clumps would have escaped untouched; thus, random distribution of clumps would have its protective effect.

All these plant defenses stood in marked contrast to what I observed in the fertile hills of the Alhambra at Granada, which resembled eastern hardwood forests. Here, the plants were profuse and lush, seemingly "unprotected." It may be that their very abundance, due to favorable conditions of growth, did not call for a protective device. Their existence was apparently unthreatened by herbivores. Perhaps in such cases the herbivores are overwhelmed by the mass of plant food, and perhaps they never approach the "food barrier" because they are held in check by

Figure 28.11 Two plants that protect themselves by producing toxins against herbivores are the foxglove (left) and oleander (right).

some other factor, such as predators. It may be, then, that protective devices, such as plant poisons, are only necessary in those areas where plants must struggle to survive and where even limited herbivore pressure could not be tolerated. All this is offered simply as suggestions for further research.

Plants may produce toxins that act somewhat directly, such as by causing convulsions and death in the animal ingesting them. Thus, the storybook picture of a cow contentedly chewing a buttercup is misleading, since buttercups (*Ranunculus* spp.) are distasteful and can kill cattle; cattle on western ranges also frequently fall victim to larkspurs (*Delphimium* spp.). Steroids in foxglove (*Digitalis purpurea*) can cause heart attacks in vertebrates, and a single leaf of oleander (*Nerium oleander*) may be lethal to man (Figure 28.11).

The actions of plant toxins are sometimes less direct, perhaps even subtle and devious. Tannins make proteins in plants indigestible for vertebrates, but they also inhibit growth in other plants and interfere with fungus and virus transmission, thus providing certain plants with a broad-spectrum defense against both animal and microbial enemies.

Several plant species, mostly gymnosperms, even contain important insect hormones that function in molting. If the insect eats too much of these plants, its metamorphosis is fatally accelerated. Some plants also produce a juvenile hormone. For example, a chemical called the "paper factor," present in these very pages, has been isolated from the balsam fir. Applied to bugs of the family Pyrrhocoridae, the substance stops development at an immature stage and prevents the formation of a normal adult.

The forces of mutation and natural selection are operating in animals too, of course, so we can expect chemical defenses of plants, in some cases, to be obviated. Hypericin is a red secretion of plants of the genus *Hypericum.* Hypericin somehow causes extreme sensitivity to light, skin irritation, and blindness in animals that eat these plants. Some beetles, though, have evolved a mechanism for detoxifying hypericin and thus they have a food supply almost untouched by other herbivores. At least one beetle species even uses the repellent as a cue to locate the plant!

The field of chemical ecology is rapidly attracting the interest of increasing numbers of researchers. The potential subtleties and possible interactions made possible by the specific nature of chemicals promise to pose intriguing and complex questions.

## Coevolution: When Trees Get After Squirrels

*Coevolution* refers to the joint evolution of two or more kinds of organisms that have close ecological relationships, but do not exchange genes, and in which reciprocal selection operates to make either group partially dependent upon the evolution of the other (Ehrlich and Raven 1964). At present, the term *coevolution* is often used to refer to the interdependency of evolutionary actions between plants and herbivorous animals. We just referred to one example of coevolution—when plants manufacture toxins to ward off herbivores, and certain herbivores then develop a tolerance, or even affinity, for the toxin. Earlier, we considered *Drosophila pachea,* which is the only species of fruit fly able to live on the alkaloid-producing senita cacti.

In some cases, plants defend themselves against herbivores by bringing in hired guns. Some species of acacia are highly palatable to plant-eaters. But the plants produce nectar-forming areas and swollen thorns that attract certain ants. The ants swarm all over the acacia leaves and attack any herbivore that makes the mistake of trying to eat the plant (Janzen 1971b).

One of the more interesting examples of such coevolution comes from the relationships between pine squirrels (*Tamiasciurus*) and their coniferous food trees in the Pacific northwest (Smith 1970). These squirrels can literally strip a pine tree of most of its cones. The trees, in turn, have developed a number of protective devices such as:

*(1)* Producing cones that are difficult for squirrels to get at, open, and/or carry

*(2)* Putting fewer seeds into each cone, thus causing the squirrels to work for less

*(3)* Increasing the thickness of seed coats, thus requiring the squirrels to work harder

*(4)* Putting less energy/nutrients into each seed (reducing the value of the seed)

*(5)* Shedding seeds from the cones before that year's crop of young squirrels begins to forage on its own

*(6)* Periodically producing cone "failures" that decimate the squirrel population

Fluctuations in cone crops from year to year may be pronounced and the trees may produce few cones even in an optimal growth year. The trees are thus able to withhold their materials for seed production until another year, while the squirrels starve. Also, cone failures are often synchronized among different tree species in a given area, evidence that their behavior is directed at herbivores. Different conifers have *diverged* from each other in cone anatomy, size, location, and time of seed shedding, and this makes it hard for the squirrel, which is forced to become a jack-of-all trades. At the same time, the trees have *converged* in their fluctuations of cone crops, preventing food switching in the squirrels. The squirrels, as we see, then, have been forced to become "cone generalists," to choose from among various sorts of cones carefully and to stockpile them so that the pines are less successful at starving them out.

As we have seen in this chapter, animal aggression may arise from stimulation of certain neural centers and, in such cases, is not dependent on the background or values of the individual. The individual simply responds to the stimulation. We know that a number of factors may act as neural stimuli, from pressure, to electricity, to chemicals. Can one argue, then, that aggressiveness is a response to some sort of stimulation of specific brain centers? Can one argue that aggression in the sexes is different either because of differences in the brains or in the kind of stimuli the brain is exposed to (that is, blood chemicals)? Obviously, the questions are complex, but the prevalence of aggression in human society indicates not only a high historical adaptiveness, but demands increasing attention to the mechanisms of its development.

# 29

# ANIMAL SOCIETIES

We have all encountered intriguing tales of animal societies. Many of these stories are patently fiction, but have turned out to be based in fact. The evidence, however, is not enough to convince all the doubters. For example, who could be blamed for refusing to believe that some ants sew leaves together by passing a silk-spinning larva back and forth between them as a shuttle (Figure 29.1)?

Other "unlikely" stories involve one species "tending" another one. Do some ants keep fungus gardens? Do others tend "cattle"? And what about those fascinating stories of one animal risking—or even sacrificing—its life for another? We grew up with stories of valiant dogs, Lassies, given to racing onrushing trains to save baby chickens. Although the Lassie stories may be dismissed out of hand, other anecdotes persist, sometimes cross-culturally, and sometimes even the most hard-nosed skeptics have their unsettling doubts. Do porpoises really sometimes push men through shark-infested waters to the safety of shore?

Before going on, I should make one point. It may seem that after discussing competition and aggression among animals, we are now considering the other end of the behavioral spectrum—that we have gone, in a sense, from "hate" to "love." However, it will become apparent that we have done no such thing. The type of behavior we will consider now is simply another way of *increasing individual fitness.* "Cooperation . . . seems often to be a subterfuge whereby an individual is enabled to gain or maintain that degree of social control of others at which his or her own behavior is relatively unconstrained" (Crook 1970a). As cynical as it sounds, social cooperation may simply be a way of advancing one's self at the expense of one's fellows.

Although the end results of competition and cooperation may be the same, it should be made clear that the kind of behavior that spells success in competitive encounters does not necessarily operate in the development of cooperation. For example, it has been pointed out that cooperative tendencies may be *complicated* by the "problem of the male" (Crook 1970a). His more intense aggressiveness, so adaptive in competitive encounters, could easily be a disruptive influence on the harmony of an animal group, resulting in unsuccessful bearing or rearing of offspring.

Since probably no area of behavioral studies is so rife with anecdote and misinterpretation as that relating to social behavior, cooperation, and self-sacrifice among animals, we should review a few general problems in interpreting animal behavior. First of all, the occurrence of a pattern does not necessarily mean that the pattern is adaptive. Rather, it may be a not-too-detrimental side effect of another behavioral characteristic that *is* adaptive. For example, we have learned that the highly adaptive aggressive behavior of some male animals may sometimes be turned toward prospective mates, thus reducing its overall advantage; but still, in sum, it is adaptive. It is therefore naive to assume that a behavioral pattern is adaptive, even if it appears repeatedly. Also, when a behavioral pattern of any animal directly benefits another individual, even of a different species, the advantage bestowed may be inadvertent. A bird taking flight to avoid a predator may incidentally alarm other species in the area, or a parent's alarm

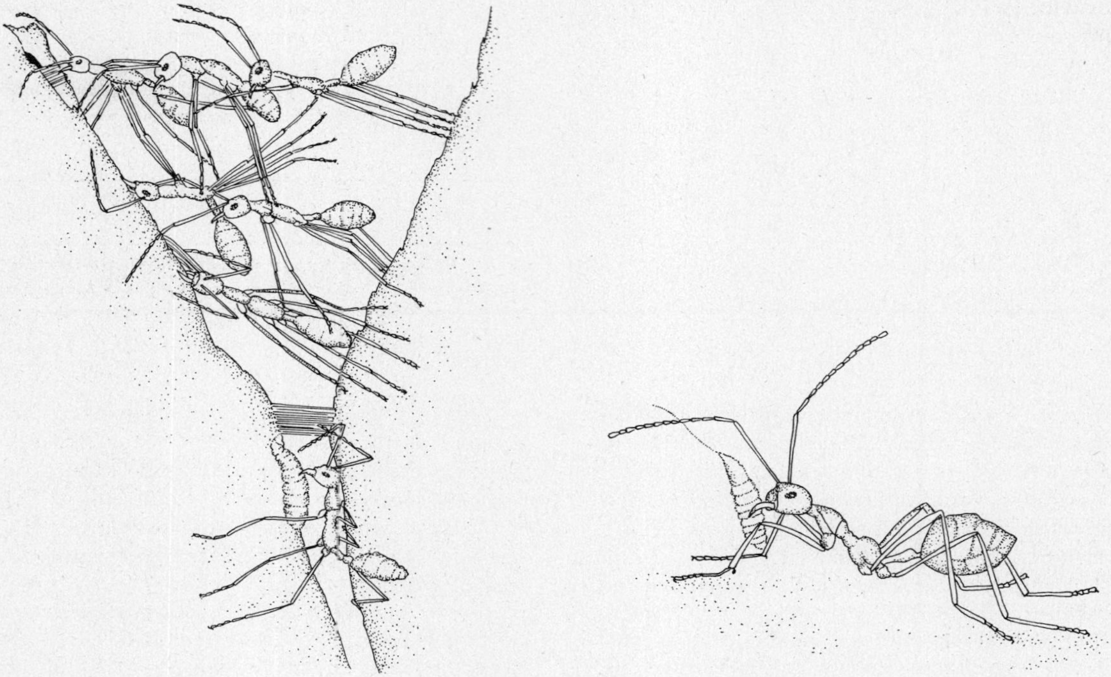

Figure 29.1 An example of highly coordinated cooperative behavior in the weaver ant (*Oecophylla longinoda*). A group of worker ants pull the two leaves close so that others may sew them together. The sewing is accomplished by pressing the spinning glands of larvae against the edges and passing them back and forth.

call to its offspring may cause a reaction in other nearby animals that are also in danger. Since a number of factors might account for any observed pattern, the most apparent explanation may not be valid.

It seems that the search for adaptiveness of any behavior is simplified if the animal being considered is one with relatively stereotyped behavior. In the case of mammals, a pattern may have developed through learning and may even be, to one degree or another, a function of insight. If the insight of the man and the animal are at variance, the problems of explaining a pattern are increased. There are those who say that porpoises push *anything* afloat ashore, perhaps to clear their water of obstructions, and that men are simply considered flotsam. Others, in explaining the recurring tales of such assistance believe that, if it indeed occurs, it may be a conscious effort on the part of the intelligent porpoise to help. It *is* known, for example, that they will help a wounded comrade reach the surface so it can breathe (Figure 29.2).

## GROUPING BEHAVIOR

Some animals are usually found in groups—groups that may be composed of one or many species. So, we may ask the deceptively simple question, "Why do animals come together?" The answers are varied. Animals may be brought together passively by physical factors, such as the movement of wind or water. Herpetologist Chaim Kropach tells me the great rifts of deadly sea snakes that have been reported in drift lines in the Pacific Ocean are probably due to current convergence that brings these surface swimmers together.

Also, animals may aggregate by some form of taxis, or simply as a result of a common search for favorable conditions. In other words, a number of animals may be attracted to the same place, but not necessarily to each other—as insects are attracted to a light. The different scents of various species of pine trees function in attracting the parasitic fly *Drino bohemica* to particular species of larvae upon which they lay their eggs (Monteith 1958). Often, specific chemical stimuli attract parasites to their hosts. The odor of humans attracts the louse *Pediculus* (Wigglesworth 1953). Carbon dioxide, an animal byproduct, and other factors, attract female mosquitoes to areas where animals may be (Laarman 1960). Animals may also come together for hibernation, breeding, or migration.

Basically, we will concern ourselves in this chapter with social groups that are formed as a result of individuals being attracted to each other. Animals may be classified as *solitary* (such as the mountain lion, a species in which individuals live alone except when mating or caring for offspring) or *gregarious.* Generally, the term gregarious includes all nonhostile, nonpredatory reactions between two or more individuals, excluding sexual or parent-offspring interactions (Eibl-Eibesfeldt 1970). (See Moynihan's 1960 paper for an extended bibliography on early work relating to the subject.) As an aid to background reading, perhaps it should be pointed out here that to people who study insects, "social" usually refers to animals that live in communities in which there is more than one female, one or more being sterile, that care for the young. This definition does not apply to vertebrate social organization, as we shall see.

Actually, gregariousness does not satisfactorily indicate the *level of social interaction.* For some purposes of description, perhaps a "social index" should be employed. Such an index might reflect the relationship between group size and the frequency of the dyad (two-animal) interaction in the sense of Lorenz's classical descriptions. For example, three tomcats living in the same neighborhood might have a higher social index than would an anonymous flock of a thousand starlings, but a lower figure than a dominant clique of three highly interactive baboons.

Among gregarious or social animals we find *distance animals* such as starlings, which appear in great flocks but maintain discrete individual distances, and *contact animals.* Contact animals are represented by the goldcrest (*Regulus regulus*), a kinglet of Europe and Asia, which keeps together as a family even after the young are fledged, the little birds huddling through the cold nights. Contact animals even show a definite appetitive behavior, or a "drive" (Eibl-Eibesfeldt 1970) for contact. Chimpanzees, gorillas, and other primates in zoos need to be touched and scratched by their keepers if the relationship is to remain on good terms. One may observe primates, including humans, sitting with their arms around each other; and among apes, higher ranking animals reassure subordinates by touching or clasping them (Figure 29.3).

Many contact animals *groom* each other. The pattern is common in Norway rats, agoutis, house mice, and many primates such as vervet monkeys, baboons, and chimpanzees. The effect of such behavior is twofold. It frees the "groomee" of parasites, hardened skin secretions, and debris in the skin, and it enables the animals to touch and be touched. The touching itself confers some degree of satisfaction to the participants and may facilitate group ties via the reinforcement mechanism (Eibl-Eibesfeldt 1970). Jane Goodall noted that wild chimpanzee mothers

Figure 29.2 An injured porpoise may be carried to the surface for air by its comrades.

Figure 29.3 Among contact animals, such as the chimpanzee, touching may be a form of reassurance.

are almost always in contact with their very young offspring and frequently touch them as the young gain their independence, possibly reassuring both mother and infant.

Harry F. Harlow (1958), working at the University of Wisconsin, described infant monkeys' love for their mothers being derived, not from the satisfaction of feeding, but from close bodily contact. Anthropologists have described very primitive tribes, such as the recently discovered stone-age forest men of the Philippines, who almost constantly hold or touch their small children. In these tribes, the young often radiate an unusual degree of self-assurance.

In any group of animals that interacts socially, there is the problem of coordinating the activities of the group. The problem may actually be considered an aspect of the greater area of communication. Coordination through communication could take place by several means: (1) through social feeding or trophallaxis (see Tavolga 1970)—as found, for example, in insects that transfer hormone-containing food between them in a form of chemical communication; (2) social facilitation—the activity of one animal increasing the probability of another animal performing the same act; (3) tradition learning; and (4) through a chain of command or hierarchy (Dimond 1970). The observer must keep in mind that animals may react similarly but independently to common environmental stimuli, such as stampeding at the appearance of a predator, and thus could give the semblance of coordinated behavior.

It will become apparent as we continue that there are incredibly complex and highly coordinated social systems in animals ranging from insects to primates. In fact, in a sense, we are only now beginning to see the extent of the ecological importance of social behavior and what we see has caused us to reconsider certain ideas and to redefine many of our terms. The field of *sociobiology* has been best and most extensively defined in E. O. Wilson's book of the same name.

Before embarking on the study of higher animals, especially birds and mammals, we should first take a look at the astounding complexity of social systems in the hymenopteran insects and termites.

## INSECT SOCIETIES

Some species of ants, bees, and termites show extremely complex and coordinated social patterns. Regimentation is exceedingly precise—enhanced by a caste system in which each individual performs a carefully programmed series of roles in which there is little room for variation.

Although behavioral variation within a caste is usually minimized, there are exceptions to the rule. For example, in winter, some species of bees dig a hibernation chamber over a yard deep where the temperature is about 10°C, amassing themselves into a great ball to conserve heat. The workers show great variation in their resistance to low temperatures. When the temperature in the hive reaches about 14°C, they form their winter cluster, a huge ball, with the queen in the center. The bees rotate their positions so that those at the outside can move in and have a chance to warm up. Those most resistant to cold wander between the surface of the ground and the chamber all winter. If the day is relatively warm, they crawl outside, catch a few rays, and go back down with the extra calories they have absorbed to add to the average heat of the chamber. As spring advances, other members of the group are activated by the heat and these, too, join in the "heat transport" parade, until the entire population is activated (Chauvin 1963). While such a system could best operate when there is some physiological and behavioral variation, generally, any single caste within an insect society is probably rather homogeneous for most traits.

### Roles Among Honeybees

Among the most intensively studied of the social insects is the honeybee (*Apis mellifera*). In this group, the queens lay eggs, but other duties are performed by the workers, which are sterile females. As a worker matures, she changes jobs. Newly emerged workers prepare cells to receive eggs and food; after a day or so, their brood food glands develop and they begin to feed larvae. After their nursing duty, they begin to accept nectar from the field workers and to pack pollen loads into cells. About this time, their wax glands develop and they also begin to build combs. Toward the end of their house duty, some become guards and continually patrol the hive. Lastly, they become foragers, as their brood food glands atrophy; they then fly afield and begin to collect nectar, pollen, propolis, or water according to the immediate needs of the hive. Those needs may be communicated to the field workers by the "eagerness" with which different loads are accepted.

Under abnormal circumstances, such as the deaths of a large number of bees with a particular duty, the normal sequence of duties can be altered. Young bees may shorten or omit house duties and quickly go to foraging, or foragers may even redevelop their brood food and wax glands and return to house duties.

In the hive, the emphasis is on efficiency. In the more feminist species, after the queen has been fertilized, the rest of the drones are quickly killed by the workers. Drones are treated as sex objects, useful only in copulation. The females live, but they live only to work. They tend the queen, rear the young, and maintain and defend the hive. And they literally wear themselves out in their service. When their wings are so torn and battered they can no longer fly, they die or are killed. But the hive goes on.

Immediately, there may come to mind questions regarding the concept of fitness. Since fitness is related to reproductive success, how have these devoted but sterile workers been selected for? In spite of the fact that they cannot pass along their *own* genes, they can see to it that their *kinds* of genes appear in the next generation. And that, of course, is what reproduction is all about. The workers, of course, must continue to care for the egg-laying machine that is the queen, which is their sole hope for continuing their own genotype. The problem is discussed in more detail in the section on altruism, but for fuller descriptions of bee societies, see Wilson (1971a); Michener (1974); Alexander (1974).

### Shepherds and Farmers

Ants provide us with fascinating examples of social behavior. Some ants, it turns out, are shepherds, while others are farmers. Certain ants of the genera *Formica* and *Larius* care for aphids, or plant lice, in the true sense of husbandry (Figure 29.4). The aphids are kept, not for their flesh, but for their "milk"—honeydew, a secretion high in sugar and amino acids. The honeydew is released when the ants stroke the aphids with their antennae. The ants tend to their herds in a variety of ways, such as by driving away predators and building sheds of earth, "paper," or silk for their charges. Also, they may rush them to safety if danger threatens. In the fall, they may gather aphid eggs and keep them safe through the winter. In spring, the young aphids are carried out and placed back on the stems or roots of plants to feed (Trager 1970).

The attine ants are the farmers. Their crops are fungi, grown underground in their dark nests. For these ants, fungus serves as a primary, if not sole, source of food. The substrate (ground) for the fungus depends on the genus of the farmer. The "simpler" genera use insect feces and carcasses. The "intermediate" genera use plant debris and fragments cut from live plants. The most "advanced" genera culture their fungus almost entirely on bits of leaves and flowers cut from large plants. The fungus of the ants is a poor competitor and doesn't do very well on its own. It is easily overcome by wild fungi unless tended by the ants. The ants use their own feces as a fertilizer, and it has been found to contain proteolytic enzymes that compensate for those lacking in the fungus. Also, the nitrogenous components of the feces act as a true nitrogen fertilizer, facilitating the initial growth of the fungus.

Figure 29.4 Ants tending their "herd" of plant-eating insects. The adults of the herd secrete droplets of fluid, which the ants consume.

In the preceding two examples, both the ants and their symbiotes benefited from the relationship and therefore the association may be labeled mutualistic. In the case of the ant-fungus relationship, however, we can see that the definition may be somewhat strained, since the fungus is actually eaten. However, without the ant the fungus apparently would not survive at all.

E. O. Wilson (1971a) emphasized the functional similarity between insects and vertebrate societies. He noted that both may occupy territories and communicate hunger, alarm, hostility, caste, status or rank, and reproductive condition by nonsyntactical signals. Also, he pointed out that in both groups individuals are aware of the distinction between group members and nonmembers; that kinship is important in group structure; and that it is probably the ultimate force behind sociality. We are reminded that certain groups, among both insects and vertebrates, share the advantages of division of labor and that both have been subject to evolutionary forces where fitness is increased through cooperative tendencies—at least between relatives. In spite of these similarities, however, there are important differences that do not lend themselves to easy generalizing from one group to the other, as we shall see.

## VERTEBRATE SOCIETIES

Most of the vertebrate social research thus far has been related to birds and mammals, but for discussions of fish, see Barlow (1974) and Myrberg and Thresher (1974); for amphibia, Blair (1963) and Lanyon and Tavolga (1960); and for reptiles, Brattstrom (1974). See also Wallace and Klopfer (1979).

Cooperative behavior may occur in a number of ecological contexts, but especially in relation to food storage, territorial defense, predator defense, group hunting or group foraging, parental care and kin care, and, at its most complex, the cooperative control of social interaction (such as we see in the dominance cliques of some primates). The advantages of sociality will be considered primarily in terms of birds and mammals because of the data available and because of the apparent ease of deducing the adaptiveness of their behavior. The discussion is intended to give some idea of the variety of adaptive effects of sociality in vertebrates. However, extrapolation of the adaptiveness of any pattern from one species to another may not be warranted, since social patterns so easily reflect other facets of the animals' histories.

## Protection

It is axiomatic that there is safety in numbers. The "safety" element may be due to the superior fighting strength of large groups, the "confusion effect" of large numbers of prey animals on predators, and the increased likelihood of another individual being between the predator and any potential prey animal. The protective benefits of groups are high and, in some cases, it is even dangerous for animals to live in groups smaller than some critical size (Collias 1944b). For example, individuals in guanay colonies below a certain size do not reproduce. It has been suggested that in the smaller colonies the ratio of circumference to area is large and, therefore, the likelihood of predation is increased due to a greater proportion of individuals being exposed at the periphery (Hutchinson and Dewey 1949). Group size may also be important in ways other than as a passive factor. For example, in some smaller ungulate species, herds composed of fewer than fifteen individuals are subject to decimation by wolves, since they are not numerous enough to fight off the predators (Leopold 1933).

The importance of the group is clearly demonstrated in those cases in which animals that find themselves temporarily isolated from their groups are unable to perform behavioral sequences in an adaptive manner. For example, if a single bird in a group takes flight, he provides strong stimulus for others to follow, and a few almost invariably will. But if a spice finch (*Lonchura punctulata*) leaves the flock in order to feed or drink, and no other birds follow, it rushes its activities, performs them in a disorganized manner, or abandons them altogether and returns to the group (Moynihan and Hall 1953).

Although some groups such as flocks of birds increase the likelihood of detecting, mobbing, confusing, or distracting predators (Tinbergen 1951; Kruuk 1964), they may also attract predators. Visual and auditory signals are necessary for maintaining flock cohesion (Moynihan 1962b) and therefore may make the group conspicuous to predators. The evolutionary aspects of the resulting compromise are discussed in the section on social mimicry in Chapter 30. So, paradoxically, animals may, in a sense, show cooperation by spacing out, thereby making themselves inconspicuous to predators; they may also band together for mutual protection.

One advantage to grouping is encompassed in the *you-first principle* (Hamilton 1971). Animals may group together for protection, but the protection may lie in an increased likelihood of another animal being taken first. Imagine a pond that harbors one snake and a number of frogs. Suppose that the frogs sit around the edge of the water and that once each day the snake appears at some random point around the pond and eats the nearest frog. The frogs obviously have a problem—how *not* to be the one. It works out so that the risk of any frog's number being up on a given day is equal to the proportion of the circumference of the circle that can be drawn by a line running from a point halfway between that frog and his neighbor on the right to a point halfway between him and his neighbor on the left. So, to be on the safe side, the frog should minimize the distance between himself and his neighbors. Thus, by moving between the two frogs that are sitting closest to each other, it may appear that he is demonstrating his love for his neighbors, but he is actually increasing the likelihood of one of them being taken by the snake instead of himself. The parable points out once again that often what appears to be "social" behavior may be, in the final analysis, self-serving. The "you-first" strategy is, of course, a resort to statistics. (But much of what we know about areas of animal behavior—such as communication—can be boiled down to statistics—information about the *probability* of something happening.) Prey animals, by the way, are most likely to resort to probabilistic defense when the predator is faster or more agile than themselves. In such cases, the predator is likely to catch *someone* in the crowd—so better someone else.

**Levels of Protective Cooperation** The protection conferred by grouping may be rather "impersonal," such as when a flock of starlings closes ranks in the presence of a hawk (Tinbergen 1951), or perhaps when coots attacked by an eagle splash up the water with their wings and escape in the confusion they manufactured themselves. A higher level of cooperation is shown in defense of a group territory (King 1955; Southwick, Beg, and Siddiqi 1965) where a group attacks a threatening interloper, and the protective advantage is derived from what is apparently sheer multiples of individual territorial behavior—each animal protecting its own interest in the territory. In other cases, the effort may be much more coordinated, such as we see in musk oxen that form a circle of outwardly facing adults, with their massive heads lowered to meet the danger. Such behavior is effective in protecting the less powerful females and young inside the circle from wolves. (Unfortunately, the hapless beasts tend to remain in this defensive position as they are felled one by one by the rifles of men.)

Probably, one of the most advanced forms of cooperative protection is found in animals that actually try to rescue any members of the group in distress. For example, Lorenz (1935) found that his tame jackdaws would attack anyone holding another jackdaw, or even a pair of black bathing trunks, in his hand. Rhesus monkeys in a zoo will attack their caretaker if he catches one of them (Eibl-Eibesfeldt 1970). Porpoises circle a female that is giving birth and protect her against any sharks attracted by her labor (Siebenaler and Caldwell 1956). It would seem, on the surface, that such rescue behavior involves at least a measure of insight.

**Lek Species** Although we are considering the adaptiveness of group behavior for widely different groups of animals, there is no intention to imply that they all benefit in the same ways. For example, the rescue behavior of the jackdaws and the porpoises may both show insight, but the advantages in attacking the threatening individual may be quite different. And the neural mechanisms involved may have little in common. The problems of synthesizing unitary rules are enormous.

When grouping has more than one advantage, it may be difficult to tell which one developed first and which were added secondarily. For example, in some birds such as grouse, males display on small territories or arenas (leks) on grassland, well within sight of each other, and the females choose from among them (Wiley 1978) (Figure 29.5). Lack (1968) figured that leks have evolved separately nine times in birds, a theory that is supported by the marked diversity among the various lek species. For example, males of some species display silently, while others call or drum, and others rely on flashy colors and patterns. Although the leks must be conspicuous to attract females, risk of predation is reduced by the males taking certain precautions. For example, they may display early in the morning or late afternoon, when the cooled earth does not furnish upward drafts that could carry predatory birds. Also, most leks are on flat ground where visibility is good and mammals cannot approach undetected. Some even scratch fallen leaves from the area, removing cover for snakes, while others display safely in the tops of trees (Lack 1968). Also, leks are formed each year on traditional sites that have proved safe in the past.

Leks may offer protection by accelerating and coordinating breeding. The cumulative effect of a group of displaying males might release reproductive behavior in females more quickly than would the efforts of a single male. This would be particularly important where the breeding season is abbreviated, but also the shorter the time spent on the lek, the fewer opportunities exist for attracting prowling predators. In fact, some researchers (Koivisto 1965) believe predator defense was the evolutionary basis of lek formation and that the reproductive advantage appeared secondarily.

Actually, there are many advantages of leks in terms of predator defense. For example, if predators are drawn to leks and then detected by any of the numerous watching eyes, at least the birds then know the whereabouts of the local predators, even if it is at the expense of a temporary disruption of the mating process. Also, the sight of so many dancing birds may overexcite a predator, causing him to rush them prematurely, thereby reducing his general efficiency.

It is fortunate, and not entirely coincidental, that most of what we know about the social behavior of vertebrates comes primarily from studies of two groups. On the one hand are birds—animals with relatively stereotyped behavior, probably related to exceedingly powerful selection operating through specialized food supplies, high predation, and stringent habitats. On the other hand are the primates—intelligent and exploratory animals that inhabit tropical areas of relatively stable food supply and that have the facility for exceedingly complex, subtle, and malleable social interactions.

**Primates** A troop of hamadryas baboons (*Papio hamadryas*) of Ethiopa usually consists of only one adult male travelling with several females and adolescents of both sexes. The group makeup suggests a kinship breakdown of a male, his consorts, and their young. During the day they feed together, but remain apart from other groups. At night, the groups may converge to sleep on steep and relatively safe cliffsides. The slumber parties may number several hundred individuals. There is no experimental evidence with which to test any hypothesis, but it has been suggested that the pattern of daytime dispersion and nighttime congregating may serve to spread out the population during the day to best utilize certain types of food distribution, to ensure regular mating (females are not allowed to stray from the group, and wanderers are retrieved by the dominant male), and to best utilize the protection offered by the male. Baboons are prey for lions and leopards, but the cats usually go for isolated individuals and, even then, they are often unwilling to confront a formidable full-grown male (Figure 29.6). On the other hand, at night the rigorous duties of the male can be laid aside, since the group now has the protection of numbers and an inaccessible resting place (Barnett 1971).

Most baboon species live in groups larger than the family unit just described (Figure 29.7). When a baboon spots a predator, it gives a loud bark and the troop closes ranks. The alarm is usually sounded by a young male that "walks point," ahead of the rest. Chauvin (1963) reports that females may also give the alarm at times, but that the real sentinels are the big males that sit scanning the countryside as the troop passes, taking their job so seriously that they even ignore estrous females at such times. The quick barks, which are the alarm, cause quite a hubbub within the troop as everyone prepares to flee or give chase. If the troop attacks, the pattern is carried out in a rather well-coordinated manner.

Figure 29.5 Males of the ruff (*Philomachus pugnax*) displaying on their lek (female is at right). These birds are unusual in that the males are highly variable in plumage.

Figure 29.6 A male baboon displaying his large canine teeth as a threat signal. These males are extremely powerful and dangerous.

Not only does the social structure of some primates protect them from predators, but in some cases from each other. For example, among gorillas the large and formidable silver-backed males rarely fight, but if two females begin a conflict and scream and wrestle, a male has only to draw near and grunt to break it up (Barnett 1971). Thus, the stability of the social system is maintained and the females are not allowed to hurt each other.

And then in some species we find dominance cliques—small groups at the top of the hierarchy in which each group member aids other group members against outsiders. The cliques are difficult to disrupt and such power at the top helps maintain the general stability of the troop structure. In sheep (Geist 1966a) and baboons (Hall and DeVore 1965), two or three males may join forces to occupy the top of the hierarchy. A challenge to any member of the clique is met by a unified front of all the members. Dominance cliques are usually thought of as a means of defense against other group members, but cliques may repel nongroup members in some cases. There are reports of lion prides in which several adult males join forces to maintain the integrity of their area against other lions (Schaller and Lowther 1969) (although some researchers note that lion prides usually have only one adult male [Ewer 1968]). In all cases in which dominance cliques exist, however, as "authority" increases, the social system can be considered to be more stabilized and secure.

**Reproduction and Social Units**

In many species, the density of the reproducing segment of the population is exceedingly critical. That density can be influenced by the social structure of the species; it can also be influenced by more passive factors that simply affect aggregation.

The reproductive advantage of aggregation in some simpler animals may be as uncomplicated as merely increasing the likelihood of a chance union of male and female gametes. In sea urchins, for example, sperm released into the sea may not reach sparsely scattered eggs, so the fewer reproductive individuals there are in any area, the more remote is the possibility of fertilization occurring.

In other animals, the reproductive advantage of grouping may be physiological. Allee, probably the person most responsible for recent interest in animal sociality, cited some important pilot studies showing that even a lowly protozoan will reproduce (through dividing itself) more frequently when it is in the company of others of its species than when it is alone. The same effect has been noticed in vertebrates. For instance, a sow's ovaries do not develop normally if she is prevented from hearing and smelling a boar (Chauvin 1963). A pair of mice alone may be unable to breed successfully. Unless several breeders are present, a female may start a gestation and then reabsorb the growing fetus. The physiological mech-

Figure 29.7 A gelada baboon troop on the move. Most baboons live in large troops and when they stay together, they are relatively safe from predators.

anisms of such behavior are not entirely understood.

Although vertebrate breeding behavior is influenced by environmental factors such as daylength, it may also be strongly affected by the behavior of other pairs. The phenomenon is not very well understood. One type of possible influence has been labelled *social stimulation* or *facilitation* (see Chapter 11). Facilitation exists when the performance of an action by some member of a group increases the probability of others in the group performing the same act. For example, an ant that works at an intermediate rate will slow down if placed with a slower worker, and speed up if placed with a faster worker. Chickens in a group eat more food than those kept singly and even those that are not hungry can be induced to eat by the sight of their fellows eating (Harlow 1932). Also, Lorenz has noted that if a heron, after eating, strikes the typical well-fed and lethargic pose, some of his hungry companions will stop eating and strike the same pose.

Facilitation may be important in breeding behavior in that it would tend to cause animals to breed synchronously. Fraser Darling, in 1938, reported that larger colonies of seabirds showed earlier and more synchronous breeding than smaller ones and he attributed this phenomenon to the "contagion" of social stimuli that altered the birds in preparation for nesting. He assumed that any such social contagion would be stronger in large groups than smaller ones.

In fact, he proposed that the chief value of colonial nesting is mutual sexual stimulation. His ideas were criticized, and alternate explanations of his observations were proposed (Fisher 1954; Coulson and White 1960). Lack (1970) stated simply that there is no reason to suppose that social facilitation is operating, at least on the basis of Darling's observations, although he agreed that social facilitation in colonial species may well exist. Hailman (1964) delineated the possible ways the correlation of colony size and breeding synchrony could arise: (1) it could be due to a common reaction of all pairs to environmental cues; (2) it could be due to a result of normal behavioral responses, such as the approach of an individual in reproductive condition to conspecifics; (3) it could be due to increased recruitment of individuals to places occupied by birds in a similar reproductive condition; and (4) that Darling's "contagion" hypothesis was correct.

But what *are* the advantages of synchrony in reproduction? First, it is obviously advantageous for males to be in a reproductive state at the same time as females are. Second, synchrony in a colony would reduce the period that chicks and eggs would be available to attract predators. Synchrony, if regulated by physical environmental conditions to some degree, might also effectively permit the exploitation of a seasonally limited food supply.

In some cases, group display may accelerate the sexual development of a local population (Emlen and Lorenz 1942). In one experiment, testosterone was injected into two California quail (*Lophortyx californicus*), causing copulatory behavior to appear two days later. Strangely enough, eight untreated birds reared with this pair began to show the same behavior within two weeks—two months before the normal date. Crook (1965) stressed that synchrony would be especially important for birds with relatively limited breeding seasons. Synchrony through group behavior has been described in studies by Orians (1961); Ashmole (1963b); Crook (1964).

In studies of social effects on weaverbirds (Collias, Brandman et al. 1971; Collias, Victoria et al. 1971), it was found that the effects of increased crowding on breeding behavior was largely dependent on the sex ratio of the group. When the relative number of birds of one sex was experimentally increased, the result was an increased stimulation of members of the opposite sex rather than the same sex. When relative numbers of the sexes were not altered, increased crowding resulted in increased singing of males to each other as well as to females. There was no increase in clutch size with added numbers, but there was an extension of the breeding and nest-building seasons. The extension of these periods was considered to be favorable to the establishment of new colonies when conditions were crowded.

There are apparently advantages in nesting colonially that are not yet completely understood. Observations of nesting in the kittiwake (*Rissa tridactyla*) reveal interesting behavioral patterns, but add a few more unknowns to the Great Puzzle (Coulson 1968). In the kittiwake, there is more intense competition for nests in the center of the colony than at the periphery. Birds removed from the center will immediately be replaced—perhaps by a member of the nonbreeding population (comprised of those individuals previously unable to secure a site). Should a vacancy at the periphery of a colony occur, however, it may not be filled. There is less demand for these places. Annual mortality rates show higher death rates for males at the periphery, but not for females. The high death rate, of course, might not be due to any effect of the peripheral area per se, but to its inhabitation by inferior males—those unable to secure a central position—who might be expected to succumb more easily than the more vigorous, centrally located males. Stress, it is suggested, is apparently not greater at the edge, but is, in fact, greater in the center. It would follow that since the center is occupied by competitively successful males, the level of aggression would be higher there. Coulson also found physical differences in the males of the two areas. The males at the periphery were slightly smaller and, he believes, differed also from the central males in the quality and quantity of certain body tissues. Finally, he found a larger mean clutch size, higher hatching success, and more young fledged per pair in the center of the colony.

*Cooperative breeding* exists when more than one pair share in building a nest, or lay eggs in a single nest, or help to feed one brood (Lack 1968). In a number of species of perching birds (and one parrot), several pairs may build a large common nest together in which each pair has its own chamber. Since cooperative breeding occurs in several groups of not-too-distantly related species, we may search for the ecological variables that could select for such behavior. Lack (1968) pointed out that diet seems to have had no influence on the evolution of cooperative breeding. Crook (1965) surmised that communal behavior involving cooperation in reproductive activities in birds appears to have evolved under two kinds of conditions. The first is where crowding into protected areas has reduced territorial sizes. Communal nesting could begin by one bird "extending" the nest of another—as long as it didn't infringe upon the entrance of the first one. In such "extensions," agonistic behavior and territorialism around entrances are not lost. The second condition is where parent-young bonds extend into the adulthood of the offspring. The advantage here would be in kinship selection and an increased likelihood that any young would be in a position to breed later. In such cases, group territoriality may arise. It has been noted that a high proportion of cooperative breeders occurs in the warmer climes of Australia, and perhaps at higher latitudes migration might make it difficult for young to stay with their parents. Because of this, we well might look for cooperative breeding primarily in warmer regions (Lack 1968).

Virtually all animals, even cooperative ones, tend to concentrate their care, if they care at all, on their own offspring, not someone elses. Most female mammals, for example, are very selective regarding which young they will nurse, usually rejecting all but their own. This selectivity means that the energy of the parent goes into its own offspring—the advancement of its own kinds of genes. However, it has been suggested that if, in a group of mammals, the young are all about the same age, if they remain close together, and if feeding times are roughly synchronized, a more effective system would be for a lactating female to nurse any infant she encountered, since the female would not have to waste any time seeking out her own offspring. The system would work particularly well if the females only fed the young and did not have to defend them (Ewer 1968).

As a case in point, consider the Mexican guano bat *Tadarida brasiliensis* (Davis, Herreid, and Short, 1962). The young are born at about the same time, around mid-June, in the dark caves that serve as summer roosts. The females do not carry their young with them when they forage. Instead, all the baby bats are "hung" in a special nursery area on the roof of the cave. When the females return from the hunt, they all converge at the nursery where they are set upon by every youngster within reach. Each female has only one young, but she feeds the first two to reach her teats. Thus, the feeding is a very communal affair. Probably a few weaklings get pushed aside and die, but the system assures that the majority will be fed.

### Foraging and Hunting

The distribution, abundance, and periodicity of food will strongly influence the social and reproductive structure of any species, as we have seen. We know that in those richer savannahs, abundant food is able to support large numbers of baboons and macaques. The food level, then, permits the primates to form large groups, to avail themselves of the protective and reproductive advantages of grouping. The smaller troops of hamadryas baboons utilize what is probably a scattered or sparse food supply, one that is best exploited by smaller, wide-ranging groups.

Any casual observer has probably noticed that some birds that are territorial in spring may flock together in the winter. Such flocks may even be composed of a number of species, several kinds of finches being particularly good "joiners." Apparently, localized and patchily abundant food in the nonbreeding season is easier to find for groups, and competition is minimized when it is discovered. Flocks may be formed, then, not only as a means of efficiently utilizing food sources, but for reasons discussed previously, such as mutual surveillance for predators (Marler and Hamilton 1966).

Among birds, insectivores are particularly gregarious, as are those whose food usually occurs in large, irregularly scattered masses (for example, fish, carrion, or fruit) (Moynihan 1960). Although gregariousness in birds is more characteristic of the nonbreeding season, we know that many species also nest colonially. (In addition to seabirds, there are colonial herons, swifts, weavers, hirundines, and a few other species, such as some crows and woodpeckers [Lack 1970].) Food must be abundant near any colony, of course, but food does not seem to be the primary cause of nesting groups, except in a few species, such as the rosy pastor and wattled starling, which breed alongside locust swarms (Schenk 1934; Roberts 1940; Lack 1954).

There are also instances of animals not only associating because of food factors but actually working together to more efficiently exploit that food. For example, acorn woodpeckers establish stores (Figure 29.8) to tide them through periods of scarcity, and these stores are then used by the community as a whole (Ritter 1938; Morris 1962). Primates may also store food, but not in a coordinated manner (Crook 1966).

Hunting animals that live in groups may catch their prey by working together. Groups of predatory jackfish sometimes circle swarms of small prey species apparently to contain their prey as they make strikes into the midst of the swarms. Wolves chasing an intended victim may split up, some continuing to chase at the heels, while others run ahead, apparently in an effort to head it off (Murie 1944).

In the same vein, we might consider the hunting dogs (*Lycaon*), of South Africa. They rush as a group into a herd of gazelle, each chasing after the nearest animal. The chances of catching a sprinting gazelle are probably not particularly good, and a single dog may be too small to make a kill alone. But each dog keeps an eye on how the others are doing. If a pack member seems to be gaining on his gazelle, the others leave off their chase and join him or her for the kill (Kuhme 1965).

Figure 29.8 Acorn woodpecker storage area. The birds excavate small holes and then firmly wedge green acorns into them. The stores will be utilized in the leaner seasons and are protected against interspecific competitors such as jays.

It has long been known that some fishing birds, such as certain pelicans, often fish together by swimming in a specific formation and diving in a rather coordinated manner. It was assumed that they caught more fish that way than by feeding alone, but the importance of group feeding was not really understood. Then it was found that the double-crested cormorants (*Phalacrocorax auritus*) of San Francisco Bay also coordinate their fishing efforts. After spending the night on the telephone wires of San Francisco, they fly to the bay in small groups. The birds usually don't begin serious fishing immediately, but individuals may make a few dives, while incoming flocks join the growing group on the water. When the group has swelled to perhaps several hundred individuals, the birds line up side by side and all begin to swim in the same direction. A few birds may follow the long line, but they usually aren't actively fishing. The larger fishing parties primarily pursue schools of fish, while smaller flocks may hunt individual prey. When a flock turns sharply, it is apparently following the school of fish below. Birds in the line splash vigorously, apparently discouraging the fish from attempting to slip beneath the line, while individual divers disappear beneath the water, the divers showing no synchronization or coordination with others in the flock. The flock appears much smaller than it really is because so many divers are beneath the surface at any one time. It is believed that, as a rule, fewer fish are caught by members of small flocks than by those of large flocks, probably because the former have to hunt for scattered prey, while the latter need merely to swim into a herded school and seize a fish (Bartholomew 1942).

Occasionally, coordinated foraging involving two species is reported. In one case, snowy egrets (*Leu-*

*cophoyx thula*) and pied-billed grebes (*Podilymbus podiceps*) were seen catching small fish in a man-made canal (Leck 1971a). Apparently (again hard data are needed), the grebes diving in the center of the canal drove fish toward the egrets waiting near the edge. The egrets, for their part, kept the fish from hiding along the shallow banks. Significantly, the grebes did not fish the other side of the canal where fish were undoubtedly hiding, but where no egrets stalked about. Of course, it can be argued that each species was simply fishing where it was finding the most fish and was oblivious to the other species. It is not unreasonable to assume, though, that the other species may have been noticed and, in time, it could have come to be associated with the presence of fish so that a form of cooperation could be developed through association learning.

## Sociophysiology

We already know that an animal can have certain physiological effects on other animals. In some species, for example, an individual will not reach its reproductive condition unless it is in the company of the opposite sex. In rats, a female will abort or reabsorb developing fetuses if she detects a strange male. In yet other species, crowding can cause changes in the organs and behavior of an individual. There are other examples of physiological effects of animals upon other members of the species that indicate a research area ripe with possibilities—an area that might be termed *sociophysiology*.

The simplest sort of physiological effects of groups are seen in animals that huddle together to conserve moisture or heat. Normally, solitary land isopods bunch up into balls in dry periods and conserve their moisture (Allee 1926). Many warm-blooded animals avail themselves of heat emanating from other animals. Pigs, in winter, may sleep in great restive heaps, squealing and clambering about all night long, as those on the perimeter attempt to move to the middle. I have been told by one in whom I have great confidence in such matters that a strict hierarchy exists among the itinerants who sleep under the bridges of Paris in the winter. As the men lie huddled side by side, the dominant or more revered among them have the warmer central positions and newcomers must take their place at either end.

More elaborate forms of social thermoregulation may also exist. In Antarctic storms, penguins form large crowds that rotate slowly as the birds walk in concentric circles. The great masses of birds move continuously leeward, with the wind, and thus apparently conserve heat (Chauvin 1963).

Allee (1951) found that the small marine flatworm *Procerodes* swells and bursts due to osmotic effects if placed into fresh water by itself. However, if a *group* of worms is placed into the fresh water, most of them last longer. It was found, though, that the first worms to die in any group did so in about the same time as isolated worms. The benefit to the rest of the worms of the rupture of the first individuals, however, was greater than could be accounted for by changes in osmotic gradient alone. It was then found that the ruptured worms "conditioned" the water by releasing calcium into the medium, and that calcium, for some reason, has a protective effect for animals placed into new osmotic conditions.

Later Allee and his students (1951) found that if aquatic animals, such as goldfish or flatworms, are grown in unfavorable waters (for example, waters in which small amounts of harmful chemicals have been added), they do better in small groups than as individuals, perhaps because their body secretions tend to neutralize the harmful additives (Etkin 1964).

As other examples, newly hatched larvae of the sheep blowfly (*Lucilia sericata*) cannot survive on a sheep long enough to penetrate the skin unless the relative humidity is above 90° for several hours. In dry weather, practically the only larvae that live are those from aggregations of several egg-masses laid together (Cragg 1955). In another experiment, it was found that the oxygen consumption of mice at 26°C, a temperature at which much of the mouse's energy must be used to maintain its proper temperature, is 1.82 times higher in isolated mice than in those huddled together.

In 1944, Collias made an interesting observation on quail, the implications of which have unfortunately been rather ignored since that time. He found that the average covey size of quail is larger in North-Central states than in the Gulf states. He called this "a social application of Bergmann's rule" (1941a,p. 41), which states that warm-blooded animals attain a larger size in colder regions than in warmer ones. Larger size of *groups* in colder areas may also be a mechanism for conserving heat if the animals pack together at certain times such as when sleeping.

Other physiological effects of animals upon each other have been observed and need explanation. For example, many body processes accelerate for animals in groups. To illustrate, up to a point, the rate of cleavage in sea urchin eggs increases with the density of individuals. The rule is not universal, however, because when snakes are in groups, their metabolism decreases (Etkin 1964).

It should be apparent that the physiological advantages of grouping together can only be realized if behavioral characteristics permit the grouping. For example, grouping may demand a small individual distance. Also, behavioral traits may change *because* of aggregating. Grouping on a physiological basis may set the stage for more intense social interplay and subsequently more complex social structures.

# 30

# INTERSPECIFIC ASSOCIATIONS

Most of the groups we have discussed have been comprised of individuals of a single species, but interspecific, or mixed-species, groups are far from uncommon. Adaptations for interspecific socializing often provide fascinating glimpses into the array of directions open to natural selection.

## KINDS OF ASSOCIATIONS

In the United States, *symbiosis* has come to refer to many types of interaction between species, including parasitism, commensalism, and mutualism. In *parasitism,* as we know, one of the species is harmed in some way. *Commensalism* refers to the sort of interaction in which one of the species is benefited and the other largely unharmed. Certain species of fish such as jackfish (*Caranx*) may follow a barracuda, swimming just above its back. They are relatively safe there from him and they are also protected from other predatory fish that stay clear of the barracuda. It should be noted that the barracuda may thus be more conspicuous because of its entourage and may, therefore, suffer some small disadvantage. So, the difference between parasitism and commensalism may be a quantitative one. It has been suggested that, evolutionarily, perhaps both mutualism and parasitism usually begin with commensalism.

In *mutualism,* both species derive some benefit from the association. We are all familiar with the soft-bodied hermit crab that lives in an empty snail shell and carries the protective shell about with it. Sometimes, sea anemones settle on the shell and therefore also travel with the crab (Figure 30.1). The anemones are equipped with stinging cells and are avoided by many predators that would otherwise make a meal of the crab, shell or no. The anemone derives its benefit from the association by eating the scraps of food that fall from its companion's table. When the crab outgrows its old shell, it often prods and pinches the anemone to induce its comrade to detach itself and move to a new shell (von Frisch 1964).

Other mutualistic animals are much less closely associated. Mixed groups of baboons and antelopes may largely ignore each other until danger threatens, then they become extremely sensitive to each other's warning signals. Such a group is indeed difficult to approach. The keen noses of the antelope are complimented by the sharp eyes of the baboons. Also, antelope moving with a troop of baboons often do not flee from a cheetah, but rely on the baboons to chase it away (Chauvin 1963).

Often, animal associations provide no clear indication of their adaptive basis. In some cases, they are apparently an unadaptive side effect of a misdirected general gregarious tendency. The highly social springbok antelope, for example, readily joins up with members of many species—even ostriches. There seems to be no advantage in such associations; however, it cannot be denied that they may confer some unrecognized benefit. Perhaps the springbok is able to react appropriately to a wide variety of warning systems. The apparent unadaptiveness of any association, of course, may be simply due to our inability to recognize signals or our ignorance of the myriad and subtle ways animals can interact.

There is one rather unusual interspecific reaction that involves man and deserves mentioning. The black-throated honeyguide (*Indicator indicator*) may attract humans or other mammals, such as the ratell or honeybadger (*Mellivora*) to beehives. The bird likes honey, but it has problems reaching it. If it happens to find a beehive, then, it sets out to find help. When it finds a man, it flutters and darts around him in a most peculiar manner; then it flies away in the direction of the hive, calling loudly and darting back occasionally, apparently just to be sure it is being followed. When the man arrives at the area of the hive, the bird reaches a frenzy of excitement. If the man robs the nest, there is always enough wax and honey left to reward the bird. It must be admitted that such a description involves perhaps unwarranted anthropomorphizing, but according to those who have observed it, there is little room for doubt that the bird is purposefully leading. Such a system undoubtedly requires several facets of learning behavior as we understand it, and important questions arise regarding the pattern's ontogenetic and phylogenetic development.

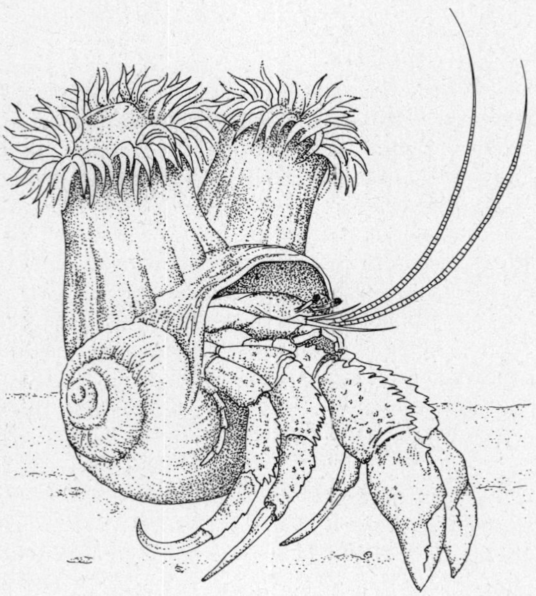

Figure 30.1 An example of mutualism between a sea anemone and a hermit crab. The anemone has attached itself to the crab's shell. The crab is protected by the anemone, which has stinging tentacles, and the anemone benefits by gathering bits of food from the crab's meals.

Because of the incredible complexity of ecological systems and the level of our understanding of them, we are often forced to deal with the problem of cooperative interaction at very simplistic levels—usually by describing how some specific behavioral pattern results in some broad benefit to two species. But studies are now being attempted that seek to correlate a wide variety of ecological interaction at several levels simultaneously. These studies employ what is sometimes called a "systems" approach. While the new light often reveals more puzzles than answers, the systems approach shows renewed promise of our one day understanding a bit more about what sort of world we live in. As an example of such a study, consider grazing animals on the African plains (Bell 1971). Great herds move across the plains in waves, one species following another in what is now known to be a certain sequence. The principal grazers are wildebeest (Figure 30.2), zebra, and Thomson's gazelle, and their movement is in response to annual shifts in rainfall. Each species is physiologically adapted to eat specific types of grasses and herbs (leaves, sheaths, or stems). As each group begins to run out of its preferred food, it moves on—the zebra first, then wildebeest, then Thomson's gazelle. The first species to enter a new area actually alters it in such a way as to prepare it for the species that follow. Therefore, a reduction in the numbers of one species is likely to reduce the numbers of other species.

Another type of behavior in birds *seems* to be the antithesis of mutualism, but if the definition were sufficiently flexible, it could be covered under the term. In this case, instead of the advantage being a result of coming together, the benefit apparently lies in species actively and specifically avoiding each other to their mutual advantage. Four types of honeycreepers in the Quito region of central Ecuador ignore other species of birds except those in the "complex" (three *Diglossa* species and one *Conirostrum*) (Moynihan 1963). Each species of the complex manages to avoid every other species of the complex, although they share the same area and even sing from the same perches at different times. Normally,

Figure 30.2 Wildebeest migration in Kenya. The great ungulate herds that seasonably travel across the vast African plains move in a certain sequence of species, each, in a sense, capitalizing on the changes brought about by the preceding group.

birds species show some interest in each other—a fact that makes this avoidance behavior seem even more unusual. The birds in the complex show almost no outright hostility to other complex members, and within the complex, there are no mutual appeasement gestures. Apparently, for these birds, avoidance offers more benefits than aggression. The social structure of the complex is elaborate and stereotyped. Each species is belligerent toward its own species, and to a lesser degree, to all other species outside the complex. Members of the complex do not hesitate to sing when one of their own species is singing, but fall silent at the song of another species of the complex. The songs of other species that share the habitat with these four are, for the most part, ignored.

## Helpful Parasitism

The very word "parasite" immediately connotes images of revolting and harmful organisms. Worms! Ugh! Perhaps this reaction is unfortunate. Maybe parasitism can be "good." Perhaps parasitism possesses a quality of goodness that has been largely overlooked because man has been so possessed with its disease aspects (Linicome 1971).

There are those who have investigated potential beneficial physiological effects of parasites, but Neal

Smith (1968) presented interesting ecological data extolling its benefits. He found that under certain conditions, it is advantageous for birds to accept the offspring of brood parasites. In fact, he reported that in one host species, more offspring reach breeding age in parasitized nests than in parasite-free nests. He studied Central American colonies of oropendolas and caciques, species that were parasitized by the giant cowbird *Scaphidura oryzivora.* Strangely, he found that the host groups differed in their behavior. In some colonies, the host species tossed out the eggs of the nest parasite, in other colonies the cowbird egg was allowed to remain. Also, he found that some colonies were built around huge bee or wasp nests. The birds' nests were generally safe from vertebrate marauders, since they were placed far out at the tips of branches. But they weren't safe from insects. In fact, the chief source of mortality was a parasitic botfly—a fly that deposits its eggs directly on the bodies of the nestlings. The hatched larvae later would bore into their young hosts' soft bodies, killing them. It turns out that in colonies with no bees and wasps, the birds were heavily parasitized by the botfly, but in nests built near stinging insects, the young escaped unharmed. Also (and this was the puzzling part), if the nests were parasitized by the giant cowbird, then no botflies occurred. Then it was discovered that the cowbird young preened their nestmates, thus removing the eggs and larvae of the botfly. This, then, is why those colonies lacking the protection of bees and wasps were the ones that did not discriminate against cowbirds.

## Mixed Species Flocks

The careful and extensive field studies of mixed flocks of birds have furnished us with new information regarding adaptive characters related to competition, communication, population density, and social behavior.

We have been aware of the existence of mixed-species flocks for some time (Chapin 1932; Davis 1946; Murphy 1936; Odum 1942; Wing 1946). Take a closer look at those winter flocks of "sparrows" where you live. How many species do you find? While mixed flocks occur in temperate areas, they are more prevalent in the tropics. In the tropics, in fact, more species form mixed flocks than form single-species flocks. (The reverse is true in temperate areas.) In the tropics, most mixed flocks are composed of anywhere from two to about twelve species of passerines. The flocks may be transitory or semi-permanent and may be dissolved and reformed at intervals. They may also have tight or loosely knit social structures.

As might be expected, mixed flocks are more likely to form in winter when territorial defenses are down and many birds tend to be gregarious. In northern Europe, for example, tits of several species travel with tree creepers, nuthatches, and certain woodpeckers through the winter woodland.

Attempts to show that the more gregarious species of birds are less belligerent than the solitary ones have been unsuccessful (Moynihan 1960). It is theorized that perhaps gregarious species are not necessarily less aggressive, but merely less likely to actually attack each other physically, relying more on threat and appeasement gestures instead. Moderately gregarious green honeycreepers, for example, frequently supplant each other, indicating a belligerent spirit, but actual contact of any sort is very rare. It must be realized that aggressiveness may vary between species in a mixed flock and the differences may help to determine the flock's composition (Moynihan 1960; Wiley 1971).

Certain of the advantages in forming mixed-species flocks are readily apparent and, in fact, may be the same as those for single-species flocks. There is, for example, the simple factor of safety in numbers. Predators might be confused by a mixed group just as by a single-species group. (There may be less of such advantage in mixed flocks, since it is possible that a predator would be less distracted by the sight of several individuals of different types than he would by several individuals of a type for which he had formed a "search image.") In addition, a predator might simply be intimidated by a large flock.

Also, many watchful eyes are better than a few. If the birds in a mixed flock were "tuned in" to each other's warning signals, they would avoid predators more easily. Such alarm recognition could be specific, such as when one species reacts to the specific warning cry of another. (Remember, there may be marked convergence of such signals due to need for sound carriage and crypticity.) Alarms may also be less specific. For example, the sight of any bird taking flight might alert other birds, possibly inducing them to follow.

The protective advantages of mixed flocks may be unmistakable in certain cases. As an example, in some mixed flocks the species will join together to mob a common predator, such as an owl (Lack 1970). It must be pointed out, however, that there are as yet little hard data that clearly demonstrate that flocking furnishes protection from predators.

When any animals forage in the same area simultaneously, they may, of course, begin to compete for food, thus grouping could have its disadvantages. If birds tended to group together for protection from predators, any resulting competition for food could be reduced in a number of ways. In species of different adaptive types, such as tits and woodpeckers, little foraging overlap would be expected as each went its way. Some might be foraging on leaves and buds, others on twigs and bark, and others might be catching insects in flight. Even if the species take similar sorts of food (for example, insects) competition can be reduced by each species foraging in a particular fashion (Willis 1966a; Vuilleumier 1967; Morse 1969). In this light, it is not surprising that mixed flocks are more common in the tropics, since many tropical species, as we know, have highly specialized foraging adaptations, and specialists would compete less. Also, in the tropics, a single species may be represented by relatively few individuals, thus further reducing competition. And then some displacement of foraging behavior could occur, at least temporarily, in flocks composed of species of similar adaptive types.

For many species there is an actual foraging *advantage* in forming mixed flocks. It has been suggested that mixed flocks are primarily "foraging associations" in which each bird gets more food than it would by foraging alone (Rand 1954; Short 1961; Brosset 1969; Moynihan 1962b). For insectivorous (insect-eating) birds, the flock may serve as a "beater" to arouse insects, causing them to stir about or to flee, thus giving their positions away. In flocks in which insectivorous birds mix with those of other dietary specializations, the main recipients of the grouping advantage may be the insect eaters (Lack 1970).

But grouping may be important to birds with other kinds of diets. Frugivorous (fruit-eating) or nectarivorous (nectar-eating) birds are forced to search for widely dispersed food sources such as fruit-bearing or flowering trees—trees that yield a rich supply once they are found. But they may not be easy to find. The probability of finding these kinds of irregularly dispersed food supplies may be increased by flocking behavior. In fact, if the food is regularly dispersed and easy to locate, there may be little "food advantage" in flocking, and therefore any association formed under such conditions can be considered due to other advantages—such as protection (Moynihan 1962b).

It seems that interspecific bonding (or mutual attraction of the species) of mixed flocks actually at fruiting trees is weaker than the bonds that exist within flocks moving through forests in search of food (Diamond and Terborgh 1967). It has been suggested that the reasons for such differences are that species already at a stationary food source would, of course, not need assistance in locating food (Lack 1971). The interspecific bonds, however, might remain strong within a flock that had located a moving food source, such as an insect swarm (Willis 1967). Frugivorous birds also don't need help flushing prey, and once food is found, they all tend to exploit the same food source, and thus a weakening of their bonds might be expected, at least temporarily. For a discussion of the variation and distribution of flock-forming species, see Moynihan (1962b).

In mixed flocks of finches on the Mojave desert that feed on small dispersed food items, winter flocks may help to regulate "return time" to a food site (Cody 1973). In other words, the flocking ensures that birds don't continually revisit a food site before it has time to recover from previous visits. The idea is that the habitat must have time to replenish its stock, and a group of birds can move in such a way as to give it this time. On the other hand, individuals foraging alone would have no knowledge of previous visitors at any one site and hence could only determine whether any area held food by trial and error—an inefficient process.

In the light of what we know about interspecific social patterns, we might ask what kinds of signals contribute to the formation of mixed flocks.

**Signaling in Mixed Flocks** Mixed flocks may be comprised of rather similar species. For example, in the nonbreeding season, several species of finches may congregate into one flock. The advantage in species similarity may lie in the similarity of signals in related groups. In mixed flocks, signaling would, of course, be adaptive in aiding group coordination. In mixed flocks, the vocalizations of the flock members are much less distinct than in the breeding season, a time when it would not be advantageous for signals of one species to become confused with those of another (Marler 1957a).

Coloration may play a more important part in mixed-flock formation than vocalization does. It is interesting that many gregarious species of birds have developed conspicuous markings and color patterns. Many are all black, such as crows, blackbirds, starlings, anis, cormorants, and vultures; or all white, such as herons, spoonbills, swans, and geese; or pied, such as magpies, gulls, and terns (Moynihan 1960). The apparent advantage is that they can see each other more clearly, thus facilitating signaling and group action. Conspicuousness to predators, it is reasoned, is countered by the increased advantages of group defense. This kind of strong coloration has been termed "episematic" (Huxley 1938). It has been pointed out that the glaring appearance may, in some cases, be identical with aposematic, or warning, coloration and may serve both purposes simultaneously (Figure 30.3).

Figure 30.3 Strong markings may aid in group cohesiveness and may have episematic functions.

Not all the species in mixed flocks are so gaudy. In fact, some species are quite drab and indistinct. It has been suggested that in some circumstances the "nucleus" of the flock may be, in fact, the drab individuals. The dull plain-colored tanager, for example, forms the nucleus of certain mixed flocks in the tropics. Its drabness is believed to increase its attractiveness (perhaps by presenting no threat) to other, brighter colored species. Dull plumage, then, might well minimize aggressive signaling and so promote the formation of the mixed flocks (Hamilton and Barth 1962); so, the very neutrality of drab species adds to their appeal. In a sense, then, although the drab tanagers may not turn on their brighter fellows, they don't turn them off either.

**Social Mimicry** Moynihan (1960, 1962b, 1963) noted that mixed flocks (especially in the tropics) tend to have similar plumage types. Such similarity has actually been found in a number of characters. For example, in seven species of Australian ground finches that form mixed flocks, each species associates only with other species of ground finches that have the same bill color (Immelmann 1962). Mountain-dwelling bush flocks in Western Panama are mostly composed of species with black and/or yellow markings. The humid temperate forests of the northern Andes are composed of mixed flocks whose members tend to be either blue or blue and yellow. And further south, such flocks are commonly bluish above with chestnut or buff underparts.

There are several routes by which physical similarity in associated species could develop, of course, such as through common descent or convergence due to common environmental pressures, but perhaps, in some cases, it is due to a specific *social mimicry* where selection moves one species toward the appearance of another species (Moynihan 1968). There would certainly be restrictions on such selection. Obviously, selection would never permit any two species to become identical, because of the reproductive importance of species distinctiveness. Also, the types of species in which social mimicry might play an important role may be rather limited. Social mimicry, it seems, may be most likely to develop in species in which discrete sign stimuli (rather than gestalt perception) elicit and direct adaptive responses. Thus, the system could be expected to function best in species of rather stereotyped behavior—such as in birds. Social mimicry is certainly not associated with all species that have relatively stereotyped behavior, such as crocodiles and sharks, since they normally do not have very complex social systems.

In theory, when related forms come into "secondary" contact (subgroups rejoining after a division of the parent stock), any divergence caused by character displacement must be checked and reversed in order for social mimicry to occur. If this is not true, then character displacement and social mimicry must operate on different physical characteristics. For example, two bird species could possibly converge in plumage while diverging in bill length, but such an evolutionary dichotomy has yet to be demonstrated.

**Division of Labor in Flock Cohesion** An interesting sort of association in three species of antwrens has been described (Wiley 1971). Two were of the genus *Myrmotherula* and a third, *Microrhopias*. Apparently, the dissimilarities in the species permit them to interact in a complementary manner. One species of the congeners was more visually conspicuous, but the other was much louder (usually, though, all vocalizations were relatively subdued). The loud species, *Myrmotherula fulviventris,* was the more visually cryptic, and members of this species would immediately congregate around a human observer in the woods, setting up a raucous clamor. Although its alarm could be heard as it darted about the foliage, its plumage made it difficult to see. The male *M. axillaris,* with its bold plumage and quieter manners, probably functioned, not in warning, but in keeping the flock together by serving as a signpost. The call of this species was never loud and was difficult to pinpoint. It is assumed that flock cohesion is better maintained by visual signals, since loud calls might attract predators from afar. The third species probably had a less critical function in terms of group function. Both sexes of *Microrhopias* had bold plumage, but they frequented the densest foliage so their patterns functioned for shorter distances, perhaps primarily between mates.

We see, then, that species need not interact in incidental or antagonistic ways. Some species need to associate with each other and have developed subtle and complex methods for doing so. Communication between species may be an important coordinating mechanism, and the constraints placed upon interspecific communication have provided invaluable insight into the options of natural selection.

# 31

# THE DEVELOPMENT OF SOCIAL BEHAVIOR

One of the primary questions in sociobiology is, How does social behavior develop? How greatly are animals constrained or compelled by that mysterious complex called the "old brain"? To what extent can its directives be overridden by the cerebrum of those animals who possess this vaunted "new brain"? How can experience alter behavior toward other animals? How important is genetics (or evolutionary learning)? Other questions center over increasingly complex social acts. How does intricate cooperative behavior arise? Do weaver ants get better at sewing as they practice? What causes some baboons to take upon themselves the mantle of sentinel? We can't hope to answer any of these questions in definitive terms, but perhaps at this stage we should satisfy ourselves with answers to smaller questions and clarification of the larger ones.

## THE DEVELOPMENT OF SOCIAL PATTERNS BETWEEN INDIVIDUALS

We should be able to discuss the ontogenetic, or individual, development of social patterns more meaningfully than the evolution of social behavior. After all, it is possible to observe the development of social patterns in an individual and to experiment. It should be pointed out, however, that the picture revealed by experimentation is not, as yet, a clear one. One reason is that we're still especially fuzzy regarding the *adaptiveness* of social behavior. Certain laboratory animals have been studied in great detail in experimental situations, but the experiments are often not geared toward revealing how a pattern operates in nature. As an example, we now know quite a bit about the aberrant behavior of laboratory monkeys that have been isolated at birth. But how important is such information when we consider animals that are never likely to be isolated? I know of only one field report related to such work, that of unusual social behavior in an orphaned chimpanzee—but even this unfortunate animal continued to live with the group (Goodall 1971). Admittedly, though, other laboratory investigations have more ecological relevance, such as those that consider the ramifications of chimpanzee adolescence. For chimpanzees, adolescence is a period of adjustment (just as in humans) that may have direct bearing on the animal's later status and, hence, on his general success.

### Incidental Relationships

There are a number of ways in which social patterns could develop in any species. For example, we know that some animals may associate because they are simply exploiting the same parts of an environment, such as we see in insects that respond independently to taxic influences. Animals may also show positive affiliation on an anonymous and nondirected basis, such as do schooling fish. However, we are not particularly concerned here with such species; of greater interest are those that recognize each other and interact on a truly social basis. At the same time, we must quickly admit that these more interesting forms of social behavior could have had their origin in association based on simple aggregation. In a hypothetical sequence, animals could associate incidentally, come to recognize each other, and then to rely on each other—all this springing from common utilization of the habitat. Perhaps, for example, the sleeping parties of the gelada baboons, which form as they congregate at safe sleeping sites, provided the original association upon which their complex social structure has developed.

### Pair Relationships

Complex social patterns may also arise as an extension of reproductive patterns. For many species, the basis of the group itself may be an extension of pair bonds over increasingly greater portions of the life span (Allee 1951). In other words, animals may affiliate as a result of pairing. The affiliations may be temporary (such as those that extend through the reproductive season or until the young become independent) or relatively permanent (such as those that may span several seasons or a lifetime).

The importance of sexual bonds in social patterns is so marked that in some cases the real basis of a social group may be masked temporarily by the affiliation imposed by the demands of mating. For example, during the mating season, a male deer may appear to "lead" a harem. However, when the group is alarmed, the buck may take off in one direction, while the *actual* leadership of the herd is revealed as an old female leads the group in another direction. Obviously, the basis of the group was not as strongly rooted in mating affiliation as might at first be presumed.

Reproductive behavior may also influence social structure in other ways. Grouping among primates, for example, may be due to the likelihood that some females in the group may be sexually receptive at any given time, since many primates do not show pronounced estrous seasons. Males, then, tend to stay with the group for sexual purposes. Social structure may also be based on females seeking the protection of males (Hall 1965b).

We really don't know much about the nature of the pair bond—that mysterious "force" that holds mates together as a cooperating unit. We don't even know what *love* is, although every poet, drunk, preacher, and psychologist has his opinion. Animals that come together only to copulate present no great conceptual problem. It is apparent why they are together—to us and to them. Animals that mate for longer periods, such as through the breeding season, are ultimately held together by the fitness advantage of joint efforts in rearing young. But those animals that form lifelong pairs present fascinating realms for speculation regarding the "mind" of the animal. What is the nature of the attraction between them outside the breeding (or sexual) season? The clues are meager. It has been theorized that perhaps personal bonds are somehow derived from the same motivation that results in aggressiveness. (Among humans, love is sometimes described as a love-hate relationship, the smile may be an intention movement to bite, and we know of the aggressiveness that the males of a number of species show toward their prospective mates when pairing and, sometimes, even after the pair is established.) It is apparent that the pair bond is not entirely a result of some kind of aggressiveness. Aggressive behavior waxes and wanes in some species that remain firmly mated for long periods. Jane Goodall (1971) noted that when certain female chimpanzees became sexually receptive, they tended to leave the group in the company of specific males. The pair would travel alone until the female lost her sexual swelling. Then the pair would return to the group. However, there was never an indication of any feeling of "love" or even affection between them. When the mating period was over, they seemed to lose interest in each other and returned to the group separately. It must be admitted that the attempt to look into the mind of animals is not a promising one. Who can say that the greatest expression of chimpanzee love is not to ignore each other? To emphasize the problem, consider love in humans. People who do not get along often spend decades together, chipping away at each other and professing love all the while. Why do they do it?

### Parent-Offspring Relationships

Animal sociality, in some cases, may be founded in extended parent-offspring relations. The key relationship in group cohesiveness may be between the mother and her offspring. Goodall noted that adult chimpanzees often showed great respect for their old mothers in that they sought to be near them and showed a general reluctance to enter into any conflict with them. A family unit in chimpanzees is comprised of a female and her young (an estrous female accepts a number of males, and fatherhood can never be determined). In the macaques, a female may be followed closely by her juvenile, adolescent, and even young adult offspring, and the social interaction among them is generally higher than for the group as a whole (Figure 31.1).

Lambs also tend to follow their mothers closely. As a result, the oldest females emerge as the leaders of the flock, the matriarch of offspring from successive seasons (Scott 1945).

Red deer are matriarchal as a result of the continuing association of the young with their mothers. The females usually stay together the year-round, are long-lived, and have a strong social system, with the old females leading. The yearling females pay close attention to the older females and thus learn by tradition. Yearling males, on the other

Figure 31.1 Female macaques may associate with several generations of their young at once.

hand, are less attentive to their mothers and they leave the group when they become adults, forming bachelor herds with a loose social structure and no hierarchy. Each rutting season, the male herds disband as each seeks to round up his own group of estrous hinds, every male collecting as many females as he is able. As harem-master, he services all the females and defends those rights against other males. The demands of such duties are so rigorous that he is only able to last about a week, when he is replaced by a fresh male. The females make no effort to help any harem-master defend the group from other males. At the end of the rutting season, all the males depart and the females continue on as a group. The leader females must direct the herd over their migratory route through the mountains and valleys as winter draws on, a route they most likely learned by following an older leader when they were young (Etkin 1964). The red deer, then, exhibits a flexible behavior in which learning by females plays an important role in the functioning of a social unit.

In other species, both parents may be involved in the formation of the "core" relationship with the young. A wolf pack is apparently a family unit composed of a father (alpha male), mother (alpha female), and the offspring of several years. Mating is apparently for life. In some cases, two or more such packs may join forces for a time. When a female is pregnant and unable to hunt, her mate, or other members of the pack, may bring her food. Members of the pack also bring food to the wolf pups, either in chunks or by regurgitation. The wolves' social unit is exceptional in that a pack includes a dominant male that may carry out almost all the breeding (Mech 1970). Dominance hierarchies exist, but little agonistic behavior appears, since the structure is formed when the wolves are very young and remains rather stable through life (Etkin 1964).

### Sibling Relationships

I have alluded to several instances of hierarchies and affiliations being formed between siblings when they are young. Probably not many species are viscous enough for adults to interact on the basis of youthful associations with brothers and sisters, but such instances are probably most common among the primates and social carnivores. Among wolves, the group affiliation is strengthened by the association of the pups. They show a high level of interaction through their romping, licking, touching, play-fighting, and chasing. In fact, it has been suggested that in wolves and dogs the strongest social relationships are developed with littermates, not parents (Scott 1945). The parents are likely to be off hunting during the time when the young are being socialized. There is a current revival of interest in our diminishing populations of wolves that may lead to important new evidence.

### Male-Male Relationships

Males of many species, from grouse to red deer, may associate for a number of reasons. But we are primarily concerned with male-male bonding—which may be entirely different from foraging or protective associations. The strongest bonds between males are probably found in the highly social mammals, especially primates.

Wild adult male chimpanzees show particularly strong bonding and affiliation. They often converge in all-male grooming sessions, sessions to which adolescent males are attracted, but which they are afraid to enter. No female is likely to even approach such a group. In general, Goodall has described a higher level of social interaction among males than among females. If females interacted with others it was usually with young or, rather timorously, with adult males.

We have already considered all-male dominance cliques and I have referred to Tiger and Fox's (1971) contention that political organizations and secret societies in humans are primarily male-oriented as a result of the male bonding necessary to carry out a successful hunt.

### The Social Imperative

There has been some argument regarding whether social tendencies are innate or largely acquired through learning. In some cases, social behavior must certainly be founded in the animal's genetic heritage. Bees, for example, certainly do not learn complex social patterns in their brief lifetimes. However, social interaction may be largely the result of experience in monkeys, as is suggested by the aberrant social behavior of isolates.

If social behavior is not learned in some species, then we may ask to what degree the patterns are based in instinct in the classical sense. Tinbergen (1951) thought that social "instinct" is really due to the operation of other instincts and that there is no social instinct per se. Others disagreed (Moynihan and Hall 1953; Keenleyside 1955). However, one argument was based on rather unrigorous observations of members of a species (the spice finch, *Lonchura punctulata*) "wanting" to be together, presumably in the appetitive sense. The other argument was built on the species-specific nature of schooling in fish, but was open to criticism, since there remained the possibility that fish could learn the patterns. Although there is some evidence that indicates a social instinct, perhaps we should not consider it instinctive in the classical sense. Perhaps, instead, we should call it a "social tendency" (Crook 1961). But what is the nature of the evidence?

There is indeed a "need" (appetitive?) for social contact in gregarious animals. A fish separated from its group swims rapidly back and forth apparently seeking its fellows (Eibl-Eibesfeldt 1970). Gregarious mammals, if isolated, become apathetic and show clear signs of loneliness (Meyer-Holzapfel 1958). Lorenz (1931) reported that his tame jackdaws would not even *allow* members of the group to leave. In one case, a large flock of migrating crows and jackdaws settled near his tame jackdaws, and it looked for a time as if much of Lorenz's flock would travel with the wild group when it took to the wing. However, two older birds of the tame flock went among the wild flock, rounded up the other tame birds, and headed them home, thus retrieving all but two in two hours. Of course, the real basis of this retrieval behavior is unknown, but it could be the expression of an unwillingness of the retrievers to become isolated.

The problems of nullifying the effects of learning in any experiment designed to investigate the innateness of social behavior are formidable. For example, there may be simply an innate tendency of species to associate with their own kind (the "social tendency"); then once drawn together, they might learn complex social patterns.

No matter what the opulence of the environmental argument, the importance of inheritance in social behavior must be acknowledged. The problem in ethology is that inheritance has somehow been viewed too long as the overriding determinant of social interaction. Now it seems that the effects of tradition and other forms of learning have been underestimated. Crook (1970a,b) repeatedly emphasized the importance of learning as a determinant of social behavior. He even referred to the habitat, physical and social, as "nature's Skinner box." As indirect evidence of the environmental shaping of behavioral patterns, he cited the lability of social structures as they appear across a species' range, together with their local stability. Such a mosaic of behavioral patterns across a species' range is considered to be an effect of local differences in the learning situation, whether that situation is a construct of environmental differences, local tradition, or a mixture of both.

### The Development of Communal Breeding in Jays

There is an interesting theory of how communal breeding (where several adults may help to feed one brood, as defined by Lack 1968) could have evolved in jays (Brown 1974). According to the model, there are three stages in the development of the behavior. A review of the concepts might provide us with additional insight into how sociality can develop among animals.

The first step is the development of a *K-selection phase* in which there is strong competition for space as a population continuously presses against the environment's carrying capacity. Such a phase would most likely appear in nonmigratory animals inhabiting an area with stable, mature type of vegetation. Thus older, more experienced animals should be better competitors than young ones, and in such cases, older birds would be doing most of the breeding. With competition having forced the population into all available niches, there would be nowhere for the young to go, so it might be to their advantage not to attempt to disperse, but to gain a year's apprenticeship, to "age" while remaining on the parent's territory. This viscosity sets the stage for the next phase.

The *kin-selection phase* would appear because of the close genetic relationship between the offspring of different broods. It is to a young bird's genetic advantage to help its parents raise the next brood. The young bird thus helps to get more of its own kinds of genes into the population and as it ages, it gains experience through its apprenticeship. Not incidentally, it also stands a good chance of inheriting the territory should the parent of its own sex pass on. It is estimated that if the probability of a young jay finding a vacant territory of its own equals the probability of the death of its parent of the same sex, then natural selection would dictate that about half the young birds leave and half remain as inheritors. We could call this one the old Cain and Abel decision. On the other hand, if a young bird has a lower chance of finding a territory than inheriting one, all the young will tend to stay on the family premises. With the prolonged retention of young in the group, very large families could develop and these could be expected to dominate smaller families. Thus, there is an added advantage in remaining to give weight to the family unit. Inheritors could expect to receive more if their family is large.

Then the third phase can begin—the *kin-group-selection phase.* As territories become larger through the influence of more and more offspring, finally, more than one breeding pair might be tolerated within the territory. Inbreeding would become more intense, and flock members would become more genetically uniform. Thus, by helping another member of the flock, one would be advancing his own kinds of genes—and this, after all, is what natural selection is all about. The flock at this point would become an important unit in natural selection. As flocks (each with their own inbred genotype) competed, selection would favor the one with the "best" type of genes. Individual genotypes would be selected according to how well they functioned within the flock. In essence, then, what I'm saying is that it is possible for vertebrates to arrive at social systems that are reminiscent of those of the social insects. An uncritical reading of the material might leave one with the impression that we had arrived at a mechanism for group selection. However, again we see that each animal is behaving in such a way as to advance its own kinds of genes.

## THEORIES OF SOCIOBIOLOGY

Since Tinbergen and Lorenz first began to describe social behavior in animals, there has been a divergence of opinion as to just how animal relationships should be defined in operational terms. The question is important to setting up any study of animal interaction. There are those who continue to describe animal interaction in terms of the classical dyad relationship (how two animals interact); but, rather recently, a new school of thought has developed that more fully incorporates principles of ecology, population dynamics, and social behavior. The object of the new school is to attempt to establish unifying principles of social behavior by presenting it as an ecological construct that reflects the net effect of environmental conditions, innate tendencies, and acquired behavior. The task will be difficult, since we can only broadly generalize regarding the physical influences of an environment and we know even less about the predisposition and intelligence of wild animals. (You would probably be surprised at how little we know about the behavior of wild animals. Unfortunately, "natural history," or descriptive biology, fell out of fashion while still in its infancy. Field researchers such as Jane Goodall, Patty Moehlman, Valerius Geist, Ian Hamilton-Smith, and George Schaller are currently helping to refocus our attention on such matters.) Much of the best integration of behavioral constructs is derived from field studies of primates, and the most articulate chronicler of the method is probably John H. Crook of the University of Bristol.

### Social Ethology

Crook (1970a) calls the broader study "social ethology" and he bases his approach on several points. He noted that, first, variation within a group and relations between individuals can often be correlated with the environments of the subgroups within the species. Second, social structure is not subject to the same evolutionary pressures as morphological characters and cannot be dealt with on the same conceptual basis. Instead, social structures are much more dynamic and labile—within the behavioral limitations of the species. Third, since individuals are dependent upon the group, they must conform to its norms or be selected out. For example, if a bird chooses to feed apart from the flock, it may be taken by a predator; if it does not read the signals of its fellows, it may be attacked and killed by them. Fourth, the problem should be studied at the level of the social *process* itself rather than by asking questions about sources (that is, the relative significance of genetics, traditions, environmental programming, and so forth). It is hoped that such an approach would avoid the persistent nature-nurture controversy. And, lastly, Crook maintains that historical change in social structure appears as the net result of changes occurring at different rates in various processes. Thus, the environment may change a social structure quickly, but its indirect effect on tradition occurs slowly, and its effect on the genetics of a population more slowly still.

We must bear in mind that selection pressures should not be conceived as acting directly on social organization (a statement reminiscent of Tinbergen's (1951) argument that social behavior in the dyad sense is not the result of a social *instinct*). Rather, selection acts upon other tendencies (nest-site selection, escape tendencies, sexual behavior), which then alter the social organization accordingly (Crook 1965). So, while "social organization" may not be a subject of selection in and of itself, it reflects the sum of a series of other pressures and is useful in description and comparison.

The term *grade* has been coined to describe characteristic lifestyles of any species adapted to a particular ecology (Huxley 1959). Crook (1970a) pointed out that species belonging to the same grade often show similar social organization. For example, grassland sciurid and cricetid rodents living in comparable niches show similar social structure. Also, carnivores are often unsociable whether they are marsupial or placental (possible reasons are discussed in Chapter 21). Other such correlations have been demonstrated by Pitelka (1942); Orians (1961); Stonehouse (1960); Crook (1964, 1965, 1970a,b); Cullen (1960); Nelson (1967); Tinbergen (1967). See Lack (1968) and Wilson (1975b) for reviews.

Again, we must remind ourselves that any comprehensive statement regarding the molders of social patterns certainly must reflect our knowledge of both heritable and acquired characteristics. The principal goal of a general theory of sociobiology should be the ability to predict features of social organization from a knowledge of population parameters, combined with information on the behavioral constraints on a species imposed by its genetic constitution (Wilson 1971b).

## ECOLOGICAL EFFECTS ON SOCIAL PATTERNS

Let's now attempt to summarize some of the findings that illustrate ecological effects on social patterns. The principle was discussed earlier with a different emphasis and we will simply add to some of those findings and investigate the possibility of developing unitary rules.

Tinbergen's group (see the discussion of Cullen's 1960 work in Chapter 4) at Oxford demonstrated that many features of the behavior of the kittiwake could be correlated with features of the environment. Other work, primarily involving field studies of birds and mammals, also indicated a strong interrelationship between social systems and the environment. The studies reiterated the concept that social structures could be interpreted or described on the basis of their adaptive significance, an idea that had, until the 1960s, been largely ignored. Earlier, social behavior was considered primarily or entirely the result of natural selection and therefore innately determined. This idea persisted in hard-headed ethological circles until certain studies clearly implicated learning as a co-determinant. Some of these studies, done on primates, showed that primate social organizations were highly variable from one population to the next and that the different social systems could be correlated with the habitat type, a relationship that could most easily be explained by different experiences of groups subjected to different kinds of environments.

### Bird Socioecology

But we're getting ahead of our story. Let's first review some of the evidence from field studies of birds, then consider primates. We continue to note certain correlations between the environment and the social behavior of its denizens. For example, among birds, colonial nesting increases foraging efficiency and predator avoidance, but it is only likely to appear where there are suitable nesting sites surrounded by patchily distributed food (abundant in separated areas). In Crook's extensive studies of weaverbirds of the subfamily Ploceinae (1961, 1964), he emphasized the principle that their social systems are strongly correlated with ecology. He found, for example, that dispersed breeding, active pursuit in courtship, and monogamy could be correlated with insect eating and low-light values in the forests of tropical Africa, while flocking, static courtship with visual advertisement of colonial nests, and polygamy were characteristic of

Figure 31.2 Birds that are primarily herbivorous, such as evening grosbeaks (left), are more likely to be gregarious than are predatory species like the broad-winged hawk (right). Predators seek sparse and elusive prey and cannot afford company on the hunt.

well-lit savannahs. Aerial advertisements, furthermore, were performed only by those populations that dispersed their hidden nests in grassland. In order to clearly establish the adaptive effect of the behavioral correlates, one must keep in mind such variables as food dispersal and food depth (biomass per unit area), considering both its presence and its locatability.

So, how has the environment imposed a particular social system on these birds? Solitary or territorial nesters are likely to have nests that are accessible to a predator should they be discovered. The birds avoid attracting attention to the nests by reducing the level of activity around them (dispersed nesting) and by approaching them discreetly. A dispersed nesting pattern also permits foraging without social interference, but this advantage is only likely to be important if food is not abundant. And then you may recall that omnivorous and herbivorous birds have smaller spatial needs as opposed to predatory species that hunt sparse or elusive prey (Schoener 1968).

Patchiness of the food of primarily herbivorous birds may contribute to the fact that they are much more frequently colonial or gregarious than are predatory birds, such as hawks (Figure 31.2). We know that in a situation where food is patchy, grouping may increase the likelihood of finding food. If grouping also results in colonial nesting, the nests are usually made inaccessible to predators, since the increased activity around them would make them conspicuous. Crook (1965) maintained that there must be sufficient food around the colony for some recruitment to occur; that is, there must be enough food to support a nonbreeding population so that at some time they may take their place among the breeders. (We should also remember that the *historical* availability of food may strongly affect the sort of mating system that we see today.)

Under conditions of low food abundance, monogamy might ensure enough food for the offspring through the cooperative efforts of both parents. In birds, then, population dispersion and, to some degree, social systems are considered to be functions of (1) the preferred food, its abundance, accessibility, and distribution, and (2) nest-site preference in relation to predation. Let's now examine some of the evidence from primates, where again we will see the importance of food, predation, and habitat type as determinants of social systems.

### Primate Socioecology

We noted in earlier discussions the effects of environment upon the social structure of some primate species. For example, in those savannah areas that have seasonal periods of low food supply, certain species of baboons move about in one-male harem units, or bachelor groups. Predation is generally low, so the added strength of numbers of males is not needed for protection, and the food supply wouldn't support a multimale group anyway. In ecologically more stable woodland-type savannah and sparsely forested areas, there is less climatic fluctuation and a more dependable food supply. Here, multimale groups may exist as a protection against higher predator risk. Less is known about species of heavy tropical and semitropical forests, but it seems that in these areas, too, social structures are often correlated with the habitat type.

Goodall's remarkable and loving study of the chimpanzees of the Gombe Stream area of East Africa is the most complete account yet produced of the behavior of wild primates. She has described large and powerful males that move with smaller, subordinate females through the deep forest. The size of the males apparently does not function in sexual selection, since both sexes are highly promiscuous, the females even copulating with adolescent males. Size in males is probably more important in the defense of the group, although the chimpanzees are not often threatened by any predator in an Africa being decimated of its large cats, their skins reappearing as status symbols, albeit low, in human wardrobes.

Goodall was able to discern specific loosely defined, wide-ranging groups that sometimes converged or interchanged membership. The resourceful and omnivorous primates probably never face severe food shortage, but they converged at sites of preferred food, such as fruiting trees and termite mounds, as well as Goodall's banana supply. In nature, the food sources of chimpanzees are likely to be patchily distributed; thus, it would pay for groups to form and then to cover a lot of ground. (It will be most interesting to compare Goodall's study of chimpanzees to Diane Fossey's continuing study of wild gorillas.)

A comparison of several species of monkeys revealed that an arboreal habitat imposes a certain uniformity on group size, population density, and home range for all the primate species living in that habitat (Aldrich-Blake 1970). Critics have seen fit to mention that since food, sleeping sites, and predation would be similar for any species living in the area, the adaptive similarities are not unexpected. But this is precisely the point. Can we generalize regarding social systems if the ecological conditions are known? The answer can only be deduced from extensive fieldwork. It is often a giant step between armchair predictions and real proof acquired in the field, and in fact, the field data have time and time again shown a disconcerting lack of respect for hypothesis.

The high population density and small range of forest monkeys is presumably related to the greater year-round productivity of forest as compared to savannah or grassland. Also, forests are "three dimensional" as opposed to "two dimensional" open country. The added dimension provided by tree height increases the habitat complexity and therefore presumably enhances the area's stability and environmental carrying capacity.

What are the other social differences that have so far been found between forest and open-area primates? Not only are home ranges of savannah primates usually larger than those of forest species, but the savannah groups are generally larger than forest groups (although more species may be encountered in the forests). It is thought that since forest monkeys may escape from terrestrial predators by running along small branches to another tree, or from eagles by dropping into the understory, large males are unnecessary, and so these species may show reduced sexual dimorphism and few males in each reproductive group. In grassland or sparsely forested areas, primates are left with two alternatives: (1) to fight (thus selection would favor a number of large males within the same group), or (2) run away (thus we see great speed in some species such as patas). But predation pressure can be expected to select for large size in males or multimale groups only if the food supply permits, abundant food being a prerequisite for body growth and for the maintenance of large groups. Of course, once large males were selected for on any single basis, a change in social structure might result, as we learned in the last chapter.

Anyone who has attempted to observe animals in deep woodland knows that it is often hard to see what an animal is doing, to observe an individual for any extended period, or to find the animal to begin with. Obviously, the heavy vegetation presents similar problems for its inhabitants. Thus, in deep forest, reduced levels of visual communication might select against large coordinated social units. Vocal communication would probably generally not replace visual signaling because of the likelihood of attracting predators from long distances. So, communication problems may also select for smaller groups in woodlands.

Because of the problems inherent in forest study, many researchers have been discouraged from chasing their animals through the woods. However, more data are sorely needed if we are to be able to make any concrete generalizations regarding ecological effects on social structure. It has been suggested that social diversity among forest monkeys is as great as the differences between forest and open-country species (Aldrich-Blake 1970). For example, some forest species have the one-male groups, formerly considered an adaptation to seasonally arid conditions of the savannah (Crook and Gartlan 1966). The question is whether the tendency toward one-male groups is adaptive or is a side effect of an adaptive response to a greater problem. Could the forest support the heavy demands of multimale groups? The one-male structure in forest monkeys is apparently permitted by their being able to flee from predators through the complex habitat, thus alleviating much of the necessity for protective males. But is the system simply *permitted,* or is it *maintained at some cost?* These are but a few of the problems regarding the attempt to correlate ecology and social behavior.

## Adaptive Grades Among Primates

Finally, with this background in mind, let's review one of the more important studies done so far that was aimed at developing a unifying theory of sociobiology. In this study, Crook and Gartlan (1966) correlated environmental pressures with the social patterns of several primate species (Table 31.1). Their object was to try to understand some of the selective pressures responsible for adaptive change in social structure and thereby to be able to construct evolutionary hypotheses regarding the phyletic history of some primates. They theorized that the adaptive radiation of primates has given us a gamut of types that run from forest-dwelling insectivores to larger, open-country vegetarians. The table shows a shift from insectivorous, nocturnal, solitary forest animals with a population dispersal based on aggressive contacts; through leaf-eating and fruit-eating forms that move in small family groups or larger parties and show many types of territorial defense (but little intraspecific aggression, sexuality, or intermale competition); culminating in largely vegetarian browsers of open country that live in well-structured groups, usually within a home range, and that are highly sexual (often seasonal), with high intermale competition.

I mentioned that there are more arboreal species than open-country species. The probable reason is because there are more niches in forests, and populations have specialized in their diet, feeding area, daily rhythms, and so forth. The greater complexity of the forest should also increase population stability, thus increasing the number of species that can be maintained there. While the species-packed forests might increase the specialization of each species as it exploits a narrow niche, the opposite forces would operate on the open-area species. They would show very flexible behavior and great adaptability in exploiting a range of habitats in their relatively low competitive milieu.

Table 31.1 shows that coadaptations of Grade I primates are related to their nocturnal insect-hunting behavior (except for two leaf-eating forms). The switch to frugivorous and leaf-eating behavior in II and III is associated with diurnal activity and the formation of larger groups. One-male or multimale groups with territorial inclinations in II and III may not be able to reserve a specific territory, but such behavior at least serves to disperse the groups and therefore achieves one of the effects of successful territoriality.

Small size of the groups in many forest frugivores is probably related to a limited food supply due to high competition in the tropical forest. Under such

Table 31.1 Adaptive grades of primates.

| SPECIES, ECOLOGICAL, AND BEHAVIORAL CHARACTERISTICS | GRADE I | GRADE II | GRADE III | GRADE IV | GRADE V |
|---|---|---|---|---|---|
| *Species* | Microcebus *sp.*<br>Chierogaleus *sp.*<br>Phaner *sp.*<br>Daubentonia *sp.*<br>Lepilemur<br>Calago<br>*Aotus* trivirgatus | Hapelemur griseus<br>Indri<br>Propithecus *sp.*<br>Avahi<br>Lemur *sp.*<br>Callicebus *moloch*<br>Hylobates *sp.* | Lemur macaca<br>Alouatta palliata<br>Saimiri sciureus<br>Colobus *sp.*<br>Cercopithecus ascanius<br>Gorilla | Macaca mulatta, *etc.*<br>Presbytis entellus<br>Cercopithecus aethiops<br>Papio cynocephalus<br>Pan satyrus | Erythrocebus patas<br>Papio hamadryas<br>Theropithecus gelada |
| *Habitat* | *Forest* | *Forest* | *Forest—forest-fringe* | *Forest-fringe, tree savannah* | *Grassland or arid savannah* |
| *Diet* | *Mostly insects* | *Fruit or leaves* | *Fruit or fruit and leaves, stems, etc.* | *Vegetarian-omnivore Occasionally carnivorous in* Papio *and* Pan | *Vegetarian-omnivore* P. hamadryas *occasionally also carnivorous* |
| *Diurnal activity* | *Nocturnal* | *Crepuscular diurnal* | *Diurnal* | *Diurnal* | *Diurnal* |
| *Size of groups* | *Usually solitary* | *Very small groups* | *Small to occasionally large parties* | *Medium to large groups.* Pan *groups inconstant in size* | *Medium to large groups, variable size in* T. gelada *and probably* P. hamadryas |

| SPECIES, ECOLOGICAL, AND BEHAVIORAL CHARACTERISTICS | GRADE I | GRADE II | GRADE III | GRADE IV | GRADE V |
|---|---|---|---|---|---|
| *Reproductive units* | *Pairs where known* | *Small family parties based on single male* | *Multimale groups* | *Multimale groups* | *One-male groups* |
| *Male mobility between groups* | — | *Probably slight* | *Yes—where known* | *Yes in* M. fuscata *and* C. aethiops, *otherwise not observed* | *Not observed* |
| *Sex dimorphism and social role differentiation* | *Slight* | *Slight* | *Slight—Size and behavioral dimorphism marked in* Gorilla. *Color contrasts in* Lemur | *Marked dimorphism and role differentiation in* Papio *and* Macaca | *Marked dimorphism Social role differentiation* |
| *Population dispersion* | *Limited information suggests territories* | *Territories with display, marking, etc.* | *Territories known in* Aloutta, Lemur. *Home ranges in* Gorilla *with some group avoidance probable* | *Territories with display in* C. aethiops. *Home ranges with avoidance of group combat in others. Extensive group mixing in* Pan | *Home ranges in* E. patas. P. hamadryas *and* T. gelada *show much congregation in feeding and sleeping.* T. gelada *in poor feeding conditions shows group dispersal* |

*Source: Crook and Gartlan (1966).*

conditions, populations of similar adaptive types may develop, each efficiently exploiting its narrow niche. Recruitment in such groups may be a highly competitive process with most efforts to join the reproductive group meeting failure. Another advantage of small reproductive groups is that they may breed on a nonseasonal basis, since they probably do not suffer periods of drastic food reduction.

Grades IV and V, on the other hand, inhabit dry forest, savannah, and steppes where the seasons bring harsh, dry periods of food shortage. These grades probably show high mortality rates in older and infirm animals at such times. But in the rainy season food is abundant, and those animals that have survived the dry season embark on a program of vigorous breeding, which results in a replenishment of the population. As with birds, it is not clear whether the timing of the breeding season is correlated with the demands of "childbearing" or "childrearing."

Woodland savannah primates are subject to harsh predation pressures, and large groups probably provide a form of defense. Grouping, as we know, is not possible under all food supply conditions. Some baboons (for example, *Papio*) and other primates such as rhesus monkeys are able to congregate because they largely utilize patchy food sources that are not subject to overexploitation.

In the larger groups that occur in open country, males compete strongly for females, and therefore the animals show the results of sexual selection. Actually, however, their sexual dimorphism in size, appearance, and behavior is probably the result of a number of selective influences. For example, the animals live in close proximity in open country and therefore can easily see and respond to each other, thus increasing the likelihood of visual signaling. There is also stronger seasonality in savannah areas than in forests, and the females may all come into estrus at about the same time, thereby seasonally elevating the level of intermale competition. It has been suggested that the sex ratio is unbalanced, that there are fewer males than females, thus further heightening competition, and that open-country females, when they do come into heat, may be sexually attractive for longer periods than their forest-dwelling counterparts. All these factors could increase intermale competition and thus produce strong sexual dimorphism. The higher aggressive behavior that is correlated with the increased size of the male has an effect on the selectivity of the males involved in mating at any time in the mating season. The estrous females of open-country groups are promiscuous until the time of their greatest receptivity when they are likely to mate only with the troop leader. Crook and Gartlan believed that the sexual dimorphism arose as a response to sexual selection, that size and aggressiveness in males did not develop directly as a response to the need for troop defense, but that this need played a direct role in the organization of the troop and the male cooperative tendencies.

Grade IV encompasses species of rather varied habitats including not only typical forest-fringe and savannah-woodland species, but also a number of forest forms. These latter, in secondary habitats provided by man's destruction of their natural woodland home, show social organization similar to that of the savannah and forest-fringe species. All these species are characterized by a marked adaptability and flexibility of behavior. For example, where *Cercopithecus aethiops* exists on the rich little island of Louli, it is territorial, but it is not territorial where it exists in more expansive and impoverished areas. Such flexibility of behavior is undoubtedly highly adaptive in the variable forest-fringe environments.

The change from Grade IV to Grade V is characterized by a shift to grassland or dry savannah where food is even less abundant and, at least seasonally, more sparsely distributed. In such areas, it would be best for groups to split up into smaller more widely ranging parties to avoid local overexploitation of food. In such hard-pressed groups, there is no longer a place for large males that function only in mating and do not help rear the young. Such males only consume large amounts of food, which then is not used in maintaining the population. Instead, these species have developed another system. One male that can mate with all the females, and doesn't eat much, is about all the group can afford. In small groups, the protective advantage of males is reduced anyway, since even powerful males derive an important advantage from numbers. One-male groups (males with harems) necessitate the exclusion of other males from the group, and thus the intermale competition is intensified. The occurrence of all-male groups indicates strong forces that are able to exclude potentially reproductive individuals. In some

Figure 31.3 A colobus monkey. The striking cape clearly identifies this animal as a male, but the secondary sexual character is not particularly metabolically expensive, as increased size would be.

species, the strong sexual selection has resulted in such embellishments as lengthened and strongly marked hair on the back, forming dramatic capes that increase the visual impact and apparent size of the animals without great metabolic expense (Figure 31.3).

The differences in the coadapted characters of the three species of baboons represented in Grade V suggest the possible variation in the interaction of such environmental factors as food shortage, predation, and habitat structure. Thus *Theropithecus gelada*, which roams the Ethiopian mountains, forms herds under good conditions, but they split up as conditions become more rigorous. There is never a shortage of sleeping sites on the cliffs, no matter how large the group.

On the other hand, *Papio hamadryas* forms one-male groups, but tends to disperse less, apparently because they must remain near their limited number of sleeping sites. Such groups may be forced to converge at night.

The patas of open savannah grassland form one-male groups that keep well apart. There are no cliffs to serve as sleeping sites, and plenty of predators are about. At night, the animals disperse into separate trees and reassemble each morning. The male does not defend members of his group, but acts as a watchdog, sometimes risking himself as a diversion for predators. Such males show a rather low level of aggression in other ways also; for example, they do not show dominance over other members of the group. It seems likely that aggressiveness in these

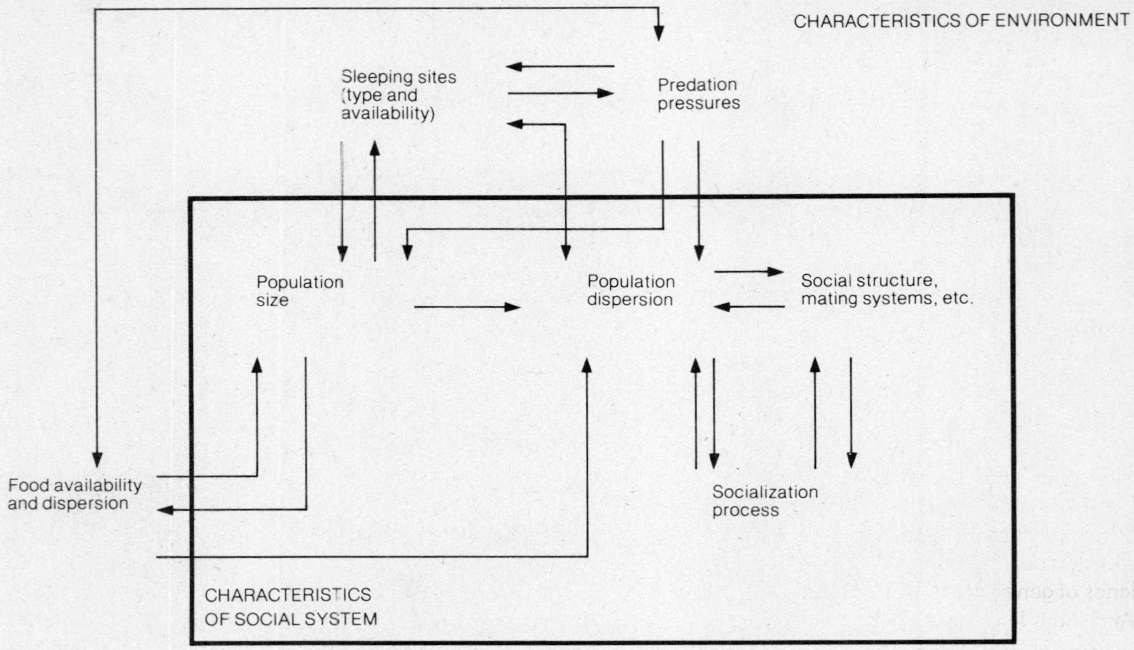


Figure 31.4 The theoretical relationship between habitat and specific social system through time in a given environment. The characteristics of the social system are shown inside the rectangle. The arrows show the direction of influence of one factor upon another.

males would be strongly selected against under the heavy predation pressures.

For each grade, the environmental variables such as food availability, predation, and availability of sleeping sites have a different value. Any environmental factor may be more important to one species than another. Crook and Gartlan prepared a general flow chart (Figure 31.4) that relates these environmental characters to the adaptive response of any of these grades. The study is an important one in that it most clearly demonstrates the importance of the interrelationship of social structure to environmental variables. Such studies, as they grow in number, should ultimately be useful to many disciplines including some areas of anthropology and psychology where there is too often a tendency to regard human social structures as cultural—divorced from any environmental, ecological, or evolutionary effects. This study has fallen under some criticism and certain parts have been discounted, but it remains a landmark study because of the discussion it generated (see Wallace and Klopfer 1979).

Social behavior, then, presents us with intense challenges as we seek to explain its adaptiveness and evolution. Obviously, it may arise on a number of selective bases, but our problems do not end there. Conceptually, we must decide at what point true social behavior begins, both individually and phylogenetically, and how it may shift in time, either in the normal course of development or in response to changing ecological pressures. It is obviously a powerful evolutionary force, as we will see in the next chapter.

# 32

# ALTRUISM AND WARFARE

Does the selfless act exist? Does any animal ever do anything for another one at a cost to itself? When a mother defends her young, is she doing it because she loves them or is she really only protecting her own genetic investment? When an uncle risks his life to save a nephew, is he only acting to increase his own kinds of genes in the population? Does altruism exist? And can it lead to war?

Anyone approaching the problem of the development of social behavior must eventually encounter the prickly question of altruism. It must be admitted that, so far, some of the answers have been unsettling. After all, we are attempting to dissect some of our most cherished values. One doesn't discuss mother's love lightly. But before we can decide if mother's love is altruistic, we must define the term.

## ALTRUISM

*Altruism* may be defined as any behavior that is detrimental in some way to the immediate well-being of the animal performing it, but beneficial to another organism. Altruistic behavior appears in very different sorts of animals from insects to vertebrates. It is common among the social insects (Hymenoptera and Isoptera), but among vertebrates it is most common in birds and mammals.

### Altruism and Fitness

We can account for a certain level of altruism according to our concept of fitness. A pregnant animal is at a disadvantage as she carries a fetus that drains her energy and may increase the likelihood of her being taken by a predator. However, the genes of animals that do not make this sacrifice are not represented in the next generation. Put another way, the species at any one time is composed of individuals whose mothers made that sacrifice. If the trait is inherited, or even culturally transmitted, the offspring would be prone to make the same sacrifice. An old one-liner states that if your parents didn't have any children, you probably won't either.

Males also may place themselves at a disadvantage for the sake of their offspring, although there is probably more variety in the levels and design of their contributions. Among polygynous species, a copulating male may reduce his vigilance for predators—or anything else—for a time, and so he is potentially altruistic as he seeks to produce offspring (albeit unwittingly). Males of monogamous species may spend their energy in bringing food to the young and they may even defend their family to the death. In each case, the net effect of male altruism is to increase the number of offspring that live to adulthood, each of which would then be themselves disposed to behave altruistically.

Altruistic behavior may also take forms that are not so easily explained. We have mentioned uncles rescuing nephews, but there are other, more peculiar, instances. Why would an individual feed the offspring of another animal, occasionally even of another species? In some cases, we must admit our ignorance. In other cases, we weakly explain, "The pattern is probably unadaptive." We are not forced to shrug before all such peculiar behavior, however. We are now armed with mathematical models that indicate the adaptiveness of "misplaced" altruism. For references, refer to the work of W. D. Hamilton (1963, 1964 a, b).

Problems in explaining altruism arise when behavior cannot be readily interpreted in line with the classical view of natural selection as it pertains to fitness. It seems that unless one wishes to accept the idea of group selection (Lack 1966; Maynard-Smith 1964), the fitness concept must be expanded. With very few exceptions, the mathematical explanation of the theory of natural selection does not account for the possibility of the evolution of any characters that are disadvantageous to the bearer. If natural selection actually operated as depicted in the classical models, species would show no behavior any more social than mating behavior or parental care. So, how can the classical model be expanded? The problem must be approached through describing the likelihood of any behavioral pattern resulting in an increase in the frequency of a certain kind of gene in a population. One must keep in mind that the gene that is thus "benefited" need not be directly descended from the individual performing the act. After all, we're talking about propagating *kinds* of genes in a population (Figure 32.1), and any individual shares *kinds* of genes with relatives, as we saw in our discussion of kin selection.

## A Model for Altruism

Haldane (1932) proposed a model by which altruism could be selected for if the starting gene frequency was high enough and the advantage to the group outweighed the individual disadvantage borne by the altruist. Hamilton expanded the idea to say that if the individual that is benefited is a relative of the altruist, then there is a strong likelihood of that recipient also carrying the gene for altruism, and so, given Haldane's requirements, altruism would be selected for.

So, what must be the relative advantages and disadvantages of altruistic behavior in order for a hypothetical "gene" for altruism to increase in a population? Hamilton determined that the relationship depends on how closely the altruist and the benefiter are related. The advantage must increase with the distance of the kinship or the behavior will disappear. To illustrate, altruism toward brothers and sisters will be selected for if the resulting gain is *more* than twice the loss, for half-brothers, four times the loss, and so on. This means that the altruistic animal

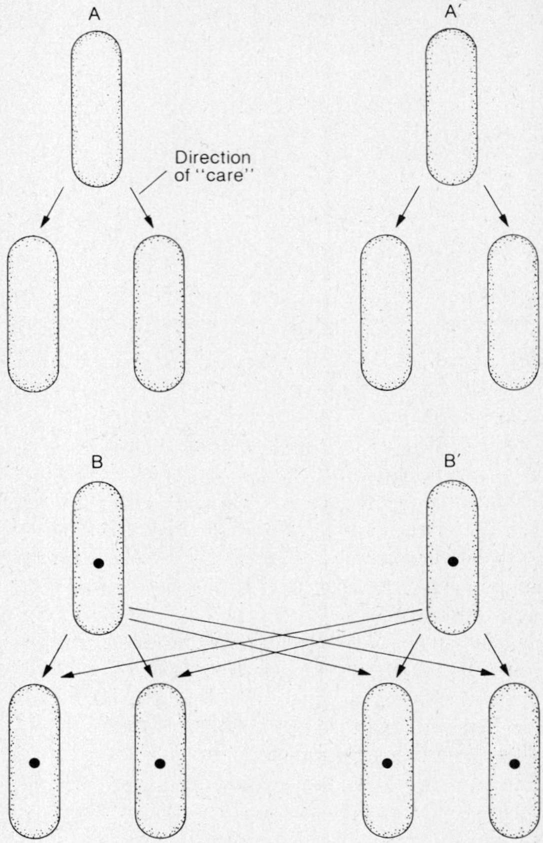


Figure 32.1 In this population, *A* and *A′* are nonaltruists. They behave in such a way as to maximize their own reproductive success, but do nothing to benefit the offspring of other individuals. In another segment of the population (*B* and *B′*), a gene for altruism has appeared that results in individuals benefiting the offspring of others in some way. It can be seen that, assuming the altruistic behavior is only minimally disadvantageous to the altruist, generations springing from *B* and *B′* are likely to increase in the population over those from *A* and *A′*. It should be apparent that the altruistic behavior is likely to be greatest where *B* and *B′* are most strongly related, so that *B* shares a maximum number of genes in common with the offspring of *B′* and vice versa. The idea is that *B*, for example, can increase its own reproductive success by caring for the offspring of a relative with whom it has some genes in common. After all, reproduction is simply a way of continuing one's own *kinds* of genes.

would sacrifice its life for more than two brothers, but not for less, for more than four half-brothers, but not for less, and so on. Therefore, we may deduce that a bird seeing a hawk will give a warning cry when the chance of attracting attention to itself is not too great and when the average neighbor is not too distantly related.

In keeping with our broadened concept of fitness, it seems that mutations toward altruistic behavior could best survive in closed groups such as families, demes, clans, and tribes, since individuals of such groups are most likely to be related (Eibl-Eibesfeldt 1970) (Figure 32.2). Hamilton (1964b) discussed the probability of individuals in colonies or clones being related.

Individual recognition may be an important factor in the development of altruism. Animals might conceivably be more inclined to behave altruistically in the presence of individuals they recognize, since recognition might reflect a kinship. The recognition might be a simple correlate of altruistic behavior or, in some mammals, such behavior might be predicated upon recognition in a purposeful sense. However, as yet there are almost no data to support such theorizing. On the same basis, altruism might also be expected in "viscous" groups. Species that do not move about much or those that return to or frequent the same areas would be likely to be more closely related than those that disperse randomly and freely.

## Reciprocal Altruism

The good Samaritan was perhaps a reciprocal altruist. He helped a fellow human in need, but was this a selfless act? Not if his behavior results in his receiving the same treatment should he be mugged someday himself. So, again, the selfless act appears to be shadowed by the umbrella of Darwinian selection. Let's take a closer look.

A man is drowning, and another man risks himself by jumping in to save him. The two are not related, so the rescuer was not saving his own kind of genes. It seems to be a case of pure altruism. However, suppose that the drowning man has $\frac{1}{2}$ a chance of drowning if left unassisted, and the rescuer has $\frac{1}{20}$ of a chance of drowning if he helps. Also suppose that if the rescuer drowns, the victim also drowns, but if the rescuer lives, he almost always saves the victim. If it were true that almost no one ever needed to be saved from drowning, there would be very little advantage in saving someone in the water. But if it happens frequently enough that the rescuer may well find himself in over his head sometime, then it benefits both men to have played the role of rescuer. The calculus shows us that each man would have traded a $\frac{1}{2}$ chance of drowning for about a $\frac{1}{10}$ chance. A population of reciprocal altruists, then, will be comprised of individuals with an increased fitness.

Now, if you're following all this, it may have occurred to you that you can beat the system. Cheat. Pass as an altruist, but never really risk your life. The problem is, since we are able to recognize each other individually and we have pretty good memories, people will sooner or later catch on and brand you as a cheater. Vigilant and suspicious citizens, then, will make sure that crime doesn't pay. They will remember you didn't play by the rules and might not save you should the occasion arise.

Perhaps, then, righteous indignation is selected for in humans. Aggressive moralists keep the system straight. In addition, guilt feelings may be adaptive in compelling people to behave altruistically (so that the guilt-ridden person may stand to benefit next time). Can guilt be selected for (Wilson 1975b)?

Obviously, if we're talking about morals, righteousness, and guilt, we're talking about humans. Since these traits are necessary to make reciprocal altruism work, it should not be surprising that the behavior is probably absent in the rest of the animal world. Perhaps animals are not smart enough to develop empathy or perhaps their relationships and their memory are too short to permit this sort of highly personal contract. Behavior that may be based in reciprocal altruism may be found in rhesus macaques, baboons, and anthropoid apes in which members of dominance cliques may aid each other reciprocally. And then chimpanzees, gibbons, African hunting dogs, and wolves may beg food from each other reciprocally (Wilson 1975b). The details of reciprocal altruism were described by Trivers (1971), and the genetic mechanism elaborated by Boorman and Levitt (1973).

## Altruism in Insects—A Special Case

Insects and termites present special cases of altruistic behavior. The reproductive specializations of these insects have rendered them so distinct from most

Figure 32.2 An animal will risk its own welfare to protect others if it is related to them, in other words, if they have genes in common. The protective animal will take greater risks tne more closely they are related. Here a leopard has been treed by three male baboons from a troop acting together.

other animals that generalizing from their behavior to that of other species is not likely to be valid. In some of these insects, as we know, there may be large segments of the population in which individuals labor valiantly and defend the group with their lives; but they are not involved in reproduction. The occurrence of such individuals may be an evolutionary result of deferred maturation culminating in permanently neotenous (sexually developed at immature stages) individuals. Among some of these social insects, the workers are strongly related as a result of all having descended from a single queen. Under such conditions, any genes present in an individual are very likely represented in the other individuals as well, so altruistic behavior may be expected. The key lies in whether the queen carries genes for altruism so that a worker bee sacrificing herself for a queen (or a soldier termite sacrificing himself) ensures the transmission of genes for such behavior. It will be recalled that the recipient must carry the altruistic gene expressed by the altruist. (The term "altruistic gene" is used advisedly and refers to any genetically-based mechanism that encourages altruistic behavior.)

Hamilton provided us with an elegant description of how such a system could have arisen in the hymenoptera and termites, animals in which an altruistic system has apparently evolved independently at least ten times. The system works for them because of their unusual system of sex determination. Males are produced from unfertilized eggs and so they have half the number of chromosomes as the females; in other words, only one of each type. The result is that all the sperm from any male contain identical chromosomes, since they do not undergo meiosis. So, all the daughters have at least fifty percent of their genes in common as a result of one parent producing only one type of gamete. Furthermore, on the average, the daughters share another twenty-five percent of their genes in common from their mother so that the total genetic affinity among daughters is seventy-five percent. Now, each daughter has only the normal fifty percent genetic affinity with her mother, so hymenopteran sisters are more strongly related to each other than to their mother. In the context, then, of what we know about kin selection, we can see that it is more advantageous for each daughter to help rear her sisters than for her to have offspring of her own. We can also see the role of sex determination in the development of altruism in the hymenoptera.

Another aspect of altruism in insects involves the evolution of distastefulness (Hamilton 1964b). The concept is an important one since it involves not only the basis for warning coloration but also for both Batesian and Mullerian mimicry. The problem is, How could warning coloration have evolved, since at every stage, the more conspicuous individuals would be selected out? An early geneticist, R. A. Fisher (1930), suggested that nearby siblings would be benefited the most and theorized that perhaps such warnings are correlated with gregariousness. To illustrate, if a noxious larva with warning coloration (Figure 32.3) were taken by a bird, the bird would not be likely to eat another one. So, while the selective advantage of saving a brother is only half as great as if the individual itself remained protected, the possibility of saving an entire brood increases the likelihood of conspicuousness developing. This means that it is best for a conspicuously colored moth to continue to live at least through the egg-laying period of the species. This way, the number of noxious individuals is increased and the probability that any specific individual will be taken is reduced.

There is another way altruism can manifest itself—through the suicidal behavior of those individuals who have passed their reproductive period. A postreproductive animal who can offer its offspring nothing more might be expected to be entirely altruistic. Any kinship at all with a neighbor would be sufficient to select for all-out sacrifice. Certain postreproductive moths exhibit such selflessness (Blest 1957). It was noted that in cryptic species the discovery of any member by a predator increases the likelihood of others also being taken, as predators form search images. The predator that captures a cryptic individual, in effect, will take a step toward recognizing the crypsis. (The operational principle is the opposite of that for species with warning coloration where the discovery of any individual benefits the rest.) Therefore, with the edible, cryptic moth it is best to die directly after reproduction. It was found that some postreproductive cryptic moths behave in such a way as to destroy their camouflage. They fly about conspicuously, increasing their chances of being caught and using up the remains of their vital reserves. They fray their wings beyond recognition, thus no bird will be able to cue in on their physical appearance to find more like those. Hamilton pointed out that the selective forces operating on the postreproductive life span are doubtless weak and will generally be strongest where the relationship of neighbors is highest (that is, in the more viscous populations). (For an interesting and readable account of how the system could operate, see Dawkins 1976.)

### Continuing Puzzles

Hamilton (1964b) noted that the literature is scattered with records of behavior that do not conform (or are antithetical) to the principles we have just discussed. In some cases, kinship is blatantly disregarded, and in other cases, animals are not harmed or benefited at all according to our expectations. For example, how can one explain the behavior of birds feeding the young of other species (Summers-Smith 1963)? Actually, such behavior occurs so seldom that one can say it is undoubtedly unadaptive behavior—a mistake. (But, then, perhaps the bird is "practicing.")

Other puzzling cases are much more regularly reported. As examples, some beehives pool their broods in a common nest, as do some birds; some hymenoptera are unable to recognize their own nests; some birds are unable to recognize their own young in a group with other offspring; and penguins leave their young with baby-sitters while they go off on fishing expeditions. Hamilton suspects in such regularly occurring cases that there is a strong likelihood of misinterpretation by the human observer or that some animals are deceiving other animals.

For general reviews of social behavior with extensive references to the literature, see Allee (1951); Eibl-Eibesfeldt (1970); Ewer (1968); Tinbergen (1953a); Crook (1970a,b); Wilson (1975b); Barash (1977).

## COHESIVENESS, ORDER, AND WAR

It might seem strange that in a chapter on altruism we should discuss war. One might think, if it is going to be considered at all, it should be included in the discussion of aggression. However, war is a strange phenomenon, to say the least. It may well be that its origins are steeped in a community spirit as well as xenophobia. Perhaps, in fact, warfare can be explained on the basis of group cohesiveness in an orderly system, coupled with out-group aggression.

Among warfare's most interesting qualities are its paradoxes. I have only met one person who said that he supported the notion of war. He was a second lieutenant, just out of ROTC. Almost everyone seems to abhor the idea. Politicians uniformly agree it is a tragic endeavor. Young men go to war either reluctantly or saying something about "fighting for peace." But in any case, the decisions are made and the young men bundle their belongings and go off. Of course, they are risking their lives for the lives,

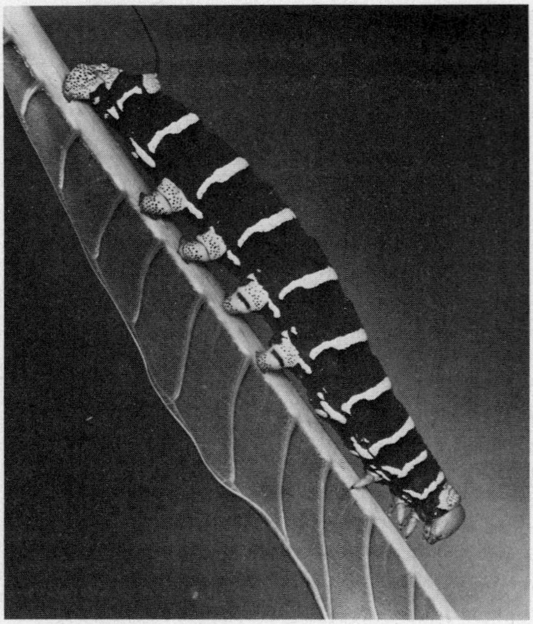

Figure 32.3 Warning coloration in poisonous species could have arisen if a strongly marked organism saved a large number of siblings by being taken by a predator. Shown here is a *pseudosphynx tetrio*.

businesses, or commodities of others, so perhaps they are behaving as altruists. But the point is, no one seems to like war. So what is the nature of our character, as a species, that enables us to continue, time and again, to do something we know is wrong?

## The Value of War

There's a catch. *Is* it "wrong"? Or has warfare served a valuable service in the evolution of our species? As men roamed the earth in small, intrarelated groups during those first cooperative efforts at bringing down big game, they might have reacted negatively to the presence of competing bands. After all, big game may be sparsely distributed, or seasonal in its appearance, so that meat was likely to be scarce at times. It would then have been to any group's advantage to exclude other groups from sharing its resources.

By the time men had coordinated their hunting efforts and become skilled at hurling weapons at large game, they would also have been in a position to effectively kill other men. The same skills that went into making man an efficient predator could also be used in making him a formidable warrior in those skirmishes over game or hunting areas. Of course, the goods of the vanquished would also become available to the victor. So, successful warriors would reap immediate rewards, such as booty, food, and women. Thus, the practice of men attacking other men to drive them from the hunting area might have been expanded as they began to actively seek out other men in order to conquer them, thereby nurturing the development of warfare.

One quality that is apparently extremely important in our permitting war is a sense of belonging. We are acutely aware of our affiliations. The breakdown of human families and tribes is apparently a very recent evolutionary phenomenon. Traditionally, we have "belonged." Even if we consider ourselves rebels against the established social system, we identify with other rebels in a reciprocated dislike of the establishment types. Often, all that is necessary to cause a feeling of dislike is to label someone as one of "them," whoever "them" is. People who don't know chess from Chinese checkers will devotedly follow a championship chess match if one of "ours" is playing one of "theirs."

## The Bull Dogs Versus the Red Devils

In one well-known experiment, two social psychologists (Sherif and Sherif 1953) found how important the concepts of "we" and "they" can be in aggression. At a boy's camp, after watching natural friendships develop spontaneously, the boys were divided into two groups, so as to split up the friends. The boys were initially disappointed at being separated from their friends, but quickly established new friendships within their subgroups. The groups were given names with which the boys could identify themselves and each other: the Bull Dogs and the Red Devils. Competition between the groups was encouraged and quickly grew intense. After a long series of tournaments, the winners, the Bull Dogs, were awarded prizes. Then, the "counselors" staged a smash-up of the Bull Dogs' refreshments, which were to be enjoyed at a party, and blamed the Red Devils.

At that point, the groups clashed so severely that the researchers tried to call off the experiment. They found that this was impossible and the two groups continued nighttime raids on each others quarters. The psychologists were finally able to restore a measure of order by scheduling a softball game against a team from a nearby town. The two groups united to meet this outside threat and remained friendly enough thereafter. This, of course, is a ploy still used by political leaders who feel their position threatened and who therefore initiate hostilities with an adjoining nation or some other out-group in an effort to close their own ranks. It has even been suggested that nations had their origins in the unions of separate tribes so that limited resources such as fertile river valleys could be better defended (Carneiro 1970).

### Social Facilitation and Warfare

Social facilitation may figure importantly in the behavior of warring humans. It has been well known for centuries that ceremonies, war dances, parades, and other grandiose affairs can bring men who aren't particularly keen about fighting to the point that they are anxious to enter the conflict (Figure 32.4). As the mood picks up, there may even be an exuberance, expressed at the "give 'em hell" sendoffs of the U.S. troopships in World Wars I and II. And, once the fighting actually starts, the social contagion becomes even more apparent. The sight of some men charging over the hill is likely to serve as an impetus for others surging forward. And, as we saw in the fall of Thieu's South Vietnam, when soldiers begin witnessing others retreating, even brave men may throw down their guns and run. Social facilitation may also contribute to the slaughters, burnings, and rapes that are so common to rampaging armies. Keep in mind that the general phenomenon of social facilitation may be highly advantageous in the coordination of less drastic forms of group behavior and could have been maintained in our behavioral repertoire on some other adaptive basis.

It has been suggested that obedience to authority may also be highly adaptive in that it permits young individuals to profit from their elders' experiences while they learn the patterns that will enable them to function competently as individuals (Waddington 1960). Of course, it is just such obedience that compels young men to go to war when they would rather be at the beach. Obedience to authority may have been an important quality in the success of our species (and still can be, depending on who is in authority). As long as we have this tendency, then, it is exceedingly important whom we elect to positions of power. Perhaps we should look more carefully at our political leaders and ask if this is *really* the best we can do. Are these the best and wisest from among us? If not, perhaps we should reanalyze our political processes.

### Response to Authority

Leaders are important in maintaining group harmony, and systems involving leaders and followers have undoubtedly been exceedingly important in our own evolution (Tiger and Fox 1969), but our tendency to blindly obey anyone in authority can sometimes be startling.

Stanley Milgram performed an experiment with a group of men who were told that they were going to be taking part in a learning "experiment." They would be the "teachers," they were told, and a stranger (actually an accomplice of Milgram's) would be the "learner." The teacher watched the learner being taken into another room and strapped into a chair fitted with electrodes from which he could receive shocks ranging from 15 to a powerful 450 volts. Actually, no shock was possible at all. The teacher then sat at a control panel in the next room and administered increasingly stronger "shocks" each time the learner made a mistake in memorizing a list of words. The teacher could not see the learner, but he could hear him, or so he thought, in the next room. On the panel before the teacher the voltages were labelled from "low shock" to "danger—heavy shock."

Four sets of teachers were tested. In each set were forty blue-collar workers, forty white-collar workers, and twenty professional people. In the first group, the learner objected by pounding against the wall when the 300 volt lever had been pressed. At 315 volts and above, he ceased to respond at all. In the second series, a protesting recorded voice was played to each teacher. The teacher heard mumbling at 75 volts, at 120 volts the learner shouted at the pain and demanded to be released. At 180 volts, he screamed that he couldn't stand it. From 315 volts on, the teacher heard only groans each time he administered

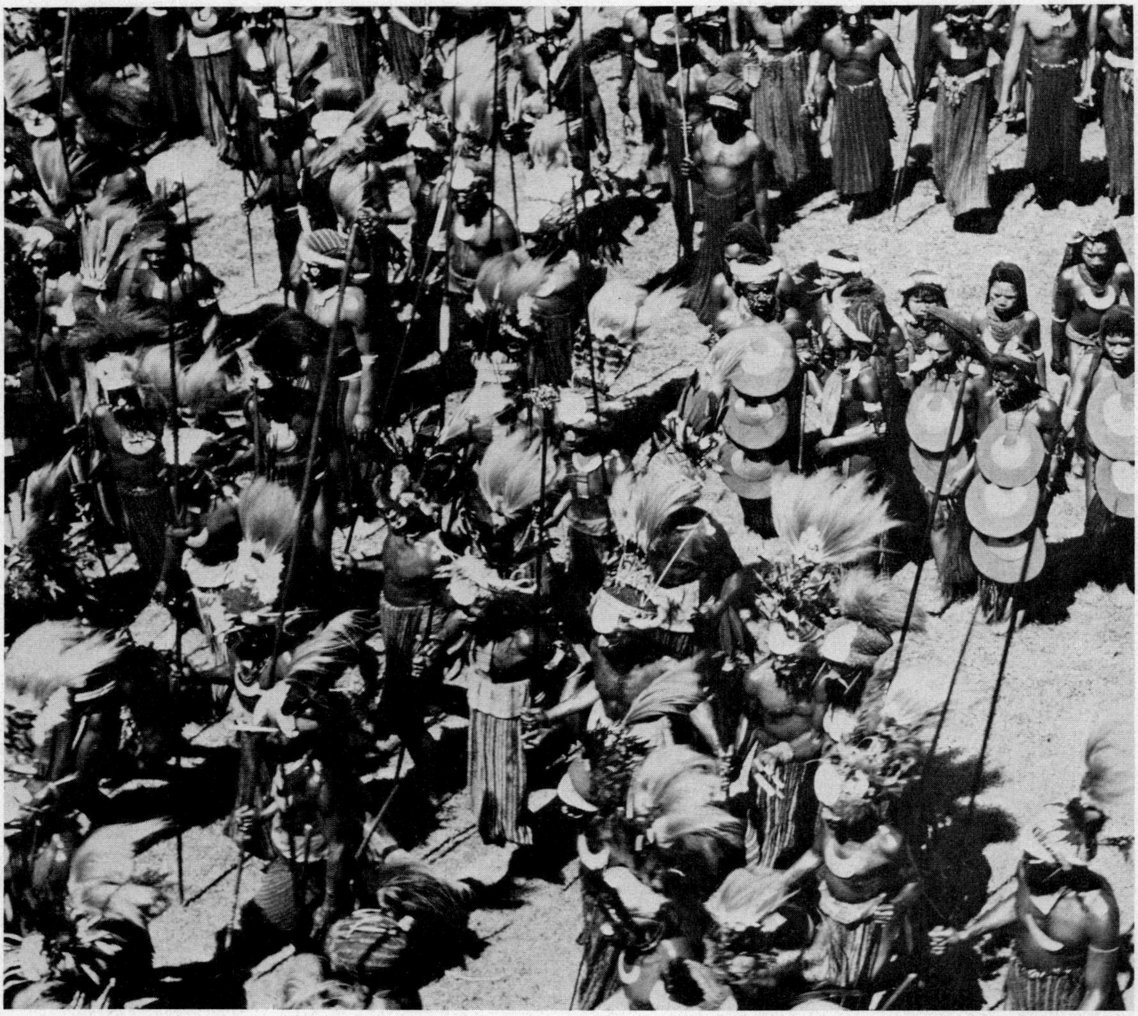

Figure 32.4 There are certain elements in common in human warfare, no matter what social differences exist. One such element is a form of social contagion. Men begin to inflame each other in a New Guinea tribe (above), and emotions may run strong as a result of military parades in industrial countries (opposite page).

a shock. In the next set of experiments, the teacher and learner were in the same room, about two feet apart, so that the teacher could see the reactions of the learner, an accomplished actor.

And in the fourth series, forty men were tested in each social group. But here the learner was able to remove his hand from the "electrified" metal plate. The teacher in this case was ordered to hold the hand of the learner on the plate if he removed it, so that he could receive his shock. In any of the experiments, if the teacher (men twenty to forty years old) objected to administering the shock, they received a stereo-

typed answer from the experimenter: "You have no choice. You must continue."

In the first two experiments, over sixty percent of the teachers continued to give the shocks in obedience to the psychologist's orders. When they were sitting next to the learner, forty percent carried out the orders. And when they were forced to touch the learner, "only" thirty percent would obey.

The results showed that a large number of men would not oppose authority (in the form of the experimenter). Under the vocal feedback conditions, they administered shocks to the victims in 62.5 percent of the cases when, in reality, they would have killed or seriously injured these people. And another thing. This took place in the United States, in a culture that more-or-less discourages blind obedience and prides itself on fostering independent thinking.

Before the experiments were started, forty leading psychiatrists were asked to predict the outcome. Most predicted that the "teachers" would not go beyond

150 volts and that only one in a thousand would follow all the steps. How wrong they were. Human behavior again failed to meet the cultural ideal.

Milgram attached a postscript to his report that provides some food for thought:

*With a numbing regularity we saw good people submit to the demands of authority and commit actions that were without feeling and cruel. Persons who in their daily lives were responsible and decent were led to commit cruel acts by the pretension of authority and the uncritical acceptance of the experimenter's definition of the situation.*

*Where is the boundary for such obedience? In many instances we tried to provide such limits. Screams of the victims were used; they were not enough. The victim complained about heart trouble; still some subjects shocked them when they were ordered to do so. The victim pleaded to be released and his answers were not registered by the signal apparatus; the subjects continued to shock them. Initially we had not expected that such drastic measures would be needed to obtain a refusal to cooperate, and each step in the experimental procedure was only added to the degree to which the ineffectiveness of the previous condition became revealed. The last attempt to erect a barrier was the condition in which the subject had to touch the victim. However, the very first subject already used force against the victim and proceeded to the highest shock intensity. One quarter of the subjects behaved similarly under these conditions.*

*The results—as they were observed and felt in the laboratory—cause great concern to this author. They allow for the possibility that human nature or—more specifically—the kind of person produced by American society would not afford much protection to its citizens against brutal and inhuman treatment at the behest of an evil authority. These people did to a large percentage what they were told, irrespective of the nature of their activity and without conscientious objections, as long as they saw that the order came from a legitimate authority. When, as in this study, an anonymous experimenter could successfully order adults to force a 50-year-old man into submission and administer painful electric shocks to him despite his protests, then one can only be apprehensive about what a government—with much more authority—could order its subject to do.* [*Miligram* 1966, *pp.* 460 *ff.*]

Again, however, let us recall that blind obedience to authority has not traditionally been necessarily evil. In fact, it may have been highly advantageous in other times and in other places. It may still be an essential component of smoothly running societies. Obedience to authority certainly diminishes dissonance within the ranks. And this may have been critically important when men moved over the plains in small groups in search of large animals to kill. Killing demanded organization and organization demanded a system of leaders and followers. If everyone is making the decisions, everyone starves to death. It may have been just that simple.

## How We Ended Up Here—A Scenario

So, let's see if we can pull a few of these admittedly oversimplified ideas together and try to reconstruct the series of events that led us to the point that we are capable of, and continually threatening to, annihilate ourselves. Why was a large, strange-looking ship, reportedly loaded with nuclear weapons, anchored in front of my house in the Bay of Ibiza in the Balearic Islands for several days? Why did Spanish newspapers that very day carry reports of a huge nuclear blast conducted in the United States? Are we completely insane? Maybe a case could be made for that, but the admission puts us "educators" out of business. So, let's assume that we're sane and that we are acting quite rationally, but, to some unknown degree, under the influence of our heritage—a history of selection for traits that have faithfully brought us to whatever success we enjoy today, but which has perhaps not had time to accommodate itself to our new kind of world.

It may have all started with those primates that began to regularly eat meat possibly as long as five million years ago. These animals, in their quest for highly proteinaceous game, then moved from the forests out into the flat, grassy savannahs where game was more plentiful (where remnants of the great herds still remain). The shift to open grassland necessitated two important changes in the predatory primates. If we assume that they were disposed toward taking larger game rather than searching for small mouselike forms or insects, we can see there may have been selection for bipedalism. On two legs they could see farther across the land as they searched for dispersed prey. The tendency toward bipedalism would have been associated with changes in the hip musculature and the development of longer legs. The result would be an animal that could run longer and faster distances in pursuit of prey. Even today, trained human runners can outdistance most four-legged animals, including canids, ungulates, and many members of the horse family.

The bipedalism and erect posture would have then freed the hands from their locomotor functions. This would have been an exceedingly critical development, since the hand would have been able to specialize. The result is a structure that can grip powerfully or hold delicately and precisely. This kind of hand would then become capable of manufacturing and using devices to bring down ever-larger (and hence more economical) game (see Figures 4.13 and 4.15 in Chapter 4).

However, the freed hand would have gone nowhere without a nervous system that could take advantage of its potential. Thus, a larger brain would have developed in conjunction with an ever-more-talented hand.

At the same time, a larger brain would have been selected for on another basis. As the savannah primates brought down larger, fleeter, or more dangerous game, a premium would have been placed on coordinated group effort. Thus, the need for social organization would have selected for intelligence. As leaders emerged, of course, they would have had to be able to communicate with their followers, and group coordination would have demanded that the members of the hunting parties be able to communicate with each other. Thus, speech would have become more highly developed. And, here, we're not only referring to the ability to talk, but the ability to communicate abstract ideas, such as, "Tomorrow we will hunt there unless. . . ." or "You get his attention and I'll hit him with a rock." So, the twin forces of freed hands and the need for language would have rapidly increased intelligence. (Of course, a certain level of social hunting can arise without language as demonstrated in carnivores such as wolves and African hunting dogs.)

Social bonding would have been strong by this time. Anthropologists Lionel Tiger (1969) with Robin Fox (1971) argued that the necessity for strong bonds between males hunting as a group has resulted in men being more "political" than women, who never had to develop such a strong social structure in their roles of mothers and gatherers of plant food. (The importance of plant food in primitive societies should not be underemphasized. For example, the diet of at least one modern hunter-gatherer group is composed of about eighty percent plant material (Lee and DeVore 1968).) Such bonds, Tiger and Fox maintain, help to explain the preponderance of males in political circles throughout history, and the stronger tendency of males to join all-male secret societies or other all-male social groups.

And this brings up another point. With hunting, there would have developed a strong difference in the roles of the sexes. Women were undoubtedly generally slower, smaller, and weaker than men because their roles as mothers and plant-gatherers did not call for size, strength, or speed. Since women would have been forced into close association with their young through their nursing duties, brash or aggressive women might have left fewer offspring in each generation as they drew attention to themselves or failed to seek out the protection of males. Also, the body that was adapted for childbearing (for example, wide hips) and nursing would be incompatible with hunting. On the other hand, women were probably closer to optimal size than men and tended to live longer. Women were physically and psychologically suited to their duties, and men to theirs (Tiger and Fox 1971). As greater premium was placed on intelligence, babies would have been born with increasingly large heads to accommodate their complex brains. Thus, the pelvis of the woman would have had to be wider to allow free passage of the cerebrated baby. The wider hips would have resulted in a shift in the joints of the arms and legs of the woman, which would make them even less efficient at throwing and running (Napier 1967), increasingly dependent on the male for such tasks, as sexual roles became increasingly defined.

Apparently associated with a change in social behavior and the roles of the sexes, was a loss of estrus in human females. They no longer came into "heat" periodically, copulating only when the baby could be expected to arrive at a propitious season. This meant that women were ready to copulate at any time, and it may have been just this willingness that kept males around.

Pairing choice is *another* question. Why a male and female would develop a strong attachment to each other to the exclusion of others, more or less, is not well understood, but it obviously resulted in each leaving more offspring than if they were promiscuous—as are most primates. However, with the female capable of copulating at any time, it would have been to the male's advantage to be vigilant (perhaps suspicious and jealous) and to see to it that his hunting and protective prowess were not expended in rearing another male's offspring. A strong feeling of attachment ("love") would have increased the likelihood of his remaining near his mate and decreased the likelihood of her copulating with another male. Continual sexual receptivity would have aided the woman's reproductive efforts in the same way. She would have been able to keep her own male with her to avail himself of her favors and to see to it that other males were denied access to her. In addition, she could immediately avail herself of the services of a superior male, should one appear, thus gaining superior genes for her offspring and allowing her own mate to assist her in rearing them.

The loss of estrus may also have been important in the social cohesion of the group (Jay 1968). Among social primates, when many of the females enter estrus at about the same time there is a sudden increase in tension and aggressiveness within the group as a result of the onset of sexual activity. Asynchronous breeding would have reduced such intensely disruptive periods. Also, the loss of traumatic aggressive interactions, as mothers consorted with males, would have resulted in a calmer period of early development of offspring and hence strengthened mother-offspring bonds, the real basis of many kinds of social units. The idea, then, is that it would have been difficult for early man to maintain a small interrelated group if females underwent estrus cycles. There is another upshot to all this. The fact that children were being born at all times of the year meant that a great deal of food was constantly needed. This could have placed a greater premium on real estate for early humans. Arable land or land with a standing biomass held food needed for the young. The more land available, the greater the likelihood of some part of it bearing food at any given time. Thus, lack of estrus could have resulted in a greater interest in space in early humans. Women, then, may have been more likely to accept a man as a mate who showed promise of being a holder of land as well as a good hunter and protector.

Protector from what? Perhaps other carnivores such as the great cats that roamed the earth. Or perhaps from other humans behaving altruistically, risking their personal well-being to bring home, to their own group, commodities. Or perhaps from other humans seeking land. Those "others" may have been acting out of greed or aggression or may have simply been responding to orders, not sure at all why they were there except that they were told to be. In any case, however, confrontation produced goods that benefited either one's own descedant genes or one's kind of genes, perhaps in the long run, perhaps immediately.

Much of this material is obviously sheer conjecture. Our generalities are compromised by the exceptions. If we are genetically driven, then we must admit that the Hopis and other peaceful groups are either overriding their natural aggressive tendencies or that they fall at one tail of the aggressive curve. We are perhaps far afield in considering human aggression in this way at all. There are those who say we have no *right* to do so, that such discussion, or the fruits of such discussion, can do us no good; that the consideration of human behavior in Darwinian terms breeds or justifies social darwinism. And, besides, if we are genetically driven, what good are our efforts at social reform? (For an account, see *Time*, December 23, 1976.) However, we did not overcome gravity by denying its existence. The biological study of social behavior does not (should not) place values on that behavior (Wallace 1977). It is but another way to see—a paradigm within which to frame our questions. It cannot be denied that there are ill-motivated and biased souls who would seek to derive some solace from this kind of discussion, but we can ignore their efforts and look now to the greater good. It is important that we do. Natural selection is a powerful and pervasive force, and we are but one of its products. To set ourselves aside from the other species, pretending to be exempt from its influences, is to deny ourselves yet another window into our own nature. The windows we have are few enough and besmudged at best, and some that appear to be windows, it turns out, let no light pass. In these times, we can't afford to ignore any opportunity to better understand ourselves. If the effort is unproductive, we will know soon enough; our methods of discovering these things are becoming more effective. We can no longer afford the luxury of smugly denouncing paradigms that seem not to fit our political or personal preconceptions. If we are to ever see the kind of world we want, it is foolish to pretend it now exists. Our position grows exceedingly precarious as we add seventy million new people to our small planet each time it circles its star, even as our technological abilities place increasingly awesome powers into the hands of so few people. We indeed need all the wisdom and insight we can muster, and ethology is at last beginning to yield its promise as a means of helping us know what manner of creature we really are.

# BIBLIOGRAPHY


Adler, H. E. 1963. Psychophysical limits of celestial navigation hypotheses. *Anim. Behav.* 11:566–577.

Adler, K. 1976. Extra-ocular photoreception in amphibians. *Photochem. Photobiol.* 23:275–298.

Adriaanse, A. 1947. *Ammophila campestris* Ltr. und *Ammophila adriaansei* Wilcke: Ein Beitrag zur vergleichenden Verhaltensforschung. *Behaviour* 1:1–34.

Agranoff, B. W. 1967. Memory and protein synthesis. *Sci. Amer.* 216:115–122.

Ahrens, R. 1953. Beitrag zür Entwicklung des Physiognomie und Mimikerkennens. *Z. Exptl. Angew. Psychol.* 2:412–454, 599–633. Cited in Eibl-Eibesfeldt 1970.

Albrecht, H. 1966. Zür Stammesgeschichte eineger Bewegungsweisen bei Fischen, untersucht am Verhalten von Haplochronis (Pisces, Chichlidae). *Z. Tierpsychol.* 23:270–301.

Aldrich-Blake, F. P. G. 1970. Problems of social structure in forest monkeys. In *Social Behaviour in Birds and Mammals,* ed. J. H. Crook, pp. 79–99. Academic Press, New York.

Alexander, R. D. 1957. Sound production and associated behavior in insects. *Ohio Sci.* 57:101–113.

———. 1960. Sound communication in Orthoptera and Cicadidae. In *Animal sounds and communications,* ed. W. E. Lanyon and U. N. Tavolga, pp. 38–92. AIBS, Washington, D.C.

———. 1961. Aggressiveness, territoriality and sexual behavior in field crickets (Orthoptera: Gryllidae). *Behavior* 17:130–223.

———. 1966. The evolution of cricket chirps. *Natur. Hist.* 75:26–31.

———. 1974. The evolution of social behavior. *Ann. Rev. Ecol. Syst.* 5:325–383.

Alexander, R. D., and W. L. Brown, Jr. 1963. Mating behavior and the origin of insect wings. Occasional Papers of Museum of Zoology of the Univ. of Michigan. No. 628., pp. 1–16.

Allee, W. C. 1926. Studies in animal aggregations: causes and effects of bunching in land isopods. *J. Exper. Zool.* 45:255–277.

———. 1951, reprint of 1938 edition. *The social life of animals.* Beacon Press, Boston.

Allee, W. C., N. Collias, and C. Z. Lutherman. 1939. Modification of the social order among flocks of hens by injection of testosterone propionate. *Physiol. Zool.* 12:412–420.

Altman, J. 1966. *Organic foundations of animal behavior.* Holt, Rinehart & Winston, New York.

Amadon, D. 1950. The Hawaiian honeycreepers (Aves, Drepaniidae). *Bull. Amer. Mus. Natur. Hist.* 95:151–262.

———. 1959. The significance of sexual differences in size among birds. *Proc. Amer. Philos. Soc.* 103: 531–536.

Andersson, B. and S. M. McCann. 1955. Drinking, antidiuresis and milk ejection from electrical stimulation within the hypothalamus of the goat. *Acta Physiol. Scand.* 35:191–201.

Andrew, R. J. 1956. Intention movements of flight in certain passerines, and their use in systematics. *Behaviour* 10:179–204.

Andrewartha, H. G. 1961. *Introduction to the study of animal populations.* Univ. of Chicago Press, Chicago.

Andrewartha, H. G., and L. C. Birch. 1954. *The distribution and abundance of animals.* Univ. of Chicago Press, Chicago.

Anthoney, T. R. 1968. The ontogeny of greeting, grooming and sexual motor patterns in captive baboons (supersp. *Papio cynocephalus*). *Behaviour* 31:358–372.

Antonovics, J. 1971. The effects of a heterogeneous environment on the genetics of natural populations. *Amer. Sci.* 59:593–599.

Archer, J. 1970. Effects of population density on behavior in rodents. In *Social behavior in birds and mammals,* ed. J. H. Crooke. Academic Press, London.

Ardrey, R. 1962. *African genesis.* Collins, London.

———. 1966. *The territorial imperative.* Dell, New York.

———. 1970. *The social contract.* Atheneum, New York.

Armstrong, E. A. 1955. *The wren.* Collins, London.

———. 1965. (first published 1947). *Bird display and behavior.* Dover, New York.

Aronson, L. R. 1965. *Environmental stimuli altering the physiological condition of the individual among lower vertebrates.* In *Sex and behavior,* ed. F. A. Beach. Wiley & Sons, New York.

Aschoff, J. 1963. Comparative physiology; diurnal rhythms. *Ann. Rev. Physiol.* 25:581–600.

———. 1965a. *Circadian clocks.* North Holland Publishers, Amsterdam.

———. 1965b. Circadian rhythms in man. *Science* 148:1427–1432.

Ashmole, N. P. 1963a. The biology of the wide-awake or sooty tern *Sterna fuscata* on Ascension Island. *Ibis* 103b:297–364.

———. 1963b. The regulation of numbers of tropical oceanic birds. *Ibis* 103b:458–473.

———. 1967. Sexual dimorphism and colonial breeding in the woodpecker, *Centurus striatus, Amer. Naturalist* 101:353–356.

Atz, J. W. 1970. The application of the idea of homology to behavior. In *Development and evolution of behavior,* ed. L. R. Aronson, E. Tobach, D. S. Lehrman, J. S. Rosenblatt. W. H. Freeman, San Francisco.

Ayala, F. J. 1972. Competition between species. *Amer. Sci.* 60:348–357.

Azrin, N. H., R. R. Hutchinson, and R. McLaughlin. 1965. The opportunity for aggression as an operant reinforcer during aversive stimulation. *J. Exper. Anim. Behav.* 8:171–180.

Baerends, G. P. 1941. Fortplanzungsverhalten und orientierung der Grabwespe, *Ammophila campestris. Tijdschrift voor de Entomologie* 84:68–275.

———. 1950. Specializations in organs and movements with a releasing function. *Symp. Soc. Exper. Biol.* 4:337–360.

Bainbridge, R. 1961. Migrations. In *The physiology of crustacea,* vol. 2, ed. T. H. Waterman, pp. 431–463. Academic Press, New York.

Baker, H. G., and G. L. Stebbins, eds. 1965. *The genetics of colonizing species.* Academic Press, New York.

Baldwin, J., and J. I. Baldwin. 1972. The ecology and behavior of squirrel monkeys (*Saimiri oerstedi*) in a natural forest in Western Panama. *Folia Primatol.* 18:161–184.

Baldwin, P. H. 1953. Annual cycle. Environment and evolution in the Hawaiian Honeycreepers (Aves, Drepaniidae). *Univ. Calif. Publ. Zool.* 52:285–398.

Barash, D. P. 1973. Human ethology: personal space reiterated. *Envir. and Behav.* 5:67–72.

———. 1974. Neighbor recognition in two "Solitary" carnivores: the raccoon (*Procyon lotor*) and the red fox (*Vulpes fulva* ). *Science* 185:794–796.

———. 1977. *Sociobiology and behavior.* Elsevier, New York.

Barber, H. S. 1951. North American fireflies of the genus *Photuris.* Smithsonian Institution's Miscellaneous Collection 17:1–58.

Barlow, G. W. 1968. Ethological units of behavior. In *The central nervous system and fish behavior,* ed. D. Ingle. Univ. of Chicago Press, Chicago.

———. 1974. Contrasts in social behavior between Central American cichlid fishes and coral-reef surgeon fishes. *Amer. Zool.* 14:9–34.

Barlow, H. B. 1953. Summation and inhibition in the frog's retina. *J. Physiol.* 119:69–88.

Barlow, H. R., and R. Hill. 1963. Selective sensitivity to direction of movement in ganglion cells of the rabbit retina. *Science* 139:412–414.

Barnett, S. A. 1963. *A study in behavior.* Methuen, London.

———. 1964. The concept of stress. In *Viewpoints in biology,* pp. 170–218. 3, Butterworth, London.

———. 1971. Apes and monkeys: social behavior. In *Topics in the study of life.* Harper and Row, New York.

Barth, R. H., Jr. 1961. Hormonal control of sex attractant production in the Cuban cockroach. *Science* 133:1598–1599.

———. 1962. The endocrine control of mating behavior in the cockroach *Byrsotria fumigate* (Guerin). *Gen. Comp. Endocrin.* 2:53–69.

Bartholomew, G. A., Jr. 1942. The fishing activities of double-crested cormorants on San Francisco Bay. *Condor* 44:13–21.

———. 1970. A model for the evolution of pinniped polygyny. *Evolution* 24:546–559.

Bastock, M. 1956. A gene mutation which changes a behavior pattern. *Evolution* 10:421–439.

———. 1967. *Courtship, an ethological study.* Aldine, Chicago.

Bateson, P. P. G. 1971. Imprinting. In *The ontogeny of vertebrate behavior,* ed. H. Mottz, pp. 369–387. Academic Press, New York.

Beach, F. A. 1944. Responses of captive alligators to auditory stimulation. *Amer. Naturalist* 78:481–515.

———. 1945. Current concepts of play in animals. *Amer. Naturalist* 79:523–541.

———. 1955. The descent of instinct. *Psychol. Rev.* 62:401–410.

Beer, G. R. de. 1938. Embryology and evolution. In *Evolution,* essays on *Aspects of evolutionary biology,* ed. G. R. de Beer, pp. 57–78. Oxford Univ. Press, London.

———. 1958. *Embryos and ancestors,* 3rd ed. Oxford Univ. Press, London.

Beidler, L. M. 1961. The Chemical Senses. *Ann. Rev. Psychol.* 12:363–388.

Bekoff, M. 1972. The development of social interaction, play and metacommunication in mammals: an ethological perspective. *Quart. Rev. Biol.* 47:412–433.

———. 1974a. Social play and play-soliciting by infant canids. *Amer. Zool.* 14:323–340.

———. 1974b. Social play in coyotes, wolves and dogs. *Bioscience* 24:225–230.

———. 1974c. Social play in mammals. *Amer. Zool.* 14:265–436.

Bell, R. H. V. 1971. A grazing system in the Serengeti. *Sci. Amer.* 225:86–93.

Bellugi, U. 1970. Learning the language. *Psychol. Today* (Dec.):32 ff.

Bendell, J. F., and P. W. Elliott. 1967. Behaviour and the regulation of numbers in Blue Grouse. *Canad. Wildlife Service Rep. Ser.* 4.

Bennett, M. A. 1940. Social hierarchy in ring doves. II. The effects of treatment with testosterone propionate. *Ecology* 21:148–165.

Benson, C. W. 1948. Geographical voice-variation in African birds. *Ibis* 90:48–71.

———. 1963. The breeding seasons of birds in the Rhodesias and Nyasaland. *Proc. XIII Int. Orn. Congr.,* pp. 623–639.

Bent, A. C. 1926. Life histories of North American marsh birds. *U.S. Nat. Mus. Bull.* 142.

———. 1939. Life histories of North American woodpeckers. *Bull. Amer. Mus. Natur. Hist.* 45:271–331.

Bentley, D., and R. R. Hoy. 1974. The neurobiology of cricket song. *Sci. Amer.* 231:34–44.

Benzer, S. 1967. Behavioral mutants of *Drosophila* isolated by counter-current distribution. *Proc. Nat. Acad. Sci.* 58:1112–1119.

Berkowitz, L. 1962. *Aggression. A social psychological analysis.* McGraw-Hill, New York.

Bernstein, I. S. 1966. Analysis of a key role in a capuchin (*Cebus albifrons*) group. *Tulane Studies in Zool.* 13:49–54.

Bernstein, I. S., and L. G. Sharpe. 1966. Social roles in a rhesus monkey group. *Behaviour* 26:91–104.

Billings, S. M. 1968. Homing in Leach's petrel. *Auk* 85:36–43.

Binford, S. R., and L. R. Binford. 1969. Stone tools and human behavior. *Sci. Amer.* 220:70–84.

Bingham, H. C. 1922. Visual perception of the chick. *Behav. Monogr.* 4:104.

Birch, L. C. 1957. The meanings of competition. *Amer. Naturalist* 91:5–18.

Bitterman, M. E. 1965. The evolution of intelligence. *Sci. Amer.* 212:92–100.

———. 1975. The comparative analysis of learning. *Science* 188:699–709.

Blair, W. F. 1950. Ecological factors in the speciation of *Peromyscus. Evolution* 4:253–275.

———. 1951. Population structure, social behavior, and environmental relations in a natural population of the beach mouse (*Peromyscus polionotus leucocephalus*). Univ. of Michigan, *Contr. Lab. Vert. Biol.* 48:1–47.

———. 1963. Acoustic behavior of Amphibia. In *Acoustic behavior of animals,* ed. R. G. Busnel, pp. 694–708 Elsevier, Amsterdam.

Blank, T. H., and J. S. Ash. 1956. The concept of territory in the partridge *Perdix p. perdix. Ibis* 98:379–389.

Blest, A. D. 1957. The evolution of protective displays in the Saturniodea and Sphingidae (Lepidoptera). *Behaviour* 11:257–309.

———. 1961. The concept of 'ritualisation.' In *Current problems in animal behaviour,* ed. W. H. Thorpe and O. L. Zangwill, pp. 102–124. Cambridge Univ. Press, New York.

Blurton-Jones, N., ed. 1972. *Ethological studies of child behavior.* Cambridge Univ. Press, New York.

Bogert, C. M. 1960. The influence of sound on the behavior of amphibians and reptiles. In *Animal sounds and communications,* ed. W. E. Lanyon and W. N. Tavolga, pp. 137–320. American Institute of Biological Sciences, Publ. No. 7. Washington, D.C.

Bonner, J. T. 1965. *Size and Cycle.* Princeton Univ. Press, Princeton.

Boorman, S. A., and P. R. Levitt. 1973. A frequency-dependent natural selection model for the evolution of social cooperation networks. *Proc. Nat. Acad. Sci., U.S.A.* 70:187–189.

Borror, D. J. 1959. Variation in the songs of the rufous-sided towhee. *Wilson Bull.* 71:54–72.

———. 1960. The analysis of animal sounds. In *Animal sounds and communication,* ed. W. E. Lanyon and W. N. Tavolga, pp. 26–37. American Institute of Biological Sciences, Washington, D.C.

———. 1961. Intraspecific variation in passerine bird songs. *Wilson Bull.* 73:57–78.

Bower, T. G. R. 1966. Heterogeneous summation in human infants. *Anim. Behav.* 14:395–398.

Bradt, G. W. 1938. A study of beaver colonies in Michigan. *J. Mammal.* 19:139–162.

Braemer, W. and H. O. Schwassmann. 1960. Von Rhythmus der Sonnenorientierung am Aquator (bei Fischen). *Ergeb. Biol.* 26:182–201.

Brain, P. F. 1971. The physiology of population limitation in rodents—a review. *Comm. Behav. Biol.* 6:115–123.

Brand, A. R., and P. P. Kellogg. 1939. Auditory responses of starlings, English sparrows and domestic pigeons. *Wilson Bull.* 51:38–41.

Brattstrom, B. H. 1974. The evolution of reptilian social behavior. *Amer. Zool.* 14:35–49.

Bronson, F. H. 1968. *Pheromonal influences on mammalian reproduction and sexual behavior,* ed. M. Diamond. Univ. of Indiana Press, Bloomington.

Bronson, F. H., and C. Desjardins. 1968. Aggression in adult mice: modification by neonatal injections of gonadal hormones. *Science* 161:705–706.

Brossett, A. 1969. La vie sociale des oiseaux dans une forêt equitoriale du Gabon. *Biol. Gabonica* 5:29–69.

Brower, L. P. 1969. Ecological chemistry. *Sci. Amer.* 220:22–29.

Brown, F. A. 1970. In *The biological clock: two views,* ed. F. A. Brown, J. W. Hastings, J. D. Palmer, pp. 15–59. Academic Press, New York.

Brown, J. L. 1963. Social organization and behavior of the Mexican jay. *Condor* 65:126–153.

———. 1964. The evolution of diversity in avian territorial systems. *Wilson Bull.* 76:160–169.

———. 1969a. Territorial behavior and population regulation in birds. *Wilson Bull.* 81:293–329.

———. 1969b. The buffer effect and productivity in tit populations. *Amer. Naturalist* 103:347–354.

———. 1974. Alternate routes to sociality in jays—with a theory for the evolution of altruism and communal breeding. *Amer. Zool.* 14:63–80.

Brown, J. L., and G. H. Orians. 1970. Spacing patterns in mobile animals. *Ann. Rev. Ecol. Syst.* 1:239–262.

Brown, R., and U. Bellugi. 1964. Three processes in the child's acquisition of syntax. In *New directions in the study of languages,* ed. E. H. Lenneberg, pp. 131–162. M. I. T. Press, Cambridge, Mass.

Brown, R. G. B. 1967. Species isolation between the herring gull, *Larus argentatus,* and the lesser black-backed gull, *Larus fuscus. Ibis* 109:310–317.

Brown, R. Z. 1953. Soçial behaviour, reproduction and population changes in the house mouse. *Ecol. Monog.* 23:217–40.

Brown, W. L., and E. O. Wilson. 1956. Character displacement. *Syst. Zool.* 5:49–64.

Buettner-Janusch, J. 1963., Hemoglobins and transferrins of baboons. *Folia Primatol.* 1:73–87.

Bullock, T. H. 1965. Physiological bases of behavior. In *Ideas in Modern Biology,* ed. J. A. Moore, pp. 449–482. Natural History Press, Garden City.

Bunning, E. 1967. *The physiological clock.* Springer, Berlin.

Burkhardt, D., W. Schleidt, and H. Altner. 1967. *Signals in the animal world.* McGraw-Hill, New York.

Burns, R. D. 1968. The role of agonistic behavior in regulation of density in Uinta ground squirrels, *Citellus armatus.* M.S. thesis, Utah State Univ.

Burrell, H. 1927. *The platypus.* Angus and Robertson, Ltd., Australia.

Bush, G. L. 1969. Sympatric host race formation and speciation in frugivorous flies of the genus *Rhagoletis* (Diptera, Tephritidae). *Evolution* 23:237–251.

Bustard, R. H. 1970. The role of behavior in the natural regulation of numbers in the gekkonid lizard *Gehyra variegata. Ecology* 51:724–728.

Buxton, P. A. 1923. *Animal life in deserts.* Edward Arnold, London.

Cade, T. J., E. J. Willoughby, and G. Maclean. 1966. Drinking behavior of sandgrouse in the Namib and Kalahari deserts, Africa. *Auk* 83:124–126.

Cain, A. J. 1964. The perfection of animals. In *Viewpoints in biology,* ed. J. Carthy and R. Duddington, pp. 36–62, Vol. 3. Butterworth, London.

Calhoun, J. B. 1962a. A behavioral sink. In *Roots of Behavior,* ed. E. L. Bliss. Harper and Bros., New York.

———. 1962b. Population density and social pathology. *Sci. Amer.* 206:139–146.

———. 1963. The ecology and sociology of the Norway rat (cited in Watson and Moss 1970).

Carl, E. 1971. Population control in arctic ground squirrels. *Ecology* 52:395–413.

Carlquist, S. 1965. *Island life. A natural history of islands of the world.* Natural History Press, New York.

———. 1972. *Island biology: we've only just begun. Bioscience* 22:221–225.

Carneiro, R. L. 1970. A theory of the origin of the state. *Science* 169:733–738.

Carpenter, C. R. 1934. A field study of the behavior and social relations of howling monkeys (*Alouatta palliata*). *Comp. Psychol. Monog.* 10:1–168.

———. 1958. Territoriality: a review of concepts and problems. In *Behavior and evolution*, ed. A. Roe and G. G. Simpson, pp. 224–250. Yale Univ. Press, New Haven, Conn.

Carr, A. 1965. The navigation of the green turtle. *Sci. Amer.* 212 (5):79–86.

Carrington, R. 1958. *Elephants. A short account of their natural history, evolution and influence on mankind.* Basic Books, New York.

Carthy, J. D. 1958. *An introduction to the behavior of invertebrates.* MacMillan, New York.

Case, J. 1966. *Sensory mechanisms.* Macmillan, New York.

Chapin, J. P. 1932. The birds of the Belgian Congo. Part I. *Bull. Amer. Mus. Nat. Hist.* 75.

Chapman, D. W. 1966. Food and space as regulators of salmoniad populations in streams. *Amer. Naturalist* 100:345–357.

Chauvin, Remy. 1963. Translation 1968. *Animal societies: from the bee to the gorilla.* Hill and Wang, New York.

Chitty, D. 1952. Mortality among voles (*Microtus agrestis*) at Lake Vyrnwy, Montgomeryshire in 1936–1939. *Philos. Trans. Roy. Soc. Lond.* B, 236: 505–552.

———. 1957. Self-regulation in numbers through changes in viability. *Symp. Quant. Biol.* 22:277–280.

———. 1960. Population processes in the vole and their relevance to general theory. *Canad. J. Zool.* 38: 99–113.

———. 1967. What regulates bird population? *Ecology* 48:698–701.

Christian, J. J. 1961. *Proc. Nat. Acad. Sci.* 47:428–449.

———. 1963. Endocrine adaptive mechanisms and the physiologic regulation of population growth. In *Physiological mammalogy*, ed. W. V. Mayer and R. G. van Gelder, pp. 189–353. Academic Press, New York.

———. 1968. The potential role of the adrenal cortex as affected by social rank and population density on experimental epidemics. *Amer. J. Epidem.* 87: 255–266.

Christian, J. J., and D. E. Davis. 1964. Endocrines, behavior and population. *Science* 146:1550–1560.

Christian, J. J., J. A. Lloyd, and D. E. Davis. 1965. The role of endocrines in the self-regulation of mammalian populations. *Recent Progr. Horm. Res.* 21: 501–578.

Clark, R. B. 1960. Habituation of the polychaete *Nereis* to sudden stimuli; 2. Biological significance of habituation. *Anim. Behav.* 8:92–103.

Clough, G. C. 1965. Variability of wild meadow voles under various conditions of population, density, season, and reproductive activity. *Ecology* 46:119–134.

Clulow, F. V., and J. R. Clarke, Jr. 1968. Pregnancy block in *Microtus agrestis* and induced ovulation. *Nature* 219:511.

Cody, M. L. 1966. A general theory of clutch size. *Evolution* 20:174–184.

———. 1969. Convergent characteristics in sympatric species: A possible relation to interspecific competition and aggression. *Condor* 71:222–239.

———. 1973. Character convergence. In *Annual review of ecology and systematics*, ed. R. F. Johnston, pp. 189–211. Annual Reviews, Inc., Palo Alto, Calif.

Cody, M. L. and J. H. Brown. 1970. Character convergence in Mexican finches. *Evolution* 24:304–310.

Colbert, E. H. 1955. *Evolution of the vertebrates: a history of backboned animals through time.* Wiley, New York.

Collias, E. C., and N. E. Collias. 1964. The development of nest-building behavior in a weaver bird. *Auk* 81:42–52.

Collias, N. E. 1944a. Aggressive behavior among vertebrate animals. *Physiol. Zool.* 17:83–123.

———. 1944b. Aggressive behavior and evolution. Excerpted in *Animal aggression*, ed. C. H. Southwick (1970). Van Nostrand Reinhold Co., New York.

———. 1960. An ecological and functional classification of animal sounds. In *Animal sounds and communication*, ed. W. E. Lanyon and W. N. Tavolga, pp. 368–391. American Institute of Biological Sciences Publication No. 7, Washington, D.C.

———. 1963. A spectrographic analysis of the vocal repertoire of the African village weaver bird. *Condor* 65:517–527.

Collias, N. E., M. Brandman, K. Victoria, L. F. Kiff and C. E. Rischer. 1971. Social facilitation of nest-building in experimental colonies of a weaver bird: effects of varying the sex ratio. *Ecology* 52:829–836.

Collias, N. E., and L. R. Jahn. 1959. Social behavior and breeding success in Canada geese (*Branta canadensis*) confined under semi-natural conditions. *Auk* 76:478–509.

Collias, N. E., J. K. Victoria, and R. J. Shallenberger. 1971. Social facilitation in weaverbirds: importance of colony size. *Ecology* 52:823–828.

Collias, N. E., and E. C. Collias. 1962. An experimental study of the mechanisms of nest-building in a weaverbird. *Auk* 79:568–595.

Collins, C. W., and S. F. Potts. 1932. Attractants for the flying gypsy moth as an aid in locating new infestations. *U.S. Dept. Agr. Tech. Bull.* 336:1–43.

Connell, J. H. 1961. The influence of interspecific competition and other factors on the distribution of the barnacle *Chthamalus stellatus*. *Ecology* 42:710–723.

———. 1963. Territorial behavior and dispersion in some marine invertebrates. *Res. Popul. Ecol.* 2: 87–101.

———. 1973. *Some mechanisms producing structure in natural communities: a model in evidence from field experiments in ecology and evolution of communities*, ed. M. L. Cody and J. M. Diamond, pp. 460–490. Belknapp Press, Harvard Univ. Press, Cambridge, Mass.

Cook, L. M., R. R. Askew, and J. A. Bishop. 1970. Increasing frequency of the typical form of the peppered moth in Manchester. *Nature* 227:1155.

Corning, W. C. and E. R. John. 1961. Effect of ribonuclease on retention of conditioned response in regenerating planaria. *Science* 134:1363–1365.

Corso, J. F. 1967. *The experimental psychology of sensory behavior.* Holt, Rinehart and Winston, New York.

Cott, H. B. 1957. *Adaptive coloration in animals.* Methuen, London.

Coulson, J. C. 1968. Differences in the quality of birds nesting in the centre and on the edges of a colony. *Nature* 217:478–479.

Coulson, J. C., and E. White. 1960. The effect of age and density of breeding birds on the timing of breeding of the kittiwake *Rissa tridactyla*. *Ibis* 102:71–86.

Cowan, I. McT. 1946. Parasites, diseases, injuries, and anomalies of the Columbian black-tailed deer, *Odocoileus hemionus columbianus.* (Richardson), in British Columbia. *Canad. J. Res.* 24:71–103.

———. 1947. The timber wolf in the Rocky Mountain national parks of Canada. *Canad. J. Res.* 25: 139–174.

Cowley, J. J. 1970. Food intake and the modification of behaviour. In *Animal populations in relation to their food resources,* ed. A. Watson. Blackwell, Oxford.

Cracraft, J. 1972. The relationship of the higher taxa of birds: problems in phylogenetic reasoning. *Condor* 74:379–392.

Cragg, J. B. 1955. The natural history of sheep blowflies in Britain. *Ann. Appl. Biol.* 42:197–207.

Craig, W. 1918. Appetites and aversions as constituents of instincts. *Biol. Bull.* 34:91–107.

———. 1928. Why do animals fight? *Int. Ethics* 31: 264–278.

Crane, J. 1952. A comparative study of innate defensive behavior in Trinidad mantids (Orthoptera, Mantoidea). *Zoologica* 37:259–293.

———. 1955. Imaginal behavior of a Trinidad butterfly, *Heliconius erato hydara* Heivitson, with special reference to the social use of color. *Zoologica* 40: 167–196.

———. 1957. Basic patterns of display in fiddler crabs (Ocypodidae, genus *Uca*). *Zoologica* 42:69–82.

———. 1966. Combat, display and ritualization in fiddler crabs. *Philos. Trans. Roy. Soc. Lond. B* 251: 459–472.

Crescitelli, F., and J. D. Pollack, 1965. Color vision in the antelope ground squirrel. *Science* 150:316–318.

Crook, J. H. 1961. The basis of flock organization in birds. In *Current problems in animal behaviour,* ed. W. H. Thorpe and O. L. Zangwill, pp. 125–129. Cambridge Univ. Press, New York.

———. 1963. Comparative studies on the reproductive behaviour of two closely related weaverbird species (*Ploceus cucullatus* and *Ploceus nigerrimus*) and their races. *Behaviour* 21:177–232.

———. 1964. The evolution of social organization and visual communication in the weaver birds (Ploceinae). *Behaviour* supplement No. 10.

———. 1965. The adaptive significance of avian social organizations. Social organizations of animal communities. *Symp. Zool. Soc. Lond.* 14:181–218.

———. 1966. Gelada baboon herd structure and movement: a comparative report. *Symp. Zool. Soc. Lond.* 18:237–258.

———. 1969. Functional and ecological aspects of

vocalization in weaver birds. In *Bird vocalizations,* ed. R. A. Hinde, pp. 265–290. Cambridge.

———. 1970a. The socio-ecology of primates. In *Social behaviour in birds and mammals,* ed. J. H. Crook, pp. 103–159. Academic Press, New York.

———. 1970b. Social organization and the environment: aspects of contemporary social ethology. *Anim. Behav.* 18:197–209.

———. 1970c. Sources of cooperation in animals and man. *Smithsonian Soc. Sci. Infor.* 9:27–48.

Crook, J. H. and J. S. Gartlan. 1966. Evolution of primate societies. *Nature* 210:1200–1203.

Crowell, K. 1961. The effects of reduced competition in birds. *Proc. Nat. Acad. Sci.* 47:240–243.

———. 1962. Reduced interspecific competition among the birds of Bermuda. *Ecology* 43:75–88.

Cruz, A. 1977. Ecology and behavior of the Jamaican woodpecker. *Bull. Flor. State Mus. Biol. Sci.* 22(4):149–204.

Cullen, E. 1957a. Adaptations in the kittiwake in the British Isles. *Bird Study* 10:147–179.

———. 1957b. Adaptations in the kittiwake to cliff nesting. *Ibis* 99:275–302.

———. 1960. Experiments on the effects of social isolation on reproductive behaviour in the three-spined stickleback. *Anim. Behav.* 8:235.

Cullen, J. M. 1959. Behaviour as a help in taxonomy. *Pub. Syst. Assoc.* 3:131–140.

———. 1962. The pecking response of young wide-awake terns, *Sterna fuscata. Ibis* 103:162–170.

Cullen, J. M., and N. P. Ashmole. 1963. The black noddy *Anous tenuirostris* on Ascension Island. Part 2, Behaviour. *Ibis* 103b:423–446.

Daanje, A. 1950. On locomotory movements in birds and the intention movements derived from them. *Behaviour* 3:48–99.

Darling, F. F. 1938. *Bird flocks and the breeding cycle: a contribution to the study of avian sociality.* Cambridge Univ. Press, Cambridge.

Darlington, P. J. 1957. *Zoogeography: the geographical distribution of animals.* Wiley & Sons, New York.

———. 1959. Area, climate and evolution. *Evolution* 13:488–510.

Darwin, C. R. 1859. *The origin of species.* John Murray, London.

———. 1871. *The descent of man and selection in relation to sex.* John Murray, London.

———. 1872. *Expression of the emotions in man and animals.* Murray, London.

———. 1890. *The descent of man,* 2nd ed. Murray, London.

Davidson, J., and H. G. Andrewartha. 1948a. Annual trends in a natural population of *Thrips imaginis* (Thysanoptera). *J. Anim. Ecol.* 17:193–199.

———. 1948b. The influence of rainfall, evaporation and atmospheric temperature on fluctuations in size of a natural population of *Thrips imaginis* (Thysanoptera). *J. Anim. Ecol.* 17:200–202.

Davis, D. E. 1942. The phylogeny of social nesting habits in the Crotophaginae. *Q. Rev. Biol.* 17:115–134.

———. 1946. A seasonal analysis of mixed flocks of birds in Brazil. *Ecology* 27:168–181.

Davis, H. 1961. Peripheral coding of auditory information. In *Sensory communication,* ed. W. A. Rosenblith, pp. 119–141. M. I. T. Press, Cambridge, Mass.

Davis, R., and E. L. Cockrum. 1962. Repeated homing experiments exhibited by a female pallid bat. *Science* 137:341–342.

Davis, R. B., C. F. Herreid, and H. L. Short. 1962. Mexican free-tailed bats in Texas. *Ecol. Monogr.* 32:311–346.

Dawkins, R. 1976. *The selfish gene.* Oxford Univ. Press, New York.

Dayhoff, M. O. 1972. *Atlas of protein sequence and structure,* V. National Biomedical Research Foundation, Washington, D.C.

De Bach, P. 1965. Some biological and ecological phenomena associated with colonizing entomophagous insects. In *The genetics of colonizing species,* ed. H. G. Baker and G. L. Stebbins. Academic Press, New York.

Deets, A. C., and H. F. Harlow. 1971. Early experience and the maturation of aggression. Presented at the American Association for the Advancement of Science symposium: Value and Knowledge Requirements for Peace. Chicago.

Delius, J. D. 1965. A population study of skylarks, *Alauda arvensis. Ibis* 107:465–492.

Delong, K. T. 1967. Population ecology of feral house mice. *Ecology* 48:611–634.

Denenberg, V. H., and K. M. Rosenberg. 1967. Nongenetic transmission of information. *Nature* 216: 549–550.

De Ruiter, L. 1955. Countershading in caterpillars. *Arch. Neerlandaises Zool.* 11:1–57.

DeSisto, M. J. 1970. Hypothalamic mechanisms of killing behavior in the rat. Ph.D. thesis, Tufts Univ. Cited in R. N. Johnson 1972.

Dethier, V. G., and E. Stellar. 1961. *Animal behavior.* Prentice-Hall, Inc., Englewood Cliffs, N. J.

DeVore, I., ed. 1965. *Primate behavior: field studies of monkeys and apes.* Holt, Rinehart and Winston, New York.

DeVore, I., and K. R. L. Hall. 1965. Baboon ecology. In *Primate behavior,* ed. I. DeVore. Holt, Rinehart and Winston, New York.

Diamond, J. M. 1969. Avifaunal equilibria and species turnover rates on the Channel Islands of California. *Proc. Nat. Acad. Sci.* 64:57–63.

———. 1974. Colonization of exploded volcanic islands by birds: the supertramp strategy. *Science* 184: 803–806.

———. 1978. Niche shifts and the rediscovery of interspecific competition. *Amer. Sci.* 66:322–331.

Diamond, J. M., and J. Terborgh. 1967. Observations on bird distribution and feeding assemblages along the Rio Callaria, Department of Loreto, Peru. *Wilson Bull.* 79:273–282.

DiCara, L. V. 1970. Learning in the autonomic nervous system. *Sci. Amer.* 222:30–39.

Dietz, R. S., and J. C. Holden. 1970. The breakup of Pangaea. *Sci. Amer.* (Oct):30–41.

Dilger, W. C. 1956. Hostile behavior and reproductive isolating mechanisms in the genera *Catharus* and *Hylocichla. Auk* 73:313–353.

———. 1960. The comparative ethology of the African parrot genus *Agapornis. Z. Tierpsychol.* 17:649–685.

———. 1962. The behavior of love birds. *Sci. Amer.* 206:88–98.

Dimond, S. J. 1970. *The social behaviour of animals.* B. T. Batsford, Ltd., London.

Dingle, H., and R. L. Caldwell. 1968. The aggressive and territorial behavior of the mantis shrimp *Gonodactylus bredini* Manning (Crustacea: Stomatopoda). *Behaviour* 33:115–136.

Dixon, K. 1961. Habit distribution and niche relationships in North American species of *Parus.* In *Vertebrate speciation,* ed. W. F. Blair, pp. 179–216. Texas Univ. Press, Austin.

Dobzhansky, T. 1950. Evolution in the tropics. *Amer. Sci.* 38:209–221.

———. 1951. *Genetics and the origin of the species.* Columbia Univ. Press, New York.

———. 1962. *Mankind evolving.* Yale Univ. Press, New Haven, Conn.

Dollard, J., L. Doob, N. Miller, O. Mowrer, and R. Sears. 1939. *Frustration and aggression.* Yale Univ. Press, New Haven, Conn.

Dorst, J. 1962. *The migration of birds.* Houghton Mifflin Co., Boston.

Doty, R. W., and J. F. Bosma. 1956. An electromyographic analysis of reflex deglutination. *J. Neurophysiol.* 19:44–60.

Dufay, C. 1957. Sur l'attraction sexuelle chez *Lasiocampa quercus* L. (Sep., Sasio-campidae). *Bull. Société de' Entomologië* 62:61–64.

Douglas-Hamilton, I., and O. Douglas-Hamilton. 1975. *Among the elephants.* Viking Press, New York.

Earhart, C. M., and N. K. Johnson. 1970. Size dimorphism and food habits of North American owls. *Condor* 72:251–264.

Eccles, J. C. 1953. *The neurophysiological basis of mind: the principles of neurophysiology.* Oxford Univ. Press, London.

Edminster, F. C. 1939. The effect of predator control on ruffed-grouse populations in New York. *J. Wildl. Mgmt.* 3:345–352.

Edney, E. B. 1954. Woodlice and the land habitat. *Biol. Rev.* 29:185–219.

Ehrenfeld, D. W. 1968. The role of vision in the sea-finding orientation of the green turtle (*Chelonia mydas*), 2. Orientation mechanism and range of spectral sensitivity. *Anim. Behav.* 16:281–287.

Ehrlich, P. R., and L. C. Birch. 1967. The "balance of nature" and "population control." *Amer. Naturalist* 101:97.

Ehrlich, P. R., and A. H. Ehrlich. 1970. *Population, resources, environment.* W. H. Freeman, San Francisco.

Ehrlich, P. R., and P. H. Raven. 1964. Butterflies and plants; a study in coevolution. *Evolution* 18:586–608.

Eibl-Eibesfeldt, I. 1950. Über die Jungendentwicklung des Verhaltens eines männlichen Dachses (*Meles meles* L.) unter besonderer Berücksichtigung des spieles. *Z. Tierpsychol.* 7:327–355.

———. 1959. Der Fisch *Aspidontus taeniatus* als Nachahmer des Putzers *Labroides dimidiatus. Z. Tierpsychol.* 16:19–25.

———. 1961a. The fighting behavior of animals. *Sci. Amer.* 205:112–121.

———. 1961b. The interactions of unlearned behavior patterns and learning in mammals. In *Brain mechanisms and learning,* ed. J. F. Delafresnage, pp. 53–73. Blackwell.

———. 1963. Angeborenes und Erworbenes im Verhalten einiger Säuger. *Z. Tierpsychol.* 20:705–754.

———. 1967. Concepts of ethology and their significance for the study of human behavior. In *Early behavior: comparative and development approaches,* ed. H. W. Stevenson, pp. 127–146. Wiley, New York.

———. 1970. *Ethology, the biology of behavior.* Holt, Rinehart and Winston, New York.

Einarsen, A. S. 1945. Some factors affecting ring-necked pheasant population density. *Murrelet* 26:39–44.

Eisenberg, J. F. 1966. The social organizations of mammals. *Handb. Zool.* 10:1–92.

———. 1967. A comparative study in rodent ethology with emphasis on the evolution of social behavior. *Proc. U.S. Nat. Mus.* 122:1–51.

Eisner, T. 1970. Chemical defense against predation in arthropods. In *Chemical ecology,* ed. E. Sondheimer and J. B. Simeone, pp. 157–217. Academic Press, New York.

Eisner, T., and F. C. Kafatos. 1962. Defense mechanisms of arthropods. X. A pheromone promoting aggregation in an aposematic distasteful insect. *Psyche,* Cambridge, 69:53–61.

Elkin, E. 1960. The common toad (*Bufo bufo L.* ) in the laboratory. *Brit. Herpetol.* 2:177–182.

Elson, P. F. 1962. Predator-prey relationships between fish-eating birds and Atlantic salmon. *Bull. Fish. Res. Bd. Can.,* No. 133.

Elton, C. 1927. *Animal ecology.* Sidgwick and Jackson, London.

Elton, C. S. 1958. *The ecology of invasions by animals and plants.* Methuen, London.

Emlen, J. M. 1973. *Ecology: an evolutionary approach.* Addison Wesley, Reading, Mass.

Emlen, J. T., and R. L. Penney. 1964. Distance navigation in the Adélie penguin. *Ibis* 106:417–431.

Emlen, J. T., and F. W. Lorenz. 1942. Pairing responses of free-living valley quail to sex hormone pellet implants. *Auk* 59:369–378.

Emlen, S. T. 1967. Migratory orientation in the indigo bunting, *Passerina cyanea.* Part I. Evidence for use of celestial cues; Part II. Mechanism of celestial orientation. *Auk* 84:309–342, 463–489.

———. 1968. Territoriality in the bullfrog, *Rana Catesbeiana. Copeia* 1968:240–243.

———. 1971. Adaptive aspects of coloniality in the bank swallow. *Amer. Zool. 11.*

———. 1974. Bird migration. In *Avian biology,* vol. 4, ed. D. Farner and J. King. Academic Press, New York.

———. 1975. The stellar-orientation system of a migratory bird. *Sci. Amer.* (Aug.) 233:102–111.

Errington, P. L. 1943. An analysis of mink predation upon muskrats in north central United States. *Res. Bull. Iowa Agr. Exp. Sta.* 320:797–924.

———. 1945. Some contributions of a fifteen year local study of the northern bobwhite to a knowledge of population phenomena. *Ecol. Monog.* 15:1–34.

———. 1946. Predation and vertebrate populations. *Q. Rev. Biol.* 21:145–177, 221–245.

Erulkar, S. D. 1972. Comparative aspects of spatial localization of sound. *Physiol. Rev.* 52:237–360.

Estes, W. K., S. Koch, K. McCorquodale, P. E. Meehl, C. G. Mueller, W. N. Schoenfeld and W. S. Verplanck. 1954. *Modern learning theory.* Appleton-Century-Crofts, New York.

Etkin, W. 1964. Cooperation and competition in social behavior. In *Social behavior and organization among vertebrates,* ed. W. Etkin, pp. 1–34. Univ. of Chicago Press, Chicago.

———. 1967. *Social behavior from fish to man.* Univ. of Chicago Press, Chicago.

Ettlinger, G. 1960. Cross-modal transfer of training in monkeys. *Behaviour* 16:56–64.

Evans, H. E. 1953. Comparative ethology and the systematics of spider wasps. *Syst. Zool.* 2:155–172.

Evans, S. M., and G. R. Patterson, 1971. The synchronization of behavior in flocks and estrildine finches. *Anim. Behav.* 19:419–428.

Ewer, R. F. 1968. *Ethology of mammals.* Plenum Press, New York.

Ewert, J. P. 1974. The neural basis of visually guided behavior. *Sci. Amer.* (March) 231:34–42.

Farner, D. S., and R. A. Lewis. 1971. Photoperiodism and reproductive cycles in birds. *Photophysiol.* 6:325–370.

Felipe, N., and R. Sommer. 1966. Invasions of personal space. *Social Problems* 14:206–214.

Festetics, A. 1959. Okologische Untersuchungen an Brutvögeln des Saser. *Vogelwelt* 80:1–21.

Fisher, J. 1954. Evolution and bird sociality. In *Evolution as a process,* ed. J. Huxley, A. C. Hardy, and E. B. Ford, pp. 71–83. Allen and Unwin, London.

Fisher, R. A. 1930. *The genetical theory of natural selection.* Dover, New York.

Fisher, R. A., and R. T. Peterson. 1964. *The world of birds.* Doubleday and Co., Inc., Garden City, N.Y.

Fitch, H. S. 1948. Ecology of the California ground squirrel on grazing lands. *Amer. Midland Naturalist* 39:513–596.

Flickinger, G. L. 1966. Responses of the testes to social interaction among grouped chickens. *Gen. Comp. Endocrin.* 6:89–98.

Flynn, J. P. 1967. The neural basis of aggression in cats. In *Neurophysiology and emotion,* ed. D. C. Glass. Rockefeller Univ. Press and Russell Sage Foundation, New York.

Fox, M. W. 1970. A comparative study of the development of facial expressions in canids: wolf, coyote, and foxes. *Behaviour* 36:49–73.

———. 1972. Socio-ecological implications of individual differences in wolf litters: a developmental and evolutionary perspective. *Behaviour* 41:298–313.

Fraenkel, G. S. and D. L. Gunn. 1940. *The orientation of animals.* Oxford Univ. Press, London.

Free, J. B. 1965. The allocation of duties among worker honeybees. Social organization of animal communities. *Symp. Zool. Soc. Lond.* 14:173–178.

Fretwell, S. D. 1968. Habitat distribution and survival in the field sparrow (*Spizella pusilla*). *Bird Banding* 39:293–306.

———. 1969a. On territorial behavior and other factors influencing habitat distribution in birds. I-III *Acta Biotheor.* 19:16–52

———. 1969b. The adjustment of birth rate to mortality in birds. *Ibis* III:624–627.

Fretwell, S. D., and H. L. Lucas. 1969. On territorial behavior and other factors influencing habitat distribution in birds. I. Theoretical development. *Acta biotheor.* 19:16–36.

Friedmann, H. 1929. *The cowbirds.* Thomas, Baltimore.

Frings, H., M. Frings, J. Jumber, R. Busnel, J. Giban and P. Gramet. 1958. Reactions of American and French species of *Corvus* and *Larus* to recorded communication signals tested reciprocally. *Zoology* 39:126–131.

Frisch, K. von. 1950. *Bees: Their chemical senses, vision, and language.* Cornell Univ. Press, Ithaca, New York.

———. 1953. *The dancing bees.* Harcourt Brace Jovanovich, New York.

———. 1964. *Biology.* Harper and Row, New York.

Fromme, H. G. 1961. Untersuchungen über das Orientierungsvermögen nächtlich ziehender Kleinvögel (*Erithacus rubecula, Sylvia communis*). *Z. Tierpsychol.* 18:205–220.

Fryer, G. 1959. Some aspects of evolution in Lake Nyasa. *Evolution* 13:440–451.

Fuller, J. L., and W. R. Thompson. 1960. *Behavior genetics.* John Wiley and Sons, New York.

Gadgil, M., and O. T. Solbrig. 1972. The concept of r and K selection: evidence from wild flowers and some theoretical considerations. *Amer. Naturalist* 106: 14–31.

Garattini, S., and E. B. Sigg, eds. 1969. *Aggressive behavior.* Wiley Interscience, New York.

Gardner, R. A., and B. T. Gardner. 1969. Teaching sign language to a chimpanzee. *Science* 165:664–672.

Gartlan, J. S. 1968. Structure and function in primate society. *Folia Primatol.* 8:89–120.

Gause, G. F. 1934. *The struggle for existence.* Williams and Wilkins, Baltimore.

Geist, V. 1966a. The evolutionary significance of mountain sheep horns. *Evolution* 20:558–66.

———. 1966b. Evolution of horn-like organs. *Behaviour* 27:175–214.

Gerard, R. W. 1961. The fixation of experience. In *Brain mechanisms and learning,* ed. J. F. Delafresnaye, pp. 21–35. Blackwell, Oxford.

Gerdes, K. 1962. Richtungstendenzen vom Brutplatz verfrachteter Lachmöven (*Larus ridibundus* L.) unter Ausschluss visueller Gelände und Himmelsmarken. *Z. Wiss. Zool.* 166:352–410.

Gerking, S. D. 1959. The restricted movement of fish populations. *Biol. Rev.* 34:221–242.

Gershenson, S. 1945. Evolutionary studies on the distribution and dynamics of melanism in the hamster (*Cricetus cricetus* L.) *Genetics* 30:207–251.

Geschwind, N. 1964. The development of the brain and the evolution of language. *Monogr. Ser. Lang. and Ling.* 17:155–169.

Gibb, J. A. 1950. The breeding biology of the great and blue titmice. *Ibis* 92:507–539.

———. 1954. Feeding ecology of tits, with notes on treecreeper and goldcrest. *Ibis* 96:513–543.

———. 1956. Food, feeding habits, and territory of the Rock Pipit, *Anthus spinoletta. Ibis* 98:506–530.

———. 1960. Population of tits and goldcrests and their food supply in pine plantations. *Ibis* 102:163–208.

Gibb, J. A., and M. M. Betts. 1963. Food and food supply of nesting tits (Paridae) in Breckland pine. *J. Anim. Ecol.* 32:489–533.

Gibson, E. J., and R. D. Walk. 1960. The "visual cliff." *Sci. Amer.* 202(4):64–71.

Gilbert, L. I. 1963. Hormone controlling reproduction and molting in invertebrates. In *Comparative endocrinology,* vol. 2, ed. C. von Euler and H. Heller, pp. 1–46. Academic Press, New York.

Gilliard, E. T. 1963. The evolution of bowerbirds. *Sci. Amer.* 209:38–46.

Goethe, F. 1954. Vergleichende Biobachtungen zum Verhalten der Silbermowe (*Larus argentatus*) und der Heringsmowe (*Larus fuscus*). *Proc. XI Int. Orn. Congr.,* pp. 577–582.

Goldman, P. S., H. T. Crawford, L. P. Stokes, T. W. Galkin, and E. H. Rosvold. 1974. Sex-dependent behavioral effects of cerebral cortical lesions in the developing rhesus monkey. *Science* 186:540–542.

Gompertz, T. 1961. The vocabulary of the great tit. *Brit. Birds* 54:369–394, 404–418.

Goodall, Jane. *See also* Lawick, Hugo van.

Goodall, J. 1968. A preliminary report on expressive movements and communication in the Gombe Stream chimpanzees. In *Primates,* ed. P. Jay, pp. 313–419. Holt, Rinehart and Winston, New York.

———. 1971. *In the shadow of man.* Houghton Mifflin Co., Boston.

Goodall, J., and H. van Lawick. 1966. Use of tools by Egyptian vultures. *Nature* 212:1468–1469.

Goodhart, C. B. 1960. The evolutionary significance of human hair patterns and skin coloring. *Adv. Sci.* 17:53–59.

Goodman, J. D., and J. M. Goodman. 1976. Possible mimetic behavior of the twig snake, *Thelotornis Kirtlandi Kirtlandi* (Hallowell). *Herpetologica* 32:148–150.

Goodwin, B. 1963. *Temporal organization in cells.* Academic Press, New York.

Goss-Custard, J. D. 1970. Feeding dispersion in some overwintering wading birds. In *Social behavior in birds and mammals,* ed. J. H. Crook, pp. 3–34. Academic Press, New York.

Gould, J. L., M. Henery, and M. C. Macleod. 1970. Communication of direction by the honeybee. *Science* 169:544–554.

Goy, R. W. 1970. Early hormonal influences on the development of sexual and sex-related behavior. In *The neurosciences,* ed. F. O. Schmitt. Rockefeller Univ. Press, New York.

Goy, R. W., and C. H. Phoenix. 1963. Hypothalamic regulation of female sexual behavior: establishment of behavioral estrus in spayed guinea pigs following hypothalamic lesions. *J. Reprod. Fert.* 5:23–40.

Grant, P. R. 1965. The adaptive significance of some size trends in island birds. *Evolution* 19:355–367.

———. 1966a. Ecological compatability of bird species on islands. *Amer. Naturalist* 100:451–462.

———. 1966b. The density of land birds on the Tres Marias Islands in Mexico. I. Numbers and biomass. *Canad. J. Zool* 44:391–400.

———. 1967. Bill length variability in birds of the Tres Marías Islands, Mexico. *Canad. J. Zool.* 45:599–614.

———. 1968. Bill size, body size, and the ecological adaptations of bird species to competitive situations on islands. *Syst. Zool.* 17:319–333.

———. 1969. Community diversity and the coexistence of congeners. *Amer. Naturalist* 103:552–556.

———. 1971a. Comment on Simberloff's letter. *Amer. Naturalist* 105:194–197.

———. 1971b. Variation in the tarsus lengths of birds in island and mainland regions. *Evolution* 25:599–614.

Grant, V. 1963. *The origin of adaptations.* Columbia Univ. Press, New York.

Grant, W. C., Jr. 1955. Territorialism in two species of salamanders. *Science* 121:137–138.

Griffin, D. R. 1955. Bird navigation. In *Recent studies in avian biology,* ed. A. Wolfson, pp. 154–172. Univ. of Illinois Press, Urbana.

———. 1958. *Listening in the dark.* Yale Univ. Press, New Haven, Conn.

———. 1964. *Bird migration.* Doubleday & Company, Inc., Garden City, New York.

Grimes, L. G. 1974. Dialects and geographical variation in the song of the splendid sunbird *Nectarinia coccinigaster. Ibis* 116:314–329.

Gross, A. O. 1940. The migration of Kent Island herring gulls. *Bird Banding* 11:129–155.

Grossman, S. P. 1967. *A textbook of physiological psychology.* Wiley and Sons, New York.

———. 1968. The physiological basis of specific and

non-specific motivational processes. *Nebr. Symp. Motiv.* 16:1–46.

Guhl, A. M. and L. L. Ortman. 1953. Visual patterns of recognition of individuals among chickens. *Condor* 55:287–298.

Guillion, G. W. 1960. The ecology of Gambel's quail in Nevada and in the arid southwest. *Ecology* 41: 518–536.

Guthrie, R. D. and R. G. Petocz. 1970. Weapon automimicry among mammals. *Amer. Naturalist* 104:585–588.

Hahn, W. E., and D. W. Tinkle. 1965. Fat body cycling and experimental evidence for its adaptive significance to ovarian follicle development in the lizard *Uta stansburiana. J. Exper. Zool.* 158:79–86.

Hailman, J. P. 1964. Breeding synchrony in the equatorial swallow-tailed gull. *Amer. Naturalist* 98:79–83.

———. 1965. Cliff-nesting adaptations of the Galápagos swallow-tailed gull. *Wilson Bull.* 77:346–362.

———. 1969. How an instinct is learned. *Sci. Amer.* 221:98–106.

———. 1977. *Optical Signals.* Indiana Univ. Press, Bloomington.

Hairston, N. G., F. E. Smith, and L. B. Slobodkin. 1960. Community structure, population control, and competition. *Amer. Naturalist* 94:421–425.

Haldane, J. B. S. 1932. *The causes of evolution.* Longmans, Green, London.

———. 1957. The cost of natural selection. *J. Genet.* 55:511–524.

Hall, E. 1969. *The hidden dimension.* Doubleday, Garden City, New York.

Hall, J. R. 1970. Synchrony and social stimulation in colonies of the blackheaded weaver *Ploceus cucullatus* and Vieillot's black weaver *Melanopteryx nigerrimus. Ibis.* 112:93–103.

Hall, K. R. L. 1964. Aggression in monkey and ape societies. In *The natural history of aggression,* ed. J. D. Carthy and F. J. Ebling, pp. 51–64. Academic Press, New York.

———. 1965a. Ecology and behavior of baboons, patas and vervet monkeys. *Proceedings of the 1st International Conference on the Baboon as an Experimental Animal,* San Antonio, Texas.

———. 1965b. Social organization of the old-world monkeys and apes. Social organization of animal communities. *Symp. Roy. Soc. Lond.* 14:265–289.

Hall, K. R. L., and I. DeVore. 1965. Baboon social behavior. In *Primate behavior,* ed. I. DeVore, pp. 53–110. Holt, Rinehart and Winston, New York.

Hall, M. F. 1962. Evolutionary aspects of estrildial song. *Symp. Zool. Soc. Lond.* 8:3.

Hall-Craggs, J. 1962. The development of song in the blackbird (*Turdus merula*). *Ibis* 104:277–300.

Hamilton, T. H. 1962. Species relationships and adaptations for sympatry in the avian genus *Vireo. Condor* 64:40–68.

Hamilton, T. H., and R. H. Barth, Jr. 1962. The biological significance of season change in male plumage appearance in some new world migratory bird species. *Amer. Naturalist* 96:123–144.

Hamilton, T. H., and G. H. Orians. 1965. The evolution of brood parasitism in altricial birds. *Condor* 67:361–382.

Hamilton, W. D. 1963. The evolution of altruistic behavior. *Amer. Naturalist* 97:354–356.

———. 1964a. The genetic evolution of social behavior. I. J. Theor. Biol. 7:1–16.

———. 1964b. The genetic evolution of social behaviour. II. *J. Theor. Biol.* 7:17–52.

———. 1966. The moulding of senescence by natural selection. *J. Theor. Biol.* 12:12–45.

———. 1971. Geometry for the selfish herd. *J. Theoret. Biol.* 31:295–311.

Hamilton, W. J., Jr. 1937. The biology of microtine cycles. *J. Agr. Res.* 54:779–790.

Hamilton, W. J. III. 1962. Reproduction adaptations of the red tree mouse. *J. Mammal.* 43:486–504.

Hansen, E. W. 1966. The development of maternal and infant behavior in the rhesus monkey. *Behaviour* 27:107–149.

Hardin, G. 1960. The competitive exclusion principle. *Science* 131:1291–1297.

Harger, J. R. E. 1972. Competitive coexistence among intertidal vertebrates. *Amer. Sci.* 60:600–607.

Harker, J. E. 1958. Diurnal rhythms in the animal kingdom. *Biol. Rev.* 33:1–52.

Harlow, H. F. 1932. Social facilitation of feeding in the albino rat. *J. Gen. Psychol.* 41:211–221.

———. 1949. The formation of learning sets. *Psychol. Rev.* 56:51–65.

———. 1958. The nature of love. *Amer. Psychologist* 13:673–685.

———. 1963. Basic social capacity of primates. In *Primate social behavior,* ed. C. H. Southwick. D. Van Nostrand, Princeton, New Jersey.

Harper, J. L., et al. 1961. The evolution and ecology of closely related species living in the same area. *Evolution* 15:209–227.

Harrington, R. W., Jr. 1961. Oviparous hermaphroditic fish with internal self-fertilization. *Science* 134: 1749–1750.

Harris, V. T. 1952. An experimental study of habitat selection by prairie and forest races of the deer mouse, *Peromyscus maniculatus. Contrib. Lab. Vert. Biol.,* Univ. of Michigan, 56:1–53.

Hartley, P. H. T. 1949. The biology of the mourning chat in white quarters. *Ibis* 91:393–413.

———. 1953. An ecological study of the feeding habits of the English titmice. *Anim. Ecol.* 22:261–288.

Hartry, A. L., P. Keith-Lee, and W. D. Morton. 1964. Planaria, memory transfer through cannibalism reexamined. *Science* 146:274–275.

Hartshorne, C. 1958. The relation of bird song to music. *Ibis* 100:421–445.

Haskell, P. T. 1957. Stridulation and associated behavior in certain Orthoptera. I. Analysis of the stridulation of, and behaviour between males. *Brit. J. Anim. Behav.* 5:139–148.

Hasler, A. D. 1960. Guideposts of migrating fishes. *Science* 132:785–792.

Hastings, J. W. 1970. In *The biological clock: two views,* ed. F. A. Brown, J. W. Hastings, J. D. Palmer, pp. 63–94. Academic Press, New York.

Hatch, J. 1966. Collective territories in Galapagos mockingbirds, with notes on other behavior. *Wilson Bull.* 78:198–207.

Hays, H. 1972. Polyandry in the spotted sandpiper. *Living Bird* 11:43–58.

Hebb, D. O. 1953. Heredity and environment in animal behavior. *Brit. Anim. Behav.* 1:43–47.

Hediger, H. 1948. Die Zucht des Feldhasen (*Lepus europaeus* Pallus) in Gefangenschaft. *Physiologia Comparata et Oecologia* 1:46–62.

———. 1955. *Studies of the psychology and behaviour of captive animals in zoos and circuses* Butterworth & Co., London.

Heinroth, O. 1910. Beiträge zur Biologie, näm entlich Ethologie und Physiologie der Anatiden. *Verhandlungen der V Internazionalen Ornothologischen Kongres,* pp. 589–702.

———. 1930. Uber bestimmte Bewegungsweisen bei Wirbeltieren. *S. B. Ges. naturf.* Fr. Berl. Feb.

Hensley, M. M., and J. B. Cope. 1951. Further data on removal and repopulation of the breeding birds in a spruce-fir forest community. *Auk* 68:483–493.

Hespenheide, H. A. 1966. The selection of seed size by finches. *Wilson Bull.* 78:191–197.

———. 1971. Food preference and the extent of overlap in some insectivorous birds, with special reference to the Tyrannidae. *Ibis* 113:59–72.

———. 1973. Ecological inferences from morphological data. In *Annual Review of Ecology and Systematics,* ed. R. F. Johnston. Annual Reviews, Inc., Palo Alto, Calif.

Hess, E. H. 1956. Space perception in the chick. *Sci. Amer.* 195:71–80.

———. 1962. Ethology: an approach toward the complete analysis of behavior. In *New directions in psychology,* pp. 159–266. Holt, Rinehart and Winston, New York.

———. 1964. Imprinting in birds. *Science* 146:1128–1139.

———. 1973. *Imprinting.* Van Nostrand Reinhold, New York.

Hess, W. R. 1943. Das Zwischenhirn als Koordinationsorgan. *Helv. Physiol. Acta.* 1:549–565.

Heston, L. 1970. The genetics of schizophrenic and schizoid disease. *Science* 167:248–255.

Hilgard, E. R. 1956. *Theories of learning.* Appleton-Century-Croft, New York.

Hilgard, E. R., and D. G. Marquis. 1961. *Conditioning and learning.* Revised G. A. Kimble. Methuen, London.

Hinde, R. A. 1952. The behavior of the great tit (*Parus major*) and some other related species. *Behaviour Suppl.* 2:1–201.

———. 1956a. Ethological models and the concept of drive. *Brit. J. Philos. Sci.* 6:321–331.

———. 1956b. Territories of birds. *Ibis* 98:340–369.

———. 1956c. The biological significance of the territories of birds. *Ibis* 98:340–369.

———. 1959. Unitary drives. *Anim. Behav.* 7: 130–141.

———. 1960a. Energy models of motivation. *Symp. Soc. Exper. Biol.* 14:199–213.

———. 1960b. Factors governing the changes in strength of a partially inborn response, as shown by the mobbing behaviour of the chaffinch (*Fringilla coelebs*): III. The interaction of short-term and long-term incremental and decremental effects. *Proc. Roy. Soc. Ser. B* 153:398–420.

Hinde, R. A., ed. 1969. *Bird vocalizations.* Cambridge Univ. Press, Cambridge.

———. 1970. (2nd ed.) *Animal behaviour: a synthesis of ethology and comparative psychology.* McGraw-Hill, New York.

Hinde, R. A., and J. Fisher. 1951. Further observations on the opening of milk bottles by birds. *Brit. Birds* 44:393–396.

Hinde, R. A., and N. Tinbergen. 1958. The comparative study of species-specific behavior. In *Behavior and evolution,* ed. A. Roe and G. G. Simpson, pp. 251–268. Yale Univ. Press, New Haven, Conn.

Hirsch, J., ed. 1967a. *Behavior-genetic analysis.* McGraw-Hill, New York.

———. 1967b. Behavior-genetic, or "experimental," analysis: the challenge of science versus the lure of technology. In *Readings in animal behavior,* ed. T. E. McGill, pp. 58–69. Holt, Rinehart and Winston, New York.

Hoar, W. S. 1966. *General and comparative physiology.* Prentice-Hall, Englewood Cliffs, New Jersey.

Hodos, W., and C. B. G. Campbell. 1969. Scala Naturae: Why there is no theory in comparative psychology. *Psychol.* Rev. 76:337–350.

Hoffman, K. 1954. Versuche zu der im Richtungsfinden der Vögel enthaltenen Zeitschatzung. *Z Tierpsychol.* 11:453–475.

———. 1965. Overt circadian frequencies and circadian rule. In *Circadian clocks,* ed. J. Aschoff. North-Holland, Amsterdam.

Hoffman, R. S. 1958. The role of reproduction and mortality in population fluctuations of voles (*Microtus*). *Ecol. Monogr.* 28:79–109.

Hokanson, J. E., and M. Burgess. 1962. The effects of three types of aggression on vascular processes. *J. Abnor. Soc. Psychol.* 64:446–449.

Hokanson, J. E., and S. Shetler, 1961. The effect of overt aggression on physiological tension level. *J. Abnor. Soc. Psychol.* 63:446–448.

Holberg, F. 1960. Temporal coordination of physiologic function. *Cold Spring Harbor Symp. Quant. Biol.* 25:289–308.

Holling, C. S. 1959.The components of predation as revealed by a study of small-mammal predation on the European sawfly *Canad. Entom.* 5:293–320.

———. 1964. The analysis of complex population processes. *Canad. Entom.* 96:335–347.

———. 1968. Tactics of a predator. *Insect Abundance Symp. Roy. Entom. Soc. Lond.* 4:47–58.

Holmes, R. T. 1970. Differences in population density, territoriality and food supply of dunlin on arctic and subarctic tundra. In *Animal populations in relation to their food resources,* ed. A. Watson, pp. 303–322. Blackwell, Oxford.

Holst, E. von. 1935. Über den Lichtrückenreflex bei Fischen. *Publ. Staz. Zool. Napoli.* 15:143–158.

Hooff, J. A. R. A. M. van. 1967. The facial displays of the catarrhine monkeys and apes. In *Primate ethology,* ed. D. Morris, pp. 7–68. Weidenfield and Nicolson, London.

Hoogland, J. L., and P. W. Sherman. 1976. Advantages and disadvantages of bank swallow (*Riparia riparia*) coloniality. *Ecol. Monog.* 46:33–58.

Hoogland, R., D. Morris, and N. Tinbergen. 1957. The spines of sticklebacks (*Gasterosteus,* and *Pygosteus*) as means of defense against predators (*Perca* and *Esox*) *Behaviour* 10:205–236.

Hook, E. B. 1973. Behavioral implications of the human XYY genotype. *Science* 179:139–150.

Horn, H. S. 1968a. The adaptive significance of colonial nesting in the Brewer's blackbird (*Euphagus cyanocephalus*). *Ecology* 49:682–694.

———. 1968b. Regulation of animal numbers: a model counterexample. *Ecology* 49:776–778.

Howard, H. E. 1920. *Territory in bird life.* Antheneum, New York.

———. 1929. *An introduction to the study of bird behavior.* Cambridge Univ. Press, Cambridge.

Hüchtker, R., and J. Schwartzkopff. 1958. Soziale Verhaltensveisen bei ho renden and gehörlosen Compfaffen (*Pyrrhula pyrrhula* L.). *Experientia* 14:106–107.

Hunsaker, D. 1962. Ethological isolating mechanisms in the *Sceloporus torquatus* group of lizards. *Evolution* 16:62–74.

Hunsaker, D., and B. R. Burrage. 1969. The significance of interspecific social dominance in iguanid lizards. *Amer. Midland Naturalist* 81:500–511.

Hunter, I. M. L. 1964. *Memory.* Penguin Press, Harmondsworth, England.

Hunter, J. L. 1965. *Acoustics.* Prentice-Hall, Inc., Englewood Cliffs, New Jersey.

Hutchinson, G. E. 1957. Concluding remarks. *Cold Spring Harbor Symp. Quant. Biol.* 22:415–427.

———. 1959. Homage to Santa Rosalia or why are there so many different kinds of animals? *Amer. Naturalist* 93:145–159.

Hutchinson, G. E. and E. S. Deevey, Jr. 1949. Ecological studies on populations. *Surv. Biol. Prog.* 1:325–359.

Huxley, J. S. 1942. *Evolution: the modern synthesis.* George Allen and Unwin, London.

———. 1934. A natural experiment on the territorial instinct. *Brit. Birds* 27:270–277.

———. 1938. Threat and warning coloration in birds, with a general discussion of the biological functions of color. *Proc. 8th Int. Orn. Cong.*, Oxford: 430–455.

———. 1959. *Clades and grades. Syst. Assn. Publ.* No. 3:21–22.

Hyden, H. 1969. Biochemical aspects of learning and memory. In *On the biology of learning,* ed. K. Pribram. Harcourt Brace Jovanovich, New York.

Iersel, J. J. A. van. 1953. An analysis of the parental behavior of the male three-spined stickleback. *Behav. Suppl.* 3:1–159.

Iersel, J. J. A. van, and A. C. A. Bol. 1958. Preening of two tern species. A study on displacement activities. *Behaviour* 13:1–88.

Immelmann, K. 1962. Biologische Bedeutung optischer und akustischer Merkmale bei Prachtfinken (Aves: Spermestidae). *Verhandlungen der Deutschen Zoologischen Gesellschaft,* Zoologischer Anzeiger Supplement, 25:369–374.

Immelmann, K. 1972. The influence of early experience upon the development of social behavior in estrildine finches. *Proc XV Int. Orn. Congr.,* pp. 316–338.

Itani, J. 1958. On the acquisition and propagation of a new food habit in the troop of Japanese monkeys at Takasakiyama. *Primates* 1:84–98.

———. 1959. Paternal care in the wild Japanese monkey, *Macaca fuscata. Primates,* Japan, 2:61–93.

Jacklet, J. W. 1969. Circadian rhythm of optic nerve impaired. Recorded in darkness from isolated eye of *Aplysia. Science* 164:562–563.

Jackson, J. B. C. 1968. Bivalves: spatial and size-frequency distributions of two intertidal species. *Science* 161:479–480.

Jacobs, W. 1953. Verhaltensbiologische Studien an Feldheuschrecken. Beiheft 1: *Z. Tierpsychol.*

Jacobson, M. 1972. *Insect sex pheromones.* Academic Press, New York.

Janowitz, H. D., and M. I. Grossman. 1949. Some factors affecting the food intake of normal dogs and dogs with esophagotomy and gastric fistulas. *Amer. J. Physiol.* 159:143–148.

Janzen, D. H. 1971a. Escape of juvenile *Dioclea megacarpa* (leguminosae) vines from predators in a deciduous tropical forest. *Amer. Naturalist* 105:97–112.

———. 1971. Seed predation by animals. *Ann. Rev. Ecol. Syst.* 2:465–492.

Jarman, P. 1968. The effect of the creation of Lake Kariba upon the terrestrial ecology of the Middle Zambezi valley, with particular reference to the large mammals. Ph. D. thesis, Manchester Univ.

Jay, P. 1968. Primate field studies and human evolution. In *Primates: studies in adaptation and variability,* ed. P. Jay. Holt, Rinehart and Winston, New York.

Jehl, J. R., Jr. 1970. Sexual selection for size differences in two species of sandpipers. *Evolution* 24:311–319.

Jenkin, P. M. 1962. *Animal hormones.* Pergamon Press, London.

Jenkins, D. 1961. Population control in protected partridges (*Perdix perdix*). *J. Anim. Ecol.* 30:235–258.

———. 1963. Population control in red grouse (*Lagopus lagopus scoticus*). *Proc. XIII Int. Orn. Congr.,* pp. 690–700.

———. 1964. Predation and red grouse populations. *J. Appl. Ecol.* 1:183–195.

Jenkins, D., A. Watson, and G. R. Miller. 1963. Population studies on red grouse, *Lagopus l. scoticus. J. Anim. Ecol.* 36:97–122.

Jespersen, P. 1929. On the frequency of birds over the high Atlantic Ocean. *Verhandlung der Internationalen Ornithologischen Kongres* 6:163–172.

Johnels, A. B. 1956. On the peripheral autonomic nervous system of the trunk region of *Lampetra planeri. Acta Zool.* 37:251–286.

Johnsgard, P. A. 1965. *Handbook of waterfowl behaviour.* Constable, London.

———. 1967. *Animal behavior.* Wm. C. Brown, Dubuque, Iowa.

Johnson, C. G. 1964. Migration. In *The physiology of insecta,* vol. 2, ed. M. Rockstein, pp. 188–226. Academic Press, New York.

Johnson, N. K. 1963. Biosystematics of sibling species of flycatchers in the *Empidonax hammondii-oberholseri-wrighti* complex. *Univ. Calif. Publ. Zool.* 66: 79–238.

Johnson, R. G. 1959. Spatial distribution of *Phoronopsis viridis* Hilton. *Science* 129:1221.

Johnson, R. N. 1972. *Aggression in man and animals.* Saunders, Philadelphia.

Jolly, A. 1972. *The evolution of primate behavior.* Macmillan, New York.

Kaada, B. 1967. Brain mechanisms related to aggressive behaviour. UCLA Forum in Medical Science. In *Aggression and defense. Neural mechanisms and social pattern,* ed. C. D. Clemente and D. B. Lindsley. Univ. of California Press, Berkeley.

Kagan, J., and H. A. Moss. 1962. *Birth to maturity.* John Wiley, New York.

Kahn, M. W. 1951. The effect of severe defeat at various age levels on the aggressive behavior of mice. *J. Gen. Psychol.* 79:117–130.

Kalela, O. 1957. Regulation of reproduction rate in subarctic populations of the vole *Clethrionomys rufocanus* (Sund.) *Ann. Acad. Sci. Fennicae Ser. A.,* Sect. IV (Biol.), No. 34:1–60.

Kalleberg, H. 1958. Observations in a stream tank of territoriality and competition in juvenile salmon and trout (*Salmo salar* L. and *S. trutta* L.). *Rep. Inst. Freshwater Res.* Drottningholm, 39:55–98.

Kalmus, H. 1955. The discrimination by the nose of the dog of individual human odours and in particular the odours of twins. *Brit. J. Anim. Behav.* 3:25–31.

Kawai, M. 1965. Newly acquired pre-cultural behavior of the natural troop of Japanese monkeys on Koshima Islands. *Primates,* Japan, 6:1–30.

Kear, J. 1961. Food selection in finches with special reference to interspecific differences. *Proc. Zool. Soc. Lond.* 138:163–204.

———. 1966. The pecking responses of young coots *Fulica atra* and moorhens *Gallinula chloropus. Ibis* 108:118–122.

Keenleyside, M. H. A. 1955. Some aspects of the schooling behaviour of fish. *Behaviour* 8:183–249.

Keenleyside, M. H. A., and F. T. Yamamoto. 1962. Territorial behaviour of juvenile Atlantic salmon (*Salmo salar* L.) *Behaviour* 19:139–169.

Keeton, W. 1967. *Biological science.* Norton, New York.

———. 1969. Orientation by pigeons: is the sun necessary? *Science* 1965:922–928.

Kellogg, W. N. 1961. *Porpoises and sonar.* Univ. of Chicago Press, Chicago.

Kemp, G. A., and L. B. Keith. 1970. Dynamics and regulation of red squirrel (*Tamiasciurus hudsonicus*) populations. *Ecology* 51:763–779.

Kendeigh, S. C. 1941. Territorial and mating behavior of the house wren. *Illinois Biol. Monogr.* 10:1–20.

———. 1945. Community selection by birds on the Helderberg plateau of New York. *Auk* 62:418–436.

Kennedy, J. S. 1956. Phase transformation in locust biology. *Biol. Rev.* 31:349–370.

Kenney, N. T. 1968. Sharks: wolves of the sea. *Nat. Geog.* 133(2):222–257.

Kettlewell, H. B. D. 1956. Further selection experiments on industrial melanism in the Lepidoptera. *Heredity* 10:287–301.

Kiester, A. R. 1971. Species density of North American amphibians and reptiles. *Syst. Zool.* 20:127–137.

Kilham, L. 1958a. Territorial behavior of wintering red-headed woodpeckers. *Wilson Bull.* 70:347–348.

———. 1958b. Sealed-in winter stores of redheaded woodpeckers. *Wilson Bull.* 70:107–113.

———. 1965. Differences in feeding behavior of male and female hairy woodpeckers. *Wilson Bull.* 77: 134–145.

———. 1970. Feeding behavior of downy woodpeckers. I. Preference for paper birches and sexual differences. *Auk* 87:544–556.

King, J. 1955. Social behavior, social organization and population dynamics in a black-tailed prairie dog town in the Black Hills of South Dakota. *Contr. Lab. Vert. Biol.* No. 67. University of Michigan, Ann Arbor.

King, M., and A. C. Wilson. 1976. Evolution at two levels in humans and chimpanzees. *Science* 188:107–116.

Kitching, J. A., and F. J. Ebling. 1967. Ecological studies at Lough Inc. *Adv. Ecol. Res.* 4:197–291.

Klomp, H. 1954. De terreinkeus van de Kievit, *Vanellus vanellus* (L). *Ardea* 43:1–140.

———. 1972. Regulation of the size of bird populations by means of territorial behaviour. *Netherlands J. Zool.* 22:456–488.

Klopfer, P. 1959. An analysis of learning in young Anatidae. *Ecology* 40:90–102.

———. 1961. Observational learning in birds: the establishment of behavioral modes. *Behaviour* 17: 1–80.

———. 1962. *Behavioral aspects of ecology.* Prentice Hall, Englewood Cliffs, New Jersey.

———. 1969a. *Habitats and territories.* Basic Books, New York.

———. 1969b. Instincts and chromosomes: what is an "innate" act? *Amer. Naturalist* 103:556–560.

———. 1971. Mother love: what turns it on? *Amer. Sci.* 59:404–407.

———. 1973. Does behavior evolve? *Ann. N.Y. Acad. Sci.* 223:113–119.

———. 1976. Evolution, behavior and language. *Communicative behavior and evolution.* Academic Press, New York.

Klopfer, P. K., and J. P. Hailman. 1965. Habitat selection in birds. In *Advances in the study of behavior,* ed. D. S. Lehrman, R. A. Hinde, E. Shaw, pp. 279–304. Academic Press, New York.

———. 1967. *Ethology's first century—An introduction to animal behavior.* Prentice-Hall, Englewood Cliffs, New Jersey.

Klopfer, P. K., and R. H. MacArthur. 1961. On the causes of tropical species diversity: niche overlap. *Amer. Naturalist* 95:223–226.

Kluyver, H. N. 1951. The population ecology of the great tit *Parus m. major* L. *Ardea* 39:1–135.

———. 1963. The determination of reproductive rates in Paridae. *Proc. Int. Orn. Congr.* 13:706–716.

———. 1966. Regulation of a bird population. *Ostrich* 6:389–396.

Kluyver, H. N., and P. C. Bucy. 1937. Psychic blindness and other symptoms following bilateral temporal lobectomy in Rhesus monkeys. *Amer. J. Physiol.* 119:352–353.

Kluyver, H. N., and N. Tinbergen. 1953. Territory and the regulation of density in titmice. *Arch. Neerlandaises Zool.* 10:265–289.

Knudson, M. 1971. Sex differences in dominance behavior of young human primates. Paper presented at 70th Annual Meeting Amer. Anthrop. Assoc., New York City.

Koch, S. 1959. *Psychology, a study of science. Study I, conceptual and systematic.* Vols. 1 and 2. McGraw-Hill, New York.

Köhler, W. 1927. *The mentality of apes,* 2nd ed. Kegan Paul, Trench, Trubner and Co., London.

Koivisto, I. 1965. Behaviour of the black grouse *Lyrurus tetrix* (L.) during the spring display. *Finnish Game Reserve* 26:60.

Kok, O. B. 1972. Breeding success and territorial behavior of male boat-tailed grackles. *Auk* 89: 528–540.

Kolata, G. B. 1977. Sexual dimorphism and mating systems: how did they evolve? *Science* 195: 382–383.

Konishi, M. 1965. The role of auditory feedback in the control of vocalization in the white-crowned sparrow. *Z. Tierpsychol.* 22:770–783.

———. 1970. Evolution of design features in the coding of species specificity. *Amer. Zool.* 10:67–72.

Konishi, M., and F. Nottebohm. 1969. Experimental studies on the ontogeny of avian vocalization. In *Bird vocalization,* ed. R. A. Hinde. Cambridge Univ. Press, New York.

Kormondy, E. J. 1969. *Concepts of Ecology.* Prentice-Hall, Englewood Cliffs, New Jersey.

Kortlandt, A. 1938. De uitdrukkingsbewegingen en -geluiden van *Phalacrocorax carbo sinensis* (Shaw and Nodder). *Ardea* 27:1–40.

———. 1940a. Wechselwirkung zwischen Instinkten. *Arch. Neerlandaises. Zool.* 4:442–520.

———. 1940b. Eine u bersicht der angelborenen Verhaltungsweisen des Mittel-Europa ischen Kormorans (*Phalacrocorax carbo sinensis* Shaw and Nodd), ihre Funktion, ontogenetische Entwicklung und phylogenetische Herkunft. *Arch. Neerlandaises Zool.* 4:401–442.

Koyama, N. 1967. On dominance rank and kinship of a wild Japanese monkey troop in Arashiyama. *Primates* 8:189–216.

Kozhov, M. 1963. *Lake Baikal and its life.* W. Junk, the Hague.

Kramer, G. 1949. Über Richtungstendenzen bei der nächtlichen Zugunruhe gekäfigter Vögel. In *Ornithologie als biologische Wissenschaft.* Heidelberg.

———. 1957. Experiments in bird orientation and their interpretation. *Ibis* 99:196–227.

———. 1961. Long-distance orientation. In *Biology and comparative physiology of birds,* ed. A. J. Marshall, pp. 341–371. London.

Krebs, C. J. 1972. *Ecology: the experimental analysis of distribution and abundance.* Harper and Row, New York.

Krebs, J. R. 1970. Regulation of numbers in the great tit (Aves: Passeriformes). *J. Zool.,* London, 162: 317–333.

———. 1971. Territory and breeding density in the great tit, *Parus major* L. *Ecology* 52:2–22.

Krefting, L. W. 1951. What is the future of the Isle Royale moose herd? *Trans. N. Amer. Wildl. Conf.* 16: 461–470.

Kress, S. W. 1967. A robin nests in winter. *Wilson Bull.* 79:245–246.

Kruuk, H. 1964. Predators and anti-predator behavior of the black-headed gull (*Larus ridibundus* L.) *Behaviour Suppl.* 11:1–129.

Kuhme, W. 1965. Freilandstudien zur Sociologie des Hyanenhundes. *Z. Tierpsychol.* 22:495–541.

Kullenberg. B. 1961. Studies in *Ophrys* pollination. *Zool. Bridrag*, Uppsala, 34:1–340.

Kummer, H. 1971. *Primate societies.* Aldine-Atherton, Chicago.

Kuo, Z. Y. 1960. Studies on the basic factors in animal fighting. *J. Gen. Psychol.* 96:201–239.

———. 1961. Studies on the basic factors in animal fighting. *J. Gen. Psychol.* 97:181–209.

Kurten, B. 1969. Continental drift and evolution. *Sci. Amer.* 220:54–64.

Kusnezov, M. 1957. Numbers of species of ants in fauna of different latitudes. *Evolution* 11:298–299.

Laarman, J. J. 1960. The plasticity of response patterns in host-seeking mosquitoes. *Proc. XI Int. Congr. Entom.* Vol. 3. Symp. 5:60–61.

Lack, D. 1933. Habitat selection in birds. *J. Anim. Ecol.* 2:239–262.

———. 1939. The display of the blackcock. *Brit. birds.* 32:290–303.

———. 1943. *The life of the robin.* Pelican Books, New York.

———. 1944. Ecological aspects of species-formation in passerine birds. *Ibis* 86:260–286.

———. 1945. The ecology of closely related species with special reference to Cormorant (*Phalacrocorax carbo*) and Shag (*P. aristotelis*). *J. Anim. Ecol.* 14:12–16.

———. 1947. *Darwin's finches.* Harvard Univ. Press, Cambridge, Mass.

———. 1948. Selection and family size in starlings. *Evolution* 2:95–110.

———. 1953. *The life of the robin.* Pelican Books, New York.

———. 1954. *The natural regulation of animal numbers.* Clarendon Press. Oxford.

———. 1958. A quantitative breeding study of British tits. *Ardea* 46:91–124.

———. 1964. A long-term study of the great tit (*Parus major*). *J. Anim. Ecol.* 33 (Jubilee suppl.):159–173.

———. 1966. *Population studies of birds.* Clarendon Press, Oxford.

———. 1968. *Ecological adaptations for breeding in birds.* Methuen & Co., Ltd., London.

———. 1969. Subspecies and sympatry in Darwin's finches. *Evolution* 23:252–263.

———. 1970. Introduction. In *Animal populations in relation to their food resources,* ed. A. Watson. Blackwell, Oxford.

———. 1971. *Ecological isolation in birds.* Harvard Univ. Press, Cambridge, Mass.

———. 1976. *Island biology.* Univ. of California Press, Berkeley.

Lack, D. and E. Lack. 1958. The nesting of the long-tailed tit. *Bird Study* 5:1–19.

Lancaster, J. B. 1968. Primate communication systems and the emergence of human language, In *Primates,* ed. P. Jay, pp. 439–457. Holt, Rinehart and Winston, New York.

Lanyon, W. E. 1957. The comparative biology of the meadowlarks (*Sturnella*) in Wisconsin. *Publ. Nuttall Orn. Club* 1:1–67.

———. 1960a. The middle American populations of the crested flycatcher *Myiarchus tyrannulus. Condor* 62:341–350.

———. 1960b. The ontogeny of vocalization in birds. In *Animal sounds and communication,* ed. W. E. Lanyon and W. N. Tavolga. Publication no. 7. American Institute of Biological Sciences, Washington, D.C.

———. 1963. Experiments on species discrimination in *Myiarchus* flycatchers. *Amer. Mus. Novitates,* No. 2126:1–16.

Lanyon, W. E. and W. N. Tavolga, eds. 1960. *Animal sounds and communication.* American Institute of Biological Sciences, Washington, D.C.

Lawick, Hugo van, and Jane Goodall. 1970. *Innocent killers.* Houghton Mifflin, Boston.

Leck, C. F. 1971a. Cooperative feeding in *Leucophoyx thula* and *Podilymbus podiceps* (Aves). *Amer. Midland Naturalist* 86:241–242.

———. 1971b. Measurement of social attractions between tropical passerine birds. *Wilson Bull.* 83: 278–283.

Lee, R., and I. DeVore, eds. 1968. *Man the hunter.* Aldine, Chicago.

Lehrman, D. S. 1953. A critique of Konrad Lorenz's theory of instinctive behavior. *Q. Rev. Biol.* 28: 337–363.

———. 1955. The physiological basis of parental feeding behaviour in the ringdove (*Streptopelia risoria). Behaviour* 7:241–286.

———. 1970. Semantic and conceptual issues in the nature-nurture problem. In *Development and evolution of behavior,* ed. L. R. Aronson, E. Tobach, D. S. Lehrman, J. S. Rosenblatt, pp. 17–52. W. H. Freeman, San Francisco.

Leibman, M. 1970. The effects of sex and race norms on personal space. *Envir. and Behav.* 2:208–246.

Le magnen, J. 1952. Les phénomenes olfacto-sexuels chez l'homme. *Arch. Sci. Physiol.* 6:125–160.

LeMaire, L. 1956. Danish experiences regarding the castration of sexual offenders. *J. Crim. Law and Criminol.* 47:294–310.

Leopold, A. 1933. *Game management.* Charles Scribner's Sons, New York.

Levene, H. 1953. Genetic equilibrium when more than one ecological niche is available. *Amer. Naturalist* 87:331–333.

Levick, W. R. 1967. Receptive fields and trigger features in the visual streak of the rabbit's retina. *J. Physiol.* 188:285–307.

Levins, R. 1968. *Evolution in changing environments. Some theoretical explorations.* Princeton Univ. Press, Princeton, New Jersey.

Levy, I. V., and I. A. King. 1953. The effects of testosterone propionate on fighting behavior in young C57BL/ID mice. *Anat. Rec.* 117:562.

Lewontin, R. C. 1965. Selection for colonizing ability. In *The genetics of colonizing species,* ed. H. G. Baker and G. L. Stebbins, pp. 77–91. Academic Press, New York.

Leyhausen, P. 1956. Verhaltensstudien an Katzen. Beiheft 2 zur *Z. Tierpsychol.*

———. 1965a. The communal organization of solitary mammals. *Symp. Zool. Soc. Lond.* 14:249–263.

———. 1965b. Über die Funktion der relativen Stimmungshierarchie (dargestellt am Beispiel der phylogenetischen und ontogenetischen Entwicklung des Beutefangs von Raubtieren). *Z. Tierpsychol.* 22:412–494.

Libassi, P. T. 1974. Where the past is present. *The Sciences* 14:17–23.

Ligon, J. D. 1968. Sexual differences in two species of *Dendrocopos* woodpeckers. *Auk* 85:203–215.

Lin, N. 1963. Territorial behavior in the cicada killer wasp, *Sphecius speciosus* (Drury) (Hymenoptera: Spheciadae). Part I. *Behaviour* 20:115–133.

Lindauer, M. 1961. *Communication among social bees.* Harvard Univ. Press, Cambridge, Mass.

———. 1964. Allgemeine Sinnesphysiologie: Orientierung im Raum. *Fortschr. der Zool.* 16:58–140.

———. 1965. Social behavior and mutual communication. In *The physiology of insecta,* ed. M. Rockstein, vol. 2, pp. 124–187. Academic Press, New York.

Lindsey, C. C. 1966. Body sizes of poikilotherm vertebrates at different latitudes. *Evolution* 20:456–465.

Linicome, D. R. 1971. The goodness of parasitism: a new hypothesis. In *Aspects of the biology of symbiosis,* ed. T. C. Cheng, pp. 139–227. Univ. Park Press, Baltimore.

Lissmann, H. W. 1951. Continuous electrical signals from the tail of a fish *Gymnarchus niloticus. Nature,* London, 167:201–202.

Lloyd, J. E. 1975. Reported in *Time* International, Feb. 17, 1975, p. 49.

Lockard, R. B. 1971. Reflections on the fall of comparative psychology: is there a message for us all? *Amer. Psychol.* 26:168–179.

Lockard, R. B., and D. H. Owings. 1974. Moon-related surface activity of bannertail (*Dipodomys spectabilis*) and Fresno (*D. nitratoides*) Kangaroo rats. *Anim. Behav.* 22:262–273.

Lockie, J. D. 1966. Territory in small carnivores. *Symp. Zool. Soc. Lond.* 18:143–165.

Lockley, R. M. 1967. *Animal navigation.* Hart Publishing Co. Inc., New York.

Loeb, J. 1901. *Comparative physiology of the brain and comparative psychology,* London.

Lorenz, K. Z. 1931. Beiträge zur Ethologie sozialer corviden. *J. f. Ornith.* 79:67–127.

———. 1935. Der Kumpan in der Umwelt des Vogels. *J. f. Ornith.* 83:37–213, 289–413.

———. 1937a. Über die Bildung des Instinktbegriffes. *Naturwiss.* 25:289–300, 307–318, 324–331. (English translation in C. H. Schiller, 1957).

———. 1937b. The companion in the bird's world. *Auk* 54:245–273.

———. 1939. Vergleichende Verhaltensforschung. *Zoo. Anz. Suppl. Bd.* 12:69–102.

———. 1941. Vergleichende Bewegungstudien an Anatinen. *J. f. Ornith.* Leipzig, 89:94–293.
———. 1950. The comparative method in studying innate behaviour patterns. *Symp. Soc. Exper. Biol.* 4:221–268.
———. 1952. *King Solomon's ring,* trans. M. K. Wilson. Thomas Y. Cromwell, New York.
———. 1957. Companionship in bird life and other essays. In *Instinctive behavior,* ed. C. H. Schiller. International Universities Press, New York.
———. 1958. The evolution of behavior. *Sci. Amer.* 199:67–78.
———. 1961. Phylogenetische Anpassung und adaptive Modifikation des Verhaltens. *Z. Tierpsychol.* 18: 139–187.
———. 1963. *On aggression.* (transl. 1966). Harcourt, Brace and World, New York.
———. 1965. *Evolution and modification of behavior.* Univ. of Chicago Press, Chicago.
———. 1969. Innate bases of learning. In *On the biology of learning,* ed. K. H. Pribram, pp. 13–93. Harcourt, Brace and World, New York.
Lorenz, K. Z., and N. Tinbergen. 1938. 1957. Taxis and Instinct. In *Instinctive behavior,* ed. C. H. Schiller, pp. 176–208. International Universities Press, New York.
———. 1939. Taxis und Instinkthandlung in der Eirollbewegung der Graugans I. *Z. Tierpsychol.* 2:1–29.
Lotka, A. J. 1925. *Elements of physical biology.* Baltimore.
MacArthur, R. H. 1959. On the breeding distribution pattern of North American migrant birds. *Auk* 76:318–325.
———. 1961. Population effects of natural selection. *Amer. Naturalist* 95:195–199.
———. 1965. Patterns of species diversity. *Biol. Rev.* 40:510–533.
———. 1967. *The theory of island biogeography.* Princeton Univ. Press, Princeton, New Jersey.
———. 1972. *Geographical ecology: patterns in the distribution of species.* Harper and Row. New York.
MacArthur, R. H., and R. Levins. 1964. Competition, habitat selection and character displacement in a patchy environment. *Proc. Nat. Acad. Sci.,* U.S.A., 51:1207–1210.
MacArthur, R. H., and J. W. MacArthur. 1961. On bird species diversity. *Ecology* 42:594–598.
MacArthur, R. H., and E. O. Wilson. 1963. An equilibrium theory of insular zoogeography. *Evolution* 17: 373–387.
———. 1967. *The theory of island biogeography.* Princeton Univ. Press, Princeton, New Jersey.
McBride, G. 1964. *A general theory of social organization.* Univ. of Queensland Press, St. Lucia.
McClearn, G. E., and J. C. DeFries. 1973. *Introduction to behavioral genetics.* Freeman, San Francisco.
McConnell, J. V. 1962. Memory transfer through cannibalism. *J. Neuropsychiatry* 3:542–548.
McConnell, J. V., A. L. Jacobson, and D. P. Kimble. 1959. The effects of regeneration upon retention of a conditioned response in the planarian. *J. Comp. Physiol. Psychol.* 52:1–5.
McDougall, W. 1905. *Physiological psychology.* London.
McIhenny, E. A. 1959. *The alligator's life history.* Christopher Publishing House, Boston.
Mackie, G. O. 1960. The structure of the nervous system in *Velella. Q. J. Microscop. Sci.* 101:119–132.
MacKinnon, J. 1974. *In search of the red ape.* Ballantine Books, New York.
MacLean, P. D., and D. W. Ploog. 1962. Cerebral representation of penile erection. *J. Neurophysiol.* 25:29–55.
McSeen, D. S. 1972. Cowbird-host relationships. *Auk* 89:360–380.
Magnus, D. 1958. Experimentalle Untersuchungen zur Bionomie und Ethologie des Kaisermantels *Argynnis paphia* L. (Lep. Nymph). I. Über optische Auslöser von Angliegeralktionen und ihre Bedeutung für das Sichfinden der Geschechter. *Z. Tierpsychol.* 15: 397–426.
Magnuson, J. J. 1962. An analysis of aggressive behavior, growth and competition for food and space in medaka (*Oryzias latipes*) (Pisces, Cyprinodontidae). *Can. J. Zool.* 40:314–363.
Maier, R. A., and B. M. Maier. 1970. *Comparative animal behavior.* Brooks/Cole, Belmont, Calif.
Makkink, G. F. 1942. Contribution to the knowledge of the behavior of the oyster catcher (*Haematopus ostralegus* L.) *Ardea* 31:23–74.
Manning, A. 1959. The sexual behaviour of two sibling *Drosophila* species. *Behaviour* 15:123–145.
———. 1961. The effects of artificial selection for mating speed in *Drosophila melanogaster. Anim. Behav.* 9:82–92.
———. 1964. Evolutionary changes and behavior ge-

netics. *Proc. XI Int. Congr. Genet.* The Hague, Netherlands, September 1963. Pergamon Press, New York.

———. 1965. *Drosophila* and the evolution of behaviour. In *Viewpoints in Biology,* 4, ed. J. D. Carthy and C. L. Duddington, pp. 125–169. Butterworths, London.

———. 1967. *An introduction to animal behavior.* Addison-Wesley, Reading, Mass.

———. 1971. Evolution of behavior. In *Psychobiology,* ed. J. L. McGaugh, pp. 1–52. Academic Press, New York.

Mark, V. H. and F. R. Ervin. 1970. *Violence and the brain.* Harper and Row, New York.

Marler, P. 1956. The voice of the chaffinch and its function as a language. *Ibis* 98:231–261.

———. 1957a. Specific distinctiveness in the communication signals of birds. *Behaviour* 11:13–39.

———. 1957b. Studies of fighting in chaffinches (4). *Appet. Consum. Behav.* 5:29–37.

———. 1959. Developments in the study of animal communication. In *Darwin's Biological work,* ed. P. R. Bell, pp. 150–206. Cambridge Univ. Press, Cambridge.

———. 1961a. The evolution of visual communication. In *Vertebrate speciation,* ed. W. F. Blair, pp. 96–121. Univ. of Texas Press, Austin.

———. 1961b. The filtering of external stimuli during instinctive behavior. In *Current problems in animal behavior,* pp. 150–166. Cambridge Univ. Press, New York.

———. 1965. Communication in monkeys and apes. In *Primate behavior: field studies of monkeys and apes,* ed. I. DeVore, pp. 544–584. Holt, Rinehart and Winston, New York.

———. 1969a. *Colobus guereza:* territoriality and group composition. *Science* 163:93–95.

———. 1969b. Vocalizations of wild chimpanzees. In *Proceedings of the Second International Congress of Primatology,* vol. 1: *Behaviour,* ed. C. R. Carpenter, pp. 94–100. Karger, Basel, Switzerland.

———. 1970. Birdsong and speech development: could there be parallels? *Amer. Sci.* 58:669–673.

Marler, P., and W. J. Hamilton, III. 1966. *Mechanisms of animal behavior.* John Wiley and Sons, Inc. New York.

Marler, P., M. Krieth, and M. Tamura. 1964. Song development in hand-raised Oregon juncos. *Auk* 79:12–30.

Marler, P., and M. Tamura. 1962. Song "dialects" in three populations of white crowned sparrows. *Condor* 64:368–377.

———. 1964. Culturally transmitted patterns of vocal behavior in sparrows. *Science* 146:1483–1486.

Marshall, A. J. 1959. Internal and environmental control of breeding. *Ibis* 101:456–478.

Martin, M. M. 1970. The biochemical basis of the fungus-attine ant symbiosis. *Science* 169:16–20.

Mason, W. A. 1965. The social development of monkeys and apes. In *Primate behavior,* ed. I. DeVore. Holt, Rinehart and Winston, New York.

Matthews, G. V. T. 1955. *Bird navigation.* Cambridge Univ. Press, Cambridge.

———. 1963. The orientation of pigeons as affected by the learning of landmarks and by the distance of displacement. *Anim. Behav.* 11:310–317.

———. 1968. *Bird navigation,* 2nd ed. Cambridge Univ. Press, Cambridge.

Mayersbach, H. von. 1967. *The cellular aspects of biorhythms.* Springer, New York.

Maynard-Smith, J. M. 1964. Group selection and kin selection. *Nature* 201:1145–1147.

Maynard-Smith, J. 1966. Sympatric speciation. *Amer. Naturalist* 100:637–650.

Maynard-Smith, J., and M. G. Ridpath. 1972. Wife sharing in the Tasmanian native hen, *Tribonex mortierii:* A Case of Kin Selection? *Amer. Naturalist* 106:447–452.

Mayr, E. 1942. *Systematics and the origin of the species.* Columbia Univ. Press, New York.

———. 1948. The bearing of the new systematics on genetical problems. The nature of species. *Adv. Genet.* 2:205–237.

———. 1958. Behavior and systematics. In *Evolution and behavior,* ed. A. Roe and G. G. Simpson, pp. 341–362. Yale Univ. Press, New Haven, Conn.

———. 1963. *Animal species and evolution.* Harvard Univ. Press, Cambridge.

———. 1965. The nature of colonization in birds. In *Genetics of colonizing species,* ed. H. G. Baker and G. L. Stebbins. Academic Press, New York.

———. 1974. Behavioral programs and evolutionary strategies. *Amer. Sci.* 62:650–659.

Mech, L. D. 1970. *The wolf.* Natural History Press, Garden City, New York.

Medawar, P. B. 1957. *The uniqueness of the individual.* Methuen, London.

Meisenheimer, J. 1921. *Geschlecht und Geschlechter im Tierreich.* Jena.

Menaker, M. 1969. Biological clocks. *Bioscience* 19:681–689.

Metzgar, L. H. 1967. An experimental comparison of screech owl predation on resident and transient white-footed mice (*Peromyscus leucopus). J. Mammal.* 48:387–391.

Meyer, M. E. 1964. Discriminative basis for astronavigation in birds. *J. Comp. Physiol. Psychol.* 58:403–406.

Meyer-Holzapfel, M. 1956. Das Speil bei Säugetieren. *Handb. Zool. Berl.* 8(10) No. 2:1–36.

———. 1958. Soziale Beziehungen bei Saugetieren. In *Gestaltungen sozialen Lebens bei Tier und Mensch.* ed. F. E. Lehmann, pp. 86–109. Franke, Bern.

Michener, C. D. 1974. *The social behavior of the bees.* Harvard Univ. Press, Cambridge, Mass.

Milgram, S. 1965. Some conditions of obedience and disobedience to authority. In *Current studies in social psychology,* ed. I. D. Steiner and M. Fishbein. Holt, Rinehart and Winston, New York.

———. 1966. Einige Bedingungen von Autoritätsgehorsam und Seiner Verweigerung. *Z. Exptl. Angew. Psychol.* 13:433–463.

Miller, N. E. 1941. The frustration-aggression hypothesis. *Psychol. Rev.* 48:337–342.

———. 1957. Experiments on motivation. Studies combining psychological, physiological and pharmacological techniques. *Science* 126:1271–1278.

———. 1961. Integration of neurophysiological and behavioral research. *Ann. N. Y. Acad. Sci.* 92: 830–839.

———. 1969. Learning of visceral and glandular responses. *Science* 163:434–445.

Miller, N. E., and A. Carmona. 1967. Modification of a visceral response, salivation in thirsty dogs, by instrumental training with water reward. *J. Comp. Physiol. Psychol.* 63:1–6.

Miller, N. E., and L. V. DiCara. 1967. Instrumental learning of heart rate changes in curarized rats: shaping, and specificity to discriminative stimulus. *J. Comp. Physiol. Psychol.* 63:12–19.

Miller, R. E. 1967. Experimental approaches to the physiological and behavioral concomitants of affective communication in rhesus monkeys. In *Social communication among primates,* ed. S. A. Altmann, pp. 125–134. Univ. of Chicago Press, Chicago.

Miller, R. S. 1969. Competition and species diversity. *Brookhaven Symp. Biol.* 22:63–70.

Miller, R. S., and W. J. D. Stephan. 1966. Spatial relationships in flocks of sandhill cranes (*Grus canadensis*). *Ecology* 47:323–327.

Milstead, W. W., ed. 1967. *Lizard ecology.* Univ. of Missouri Press, Columbia.

Missakian, E. A. 1972. Genealogical and cross-genealogical dominance relations in a group of free-ranging rhesus monkeys (*Macaca mulatta*) on Cayo Santiago. *Primates* 13:169–180.

Mitchell, D. 1978. The psychological versus the ethological rat: two views of the poison avoidance behavior of the rat compared. *Anim. Learn. Behav.* 6: 121–124.

Mitchell, D., E. T. Beatty, and P. K. Cox. 1977. Behavioral differences between two populations of wild rats: Implications for domestication research. *Behav. Biol.* 19:206–217.

Mitchell, D., E. H. Kirschbaum, and R. L. Perry. 1975. Effects of neophobia and habituation on the poison-induced avoidance of exteroceptive stimuli in the rat. *J. Exper. Psychol.: Anim. Behav. Proc.* 1:47–55.

Mitchell, D., L. F. Parker, and R. Johnson. 1976. Absence of a generalization decrement in the poison-induced avoidance of interoceptive stimuli in the rat. *Physiol. Psychol.* 4:121–123.

Mitchell, D., D. W. Scott, and K. D. Williams. 1973. Container neophobia and the rat's preference for earned food. *Behav. Biol.* 9:613–624.

Moment, G. 1962. Variation in brittle stars. *Science* 136:262–263.

Montagu, M. F. 1962. *Culture and the evolution of man.* Oxford Univ. Press, New York.

———. 1968. *Man and aggression.* Oxford Univ. Press, London.

Monteith, L. G. 1958. Influence of food plant of host on attractiveness of the host to tachinid parasites with notes on preimaginal conditioning. *Canad. Entom.* 90:478–482.

Moreau, R. E. 1950. The breeding seasons of African birds: I. Land Birds. *Ibis* 92:223–267.

———. 1966. *The bird faunas of Africa and its islands.* Academic Press, New York.

Morgan, C. T. 1965. *Physiological psychology,* 3rd ed. McGraw-Hill, New York.

Morgan, L. C. 1894. *Introduction to comparative psychology.* Arnold, London.

Morris, D. 1957. "Typical intensity" and its relation to the problem of ritualisation. *Behaviour* 11:1–12.

———. 1962. The behavior of an experimental colony of wild rabbits, *Oryctalagus cuniculus* (L), III Second breeding season. *Wildl. Res.* 5:120.

———. 1967. *The naked ape.* McGraw-Hill, New York.

Morse, D. H. 1968. The use of tools by brown-headed nuthatches. *Wilson Bull.* 80:220–224.

———. 1969. Ecological aspects of some mixed-species foraging flocks of birds. *Ecol. Monogr.* 40:119–168.

———. 1970. Ecological aspects of some mixed-species foraging flocks of birds. *Ecol. Monogr.* 40:119–168.

———. 1971. The foraging of warblers isolated on small islands. *Ecology* 52:216–228.

Morton, E. S. 1970. Ecological sources of selection on avian sound. *Diss. Abstr. Int.*, vol. 31, No. 12, 1971.

Moulton, J. M. 1956. Influencing the calling of sea robins (*Prionotus* spp.) with sound. *Biol. Bull.* 111:393–398.

Moyer, K. E. 1969. Internal impulses to aggression. *Trans. N.Y. Acad. Sci.* 31:104–114.

———. 1971. *The physiology of hostility.* Markham, Chicago.

Moynihan, M. 1960. Some adaptations which help to promote gregariousness. *Proc. 12th Int. Orn. Congr.*, pp. 523–541.

———. 1962a. Hostile and sexual behavior patterns of South American and Pacific Laridae. *Behaviour Suppl.* 8:1–365.

———. 1962b. The organization and probable evolution of some mixed species flocks of neotropical birds. *Smithsonian Misc. Coll.* 143:1–140, No. 7.

———. 1963. Inter-specific relations between some Andean birds. *Ibis* 105:327–339.

———. 1967. Comparative aspects of communication in New World primates. In *Primate ethology,* ed. D. Morris, pp. 236–266. Wiedenfield and Nicolson, London.

———. 1968. Social mimicry: character convergence versus character displacement. *Evolution* 22: 315–331.

———. 1971. Successes and failures of tropical mammals and birds. *Amer. Naturalist* 105:371–383.

Moynihan, M., and F. Hall. 1953. Hostile sexual and other social behaviour patterns of the spice finch (*Lonchura punctulata*) in captivity. *Behaviour* 7: 33–77.

Mueller, H. C., and D. D. Berger. 1970. Prey preferences in the sharp-shinned hawk: the roles of sex, experience and motivation. *Auk* 87:452–457.

Muir, R. C. 1954. Calling and feeding rates of fledged tawny owls. *Bird Study* 1:111–117.

Muntz, W. R. A. 1971. Sensory processes and behavior. In *Psychobiology,* ed. J. L. McGaugh, pp. 73–98. Academic Press, New York.

Murdoch, W. W. 1966. Community structure, population control, and competition: a critique. *Amer. Naturalist* 100:219–226.

Murdoch, W., S. Avery, and M. E. B. Smyth. 1975. Switching in predatory fish. *Ecology* 56:1094–1105.

Murie, A. 1934. The moose of Isle Royale. *Univ. Mich. Mus. Zool. Misc. Publ.* 25.

———. 1944. The wolves of Mount McKinley. *Faune of the national parks of the U.S.A.* Fauna Series, 5.

Murphy, R. C. 1936. *Oceanic birds of South America.* Macmillan Co., New York.

Murray, B. G. 1971. The ecological consequences of interspecific territorial behavior in birds. *Ecology* 52:414–423.

Murton, R. K., A. T. Isaacson, and K. T. Westwood. 1964. The relationships between woodpigeons and their clover food supply and the mechanism of population control. *J. Appl. Ecol.* 3:55–96.

Myers, K. 1966. The effects of density on sociality and health in mammals. *Proc. Ecol. Soc. Australia* 1: 40–64.

Myrberg, A. A., and R. E. Thresher. 1974. Interspecific aggression and its relevance to the concept of territoriality in reef fishes. *Amer. Zool.* 14:81–96.

Napier, J. R. 1967. The antiquity of human walking. *Sci. Amer.* 216:56–66.

Nelson, J. B. 1967. Etho-ecological adaptations in the Great Frigate-bird. *Nature* 214:318.

Newell, N. D. 1949. Phyletic size increase: an important trend illustrated by fossil invertebrates. *Evolution* 3:103–124.

*Newsweek* (International edition). June 9, 1975, p. 5.

Nice, M. M. 1937. Studies in the life history of the song sparrow. I. *Trans. Linnaean Soc.* New York, 4.

———. 1943. Studies in the life history of the song sparrow, I, II. *Trans. Linnaean Soc., New York,* 4:1–247, 6:1–328.

———. 1957. Nesting success in altricial birds. *Auk* 74: 305–321.

Nicholson, A. J. 1933. The balance of animal populations. *J. Anim. Ecol.* 2:132–178.

———. 1954. An outline of the dynamics of animal populations. *Austral. J. Zool.* 2:9–65.

———. 1957a. Comments on paper of T. B. Reynoldson. *Cold Spring Harbor Symp. Quant. Biol.* 22:326.

———. 1958. The self-adjustment of populations to change. *Cold Spring Harbor Symp. Quant. Biol.* 22:153–173.

Nicholson, A. J., and V. A. Bailey. 1935. The balance of animal populations. *Proc. Zool. Soc. Lond.* 1935:551–598.

Noble, G. K. 1936. Courtship and sexual selection of the flicker (*Colaptes auratus lutus*). *Auk* 53:269–282.

Nordsiek, F. W. 1972. The sweet tooth. *Amer. Sci.* 60:41–45.

Norris, K. S. 1967. Some observations on the migration and orientation of marine mammals. In *Animal orientation and navigation,* ed. R. M. Storm, pp. 101–125. Oregon State Univ. Press, Corvallis.

———. 1969. The echolocation of marine mammals. In *The biology of marine mammals,* ed. H. T. Andersen. Academic Press, New York.

Nottebohm, F. 1970. Ontogeny of bird song. *Science* 167:950–956.

Odum, E. P. 1942. Annual cycle of the black-capped chickadee. *Auk* 59:499–535.

———. 1971. *Fundamentals of ecology,* 3rd ed. W. B. Saunders, Philadelphia.

Orenstein, R. I. 1972. Tool-use by the New Caledonian crow (*Corvus moneduloides*). *Auk* 89:674–676.

Orians, G. H. 1961. Social stimulation within blackbird colonies. *Condor* 63:330–337.

———. 1962. Natural selection and ecological theory. *Amer. Naturalist* 96:257–263.

———. 1966. Food of nestling yellow-headed black birds, Caribou Parklands, British Columbia. *Condor* 68:321–337.

———. 1969a. On the evolution of mating systems in birds and mammals. *Amer. Naturalist* 103:589–603.

———. 1969b. The number of bird species in some tropical forests. *Ecology* 50:783–801.

———. 1971. Ecological aspects of behavior. In *Avian biology,* vol. 1, ed. D. S. Farner and J. King. Academic Press, New York.

Orians, G. H., and G. Collier. 1963. Competition and blackbird social systems. *Evolution* 17:449–459.

Orians, G. H., and M. F. Willson. 1964. Interspecific territories of birds. *Ecology* 45:736–745.

Oring, L. W., and M. L. Knudson. 1972. Monogamy and polyandry in the spotted sandpiper. *Living Bird* 11:59–73.

Osche, G. 1962. Okologie des Parasitismus und der Symbiose. *Fortschr. der Zool.* 15:125–164.

———. 1966. Amniotic contractions and embryonic motility in the chick embryo. *Science* 152: 528–529.

Paine, R. T. 1966. Food web complexity and species diversity. *Amer. Naturalist* 100:65–76.

Palmer, J. D. 1970. In *The biological clock: two views,* ed. F. A. Brown, J. W. Hastings, J. D. Palmer, pp. 3–12. Academic Press, New York.

———. 1974. *Biological clocks in marine organisms: the control of physiological and behavioral tidal rhythms.* Wiley, New York.

Papi, F., L. Fiore, V. Fiaschi, and S. Benvenuti. 1972. Olfaction and homing in pigeons. *Monitore Zool. Ital.* n. s. 6:85–95.

Parrack, J. 1973. *The naturalist in Majorca.* David and Charles Newton Abbott, England.

Parsons, P. A. 1967. *The genetic analysis of behavior.* Methuen, London.

Passano, L. M., and C. F. A. Pantin. 1955. Mechanical stimulation in the sea anemone *Callicatus parasitica. Proc. Roy. Soc.* Lond. 143:226–238.

Patterson, I. J. 1965. Timing and spacing of broods in the blackheaded gull (*Larus ridibundus*). *Ibis* 107:433–459.

Paulson, D. R. 1969. Commensal feeding in grebes. *Auk* 86:759.

Pavlov. I.P. 1927. Conditioned reflexes: an investigation of the activity of the cerebral cortex. *Trans. G. V. Anrep,* London.

Payne, A. P., and H. H. Swanson. 1971. Hormonal control of aggressive dominance in the female hamster. *Physiol. Behav.* 6:355–357.

Payne, R. B. 1962. How the barn owl locates its prey by hearing. *Living Bird* 1:151–159.

———. 1965. Clutch size and numbers of eggs laid by brown-headed cowbirds. *Condor* 67:44–60.

Pearson, O. P. 1947. The rate of metabolism of some small mammals. *Ecology* 28:127–145.

Peckham, G. W., and E. G. Peckham. 1898. On the instincts and habits of the solitary wasps. *Wisc. Geol. Natur. Hist. Surv. Bull.* No. 2.

Pelkwijk, J. J. ter, and N. Tinbergen. 1937. Eine reizbiologische Analyse einiger Verhaltensweisen von *Gasterosteus aculeatus* L. *Z. Tierpsychol.* 1:193–204.

Penfield, W. 1959. The interpretive cortex. *Science* 129:1719–1725.

Penney, R. L., and J. T. Emlen. 1967. Further experiments on distance navigation in the Adélie penguin. *Ibis* 109:99–109.

Perdeck, A. C. 1958. Two types of orientation in migrating starlings, *Sturnus vulgaris* L., and chaffinches, *Fringilla coelebs* L., as revealed by displacement experiments. *Ardea* 46:1–37.

Perrins, C. M. 1965. Population fluctuations and clutch-size in the great tit, *Parus major* L. *J. Anim. Ecol.* 34:601–647.

Pfaffman, C. 1959. The afferent code for sensory quality. *Amer. Psychol.* 14:226–232.

Pfeiffer, J. 1970. *The emergence of man.* Harper and Row, New York.

Phillips, R. E. 1964. "Wildness" in the mallard duck: effects of brain lesions and stimulation on "escape behavior" and reproduction. *J. Comp. Neurol.* 122:139–155.

Phoenix, C. H., R. W. Goy, A. A. Gerall, and W. C. Young. 1959. Organizing action of prenatally administered testosterone propionate on the tissues mediating behavior in the female guinea pig. *Endocrinology* 65:369–382.

Pianka, E. R. 1966. Convexity, desert lizards and spatial heterogeneity. *Ecology* 47:1055–1059.

———. 1970. On r- and K- selection. *Amer. Naturalist* 104:592–597.

———. 1972. r and K selection or b and d selection? *Amer. Naturalist* 106:581–588.

———. 1974. *Evolutionary ecology.* Harper and Row, New York.

Pielowski, Z. 1961. Über die Vertikalverteilung der Vögel in einem Pineto-Quercetum Biotop. *Ekologia Polska Seria A* 9:1–23.

Pilbeam, D. 1972. *The ascent of man.* Macmillan, New York.

Pilbeam, D., and S. J. Gould. 1974. Size and scaling in human evolution. *Science* 186:892–901.

Pimental, D., W. P. Nagel, and J. L. Madden. 1963. Space-time structure of the environment and the survival of parasite-host systems. *Amer. Naturalist* 97:141–167.

Pimlott, D. H. 1967. Wolf predation and ungulate populations. *Amer. Zool.* 7:267–278.

Pitelka, F. A. 1941. Distribution of birds in relation to biotic communities. *Amer. Midland Naturalist* 25:113–137.

———. 1942. Territoriality and related problems in North American hummingbirds. *Condor* 44: 189–204.

———. 1964. The nutrient-recovery hypothesis for arctic microtine cycles. I. Introduction. In *Grazing in terrestrial and marine environments. Brit. Ecol. Soc. Symp.*, ed. D. J. Crisp, pp. 55–56.

Pittendrigh, C. S., and D. H. Minis. 1964. The entrainment of circadian oscillations by light and their role as photoperiodic clocks. *Amer. Naturalist* 98:261–294.

Plotnik, R. 1968. Changes in social behavior of squirrel monkeys after anterior temporal lobectomy. *J. Comp. Physiol. Psychol.* 66:369–377.

Poirier, F. E., and E. O. Smith. 1974. Socializing functions of primate play. *Amer. Zool.* 14:275–287.

Prechtl, H. F. R., and W. M. Schleidt. 1950. Auslösende und steuernde Mechanismen des Saugaktes. I. *Z. Vergleich. Physiol.* 32:252–262. Cited in Eibl-Eibesfeldt 1970.

Premack, A. J., and D. Premack. 1972. Teaching language to an ape. *Sci. Amer.* 227:92–99.

Pribram, K. H., ed. 1969. *On the biology of learning.* Harcourt, Brace and World, New York.

Pritchatt, D. 1968. Avoidance of electric shock by the cockroach *Periplaneta americana. Anim. Behav.* 16:178–185.

Prosser, C. L., and F. A. Brown. 1961. *Comparative Animal Physiology,* 2nd ed. Saunders, Philadelphia.

Ramp, W. K. 1965. The auditory range of a hairy woodpecker. *Condor* 67:183–185.

Rand, A. L. 1954. Social feeding behavior of birds. *Fieldiana: Zoologica* 36:1–71.

Rasa, O. A. E. 1969. The effect of pair isolation on reproductive success in *Etroplus maculatus* (Cichlidae). *Z. Tierpsychol.* 19:645–651.

Recher, H. F., and J. A. Recher. 1969. Some aspects of the ecology of migrant shorebirds. II: Aggression. *Wilson Bull.* 81:141–154.

Remane, A. 1961. Gedanhen zum Problem: Homologie und Analogie, Praeadaptation und Parallelitat. *Zoo. Anz.* 166:447–470.

Rheingold, H. L. 1963. *Maternal behavior in mammals.* John Wiley & Sons, New York.

Rice, J. O., and W. L. Thompson. 1968. Song development in the indigo bunting. *Anim. Behav.* 16: 462–469.
Ricklefs, R. E. 1973. *Ecology.* Chiron, Newton, Mass.
Ripley, S. D. 1961. Aggressive neglect as a factor in interspecific competition in birds. *Auk* 78:366–371.
Ritter, W. E. 1938. *The California woodpecker and I.* Univ. of California Press, Berkeley.
Roberts, A. 1940. *The birds of South Africa.* London.
Roberts, B. B. 1934. Notes on the birds of central and south-east Iceland. *Ibis* 4:239–264.
Roberts, W. W., and E. H. Berquist. 1968. Attack elicited by hypothalamic stimulation in cats raised in social isolation. *J. Comp. Physiol. Psychol.* 66:590–595.
Roberts, W. W., and A. O. Kiess. 1964. Motivational properties of hypothalamic aggression in cats. *J. Comp. Physiol. Psychol.* 58:187–193.
Robertson, D. R. 1972. Social control of sex reversal in a coral-reef fish. *Science* 177:1007–1009.
Roeder, K. D. 1963. Nerve cells and insect behavior. Harvard Univ. Press, Cambridge, Mass.
Romer, A. S. 1941. *Man and the vertebrates* 3rd ed. Univ. of Chicago Press, Chicago.
———. 1958. Phylogeny and behavior with special reference to vertebrate evolution. In *Behavior and evolution,* ed. A. Roe and G. G. Simpson, pp. 48–75. Yale Univ. Press., New Haven, Conn.
———. 1970. *The vertebrate body.* W. B. Saunders, Philadelphia.
Rooke, K. B. 1947. Notes on robins wintering in North Algeria. *Ibis.* 89:204–210.
Rosen, D. E., and M. Gordon. 1953. Functional anatomy and evolution of male genitalia in poecilid fishes. *Zoologica,* New York, 38:1–47.
Rosenbaum, L. A., and R. W. Cooper. 1968. *The squirrel monkey.* Academic Press, New York.
Roth, L. M., and E. R. Willis. 1952. A study of cockroach behaviour. *Amer. Midland Naturalist* 47:66–129.
Rothenbuhler, W. C. 1964a. Behaviour genetics of nest cleaning in honeybees. I. Responses of four inbred lines to disease-killed brood. *Anim. Behav.* 12: 578–583.
———. 1964b. Behavior genetics of nest cleaning in honeybees. IV. Responses of $F_1$ and backcross generations to disease-killed brood. *Amer. Zool.* 4: 111–123.
Rothstein, S. I. 1970. An investigation of the defenses of the hosts of the parasitic brownheaded cowbird (*Molothrus ater*). Ph.D. dissertation, Yale Univ., New Haven, Conn.
———. 1974. Mechanisms of avian egg recognition: possible learned and innate factors. *Auk* 91: 796–807.
———. 1975. Mechanisms of avian egg-recognition: do birds know their own eggs? *Anim. Behav.* 23: 268–278.
———. 1976a. Cowbird parasitism of the cedar waxwing and its evolutionary implications. *Auk* 93: 498–509.
———. 1976b. Experiments on defenses cedar waxwings use against cowbird parasitism. *Auk* 93: 675–691.
Rowell, T. E. 1966. Hierarchy in the organization of a captive baboon group. *Anim. Behav.* 14:420–443.
Rowley, I. 1965. The life history of the superb blue wren. *Malarus cyaneus. Emu* 64:251–297.
Royama, T. 1970. Factors governing the hunting behavior and selection of food by the great tit (*Parus major* L). *J. Anim. Ecol.* 39:619–668.
Sackett, G. P. 1966. Monkeys reared in visual isolation with pictures as visual input. Evidence for an innate releasing mechanism. *Science* 154:1468–1472.
Sade, D. S. 1967. Determinants of dominance in a group of free-ranging rhesus monkeys. In *Social communication among primates,* ed. S. A. Altmann, pp. 99–114. Univ. of Chicago Press, Chicago.
Sargent, T. D. 1966. Background selections of geometrid and noctuid moths. *Science* 154:1674–1675.
———. 1968. Cryptic moths: effects on background selections of painting the circumocular scales. *Science* 159:100–101.
Sauer, F. 1955. Über Varcatonen der Artgesange bie Grasmücken. *J. f. Ornith.* 96:129–146.
———. 1957. Die Sternenorientierung nachtlich ziehender Grasmücken (*Sylvia atricapilla, borin* und *curruca*). *Z. Tierpsychol* 14:29–70.
Sauer, F., and E. M. Sauer. 1955. Zur Frage der nächtlichen Zugorientierung von Grasmücken. *Rev. Suisse Zool.* 62:250–259.
Saunders, J. W., and M. W. Smith. 1962. Physical alteration of stream habitat to improve trout production. *Trans. Amer. Fish. Soc.* 91:185–188.
Schaller, G. B. 1963. *The mountain gorilla,* Univ. of Chicago Press, Chicago.
Schaller, G. B., and G. R. Lowther. 1969. The relevance

of carnivore behavior to the study of early hominids. *Southwest. J. Anthropol.* 24:307–341.

Schenk, J. 1934. Die brutinvasionen des rosenstares in ungarn in den jahren 1932 und 1933. *Aquila* 38–41:136–153.

Schenkel, R. 1966. Zum Problem der Territorialität und des Markierens bei Säugern—am Beispiel des Schwarzen Nashorns und des Löwen. *Z. Tierpsychol.* 23:593–626.

Schiller, C. H. 1957. *Instinctive behavior: the development of a modern concept.* International Universities Press, New York.

Schmidt-Koenig, K. 1965. Current problems in bird orientation. In *Advances in the study of behavior,* ed. E. Hinde, K. Lehrman, and E. Shaw, pp. 217–278. New York.

———. 1975. *Migration and homing in animals.* Zoophysiology and Ecology 6. Springer-Verlag, New York.

Schmidt-Koenig, K., and H. J. Schlichte. 1972. Homing in pigeons with impaired vision. *Proc. Nat. Acad. Sci.* 69:2446–2447.

Schneider, D. 1962. Electrophysiological investigation on the olfactory specificity of sexual attracting substances in different species of moths. *J. Insect Physiol.* 8:15–30.

Schneider, P. 1969. Insect olfaction: deciphering system for chemical messages. *Science* 163:1031–1036.

Schneirla, T. C. 1956. The interrelationships of the "innate" and the "acquired" in instinctive behavior. In *L'instinct dans le comportement des animaux et de l'homme,* pp. 387–452. Masson, Paris.

———. 1959. An evolutionary and developmental theory of biphasic processes underlying approach and withdrawal. *Nebr. Symp. Motiv.* 1–42.

———. 1966. Behavioural development and comparative psychology. *Q. Rev. Biol.* 41:283–302.

Schoener, T. W. 1965. The evolution of bill size differences among sympatric congeneric species of birds. *Evolution* 19:189–213.

———. 1968. Sizes of feeding territories among birds. *Ecology* 49:123–141.

———. 1969a. Models of optimal size for solitary predators. *Amer. Naturalist* 103:277–313.

———. 1969b. Optimal size and specialization in constant and fluctuating environments: an energy-time approach. *Brookhaven Symp. Biol.* 22:103–114.

———. 1970. Size patterns in West Indian *Anolis* lizards. II. Correlations with the sizes of particular sympatric species-displacement and convergence. *Amer. Naturalist* 104:155–174.

———. 1974. Resource partitioning in ecological communities. *Science* 185:27–39.

Schultz, R. J. 1961. Reproductive mechanism of unisexual and bisexual strains of the viviparous fish *Poeciliopsis. Evolution* 15:302–325.

Schutz, F. 1965. Sexuelle Prägung bei Anatiden. *Z. Tierpsychol.* 22:50–103.

Schüz, E. 1949. Die Spät-Auflassung ostpreussischer Jungstörche in West-Deutschland durch die Vogelwaste Rossitten 1933. *Vogelwarte* 15:63–78.

Schwartz, E. 1965. Bau und funktzion der Seitenlinie des Streifenhechtlings (*Aplocheilus lineatus* Cuv. u. Val.) *Z. Vergl. Physiol.* 50:55–87.

Schwartzkopff, J. 1955. On the hearing of birds. *Auk* 72:340–347.

Schwieger, H. G. and J. W. Hastings. 1976. *The molecular basis of circadian rhythms.* Dahlem Conference, Berlin.

Schwink, I. 1958. A study of olfactory stimuli in the orientation of moths. *Proc. Xth Int. Congr. Entom.* 2:577–582.

Scott, J. P. 1945. Social behavior, organization and leadership in a small flock of domestic sheep. *Comp. Psychol. Monog.* No. 96, 18:1–29.

———. 1960. *Aggression.* Univ. of Chicago Press, Chicago.

———. 1966. Agonistic behavior of mice and rats: a review. In *Animal aggression: selected readings,* ed. C. H. Southwick, pp. 103–128. Van Nostrand Reinhold, New York.

———. 1970. Agonistic behavior of mice and rats: a review. In *Animal aggression,* ed. C. H. Southwick. Van Nostrand Reinhold, New York.

Sechzer, J. A. 1970. Prolonged learning and split-brain cats. *Science* 169:889–892.

Sechzer, J. A., and J. L. Brown. 1964. Color discrimination in the cat. *Science* 144:427–429.

Seilacher, A. 1967. Fossil Behavior. *Sci. Amer.* 217(2):72–80.

Seitz, A. 1940. Die Paarbildung bei einigen Cichliden. I. Die Paarbildung bei *Astatotilapia strigigena* (Pfeffer). *Z. Tierpsychol.* 4:40–84.

———. 1942. Die Paarbildung bei einigen Cichliden. II.

Die Paarbildung bei *Hemichromis bimaculatus. Z. Tierpsychol.* 5:74–101.

Selander, R. K. 1965. On mating systems and natural selection. *Amer. Naturalist* 99:129–141.

———. 1966. Sexual dimorphism and differential niche utilization in birds. *Condor* 68:113–151.

———. 1970. Behavior and genetic variation in natural populations. *Amer. Zool.* 10:53–66.

———. 1972. Sexual selection and dimorphism in birds. In *Sexual selection and the descent of man,* ed. B. Campbell, pp. 180–230. Aldine, Chicago.

Selander, R. K., and D. R. Giller. 1963. Species limits in the woodpecker genus *Centurus* (Aves). *Bull. Amer. Mus. Nat. Hist.* 124:213–274.

Serie, P. 1936. Distribution Geographica de los Ofidios Argentinos. *Obra Cincuentenario Mus. Plata* 2: 33–61.

Severaid, J. H. 1942. The snowshoe hare, its life history and artificial propagation. *Maine Dept. Inland Fish and Game.*

*Shapiro, S., and K. R. Vukovich. 1970. Early experience effects upon cortical dendrites: a proposed model for development. Science* 167:292–294.

Shaw, C. E. 1948. The male combat "dance" of some crotalid snakes. *Herpetologica* 4:137–145.

Sherif, M., and C. W. Sherif. 1953. Groups in harmony and tension. In *The dynamics of aggression,* ed. E. I. Megaree and J. E. Hokanson. Harper and Row, New York.

Sherrington, C. S. 1906. *The integrative action of the nervous system.* Cambridge Univ. Press, New York.

Shields, O. 1967. Hilltopping: an ecological study of summit congregation behavior of butterflies on a southern California hill. *J. Res. Lepidoptera.* 6: 69–178.

Short, L. L., Jr. 1961. Interspecies flocking of birds of montane forest in Oaxaca, Mexico. *Wilson Bull.* 73:341–347.

———. 1970. The habits and relationships of the Magellanic woodpecker. *Wilson Bull.* 82:115–129.

Sibley, C. G. 1957. The evolutionary and taxonomic significance of sexual dimorphism and hybridization in birds. *Condor* 59:166–191.

———. 1961. Hybridization and isolating mechanisms. In *Vertebrate speciation,* ed. W. F. Blair, pp. 69–88. Univ. of Texas Press, Austin.

Siebenaler, J. B., and D. K. Caldwell. 1956. Cooperation among adult dolphins. *J. Mammal.* 37:126–128.

Simberloff, D. 1970. Taxonomic diversity of island biotas. *Evolution* 24:22–47.

———. 1971. Population sizes of congeneric bird species on islands. *Amer. Naturalist* 105:190–193.

Simberloff, D. S., and E. O. Wilson. 1969. Experimental zoogeography and islands: the colonization of empty islands. *Ecology* 50: 278–296.

Simmons, K. E. L. 1951. Interspecific territorialism. *Ibis.* 93:407–418.

———. 1970. Ecological determinants of breeding adaptations and social behaviour in two fish-eating birds. In *Social behaviour in birds and mammals,* ed. J. H. Crook, pp. 37–77. Methuen, London.

Simpson, G. G. 1958. Behavior and evolution. In *Behavior and evolution,* ed. A. Roe and G. G. Simpson, pp. 7–26, 507–535. Yale Univ. Press, New Haven, Conn.

———. 1961. *Principles of animal taxonomy.* Columbia Univ. Press, New York.

Singer, J. J. 1968. Hypothalamic control of male and female sexual behavior in female cats. *J. Comp. Physiol. Psychol.* 66:738–742.

Sinsheimer, R. L. 1971. The brain of Pooh: an essay on the limits of mind. *Amer. Sci.* 59:20–28.

Skinner, B. F. 1938. *The behavior of organisms.* Prentice-Hall, Englewood Cliffs, New Jersey.

———. 1971. *Beyond freedom and dignity.* Alfred A. Knopf, New York.

Skutch, A. F. 1950. The nesting seasons of Central American birds in relation to climate and food supply. *Condor* 59:166–191.

———. 1961. Helpers among birds. *Condor* 63: 198–226.

Slobodkin, L. B. 1968. How to be a predator. *Amer. Zool.* 8:43–51.

Smith, C. C. 1968. The adaptive nature of social organization in the genus of tree squirrels *Tamiasciurus. Ecol. Monogr.* 38:31–63.

———. 1970. The coevolution of pine squirrels. (*Tamiasciurus*) and conifers. *Ecol. Monogr.* 40:349–371.

Smith, D. E., M. B. King, and B. C. Hoebel. 1970. Lateral hypothalamic control of killing: evidence for a cholinoceptive mechanism. *Science* 167:900–901.

Smith, F. E. 1961. Density dependence in the Australian thrips. *Ecology* 42:403–407.

Smith, H. S. 1935. The role of biotic factors in the

determination of population densities. *J. Econ. Entomol.* 28:873–898.

Smith, J. 1977. *The behavior of communication.* Harvard Univ. Press, Cambridge, Mass.

Smith, N. G. 1968. The advantage of being parasitized. *Nature* 219:690–694.

Smythe, N. 1970. On the existence of "pursuit invitation" signals in mammals. *Amer. Naturalist* 104:491–494.

Solomon, M. E. 1957. Dynamics of insect populations. *Ann. Rev. Entomol.* 2:121–142.

Soulé, M. 1970. A comment on the letter by Van Valen and Grant. *Amer. Naturalist* 104:590–591.

Soulé, M., and B. R. Stewart. 1970. The "niche-variation" hypothesis: a test and alternatives. *Amer. Naturalist* 104:85–97.

Southwick, C. H. 1955. The population dynamics of confined house mice supplied with unlimited food. *Ecology* 36:212–225.

———. 1970. *Animal aggression: selected readings.* Van Nostrand Reinhold Co., New York.

Southwick, C. H., M. A. Beg, and M. R. Siddiqi. 1965. Rhesus monkeys in North India. In *Primate behavior,* ed. I. DeVore. Holt, Rinehart and Winston, New York.

Sowles, L. K. 1966. Reproduction in the collared peccary. *Symp. Zool. Soc. Lond.* 15:155–172.

Sperry, R. W. 1958. Developmental basis of behavior. In *Behavior and evolution,* ed. A. Roe and G. G. Simpson, pp. 128–139. Harvard Univ. Press, New Haven, Conn.

———. 1961. Cerebral organization and behavior. *Science* 133:1749–1757.

Spieth, H. T. 1952. Mating behavior within the genus *Drosophila* (Diptera). *Bull. Amer. Mus. Nat. Hist.* 99:401–474.

———. 1966. Courtship behavior of endemic Hawaiian *Drosophila. Univ. of Texas Publ.* 6615:245–313.

———. 1968. Evolutionary implications of sexual behavior in *Drosophila.* In *Evolutionary biology,* ed. T. Dobzhansky, M. K. Hecht, and W. C. Steere, pp. 157–193. Appleton-Century-Crofts, New York.

Spitz, R. A., and K. M. Wolf. 1946. The smiling response: a contribution to the ontogenesis of social relations. *Gen. Psychol. Monogr.* 34:57–125.

Stehli, F. G., R. G. Douglas, and N. D. Newton. 1969. Generation and maintenance of gradients in taxonomic diversity. *Science* 164:947–949.

Stenger, J. 1958. Food habits and available food of ovenbirds in relation to territory size. *Auk* 75: 335–346.

Stettner, L. J., and K. A. Matyniak. 1968. The brain of birds. *Sci. Amer.* 218(6):64–76.

Stewart, R. E., and J. W. Aldrich. 1951. Removal and repopulation of breeding birds in a spruce-fir forest community. *Auk* 68:471–482.

Stiles, F. G., and L. L. Wolf, 1970. Hummingbird territoriality at a tropical flowering tree. *Auk* 87: 467–491.

Stoddard, H. L. 1931. *The bobwhite quail: its habits, preservation and increase.* Scribner's, New York.

Stokes, A. W. 1962a. Agonistic behaviour among blue tits at a winter feeding station. *Behaviour* 19: 118–138.

———. 1962b. Comparative ethology of great, blue, marsh and coal tits at a winter feeding station. *Behaviour* 19:208–218.

———. 1969. Aggressive man and aggressive beast. 39th Honor Facility Lecture, Utah State Univ.

Stonehouse, B. 1960. The King Penguin *Aptenodytes pategonica* of South Georgia. I. Breeding behaviour and development. *Sci. Rep. Falkland Islands Dependencies Surv.* 23:1–181.

Stoner, E. A. 1947. Anna hummingbirds at play. *Condor* 49:36.

Storer, R. W. 1952. Variation in the resident sharp-shinned hawks of Mexico. *Condor* 54:283–289.

Stout, J. F. 1963. The significance of sound production during the reproductive behaviour of *Notropis analostanus* (Family Cyprinidae). *Anim. Behav.* 11: 83–92.

Strecker, R. L., and J. T. Emlen. 1953. Regulatory mechanisms in house mouse populations: the effect of limited food supply on a confined population. *Ecology* 34:375–385.

Streseman, E. 1927–1934. *Aves in Kükenthal-Krumbäch,* Handbüch der Zoologie, Vol. VIII, Berlin and Leipzig.

Strickberger, M. 1968. *Genetics.* Macmillan, New York.

Stride, G. O. 1958. On the courtship behavior of a tropical mimetic butterfly, *Hypolimnas misipus* L. (Nymphalidae). *Proc. Xth Int. Congr. Entom.* (1956) 2:419–424.

Stromeyer, C. F. 1970. Eidetikers. *Psychol. Today* 77:80.

Sudd, J. H. 1967. *An introduction to the behavior of ants.* St. Martin's Press, New York.

Summers-Smith, D. 1963. *The house sparrow.* Collins, London.

Svärdson, G. 1949. Competition and habitat selection in birds. *Oikos* 1:157–174.

Sweeney, B. M. 1969. *Rhythmic phenomena in plants.* Academic Press, New York.

Sweet, W. H., F. Ervin, and V. H. Mark. 1969. The relationship of violent behavior to focal cerebral disease. In *Aggressive behavior,* ed. S. Garattini and E. Sigg. Wiley, New York.

Tavolga, W. N. 1956, Visual, chemical and sound stimuli as cues in sex discriminatory behavior of the Gobiid fish. *Bathygobius sporator. Zoologica* 41:49–64.

———. 1960. Sound production and underwater communication in fishes. In *Animal sounds and communication,* ed. W. E. Lanyon and W. N. Tavolga, pp. 93–136. AIBS, Washington, D. C.

———. ed. 1964. Sonic characteristics and mechanisms in marine fishes. In *Marine bioacoustics.* Macmillan, New York.

———. 1970. Levels of interaction in animal communication. In *Development and evolution of behavior,* ed. Aronson, Tobach, Lehrman and Rosenblatt, pp. 281–302. W. H. Freeman, San Francisco.

Taylor, D. H., and D. E. Ferguson. 1970. Extraoptic celestial orientation in the southern cricket frog *Acris gryllus. Science* 168:390–392.

Teitelbaum, P., and A. N. Epstein. 1962. The lateral hypothalamic syndrome: recovery of feeding and drinking after lateral hypothalamic lesions. *Psychol. Rev.* 69:74–90.

Telle, H. J. 1966. Beitrag zur Kenntnis der Verhaltensweise bei Rattan, vergleichend dargestellt bei *Rattus norvegicus* und *Rattus rattus. Angewandte Zool.* 9:129–196.

Test, F. H. 1956. Social aggressiveness in an amphibian. *Science* 120:140–141.

Thayer, G. H. 1909. *Concealing-coloration in the animal kingdom.* Macmillan, New York.

Thielcke, G. 1961. Stammesgeschichte und geographische Variation des Gesanges unserer Baumlaufer (*Certhis familiaris* L. und *Certhis brachydactyla* Brehm). *Z. Tierpsychol.* 18:188–204.

———. 1969. Geographical variation in bird vocalizations. In *Bird vocalizations,* ed. R. A. Hinde. Cambridge Univ. Press, Cambridge.

Thiessen, D. D. 1964. Population density and behavior: a review of theoretical and physiological contributions. *Texas Rep. Biol. Med.* 22:266–314.

———. 1971. *Gene organization and behavior.* Random House, New York.

Thiessen, D. D., and D. A. Rodgers, 1961. Population density and endocrine function. *Psychol. Bull.* 58:441–451.

Thompson, R. F. 1967. *Foundations of physiological psychology.* Harper and Row, New York.

Thompson, T. I. 1963. Visual reinforcement in Siamese fighting fish. *Science* 141:55–57.

———. 1964. Visual reinforcement in fighting cocks. *J. Exper. Anim. Behav.* 7:45–49.

Thompson, W. R. 1939. Biological control and the theories of the interactions of populations. *Parasitology* 31:299–388.

Thorndike, E. L. 1911. *Animal intelligence.* New York.

Thorpe, W. H. 1945. The evolutionary significance of habitat selection. *J. Anim. Ecol.* 14:67–70.

———. 1961. *Bird song: the biology of vocal communication and expression in birds.* Cambridge Univ. Press, Cambridge.

———. 1963. *Learning and instinct in animals,* 2nd ed. Harvard Univ. Press, Cambridge, Mass.

———. 1973. Duet-singing birds. *Sci. Amer.* 229: 70–79.

Thorpe, W. H., and O. L. Zangwill. 1961. *Current problems in animal behaviour.* Cambridge Univ. Press, Cambridge.

Tiger, L. 1969. *Men in groups.* Random House, New York.

Tiger, L., and R. Fox. 1971. *The imperial animal.* Holt, Rinehart and Winston, New York.

*Time.* Dec. 23, 1976. Genes über alles. P. 86.

Tinbergen, L. 1960. The natural control of insects in pine woods. I. Factors influencing the intensity of predation by songbirds. *Arch. Neerlandaises Zool.* 13:266–336.

Tinbergen, N. 1940. Die Übersprungbewegung. *Z. Tierpsychol.* 4:1–40.

———. 1942. An objectivistic study of the innate behaviour of animals. *Biblioth. Biotheor.,* Leiden, 1:37–98.

———. 1948a. Social releasers and the experimental method required for their study. *Wilson Bull.* 60: 6–52.

———. 1948b. Dierkundeles in het meeuwendiun. *De Levende Natuur* 51:49–56.

———. 1951. *The study of instinct.* Oxford Univ. Press, London.

———. 1952a. Derived activities: their causation, function and origin. *Q. Rev. Biol.* 27:1–32.

———. 1952b. On the significance of territory in the herring gull. *Ibis* 94:158–159.

———. 1952c. The curious behavior of the stickleback. In *Psychobiology. Readings from Scientific American* (1966), pp. 5–9. W. H. Freeman, San Francisco.

———. 1953a. *Social behaviour in animals, with special reference to vertebrates.* Methuen, London.

———. 1953b. Specialists in nest building. *Country Life,* Jan. 30:270–271.

———. 1953c. *The herring gull's world.* Collins, London.

———. 1956. On the functions of territory in gulls. *Ibis* 98:401–411.

———. 1957. The functions of territory. *Bird Study* 4:14–27.

———. 1958. *Curious naturalists.* Basic Books, New York.

———. 1959. Comparative studies of the behaviour of gulls (Laridae): a progress report. *Behaviour* 15: 1–70.

———. 1962. The evolution of animal communication—a critical examination of methods. *Symp. Zool. Soc. Lond.* 8:1–6.

———. 1963. On aims and methods of ethology. *Z. Tierpsychol.* 20:410–433.

———. 1964. *Social behaviour in animals.* Barnes & Noble, New York.

———. 1965a. *Animal behavior.* Time, Inc. New York.

———. 1965b. Behavior and natural selection. In *Ideas in evolution and behavior,* ed. J. A. Moore, vol. 6. *Proc. XVIth Int. Congr. Zool.* Natural History Press, New York.

———. 1967. Adaptive features of the Black-Headed Gull (*Larus ridibundus* L.). *Proc. XIVth Int. Orn. Congr.* Oxford, 1966, pp. 43–59. Blackwell, Oxford.

———. 1968. On war and peace in animals and men. *Science* 160:1411–1418.

———. 1974. Ethology and stress diseases. *Science* 185:20–27.

Tinbergen, N., G. J. Brockhuysen, F. Feekes, J. C. W. Houghton, H. Kruuk and E. Szulc. 1962. Eggshell removal by the black-headed gull, *Larus ridibundus* L.: a behavior component of camouflage. *Behaviour* 19:74–118.

Tinbergen, N., M. Impekoven and D. Franck. 1967. An experiment in spacing-out as a defense against predation. *Behaviour* 28:307–321.

Tinbergen, N., and D. J. Kuenen. 1939. Über die auslosenden und die richtunggebenden Reizsituationen der Sperrbewegung von jungen Drosseln (*Turdus m. merula* L. und *T. e. ericetorum* Turton). *Z. Tierpsychol.* 3:37–60.

Tinbergen, N., and A. C. Perdeck. 1950. On the stimulus situation releasing the begging response in the newly hatched herring gull chick (*Larus argentatus argentatus* Pont). *Behaviour* 3:1–39.

Todd, J. H. 1971. The chemical language of fishes. *Sci. Amer.* 224:98–108.

Tolman, E. C. 1937. The acquisition of string-pulling by rats—conditioned response or sign gestalt? *Psychol. Rev.* 44:195–211.

Tompa, F. S. 1962. Territorial behavior: the main controlling factor of a local song sparrow population. *Auk* 79:687–697.

———. 1964. Factors determining the numbers of song sparrows *Melospiza melodia* (Wilson) on Mandarte Island, B. C., Canada. *Acta Zool. fenn.* 109:1–68.

Trager, W. 1970. *Symbiosis.* Van Nostrand Reinhold Co., New York.

Trivers, R. L. 1971. The evolution of reciprocal altruism. *Q. Rev. Biol.* 46:35–57.

———. 1972. Parental investment and sexual selection. In *Sexual selection and the descent of man, 1871–1971,* ed. B. Campbell, pp. 136–179. Aldine, Chicago.

Trowill, J. A. 1967. Instrumental conditioning of the heart rate in the curarized rat. *J. Comp. Psychol.* 63:7–11.

Ulrich, J. 1938. The social hierarchy in albino mice. *J. Comp. Physiol. Psychol.* 25:373–413.

Utida, S. 1957. Population fluctuation, an experimental and theoretical approach. *Cold Spring Harbor Symp. Quant. Biol.* 22:139–151.

Valenstein, E. S., V. C. Cox, and J. W. Kakolewski. 1970. Re-examination of the role of the hypothalamus in motivation. *Psychol. Rev.* 77:16–31.

Van den Assem, J. 1967. Territory in the three-spined stickleback *Gasterosteus aculeatus.* L. *Behav. Suppl.* 16.

Vander Wall, S. B., and R. P. Balda. 1977. Coadaptations of the Clark's nutcracker and the piñon pine for efficient seed harvest and dispersal. Ecol. Monogr. 47(1):89–111.

Van Valen, L. 1965. Morphological variation and width of ecological niche. *Amer. Naturalist* 99:377–390.

Van Valen, L. and P. R. Grant. 1970. Variation and niche width reexamined. *Amer. Naturalist* 104:589–590.

Varley, G. C. 1941. On the search for hosts and the egg distribution of some chalcid parasites of the knapweed gall-fly. *Parasitology* 33:47–66.

———. 1947. The natural control of population balance in the knapweed gall-fly (*Vrothora jaceana*). *J. Anim. Ecol.* 16:139–187.

———. 1953. Ecological aspects of population regulation. *Trans. IXth. Int. Congr. Entom.* 2:210–214.

Vernberg, F. J., and W. B. Vernberg. *The animal and the environment.* Holt, Rinehart and Winston, New York.

Verner, J. 1964. Evolution of polygamy in the long-billed marsh wren. *Evolution* 18:252–261.

Verner, J., and M. F. Willson. 1966. The influence of habitats on mating systems of North American birds. *Ecology* 47:143–147.

Volterra, V. 1926. Variazioni e fluttuzaioni del numero d'individui in specie animale conviventi. *Atti dell' Accademia Nazionale dei Lincei* Memorie (6) 2: 31–113.

Voûte, A. D. 1946. Regulation of the density of the insect populations in virgin forests and cultivated woods. *Arch. Neerlandaises Biol.* 7:435–470.

Vowles, D. M. 1954. The orientation of ants. I. The substitution of stimuli. *J. Exper. Biol.* 31:341–355.

Vuilleumier, F. 1967. Mixed species flocks in Patagonian forest, with remarks on interspecies flock formation. *Condor* 69:400–404.

Waddington, C. 1960. *The ethical animal.* George Allen and Unwin, London.

Wahlert, Gerd von. 1965. The role of ecological factors in the origin of higher levels of organization. *Syst. Zool.* 14:288–300.

Walcott, C., and R. Green. 1974. Orientation of homing pigeons altered by a change in the direction of an applied magnetic field. *Science* 184:180–182.

Walcott, C., and M. Michener. 1967. Analysis of single tracks of homing pigeons. *Proc. XIVth Orn. Congr.*:311–329.

Wald, G. 1959. The photoreceptor in vision. In *Handbook of physiology,* ed. F. Field, vol. 1. Amer. Physiol. Soc., pp. 671–692. Washington, D. C.

Wall, W. van der. 1963. Bewegungstudien an anatinen. *J. f. Ornith.*, Leipzig, 104:1–15.

Wallace, B., and A. M. Srb. 1964. *Adaptation.* Prentice-Hall, Englewood Cliffs, New Jersey.

Wallace, R. A. 1966a. Cellular specialization for specific flight characteristics in birds. *Proc. Virginia J. Sci. Proc.*, September, p. 285.

———. 1966b. Nesting and defensive behavior of the black-naped tern in the Maldive Islands. *Auk* 83:138.

———. 1969. Sexual dimorphism, niche utilization, and social behavior in insular species of woodpeckers. Ph.D. dissertation, Univ. of Texas at Austin.

———. 1974a. Ecological and social implications of sexual dimorphism in five melanerpine woodpeckers. *Condor* 76:238–248.

———. 1974b. Aberrations in the tongue structures of some melanerpine woodpeckers. *Wilson Bull.* 86:79–82.

———. 1977. Genes über alles. *Time.* Letters to the editor, Jan. 10, p. 7.

———. 1978a. *Biology the world of life,* 2nd ed. Goodyear Publishing Co., Santa Monica, Calif.

———. 1978b. Social behavior on islands. In *Perspectives in ethology.* ed. P. P. G. Bateson and P. H. Klopfer, pp. 167–204. Plenum Press, New York.

———. 1978c. Why there are giants on islands. Animal Behavior Society symposium: Social behavior on islands. Univ. of Wash., Seattle.

Wallace, R. A., and P. H. Klopfer. 1979. *The modern roots of sociobiology.* Holt, Rinehart and Winston, New York.

Walls, G. L. 1942. *The vertebrate eye and its adaptive radiation.* Cranbrook Institute of Science, Bloomfield Hills, Mich.

Walther, F. R. 1958. Zum Kampf-und Paarungsverhalten einiger Antilopen. *Z. Tierpsychol.* 15:340–380.

———. 1961. Enlwickungszüge im Kampf und Paarungsverhalten der Horntiere. *Jahrbuch der Gesellschaft von Opel-Freigehege Tierforschung* 3: 90–115.

Ward, P. 1965. Feeding ecology of the blackfaced dioch *Quelea quelea* in Nigeria. *Ibis* 107:173–214.

Washburn, S. L., and V. Avis. 1958. Evolution of human behavior. In *Evolution and behavior,* ed. A. Roe and G. G. Simpson, pp. 421–436. Yale Univ. Press, New Haven, Conn.

Washburn, S. L., and D. A. Hamburg. 1965. The study of primate behavior. In *Primate behavior,* ed. I. DeVore, pp. 1–13. Holt, Rinehart and Winston, New York.

———. 1968. Aggressive behavior in old world monkeys and apes. In *Primates,* ed. P. C. Jay. Holt, Rinehart and Winston, New York.

Washburn, S. L., and C. S. Lancaster. 1968. The

evolution of hunting. In *Man the hunter*, ed. I. DeVore and R. Lee. Aldine, Chicago.

———. 1973. The evolution of hunting. In *Human evolution*, ed. N. Korn. Holt, Rinehart and Winston, New York.

Waterman, T. H., ed. 1961. *The physiology of crustacea;* vol. 2. Academic Press, New York.

Watson, A. 1964. Aggression and population regulation in red grouse. *Nature*, London, 202:506–7.

———. 1967. Population control by territorial behavior in the red grouse. *Nature* 215:1274–1275.

Watson A., and R. Moss. 1970. Dominance, spacing behaviour and aggression in relation to population limitation in vertebrates. In *Animal populations in relation to their food resources*, ed. A. Watson, pp. 167–220. Blackwell, Oxford.

Watson, G. E. 1962. Three sibling species of *Alectoris* partridge. *Ibis* 104:353–367.

Watson, J. B. 1913. Psychology as the behaviorist views it. *Psychol. Rev.* 20:158–177.

Watts, C. R., and A. W. Stokes. 1971. The social order of turkeys. *Sci. Amer.* 224:112–118.

Wecker, S. C. 1963. The role of early experience in habitat selection by the prairie deer mouse. *Peromyscus maniculatus bairdi. Ecol. Monogr.* 33:307–325.

Weeden, J. S., and J. B. Falls. 1959. Differential responses of male ovenbirds to recorded songs of neighboring and more distant individuals. *Auk* 76:343–351.

Weichert, C. K. 1959. *Elements of chordate anatomy*, 2nd ed. McGraw-Hill, New York.

———. 1965. *Anatomy of the chordates.* McGraw-Hill, New York.

Weidmann, U. 1961. The stimuli eliciting begging in gulls and terns. *Anim. Behav.* 9:115–116.

Weiss, P. A. 1941. Self-differentiation of the basic patterns of coordination. *Comp. Psychol. Monogr.* 17:1–96.

Wells, M. J. 1959. Functional evidence for neurone fields representing the individual arms within the central nervous system of *Octopus. J. Exper. Biol.* 36: 501–511.

———. 1960. Optic glands and the ovary of *Octopus. Symp. Zool. Soc. Lond.* 2:87–107.

Wells, P. H., and A. M. Wenner. 1974. How recruited bees find food. *Gleanings in Bee Culture*, April: 110–111.

Welty, J. C. 1963. *The life of birds.* Knopf, New York.

Wenner, A. 1971. *The bee language controversy.* Educ. Programs Improvement Corp., Boulder, Colo.

———. 1974. Information transfer in honey bees: a population approach. In *Nonverbal communication.* Advances in the study of communication and effect, vol. 1, ed. L. Krames, T. Alloway, and P. Pliner, pp. 13–169. Plenum Press, New York.

Wertheimer, M. 1912. Experimentelle Studien über das Sehen von Bewegung. *Z. Psychol.* 61:161–265.

Whitman, C. O. 1899. *Animal behavior.* Biological Lectures, Summer Session, 1897–1898. Wood's Hole, pp. 285–338.

———. 1919. The behavior of pigeons. Posthumous works of C. O. Whitman, 3:1–161, ed. H. A. Parr. Publ. Carneg. Instn., 257

Whittaker, R. H. 1970. *Communities and ecosystems.* Macmillan, New York.

Whittaker, R. H., and P. P. Feeny. 1971. Allelochemics: Chemical interactions between species. *Science* 171:757–770.

Whittaker, R. H., R. Levin, and P. Root. 1973. Niche, habitat, and ecotype. *Amer. Naturalist* 107:321–338.

Whorf, B. L. 1956. *Language, thought and reality.* Wiley & Sons, New York.

Wickler, W. 1961. Okologie und Stammesgeschichte von Verhaltensweisen. *Fortschr. für Zool.* 13:303–365.

———. 1968. *Mimicry in plants and animals.* McGraw-Hill, New York.

Wiewandt, T. A. 1969. Vocalization, aggressive behavior and territoriality in the bullfrog, *Rana catesbeiana. Copeia* 1969:276–285.

Wigglesworth, V. B. 1953. *The principles of insect physiology*, 5th ed. Methuen and Co., London.

Wilbur, H. M. 1971. The ecological relationship of the salamander *Ambystoma laterale* to its all-female gynogenetic associate. *Evolution* 25:168–179.

Wilbur, H. M., D. W. Tinkle, and J. P. Collins. 1974. Environmental certainty, tropic level, and resource availability in life history evolution. *Amer. Naturalist* 108:805–817.

Wilcox, R. Stimson. Communication by surface waves. *J. Comp. Physiol.* 80:255–266.

Wiley, R. H. 1971. Cooperative roles in mixed flocks of antwrens (Formicariidae). *Auk* 88:881–892.

Wiley, R. H., Jr. 1978. The lek mating system of the sage grouse. *Sci. Amer.* 238:114–125.

Wilkinson, D. H. 1952. The random element in bird "navigation." *J. Exper. Biol.* 29:532–560.

Williams, G. C. 1957. Pleiotropy, natural selection, and the evolution of senescence. *Evolution* 11: 398–411.

———. 1966a. *Adaptation and natural selection.* Princeton Univ. Press, Princeton, New Jersey.

———. 1966b. Natural selection, the costs of reproduction, and a refinement of Lack's principle. *Amer. Naturalist* 100:687–690.

———. 1974. *Sex and evolution.* Princeton Univ. Press, Princeton, New Jersey.

Williams, L. 1942. Display and sexual behavior of the Brandt cormorant. *Condor* 43:85–104.

Willis, E. O. 1966a. Interspecific competition and the foraging behavior of plain-brown wood-creepers. *Ecology* 47:667–672.

———. 1966b. The role of migrant birds at swarms of army ants. *Living Bird* 5:187–236.

———. 1967. The behavior of bicolored antbirds. *Univ. Calif. Publ. Zool.* 79:1–132.

Willson, M. F. 1966. Breeding ecology of the yellow-headed blackbird. *Ecol. Monogr.* 36:51–77.

———. 1971. Ecological overlap and bill size differences: a comment. *Amer. Midland Naturalist* 86:215–217.

Wilson, E. O. 1965. The challenge from related species. In *The genetics of colonizing species,* ed. H. G. Baker and G. L. Stebbins, pp. 7–24. Academic Press, New York.

———. 1968. Chemical systems. In *Animal Communication: techniques of study and results of research,* ed. T. A. Sebeak, pp. 75–102. Indiana Univ. Press, Bloomington.

———. 1970. Chemical communication within animal species. In *Chemical ecology,* ed. E. Sondheimer and J. B. Simeone, pp. 133–155. Academic Press, New York.

———. 1971a. *The insect societies.* Harvard Univ. Press, Cambridge, Mass.

———. 1971b. The prospects for a unified sociobiology, *Amer.* Sci. 59:400–417.

———. 1975a. Slavery in ants. *Sci. Amer.* 232:32–36.

———. 1975b. *Sociobiology.* Belknap Press, Cambridge, Mass.

Wilson, J. T., and Bossert. 1963. Chemical communication among animals. *Recent Progr. Horm. Res.* 19:673–716.

Wilson, J. T., ed. 1973. *Continents adrift.* A collection of articles from *Sci. Amer.* Freeman, San Francisco.

Wiltschko, W. 1968. Über den Einfluss statischer Magnetfelder auf die Zugorientiesung der Rotkehlchen (*Erithacus rubecula*). *Z. Tierpsychol.* 25:537–558.

Wing, L. 1946. Species association in winter flocks. *Auk* 63:507–510.

Wolf, L. L. 1970. The impact of seasonal flowering on the biology of some tropical hummingbirds. *Condor* 72:1–14.

Wolfenbarger, P. O. 1946. Dispersion of small organisms. *Amer. Midland Naturalist* 35:1–152.

Wolfson, F. H. 1977. Gray whale behavior. *Science* 195:534–535.

Woodworth, C. H. 1971. Attack elicited in rats by electrical stimulation of the lateral hypothalamus. *Physiol. Behav.* 6:345–353.

Wright, S. 1951. Fisher and Ford on "the Sewall Wright effect." *Amer. Sci.* 39:452–479.

———. 1968 et seq. *Evolution and the genetics of populations: a treatise.* Univ. of Chicago Press, Chicago.

Wynne-Edwards, V. C. 1962. *Animal dispersion in relation to social behaviour.* Oliver and Boyd, London.

———. 1965. Self-regulating systems in populations of animals. *Science* 147:1543–1548.

Yarrow, L. J. 1964. Separation from parents during early childhood. In *Review of child development research,* vol. I, ed. M. L. Hoffman and L. W. Hoffman, pp. 89–136. Russell Sage Foundation, New York.

Yerkes, R. M. 1943. *Chimpanzees: a laboratory colony.* Yale Univ. Press, New Haven, Conn.

Young, J. Z. 1961. Repeated reversal of training in Octopus. *Q. J. Exper. Psychol.* 14:206–222.

———. 1971. *An introduction to the study of man.* Oxford Univ. Press, New York.

# CREDITS

## TABLES

Table 5.1 A. Adriaanse, "*Ammophila campestris* Ltr. und *Ammophila adriaansei* Wilcke: Ein Beitrag zur vergleichenden Verhaltensforschung," *Behaviour* 1(1947):1–34

Table 5.2 E. Mayr, "Behavior and Systematics," in *Evolution and Behavior*, ed. A. Roe and G. G. Simpson (New Haven, Conn.: Yale Univ. Press, 1958), pp. 341–362

Tables 19.1, 19.2 *Ecology: The Experimental Analysis of Distribution and Abundance* by Charles J. Krebs. Copyright © 1972 by Harper & Row, Publishers, Inc. By permission of the publishers

Table 19.3 Adapted from E. V. Kormondy, *Concepts of Ecology* (Englewood Cliffs, New Jersey: Prentice-Hall, Inc., 1969), p. 63

Table 19.4 Merritt J. Emlen, *Ecology: An Evolutionary Approach*, 1973, Addison-Wesley, Reading, Mass.

Table 19.5 After E. S. Deevey, "The Probability of Death." Copyright © 1950 by Scientific American, Inc. All rights reserved.

Table 19.6 *Demographic Yearbook*, United Nations (1975).

Table 20.1 *Evolutionary Ecology* by Eric R. Pianka. Copyright © 1974 by Harper & Row, Publishers, Inc. By permission of the publishers

Table 21.1 From Audubon Field Notes, after *Ecology*, Robert E. Ricklefs, 1973. Chiron Press. Reprinted by permission.

Tables 23.1, 23.2, 23.3 *Ecology: The Experimental Analysis of Distribution and Abundance* by Charles J. Krebs. Copyright © 1972 by Harper & Row, Publishers, Inc. By permission of the publishers

Tables 25.1, 25.2, 25.3 Merritt J. Emlen, *Ecology: An Evolutionary Approach*, 1973, Addison-Wesley, Reading, Mass.

Table 28.1 From Roger N. Johnson, *Aggression in Man and Animals*, W. B. Saunders, 1972.

Table 31.1 J. H. Crook and J. S. Gartlan, "Evolution of Primate Societies," *Nature* 210(1966):1200–1203

## PHOTOS AND ILLUSTRATIONS

### Chapter 2

2.2 Australian Information Service

2.4, 2.5 From A. S. Romer, *The Vertebrate Body* (Philadelphia: W. B. Saunders, 1970)

2.6 From A. S. Romer, *Man and the Vertebrates* (Chicago: University of Chicago Press, 1941)

2.7 After E. H. Colbert, *Evolution of the Vertebrates: A History of Backboned Animals Through Time* (New York: John Wiley and Sons, 1955)

2.8 Irven DeVore, Anthro-Photo

2.9 Lincoln Brower

2.10 From John Alcock, *Animal Behavior: An Evolutionary Approach.* Copyright © 1975 by Sinauer Associates, Inc. Reprinted by permission

2.11 From J. J. ter Pelkwijk and N. Tinbergen, "Eine reizbiologische Analyse einiger Verhaltensweisen von *Gasterosteus aculeatus* L.," *Zeitschrift für Tierpsychologie* 1:193–204

### Chapter 3

3.2 The National Foundation, March of Dimes

3.4 D. J. Chivers, Anthro-Photo

3.5 *Evolutionary Ecology* by Eric R. Pianka. Copyright © 1974 by Harper & Row, Publishers, Inc. By permission of the publishers

3.6 U.S. Department of Health, Education, and Welfare

3.8 From A. Manning, "Effects of Artificial Selection for Mating Speed in *Drosophila melanogaster*," *Animal Behavior*, vol. 9, nos. 1 and 2, January–April, 1961

### Chapter 4

4.1 Natural Science Photos

4.3 Hugh M. Halliday, National Audubon Society, from Photo Researchers

4.4 Allan D. Cruickshank, National Audubon Society, from Photo Researchers

4.5 Robert C. Hermes, National Audubon Society, from Photo Researchers

4.6 Gordon S. Smith, National Audubon Society, from Photo Researchers

4.7 Stephen Dalton, N.H.P.A.

4.8 Karl W. Kenyon, National Audubon Society, from Photo Researchers

4.9 Irven DeVore, Anthro-Photo
4.14 Malcolm, Anthro-Photo

## Chapter 5

5.3 From R. W. Doty and J. F. Bosma, "An Electromyographic Analysis of Reflex Deglutination," *Journal of Neurophysiology* 19:44–60
5.5 From D. Hunsaker, "Ethological Isolating Mechanisms in the *Sceloporus torquatus* Group of Lizards," *Evolution* 16:62–74
5.6, 5.7 From K. Lorenz, *King Solomon's Ring,* trans. M. K. Wilson (New York: Thomas Y. Crowell, 1952)
5.8 From a film sequence after W. van der Wall, "Bewegungsstudien an anatinen," *Journal für Ornithologie*, Leipzig, 104:1–15
5.9 From N. Tinbergen, "Comparative Studies of the Behavior of Gulls (Laridae): A Progress Report," *Behaviour* 15:1–70
5.10 (top) After L. Williams, "Display and Sexual Behavior of the Brandt's Cormorant," *Condor* 43:85–104; (bottom) From A. Adrea, 27:1–40, *Phalacrocorax carbo sinensis.* By permission of E. J. Brill 1938

## Chapter 6

6.9 Runk/Schoenberger, Grant Heilman Photography
6.10 From *Comparative Animal Behavior* by Richard A. Maier and Barbara A. Maier. Copyright © 1970 by Brooks/Cole Publishing Co., Monterey, Calif.
6.13 From *Animals Without Backbones,* rev., 2nd ed., by Ralph Buchsbaum. Copyright © 1975 by University of Chicago Press. Reprinted by permission
6.14 From *Comparative Animal Behavior* by Richard A. Maier and Barbara M. Maier. Copyright © 1970 by Brooks/Cole Publishing Co., Monterey, Calif.
6.20 Aubrey Manning, *Introduction to Animal Behavior,* 2nd ed., 1972, Addison-Wesley, Reading, Mass.

## Chapter 7

7.1 E. S. Ross
7.2 From John Alcock, *Animal Behavior: An Evolutionary Approach.* Copyright © 1975 by Sinauer Associates, Inc. Reprinted by permission
7.3 Allan D. Cruickshank, National Audubon Society, from Photo Researchers
7.5 From *Animals Without Backbones,* rev., 2nd ed., by Ralph Buchsbaum. Copyright © 1975 by University of Chicago Press. Reprinted by permission
7.6, 7.7 From *Comparative Animal Behavior* by Richard A. Maier and Barbara M. Maier. Copyright © 1970 by Brooks/Cole Publishing Co., Monterey, Calif.
7.8 Thomas Eisner, Cornell University
7.10 From *Comparative Animal Behavior* by Richard A. Maier and Barbara M. Maier. Copyright © 1970 by Brooks/Cole Publishing Co., Monterey, Calif.
7.11, 7.12 E. S. Ross
7.14 From *Comparative Animal Behavior* by Richard A. Maier and Barbara M. Maier. Copyright © 1970 by Brooks/Cole Publishing Co., Monterey, Calif.

## Chapter 8

8.1 Drawn from a photograph in R. A. Fisher and R. T. Peterson, *The World of Birds* (Garden City, N.Y.: Doubleday and Co., Inc., 1964)
8.4 Adapted from N. Tinbergen, *The Study of Instinct* (London: Oxford University Press, 1951)
8.6 From K. Lorenz and N. Tinbergen, "Taxis and Instinkthandlung in der Eirollbewegung der Graugans I., *Zeitschrift für Tierpsychologie* 2:1–29
8.7 From N. Tinbergen, "An Objectivistic Study of the Innate Behavior of Animals," *Bibliotheca Biotheoretica,* Leiden, 1:37–98
8.8 From N. Tinbergen, *The Study of Instinct* (London: Oxford University Press, 1951)
8.9 Aubrey Manning, *Introduction to Animal Behavior,* 2nd ed., 1972, Addison-Wesley, Reading, Mass.
8.10 From N. Tinbergen, "Social Releasers and the Experimental Method Required for Their Study," *Wilson Bulletin* 60:6–52; and "Dierkundeles in het Meeuwendiun, *De Levende Natuur* 51:49–56
8.11 From N. Tinbergen, "Social Releasers and the Experimental Method Required for Their Study," *Wilson Bulletin* 60:6–52
8.12 From N. Tinbergen and D. J. Kuenen, "Über die auslosenden und die richtunggebenden Reizsituationen der Sperrbewegung von jungen Drosseln (*Turdus m. merula* L. und *T. e. ericetorum* Turton)," *Zeitschrift für Tierpsychologie* 3:37–60
8.13, 8.14 From N. Tinbergen, *The Study of Instinct* (London: Oxford University Press, 1951)
8.15 Aubrey Manning, *Introduction to Animal Behavior,* 2nd ed., 1972, Addison-Wesley, Reading, Mass.
8.16, 8.17, 8.18 From *Ethology: The Biology of Behavior,* 2nd ed., by I. Eibl-Eibesfeldt. Copyright © 1975 by Holt, Rinehart and Winston, Publishers

8.19 A. K. Tunstill

8.20 From *Animal Behavior: A Synthesis of Ethology and Comparative Psychology*, 2nd ed., by Robert A. Hinde. Copyright © 1970 by McGraw-Hill, Inc. Used with permission of McGraw-Hill Book Company

8.21 Photophile

8.22 Van Bucher, Photo Researchers

8.23 From K. Lorenz, *On Aggression*, Harcourt Brace Jovanovich, Inc., and Methuen and Co. Ltd., 1963

### Chapter 9

9.1 From M. M. Nice, "Studies in the Life History of the Song Sparrow," I, *Transactions of the Linnaean Society*, New York, 4

9.2 Aubrey Manning, *Introduction to Animal Behavior*, 2nd ed., 1972, Addison-Wesley, Reading, Mass.

### Chapter 10

10.2 Aubrey Manning, *Introduction to Animal Behavior*, 2nd ed., 1972, Addison-Wesley, Reading, Mass.

### Chapter 11

11.1 Leonard Lee Rue III, National Audubon Society, from Photo Researchers

11.3 Lincoln Brower

11.4 Courtesy Dr. B. F. Skinner, Harvard University

11.5 After L. V. DiCara, "Learning in the Autonomic Nervous System," *Scientific American*, 222:30–39

11.6 Aubrey Manning, *Introduction to Animal Behavior*, 2nd ed., 1972, Addison-Wesley, Reading, Mass.

11.7 Alfred Bailey, National Audubon Society, from Photo Researchers

11.10 C. R. Carpenter, courtesy Pennsylvania State University

11.11 Mark N. Boulton, National Audubon Society, from Photo Researchers

11.12 Timothy W. Ransom

11.13 Aubrey Manning, *Introduction to Animal Behavior*, 2nd ed., 1972, Addison-Wesley, Reading, Mass.

11.14 From John Alcock, *Animal Behavior: An Evolutionary Approach*. Copyright © 1975 by Sinauer Associates, Inc. Reprinted by permission

### Chapter 12

12.2 From H. E. Hess, "Space Perception in the Chick," *Scientific American* 195:71–80

12.3 After Etkin 1964 from J. S. Huxley, *Proceedings of the Zoological Society*, 1914

### Chapter 13

13.1 From E. von Holst, "Über den Lichtruckenreflex bei Fischen," *Publicazioni della Stazione Zoologica di Napoli* 15:143–158

13.2 U.S. Department of Agriculture, Office of Information

13.3, 13.4 From *Biology* by Karl von Frisch, translated by Jane M. Oppenheimer. Copyright © 1964 by Bayerischer Schulbuch-Verlag. By permission of Harper and Row, Publishers, Inc.

13.5 Courtesy A. F. Carr, University of Florida

13.7 Peter H. Klopfer, *Behavioral Aspects of Ecology*, 2nd ed., © 1973. Reprinted by permission of Prentice-Hall, Inc., Englewood Cliffs, New Jersey

### Chapter 14

14.1 United Press International Photo

14.2 C. Walcott

14.3 From R. M. Lockley, *Animal Navigation* (New York: Hart Publishing Co., Inc., 1967)

14.4 Karl H. Maslowski, National Audubon Society, from Photo Researchers

14.5, 14.6 From G. Kramer, "Experiments in Bird Orientation and Their Interpretation," *Ibis* 99:196–227

14.9 From G. V. T. Matthews, *Bird Navigation* (Cambridge: Cambridge University Press, 1955)

14.10 From F. Sauer, "Die Sternenorientierung nachtlich ziehender Grasmucken (*Sylvia atricapilla, borin* und *curruca*), *Zietschrift für Tierpsychologie* 14: 29–70

14.11 G. Ronald Austing, National Audubon Society, from Photo Researchers

### Chapter 15

15.2 Gordon Smith, National Audubon Society, from Photo Researchers

15.3 Mark N. Boulton, National Audubon Society, from Photo Researchers

15.4 Courtesy Nancy Moffatt, National Marine Fisheries Service

15.5 Ron Garrison, San Diego Zoological Society

15.6, 15.7 From J. D. Palmer, in *The Biological Clock: Two Views*, ed. F. A. Brown, J. W. Hastings, and J. D. Palmer (New York: Academic Press, 1970), pp. 3–12

15.8 From M. Menaker, "Biological Clocks," *Bioscience* 19 (1969):682, fig. 1

15.9 Allan D. Cruickshank, National Audubon Society, from Photo Researchers

15.10, 15.11 From John Alcock, *Animal Behavior: An Evolutionary Approach.* Copyright © 1975 by Sinauer Associates, Inc. Reprinted by permission

15.12, 15.13, 15.14 From J. W. Hastings, in *The Biological Clock: Two Views*, ed. F. A. Brown, J. W. Hastings, and J. D. Palmer (New York: Academic Press, 1970), pp. 63–94

15.15 From J. D. Palmer, in *The Biological Clock: Two Views*, ed. F. A. Brown, J. W. Hastings, and J. D. Palmer (New York: Academic Press, 1970), pp. 3–12

15.16 From Brown, "American Astronautical Society," 17 (1964):29–39

15.17, 15.18, 15.19 From F. A. Brown, in *The Biological Clock: Two Views*, ed. F. A. Brown, J. W. Hastings, J. D. Palmer (New York: Academic Press, 1970), pp. 15–59

**Chapter 16**

16.1 Robert A. Wallace

16.2 David Anderson, Primate Center, Duke University

16.3 After photographs in D. Morris, "'Typical Intensity' and Its Relation to the Problem of Ritualisation," *Behaviour* 11:1–12. By permission of N. V. Boekhandel and Drukkerij

16.4 From *Darwin's Biological Work*, ed. P. R. Bell, pp. 150–206, "Developments in the Study of Animal Communication," by P. Marler. © 1959, Cambridge University Press

16.5 Toni Angermayer, Photo Researchers

16.6 G. Ronald Austing, National Audubon Society, from Photo Researchers

16.7, 16.8 From *Darwin's Biological Work*, ed. P. R. Bell, pp. 150–206, "Developments in the Study of Animal Communication," by P. Marler. © 1959, Cambridge University Press

16.9 San Diego Zoological Society

16.11 Ron Garrison, San Diego Zoological Society

16.12 After J. Meisenheimer, *Geschlecht und Geschlechter im Tierreich* (Jena, 1921)

16.13 E. S. Ross

16.14 William E. Ferguson

**Chapter 17**

17.1 From G. Thielcke, "Stammesgeschichte und geographische Variation des Gesanges unserer Baumlaufer (*Certhis familiaris* L. und *Certhis brachydactyla* Brehm)," *Zeitschrift für Tierpsychologie* 18:188–204

17.2 Roger T. Peterson, National Audubon Society, from Photo Researchers

17.4 F. D. Schmidt, San Diego Zoological Society

17.5 Jerry L. Hout, Photo Researchers

17.6 After N. Tinbergen, *Animal Behavior* (New York: Time-Life Books, 1965)

17.7 From P. A. Johnsgard, *Animal Behavior* (Dubuque, Iowa: Wm. C. Brown Company Publishers, 1967)

17.10 From H. B. Cott, *Adaptive Colouration in Animals* (Associated Book Publishers, Ltd., 1957)

17.11 M. W. F. Tweedie, Photo Researchers

17.12 Ylla, Rapho/Photo Researchers

17.13 Thomas Eisner and D. Aneshansky, Cornell University

17.14 From *Biology* by Karl von Frisch, translated by Jane M. Oppenheimer. Copyright © 1964 by Bayerischer Schulbuch-Verlag. By permission of Harper and Row, Publishers, Inc.

17.17 Robert Hermes, National Audubon Society, from Photo Researchers

17.18 From A. D. Blest, "The Evolution of Protective Displays in the Saturniodea and Sphingidae (Lepidoptera)," *Behaviour* 11:257–309. With permission of N. V. Boekhandel and Drukkerij

17.19 Dr. J. E. Lloyd

17.20 Lincoln Brower

17.21 From R. D. Guthrie and R. G. Petocz, "Weapon Automimicry Among Mammals," *American Naturalist* 104(1970):585–588

17.22 From W. Wickler, *Mimicry in Plants and Animals* (New York: McGraw-Hill, 1968)

17.23 From B. Kullenberg, "Studies in *Ophrys* Pollination," *Zool. Bridag,* Uppsala, 34:1–340

17.24 From G. Moment, "Variation in Brittle Stars," *Science* 136:262–263

**Chapter 18**

18.1 From David Bentley and Ronald R. Hoy, "The Neurobiology of Cricket Song," *Scientific American*

231:42. Copyright © 1974 by Scientific American, Inc. All rights reserved.

18.2 Courtesy Dr. Peter Marler

18.3 From Mazukazu Konishi, *Zeitschrift für Tierpsychologie* 22(1965) 22:770–783

18.4 Vivienne, Photo Researchers

18.5 San Diego Zoological Society

18.6 From M. Bertrand, "The Behavioral Repertoire of the Stumptail Macaque, *Bibl. Primat.* No. 11 (Karger, Basel, 1969)

18.7 From John Alcock, *Animal Behavior: An Evolutionary Approach.* Copyright © 1975 by Sinauer Associates, Inc. Reprinted by permission

## Chapter 19

19.1 Stephen Dalton, N.H.P.A.

19.2, 19.3 *Ecology: The Experimental Analysis of Distribution and Abundance* by Charles J. Krebs. Copyright © 1972 by Harper & Row, Publishers, Inc. By permission of the publishers

19.4 *Evolutionary Ecology* by Eric R. Pianka. Copyright © 1974 by Harper & Row, Publishers, Inc. By permission of the publishers

19.5 Mark N. Boulton, National Audubon Society, from Photo Researchers

19.6 *Evolutionary Ecology* by Eric R. Pianka. Copyright © 1974 by Harper & Row, Publishers, Inc. By permission of the publishers

19.11 Merritt J. Emlen, *Ecology: An Evolutionary Approach,* 1973, Addison-Wesley, Reading, Mass.

19.12 Mel Konner, Anthro-Photo

19.13 *Ecology,* by Robert E. Ricklefs, 1973. Chiron Press. Reprinted by permission

## Chapter 20

20.1 *Ecology,* by Robert E. Ricklefs, 1973. Chiron Press. Reprinted by permission

20.2 *Evolutionary Ecology* by Eric R. Pianka. Copyright © 1974 by Harper & Row, Publishers, Inc. By permission of the publishers

20.3 After P. Jespersen, "On the Frequency of Birds over the High Atlantic Ocean," *Verhandlung der Internationalen Ornithologischen Kongres* 6:163–172

20.4 From G. F. Gause, *The Struggle for Existence,* Williams and Wilkins Co., Baltimore. © 1934

20.5 From a census each April and November 1930–1943, from Errington 1945. *Ecological Monographs.* By permission of the publishers

20.6 *Evolutionary Ecology* by Eric R. Pianka. Copyright © 1974 by Harper & Row, Publishers, Inc. By permission of the publishers

## Chapter 21

21.1 (left) Allan D. Cruickshank, National Audubon Society, from Photo Researchers; (right) San Diego Zoological Society

21.2 Leonard Lee Rue IV, National Audubon Society, from Photo Researchers

21.4 *Ecology,* by Robert E. Ricklefs, 1973. Chiron Press. Reprinted by permission

21.5, 21.6 E. S. Ross

21.7 *Ecology,* by Robert E. Ricklefs, 1973. Chiron Press. Reprinted by permission

## Chapter 22

22.1 Modified from J. J. Christian, in *Proceedings of the National Academy of Science* 47 (1961):428–449

22.3 Mary M. Thacher, Photo Researchers

22.4 George Holton, Photo Researchers

22.5 San Diego Zoological Society

## Chapter 23

23.1, 23.2 *Evolutionary Ecology* by Eric R. Pianka. Copyright © 1974 by Harper & Row, Publishers, Inc. By permission of the publishers

23.3, 23.4, 23.5 *Ecology: The Experimental Analysis of Distribution and Abundance* by Charles J. Krebs. Copyright © 1972 by Harper & Row, Publishers, Inc. By permission of the publishers

23.7 L. David Mech

23.8 Malcolm, Anthro-Photo

23.9 From R. H. MacArthur "Population Ecology of Some Warblers of N.E. Coniferous Forests," *Ecology* 39:599–618

23.11 E. S. Ross

23.12, 23.13 *Evolutionary Ecology* by Eric R. Pianka. Copyright © 1974 by Harper & Row, Publishers, Inc. By permission of the publishers

23.14 From *The Animal and the Environment* by F. John Vernberg and Winona B. Vernberg. Copyright © 1970 by Holt, Rinehart and Winston, Inc. Reprinted by permission of Holt, Rinehart and Winston, Inc.

23.15 From I. Eibl-Eibesfeldt, "Concepts of Ethology

and Their Significance for the Study of Human Behavior," in *Early Behavior: Comparative and Developmental Approaches,* ed. H. W. Stevenson (New York: John Wiley and Sons, 1967), pp. 127–146

23.18 After D. Lack, *Ecological Adaptations for Breeding in Birds* (London: Methuen & Co., Ltd., 1968)

**Chapter 24**

24.1, 24.2 *Evolutionary Ecology* by Eric R. Pianka. Copyright © 1974 by Harper & Row, Publishers, Inc. By permission of the publishers

24.3 From Bruce Wallace and Adrian M. Srb, *Adaptation,* © 1964, Prentice-Hall, Inc., p. 53

24.4 *Ecology,* by Robert E. Ricklefs, 1973. Chiron Press. Reprinted by permission

24.5, 24.6 After I. Eibl-Eibesfeldt, *Ethology: the Biology of Behavior* (New York: Holt, Rinehart and Winston, 1970)

24.7 Zoological Society of London

24.9, 24.10 *Evolutionary Ecology* by Eric R. Pianka. Copyright © 1974 by Harper & Row, Publishers, Inc. By permission of the publishers

24.13 *Ecology,* by Robert E. Ricklefs, 1973. Chiron Press. Reprinted by permission

**Chapter 25**

25.1 (top) Dr. Georg Gerster, Photo Researchers; (bottom) Tomas D. W. Friedmann, Photo Researchers

25.2 Zoological Society of London

25.3 Peter H. Klopfer, *Behavioral Aspects of Ecology,* 2nd ed., © 1973. Reprinted by permission of Prentice-Hall, Inc., Englewood Cliffs, New Jersey

25.5, 25.6, 25.7 *Evolutionary Ecology* by Eric R. Pianka. Copyright © 1974 by Harper & Row, Publishers, Inc. By permission of the publishers

25.8, 25.9 *Ecology,* by Robert E. Ricklefs, 1973. Chiron Press. Reprinted by permission

25.10 *Evolutionary Ecology* by Eric R. Pianka. Copyright © 1974 by Harper & Row, Publishers, Inc. By permission of the publishers

**Chapter 26**

26.1 From K. von Frisch, *Biology* (New York: Harper & Row, 1964)

26.2 Fratelli Alinari S.P.A., Editorial Photocolor Archives, Inc.

26.4 Photophile

26.5, 26.6 From *Ethology: The Biology of Behavior,* 2nd ed., by I. Eibl-Eibesfeldt. Copyright © 1975 by Holt, Rinehart and Winston, Publishers

26.8 From Edward J. Kormondy, *Concepts of Ecology,* © 1969, Prentice-Hall, Inc., p. 77

**Chapter 27**

27.1 After G. K. Noble, "Courtship and Sexual Selection of the Flicker (*Colaptes auratus lutus*)," *Auk* 53: 269–282

27.2 Robert A. Wallace

27.3 From John B. Calhoun, "Population Density and Social Pathology," *Scientific American* 206:140–141. Copyright © 1962 by Scientific American, Inc. All rights reserved.

27.4, 27.5 United Press International Photo

27.6 (left) Ron Garrison, San Diego Zoological Society; (right) United Press International Photo

27.7 Photophile

27.8 United Press International Photo

**Chapter 28**

28.3 From Roger N. Johnson, *Aggression in Man and Animals,* W. B. Saunders, 1972.

28.4 From original material courtesy M. Gerald Borniger

28.5 After photograph by Robert Hermes, National Audubon Society, from Photo Researchers

28.6 Lynwood Chace, National Audubon Society, from Photo Researchers

28.7 San Diego Zoological Society

28.8 From "The Fighting Behavior of Animals" by I. Eibl-Eibesfeldt. Copyright © 1961 by Scientific American, Inc. All rights reserved

28.9 After F. R. Walther, "Zum Kampf-und Paarungsverhalten einiger Antilopen," *Zeitschrift für Tierpsychologie* 15:340–380

28.10 After F. R. Walther, "Enlwickungszüge im Kampf und Paarungsverhalten der Horntiere," *Jahrbuch der Gesellschaft von Opel-Freigehege Tierforschung* 3:90–115

28.11 (left) Photophile; (right) E. S. Ross

**Chapter 29**

29.1 After F. Doflein 1905 from I. Eibl-Eibesfeldt, *Ethology: The Biology of Behavior* (New York: Holt, Rinehart and Winston, 1970)

29.2 After J. B. Siebenaler and D. K. Caldwell, "Cooperation Among Adult Dolphins," *Journal of Mammalogy* 37:126–128
29.3 Ylla, Photo Researchers
29.4 E. S. Ross
29.5 After D. Lack, *Ecological Adaptations for Breeding in Birds* (London: Methuen & Co., Ltd., 1968)
29.6 Irven DeVore, Anthro-Photo
29.7 Hans Kummer
29.8 H. D. Wheeler, National Audubon Society, from Photo Researchers

**Chapter 30**

30.2 Irven DeVore, Anthro-Photo
30.3 Hugh M. Halliday, National Audubon Society, from Photo Researchers

**Chapter 31**

31.1 Dr. Walter Angst
31.2 (left) Allan D. Cruickshank, National Audubon Society, from Photo Researchers; (right) Leonard Lee Rue III, National Audubon Society, from Photo Researchers
31.3 Ron Garrison, San Diego Zoological Society
31.4 From J. H. Crook and J. S. Gartlan, "Evolution of Primate Societies," *Nature* 210(1966):1200–1203

**Chapter 32**

32.2 William G. Levy, Photo Researchers
32.3 Lincoln Brower
32.4 (page 520) Photophile; (page 521) United Press International Photo

# INDEX

*Page numbers in italic indicate figures.*

## A

## B

## C

## D

## E

## F

## G

## H

# I

## J



## K



## L

## M

## N



## O

## P

## S

## T

## U



## V



## W

## Y



## Z